INDEX OF APPLICATIONS

(continued in back of book)

QUANTITATIVE LITERACY

QUANTITATIVE LITERACY
Thinking Between the Lines

Bruce C. Crauder
Oklahoma State University

Benny Evans
Oklahoma State University

Jerry A. Johnson
University of Nevada

Alan V. Noell
Oklahoma State University

W.H. Freeman and Company
New York

FOR DOUGLAS, ROBBIE, AND ESPECIALLY ANNE -BRUCE

FOR SIMON, LUCAS, ROSEMARY, ISAAC, AND BEAR -BENNY

FOR JEANNE -JERRY

FOR EVELYN, PHILIP, ALYSON, LAURA, STEPHEN, AND ESPECIALLY LIZ -ALAN

Publisher: Ruth Baruth
Executive Editor: Terri Ward
Executive Marketing Manager: Jennifer Somerville
Senior Developmental Editor: Randi Blatt Rossignol
Senior Media Editor: Laura Capuano
Associate Media Editor: Catriona Kaplan
Supplements Editor: Katrina Wilhelm
Photo Editor: Cecilia Varas
Photo Researcher: Dena Digilio Betz
Text and Cover Designer: Vicki Tomaselli
Project Editor: Kerry O'Shaughnessy
Illustrations: Emily Cooper; MPS Limited, a Macmillan Company
Illustration Coordinator: Bill Page
Production Manager: Susan Wein
Composition: MPS Limited
Printing and Binding: RR Donnelley

Library of Congress Control Number: 2011936039

ISBN-13: 978-1-4292-2328-7
ISBN-10: 1-4292-2328-6

Printed in the United States of America

First printing

W. H. Freeman and Company
41 Madison Avenue
New York, NY 10010
Houndmills, Basingstoke
RG21 6XS, England

www.whfreeman.com

ABOUT THE AUTHORS

BRUCE CRAUDER graduated from Haverford College in 1976, receiving a B.A. with honors in mathematics, and received his Ph.D. from Columbia in 1981, writing a dissertation on algebraic geometry. After post-doctoral positions at the Institute for Advanced Study, the University of Utah, and the University of Pennsylvania, Bruce went to Oklahoma State University in 1986 where he is now a Professor of Mathematics and also Associate Dean.

Bruce's research in algebraic geometry has resulted in 10 refereed articles in as many years in his specialty, three-dimensional birational geometry. He has since worked on the challenge of the beginning college math curriculum, resulting in the creation of two new courses with texts to support the courses. He is especially pleased with these texts, as they combine scholarship with his passion for teaching.

Bruce and his wife, Anne, have two sons who are now just graduating from college. He enjoys working with non-profits, playing with his dog, and dancing, especially contradancing.

BENNY EVANS received his Ph.D. in mathematics from the University of Michigan in 1971. After a year at the Institute for Advanced Study in Princeton, New Jersey, he went to Oklahoma State University where he is currently Professor of Mathematics. In his tenure at OSU, he has served as Undergraduate Director, Associate Head, and Department Head. His career has included visiting appointments at the Institute for Advanced Study, Rice University, Texas A&M, and University of Nevada at Reno.

Benny's research interests are in topology and mathematics education. His record includes 28 papers in refereed journals, numerous books and articles (some in collaboration with those three other guys), and 25 grants from the National Science Foundation, Oklahoma State Board of Regents, and private foundations.

Benny's hobbies include fishing and spending time with his grandchildren. Fortunately, it is possible to do both at the same time.

JERRY JOHNSON received his Ph.D. in mathematics in 1969 from the University of Illinois, Urbana. From 1969 until 1993 he taught at Oklahoma State University, with a year off in 1976 for a sabbatical at Pennsylvania State University. In 1993, he moved to the University of Nevada, Reno, to become the founding director of their Mathematics Center as well as director of a project called Mathematics across the Curriculum. From 1995 to 2001, he served as Chairman of the Mathematics Department.

Over the years, Jerry has published 17 research articles in prominent mathematics journals and over 35 articles in various publications related to mathematics education. He wrote and published a commercially successful animated 3-D graphing program called GyroGraphics and was a founding member of the National Numeracy Network, whose goal is to promote quantitative literacy.

Jerry's wife, Jeanne, is a folklorist with a Ph.D. from Indiana University, Bloomington. They enjoy traveling (Jerry has been to all 50 states), gardening (especially tomatoes), cooking (you fry it, I'll try it), and cats (the kind that meow).

ALAN NOELL received his Ph.D. in mathematics from Princeton University in 1983. After a postdoctoral position at Caltech, he joined the faculty of Oklahoma State University in 1985. He is currently Professor of Mathematics and director of the graduate program there. His scholarly activities include research in complex analysis and curriculum development.

Alan and his wife, Liz, have four children and one daughter-in-law. They are very involved with their church, and they also enjoy reading and cats (the same kind as for Jerry).

BRIEF CONTENTS

CONTENTS

CHAPTER **5**

INTRODUCTION TO PROBABILITY   269

PREFACE

Will Rogers, a popular humorist and actor from the 1920s and 1930s, famously quipped, "All I know is just what I read in the papers, and that's an alibi for my ignorance." Today we get our information from a wider variety of sources, including television broadcasts, magazines, and Internet sites, as well as newspapers. But the sentiment remains the same—we rely on public media for the information we require to understand the world around us and to make important decisions in our daily lives.

In today's world, the vast amount of information that is available in the form of print and images is overwhelming. Interpreting and sorting this information, trying to decide what to believe and what to reject, is ever more daunting in the face of rapidly expanding sources. In addition, the important news of the day often involves complicated topics such as economics, finance, political polls, and an array of statistical data—all of which are quantitative in nature.

Our Goals and Philosophy

The overall goal of this book is to present and explain the quantitative tools necessary to understand issues arising in the popular media and in our daily lives. Through contemporary real world applications, our aim is to teach students the practical skills they will need throughout their lives to be critical thinkers, informed decision makers, and intelligent consumers of the quantitative information that they see every day. This goal motivates our choice of topics and our use of numerous articles from the popular media as illustrations.

Being an intelligent reader and consumer requires critical thinking. This fact is what is conveyed by the book's subtitle, *Thinking Between the Lines*. For example, visual displays of data such as graphs can be misleading. How can you spot this when it happens? A poll often presents, along with its results, a "margin of error." What does this mean? Banks often use the terms APR and APY. What do these mean? In these and many other practical situations, math matters. A nice example of the role in the text of critical thinking can be found in a discussion of how to display federal defense spending.

College-level students with a background of no more than elementary algebra are the intended audience for this book. To engage these students in the study of quantitative literacy, important mathematical skills and concepts are taught in real-world contexts. For example, exponential functions are presented in the context of constant percentage growth/decay rather than as abstract formulas. Their utility is shown through ecological applications as well as the sorts of personal finance issues that we all face.

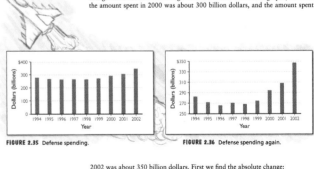

Misleading by choice of axis scale

Consider the bar graphs in **Figures 2.35 and 2.36**, which show federal defense spending in billions of dollars for the given year.

If we look carefully at the years represented and the spending reported, we see that the two graphs represent exactly the same data—yet they don't look the same. For example, Figure 2.36 gives the visual impression that defense spending doubled from 2000 to 2002, and Figure 2.35 suggests that defense spending did increase from 2000 to 2002, but not by all that much. What makes the graphs so different?

The key to understanding the dramatic difference between these two graphs is the range on the vertical axis. In Figure 2.35 the vertical scale goes from 0 to 400 billion dollars, and in Figure 2.36 that scale goes from 250 to 350 billion dollars. The shorter vertical range exaggerates the changes from year to year.

One way to assess the actual change in defense spending is to find the percentage change. Let's consider the change from 2000 to 2002. In both graphs, we see that the amount spent in 2000 was about 300 billion dollars, and the amount spent in

FIGURE 2.35 Defense spending.

FIGURE 2.36 Defense spending again.

2002 was about 350 billion dollars. First we find the absolute change:

$$\text{Change} = 350 - 300 = 50 \text{ billion dollars.}$$

Now to find the percentage change, we divide by the spending in 2000:

$$\text{Percentage change} = \frac{50}{300} \times 100\%.$$

This is about 17%. The percentage increase is fairly significant, but it is not nearly as large as that suggested by Figure 2.36, which gives the impression that spending has almost doubled—that is, increased by 100%.

Similarly, instead of simply defining logarithms by formulas, we discuss how they relate to the decibel and Richter scales. Instead of just presenting graph theory, we discuss how it relates to spell checkers and social media such as Facebook. This "in context" philosophy guides the presentation throughout the text; therefore, it should always be easy to answer a student if he or she asks the famous question, "What is this stuff good for?"

We want to do more than just help students succeed in a liberal arts math course. We want them to acquire skills and experience material that they will be able to use throughout their lives as intelligent consumers of information.

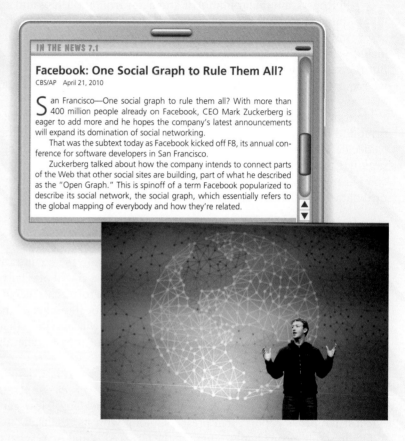

IN THE NEWS 7.1

Facebook: One Social Graph to Rule Them All?
CBS/AP April 21, 2010

San Francisco—One social graph to rule them all? With more than 400 million people already on Facebook, CEO Mark Zuckerberg is eager to add more and he hopes the company's latest announcements will expand its domination of social networking.

That was the subtext today as Facebook kicked off F8, its annual conference for software developers in San Francisco.

Zuckerberg talked about how the company intends to connect parts of the Web that other social sites are building, part of what he described as the "Open Graph." This is spinoff of a term Facebook popularized to describe its social network, the social graph, which essentially refers to the global mapping of everybody and how they're related.

Balanced Coverage

We offer students and teachers a book that strikes the right balance between reading text and doing calculations. For example, in Chapter 5 we discuss the probability of an event not occurring. The usual formula is presented, but it is also clearly described in words and in familiar contexts. This balance of concepts and calculation engages students in thinking about interesting ideas and also gives them the opportunity to apply and practice the math underlying those ideas.

Some instructors like to present derivations of formulas to their students, whereas others would prefer to avoid them. As a balance between these points of view, we have put many derivations into a feature called **Algebraic Spotlight** that may be used at the instructor's discretion. On the other hand, a feature called **Key Concept** is used to encapsulate succinct summaries of the basic ideas. These two features also illustrate the balance in the presentation. An example of this balance is seen in the Key Concept from Chapter 2 where average growth rate is described in words and is followed by the average growth rate described as a formula.

The average growth rate is always calculated over a given interval, often a time period (such as 1790 to 1800). To calculate the average growth rate, we divide the change in the function over that interval by the change in the independent variable.

Key Concept

The **average growth rate** of a function over an interval is the change in the function divided by the change in the independent variable.

We calculate the average growth rate using the formula

$$\text{Average growth rate} = \frac{\text{Change in function}}{\text{Change in independent variable}}.$$

Abundant Use of Applications: Real, Contemporary, and Meaningful

Wherever possible, we use real applications, real scenarios, and real data. For example, in Chapter 1, we use the Berkeley gender discrimination case to introduce the concept of paradox.

In Chapter 4, we discuss the pros and cons of credit card usage as the context for calculating finance charges and monthly payments. Instead of inventing examples, we use concrete examples from the news in a prominent feature called In the News.

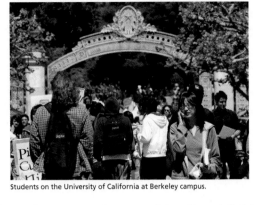

Students on the University of California at Berkeley campus.

These numbers may seem to be pretty conclusive evidence of discrimination. When we look more closely, however, we find that graduate school admissions are based on departmental selections. There are many departments at Berkeley, but to keep things simple, we will assume that the university consists of only two departments: Math and English. Here is how the male and female applicants might have been divided among these two departments.

Credit cards are convenient and useful. They allow us to travel without carrying large sums of cash, and they sometimes allow us to defer cash payments, even interest-free, for a short time. In fact, owning a credit card can be a necessity: Most hotels and rental car companies require customers to have a credit card. Some sources say that the average credit card debt per American is almost $8000.[9]

The convenience comes at a cost, however. For example, some credit cards carry an APR much higher than other kinds of consumer loans. The APR depends on various factors, including the customer's credit rating. According to the article above, credit cards intended for students may range from 10% to 19.8%. At the time of this writing, the site www.indexcreditcards.com/lowaprcreditcards.html reveals cards with APRs ranging from 7.24% to 22.9%.

In the News

Throughout the text, explanations, examples, and exercises are often motivated by everyday events reported in the popular media. Each chapter and each section opens with real articles gleaned from newspapers, magazines, and Internet sites. They are presented under the feature titled **In the News**. These lead naturally to discussions of mathematics as part of everyday life. For example, the article entitled "Being Bored Could Be Bad for Your Health" leads to a discussion of causation.

> **IN THE NEWS 6.14**
>
> ### Being Bored Could Be Bad for Your Health
> MARIA CHENG   February 10, 2010
>
> London (AP)—Can you really be bored to death? In a commentary to be published in the *International Journal of Epidemiology* in April, experts say there's a possibility that the more bored you are, the more likely you are to die early.
>
> Annie Britton and Martin Shipley of University College London caution that boredom alone isn't likely to kill you—but it could be a symptom of other risky behavior like drinking, smoking, taking drugs or having a psychological problem....
>
> Britton and Shipley said boredom was probably not in itself that deadly. "The state of boredom is almost certainly a proxy for other risk factors," they wrote. "It is likely that those who were bored were also in poor health."

The **Exercise Sets** at the end of each section are also based on realistic situations, and many are taken directly from the world in which we live. Exercise 20 on page 162 is a typical example.

An Emphasis on Solving Problems

Concepts are reinforced with worked Examples that model problem solving and are often followed by Try it Yourself exercises for students to practice on their own. The Try it Yourself exercises engage students in thinking rather than just passively reading.

Examples

Examples in the text are selected as much as possible for their relevance to student experience. For instance, Example 4.7 shows how compound interest affects savings. This example provides students with basic tools needed to deal with a myriad of financial issues. Another, Example 6.4, uses gas mileage from automobiles. It shows students how to create and interpret box plots.

EXAMPLE 4.7 Calculating present and future value: Investing for a car

You would like to have $10,000 to buy a car three years from now. How much would you need to invest now in a savings account that pays an APR of 9% compounded monthly?

SOLUTION

In this problem, we know the future value ($10,000) and would like to know the present value. The monthly rate is $r = 0.09/12 = 0.0075$ as a decimal, and the number of compounding periods is $t = 36$ months. Thus,

$$\text{Present value} = \frac{\text{Future value}}{(1+r)^t}$$
$$= \frac{\$10,000}{1.0075^{36}}$$
$$= \$7641.49.$$

Therefore, we should invest $7641.49 now.

The solution for this example includes a clear and complete explanation which students can follow and reference when solving similar types of problems.

Try It Yourself

Many worked examples are followed by a very similar problem that gives students the opportunity to test their understanding. This feature serves to actively engage students in the learning process. The answers to the Try It Yourself problems are provided at the end of the section so that students can check their solutions.

Try It Yourself 4.6: Using APY balance formula: CD balance at maturity $111,682.36.

Try It Yourself 4.7: Calculating present and future value: An account with quarterly compounding $1235.51.

---

20. A nuclear waste site. Cesium-137 is a particularly dangerous by-product of nuclear reactors. It has a half-life of 30 years. It can be readily absorbed into the food chain and is one of the materials that would be stored in the proposed waste site at Yucca Mountain (see the article opening this section). Suppose we place 3000 grams of cesium-137 in a nuclear waste site.

a. How much cesium-137 will be present after 30 years, or one half-life? After 60 years, or two half-lives?

b. Find an exponential formula that gives the amount of cesium-137 remaining in the site after h half-lives.

c. How many half-lives of cesium-137 is 100 years? Round your answer to two decimal places.

d. Use your answer to part c to determine how much cesium-137 will be present after 100 years. Round your answer to the nearest whole number.

EXAMPLE 6.4 Making and interpreting a boxplot: Gas mileage

A report on the greenercars.org Web site shows 2011-model cars that score well in terms of environmental impact. Here are the data, with mileage measured in mile per gallon (mpg) listed in order of the Web site's "green score."

Model	City mileage (mpg)	Highway mileage (mpg)
Toyota Prius	51	48
Honda Civic Hybrid	40	43
Honda CR-Z	35	39
Toyota Yaris	29	35
Audi A3	30	42
Hyundai Sonata	22	35
Hyundai Tucson	23	31
Chevrolet Equinox	22	32
Kia Rondo	20	27
Chevrolet Colorado/GMC Canyon	18	25

a. Find the five-number summary for city mileage.

b. Present a boxplot of city mileage.

c. Comment on how the data are distributed about the median. *Note:* The corresponding calculations for highway mileage are left as an exercise. See Exercise 8.

SOLUTION

a. The list for city mileage, in order from lowest to highest, is

$$18, 20, 22, 22, 23, 29, 30, 35, 40, 51.$$

The maximum is 51 mpg, and the minimum is 18 mpg. To find the median, we average the two numbers in the middle:

$$\text{Median} = \frac{23 + 29}{2} = 26 \text{ mpg.}$$

The lower half of the list is 18, 20, 22, 22, 23, and the median of this half is 22 Thus, the first quartile is 22 mpg. The upper half of the list is 29, 20, 35, 40, 51 and the median of this half is 35. Thus, the third quartile is 35 mpg.

b. The corresponding boxplot appears in **Figure 6.2.** The vertical axis is the mileage measured in miles per gallon.

c. Referring to the boxplot, we note that the first quartile is not far above the minimum, and the median is not far above the first quartile. The third quartile i well above the median, and the maximum is well above the third quartile. This

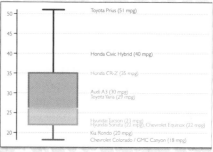

FIGURE 6.2 Boxplot for city mileage.

emphasizes the dramatic difference between the high-mileage cars (the and ordinary cars.

A Wealth of Features to Help Students Engage, Study, and Learn

The book includes many pedagogical features that will help students learn and review key concepts. These include a cutting-edge illustration program, Quick Review, Key Concepts, Chapter Summaries, and Chapter Quizzes. Each section concludes with abundant exercises that are graded from simple to more complex. We review each of the features in turn.

Artwork

Many students are visual learners, and we have used graphs, photos, cartoons, and sketches liberally throughout the text. Many illustrations are paired so that students can easily compare and contrast similar concepts.

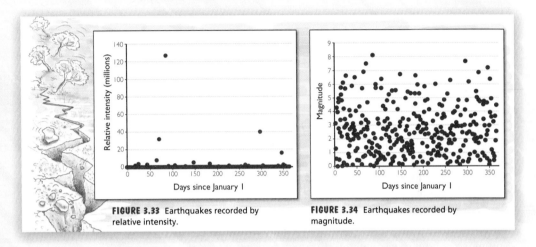

FIGURE 3.33 Earthquakes recorded by relative intensity.

FIGURE 3.34 Earthquakes recorded by magnitude.

Here is an example of the many cartoons used to lighten things up:

Key Concepts

Most mathematics books state definitions of terms formally. We have chosen to define terms in a concise and informal way and to label each as a Key Concept. We do so in order to avoid intimidating students and to focus on the concepts rather than dry terminology. Examples of this will be found on page 353 in Chapter 6 on statistics, where "outliers" and "quartiles" are explained.

Key Concept

An **outlier** is a data point that is significantly different from most of the data.

Key Concept

- The **first quartile** of a list of numbers is the median of the lower half of the numbers in the list.

- The **second quartile** is the same as the median of the list.

- The **third quartile** is the median of the upper half of the numbers in the list.

If the list has an even number of entries, it is clear what we mean by the "lower half" and "upper half" of the list. If the list has an odd number of entries, eliminate the median from the list. This new list now has an even number of entries. The "lower half" and "upper half" refer to this new list.

Summaries

Students may occasionally feel overwhelmed by new concepts, so throughout each chapter we provide brief boxed Summaries of core concepts. This is an opportunity for students to pause, collect their thoughts, and reinforce what they have learned. Summaries also provide reference points when students are solving exercises. For instance, in Section 1.2, a discussion of some common types of logical fallacies is offered. These are grouped together with brief descriptions of each in Summary 1.2.

> **SUMMARY 1.2**   **Some Common Informal Fallacies**
>
> **Appeal to ignorance**   A statement is either accepted or rejected because of a lack of proof.
>
> **Dismissal based on personal attack**   An argument is dismissed based on an attack on the proponent rather than on its merits.
>
> **False authority**   The validity of a claim is accepted based on an authority whose expertise is irrelevant.
>
> **Straw man**   A position is dismissed based on the rejection of a distorted or different position (the "straw man").
>
> **Appeal to common practice**   An argument for a practice is based on the popularity of that practice.
>
> **False dilemma**   A conclusion is based on an inaccurate or incomplete list of alternatives.
>
> **False cause**   A causal relationship is concluded based on the fact that two events occur together or that one follows the other.
>
> **Circular reasoning**   This fallacy simply draws a conclusion that is really a restatement of the premise.
>
> **Hasty generalization**   This fallacy occurs when a conclusion is drawn based on a few examples that may be atypical.

Calculation Tips

Tips are provided to help students avoid common errors in doing calculations. One kind of tip involves advice in preparing to make a calculation. For example, Calculation Tip 5.1 helps students deal with probability calculations. Others may give helpful hints for carrying out a complex calculation, such as Calculation Tip 6.1, which summarizes how one calculates a standard deviation.

> **CALCULATION TIP 5.1** **Equally Likely Outcomes**
>
> When calculating probabilities, take care that each of the outcomes considered is equally likely to occur. In cases involving identical items, such as coins or dice, it may help to label the items to distinguish them.

> **CALCULATION TIP 6.1** **Calculating Standard Deviation**
>
> To find the standard deviation of n data points, we first calculate the mean μ. The next step is to complete the following calculation template:
>
> | Data | Deviation | Square of deviation |
> |------|-----------|---------------------|
> | $\vdots$ | $\vdots$ | $\vdots$ |
> | x_i | $x_i - \mu$ | Square of second column |
> | $\vdots$ | $\vdots$ | $\vdots$ |
> | | | Sum of third column |
> | | | Divide the above sum by n and take the square root. |

Rule of Thumb Statements

One facet of quantitative literacy is the ability to make estimates. The Rule of Thumb feature provides hints on making such "ballpark" estimates without appealing to complicated formulas for exact answers. For example, Rule of Thumb 4.2 shows how to estimate certain monthly payments on loans. These estimates provide a simple and effective check to catch possible errors in exact calculations.

> **RULE OF THUMB 4.2** **Monthly Payments for Short-Term Loans**
>
> For all loans, the monthly payment is *at least* the amount we would pay each month if no interest were charged, which is the amount of the loan divided by the term (in months) of the loan. This would be the payment if the APR were 0%. It's a rough estimate of the monthly payment for a short-term loan if the APR is not large.

Algebraic Spotlights

An effort is made throughout the text to motivate complicated formulas, but some derivations require algebra skills that may be beyond the scope of a course. Such derivations are presented in a special discussion called Algebraic Spotlight. An example is the derivation of the monthly payment formula in Chapter 4. While some instructors like to present derivations of formulas to their students, others would prefer to avoid them. Therefore, these spotlights are intended to be used at the instructor's discretion.

ALGEBRAIC SPOTLIGHT 4.3 Derivation of the Monthly Payment Formula, Part II

Now we use the account balance formula to derive the monthly payment formula. Suppose we borrow B_0 dollars at a monthly interest rate of r as a decimal and we want to pay off the loan in t monthly payments. That is, we want the balance to be 0 after t payments. Using the account balance formula we derived in Algebraic Spotlight 4.2, we want to find the monthly payment M that makes $B_t = 0$. That is, we need to solve the equation

$$0 = B_t = B_0(1+r)^t - M\frac{(1+r)^t - 1}{r}$$

for M. Now

$$0 = B_0(1+r)^t - M\frac{(1+r)^t - 1}{r}$$

$$M\frac{(1+r)^t - 1}{r} = B_0(1+r)^t$$

$$M = \frac{B_0 r(1+r)^t}{((1+r)^t - 1)}.$$

Because B_0 is the amount borrowed and t is the term, this is the monthly payment formula we stated earlier.

Quick Review

Some students may need occasional reminders about basic facts. These refreshers are provided at key points throughout the text in Quick Reviews. For example, in Chapter 2 there is a Quick Review on making graphs.

Quick Review Making Graphs

Typically, we locate points on a graph using a table such as the one below, which shows the number of automobiles sold by a dealership on a given day.

Day (independent variable)	1	2	3	4
Cars sold (dependent variable)	3	6	5	2

When we create a graph of a function, the numbers on the horizontal axis correspond to the independent variable, and the numbers on the vertical axis correspond to the dependent variable. For this example, we put the day (the independent variable) on the horizontal axis, and we put the number of cars sold (the dependent variable) on the vertical axis.

The point corresponding to day 3 is denoted by (3, 5) and is represented by a single point on the graph. To locate the point (3, 5), we begin at the origin and move 3 units to the right and 5 units upward. This point is displayed on the graph as one of the four points of that graph. The aggregate of the plotted points is the graph of the function.

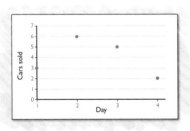

Chapter Summaries

Chapter summaries are included to wrap up the discussion and give an overview of the major points in the chapter. For example, at the end of Chapter 2, the summary brings together the major points about visualization of data.

CHAPTER SUMMARY

In our age, we are confronted with an overwhelming amount of information. Analyzing quantitative data is an important aspect of our lives. Often this analysis requires the ability to understand visual displays such as graphs, and to make sense of tables of data. A key point to keep in mind is that growth rates are reflected in graphs and tables.

Measurements of growth: How fast is it changing?

Data sets can be presented in many ways. Often the information is presented using a table. To analyze tabular information, we choose an independent variable. From this point of view, the table presents a function, which tells how one quantity (the dependent variable) depends on another (the independent variable).

A simple tool for visualizing data sets is the *bar graph*. See **Figure 2.63** for an example of a bar graph. A bar graph is often appropriate for small data sets when there is nothing to show between data points.

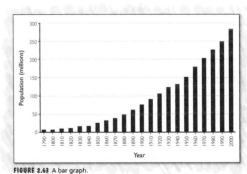

FIGURE 2.63 A bar graph.

Both absolute and percentage changes are easily calculated from data sets, and such calculations often show behavior not readily apparent from the original data. We calculate the percentage change using

$$\text{Percentage change} = \frac{\text{Change in function}}{\text{Previous function value}} \times 100\%.$$

Key Terms

Prior to each Chapter Quiz is a list of Key Terms from the chapter along with the page number on which each term is introduced. This allows a student to review vocabulary from the chapter to be sure they have not missed any important terms.

KEY TERMS

independent variable, p. 73	relative change, p. 76	scatterplot, p. 92
dependent variable, p. 73	average growth rate, p. 80	line graph, p. 93
function, p. 73	interpolation, p. 82	smoothed line graph, p. 98
percentage change, p. 76	extrapolation, p. 85	

Chapter Quiz

Each chapter ends with a Chapter Quiz, which consists of a representative sample of exercises. These are intended to give students a review of the major points in the chapter and a chance to test themselves over retention of the concepts they have learned. These exercises are accompanied by answers and references to worked examples in the text. For instance, the quiz at the end of Chapter 9 contains eight exercises.

CHAPTER QUIZ

1. We invest $2400 in an account that pays simple interest of 8% each year. Find the interest earned after five years.

Answer $960

If you had difficulty with this problem, see Example 4.1.

2. Suppose we invest $8000 in a four-year CD that pays an APR of 5.5%.
a. What is the value of the mature CD if interest is compounded annually?
b. What is the value of the mature CD if interest is compounded monthly?

Answer a. $9910.60, b. $9963.60

If you had difficulty with this problem, see Example 4.3.

3. We have an account that pays an APR of 9.75%. If interest is compounded quarterly, find the APY. Round your answer as a percentage to two decimal places.

Answer 10.11%

If you had difficulty with this problem, see Example 4.4.

4. How much would you need to invest now in a savings account that pays an APR of 8% compounded monthly in order to have a future value of $6000 in a year and a half?

Answer $5323.64

If you had difficulty with this problem, see Example 4.7.

5. Suppose an account earns an APR of 5.5% compounded monthly. Estimate the doubling time using the Rule of 72, and calculate the exact doubling time. Round your answers to one decimal place.

Answer Rule of 72: 13.1 years; exact method: 151.6 months (about 12 years and 8 months)

If you had difficulty with this problem, see Example 4.8.

6. You need to borrow $6000 to buy a car. The dealer offers an APR of 9.25% to be paid off in monthly installments over $2\frac{1}{2}$ years.
a. What is your monthly payment?
b. How much total interest did you pay?

Answer a. $224.78, b. $743.40

If you had difficulty with this problem, see Example 4.11.

7. We can afford to make payments of $125 per month for two years for a used motorcycle. We're offered a loan at an APR of 11%. What price bike should we be shopping for?

Answer $2681.95

If you had difficulty with this problem, see Example 4.10.

8. Suppose we have a savings account earning 6.25% APR. We deposit $15 into the account at the end of each month. What is the account balance after eight years?

Answer $1862.16

If you had difficulty with this problem, see Example 4.21.

9. Suppose we have a savings account earning 5.5% APR. We need to have $2000 at the end of seven years. How much should we deposit each month to attain this goal?

Answer $19.57

If you had difficulty with this problem, see Example 4.22.

10. Suppose we have a nest egg of $400,000 with an APR of 5% compounded monthly. Find the monthly yield for a 10-year annuity.

Answer $4242.62

If you had difficulty with this problem, see Example 4.24.

11. Suppose your MasterCard calculates finance charges using an APR of 16.5%. Your previous statement showed a balance of $400, toward which you made a payment of $100. You then bought $200 worth of clothes, which you charged to your card. Complete the following table:

	Previous balance	Payments	Purchases	Finance charge	New balance
Month 1					

Exercise Set 3.1

Note: In some exercises, you are asked to find a formula for a linear function. If no letters are assigned to quantities in an exercise, you are expected to choose appropriate letters and also give appropriate units.

Linear or not? In Exercises 1 through 12, you are asked to determine whether the relationship described represents a function that is linear. In each case, explain your reasoning. If the relationship can be represented by a linear function, find an appropriate linear formula, and explain in practical terms the meaning of the growth rate.

1. A staircase. Is the relationship between the height of a staircase and the number of steps linear?

2. A music club. Suppose that the cost of purchasing CDs from a music club is a flat membership fee of $30 plus $10 for each CD purchased. Is the amount of money you pay a linear function of the number of CDs you buy?

3. Another music club. Suppose that a music club charges $10 per CD but offers a 2% discount on orders larger than 15 CDs. Is the cost per order a linear function of the number of CDs you order?

Price of Amazon's Kindle **Figure 3.16** shows how the price of Amazon's Kindle 2 e-book reader has decreased over time. (The data for 2011 are projected.) For example, the price was $349 when the Kindle was launched in February 2009. The price had dropped to $299 by July 2009. The graph is a straight line, and we will assume that the relationship is linear. This information is used in Exercises 16 through 19.

16. Use the information for February and July of 2009 to determine the decrease in price over each month.

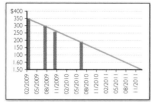

FIGURE 3.16 Price of the Kindle 2.

17. Use your answer to Exercise 16 to determine the slope of the linear function that gives the price in terms of the time in months since February 2009. Be careful about the sign of the slope.

Graded Exercises

The exercises at the end of each section begin with relatively straightforward problems. As the exercises progress, more difficult problems appear. For example, in Section 9.1, Exercises 1 and 2 are quite straightforward. By the time we get to Exercises 16 and 17, the difficulty has increased a bit.

Topics and Organization

Chapter 1 covers logic. The message is straightforward—we need to critically analyze the input we get from the popular media rather than accept it at face value. The basic tools needed to distinguish valid arguments from invalid ones are provided. The closing section of this chapter focuses on number sense, bringing home the meaning of very large numbers like the national debt and very small numbers like those used to measure the size of a bacterium. Number sense is also useful in making ballpark estimates that help us judge whether figures we encounter in daily life are accurate.

Familiar topics such as interest rates, inflation, and population growth rates are prime examples of issues that are understood in terms of rates of change. Chapter 2 discusses their general significance and shows how they are reflected in tabular and graphical presentations. The presentation and treatment of data in Chapter 2 are cast in terms of rates of change, along with the tabular and visual consequences of presenting such data. The treatment of data in terms of statistics, on the other hand, is presented in Chapter 6.

In Chapter 3, we focus on constant growth rates (linear functions) and constant percentage growth rates (exponential functions) because these are the ones most commonly encountered in the popular media. We believe that laying this foundation in Chapters 2 and 3, prior to discussing financial issues in Chapter 4, is an efficient and natural way to approach things. Furthermore, it provides students with a foundation for understanding other kinds of growth rates (both positive and negative) such as inflation (or deflation) and population (growth or decline).

The basic concept that exponential functions change at a constant percentage rate unifies the discussion of applications. Rather than treat compound interest as a topic separate from exponential population growth and other exponential models, our unified treatment emphasizes that there is one basic idea with several important applications. It also highlights the contrast between linear and exponential growth, surely one of the most important distinctions for quantitative literacy.

Chapter 4, Personal Finance, covers the basics of borrowing, saving, credit, and installment loans. Chapters 2 and 3 are recommended as a foundation for this material, but Chapter 4 is structured so that it can stand on its own. Many college students are already facing some of the complexities of financial life through their use of credit cards and loans to pay for their education. Soon enough they will be looking at large purchases such as cars or homes along with other significant financial issues which are covered in this chapter. A key feature is an emphasis on how compound interest affects both savings and debt. Illustrations are offered which show the importance of an early start to long-term savings.

Chapter 5 focuses on probability and Chapter 6 covers statistics. From weather reports to medical tests to risk management and gambling, probability plays an important role in our lives. And it lays a foundation for statistics, which is probably the single most used and misused mathematical tool employed by the popular media. We see statistics in such diverse guises as SAT scores, political polls, median home prices, and batting averages. But any attempt to distill the meaning of large amounts of data into a few numbers or phrases is a risky business, and statistics, if improperly used or understood, can mislead rather than inform. Our treatment gives students the tools they need to view the statistics dispensed by the popular media with a critical eye.

Chapter 7 provides an introduction to graph theory, where we explore connections to topics such as efficient routes and spell checkers. The most salient connection is to social networks such as Facebook, which are familiar to virtually every student these days.

Chapter 8 examines voting and social choice. Students may be surprised to learn that elections and voting are more complex than might be suspected. Multiple candidates and voting blocs lead to unexpected complications. Inheritance or divorce can lead to difficult issues involving fair division of money, property, and items of sentimental value. We look at several schemes for accomplishing this division. Finally, the issue of how members of the House of Representatives are apportioned among the states is explained.

The book closes with Chapter 9, a look at geometry. We discuss measurements, symmetry, and tilings. Proportionality arising from symmetry is key to understanding many things from the relative cost of pizzas to why King Kong is (unfortunately?) an impossibility. Geometry is important in art and architecture, but also has its share of controversy, which we explore.

For good or ill, mathematics is part of our daily lives. One need not be a professional mathematician in order to employ it effectively. A few basic mathematical ideas and the confidence to employ them can be important tools that help us understand and engage in the world we inhabit.

Media and Supplements

For Students

ⓛmathportal at www.yourmathportal.com

(Access code or online purchase required.) MathPortal is the digital gateway to *Quantitative Literacy: Thinking Between the Lines,* designed to enrich the course and enhance students' study skills through a collection of Web-based tools. MathPortal integrates a rich suite of diagnostic, assessment, tutorial, and enrichment features, enabling students to master quantitative literacy at their own pace. MathPortal is organized around three main teaching and learning components:

1. The **Interactive eBook** offers a complete and customizable online version of the text, fully integrated with all of the media resources available with *Quantitative Literacy: Thinking Between the Lines.* The eBook allows students to quickly search the text, highlight key areas, and add notes about what they're reading. Instructors can customize the eBook to add, hide, and reorder content; add their own material; and highlight key text for students.

2. **Resources** organize all of the resources for *Quantitative Literacy: Thinking Between the Lines* into one location for ease of use. Includes (for students):

  • **Student Solutions Manual** provides solutions to all the odd-numbered exercises in the text.

  • **Gradable Vocabulary Flashcards** offer additional practice to students learning key terms.

  • **MathClips** present animated whiteboard videos illuminating key concepts in each chapter by showing students step-by-step solutions to selected exercises.

  • NEW! **LEARNING***Curve* is a formative quizzing system that offers immediate feedback at the question level to help students master the course material.

  • **Interactive Applets** offer additional help to students mastering key concepts.

And for instructors:

  • **Instructor's Guide** with Solutions includes teaching suggestions, chapter comments, and detailed solutions to all exercises.

  • **Test Bank** offers hundreds of multiple choice, fill-in, and short answer questions.

  • **Lecture PowerPoint Slides** offer a detailed lecture presentation of key concepts covered in each chapter of *Quantitative Literacy: Thinking Between the Lines.*

  • **SolutionMaster** is a Web-based version of the solutions in the Instructor's Guide with Solutions. This easy-to-use tool allows instructors to generate a solution file for any set of homework exercises. Solutions can be downloaded in PDF format for convenient printing and posting. For more information or a demonstration, contact your local W. H. Freeman sales representative.

3. **Assignments** guide instructors through an easy-to-create assignment process providing access to questions from the Test Bank and Exercises from the text, including many algorithmic problems. The Assignment Center enables instructors to create their own assignments from a variety of question types for machine-gradable assignments. This powerful assignment manager allows instructors to select their preferred policies in regard to scheduling, maximum attempts, time limitations, feedback, and more.

Quantitative Literacy: Thinking Between the Lines eBook

The complete eBook is also available stand-alone, outside of MathPortal, at approximately one-half the cost of the printed textbook.

Online Study Center at www.whfreeman.com/osc/quantlit

(Access code or online purchase required.) The Online Study Center offers all the resources available in MathPortal, except the eBook and Assignment Center.

Companion Web site at www.whfreeman.com/quantlit

For students, this site serves as a FREE 24/7 electronic study guide, and it includes such features as applets, self-quizzes, and vocabulary terms.

Printed Student Solutions Manual

ISBN: 1-4292-4424-0
This printed manual provides stepped-through solutions for all odd-numbered exercises in the text.

For Instructors

The **Instructor's Web site** (www.whfreeman.com/quantlit) requires user registration as an instructor and features all of the student Web materials plus:

- **PowerPoint Slides** containing all textbook figures and tables.
- Lecture PowerPoint Slides

Printed Instructor's Guide with Solutions

ISBN: 1-4292-4423-2
This printed guide includes teaching support for each chapter and full solutions for all problems in the text.

Test Bank

ISBNs: 1-4292-4421-6 (printed version), 1-4292-4420-8 (digital [CD-ROM] version)
The Test Bank contains hundreds of multiple-choice, fill-in, and short-answer questions to generate quizzes and tests for each chapter of the text. The Test Bank is available in print as well as on CD-ROM (for Windows and Mac), where questions can be downloaded, edited, and resequenced to suit each instructor's needs.

Enhanced Instructor's Resource CD-ROM

ISBN: 1-4292-4422-4
Allows instructors to search and export (by key term or chapter):

- All text images and tables
- Instructor's Guide with Solutions
- Lecture PowerPoint Slides
- Test Bank files

Course Management Systems

W. H. Freeman and Company provides courses for Blackboard, WebCT (Campus Edition and Vista), Angel, Desire2Learn, Moodle, and Sakai course management systems. These are completely integrated solutions that you can easily customize and adapt to meet your teaching goals and course objectives. Visit www.bfwpub.com/lms for more information.

i-clicker. is a two-way radio-frequency classroom response solution developed by educators for educators. University of Illinois Physicists Tim Stelzer, Gary Gladding, Mats Selen, and Benny Brown created the i-clicker system after using competing classroom response solutions and discovering they were neither classroom-appropriate nor student-friendly. Each step of i-clicker's development has been informed by teaching and learning. i-clicker is superior to other systems from both a pedagogical and technical standpoint. To learn more about packaging i-clicker with this textbook, please contact your local sales rep or visit www.iclicker.com.

Acknowledgements

We are grateful for the thoughtful comments and insights from the reviewers of our manuscript in its many versions. Their reviews were invaluable, helping us see our materials through the eyes of others.

Reza Abbasian, *Texas Lutheran University*

Lowell Abrams, *George Washington University*

Marwan Abu-Sawwa, *University of North Florida*

Tom Adamson, *Phoenix College*

Margo Alexander, *Georgia State University*

K.T. Arasu, *Wright State University*

Kambiz Askarpour, *Essex Community College*

Wahab Baouchi, *Regis University*

Erol Barbut, *University of Idaho*

Ronald Barnes, *University of Houston—Downtown*

Wes Barton, *Southwestern Community College*

David Baughman, *University of Illinois at Chicago*

Jonathan Bayless, *Husson University*

Mary Beard, *Kapiolani Community College*

Jaromir Becan, *University of Texas at San Antonio*

Rudy Beharrysingh, *Haywood Community College*

Curtis Bennett, *Loyola Marymount University*

David Berry, *Xavier University*

Dip Bhattacharya, *Clarion University of Pennsylvania*

Phil Blau, *Shawnee State University*

Tammy Blevins, *Hillsborough Community College*

Karen Blount, *Frederick Community College*

Russell Blyth, *Saint Louis University*

Mark Brenneman, *Mesa Community College*

Pat Brislin, *Millersville University of Pennsylvania*

Albert Bronstein, *University of Illinois at Chicago*

Joanne Brunner, *Joliet Junior College*

Corey Bruns, *University of Colorado at Boulder*

Stanislaw Buchcic, *Wright College*

Gerald Burton, *Virginia State University*

Dale Buske, *St. Cloud State University*

Jason Callahan, *St. Edward's University*

Jeff Cathrall, *Colorado University—Boulder*

Christine Cedzo, *Gannon University*

Jiang-Ping Chen, *St. Cloud State University*

Lee Chiang, *Trinity Washington University*

Madeline Chowdhury, *Mesa Community College*

Lars Christensen, *Texas Tech University*

Boyd Coan, *Norfolk State University*

Shelley Cook, *Stephen F. Austin State University*

Grace Coulombe, *Bates College*

Richard Coyne, *Arizona State University*

Tony Craig, *Paradise Valley Community College*

Kenneth Cramm, *Riverside City College*

Karen Crossin, *George Mason University*

Cheryl Cunnington, *North Idaho University*

Shari Davis, *Old Dominion University*

Megan Deeney, *Western Washington University*

Dennis DeJong, *Dordt College*

Sloan Despeaux, *Western Carolina University*

Mindy Diesslin, *Wright State University*

Qiang Dotzel, *University of Missouri—St. Louis*

William Drury, *University of Massachusetts Lowell*

Mark Ellis, *Central Piedmont Community College*

Solomon Emeghara, *Bloomfield College*

Scott Erway, *Itasca Community College*

Brooke Evans, *Metropolitan State College of Denver*

Cathy Evins, *Roosevelt University*

Scott Fallstrom, *University of Oregon*

Vanessa Farren, *Seneca College of Applied Arts and Technology*

Lauren Fern, *University of Montana*

Marc Formichella, *University of Colorado at Boulder*

Monique Fuguet, *University of Massachusetts—Boston*

Joe Gallegos, *Salt Lake City Community College*

Kevin Gammon, *Cumberland University*

Stephen Gendler, *Clarion University*

Antanas Gilvydis, *City Colleges of Chicago—Richard J. Daley College*

Paul Glenn, *Catholic University of America*

Cliona Golden, *Bard College*

Marilyn Grapin, *Metropolitan Center/Empire State College*

Charlotte Gregory, *Trinity College*

Susan Hagen, *Virginia Polytechnic and State University*

C. Hail, *Union University*

Susan Haller, *St. Cloud State University*

Richard Hammond, *St. Joseph's College*

David Hartz, *College of St. Benedict*

William Haver, *Virginia Commonwealth University*

Dan Henschell, *Douglas College*

Ann Herbst, *Santa Rosa Junior College*

Carla Hill, *Marist College*

Frederick Hoffman, *Florida Atlantic University*

Fran Hopf, *University of South Florida*

Mark Hunacek, *Iowa State University*

Jerry Ianni, *CUNY—La Guardia Community College*

Kelly Jackson, *Camden County College*

Nancy Jacqmin, *Carlow University*

Benny John, *University of Houston—Downtown*

Clarence Johnson, *Cleveland State University*

Terrence Jones, *Norfolk State University*

Richard Kampf, *Great Basin College*

Robert Keller, *Loras College*

Deborah Kent, *Hillsdale College*

Jane Kessler, *Quinnipiac University*

Margaret Kiehl, *Rensselaer Polytechnic Institute*

Sung Eun Kim, *Judson College*

Jared Knittel. *Texas State University*

Marshall Kotzen, *Worcester State University*

Kathryn Kozak, *Coconino County Community College*

Karole Kurnow, *Dominican College*

Latrice Laughlin, *University of Alaska—Fairbanks*

Namyong Lee, *Minnesota State University*

Raymond Lee, *University of North Carolina at Pembroke*

Richard Leedy, *Polk Community College*

Warren Lemerich, *Laramie City Community College*

Kathy Lewis, *California State University—Fullerton*

Michael Little Crow, *Scottsdale Community College*

Antonio Magliaro, *Quinnipiac University*

Iryna Mahlay, *Cleveland State University*

Antoinette Marquard, *Cleveland State University*

Jeannette Martin, *Washington State University*

John Martin, *Santa Rosa Junior College*

Christopher Mason, *Community College of Vermont*

Kathleen McDaniel, *Buena Vista University*

Karol McIntosh, *University of South Florida*

Farzana McRae, *Catholic University of America*

Ryan Melendez, *Arizona State University*

Phyllis Mellinger, *Hollins University*

Tammy Muhs, *University of Central Florida*

Ellen Mulqueeny, *Baldwin-Wallace College*

Judy Munshower, *Clarke University*

Bishnu Naraine, *St. Cloud State University*

Jennifer Natoli, *Manchester Community College*

Lars Neises, *Spokane Falls Community College*

Donald Neth, *Kent State University*

Cao Nguyen, *Central Piedmont Community College*

Stephen Nicoloff, *Paradise Valley Community College*

John Noonan, *Mount Vernon Nazarene University*

Douglas Norton, *Villanova University*

Jackie Nygaard, *Brigham Young University*

Jon Oaks, *Henry Ford Community College*

Michael Oppedisano, *Onondaga Community College*

Anne O'Shea, *North Shore Community College*

Andrew Pak, *Kapiolani Community College*

Kenny Palmer, *Tennessee Technological University*

Donna Passman, *Bristol Community College*

Laramie Paxton, *Coconino County Community College*

Shawn Peterson, *Texas State University*

Michael Petrie, *University of Northern Colorado*

Yevgeniy Ptukhin, *Texas Tech University*

Evelyn Pupplo-Cody, *Marshall University*

Parthasarathy Rajagopal, *Kent State University*

Crystal Ravenwood, *Whatcom Community College*

Carolynn Reed, *Austin Community College*

Rochelle Ring, *City College of New York*

Nancy Rivers, *Wake Technical Community College*

Leanne Robertson, *Seattle University*

Joe Rody, *Arizona State University*

Cosmin Roman, *The Ohio State University—Lima*

Tracy Romesser, *Erie Community College*

Martha Rooney, *University of Colorado at Boulder*

M.E. Rosar, *William Paterson University*

Anita Ross, *North Park University*

Katherine Safford-Ramus, *Saint Peter's College*

Matt Salomone, *Bridgewater State University*

Paula Savich, *Mayland Community College*

Susan Schmoyer, *Worcester State University*

Alison Schubert, *Wake Technical Community College*

Kathy Schultz, *Pensacola Junior College—Pensacola*

Sal Sciandra, *Niagara County Community College*

Brian Scott, *Cleveland State University*

Stacy Scudder, *Marshall University*

Tsvetanka Sendova, *Bennett College*

Sandra Sikorski, *Baldwin-Wallace College*

Laura Smithies, *Kent State University*

Mark Snavely, *Carthage College*

Lawrence Somer, *Catholic University of America*

Sandy Spears, *Jefferson Community and Technical College*

Dori Stanfield, *Davidson Community College*

Eryn Stehr, *Minnesota State University Mankato*

Kim Steinke, *College of the North Atlantic*

Donna Stevenson, *The State University of New York—Jefferson Community College*

Robert Storfer, *Florida International University*

Sarah Stovall, *Stephen F. Austin State University*

Joseph Tahsoh, *South Carolina State University*

Susan Talarico, *University of Wisconsin—Stevens Point*

Louis Talman, *Metropolitan State College of Denver*

Julie Theoret, *Johnson State College*

Jamie Thomas, *University of Wisconsin, Manitowoc*

Mike Tindall, *Seattle Pacific University*

Karl Ting, *Mission College*

Hansun To, *Worchester State*

Emilio Toro, *University of Tampa*

Preety Tripathi, *The State University of New York at Oswego*

David Tucker, *Midwestern State University*

Ulrike Vorhauer, *Kent State University*

Walter Wallis, *Southern Illinois University*

Tom Walsh, *Kean University*

Richard Watkins, *Tidewater Community College*

Leigh Ann Wells, *Western Kentucky University*

Ed Wesly, *Harrington College of Design*

Cheryl Whitelaw, *Southern Utah University*

Mary Wiest, *Minnesota State University*

David Wilson, *The State University of New York—College at Buffalo*

Antoine Wladina, *Farleigh Dickinson University*

Alma Wlazlinski, *McLennan Community College*

Jim Wolper, *Idaho State University*

Debra Wood, *University of Arizona*

Judith Wood, *Central Florida Community College*

Bruce Woodcock, *Central Washington University*

Jennifer Yuen, *Illinois Institute of Art*

Abbas Zadegan, *Florida International University*

Paulette Zizzo, *Wright State University—Main Campus*

Cathleen Zucco-Teveloff, *Rowan University*

Marc Zucker, *Marymount Manhattan College*

We wish to thank the dedicated and professional staff at W. H. Freeman who have taken such excellent care of our project. We owe a debt of gratitude to Ruth Baruth, who believed in us and this book from the very beginning; to Terri Ward, who guided this project through to completion; and to Kerry O'Shaughnessy and Susan Wein, who deftly and patiently oversaw the entire production process. Most of all, we are entirely in debt to Randi Blatt Rossignol, whose fine editorial eye, intelligence, humor, and wisdom utterly transformed our text, refining it over and over until all involved were finally satisfied. Our text is the beneficiary of her gentle but insistent ministrations.

Bruce Crauder

Benny Evans

Jerry Johnson

Alan Noell

QUANTITATIVE LITERACY

1

CRITICAL THINKING

There is no need to make a case for the importance of literacy in contemporary society. The ability to read with comprehension and the ability to communicate clearly are universally recognized as necessary skills for the average citizen. But there is another kind of literacy, one that may be just as important in a world filled with numbers, charts, diagrams, and even formulas: *quantitative literacy*.

In contemporary society we are confronted with quantitative issues at every turn. We are required to make choices and decisions about purchases, credit cards, savings plans, investments, and borrowing money. In addition, politicians and advertisers throw figures and graphs at us in an attempt to convince us to vote for them or buy their product. If we want to be informed citizens and savvy consumers, we must be quantitatively literate.

At the heart of quantitative literacy is critical thinking, which is necessary to deal intelligently with the information and misinformation that come to us via television, computers, newspapers, and magazines.

The following article from the Web site OttawaStart shows the importance that modern educators place on critical thinking.

IN THE NEWS 1.1

21st Century Skills
Rethinking How Students Learn
October 27, 2010

Who is the 21st Century Learner and why should we care? Dr. John Barell says we should care because we need to prepare our students for a rapidly evolving global and technological world.

Dr. John Barell, an educator, author, and professor emeritus of curriculum and teaching at Montclair State University will be in Ottawa on Thursday, November 4 to share his wisdom, his experiences, and his methods of developing and nurturing curious minds. His presentation is part of the OCDSB Free Speakers' Series— Parents and Educators Helping Students Learn.

Dr. Barell will discuss ways to re-envision learning and promote innovation through critical thinking, problem solving, collaboration, and technology integration, while building on mastery of core content and background knowledge.

Dr. Barell examines the daunting challenge today's educators face: how to provide students with the skills to thrive in the 21st century. Dr. Barell believes that both knowledge and skills are required, and they are mutually dependent. He contends that effective teaching involves having students use skills to acquire knowledge.

This chapter opens with a phenomenon called *Simpson's paradox*, which illustrates in a very interesting way the need for critical thinking. We next discuss basic logic, the use of Venn diagrams, and number sense.

1.1 Public policy and Simpson's paradox: Is "average" always average?

TAKE AWAY FROM THIS SECTION

View with a critical eye conclusions based on averages.

It is a common practice today to use standardized test scores as a benchmark for measuring performance. If a school's average test scores do not compare well to statewide averages, teachers and administrators might be criticized. Penalties of some sort might be imposed on the school as a result of state or federal laws such as the No Child Left Behind Act.

IN THE NEWS 1.2

Warning: Why Average Isn't Average!

Simplistic Statistics Can Be Misleading in Measuring for Accountability in Education

DOUGLAS E. HALL   June 1998

One of the proposed measures of achievement of individual schools and school districts includes the results of the state's assessment tests of 3rd, 6th, and 10th graders. As this paper demonstrates, it is possible for the average score of students in a school to be below the statewide average, yet the average for every sub-group of students in that same school to be above the statewide average for similar sub-groups. This can occur because the demographic composition of the student population of one school can vary considerably from the state average demographic composition for all schools.

Using statewide averages of test scores or other statistics as a benchmark for measuring performance of a school or a school district can be analytically misleading. The [New Hampshire] Center [for Public Policy Studies] raises a very strong caution in this regard. While the Center enthusiastically and unequivocally supports emphasizing results, we believe measuring, reporting and interpreting these results must be done carefully.

The New Hampshire Center for Public Policy Studies strongly cautions state policymakers about moving too rapidly into an overly simplistic reporting, analysis, and use of student achievement measures.

The preceding excerpt is from an article from the New Hampshire Center for Public Policy Studies in association with the Institute for Policy and Social Science Research, University of New Hampshire.

According to the example mentioned in the article, it may be that the average score of students in a school is below the statewide average, yet, when broken down into certain subgroups, the average for each subgroup is above the statewide average. This is an example of a general phenomenon called *Simpson's paradox*. It's called a paradox because such a phenomenon seems impossible, but we will now see that it can happen.

President George W. Bush signs the No Child Left Behind Act in 2002.

Simpson's paradox and test scores

The following example illustrates how the situation in the article at the beginning of this section can occur.

EXAMPLE 1.1 Calculating averages: Test scores

Suppose a certain high school gave a math proficiency exam to its students and that the percentage who passed was below the statewide average. After examining the figures further, the school decided to report its test data by separating them into students from low-income families and students from higher-income families. (This process is referred to as *disaggregating* the data.)

The following table gives the numbers of students from each category who took the exam and the number who passed:

	Local school		Statewide	
	Students tested	Passed	Students tested	Passed
Low income	400	260	200,000	128,000
High income	700	532	1,100,000	825,000
Total	1100	792	1,300,000	953,000

Show that the local school outperformed statewide students in both the low-income and high-income categories but had a lower passing percentage overall than the statewide rate. (If you need a quick review of percentages, consult the end of this section.)

SOLUTION

First we find the local school's pass percentage for low-income families:

$$\text{Local low-income average} = \frac{\text{Number passes}}{\text{Number tests taken}} \times 100\%$$

$$= \frac{260}{400} \times 100\%$$

$$= 65.0\%.$$

We calculate the remaining percentages in a similar fashion. The results are shown in the table below. (All of the figures are rounded to one decimal place.)

	Local school % pass	Statewide % pass
Low income	260/400 or 65.0%	128,000/200,000 or 64.0%
High income	532/700 or 76.0%	825,000/1,100,000 or 75.0%
Total	792/1100 or 72.0%	953,000/1,300,000 or 73.3%

Thus for both low-income and high-income families, the pass rate of local school students is one percentage point higher than the statewide pass rate. But overall the pass rate of the local school is *lower* than the statewide rate by more than one percentage point.

TRY IT YOURSELF 1.1

The same school had an English proficiency exam with the following results:

	Local school		Statewide	
	Students tested	Passed	Students tested	Passed
Low income	400	316	200,000	156,000
High income	700	595	1,100,000	924,000
Total	1100	911	1,300,000	1,080,000

Make a table showing the pass percentage for each category and the overall pass percentage. (Round each figure to one decimal place in percentage form.) Interpret your results.

The answer is provided at the end of the section.

It is easy to see why Simpson's paradox is called a paradox. Such a result seems impossible. But the example shows that combining, or aggregating, data can mask underlying patterns.

Even though this phenomenon is called a paradox, the results of the preceding example are not so surprising if we note that the local school has a much higher percentage of students from low-income families (400 out of 1100 or about 36%) than does the state as a whole (200,000 out of 1,300,000 or about 15%). This is an important example of the type of issue that must be considered when comparing school districts with different demographics.

The Berkeley gender discrimination case

A famous but relatively simple case provides a classic illustration of Simpson's paradox. Data from a 1973 study seemed to give persuasive evidence that the University of California at Berkeley was practicing gender discrimination in graduate school admissions. The table below shows that Berkeley accepted 44.0% of male applicants but only 35.0% of female applicants. (All of the percentages in this table and the next are rounded to one decimal place.)

	Applicants	Accepted	% accepted
Male	8442	3714	44.0%
Female	4321	1512	35.0%

Students on the University of California at Berkeley campus.

These numbers may seem to be pretty conclusive evidence of discrimination. When we look more closely, however, we find that graduate school admissions are based on departmental selections. There are many departments at Berkeley, but to keep things simple, we will assume that the university consists of only two departments: Math and English. Here is how the male and female applicants might have been divided among these two departments.

	Males			Females		
	Applicants	Accepted	% accepted	Applicants	Accepted	% accepted
Math	2000	500	25.0%	3000	780	26.0%
English	6442	3214	49.9%	1321	732	55.4%
Total	8442	3714	44.0%	4321	1512	35.0%

Note that each of the two departments actually accepted a larger percentage of female applicants than male applicants: The Math Department accepted 25.0% of males and 26.0% of females, and the English Department accepted 49.9% of males and 55.4% of females. The explanation of why a larger percentage of men overall were accepted lies in the fact that the Math Department accepts significantly fewer applicants than does English, both for men (25.0% compared to 49.9%) and for women (26.0% compared to 55.4%). Most women applied to the Math Department, where it is more difficult to be accepted, but most men applied to English, where it is less difficult.

A similar type of mind-bending phenomenon occurs in sports, as the next example illustrates.[1]

EXAMPLE 1.2 Comparing averages: Batting records

The following table shows the hitting records of two major league baseball players, Derek Jeter and David Justice, in 1995 and 1996:

	1995		1996	
	At-bats	Hits	At-bats	Hits
Jeter	48	12	582	183
Justice	411	104	140	45

We calculate the batting average for a baseball player by dividing the number of hits by the number of at-bats. Which batter had the higher average in 1995? Which had the higher average in 1996? Which had the higher average over the two-year period? Round all figures to three decimal places.

SOLUTION

We calculate Jeter's average in 1995 as follows:

$$\text{Average in 1995} = \frac{\text{Hits}}{\text{At-bats}}$$
$$= \frac{12}{48}$$
$$= 0.250.$$

We calculate averages for Jeter and Justice over each of the two years in a similar fashion. The results are shown in the table below.

	1995	1996
Average for Jeter	12/48 or 0.250	183/582 or 0.314
Average for Justice	104/411 or 0.253	45/140 or 0.321

Derek Jeter at bat.

---

[1] This example is drawn from Ken Ross, *A Mathematician at the Ballpark* (New York: Pi Press, 2004).

Now we find the two-year averages. Over the two-year period, Derek Jeter got a total of $12 + 183 = 195$ hits out of $48 + 582 = 630$ trips to the plate. Therefore,

$$\text{Jeter's average over two-year period} = \frac{\text{Hits}}{\text{At-bats}}$$
$$= \frac{195}{630}.$$

This is about 0.310.

A similar calculation for David Justice gives his average over the two-year period as 149/551 or about 0.270. Hence, Justice had the higher batting average in each of the two years, but Jeter had the higher batting average over the two-year period.

TRY IT YOURSELF 1.2

In 1997 Derek Jeter got 190 hits out of 654 at-bats, and David Justice had 163 hits out of 495 at-bats. Which batter had the better average in 1997? Which batter had the better average over the three-year period from 1995 through 1997? Round all figures to three decimal places.

The answer is provided at the end of the section.

As we have seen in the other examples in this section, the explanation for this seemingly paradoxical situation with batting averages lies in the absolute numbers involved in calculating the percentages. Jeter's relatively low average of 0.250 occurred in 1995 when he had only 48 at-bats. He was at the plate 582 times in 1996, when his average was much higher. Thus, his batting average over the two years was affected more by the 1996 numbers because that was when he was most often at bat. A similar analysis applies to David Justice, who was at bat many more times in 1995 than in 1996. We will allow baseball enthusiasts to argue over who is really the better hitter, but Simpson's paradox makes the use of overall batting averages alone somewhat suspect in the debate.

The examples of Simpson's paradox presented here show the importance of critical thinking. The use of a single number such as an average to describe complex data can be perilous. Further, combining data can mask underlying patterns when some factor distorts the overall picture but not the underlying patterns. To understand information presented to us, it is important for us to know at least the basics of how the information is produced and what it means.

> **SUMMARY 1.1   Simpson's Paradox**
>
> Simpson's paradox occurs when combining, or aggregating, data masks underlying patterns. Often this occurs when a factor distorts the overall picture, but this distortion goes away when the underlying data are examined. For example, separating test scores by the economic level of the students may show that at a local school students at each economic level perform better than the statewide average for students at the same level. But if the school has more students at lower economic levels, its test scores overall may be lower than the state average. Such results can be counterintuitive. Without careful consideration, one can be led to an incorrect conclusion.

The following Quick Review is provided as a refresher for doing percentage calculations.

> ### Quick Review Calculating Percentages
>
> To calculate $p\%$ of a quantity, we multiply the quantity by $\dfrac{p}{100}$:
>
> $$p\% \text{ of Quantity} = \frac{p}{100} \times \text{Quantity}.$$
>
> For example, to find 45% of 500, we multiply 500 by $\dfrac{45}{100}$:
>
> $$45\% \text{ of } 500 = \frac{45}{100} \times 500 = 225.$$
>
> Therefore, 45% of 500 is 225.
>
> To find what percentage of a whole a part is, we use
>
> $$\text{Percentage} = \frac{\text{Part}}{\text{Whole}} \times 100\%.$$
>
> For example, to find what percentage of 140 is 35, we proceed as follows:
>
> $$\begin{aligned}
> \text{Percentage} &= \frac{\text{Part}}{\text{Whole}} \times 100\% \\
> &= \frac{35}{140} \times 100\% \\
> &= 0.25 \times 100\% \\
> &= 25\%.
> \end{aligned}$$
>
> Thus, 35 is 25% of 140.

Try It Yourself answers

Try It Yourself 1.1: Calculating averages: Test scores

	Local school % pass	Statewide % pass
Low income	79.0%	78.0%
High income	85.0%	84.0%
Total	82.8%	83.1%

For both low-income and high-income families, the pass rate of local school students is one percentage point higher than the statewide pass rate. But overall the pass rate of the local school is slightly lower than the statewide rate.

Try It Yourself 1.2: Comparing averages: Batting averages   In 1997 Jeter's average was 0.291. Justice had the higher average of 0.329. Over the three-year period, Justice had an average of 0.298. Jeter had the higher average of 0.300.

Exercise Set 1.1

In these exercises, round all answers in percentage form to one decimal place unless you are instructed otherwise.

Which hospital would you choose?   The local newspaper examined a town's two hospitals and found that, over the last six years, at Mercy Hospital 79.0% of the patients survived and at County Hospital 90.0% survived. The following table summarizes the findings.

	Lived	Died	Total	% who lived
Mercy Hospital	790	210	1000	79%
County Hospital	900	100	1000	90%

Exercises 1 through 4 refer to the information in this table.

1. **Patients in fair condition.** Patients were categorized upon admission as being in fair condition (or better) versus being in worse than fair condition. When the survival numbers were examined for these groups, the following table emerged for patients admitted in fair condition:

Patients admitted in fair condition (or better)				
	Lived	Died	Total	% who lived
Mercy Hospital	580	10	590	
County Hospital	860	30	890	

Fill in the last column for this table.

2. **Patients in worse than fair condition.** Patients were categorized upon admission as being in fair condition (or better) versus being in worse than fair condition. When the survival numbers were examined for these groups, the following table emerged for patients admitted in worse than fair condition:

Patients admitted in worse than fair condition				
	Lived	Died	Total	% who lived
Mercy Hospital	210	200	410	
County Hospital	40	70	110	

Complete the last column for this table.

3. **The head of County Hospital.** *This exercise uses the results of Exercises 1 and 2.* The head of County Hospital argued that, based on the overall figures, her hospital came out well ahead of Mercy by 90.0% to 79.0% and therefore is doing a far better job. If you were head of Mercy Hospital, how would you respond?

4. **A sick child.** *This exercise uses the results of Exercises 1 and 2.* You are a parent trying to decide to which hospital to send your sick child. On the basis of the three tables above, which hospital would you choose? Explain.

Hiring   In a recent hiring period, a hypothetical department store, U-Mart, hired 62.0% of the males who applied and 14.0% of the females. A lawsuit was contemplated because these numbers seemed to indicate that there was gender discrimination. On closer examination, it was found that U-Mart's hiring was only for two of the store's departments: hardware and ladies' apparel. The hardware department hired 60 out of 80 male applicants and 15 out of 20 female applicants. The ladies' apparel department hired 2 out of 20 male applicants and 30 out of 300 female applicants. This information is summarized in the accompanying table.

	Males applying	Males hired	Females applying	Females hired
Hardware	80	60	20	15
Ladies' apparel	20	2	300	30
Total	100	62	320	45

This information is used in Exercises 5 through 11.

5. **Male applicants in hardware.** For the hardware department, what was the percentage of male applicants hired?

6. **Female applicants in hardware.** For the hardware department, what was the percentage of female applicants hired?

7. **Males in ladies' apparel.** For the ladies' apparel department, what was the percentage of male applicants hired?

8. **Females in ladies' apparel.** For the ladies' apparel department, what was the percentage of female applicants hired?

9. **Attorney for plaintiff.** You are an attorney for a female plaintiff. How would you argue that there is gender discrimination?

10. **Attorney for U-Mart.** You are an attorney for U-Mart. How would you argue that there is no gender discrimination based on the results of Exercises 5 through 8?

11. **You decide.** How would you vote if you were on the jury based on the results of Exercises 5 through 8?

Success and poverty   The success rates of two schools were compared in two categories: students from families above the poverty line and students from families below the poverty line. The results are in the tables below.

Students from families above the poverty line				
	Passed	Failed	Total	% passing
School A	400	77	477	
School B	811	190	1001	

Students from families below the poverty line				
	Passed	Failed	Total	% passing
School A	199	180	379	
School B	38	67	105	

This information is used in Exercises 12 through 20.

12. **Students above the poverty line.** Complete the table for students above the poverty line.

13. **Students below the poverty line.** Complete the table for students below the poverty line.

14. What percentage of students from School A are from families above the poverty line?

15. What percentage of students from School B are from families above the poverty line?

16. What percentage of students from families below the poverty line in School A passed?

17. What percentage of students from families below the poverty line in School B passed?

18. All students. Complete the following table:

All students in both groups				
	Passed	Failed	Total	% passing
School A				
School B				

19. The principal. The principal of School B argued that, based on the overall figures in the table from Exercise 18, her school came out ahead and therefore is doing a better job than School A. How would you respond?

20. A parent. You are a parent trying to decide to which school to send your child. On the basis of the tables in Exercises 12, 13, and 18, which school would you choose? Explain.

Children in South Africa In 1990 scientists initiated a study of the development of children in the greater Johannesburg/Soweto metropolitan area of South Africa. They collected information on certain children at birth, and five years later they invited these children and their caregivers to be interviewed about health-related issues. Relatively few participated in the follow-up interviews, and the researchers wanted to know whether those who participated had similar characteristics to those who didn't. In the tables below, these are referred to as the *five-year group* and the *children not traced*, respectively. One factor used to compare these groups was whether the mother had health insurance at the time the child was born. Here are the results for this factor, including only participants who were either white or black.

All participants				
	Had insurance	No insurance	Total	% with insurance
Children not traced	195	979		
Five-year group	46	370		

Exercises 21 through 26 refer to this information.

21. Fill in the blanks in the table above. Round all percentages to whole numbers.

22. In the above-cited summary of the study, the author notes, "If the five-year sample is to be used to draw conclusions about the entire birth cohort, the five-year group should have characteristics similar to those who were not traced from the initial

group." Are the percentages of those with insurance similar between the two groups in the table? If not, what conclusion would you draw on the basis of the quotation above?

23. Here is the table showing the results for whites. Complete the table. Round all percentages to whole numbers.

White participants				
	Had insurance	No insurance	Total	% with insurance
Children not traced	104	22		
Five-year group	10	2		

24. Here is the table showing the results for blacks. Complete the table. Round all percentages to whole numbers.

Black participants				
	Had insurance	No insurance	Total	% with insurance
Children not traced	91	957		
Five-year group	36	368		

25. For whites, compare the percentages having insurance between the two groups: children not traced and the five-year group. For blacks, compare the percentages having insurance between the two groups: children not traced and the five-year group.

26. On the basis of your answer to Exercise 25, revisit the question at the end of Exercise 22. Do the tables for whites and blacks provide any evidence that the five-year sample cannot be used to draw conclusions about the entire group? *Note*: The author of the summary observed that in South Africa whites generally were much more likely to have health insurance than blacks. Further, relatively few whites were included in the original study. As a result, there were relatively few in the first table above who had health insurance. Another factor in the paradoxical outcome is that, in contrast to blacks, relatively few of the whites who were initially involved participated in the interviews and were included in the five-year group.

27. History of Simpson's paradox. Research the history of Simpson's paradox and describe some of the early examples given.

---

[2] Christopher H. Morrell, "Simpson's Paradox: An Example from a Longitudinal Study in South Africa," *Journal of Statistics Education* 7 (1999).

1.2 Logic and informal fallacies: Does that argument hold water?

Voters must judge the quality of arguments in political debates. For example, in 2010 the people of Illinois were challenged to sort truth from fiction in the gubernatorial debate reported in the following article (from the Web site of Chicago Public Radio).

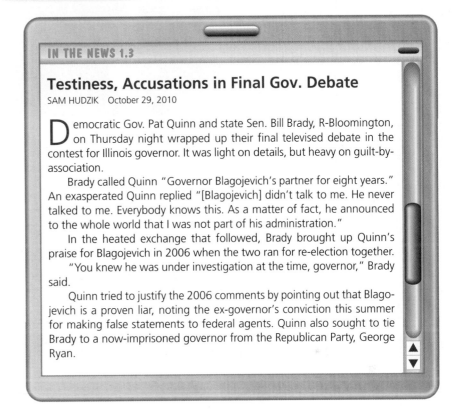

IN THE NEWS 1.3

Testiness, Accusations in Final Gov. Debate

SAM HUDZIK   October 29, 2010

Democratic Gov. Pat Quinn and state Sen. Bill Brady, R-Bloomington, on Thursday night wrapped up their final televised debate in the contest for Illinois governor. It was light on details, but heavy on guilt-by-association.

Brady called Quinn "Governor Blagojevich's partner for eight years." An exasperated Quinn replied "[Blagojevich] didn't talk to me. He never talked to me. Everybody knows this. As a matter of fact, he announced to the whole world that I was not part of his administration."

In the heated exchange that followed, Brady brought up Quinn's praise for Blagojevich in 2006 when the two ran for re-election together.

"You knew he was under investigation at the time, governor," Brady said.

Quinn tried to justify the 2006 comments by pointing out that Blagojevich is a proven liar, noting the ex-governor's conviction this summer for making false statements to federal agents. Quinn also sought to tie Brady to a now-imprisoned governor from the Republican Party, George Ryan.

Political debates and commercials are often delivered in sound bites to evoke an emotional reaction. Critical examination of what these "pitches" actually say can reveal exaggerations, half-truths, or outright fallacies.

In the face of such heated rhetoric and tenuous logic, we appeal to an older source, one that remains relevant today.

Quote from Thomas Jefferson

In a republican nation, whose citizens are to be led by reason and persuasion and not by force, the art of reasoning becomes of first importance.

In this section, we begin a study of logic and examine some informal logical fallacies.

Logical arguments

Key Concept

Logic is the study of methods and principles used to distinguish good (correct) from bad (incorrect) reasoning.

2008 presidential debate between John McCain and Barack Obama.

A logical argument involves one or more *premises* (or *hypotheses*) and a *conclusion*. The premises are used to justify the conclusion. The argument is *valid* if the premises properly support the conclusion. Here is an example.

Premises: (1) *Smoking increases the risk of heart disease*, and (2) *My grandfather is a smoker.*

Conclusion: *My grandfather is at increased risk of heart disease.*

This is a valid argument because the premises justify the conclusion. In determining whether an argument is valid, we do not question the premises. An argument is valid if the conclusion follows from the assumption that the premises are true. The truth or falsity of the premises is a separate issue. Nor is it simply the fact that the conclusion is true that makes an argument valid—the important thing is that the premises support the conclusion.

"The child is very illogical."

Consider, for example, the following invalid argument:

Premises: (1) *My grandfather smoked*, and (2) *he suffered from heart disease.*

Conclusion: *Therefore, smoking is a contributing factor to heart disease.*

Here, the conclusion may indeed be true, but the case of a single smoker is not sufficient evidence to draw the conclusion. Even if it were, this example does not show that smoking contributed to the heart disease—it is established only that the two events of smoking and having heart disease both occurred.

Key Concept

A logical argument consists of **premises** (also called **hypotheses**) and a **conclusion**. The premises are assumptions that we accept as a starting point. The argument is **valid** if the premises justify the conclusion.

EXAMPLE 1.3 Identifying parts of an argument: Wizards and beards

Identify the premises and conclusion of the following argument. Is this argument valid?

All wizards have white beards. Gandalf is a wizard. Therefore, Gandalf has a white beard.

SOLUTION

The premises are (1) *All wizards have white beards* and (2) *Gandalf is a wizard*. The conclusion is *Gandalf has a white beard*. This is a valid argument because the premises support the conclusion: If in fact all wizards have white beards and Gandalf is a wizard, it must be the case that Gandalf has a white beard.

TRY IT YOURSELF 1.3

Identify the premises and conclusion of the following argument. Is this argument valid?

All wizards have white beards. Gandalf has a white beard. Therefore, Gandalf is a wizard.

The answer is provided at the end of the section.

The term *fallacy*, as used by logicians, refers to an argument that may on the surface seem to be correct but is in fact incorrect. There are many types of fallacies. Some are easy to spot, and others are more subtle. It has been suggested that it is impossible to list all the ways in which incorrect inferences can be constructed. Fallacies fall into two general categories, however: informal and formal.

Key Concept

An **informal fallacy** is a fallacy that arises from the content of an argument, not its form or structure. The argument is incorrect because of *what* is said, not *how* it is said. A **formal fallacy** arises in the form or structure of an argument. The fallacy is independent of the content of the argument.

In this section, we focus on informal fallacies. We reserve the treatment of formal fallacies for the next section. Our discussion generally follows the development of Copi's classic work[3] but includes other sources as well. We divide informal fallacies into two categories: those mistakes in reasoning that arise from appeals to irrelevant factors and those that arise from unsupported assumptions.

---

[3]Irving M. Copi and Carl Cohen, *Introduction to Logic*, 13th ed. (Upper Saddle River, NJ: Prentice-Hall, 2008).

Fallacies of relevance

In fallacies of relevance, the premises are logically irrelevant to, and hence incapable of establishing the truth of, their conclusions. We offer several examples, but this collection is by no means exhaustive.

Appeal to ignorance   This type of fallacy occurs whenever it is argued that a statement is true or not true simply on the basis of a lack of proof. This is a fallacy of relevance because the lack of proof of a statement is not by itself relevant to the truth or falsity of that statement. The structure of this fallacy is as follows:

- A certain statement is unproven.

- Therefore, the statement must be false.

Example: *For over 75 years people have tried and failed to show that aliens have not visited Earth. So we must finally accept the fact that at least some of the UFO reports are based on actual alien visits.*

In this example, the inability to disprove alien visits is not relevant to the existence of such visits. Perhaps it would help to consider the opposite argument: *For over 75 years people have not been able to show that aliens have visited Earth. So we must finally accept the fact that no such visits have ever occurred.* The same set of facts is used to establish both that aliens have visited Earth and that they have not. The point is that an absence of proof cannot establish either that aliens have visited or that they have not.

Dismissal based on personal attack   This type of fallacy occurs when we simply attack the person making an assertion instead of trying to prove or disprove the assertion itself. This is a fallacy of relevance because the character of the person making a claim is not necessarily relevant to the claim being made. This fallacy has the following structure:

- A person presents an argument or point of view.

- The character of that person is brought into question.

- Based on the character attack, it is concluded that the argument or point of view is incorrect.

Example: *My political opponent is against government-funded health care, and she also has a reputation for being heartless. She refused to seek medical treatment for her own father when he was ill. So government-funded health care is a good idea.*

In this argument, the conclusion is that a certain health-care proposal is good, but no information is provided regarding the merits of the proposal itself. Rather, the argument is that the proposal must be good because of the claim that one of its opponents is suspect. Even if that accusation is true, it does not necessarily follow that the proposal is good. We cannot evaluate the proposal without information about what it is.

False authority   This fallacy arises whenever the validity of an argument is accepted based solely on the fact that it is supported by someone, but that person's credentials or expertise are not relevant to the argument. This is clearly a fallacy of relevance. This fallacy has the following structure:

- A person makes a claim based on his or her authority.

- The claim is outside the scope of that person's authority.

- The truth of the argument is concluded based on the authority of the claimant.

Example: *Over the past few years, I have starred in a number of the most popular movies in America, so I can assure you that Johnson and Johnson's new anti-nausea drug is medically safe and effective.*

The fallacy here is obvious—starring in movies in no way qualifies anyone to judge the efficacy of any medical product. The prevalence of commercials where sports figures or movie stars tell us which clothes to buy, which foods to eat, and which candidates to vote for speaks volumes about the effectiveness of such fallacious arguments.

Of course, *appropriate* authority does indeed carry logical weight. For example, the authority of Stephen Hawking lends credence to a claim regarding physics because he is a world-renowned physicist. On the other hand, his authority would probably add nothing to a toothpaste commercial.

Straw man   This fallacy occurs whenever an argument or position is dismissed based on the refutation of a distorted or completely different argument or position (the straw man)—presumably one that is easier to refute. This is a fallacy of relevance because the refutation does not even address the original position. The structure of this fallacy is as follows:

- A position or point of view is presented.

- The case for dismissing a distorted or *different* position or point of view (the straw man) is offered.

- The original position is dismissed on the basis of the refutation of the straw-man position.

Example: *A group of my fellow senators is proposing a cut in military expenditures. I cannot support such a cut because leaving our country defenseless in these troubled times is just not acceptable to me.*

The U.S.S. Ronald Reagan.

The "straw man" here is the proposition that our military defense program should be dismantled. It is easy to get agreement that this is a bad idea. The fallacy is that no such proposal was made. The proposal was to cut military expenditures, not to dismantle the military.

Political speech is often full of this type of blatant fallacy. Rather than engaging in honest, reasoned debate of the issues, we often see distortions of an opponent's position (the straw man) refuted. The politicians who do this are trying to appeal to emotion rather than reason. Critical thinking is our best defense against such tactics.

Appeal to common practice   This fallacy occurs when we justify a position based on its popularity. It is a fallacy of relevance because common practice does not imply correctness. This fallacy has the following structure:

- The claim is offered that a position is popular.

- The validity of the claim is based on its popularity.

Example: *It is OK to cheat on your income tax because everybody does it.*

Of course, it isn't true that everybody cheats on their income taxes, but even if it were, that is not justification for cheating. (That argument certainly will not carry much weight with the IRS if you're audited.) This type of argument is often used in advertising.

EXAMPLE 1.4 Identifying types of fallacies of relevance: Two fallacies

Identify the type of each of the following fallacies of relevance.

1. *I have scoured the library for information on witchcraft. I cannot find a single source that proves that anyone accused of witchcraft was actually capable of anything magical. This shows that there is no such thing as a real witch.*

2. *The popular actress Jane Fonda condemned America's involvement in the Vietnam War. So we know that that conflict was a civil war in which America should never have been involved.*

SOLUTION

In the first argument, we rely on a lack of proof to draw the conclusion. This is appeal to ignorance.

The second argument involves false authority. The fact that Jane Fonda was a popular actress does not lend authority to her opinion about the war.

TRY IT YOURSELF 1.4

Identify the fallacy of relevance in the following argument:

I don't see why I should vote in the election. Quite often less than half of the adult population casts a vote.

The answer is provided at the end of the section.

Fallacies of presumption

In fallacies of presumption, false or misleading assumptions are either tacitly or explicitly assumed, and these assumptions are the basis of the conclusion.

False dilemma This fallacy occurs when a conclusion is based on an inaccurate or incomplete list of choices. This is a fallacy of presumption because the (incorrect) assumption is being made that the list offered is complete. This fallacy has the following structure:

• An incomplete or inaccurate list of consequences of not accepting an argument is presented.

• A conclusion is drawn based on the best (or least bad) of these consequences.

Example: *You'd better buy this car or your wife will have to walk to work and your kids will have to walk to school. I know you don't want to inconvenience your family, so let's start the paperwork on the automobile purchase.*

The salesman is telling us we have two choices, "Buy this car or inconvenience your family." In fact, we probably have more choices than the two offered by the salesman. There may be a less expensive car that will serve my family's needs, or perhaps public transportation is an option in my neighborhood. The dilemma presumed by the salesman is not necessarily true. The conclusion that we should purchase the car is based on choosing the best of the options offered by the salesman, but in fact there are additional choices available.

An argument of this sort is valid only if the list of choices is indeed complete. This type of argument is commonly presented by public officials: *We must either increase your taxes or cut spending on education. So you must support an increase in taxes.*

False cause The fallacy of false cause occurs whenever we conclude a causal relationship based solely on common occurrence. It is an error of presumption because we are tacitly assuming that a causal relationship must exist. The structure of this fallacy is as follows:

- Two events occur together, or one follows the other.

- The fact that the events are related is used to conclude that one *causes* the other.

Example: *Studies have shown that many people on a high-carbohydrate diet lose weight. Therefore, a high-carbohydrate diet leads to weight loss.*

The premise of this argument is that the two events of losing weight and eating a high-carbohydrate diet sometimes occur together. The fallacy is that if two events occur together, then one must cause the other. It is just as reasonable to conclude that the weight loss causes the high-carbohydrate diet!

Consider this example: *Each morning my dog barks, and there are never any elephants in my yard. Therefore, my dog is keeping the elephants away from my home.* It is probably true that the dog barks and that there are no elephants in the yard, but these facts have nothing to do with each other.

When two events occur together, there are three distinct possibilities:

1. *There is no causal relationship.* Example: My dog barks and there are no elephants in my yard.

2. *One event causes the other.* Example: Ice on the sidewalk causes the sidewalk to be slippery.

3. *Both events are caused by something else.* Example: A mother observes that her child often has a fever and a runny nose at the same time. She doesn't conclude that the runny nose causes the fever. Instead, she concludes that the child has a cold, which causes both symptoms.

In general, causality is a difficult thing to establish, and we should view such claims skeptically. We will return to the topic of causality when we discuss *correlation* in Section 4 of Chapter 6.

Circular reasoning or begging the question This fallacy occurs when the repetition of a position is offered as its justification. This is a fallacy of presumption because it is assumed that a statement can prove itself. The structure of this fallacy is as follows:

- A position or argument is offered.

- The position or argument is concluded to be true based on a restatement of the position.

Example: *Establishing government-run health insurance would be a mistake, because it is just flat wrong to do it.*

This argument says essentially that government-run health insurance is wrong because it is wrong. Sometimes circular arguments may be longer and more convoluted than the simple example presented here and therefore more difficult to spot.

Another example is this: *I have listed Mary as a reference—she can vouch for me. Tom will verify her trustworthiness, and I can vouch for Tom.*

Hasty generalization The fallacy of hasty generalization occurs when a conclusion is drawn based on a few examples that may be atypical. This is a fallacy of presumption because we are assuming that the few cases considered are typical of all cases. This fallacy has the following form:

- A statement is true in several cases that may be atypical.

- The conclusion that it is generally true or always true is drawn based on the few examples.

President Obama signing the Patient Protection and Affordable Health Care Act on March 23, 2010.

Example: *I know the quarterback, the tight end, and the center on our football team, and all three are excellent students. The athletes at our university do not shirk their scholarly duties.*

Here we are making a judgment about athletes at the university based on the good qualities of three football players. The exemplary behavior of these three athletes tells us nothing about how the remaining athletes are behaving.

Finding three athletes who are scholars does not mean that all athletes are, but these examples might prompt us to explore further. Examples cannot prove the general case, but they can help us formulate ideas that may merit further investigation. For example, I might be very pleased to learn that three of my daughter's friends are honor students. Perhaps she is wise in her choice of friends. It is certainly worth further attention.

SUMMARY 1.2   Some Common Informal Fallacies

Appeal to ignorance   A statement is either accepted or rejected because of a lack of proof.

Dismissal based on personal attack   An argument is dismissed based on an attack on the proponent rather than on its merits.

False authority   The validity of a claim is accepted based on an authority whose expertise is irrelevant.

Straw man   A position is dismissed based on the rejection of a distorted or different position (the "straw man").

Appeal to common practice   An argument for a practice is based on the popularity of that practice.

False dilemma   A conclusion is based on an inaccurate or incomplete list of alternatives.

False cause   A causal relationship is concluded based on the fact that two events occur together or that one follows the other.

Circular reasoning   This fallacy simply draws a conclusion that is really a restatement of the premise.

Hasty generalization   This fallacy occurs when a conclusion is drawn based on a few examples that may be atypical.

There are other types of fallacies. Some of these will be examined in the exercises.

EXAMPLE 1.5 Identifying types of informal fallacies: Two fallacies

Classify each of the following fallacies:

1. *He says we should vote in favor of lowering the sales tax, but he has a criminal record. So I think decreasing the sales tax is a bad idea.*

2. *My dad is a professor of physics, and he says Dobermans make better watchdogs than collies.*

SOLUTION

In the first argument, we dismiss a position based on a personal attack. In the second, knowledge of physics does not offer qualification for judging dogs. This is a use of false authority.

TRY IT YOURSELF 1.5

Identify the type of fallacy illustrated in the following argument. *Do you want to buy this MP3 player or do you want to do without music?*

The answer is provided at the end of the section.

Inductive reasoning and pattern recognition

The fallacies we have looked at are usually flawed arguments in which the premises are *supposed* to provide conclusive grounds for accepting the conclusion, but in fact they do not. An argument with premises and a conclusion is called a *deductive* argument because it *deduces* a specific conclusion from the given premises. The classic example is:

Premises: (1) *All men are mortal* and (2) *Socrates is a man.*

Conclusion: *Therefore, Socrates is mortal.*

In a valid deductive argument, the premises provide *conclusive* grounds for accepting the conclusion.

An *inductive* argument, unlike a deductive argument, draws a general conclusion from specific examples. It does not claim that the premises give irrefutable proof of the conclusion (that would constitute the fallacy of hasty generalization), only that the premises provide plausible support for it.

An inductive argument related to the deductive argument above is as follows:

Premise: Socrates is mortal. Plato is mortal. Aristotle is mortal. Archimedes is mortal.

Inductive conclusion: Therefore, all men are mortal.

In this argument, the conclusion that all men are mortal is based on knowledge of a few individuals. A proper deductive argument can establish a conclusion with certainty. But an inductive argument draws a conclusion based on only partial evidence in support of it.

Key Concept

A deductive argument draws a conclusion from premises based on logic. An inductive argument draws a conclusion from specific examples. The premises provide only partial evidence for the conclusion.

In some cases inductive arguments are extremely convincing. Take this one for example:

Premise: Every human being who has ever lived has eventually died.

Inductive conclusion: All human beings are mortal.

Because the evidence is so overwhelming, some people might say that the conclusion is deduced from the premise. Technically speaking, however, this is still an inductive argument. The conclusion is based on evidence, not on deductive logic.

Here's another inductive argument with lots of evidence:

Premise: Every college football player I have ever seen is male.

Inductive conclusion: All college football players are male.

This argument may sound convincing, but in 2002 Katie Hnida at the University of New Mexico became the first woman to play in an NCAA Division I-A football game.

The weight of an inductive argument depends to a great extent on just how much evidence is in the premises. Such argumentation also has value in that it may suggest further investigation. In mathematics, this is often how we discover things—we observe patterns displayed by specific examples. This may lead us to propose a more general result, but in mathematics no such result is accepted as true until a proper deductive argument is found.

To show what we mean, let's look at the following table, which shows the squares of the first few odd integers:

Katie Hnida kicking a field goal.

Odd integer	1	3	5	7	9
Odd integer squared	1	9	25	49	81

Note that each of the squares turns out to be odd. Reasoning inductively, we propose that the square of *every* odd integer is odd. Note that the pattern in the table lends credence to this claim, but the table does not prove the claim. The table does not, for example, guarantee that the square of 163 is an odd integer. Mathematicians are indeed able to give a (relatively simple) deductive proof of this proposal. Thus, the pattern that we saw in the table is, in fact, true for all odd integers.

Inductive reasoning of this type can occasionally lead us astray. Recall that a prime number is a number (other than 1) that has no divisors other than itself and 1. So 7 is prime because its only divisors are 1 and 7. On the other hand, 15 is not prime because both 3 and 5 are divisors of 15. It has divisors other than 1 and 15. Let's consider the expression $n^2 + n + 41$. The table below shows some values for this expression.

n	1	2	3	4	5	6	7
$n^2 + n + 41$	43	47	53	61	71	83	97

After we observe that 41, 43, 47, 53, 61, 71, 83, and 97 are all prime, we might be led to propose that the expression $n^2 + n + 41$ always gives a prime number. If you explore further, you will find that the result is prime for $n = 1$ up through $n = 39$. But if we put in $n = 40$, the result is not prime:

$$40^2 + 40 + 41 = 1681 = 41 \times 41.$$

In this case, the pattern we observed in the table is not true for all numbers. Inductive reasoning can lead us to important discoveries. But as we have seen, it can sometimes lead us to wrong conclusions.

The idea of an inductive argument is closely tied to pattern recognition. We explore this connection further in the following example.

EXAMPLE 1.6 Pattern recognition: Inductive logic

A certain organism reproduces by cell division. The following table shows the number of cells observed to be present over the first few hours:

Hours	0	1	2	3	4	5	6
Number of cells	1	2	4	8	16	32	64

Describe the pattern shown by the table and suggest a general rule for finding the number of cells in terms of the number of hours elapsed.

SOLUTION

Looking closely at the table, we see a pattern. The number of cells present is always a power of 2.

Hours	0	1	2	3	4	5	6
Number of cells	$1 = 2^0$	$2 = 2^1$	$4 = 2^2$	$8 = 2^3$	$16 = 2^4$	$32 = 2^5$	$64 = 2^6$

The pattern suggests that, in general, we find the number of cells present by raising 2 to the power of the number of hours elapsed.

TRY IT YOURSELF 1.6

An Internet guide gives a table showing how much fertilizer I need to cover a yard with given square footage. Measurements are in square feet.

Yard size	100 sq ft	400 sq ft	900 sq ft	1600 sq ft	2500 sq ft
Pounds needed	1	4	9	16	25

Use this table to determine a pattern that leads to a general relationship between yard size and fertilizer needed.

The answer is provided at the end of the section.

Try It Yourself answers

Try It Yourself 1.3: Identifying parts of an argument: Wizards and beards   The premises are (1) *All wizards have white beards* and (2) *Gandalf has a white beard.* The conclusion is *Gandalf is a wizard.* This argument is not valid.

Try It Yourself 1.4: Identifying types of fallacies of relevance: Types of fallacies This is an appeal to common practice.

Try It Yourself 1.5: Identifying types of informal fallacies: Types of fallacies   This is a false dilemma.

Try It Yourself 1.6: Pattern recognition: Fertilizer   To find the number of pounds needed, we divide the yard size in square feet by 100.

Exercise Set 1.2

Premises and conclusion   In Exercises 1 through 7, identify the premises and conclusion of the given argument.

1. *All dogs go to heaven, and my terrier is a dog. So my terrier will go to heaven.*

2. *If you take this new drug, you will lose weight.*

3. *Nobody believes the president anymore because he was caught in a lie.*

4. *Today's children do not respect authority. A disdain for authority leads to social unrest. As a result, the social fabric is at risk.*

5. *A people free to choose will always choose peace.* (Ronald Reagan)

6. *He who would live in peace and at ease must not speak all that he knows or all he sees.* (Benjamin Franklin)

7. *Bad times have a scientific value. These are occasions a good learner would not miss.* (Ralph Waldo Emerson)

Valid arguments   In Exercises 8 through 13, decide whether or not the given argument is valid.

8. *All dogs go to heaven. My pet is a dog. Therefore, my pet will go to heaven.*

9. *The senate has a number of dishonest members. We just elected a new senator. Therefore, that senator is dishonest.*

10. *If I have diabetes, my blood-glucose level will be abnormal. My blood-glucose level is normal, so I do not have diabetes.*

11. *People who drive too fast get speeding tickets. Because you got a speeding ticket, you were driving too fast.*

12. *If you follow the yellow brick road, it will lead you to the Wizard of Oz. I have met the Wizard of Oz, so I must have followed the yellow brick road.*

13. *The job of every policeman is to protect and serve the public. Because I am dedicated to protecting and serving the public, I am a policeman.*

Types of fallacies   Classify the fallacies in Exercises 14 through 31 as one of the types of fallacies discussed in this section.

14. *Every time a Democrat gets elected, our taxes go up. Raising taxes is a Democratic policy.*

15. *I don't think we should believe him because I heard he dropped out of school.*

16. *My math teacher says Cheerios is the healthiest breakfast cereal.*

17. *The Republican Party has been in power for the past 8 years, and during that period the number of jobs has declined. A decrease in jobs is a Republican policy.*

18. *We can't afford the more expensive car because it will cost more.*

19. *In this war you are either for us or against us. Because you do not support the war effort, you are our enemy.*

20. *My congressman accepted large amounts of money from a political action group. All congressmen do it, so he didn't do anything wrong.*

21. *More people use Crest than any other toothpaste. So it must be the best toothpaste on the market.*

22. *I visited three state parks in California and saw redwood trees in all three. There must be redwood trees in all California state parks.*

23. *No one in my family has tested positive for HIV. So my family is free of AIDS.*

24. *My history teacher claims that accepted statistics for the population of South America before Columbus are, in fact, far too small. I have heard that he is something of a racist with regard to Native Americans. His conclusions about the Indian population cannot be valid.*

25. *My math teacher says studying can improve my grade. I don't believe I will make an "A" no matter how hard I study, so I'm skipping the review session tonight.*

26. *Everybody wants to enroll in Professor Smith's section. He must be the best teacher on campus.*

27. *If you don't go with me, you are going to miss the biggest party of the year. So hop in, and let's go.*

28. *I bought that product, and I lost 25 pounds. It is the best weight-loss product on the market.*

29. *I know that my income tax is too high because I have to pay too much.*

30. *The three smartest students in my math class are of Asian descent. Asians are better at mathematics than Americans.*

31. *The president says we need to help people in Third World countries who don't have sufficient food or medical care. Because all of our money is going overseas, I don't support foreign aid.*

Additional informal fallacies   Exercises 32 through 35 introduce additional informal fallacies. In each case, explain why the example given is a fallacy.

32. **Appeal to fear.** *You should support my point of view. I am having dinner with your boss tonight, and I might have to tell her what you said last week.*

33. **Fallacy of accident.** *All men are created equal. So you can throw the football just as well as the quarterback for the Dallas Cowboys.*

34. **Slippery slope.** *They keep cutting state taxes. Eventually, the state won't have enough revenue to repair the roads.*

35. **Gambler's fallacy.** *I have lost on 20 straight lottery tickets. I'm due for a win, so I'm buying extra tickets today.* Note: This fallacy will be explored further in Section 5 of Chapter 5.

36. **Currency conversion.** The euro is the currency used by most member states of the European Union. Here is a table of values from the Web for converting euros to American dollars.

Euro	1	2	3	4	5	6	7	8	9
Dollar	1.43	2.86	4.29	5.72	7.15	8.58	10.01	11.44	12.87

Based on this table, what would be the cost in American dollars of a bottle of wine marked 16 euros?

37. **Children's blocks.** The following chart shows the number of children's blocks that will fit in cubes of various sizes. Size measurements are in inches.

Size of cube	1 by 1 by 1	2 by 2 by 2	3 by 3 by 3	4 by 4 by 4	5 by 5 by 5
Number of blocks	1	8	27	64	125

What pattern is shown in this chart? In general, how can you find the number of children's blocks that will fit in a cube of a given size? How many blocks will fit in a cube of 50 inches on a side?

Finding patterns In Exercises 38 through 40, find a pattern suggested by the given sequence.

38. 2, 4, 6, 8, 10, . . .

39. 1, 3, 5, 7, 9, . . .

40. $1, \dfrac{1}{2}, \dfrac{1}{4}, \dfrac{1}{8}, \dfrac{1}{16}, \dfrac{1}{32}, \cdots$

41. Research. Find and report on an argument presented in the popular media that is an informal fallacy.

1.3 Formal logic and truth tables: Do computers think?

TAKE AWAY FROM THIS SECTION

Understand the formal logic used by computers.

The following article appears at the physicsworld.com Web site.

This article talks about "logic gates," which are the basic building blocks of both ordinary and quantum computers. Formal logic focuses on the structure or form of a statement. We as humans make judgments based on information we gather. But computers have a strict set of rules that they follow, and there is no room for judgment. The strict rules by which computers operate is the formal version of logic that we will present here.

IN THE NEWS 1.4

Quantum Logic Gate Lights Up

BELLE DUMÉ   August 8, 2003

Physicists in the US have taken another important step towards making a quantum computer. Duncan Steel of the University of Michigan and co-workers have created a logic gate using two electron-hole pairs—also known as "excitons"—in a quantum dot (X Li *et al.* 2003 *Science* **301** 809).

Classical computers deal with binary logic and the bits being processed must be either "0" or "1". Quantum computers, on the other hand, exploit the ability of quantum particles to be in two or more states at the same time. A quantum bit or "qubit" can therefore be "0" or "1" or any combination of the two. This means that a quantum computer could, in principle, outperform a classical computer for certain tasks. However, all the quantum computers demonstrated so far have only contained a handful of qubits.

Although qubits have been made with trapped photons, atoms and ions, it is generally thought that it should be easier to build working devices with solid-state systems. Several teams have made significant progress with the superconducting approach to solid-state quantum computing. Now Steel and co-workers at Michigan, Michigan State, the Naval Research Laboratory and the University of California at San Diego have demonstrated the first all-optical quantum gate in a semiconductor quantum dot.

Symbols and rules can be used to analyze logical relationships in much the same way they are used to analyze relationships in algebra. This formalism has the advantage of allowing us to focus on the structure of logical arguments without being distracted by the actual statements themselves.

Formal logic deals only with statements that can be classified as either true or false (but not both). To clarify what we mean, consider the following sentences:

This sentence is false. Is this sentence true or is it false? If the sentence is true, then it's telling us the truth when it says that it's false—so the sentence must be false. On the other hand, if the sentence is false, then it's not telling us the truth when it says that it's false—so the sentence must be true. This is rather paradoxical: It appears that there is no reasonable way to classify this sentence as either true or false. Therefore, the sentence does not qualify as a statement to which the rules of logic can be applied.

Please turn down the music. This sentence is a request for action. It cannot be classified as either true or false.

I am 6 feet tall. Although you may not know my height, it is the case that either I am 6 feet tall or I am not. The statement is either true or false (and not both), so it is a statement in the formal logical sense. We emphasize that it is necessary only that we know that a statement can be classified as either true or false—it is not necessary that we know which of the two applies.

Operations on statements: Negation, conjunction, disjunction, and implication

In order to focus on the formal structure of logic, we use letters to represent statements just as algebra uses letters to represent numbers. For example, we might represent the statement *He likes dogs* by p. In algebra we combine variables that represent numbers using operations such as addition (+), multiplication ($\times$), and so forth. In formal logic we use logical operations to combine symbols that stand for statements. But the logical operations are not multiplication or division. Instead, they are *negation, conjunction, disjunction,* and *implication.*

Negation   The *negation* of a statement is the assertion that means the opposite of the original statement: One statement is true when the other is false. As in common English usage, we will indicate the negation using NOT. For example, if p represents the statement that *He likes dogs*, then the statement NOT p means *It is not the case that he likes dogs* or more simply *He does not like dogs.*

$$p = \text{He likes dogs.}$$

$$\text{NOT } p = \text{He does not like dogs.}$$

By its definition, the negation of a true statement is false, and the negation of a false statement is true. For example, if the statement p that *He likes dogs* is true, then the statement NOT p that *He does not like dogs* is false. We show this below using a *truth table*. This table lists the possibilities for the truth of p and the corresponding value for the truth of NOT p.

p	NOT p	
T	F	This means that the negation of p is false when p is true.
F	T	This means that the negation of p is true when p is false.

In common usage, we sometimes encounter a *double negative*. Consider the following statement:

I enjoy dogs, but it's just not true that I don't like other animals as well.

In this statement, the speaker is telling us that, in fact, he does like other animals. If q is the statement that *I like animals other than dogs*, then the quote above is represented by NOT NOT q, which has the same meaning as q.

The Internet search engine Google will perform the NOT operation, but it uses the minus sign (preceded by a space) to indicate negation. For example, if you search on **Iraq**, you will get a number of hits, including at least one from the Central Intelligence Agency. If you want to refine your search to exclude references to the CIA, you use

Iraq -CIA. (The space before the minus sign is necessary.) The result is a list of pages about Iraq that don't mention the CIA.

Google advanced search page.

Conjunction   The *conjunction* of two statements is the assertion that *both* are true. In common English usage, *and* signifies conjunction. We will use the symbol AND to denote conjunction. The following example shows how:

Assume that

$p = $ *Baseball is 90% mental.*

$q = $ *The other half of baseball is physical.*

The conjunction of these two statements is one of the many humorous quotes of Yogi Berra. In symbols, p AND q represents *Baseball is 90% mental* **and** *the other half is physical.*

Because the conjunction of two statements is the assertion that both are true, the conjunction is true when, and only when, both statements are true. For example, Yogi Berra's quote above cannot be true because the truth of either of the two statements implies that the other is false. Hence, one or the other is false—so the conjunction is false.

The truth table for the conjunction shows the truth value of p AND q for each possible combination of the truth values of p and q.

p	q	p AND q	
T	T	T	This means that p AND q is true when p is true and q is true.
T	F	F	This means that p AND q is false when p is true and q is false.
F	T	F	This means that p AND q is false when p is false and q is true.
F	F	F	This means that p AND q is false when p is false and q is false.

The following four statements illustrate the four different cases in the conjunction truth table:

True conjunction: *A fever may accompany a cold* **and** *a headache may accompany a cold.*

False conjunction: *You are required to pay income tax* **and** *pigs often fly.*

False conjunction: *Aspirin cures cancer* **and** *water is wet.*

False conjunction: *The president's policies are always best for America* **and** *Congress never passes an unwise bill.*

Internet search engines understand the use of the AND operator. If you search for **Iran and Iraq,** you will see sites that mention both Iran and Iraq—the Iran-Iraq war, for example.

EXAMPLE 1.7 Determining truth in a conjunction: Health insurance policies

In the past, many health insurance policies did not cover preexisting conditions. That is, they did not cover illness that existed prior to the purchase of the policy. A salesman for such a policy stated, *If you buy this policy, it will cover cases of flu in your family next winter, and it will cover treatment for your wife's chronic arthritis.* Was the salesman telling the truth?

SOLUTION

The policy did not cover the preexisting arthritic condition, so that part of the conjunction was not true. Conjunctions are true only when both parts are true. The salesman did not speak truthfully.

Health insurance card.

TRY IT YOURSELF 1.7

I scored 62% on a difficult exam where 60% was a passing grade. I told my friend *The test was hard, and I passed it.* Did I speak truthfully?

The answer is provided at the end of this section.

Disjunction   The *disjunction* of two statements is the assertion that either one or the other is true (or possibly both). Compare this with the conjunction, where both statements *must* be true. Disjunction in common English usage is indicated with *or*. For symbolic representations, we use OR. Here is an example.

Assume that

$p = $ *This medication may cause dizziness.*

$q = $ *This medication may cause fatigue.*

We use the disjunction to state the warning on my medication: p OR q stands for the statement *This medication may cause dizziness or fatigue.*

In common English usage, there is no doubt as to what is meant when we combine statements using **and**. There can be a question, however, when we consider the usage of the word **or**. In the warning above, it is clearly intended that the medicine may cause either or *both* conditions. This is the *inclusive* use of **or**, which means either or both of the combined statements may be true. By way of contrast, in the spring of 2008, Democrats in Pennsylvania had the opportunity to vote for Hillary Clinton **or** Barack Obama. That definitely did not mean they could vote for both. This is an illustration of what is called the *exclusive or*, meaning that one choice or the other is allowed, but not both.

Key Concept

In mathematics and logic, the word **or** is always used in the inclusive sense: one or the other, or possibly both.

In logic, a disjunction is true when either one part is true or both parts are true. It is false only when *both* parts are false. Here is the truth table for disjunction.

p	q	p OR q	
T	T	T	This means that p OR q is true when p is true and q is true.
T	F	T	This means that p OR q is true when p is true and q is false.
F	T	T	This means that p OR q is true when p is false and q is true.
F	F	F	This means that p OR q is false when p is false and q is false.

The following four statements illustrate the four different cases in the disjunction truth table:

True disjunction: *A fever may accompany a cold* **or** *a headache may accompany a cold.*

True disjunction: *You are required to pay income tax* **or** *pigs often fly.*

True disjunction: *Aspirin cures cancer* **or** *water is wet.*

False disjunction: *The president's policies are always best for America* **or** *Congress never passes an unwise bill.*

Internet searches use OR in the inclusive sense. For example, **red or blue Toyota** will bring up information about Toyotas that have something to do with red, or blue, or *both*.

EXAMPLE 1.8 Determining the meaning of a disjunction: Menu options

In my favorite restaurant, the waiter asks if I want butter or sour cream on my baked potato. Is this the inclusive or exclusive **or**?

SOLUTION

His statement surely means you can have butter, sour cream, or both. Therefore, it is the inclusive **or**.

TRY IT YOURSELF 1.8

A waiter says, *Your meal comes with fries or a baked potato.* Is this the inclusive or exclusive **or**?

The answer is provided at the end of this section.

Implication   A statement of the form "If p then q" is called an *implication* or a *conditional* statement. We use the same terminology as in the preceding section: p is called the *premise* or the *hypothesis*, and q is called the *conclusion*. We represent the conditional symbolically as $p \rightarrow q$. Consider the following example:

Assume that

p = *Your average is 90% or more.*

q = *You get an A for the course.*

The string of symbols $p \to q$ represents the conditional **If** *your average is 90% or more*, **then** *you get an A for the course*. In this conditional, the statement *Your average is 90% or more* is the premise or hypothesis, and the statement *You will get an A* is the conclusion.

You can think of a conditional as a promise that the conclusion will happen upon the condition that the premise happens. (This is why such statements are called conditional.) It is important to note that when the teacher promised an A for an average of 90% or better, nothing was said about what might happen if your average is less than 90%. So, for example, if your average is 88%, you may still get an A or you may get a lower grade. Neither outcome would indicate that the grading promise had been broken. In fact, the only situation where the grading promise is broken is when a student gets an average of 90% or more but does not get an A for the course.

In general, we regard a conditional statement to be true as long as the promise it involves is not broken. The promise is not broken if the premise is not true, so a conditional statement is regarded as true any time the premise is false.

The fact that a conditional statement is true whenever the hypothesis is false may seem counterintuitive. Think about it this way: If the premise is not true, then the promise has not been broken for the simple reason that it has never been tested. Therefore, we cannot say the conditional statement is false. But if the conditional statement is not false, then it must be true, because a statement in logic must be either true or false.

Here's an old joke that uses the mathematical interpretation of the conditional statement:

Sue: "If you had two billion dollars, would you give me one billion?"
Bill: "Sure."
Sue: "If you had two million dollars, would you give me one million?"
Bill: "Of course."
Sue: "If you had twenty dollars, would you give me ten?"
Bill: "No!"
Sue: "Why not?"
Bill: "Because I *have* twenty dollars!"

The point of the joke is that one can promise anything based on a condition and not be accused of lying, provided that the condition is not fulfilled.

The truth table for the conditional statement is as follows:

p	q	$p \to q$	
T	T	T	This means that $p \to q$ is true when p is true and q is true.
T	F	F	This means that $p \to q$ is false when p is true and q is false.
F	T	T	This means that $p \to q$ is true when p is false and q is true.
F	F	T	This means that $p \to q$ is true when p is false and q is false.

Here are examples of the truth table entries.

True conditional: If *Earth is spherical,* **then** *you can go from Spain to Japan by traveling west.*

False conditional: If *the Moon orbits Earth,* **then** *the Moon is a planet.*

True conditional: If *Earth is the planet nearest to the Sun,* **then** *Earth orbits the Sun once each year.*

True conditional: If *Earth is the planet farthest from the Sun,* **then** *the Moon is made of green cheese.*

EXAMPLE 1.9 Determining the meaning of the conditional: Presidential promise

The new president promises, *If Congress passes my economic package, then the recession will end within two years.*

Under which of the following scenarios would we think the president's promise was kept?

Scenario 1: Congress passes the economic package, but two years later the recession persists.

Scenario 2: Congress does not pass the economic package, but two years later the recession persists.

Scenario 3: Congress does not pass the economic package, and the recession ends.

(from right) Senators Richard Durbin, Harry Reid, and Charles Schumer, with a copy of the stimulus bill that was passed by Congress on February 13, 2009.

SOLUTION

In Scenario 1, the premise is true but the conclusion is false, so the implication is false. The president's promise was not kept. In Scenarios 2 and 3, the premise of the conditional is false. Thus, the conditional is true, and the promise should be considered to be kept.

Conditional statements can be expressed in a variety of ways. Sometimes the phrase *only if* is used. For example, consider the statement *You get a ticket only if you are speeding.* This means *If you get a ticket, then you are speeding.* In general, "*p* only if *q*" means "If *p* then *q*" and is represented by $p \rightarrow q$.

EXAMPLE 1.10 Representing statements and finding the truth: Majors

Let m represent the statement *Bill is a math major*, and let c represent the statement *Bill is a chemistry major*. First, represent each of these combinations of statements in symbolic form. Second, assume that Bill is indeed a math major but not a chemistry major and determine which of the statements are true and which are false.

    a. Bill is a math major or a chemistry major.

    b. Bill is not a math major and is a chemistry major.

    c. If Bill is not a math major, then he's a chemistry major.

SOLUTION

    a. This is a disjunction, and we represent it using m OR c. This statement is true (even though c is false) because m is true.

    b. Note that NOT m means *Bill is not a math major*. So we use (NOT m) AND c to represent the conjunction. Observe that we are careful to include the parentheses here. It is always advisable to use parentheses to make sure that the use of NOT is properly understood. The conjunction is false because one of the statements is not true. (In fact, neither is true.)

    c. This is a conditional. Note that NOT m means *Bill is not a math major*. So (NOT m) → c is the symbolic statement we seek. The premise *Bill is not a math major* is false, so the implication is true (even though the conclusion is false).

TRY IT YOURSELF 1.10

Represent these combinations of statements in symbolic form. Then determine which are true under the assumption that Bill is a math major but not a chemistry major.

    a. Bill is neither a math major nor a chemistry major.

    b. If Bill is a math major, then he's a chemistry major.

The answer is provided at the end of this section.

Using parentheses

Let's look more closely at the use of parentheses in part b of the preceding example. Without explanation the expression NOT m AND c is ambiguous and could be interpreted using parentheses in at least two ways: (NOT m) AND c; NOT (m AND c).

Interpretation 1: (NOT m) AND c. This expression means, "It is not true that he is a math major. In addition, he is a chemistry major." This is the same as the statement *He is not a math major and is a chemistry major.*

Interpretation 2: NOT (m AND c). This expression means, "It is not the case that he has the two majors math and chemistry."

    These two interpretations say entirely different things. Thus, the use of parentheses is very important. If we leave them out and just write NOT m AND c, the meaning is ambiguous. It could mean either of the two interpretations above.

    This illustrates why parentheses are often necessary to avoid ambiguity. You should use parentheses when you want to view a whole collection of symbols as one single entity. When a complex statement contains one or more parts that are themselves combinations of statements, the parentheses are nearly always vital to a proper understanding of the overall meaning. *When in doubt, use parentheses!*

Truth tables for complex statements

We can use the basic truth tables presented above to analyze more complex statements. Consider, for example, the following quote from the philosopher George Santayana: "*Those who do not remember the past are condemned to repeat it.*" We can rephrase this statement as **If** *you do not remember the past,* **then** *you are condemned to repeat it.*

Assume that

$$p = \textit{You remember the past.}$$

$$q = \textit{You are condemned to repeat the past.}$$

We can express the given statement symbolically as $(\text{NOT } p) \to q$.

To make a truth table, we begin with a table that lists the possible truth values for p and q. We include columns for NOT p, and for the implication $\to$.

p	q	NOT p	$(\text{NOT } p) \to q$	
T	T			p is true, and q is true.
T	F			p is true, and q is false.
F	T			p is false, and q is true.
F	F			p is false, and q is false.

When parentheses are present, we always work from "the inside out," so the next step is to complete the column corresponding to the "inside," that is, NOT p. The truth values of this statement are the opposite of those for p.

p	q	NOT p	$(\text{NOT } p) \to q$	
T	T	F		NOT p is false.
T	F	F		NOT p is false.
F	T	T		NOT p is true.
F	F	T		NOT p is true.

We complete the table by applying the truth table for $\to$ to the last column. Remember that the conditional is true except when the premise is true but the conclusion is false.

p	q	NOT p	$(\text{NOT } p) \to q$	
T	T	F	T	NOT p is false, and q is true, so the conditional is true.
T	F	F	T	NOT p is false, and q is false, so the conditional is true.
F	T	T	T	NOT p is true, and q is true, so the conditional is true.
F	F	T	F	NOT p is true, and q is false, so the conditional is false.

The truth value of the compound statement $(\text{NOT } p) \to q$ is shown in the last column. Note, for example, that the table tells us that $(\text{NOT } p) \to q$ is true when p is true and q is true. This fact is relevant to those people who remember the past and are condemned to repeat it. In these cases, Santayana's statement is true.

EXAMPLE 1.11 Using truth tables for complex statements: Proposition 187

In 1994 California voters approved Proposition 187, which included the statement *A person shall not receive any public social services until he or she has been verified as a United States citizen or as a lawfully admitted alien.* The law was later judged unconstitutional by a federal court.

Let

$$C = \textit{Citizenship has been verified.}$$

$$A = \textit{Lawfully admitted alien status has been verified.}$$

Students protesting Proposition 187.

Then under Proposition 187, the condition that would deny services is NOT (*C* OR *A*). Use a truth table to analyze this statement.

SOLUTION

To begin making the truth table, we list the four possibilities for *C* and *A*.

C	*A*	*C* OR *A*	NOT (*C* OR *A*)	
T	T			*C* is true, and *A* is true.
T	F			*C* is true, and *A* is false.
F	T			*C* is false, and *A* is true.
F	F			*C* is false, and *A* is false.

Working from the inside out, we next fill in the column for *C* OR *A*.

C	*A*	*C* OR *A*	NOT (*C* OR *A*)	
T	T	**T**		*C* is true, and *A* is true, so *C* OR *A* is true.
T	F	**T**		*C* is true, and *A* is false, so *C* OR *A* is true.
F	T	**T**		*C* is false, and *A* is true, so *C* OR *A* is true.
F	F	**F**		*C* is false, and *A* is false, so *C* OR *A* is false.

We take the negation of this column to obtain the final answer.

C	*A*	*C* OR *A*	NOT (*C* OR *A*)	
T	T	T	**F**	*C* OR *A* is true, so NOT (*C* OR *A*) is false.
T	F	T	**F**	*C* OR *A* is true, so NOT (*C* OR *A*) is false.
F	T	T	**F**	*C* OR *A* is true, so NOT (*C* OR *A*) is false.
F	F	F	**T**	*C* OR *A* is false, so NOT (*C* OR *A*) is true.

The final truth value of the statement NOT (*C* OR *A*) is the last column in the truth table.

TRY IT YOURSELF 1.11

Make the truth table for NOT (*p* AND *q*).

The answer is provided at the end of this section.

More on the conditional: Converse, inverse, contrapositive

The conditional is the combination of statements that is most often misinterpreted, so it deserves further attention.

Assume that

$$p = \text{You support my bill.}$$
$$q = \text{You are a patriotic American.}$$

The conditional is a statement that a politician might use to garner support for a bill he or she wants to see passed:

$p \rightarrow q$: **If** *you support my bill,* **then** *you are a patriotic American.*

There are three common variations of the conditional that may lead to confusion.

The converse: $q \rightarrow p$. **If** *you are a patriotic American,* **then** *you support my bill.* The converse reverses the roles of the premise and conclusion.

The inverse: (NOT p) $\rightarrow$ (NOT q). **If** *you do not support my bill,* **then** *you are not a patriotic American.* The inverse replaces both the premise and conclusion by their negations.

The contrapositive: (NOT q) $\rightarrow$ (NOT p). **If** *you are not a patriotic American,* **then** *you do not support my bill.* The contrapositive both reverses the roles of and negates the premise and conclusion.

The original conditional $p \rightarrow q$ says that all supporters of the bill are patriotic Americans, but it allows for the possibility that some nonsupporters of the bill are also patriotic Americans. The contrapositive, (NOT q) $\rightarrow$ (NOT q), says that if you are not a patriotic citizen, then you do not support the bill. This is just another way of saying the same thing as $p \rightarrow q$, that is, that supporters of the bill are patriotic but nonsupporters may be patriotic as well. The conditional and its contrapositive are really just different ways of saying the same thing.

Similarly, the converse $q \rightarrow p$ and the inverse (NOT p) $\rightarrow$ (NOT q) of a conditional are also different ways of saying the same thing, but what they say is *not* the same as the original conditional. They say that nonsupporters are unpatriotic.

One way to compare the various forms is to look at a truth table showing all four.

				Conditional	Converse	Inverse	Contrapositive
p	q	NOT p	NOT q	$p \rightarrow q$	$q \rightarrow p$	(NOT p) $\rightarrow$ (NOT q)	(NOT q) $\rightarrow$ (NOT p)
T	T	F	F	T	T	T	T
T	F	F	T	F	T	T	F
F	T	T	F	T	F	F	T
F	F	T	T	T	T	T	T

This table shows that the conditional and its contrapositive have the same truth value. Thus, they are termed *logically equivalent.* Similarly, the converse and inverse are logically equivalent.

Key Concept

Two statements are **logically equivalent** if they have the same truth tables. Logically equivalent statements are just different ways of saying the same thing.

The fact that a conditional is logically equivalent to its contrapositive is useful whenever it may be easier to observe consequences of a phenomenon than the phenomenon itself. For example, it is not easy to determine whether heavier objects fall faster than lighter objects when air resistance is ignored. Aristotle believed they did. But if this were true, then heavier and lighter objects dropped from a high place would strike the ground at different times. Famously, Galileo observed that this was not true, and an important fact about gravity was verified.

The contrapositive arises in thinking about the results of medical tests. For example, people suffering from diabetes have abnormal blood sugar (*If you have diabetes, then you have abnormal blood sugar*). When a doctor finds your blood sugar to be normal, the contrapositive is used to conclude that you do not have diabetes (*If you do not have abnormal blood sugar, then you do not have diabetes*).

Whether intentional or not, the error of confusing a conditional statement with its converse or inverse is commonplace in politics and the popular media. The statement *The perpetrator has blood type A* can be restated as *If you are the perpetrator, then you have blood type A*. If the prosecutor then tells us that the defendant has blood type A, he or she may be hoping the jury will think the converse, namely *If you have blood type A, then you are the perpetrator*. This is not a valid conclusion because many innocent people also have blood type A.

SUMMARY 1.3   **Conditional Statements**

Consider the conditional statement **If** *statement A*, **then** *statement B*.

- The contrapositive statement is
  If *statement B is false*, **then** *statement A is false*.
A conditional is logically equivalent to its contrapositive.

- The converse statement is
  If *statement B*, **then** *statement A*.

- The inverse statement is
  If *statement A is false*, **then** *statement B is false*.
The inverse is logically equivalent to the converse.

- Replacing a conditional statement with its inverse or converse is *not* logically valid.

EXAMPLE 1.12 Formulating variants of the conditional: Politics

Formulate the converse, inverse, and contrapositive of each of the following conditional statements:

    a. *If you vote for me, your taxes will be cut.*

    b. *All Democrats are liberals.*

SOLUTION

    a. The converse of the statement is *If your taxes are cut, you voted for me.* The inverse is *If you do not vote for me, your taxes will not be cut.* The contrapositive is *If your taxes are not cut, you did not vote for me.*

    b. The statement *All Democrats are liberals* is another way of saying *If you are a Democrat, you are a liberal.* So the converse is *If you are a liberal, you are a Democrat* or *All liberals are Democrats.* The inverse is *If you are not a Democrat, you are not a liberal.* The contrapositive is *If you are not a liberal, you are not a Democrat.*

TRY IT YOURSELF 1.12

Formulate the converse, inverse, and contrapositive of the statement *Every Republican is a conservative.*

The answer is provided at the end of this section.

Logic and computers

The article that opened this section referred to *logic gates* used by computers. These gates perform functions corresponding to the logical operations such as negation and conjunction that we have studied in this section. Here is the connection: In classical computing, information is stored using the two numbers 0 and 1. The connection with logic is made by thinking of 0 as "false" and 1 as "true."

There are several types of logic gates. The logic gate NOT converts 0 to 1 and 1 to 0, just as in formal logic the negation of a false statement is true and the negation of a true statement is false. The logic gate OR takes two inputs, each of them 0 or 1. It returns 1 if at least one of the inputs is 1. This corresponds to the fact that in logic the disjunction of two statements is true (corresponding to 1) if at least one of the statements is true.

EXAMPLE 1.13 Using logic gates and logical operations: Input and output

The logic gate AND takes two inputs, each of them 0 or 1. It corresponds to the logical operation of conjunction. If the inputs to the AND gate are 1 and 1, what is the output?

SOLUTION

This situation corresponds to the conjunction p AND q when both p and q are true. In that case, p AND q is true, so the output of the logic gate is 1.

TRY IT YOURSELF 1.13

The logic gate XOR takes two inputs, each of them 0 or 1. It corresponds to the *exclusive* use of **or**, in which we ask whether one or the other, but not both, is true. If the inputs to the XOR gate are 1 and 1, what is the output?

The answer is provided at the end of this section.

Try It Yourself answers

Try It Yourself 1.7: Determining truth in a conjunction: Grade on exam   The statement is true.

Try It Yourself 1.8: Determining the meaning of a disjunction: Menu options
This is an example of the exclusive **or.**

Try It Yourself 1.10: Representing statements and finding the truth: Majors
    a. (NOT m) AND (NOT c). The conjunction is false.

    b. $m \to c$. The conditional is false.

Try It Yourself 1.11: Using truth tables for complex statements: Proposition 187

p	q	p AND q	NOT (p AND q)
T	T	T	F
T	F	F	T
F	T	F	T
F	F	F	T

Try It Yourself 1.12: Formulating variants of the conditional: Politics   The converse is *If you are a conservative, then you are a Republican.* The inverse is *If you are not a Republican, then you are not a conservative.* The contrapositive is *If you are not a conservative, then you are not a Republican.*

Try It Yourself 1.13: Using logic gates and logical operations: Input and output
The output is 0.

Exercise Set 1.3

1. Representing statements. Let p represent the statement *You drive too fast*, and let q represent the statement *You get a traffic ticket.* Express using symbols the statement *If you don't drive too fast, you will not get a traffic ticket.*

2. Conditional. Write the statement *Stop talking or I'll leave* as a conditional statement.

3. A research project using the Internet. You need to write a report on France or England in the eighteenth OR nineteenth century. Propose an Internet search phrase that will help you.

4. Converse. Consider the statement *If it rains, then I carry an umbrella.* Formulate the converse. The original statement is quite reasonable. Is the converse reasonable?

5. Inverse. Consider the statement *If I buy this MP3 player, then I can listen to my favorite songs.* Formulate the inverse. The original statement seems reasonable. Is the inverse reasonable?

6. Politicians. Congresswoman Jones says, "If we pass my bill, the economy will recover." A year later, Congressman Smith says to Congresswoman Jones, "Well, you were wrong, we didn't pass your bill and the economy has recovered just fine." Is there anything wrong here? Explain.

7. Taxing implications. Consider the statement *If you don't pay your taxes by April 15, then you owe a penalty.* A taxpayer is confused because she paid her taxes by April 15 but she still owes a penalty. Explain her confusion.

More on representing statements   Let p represent the statement *You live in Utah*, and let q represent the statement *You work in Nevada.* Use this information in Exercises 8 through 12 to express the given statement in symbolic form.

8. You do not live in Utah and you work in Nevada.

9. If you work in Nevada, then you cannot live in Utah.

10. You do not live in Utah or you do not work in Nevada.

11. Nobody who works in Nevada can live in Utah.

12. You do not live in Utah only if you work in Nevada.

Translating into symbolic form   In Exercises 13 through 19, translate the given sentence into symbolic form by assigning letters to statements that do not contain a negative. State clearly what your letters represent.

13. If we pass my bill, the economy will recover.

14. We didn't pass your bill, and the economy recovered.

15. For breakfast I want cereal and eggs.

16. I want cereal or eggs for breakfast.

17. If you don't clean your room, I'll tell your father.

18. I'll give you a dollar if you clean your room and take out the trash.

19. I'll go downtown with you if we can go to a movie or a restaurant.

Translating into English   Let p represent the statement *He's American*, and let q represent the statement *He's Canadian.* Each of Exercises 20 through 22 has two logically equivalent statements. That is, they are just different ways of saying the same thing. Write each of these statements in plain English.

20. (NOT p) $\to q$ and p OR q

21. NOT (p OR q) and (NOT p) AND (NOT q)

22. NOT (p AND q) and (NOT p) OR (NOT q)

True or false   In Exercises 23 through 31, determine whether the given statement is true or false. Be sure to explain your answer.

23. $3 + 3 = 7$ or $2 + 2 = 5$.

24. $3 + 3 = 6$ or $2 + 2 = 4$.

25. If pigs fly, then Mickey Mouse is president.

26. If George Washington was the first U.S. president, then Mickey Mouse is the current U.S. president.

27. I will be president if pigs fly.

28. If pigs can fly, then Earth is flat.

29. If Earth is flat, then pigs can fly.

30. If pigs can fly, then oceans are water.

31. If oceans are water, then pigs can fly.

32. Combined statements. Which of the following statements are true?

   a. Mickey Mouse is president of the United States and $5 + 5 = 10$.

   b. Mickey Mouse is president of the United States or $5 + 5 = 10$.

   c. If Mickey Mouse is president of the United States, then $5 + 5 = 10$.

   d. If $5 + 5 = 10$, then Mickey Mouse is president of the United States.

33. True and false. Which of the following statements are true?

   a. Mickey Mouse is a cartoon character and $5 + 5 = 10$.

   b. Mickey Mouse is president of the United States or $5 + 5 = 11$.

   c. If $5 + 5 = 12$, then Mickey Mouse is president of the United States.

The inverse In Exercises 34 through 41, formulate the inverse of the given statement.

34. *All dogs chase cats.*

35. *If you don't clean your room, I'll tell your father.*

36. *I'll give you a dollar if you clean your room and take out the trash.*

37. *I'll go downtown with you if you agree to go to a movie or a restaurant.*

38. *If I am elected, then taxes will be cut.*

39. *All math courses are important.*

40. *If you don't exercise regularly, your health will decline.*

41. *If you take your medicine, you will get well.*

The converse In Exercises 42 through 49, formulate the converse of the given statement.

42. *All dogs chase cats.*

43. *If you don't clean your room, I'll tell your father.*

44. *I'll give you a dollar if you clean your room and take out the trash.*

45. *I'll go downtown with you if you agree to go to a movie or a restaurant.*

46. *If I am elected, then taxes will be cut.*

47. *All math courses are important.*

48. *If you don't exercise regularly, your health will decline.*

49. *If you take your medicine, you will get well.*

50. Contrapositives. Formulate the contrapositive of the following statements:

   a. *If there is life on Mars, then there must be traces of water there.*

   b. *All flu cases are accompanied by high fever.*

   c. *If you are neutral in situations of injustice, you have chosen the side of the oppressor.* (Desmond Tutu)

   d. *If you have ten thousand regulations, you destroy all respect for the law.* (Winston Churchill)

The contrapositive Formulate the contrapositive of the statements in Exercises 51 through 59.

51. *If you drink and drive, you will get arrested.*

52. *If aliens had visited Earth, we would have found concrete evidence by now.*

53. *If you don't clean your room, I'll tell your father.*

54. *I'll give you a dollar if you clean your room and take out the trash.*

55. *I'll go downtown with you if you agree to go to a movie or a restaurant.*

56. *If we continue to produce greenhouse gases, then global warming will occur.*

57. *My friends all drive Porsches.*

58. *Pneumonia leads to severe respiratory distress.*

59. *If thunderstorms are approaching, they will be seen on radar.*

Truth tables In Exercises 60 through 63, make the truth table for the given compound statement.

60. p AND (NOT q)

61. $p \rightarrow (p$ OR $q)$

62. $p \rightarrow (p$ AND $q)$

63. p AND (p OR q)

Equivalence In each of Exercises 64 through 66, show using truth tables that the given two statements are logically equivalent. That is, show that the two statements have the same truth values.

64. (NOT p) $\rightarrow q$ and p OR q

65. NOT (p AND q) and (NOT p) OR (NOT q)

66. (NOT p) AND q and NOT (p OR (NOT q))

67. Logic gate AND. If the inputs to the AND gate are 0 and 1, what is the output?

68. Logic gate NAND. The logical operation NAND corresponds to negating the conjunction of two statements: p NAND q means NOT (p AND q). The corresponding logic gate NAND is one of two basic gates, in the sense that every other logic gate can be constructed using it. (The other basic gate NOR is defined in Exercise 69.) The gate NAND takes two inputs, each of them 0 or 1. If the inputs to the NAND gate are 0 and 1, what is the output?

69. Logic gate NOR. The logical operation NOR corresponds to negating the disjunction of two statements: p NOR q means NOT (p OR q). The corresponding logic gate NOR is one of two basic gates, in the sense that every other logic gate can be constructed using it. (The other basic gate NAND is defined in

Exercise 68.) The gate NOR takes two inputs, each of them 0 or 1. If the inputs to the NOR gate are 0 and 0, what is the output?

70. History of truth tables. Research the development and use of truth tables in the late nineteenth and early twentieth cen-tury. Be sure to include the contributions of Charles Lutwidge Dodgson (better known as Lewis Carroll).

71. More research. Find and report on an argument presented in the popular media that makes a fallacious argument using the converse.

1.4 Sets and Venn diagrams: Pictorial logic

TAKE AWAY FROM THIS SECTION

Produce visual displays that show categorical relationships as well as logical statements.

In this section, we study graphical representations of sets known as *Venn diagrams*. The following article from the *Boston Globe* shows that they are a common tool in business settings.

Lewis Carroll.

IN THE NEWS 1.5

Venn and the Art of Diagrams

JOSHUA GLENN   May 2, 2004

Anyone who has ever spent time in a conference room equipped with an overhead projector is familiar with the basic Venn diagram—three overlapping circles whose eight regions represent every possible intersection of three given sets, the eighth region being the space around the diagram. Although it resembled the intertwined rings familiar in Christian (and more recently, Led Zeppelin) iconography, when Cambridge University logician John Venn devised the diagram in 1880 it was hailed as an innovative way to represent complex logical problems in two dimensions.

There was just one dilemma, according to British statistician and geneticist A.W.F. Edwards, author of the entertaining new book *Cogwheels of the Mind* (Johns Hopkins): Venn's diagram didn't scale up. When a problem involves four sets, circles don't have enough possible combinations of overlaps. Ovals work for four sets, Venn found, but after that one winds up drawing unreadable messes. What to do?

In Venn's day, a rival lecturer in mathematics at Oxford by the name of Charles Dodgson—better known to us as Lewis Carroll—tried to come up with a superior logical diagram using rectangles but failed (though not before producing an 1887 board game based on his design).

As the article indicates, Venn diagrams are used for representing and solving logical problems. Thus they are an important tool for critical thinking.

Sets

Underlying the Venn diagram discussed in the preceding article is the idea of a *set*, which is nothing more than a collection of objects. This is where we begin.

Key Concept

A set is a collection of objects, which are called elements of the set.

A set is specified by describing its elements. This is sometimes done in words and sometimes simply by listing the elements and enclosing them in braces. For example, we can specify the set of all U.S. states that touch the Pacific Ocean by listing the elements of this set as

{Alaska, California, Hawaii, Oregon, Washington}.

California is an element of this set, but Nebraska is not. If we let P stand for the set of Pacific Ocean states, we would write

P = {Alaska, California, Hawaii, Oregon, Washington}.

The Pacific Ocean states.

The states in the set P of Pacific Ocean states are also elements of the larger set U of all 50 U.S. states. This is an example of the idea of a *subset*: P is a subset of U.

At the other extreme, some sets have nothing in common. For example, the set of U.S. states that touch the Atlantic Ocean shares no members with the set of Pacific states. Whenever two sets have no elements in common, they are said to be *disjoint*.

Key Concept

The set B is a subset of the set A if every element of B is also an element of A. Two sets are called disjoint if they have no elements in common.

EXAMPLE 1.14 Describing sets: The U.S. Congress

In parts a, b, and c, we let S denote the set of U.S. senators, R the set of members of the U.S. House of Representatives, and G the set of all elected U.S. government officials.

   a. How many elements are in the set S?

   b. Of the three sets, which are subsets of others?

   c. Which pairs of the three sets S, R, and G are disjoint?

   d. If V is the collection of all vowels in the alphabet, specify the set as a list of elements.

SOLUTION

   a. There are 2 senators from each of the 50 states, making a total of 100 U.S. senators. Therefore, there are 100 elements in S.

   b. Both senators and representatives are elected officials. So S is a subset of G and R is a subset of G.

   c. The Senate and House have no members in common, so S and R are disjoint.

   d. $V = \{a, e, i, o, u\}$. (Sometimes y is a vowel, depending on its usage. See Exercise 45.)

TRY IT YOURSELF 1.14

   a. Let S be the set of all U.S. states. Determine which of the following are elements of S: Hawaii, Guam, Alaska, Puerto Rico.

   b. Let O be the set of odd whole numbers from 1 through 9. Write O as a list enclosed in braces.

   c. Let N denote the set of North American countries, let S denote the set of South American countries, and let A be the set {United States, Canada, Mexico}. Which of these are subsets of others? Which pairs are disjoint?

The answer is provided at the end of this section.

Introduction to Venn diagrams

In many cases, pictures improve our understanding. A Venn diagram is a way of picturing sets.

Key Concept

A Venn diagram is a visual device for representing sets. It usually consists of circular regions inside a rectangle.

   A Venn diagram typically consists of either two or three circles inside a rectangle. The area inside the rectangle represents all the objects that are under consideration at the time, and the areas inside the circles represent certain sets of these objects.

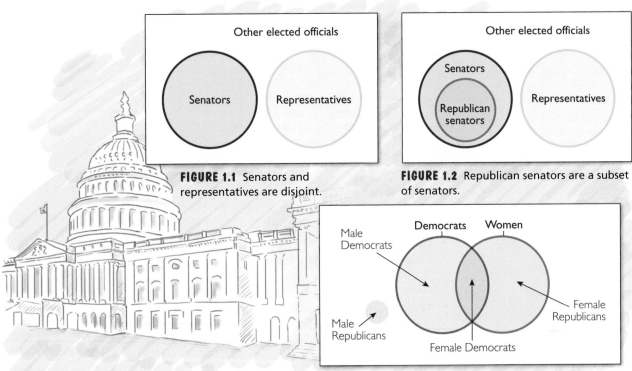

FIGURE 1.1 Senators and representatives are disjoint.

FIGURE 1.2 Republican senators are a subset of senators.

FIGURE 1.3 House of Representatives.

These diagrams can be a useful tool for answering questions about the number of elements in a set and for understanding relationships between the elements. For example, **Figure 1.1** shows a Venn diagram[4] in which the rectangle represents all elected government officials. One circle represents senators, and the other circle represents members of the House of Representatives. As we noted in the solution to part c of Example 1.14, these two sets are disjoint, and that relationship is indicated by the fact that the circles do not overlap. The region outside the two circles represents elected officials who are neither senators nor representatives. The president and vice president would go in this region.

In **Figure 1.2**, we have added a circle that represents Republican senators. This new circle is inside the "Senators" circle, indicating that the set of Republican Senators is a subset of the set of senators.

A more common example of a Venn diagram is shown in **Figure 1.3**, where the two circles overlap. In this diagram, the category under discussion is all U.S. representatives. One circle represents the set of Democratic members, and the other circle represents female members. There are four distinct regions in the diagram. (At the time of this writing, all members of the House of Representatives were either Democrats or Republicans.)

• The region common to both circles represents female Democrats.

• The region inside the "Democrat" circle but outside the "Women" circle represents male Democrats.

• The region inside the "Women" circle but outside the "Democrat" circle represents female Republicans.

---

[4]Technically, Figure 1.1 and Figure 1.2 are known as *Euler* diagrams, but we will not bother with that distinction here.

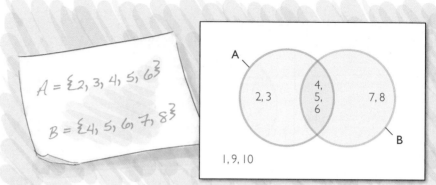

FIGURE 1.4 Parts of a Venn diagram.

• The region outside both circles represents members who are neither Democrats nor female. These are male Republicans.

To further clarify what the various regions in a Venn diagram represent, let the category be the whole numbers from 1 to 10. Let $A = \{2, 3, 4, 5, 6\}$ and $B = \{4, 5, 6, 7, 8\}$. The corresponding Venn diagram is shown in **Figure 1.4**. It shows which elements go in each part of the diagram.

Some Venn diagrams are just for fun.

EXAMPLE 1.15 Making Venn diagrams: Medical tests

Medical tests do not always produce accurate results. For this example, we use the category of all people who have undergone a medical test for hepatitis C. Let H denote the set of all patients who have hepatitis C, and let P be the set of all people who test positive:

$$H = \{\text{Patients with hepatitis C}\}.$$
$$P = \{\text{Those who test positive}\}.$$

Make a Venn diagram using these sets, and locate in it the following regions:

• **True positives:** These are people who test positive and have hepatitis C.

• **False positives:** These are people who test positive but do not have hepatitis C.

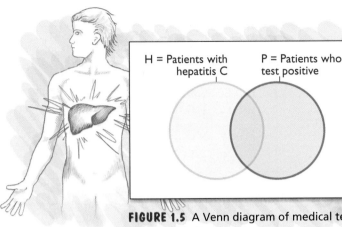

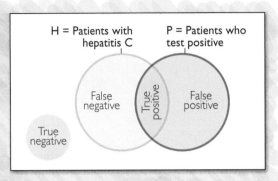

FIGURE 1.5 A Venn diagram of medical tests for hepatitis C.

FIGURE 1.6 The completed diagram.

- **True negatives:** These are people who test negative and do not have hepatitis C.

- **False negatives:** These are people who test negative but do have hepatitis C.

SOLUTION

Because the two sets C and P overlap, we make the diagram shown in **Figure 1.5**. We locate the four regions as follows:

- True positives both test positive and have hepatitis C. They are common to both sets, so they lie in the overlap between the two circles.

- False positives are inside the "Positive" circle because they test positive but outside the "Hepatitis" circle because they do not have hepatitis C.

- True negatives lie outside both circles.

- False negatives are inside the "Hepatitis" circle but outside the "Positive" circle.

The completed diagram is shown in **Figure 1.6**.

TRY IT YOURSELF 1.15

If the category is states of the United States, make a Venn diagram showing Gulf States (states that touch the Gulf of Mexico) and states west of the Mississippi River. Locate the following in your diagram: Florida, New York, Oregon, and Texas.

The answer is provided at the end of this section.

Venn diagrams can be useful in helping to count the number of items in a given category.

EXAMPLE 1.16 Making Venn diagrams with numbers: Drug trial

A double-blind drug trial separated a group of 100 patients into a test group of 50 and a control group of 50. The test group got the experimental drug, and the control group got a placebo.[5] The result was that 40% of the test group improved and only 20% of the control group improved. Let T be the test group and I the set of patients who improved:

$$T = \{\text{Test group}\}.$$

$$I = \{\text{Patients who improved}\}.$$

[5] A *placebo* is a harmless substance that contains no medication and that is outwardly indistinguishable from the real drug. We will discuss clinical trials in Section 4 of Chapter 6.

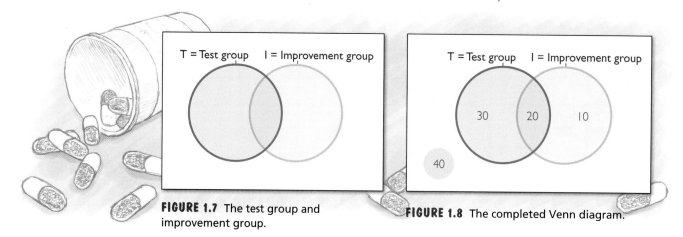

FIGURE 1.7 The test group and improvement group.

FIGURE 1.8 The completed Venn diagram.

Make a Venn diagram that shows the number of people in each of the following categories:

- Test group who improved
- Test group who did not improve
- Control group who improved
- Control group who did not improve

SOLUTION

The basic Venn diagram is shown in **Figure 1.7**. The rectangle represents all who were in the trial. The numbers and location of each group are as follows.

- **Test group who improved:** This is the overlap of the two circles. There are 40% of 50 or 20 people in this region.

- **Test group who did not improve:** This is inside the T circle but outside the I circle. There are $50 - 20 = 30$ people in this group.

- **Control group who improved:** This is in the I circle but outside the T circle. There are 20% of 50 or 10 patients in this category.

- **Control group who did not improve:** This is outside both circles. There are $50 - 10 = 40$ patients in this category.

We have entered these numbers in the completed Venn diagram shown in **Figure 1.8**.

TRY IT YOURSELF 1.16

A group of 50 children from an inner-city school and 50 children from a suburban school were tested for reading proficiency. The result was that 60% of children from the inner-city school passed and 80% of children from the suburban school passed. Make a Venn diagram showing the number of children in each category.

The answer is provided at the end of this section.

Logic and Venn diagrams

We noted at the beginning of this section that Venn diagrams can be used to analyze logical statements. Consider, for example, the conditional statement *If you are stopped for speeding, then you will get a traffic ticket.* Let S be the set of all people who are stopped for speeding and T the set of people who get traffic tickets:

$$S = \{\text{People stopped for speeding}\}.$$

$$T = \{\text{People who get traffic tickets}\}.$$

FIGURE 1.9 Venn diagram for a conditional: If you are stopped for speeding, you get a ticket.

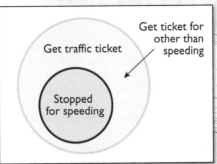

FIGURE 1.10 Why the converse may not be true.

FIGURE 1.11 Diagram for the converse: If you get a traffic ticket, then you were stopped for speeding.

Our conditional can be recast as *everyone who is stopped for speeding will get a traffic ticket*. This means that S is a subset of T. Thus to represent this statement using a Venn diagram, we put the circle for S inside the circle for T. The result is shown in **Figure 1.9**. This figure makes it clear why we cannot necessarily conclude the converse: *If you get a traffic ticket, then you were stopped for speeding.* Any driver in between the two circles is a driver who got a ticket but was not stopped for speeding. This is shown in **Figure 1.10**. The Venn diagram for the converse is shown in **Figure 1.11**. Note that the roles of the circles are reversed. The only time a statement and its converse are both true is when the two circles in the Venn diagram are the same.

We can also represent the conjunction and disjunction using Venn diagrams. For the conjunction *He is a math major and a music major*, let A denote the set of all math majors and B the set of all music majors:

$$A = \{\text{Math majors}\}.$$

$$B = \{\text{Music majors}\}.$$

To say he is a math *and* a music major means that he belongs to *both* sets. These double majors are shown in **Figure 1.12**. The shaded region in that figure is the region of the Venn diagram where the conjunction is true.

The disjunction: *Joe is a math major or a music major* means that Joe belongs to one set or the other. **Figure 1.13** illustrates the corresponding region of the Venn diagram.

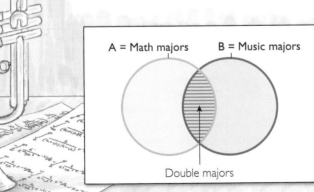

FIGURE 1.12 A conjunction is true in the overlap of the sets.

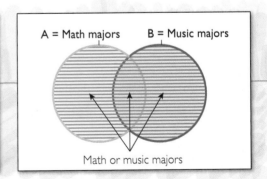

FIGURE 1.13 A disjunction is true in either one or both sets.

EXAMPLE 1.17 Matching logical statements and Venn diagrams: Farm bill

Let F denote the set of senators who voted for the farm bill and T the set of senators who voted for the tax bill:

$$F = \{\text{Senators who voted for the farm bill}\}.$$

$$T = \{\text{Senators who voted for the tax bill}\}.$$

a. Make a Venn diagram for F and T, and shade in the region for which the statement *He voted for the tax bill but not the farm bill* is true.

b. Let S denote the set of all Southern senators. Suppose it is true that *all Southern senators voted for both the tax and farm bills*. Add the circle representing Southern senators to the Venn diagram you made in part a.

SOLUTION

a. The statement is true for all senators who are in the set T but not in the set F. That group of senators is inside the "Tax" circle but outside the "Farm" circle. We have shaded this region in the Venn diagram shown in **Figure 1.14**.

b. Because all Southern senators voted for both bills, they all belong to both F and T. That makes S a subset of the overlap of the two circles. Thus, the circle representing S is inside the region common to both F and T, as shown in **Figure 1.15**.

Counting using Venn diagrams

Venn diagrams are useful in counting. To illustrate this, suppose that 17 tourists travel to Mexico or Costa Rica. Some visit Mexico, some visit Costa Rica, and some visit both. If 10 visit Mexico but not Costa Rica and 5 visit both, how many people visit Costa Rica? Let

$$M = \{\text{Tourists visiting Mexico}\}.$$

$$C = \{\text{Tourists visiting Costa Rica}\}.$$

Using these sets, we make the Venn diagram shown in **Figure 1.16**. The rectangle represents all the tourists.

The next step is to fill in the numbers we know. Now 10 tourists visit Mexico and not Costa Rica, so this number goes in the "Mexico" circle but outside the "Costa Rica" circle. Similarly, 5 people visit both countries, and this is the number that goes

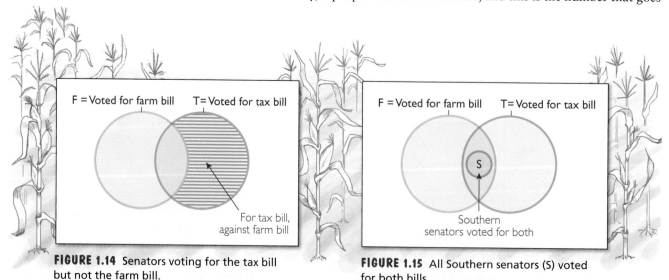

FIGURE 1.14 Senators voting for the tax bill but not the farm bill.

FIGURE 1.15 All Southern senators (S) voted for both bills.

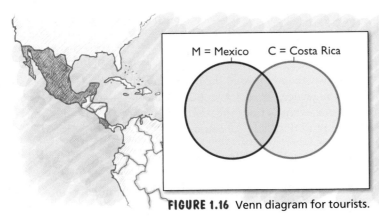

FIGURE 1.16 Venn diagram for tourists.

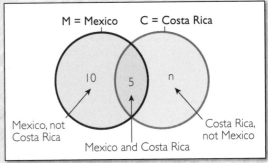

FIGURE 1.17 Entering known values.

in the overlap of the two circles. In **Figure 1.17**, we have entered these numbers and used n to denote the number who visited Costa Rica but not Mexico. Because the sum of all the numbers in the two circles gives the total number of tourists, we find

$$10 + 5 + n = 17,$$

which gives $n = 2$. So 2 tourists visited Costa Rica but not Mexico. Adding this number to the 5 tourists who visited both countries gives 7 visitors to Costa Rica.

EXAMPLE 1.18 Counting using Venn diagrams: Exam results

Suppose that 100 high school students are examined in science and mathematics. Here are the results.

- 41 students passed the math exam but not the science exam.

- 19 students passed both exams.

- 8 students failed both exams.

How many students passed the math exam, and how many passed the science exam?

SOLUTION

Let M denote those students who passed the math exam and S those students who passed the science exam. The Venn diagram for these exams is shown in **Figure 1.18**. The rectangle represents all who were examined. We have entered the data we know in **Figure 1.19**. We have put n in the "Passed science, not math" region because we don't know that number yet.

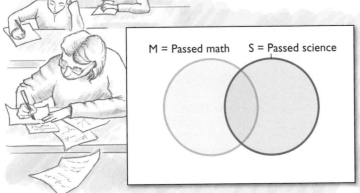

FIGURE 1.18 Venn diagram for science and math exams.

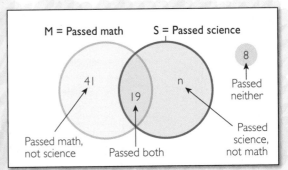

FIGURE 1.19 Entering given data.

If we add the numbers in all the regions, we get the total number of students, which is 100. Thus,

$$8 + 19 + 41 + n = 100.$$

This gives $n = 32$. We know that 41 passed the math exam but not the science exam and that 19 passed both. Hence, $41 + 19 = 60$ passed the math exam. Also, 32 students passed the science exam but not the math exam, and 19 passed both. That gives $32 + 19 = 51$ who passed the science exam.

TRY IT YOURSELF 1.18

Suppose that 100 students are given a history exam and a sociology exam. Assume that 32 passed the history exam but not the sociology exam, 23 passed the sociology exam but not the history exam, and 12 passed neither exam. How many passed the history exam?

The answer is provided at the end of this section.

Venn diagrams with three sets

A Venn diagram for three sets typically shows all possible overlaps. Consider the Venn diagram in **Figure 1.20**, where the rectangle represents all highway collisions. The three circles show the percentage of collisions resulting from one or a combination of three factors: the vehicle, human error, and the road. For example, a collision may result from faulty brakes, an impaired driver, or a curve that is too tight for the prevailing speed. In the nation, 12% of all collisions involve some factor of the vehicle traveling the roadway; 34% involve some characteristic of the roadway; and 93% are due to human factors.[6]

The various regions of the Venn diagram are labeled in Figure 1.20. For example, the number in the region labeled "Road and human, not vehicle" means that 27% of accidents are due to a combination of road conditions and human error but do not involve vehicular problems.

Having three sets makes our calculations a bit more complicated, but the underlying concept is the same as for two sets. The following example illustrates the procedure.

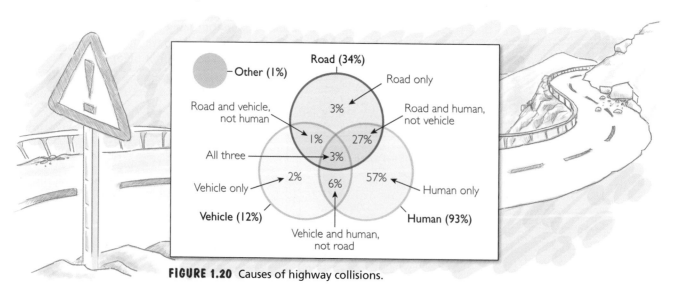

FIGURE 1.20 Causes of highway collisions.

---

[6]This information is taken from the *California 2006 Five Percent Report*. It is available at the Federal Highway Administration Web site.

EXAMPLE 1.19 Using three sets: Sports

A survey asked 110 high school athletes which sports they played among football, basketball, and baseball. It showed the following results:

Football only	24
Basketball only	13
Baseball only	12
Football and baseball, but not basketball	9
Football and basketball, but not baseball	5
Basketball and baseball, but not football	10
All three sports	6

a. How many played none of these three sports?

b. How many played exactly two of the three sports?

c. How many played football?

SOLUTION

a. In this example, we denote by F the set of those students who played football, by B the set of those who played basketball, and by S the set of those who played baseball. Using these sets, we have made the Venn diagram in **Figure 1.21** and filled in each of the regions for which we are given information. (The rectangle represents all students surveyed.) We used the following information to locate these numbers:

• Football only, 24. These athletes played football but not basketball or baseball. So this is inside the "Football" circle but outside the other two.
• Basketball only, 13. These athletes are inside the "Basketball" circle but outside the other two.
• Baseball only, 12. These athletes are inside the "Baseball" circle but outside the other two.
• Football and baseball, but not basketball, 9. These athletes played football and baseball and so are in the overlap of the "Football" and "Baseball" circles. But they did not play all three sports. So this number goes in the overlap of the "Football" and "Baseball" circles but outside the region common to all three circles.

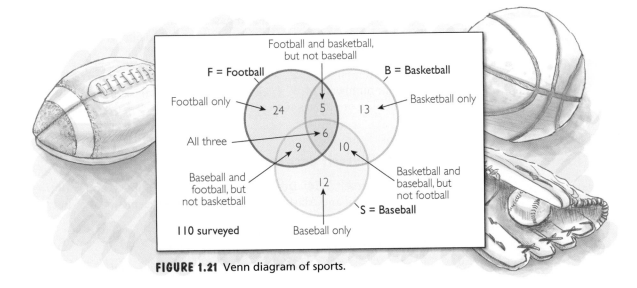

FIGURE 1.21 Venn diagram of sports.

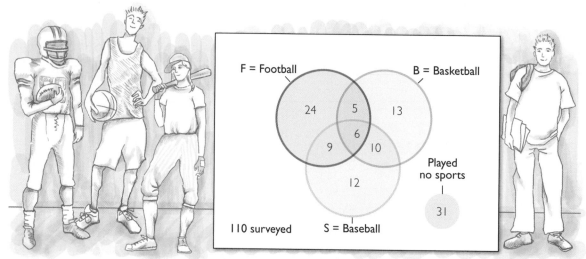

FIGURE 1.22 Adding the number who play none of these sports.

- Football and basketball, but not baseball, 5. This number goes in the overlap of the "Football" and "Basketball" circles but outside the region common to all three circles.
- Basketball and baseball, but not football, 10. This number goes in the overlap of the "Basketball" and "Baseball" circles but outside the region common to all three circles.
- All three sports, 6. These athletes participated in all three sports, so this number goes in the region common to all three circles.

Now we can answer the question in part a. There is one region that has no number in it. That is the one that is outside all three circles—the number who play none of these three sports. To find this number, we note that if we add up all the numbers, including those who play none of these sports, we get the total number surveyed: 110. So the number of students who play none of these sports is

$$110 - 24 - 13 - 12 - 9 - 5 - 10 - 6 = 31.$$

We have added this number to the diagram in **Figure 1.22**.

b. From the completed diagram in Figure 1.22, we see that the numbers 9, 5, and 10 lie in exactly two of the three sets F, B, and S. (So 9 played football and baseball, but not basketball; 5 played football and basketball, but not baseball; and 10 played basketball and baseball, but not football.) The sum of these is $9 + 5 + 10 = 24$, which is the number of students who play exactly two of the three sports.

c. From Figure 1.22, we see that the set F is divided into regions labeled by four numbers: 24, 9, 6, and 5. When we add these numbers, we obtain the number of athletes who played football: $24 + 9 + 6 + 5 = 44$.

Try It Yourself answers

Try It Yourself 1.14: Describing sets: States, numbers, and countries
a. Hawaii and Alaska are U.S. states and so elements of S, but Guam and Puerto Rico are not.

b. $O = \{1, 3, 5, 7, 9\}$.

c. Here A is a subset of N, A and S are disjoint, and N and S are disjoint.

Try It Yourself 1.15: Making Venn diagrams: U.S. states

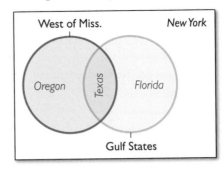

Try It Yourself 1.16: Making Venn diagrams with numbers: Reading proficiency
The rectangle represents all who were tested.

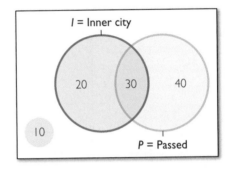

Try It Yourself 1.18: Counting using Venn diagrams: Exam results   65.

Exercise Set 1.4

Listing set elements   In Exercises 1 though 4, write the given set as a list of elements.

1. The set of whole numbers between 5 and 9 inclusive

2. The set of even whole numbers between 5 and 9 inclusive

3. The set of consonants in the alphabet

4. The set of states that border the one in which you live

Relationships among sets   Exercises 5 through 7 deal with relationships among pairs of sets. Determine which pairs are subsets and which pairs are disjoint.

5. $A = \{a, b, c, d\}$, $B = \{b, c\}$, and $C = \{d\}$.

6. A is the set of oceans, B is the set of continents, and C is the set {Asia, Pacific}.

7. A is the set of Republican governors, B is the set of governors, and C is the set of elected state officials.

8. Miscounting. A store sells only sodas and bottled water. A clerk wants to know how many items a customer wants to buy, so he asks how many sodas and how many bottles there are. The customer answers truthfully that there are 10 sodas and 10 bottles. Should the clerk conclude that there are 20 items altogether? Can the clerk determine the number of items using the given information? Use a Venn diagram to explain your answers.

9. Countries. Make a Venn diagram where the category is countries of the world. Use A to denote countries that touch the Atlantic Ocean and N to denote countries of the Northern Hemisphere. On your diagram, locate these countries: United States, Brazil, Japan, and Australia.

10. Plants. Make a Venn diagram where the category is animals. Use M for mammals and P for predators. Locate the following on your diagram: tiger, elephant, turtle, and shark.

11. Fruit. For this exercise, the category is fruit. Make a Venn diagram using the set Y of yellow fruit and the set C of citrus fruit. Locate the following fruits in your diagram: apple, banana, lemon, and orange.

Using a Venn diagram   Figure 1.23 shows a Venn diagram for students at a certain college. Let H denote the set of history majors, M the set of male students, and A the set of students receiving financial aid. Exercises 12 through 19 refer to this diagram.

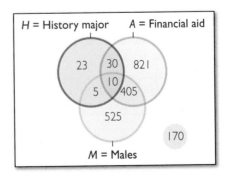

FIGURE 1.23 Students.

12. How many male history majors receive financial aid?

13. How many history majors are there?

14. How many males receive financial aid?

15. How many females receive financial aid?

16. How many history majors do not receive financial aid?

17. How many females are not history majors?

18. How many male students are there?

19. How many students are at this college?

Logic and Venn diagrams Exercises 20 through 24 refer to the Venn diagram in **Figure 1.24**, which shows the set H of history majors and the set P of political science majors at a certain college.

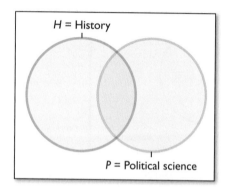

FIGURE 1.24 History and political science majors.

20. Shade in the region of Figure 1.24 where the statement *She is a political science major or a history major* is true.

21. Shade in the region of Figure 1.24 where the statement *She is a political science major but not a history major* is true.

22. Shade in the region of Figure 1.24 where the statement *She is a political science major and a history major* is false.

23. Let G be the set of all students from Greece. Add a circle for G to the Venn diagram in Figure 1.24 so that the statement *All Greek students are history majors but none are political science majors* is true.

24. Let G be the set of all students from Greece. Add a "circle" for G to the Venn diagram in Figure 1.24 so that the statement *Some Greek students are political science majors and some are history majors but none are double majors* is true. (This "circle" will not be round!)

PSA test The following table presents partial data concerning the accuracy of the PSA (Prostate Specific Antigen) test for prostate cancer.

	Has prostate cancer	Does not have prostate cancer
Test positive		2790
Test negative	220	6210

Exercises 25 through 28 refer to this table.[7]

25. Make a Venn diagram in which you locate (without numbers) the following regions. (Various choices for the sets are possible.)
- **True positives:** These are people who test positive and have prostate cancer.
- **False positives:** These are people who test positive but do not have prostate cancer.
- **True negatives:** These are people who test negative and do not have prostate cancer.
- **False negatives:** These are people who test negative but do have prostate cancer.

26. *This is a continuation of Exercise 25.* Use the table to fill in the number of people in as many of the regions of your Venn diagram as you can. For which region can you not determine a number?

27. *This is a continuation of Exercises 25 and 26.* Assume that the total number of people tested was 10,000. Fill in the number you were missing in Exercise 26.

28. *This is a continuation of Exercises 25 through 27.* The *sensitivity* of a medical test is defined to be the number of true positives divided by the number who have the disease. Find the sensitivity of the PSA test. Express your answer as a percentage. *Note*: The sensitivity is one measure of the accuracy of a medical test—accurate tests have a high sensitivity. These topics are discussed in Section 2 of Chapter 5.

29. A drug trial. In a drug trial, a test group of 50 patients is given an experimental drug, and a control group of 40 patients is given a placebo. Assume that 15 patients from the control group improved and 30 from the test group improved. Let C denote the control group and I the group of patients who improved. Make a Venn diagram using C and I. Show the number of patients in each region of the diagram.

30. Weather. Over a period of 100 days, conditions were recorded as sunny or cloudy and warm or cold. Let S denote sunny days and W warm days. Of 75 sunny days, 55 were warm. There were only 10 cold and cloudy days. Make a Venn diagram showing the numbers in each region of the diagram. How many days were warm?

31. Cars. Suppose that 178 cars were sorted by model and color. Assume that 10 were Fords but not blue, 72 were blue but not Fords, and 84 were neither blue nor Fords. Make a Venn diagram using Fords and blue cars, and include the numbers in each region of the diagram. How many blue cars were observed?

32. Employed freshmen. A certain college has 11,989 students. Assume that 6120 are unemployed sophomores, juniors, or seniors and that 4104 are employed sophomores, juniors, or seniors. There are 745 employed freshmen. Use F for freshmen and E for employed students, and make a Venn diagram marking the number of students in each region of the diagram. How many freshmen are there?

Beverage survey A survey of 150 students asked which of the following three beverages they liked: sodas, coffee, or bottled water. The results are given in the table below.

Sodas only	28
Coffee only	17
Water	51
Sodas and water	26
Coffee and water	27
Sodas and coffee but not water	9
Coffee and water, not sodas	6
All three	21

This information is used in Exercises 33 through 39.

33. Use *S* for soda, *W* for water, and *C* for coffee, and make a Venn diagram for drinks. The completed diagram should show the number in each region.

34. How many liked coffee only?

35. *This exercise uses the results of Exercise 33.* How many liked water but not coffee?

36. *This exercise uses the results of Exercise 33.* How many liked none of the three?

37. How many liked coffee and water?

38. *This exercise uses the results of Exercise 33.* How many liked exactly two of the three?

39. *This exercise uses the results of Exercise 33.* How many liked exactly one of the three?

News: A survey asked a group of people where they got their news: a newspaper, TV, or the Internet. The responses are recorded in the following table.

Newspaper and TV	46
TV only	20
Internet only	12
Newspaper and Internet	57
TV and Internet	50
Newspapers	81
TV and Internet, but not newspapers	19
All three	31
None of the three	8

This information is used in Exercises 40 through 44.

40. Using *N* for a newspaper, *T* for TV, and *I* for the Internet, make a Venn diagram for news sources. The completed diagram should give the number in each region of the diagram.

41. *This exercise uses the results of Exercise 40.* How many were surveyed?

42. *This exercise uses the results of Exercise 40.* How many got their news from only one source?

43. *This exercise uses the results of Exercise 40.* How many got some of their news from the Internet?

44. *This exercise uses the results of Exercise 40.* How many got none of their news from TV?

45. Research on vowels. In the solution to Example 1.14, it was stated that the set of vowels is $V = \{a, e, i, o, u\}$ and that sometimes *y* is a vowel, depending on its usage. Look up the meaning of the word "vowel" and write a report on what you discover.

1.5 Critical thinking and number sense: What do these figures mean?

TAKE AWAY FROM THIS SECTION

Cope with the myriad measurements the average consumer encounters every day.

The following excerpt is from an article that appeared in the *Wall Street Journal*.

An artist's depiction of 1 billion dollars. The pallets contain 10 million hundred-dollar bills.

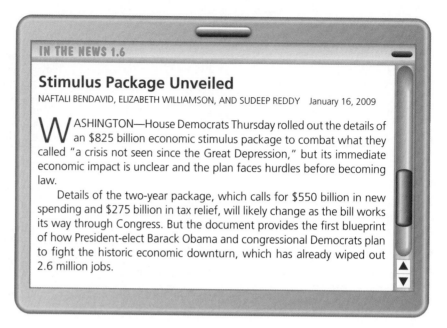

Stimulus Package Unveiled

NAFTALI BENDAVID, ELIZABETH WILLIAMSON, AND SUDEEP REDDY   January 16, 2009

WASHINGTON—House Democrats Thursday rolled out the details of an $825 billion economic stimulus package to combat what they called "a crisis not seen since the Great Depression," but its immediate economic impact is unclear and the plan faces hurdles before becoming law.

Details of the two-year package, which calls for $550 billion in new spending and $275 billion in tax relief, will likely change as the bill works its way through Congress. But the document provides the first blueprint of how President-elect Barack Obama and congressional Democrats plan to fight the historic economic downturn, which has already wiped out 2.6 million jobs.

Amounts of money like the $825 billion in the proposed stimulus package are so large that their meaning may be lost.[8] Perhaps it helps to note that if every man, woman, and child in the state of Oklahoma bought a $25,000 automobile, they wouldn't come close to spending $825 billion. In fact, there would be enough money left over to buy $25,000 automobiles for each living soul in the states of Arkansas, Iowa, Kansas, Louisiana, Minnesota, Missouri, Nebraska, New Mexico, North Dakota, and South Dakota.

Making informed decisions often depends on understanding what very large or very small numbers really mean. Putting such numbers into a personal conte us get a sense of their true scale. We can also use estimation to sort our way through otherwise intimidating calculations.

Magnitudes

Often we convey information about numbers by using relative sizes or *magnitudes* rather than specific amounts. For example, the number of dollars in your wallet may be counted in tens, the number of students on your campus may be counted in thousands, Earth's population is measured in billions, and the national debt is expressed in trillions of dollars. These examples of magnitudes are integral powers of 10. Some examples will show what powers of 10 mean.

Examples of powers of 10

Positive powers of 10

- $10^3 = 1000$ (three zeros) is a thousand.

- $10^6 = 1,000,000$ (six zeros) is a million, which is a thousand thousands.

- $10^9 = 1,000,000,000$ (nine zeros) is a billion, which is a thousand millions.

- $10^{12} = 1,000,000,000,000$ (twelve zeros) is a trillion, which is a thousand billions.

Negative powers of 10

- $10^{-2} = 0.01$ (two digits to the right of the decimal) is a hundredth.

- $10^{-3} = 0.001$ (three digits to the right of the decimal) is a thousandth.

---

[8]Everett Dirksen, a senator from Illinois several decades ago, is often quoted as saying, "A billion here, a billion there, and pretty soon you're talking real money."

- $10^{-6} = 0.000\,001$ (six digits to the right of the decimal) is a millionth.

- $10^{-9} = 0.000\,000\,001$ (nine digits to the right of the decimal) is a billionth.

The gigantic number 10^{100} is a "one" followed by 100 zeros and actually has the funny name *googol*. (It's an interesting fact that the name of the company Google is a misspelling of googol.) Physicists believe there are only about 10^{80} atoms in the universe—far fewer than a googol.

To get a better grasp of these magnitudes, we refer to the figures below. The first figure is a view of space through a frame that is 10^{21} meters on each side.[9] That frame is large enough to show the entire Milky Way galaxy, which has a diameter of about 100,000 lightyears. (A lightyear is the distance light travels in a year, about 10^{16} meters.) The pictures decrease in scale by powers of 10 until we finally view the structure of a DNA molecule through a frame that is 10^{-9} meter on a side.

The Milky Way Galaxy is about 10^{21} meters across.

The diameter of Earth is just over 10^7 meters.

San Francisco Bay is about 10^5 meters, or 100 kilometers, long.

Some lily flowers are about 0.1, or 10^{-1}, meter across.

The eye of a bee is about 10^{-3} meter across.

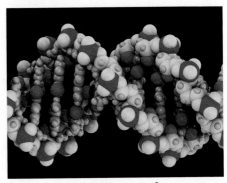

A DNA molecule is about 10^{-9} meter across.

[9]One meter is about 3.28 feet.

Dealing with powers of 10 sometimes requires a bit of algebra with exponents. The following quick review is available for those who need a refresher. Additional information about exponents, including scientific notation, is provided in Appendix 2.

Quick Review Exponents

Negative exponents: a^{-n} is the reciprocal of a^n.

Definition	Example
$a^{-n} = \dfrac{1}{a^n}$	$10^{-3} = \dfrac{1}{10^3} = \dfrac{1}{1000} = 0.001$

Zero exponent: If $a \neq 0$, then $a^0 = 1$.

Definition	Example
$a^0 = 1$	$10^0 = 1$

Basic properties of exponents:

Property	Example
$a^p a^q = a^{p+q}$	$10^2 \times 10^3 = 10^{2+3} = 10^5 = 100,000$
$\dfrac{a^p}{a^q} = a^{p-q}$	$\dfrac{10^6}{10^4} = 10^{6-4} = 10^2 = 100$
$(a^p)^q = a^{pq}$	$\left(10^3\right)^2 = 10^{3\times2} = 10^6 = 1,000,000$

A familiar example of magnitudes is provided by computer memory, which is measured in terms of *bytes*. A single byte holds one character (such as a letter of the alphabet). Modern computing devices have such large memories that we measure them using the following terms:

A *kilobyte* is a thousand $= 10^3$ bytes, usually abbreviated *KB*.

A *megabyte* is a million $= 10^6$ bytes, usually abbreviated *MB* or *meg*.

A *gigabyte* is a billion $= 10^9$ bytes, usually abbreviated *GB* or *gig*.

A *terabyte* is a trillion $= 10^{12}$ bytes, usually abbreviated *TB*.

To find out how many megabytes are in a gigabyte, we divide:

$$\text{Numbers of megs in a gig} = \frac{10^9 \text{ bytes}}{10^6 \text{ bytes}}$$
$$= 10^{9-6}$$
$$= 10^3.$$

Thus, there are 1000 megabytes in a gigabyte. Similarly, we calculate that there are a million megabytes in a terabyte.

EXAMPLE 1.20 Comparing sizes using powers of 10: Computer memory

In the 1980s, one of this book's authors owned a microcomputer that had a memory of 64 kilobytes. The computer on his desk today has 4 gigabytes of memory. Is the memory of today's computer tens, hundreds, or thousands of times as large as the old computer's memory?

SOLUTION

We express the size of the memory using powers of 10. The old computer had a memory of 64×10^3 bytes, and today's computer has 4×10^9 bytes of memory. To compare the sizes, we divide:

$$\frac{\text{New memory size}}{\text{Old memory size}} = \frac{4 \times 10^9 \text{ bytes}}{64 \times 10^3 \text{ bytes}}$$

$$= \frac{4}{64} \times 10^{9-3}$$

$$= 0.0625 \times 10^6$$

$$= 62{,}500.$$

The new computer has over 60 thousand times as large a memory as the old computer.

TRY IT YOURSELF 1.20

One kilobyte of computer memory will hold about 2/3 of a page of typical text (without formatting). A typical book has about 400 pages. Can the 9-gig flash drive I carry in my pocket hold tens, hundreds, or thousands of such books?

The answer is provided at the end of this section.

Taming large and small numbers

Getting a handle on large numbers (or very small numbers) often means expressing them in familiar terms. For example, to get a feeling for how much a billion is, let's change the distance of a billion inches into terms that are more meaningful.[10] There are 12 inches in a foot and 5280 feet in a mile. So, we change inches to feet by dividing by 12 and feet to miles by dividing by 5280. A billion inches is 10^9 inches. Therefore, a billion inches equals

$$10^9 \text{ inches} = 10^9 \times \frac{1}{12} \text{ feet}$$

$$= 10^9 \times \frac{1}{12} \times \frac{1}{5280} \text{ miles}.$$

This is about 16,000 miles. The length of the equator of Earth is about 25,000 miles, so 1 billion inches is about 2/3 of the length of the equator of Earth.

As another illustration, suppose you have a lot of dollar bills—1 billion of them. If you put your dollars in a stack on the kitchen table, how tall would the stack be? A dollar bill is about 0.0043 or 4.3×10^{-3} inch thick. A billion is 10^9, so the height of the stack would be

$$4.3 \times 10^{-3} \times 10^9 = 4.3 \times 10^{-3+9} = 4.3 \times 10^6 = 4.3 \text{ million inches.}$$

As before, we divide by 12 to convert to feet and then by 5280 to convert to miles:

$$4.3 \times 10^6 \text{ inches} = 4.3 \times 10^6 \times \frac{1}{12} \text{ feet}$$

$$= 4.3 \times 10^6 \times \frac{1}{12} \times \frac{1}{5280} \text{ miles}.$$

This is about 67.9 miles, so 1 billion dollar bills make a stack almost 68 miles high.

---

[10]Unit conversion is reviewed in Appendix 1.

EXAMPLE 1.21 Understanding large numbers: The national debt

As of January 2009, the national debt was about 10.6 trillion dollars, and there were about 305 million people in the United States. The national debt is not just an abstract figure. The American people actually owe it. Determine how much each person in the United States owes. Round your answer to the nearest $100.

SOLUTION

We use powers of 10 to express both the national debt (10.6 trillion = 10.6×10^{12}) and the population of the United States (305 million = 305×10^6):

$$\text{Debt per person} = \frac{10.6 \times 10^{12} \text{ dollars}}{305 \times 10^6}$$

$$= \frac{10.6}{305} \times 10^{12-6} \text{ dollars}$$

$$= \frac{10.6}{305} \times 10^6 \text{ dollars.}$$

This is about 0.0348×10^6 dollars, so each person owes $34,800.

TRY IT YOURSELF 1.21

The U.S. Census Bureau estimated that there were approximately 113,000,000 households in the United States in 2009. How much did each household owe on the national debt in 2009? Round your answer to the nearest $100.

The answer is provided at the end of this section.

At the other end of the scale from numbers describing the national debt or computer memory are extremely small numbers. Microchips that can be implanted in pets are about 0.0015 meter in diameter. That is 1.5 millimeters—about the size of an uncooked grain of rice. The eye of a bee, which is about 1 millimeter across (see the figure on p. 57), may provide additional context for this magnitude.

A smaller magnitude is a *micron*, which is one-millionth of a meter or 10^{-6} meter. In this scale, you could get a close-up view of a bacterium.

Even smaller are the particles used in the developing field of *nanotechnology*, which is the control of matter on an atomic or molecular scale. It may be possible to develop *nanoparticles* in the 10- to 100-*nanometer* range that can seek out and selectively destroy cancer cells. One nanometer is 10^{-9} meter, a billionth of a meter.

On this scale, we can view the structure of DNA, as we saw in the illustration on p. 57. We further develop our sense of these very small sizes in the next example.

EXAMPLE 1.22 Understanding very small numbers: Nanoparticles

Human hair can vary in diameter, but one estimate of the average diameter is 50 microns or 50×10^{-6} meter. How many 10-nanometer particles could be stacked across the diameter of a human hair?

SOLUTION

Now 10 nanometers is 10×10^{-9} meter. We divide to find the number of 10-nanometer particles that can be stacked across the diameter of a human hair:

$$\frac{\text{Diameter of hair}}{\text{Diameter of nanoparticle}} = \frac{50 \times 10^{-6} \text{ meter}}{10 \times 10^{-9} \text{ meter}}$$
$$= 5 \times 10^{-6+9}$$
$$= 5 \times 10^{3}.$$

Thus, 5000 of the nanoparticles would be needed to span the diameter of a human hair.

TRY IT YOURSELF 1.22

Suppose the universe expanded so that the 10-nanometer particle grew to the size of a 15-millimeter (15×10^{-3} meter) diameter marble. If a 15-millimeter marble grew in the same proportion, how many meters long would its new diameter be? How many miles long would the new diameter be? (Use the fact that 1 kilometer is about 0.62 mile, and round your answer to the nearest mile.)

The answer is provided at the end of this section.

Estimation

There are many occasions when a quick, ballpark figure will suffice—an exact calculation is not necessary. In fact, many of the examples already discussed above involve estimates. What we will pursue here are examples that a person would tend to encounter in everyday life.

For instance, if a salesman is showing us a car, we need to know whether it really is in our price range. Suppose we are looking at a car with a price tag of $23,344 and we plan to pay off the car in 48 monthly payments at an annual interest rate of 6%. We may need help figuring out the exact monthly payment, especially considering the interest that will be charged for the loan. But we can get a rough estimate by approximating the numbers by ones that make the calculation easier.

Suppose we estimate the cost of the car to be $25,000 and round off the 48 monthly payments to 50 monthly payments. These numbers are much easier to handle, and we should be able to figure in our heads what the monthly payments using them would be. If we ignore interest, we find

$$\text{Estimated monthly payment} = \frac{\text{Total cost}}{\text{Number of payments}} = \frac{\$25{,}000}{50} = \$500.$$

In Chapter 4, we will learn a formula for the monthly payment in such a situation. According to that formula, borrowing $23,344 at an annual interest rate of 6% and repaying in 48 monthly payments result in a monthly payment of $548.23. So our estimate was a bit low. But if we can afford (say) only $450 per month, our estimate warns us to look for a less expensive car.

EXAMPLE 1.23 Estimating costs: Buying gas in the United States and Canada

You are traveling to Canada, and you wonder whether to gas up before you cross the border. You see a sign at a gas station on the U.S. side of the border showing $3.77, and on the Canadian side of the border, you see a sign showing $1.10. One might think that the Canadian gas is much cheaper, but let's use critical thinking and look a little closer.

In Canada (and most of the rest of the world), gasoline is measured in liters rather than gallons. Also, the dollar sign refers not to U.S. dollars but to *Canadian* dollars. So the sign on the Canadian side of the border means that gasoline costs 1.10 Canadian dollars per liter.

In parts a and b, use an exchange rate of 0.94 U.S. dollar per Canadian dollar.

a. Use the fact that a quart and a liter are nearly the same and that the Canadian dollar is worth a little less than the U.S. dollar to get a quick estimate of the cost of gasoline in Canada measured in U.S. dollars per gallon.

b. There are 3.79 liters in a gallon. Use this fact to find the actual cost in U.S. dollars per gallon of gasoline at the Canadian station. Was your estimate in part a good enough to tell you where you should buy your gas?

SOLUTION

a. Because 1 Canadian dollar is a little less than 1 U.S. dollar and 1 liter costs 1.10 Canadian dollars, gas costs about 1 U.S. dollar per liter. Because a quart is about the same as a liter, there are about 4 liters in a gallon. This means that gas in Canada costs approximately 4 U.S. dollars per gallon. This estimate suggests that gas is cheaper at the station in the United States.

b. Because there are 3.79 liters in a gallon, gasoline in Canada costs

$$3.79 \frac{\text{liters}}{\text{gallon}} \times 1.10 \frac{\text{Canadian dollars}}{\text{liter}} = 4.169 \frac{\text{Canadian dollars}}{\text{gallon}}.$$

One way to check that the units are converted correctly is to include them in the computation as we did above and combine or cancel them using the usual rules of algebra. Note that the units involving liters cancel. Further information on unit conversion is in Appendix 1.

Because each Canadian dollar is worth 0.94 U.S. dollar, we multiply by 0.94 to get the cost in U.S. dollars of a gallon of gas in Canada:

$$\text{Cost in U.S. dollars} = 4.169 \times 0.94.$$

This is about 3.92, so the actual cost of gas on the Canadian side of the border is 3.92 U.S. dollars per gallon. This calculation confirms that gas is cheaper at the station in the United States.

Our estimate in part a was good enough to tell us that gas on the American side was the better buy.

Areas and volumes are important measurements for consumers. For example, carpet is priced in both square feet and square yards. There are 3 feet in a yard. How many square feet are in a square yard? In **Figure 1.25** (p. 64), we have made a picture of a square yard. It has a length and width of 3 feet. Recall that the area of a rectangle is the length times the width. So the area in square feet of a square yard is

$$\text{Area of 1 square yard} = \text{Length} \times \text{Width}$$
$$= 3 \text{ feet} \times 3 \text{ feet}$$
$$= 9 \text{ square feet}.$$

Concrete is usually priced by the cubic yard.[11] **Figure 1.26** (p. 64) shows that 1 cubic yard is a cube that is 3 feet on each side. Recall that the volume of a box is the length times the width times the height. So the volume in cubic feet of a cubic yard is

$$\text{Volume of 1 cubic yard} = \text{Length} \times \text{Width} \times \text{Height}$$
$$= 3 \text{ feet} \times 3 \text{ feet} \times 3 \text{ feet}$$
$$= 27 \text{ cubic feet}.$$

So 1 square yard is 9 square feet, and 1 cubic yard is 27 cubic feet.

The next example illustrates that savvy consumers need both the skill of estimation and the ability to measure areas and volumes.

---

[11] Often sellers of concrete quote the price "per yard." What they really mean is the price per *cubic* yard.

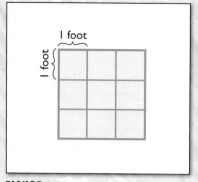

FIGURE 1.25 1 square yard is 9 square feet.

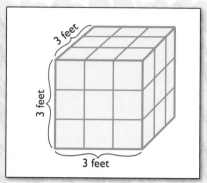

FIGURE 1.26 1 cubic yard is 27 cubic feet.

EXAMPLE 1.24 Estimating with areas and volumes: Flooring

You want to redo your living room floor. Hardwood costs $10.40 per square foot, and the carpet you like costs $28.00 per (square) yard.[12]

    a. You need to decide right away whether you should be looking at hardwood or carpet for your floor. Use the fact that the cost of hardwood is about $10 per square foot to estimate the cost of a square yard of hardwood. Use your estimate to decide how the cost of hardwood compares with the cost of carpet.

    b. Find the actual cost of a square yard of hardwood.

SOLUTION

    a. A common (and ultimately expensive) consumer error is to think that, because 1 yard is 3 feet, 1 square yard is 3 square feet. But as we noted above, 1 square yard is, in fact, 9 square feet. Because hardwood costs about $10 per square foot, 9 square feet costs about $90. Because the carpet is $28 per yard, we estimate that hardwood is over three times as expensive as carpet.

    b. To find the actual cost of a square yard of hardwood, we multiply the exact cost of $10.40 per square foot by 9 square feet:

$$\$10.40 \times 9 = \$93.60.$$

The result is a bit higher than the estimate of $90 we found in part a, and it confirms our conclusion that hardwood is much more expensive than carpet.

TRY IT YOURSELF 1.24

A German sports car advertises an engine displacement of 3500 cubic centimeters. An American sports car boasts a 302-cubic-inch engine. Use the fact that 1 centimeter is 0.39 inch to determine which displacement is larger.

The answer is provided at the end of this section.

Try It Yourself answers

Try It Yourself 1.20: Comparing sizes using powers of 10: Computer memory Thousands: The flash drive will hold about 15,000 such books.

Try It Yourself 1.21: Understanding large numbers: The national debt $93,800.

Try It Yourself 1.22: Understanding very small numbers: Nanoparticles   22,500 meters or 14 miles.

---

[12]Often sellers of carpet quote the price "per yard." What they really mean is the price per *square* yard.

Try It Yourself 1.24: Estimating with areas and volumes: Engine displacement
The American engine is larger because the German engine is about 208 cubic inches.

Exercise Set 1.5

1. Rhinoviruses. Rhinoviruses, which are the leading cause of the common cold, have a diameter of 20 nanometers (20×10^{-9} meter). Pneumococcus, a bacterium causing pneumonia, measures about 1 micron (10^{-6} meter) in diameter. How many times as large in diameter as the virus is the bacterium?

2. Trillion. How many millions are in a trillion?

3. Pennies. Some high school students want to collect a million pennies for charity. How many dollars is that?

4. Nevada. In January 2009 the state of Nevada was facing a budget deficit of $1.5 billion. The population of the state was approximately 2 million. How much would each person in the state have to contribute to pay this deficit?

A billion dollars A 1-dollar bill is 6.14 inches long, 2.61 inches wide, and 0.0043 inch thick. This information is used in Exercises 5 through 10.

5. If you had a billion dollars in 100-dollar bills, how many bills would you have?

6. If you laid a billion 1-dollar bills side by side, how many miles long would the trail be? Round your answer to the nearest mile.

7. If you laid a billion 1-dollar bills end to end, how many miles long would the trail be? (Round your answer to the nearest mile.) About how many times around Earth would this string of bills reach? (The circumference of Earth is approximately 25,000 miles.)

8. If you spread a billion 1-dollar bills over the ground, how many square miles would they cover? Round your answer to the nearest square mile.

9. How many cubic feet would a billion dollars occupy? Round your answer to the nearest cubic foot.

10. Measure your classroom. How many such rooms would a billion 1-dollar bills fill?

11. Counting to a billion. Let's say it takes about 3 seconds to count to ten out loud. How many years will it take to count out loud to a billion? Ignore leap years and round your answer to 1 decimal place.

12. Queue. Suppose we lined up all the people in the world (about 6.7 billion people). Assume that each person occupied 12 inches of the line. How many times around Earth would they reach? (Earth has a circumference of about 25,000 miles.) Round your answer to 1 decimal place.

13. Blogosphere growth. The number of blogs (or weblogs) has grown rapidly. According to one report, in July 2006 about 175,000 new blogs were created each day.[13] How many blogs were created, on average, each second? Round your answer to one decimal place.

14. Lightyear. A lightyear is the distance light travels in a year. Light travels 186,000 miles in a second. How many trillion miles long is a lightyear? Round your answer to one decimal place.

15. Mars rovers. The Mars rovers Spirit and Opportunity are remotely controlled from Earth, and the signal is delayed due to the great distance. The signal travels at the speed of light, which is 186,000 miles in a second. Assume that Mars is about 140 million miles from Earth when a controller sends a command to a rover. How long will it take for the controller to know what actually happened? In other words, how long will it take for the signal to make the 140-million-mile trip to Mars and back? Give your answer to the nearest minute.

16. Light. Light travels 186,000 miles in a second. How many feet does it travel in a nanosecond? (A nanosecond is a billionth of a second.) Round your answer to the nearest foot.

17. You try. Several ways to convey the size of a billion were explored in this section. You try to find an effective way to convey the size of a billion.

18. Loan I. You are borrowing $7250 to buy a car. If you pay the loan off in 3 years (36 months), estimate your monthly payment by ignoring interest and rounding down the amount owed.

19. Loan II. You have a $3100 student loan. Suppose you pay the loan off in two and a half years (30 months). Estimate your monthly payment by ignoring interest and rounding down the amount owed.

20. Gas. You're going on a 1211-mile trip, and your car gets about 29 miles per gallon. Gas prices along your route average $1.99. Estimate the cost of gasoline for this trip by rounding the distance to 1200 miles, the mileage to 30 miles per gallon, and the cost of gas to $2.00 per gallon. Calculate the answer without using estimation and compare the two results.

21. Foreign aid. In 2007 the U.S. government spent $31.8 billion on foreign aid. The entire budget was $2.77 trillion. Estimate the percentage of the budget spent on foreign aid by rounding the foreign aid budget to $30 billion and the entire budget to $3 trillion. Calculate the percentage without using estimation and compare the two results. Round your second answer in percentage form to two decimal places.

22. Oil. In 2008 the United States consumed 20,680,000 barrels of oil per day. A barrel of oil is 42 U.S. gallons. In 2008 there were about 305,000,000 people in the United States. Estimate how many gallons each person in the United States consumed each day in 2008 by rounding down all three numbers. Calculate the answer without using estimation and compare the two results. Round your answers to two decimal places.

---

[13] See www.sifry.com/alerts/archives/000436.html.

23. Carpet. You need to order carpet. One store advertises carpet at a price of $1.50 per square foot. Another store has the same carpet advertised at $12.00 "per yard." (What the story really means is the price per *square* yard.) Which is the better buy?

World population As of 2008, the world's population was about 6.7 billion people. The population of the United States in 2008 was 305 million. This information is used in Exercises 24 through 26.

24. In 2008 what percent of the world's population was the U.S. population? Round your answer as a percentage to one decimal place.

25. The land area of the state of Texas is about 261,797 square miles. If in 2008 all the world's people were put into Texas, how many people would there be in each square mile? Round your answer to the nearest whole number.

26. The area of New York City is 305 square miles. If in 2008 every person in the world moved into New York City, how many square feet of space would each person have? Round your answer to one decimal place. *Note*: There are 5280 feet in a mile. First determine how many square feet are in a square mile.

CHAPTER SUMMARY

Today's media sources bombard us with information and misinformation, often with the aim of selling us products, ideas, or political positions. Sorting the good from the bad is a constant challenge that requires a critical eye. Solid reasoning is a crucial tool for success.

Public policy and Simpson's paradox: Is "average" always average?

Simpson's paradox is a striking example of the need for critical thinking skills. An average based on combining categories can be skewed by some factor that distorts the overall picture but not the underlying patterns. Considering only the overall average may lead to invalid conclusions. A simple, but hypothetical, example is given by the following table, which shows applicants for medical school and law school.

	Males			Females		
	Applicants	Accepted	% accepted	Applicants	Accepted	% accepted
Medical school	10	2	20.0%	20	4	20.0%
Law school	20	5	25.0%	12	3	25.0%
Total	30	7	23.3%	32	7	21.9%

Note that the medical school accepts 20% of male applicants and 20% of female applicants. The law school accepts 25% of male applicants and 25% of female applicants. But overall 23.3% of male applicants are accepted and only 21.9% of female applicants are accepted. The discrepancy is explained by the fact that females tended to apply to the medical school and males tended to apply to the law school—and the medical school accepts fewer applicants (on a percentage basis) than does the law school. The overall average might lead one to suspect gender bias where none exists.

Logic and informal fallacies: Does that argument hold water?

Logic is the study of methods and principles used to distinguish good reasoning from bad. A logical argument involves one or more *premises* (or *hypotheses*) and a *conclusion*. The premises are used to justify the conclusion. The argument is *valid* if the premises properly support the conclusion.

People who try to sway our opinions regarding products, politics, or public policy sometimes use flawed logic to bolster their case. The well-informed citizen can recognize phony arguments. We consider two main types of fallacies: *fallacies of relevance* and *fallacies of presumption*. A classic fallacy of presumption is *false cause*. For example, a commercial may tell us that people who drive our sports cars have lots of friends. The implication is that you will have more friends if you buy their car. But, in fact, no evidence is offered to show that either event *causes* the other. The fact that two events occur together does not automatically mean that one causes the other. There are several other informal fallacies.

An argument of the form *if premises then conclusion* is a *deductive* argument. An *inductive* argument draws a general conclusion from specific examples. It does not claim that the premises give irrefutable proof of the conclusion, only that the premises provide plausible support for it. The idea of an inductive argument is closely tied to pattern recognition.

Formal logic and truth tables: Do computers think?

Logic has a formal structure that may remind us of algebra. This structure focuses on the way statements are combined, and it is the basis on which computers operate. One feature of formal logic is the *truth table*, which gives the truth or falsity of a compound statement based on the truth value of its components.

Operations on statements include *negation*, *conjunction*, and *disjunction*. Of particular interest is the *conditional statement* or *implication*. Such statements have the form "if *hypothesis* then *conclusion*." One example is, "If I am elected, then your taxes will decrease." Analyzing conditional statements in terms of the converse, the inverse, and the contrapositive can be useful. The politician who made the preceding statement may wish you to believe that if she is not elected, then your taxes will not decrease. No such conclusion is warranted: Replacing a statement by its inverse is not valid.

It is important to remember that a conditional statement is considered to be true if the hypothesis is false. For example, consider again the following statement: *If I am elected, then your taxes will decrease.* This statement is true if the politician is not elected—regardless of what happens to your taxes.

Computers use *logic gates* that perform functions corresponding to logical operations. In classical computing, information is stored using the two numbers 0 and 1. The connection with logic is made by thinking of 0 as "false" and 1 as "true."

Sets and Venn diagrams: Pictorial logic

Sets can be used, among other things, to analyze logical statements. The basic tool for this purpose is the Venn diagram. Consider this statement: *If you speed, then you will get a ticket.* We can represent it using the Venn diagram shown in **Figure 1.27**. One circle represents the collection of all speeders, and the other represents those who get a ticket. The fact that the "Speeders" circle is inside the "Ticket" circle means that all speeders get tickets.

We also use Venn diagrams to show how categories relate to each other. Suppose, for example, that our state legislature consists only of Democrats and Republicans. **Figure 1.28** shows how the House divides by party and gender: One circle represents females; the outside of that circle represents males. The other circle represents Democrats; the outside of that circle represents Republicans. So the intersection of the two circles shows female Democrats. Venn diagrams can also be used to count complicated sets.

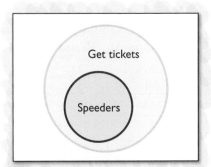

FIGURE 1.27 Venn diagram illustrating a statement.

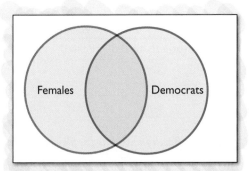

FIGURE 1.28 House members divided by sex and party.

Critical thinking and number sense: What do these figures mean?

Relative sizes of numbers are often indicated using *magnitudes* or powers of 10. For example, a city may have a population in hundreds of thousands (10^5), and world population is measured in billions (10^9).

We can get a feeling for very large and very small numbers by putting them into a more familiar context. For example, the national debt of 10.6 trillion dollars means that every man, woman, and child in the United States owes about $34,800.

One way avoid complicated computations is to estimate by rounding the given numbers to make the arithmetic easier. Another method of estimation is to ignore features that complicate the calculation.

KEY TERMS

logic, p. 13
premises, p. 15
hypotheses, p. 15
conclusion, p. 15
valid, p. 15
informal fallacy, p. 15

formal fallacy, p. 15
deductive, p. 21
inductive, p. 21
or, p. 29
logically equivalent, p. 35
set, p. 41

elements, p. 41
subset, p. 41
disjoint, p. 41
Venn diagram, p. 42

CHAPTER QUIZ

1. Students at College A and College B were allowed to take either an English exam or a history exam. The results are presented in the table below.

	College A		
	Students tested	Passed	% passed
English	500	320	
History	100	70	
Total			

	College B		
	Students tested	Passed	% passed
English	50	31	
History	60	42	
Total			

Complete the table by filling in the blank spaces. Round each percentage to one decimal place.

Answer

	College A			College B		
	Students tested	Passed	% passed	Students tested	Passed	% passed
English	500	320	64.0%	50	31	62.0%
History	100	70	70.0%	60	42	70.0%
Total	600	390	65.0%	110	73	66.4%

If you had difficulty with this problem, see Example 1.1.

2. In Exercise 1, which college performed better on the English exam? Which college performed better on the history exam? Which college performed better overall?

Answer College A performed better on English. They tied on history. College B performed better overall.
If you had difficulty with this problem, see Example 1.1.

3. Identify the premises and conclusion of the following argument. Is this argument valid?
Most basketball players are tall. Tom is tall. Therefore, Tom is a basketball player.

Answer The premises are (1) *Most basketball players are tall.* (2) *Tom is tall.* The conclusion is *Tom is a basketball player.* The argument is not valid because the premises do not support the conclusion.
If you had difficulty with this problem, see Example 1.3.

4. Identify the type of the fallacy in the following argument. *John is a smoker, so you shouldn't trust his advice about a healthy life style.*

Answer This is dismissal based on personal attack. John's habits are attacked rather than his argument.
If you had difficulty with this problem, see Example 1.5.

5. Let *p* represent the statement *You study for your math exam* and *q* the statement *You pass your math exam.* Use these letters to express symbolically the statement *If you don't study for your math exam, you fail the math exam.*

Answer (NOT *p*) → (NOT *q*).
If you had difficulty with this problem, see Example 1.10.

6. Make the truth table for *p* → (NOT *q*).

Answer

p	*q*	NOT *q*	*p* → (NOT *q*)
T	T	F	F
T	F	T	T
F	T	F	T
F	F	T	T

If you had difficulty with this problem, see Example 1.11.

7. Tests for drug use are not always accurate. Make a Venn diagram using the categories of drug users and those who test positive. Shade the area of the diagram that represents *false negatives*, that is, drug users who test negative.

Answer The region for false negatives is shaded.

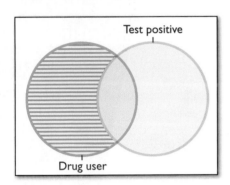

If you had difficulty with this problem, see Example 1.15.

8. The national debt in a country with a population of 1.1 million is 4.2 billion dollars. Calculate the amount each person owes. Round your answer to the nearest $100.

Answer Each person owes

$$\frac{4.2 \text{ billion dollars}}{1.1 \text{ million}} = \frac{4.2 \times 10^9 \text{ dollars}}{1.1 \times 10^6} = \frac{4.2}{1.1} \times 10^{9-6} \text{ dollars}$$

$$= \frac{4.2}{1.1} \times 10^3 \text{ dollars}.$$

This is about 3.8×10^3 dollars, so each person owes about $3800.

If you had difficulty with this problem, see Example 1.21.

9. A group of 19 people buys $207 worth of pizza. Estimate the amount each owes for lunch by rounding the number of people to 20 and the price to $200. What is the exact answer?

Answer The estimate is $200/20 = $10. The exact amount is $207/19 or about $10.89.

If you had difficulty with this problem, see Example 1.24.

CHAPTER 2

ANALYSIS OF GROWTH

2.1 Measurements of growth: How fast is it changing?

2.2 Graphs: Picturing growth

2.3 Misleading graphs: Should I believe my eyes?

The following article appears at the *Valleywag* Web site, which bills itself as *Silicon Valley's Tech Gossip Rag.*

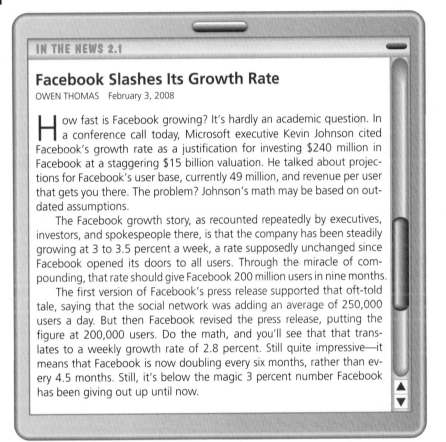

IN THE NEWS 2.1

Facebook Slashes Its Growth Rate

OWEN THOMAS   February 3, 2008

How fast is Facebook growing? It's hardly an academic question. In a conference call today, Microsoft executive Kevin Johnson cited Facebook's growth rate as a justification for investing $240 million in Facebook at a staggering $15 billion valuation. He talked about projections for Facebook's user base, currently 49 million, and revenue per user that gets you there. The problem? Johnson's math may be based on outdated assumptions.

The Facebook growth story, as recounted repeatedly by executives, investors, and spokespeople there, is that the company has been steadily growing at 3 to 3.5 percent a week, a rate supposedly unchanged since Facebook opened its doors to all users. Through the miracle of compounding, that rate should give Facebook 200 million users in nine months.

The first version of Facebook's press release supported that oft-told tale, saying that the social network was adding an average of 250,000 users a day. But then Facebook revised the press release, putting the figure at 200,000 users. Do the math, and you'll see that that translates to a weekly growth rate of 2.8 percent. Still quite impressive—it means that Facebook is now doubling every six months, rather than every 4.5 months. Still, it's below the magic 3 percent number Facebook has been giving out up until now.

This article, which is accompanied by **Figure 2.1**, makes clear that the growth rate of Facebook is crucial to an important financial decision. The crucial question is whether the graph supports the conclusions drawn in the article. It may or may not be easy to determine the degree to which the graph supports the article, but clearly this type of analysis is important not only to Microsoft executives but also to anyone who wants to understand the real facts about the growth of Facebook.

In this technological age, we are overwhelmed with information. Frequently, this information includes numerical data: the number of cases of flu each year, the variation in the inflation rate over the last decade, the growth rate of Facebook, and so forth. The media present quantitative information to have maximum visual impact; using graphics as in the *Valleywag* article to convey the information is common.

In this chapter, we illustrate how such information can be presented using graphs and tables. A key insight is that graphs and tables reflect growth rates. We will learn

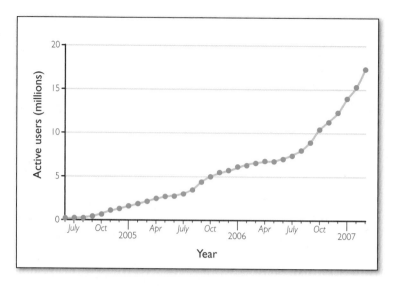

FIGURE 2.1 Growth of Facebook.

to analyze data and to make predictions. We will also examine the advantages and disadvantages of different kinds of graphical representations.

You should learn to be wary of graphical presentations and to recognize when data are misrepresented and when inaccurate conclusions are drawn. Empowering you to analyze tables and graphs critically is the ultimate goal of this chapter.

2.1 Measurements of growth: How fast is it changing?

TAKE AWAY FROM THIS SECTION

Use growth rates to analyze quantitative information.

The following article from *USA TODAY* deals with college graduation rates for athletes.

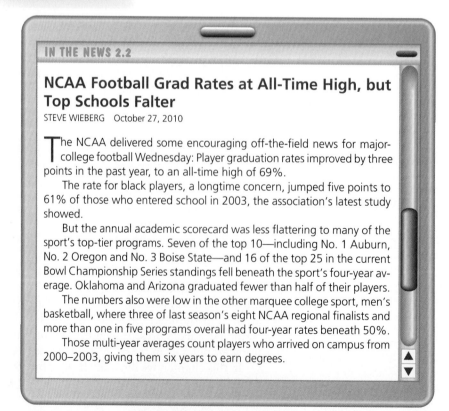

IN THE NEWS 2.2

NCAA Football Grad Rates at All-Time High, but Top Schools Falter

STEVE WIEBERG   October 27, 2010

The NCAA delivered some encouraging off-the-field news for major-college football Wednesday: Player graduation rates improved by three points in the past year, to an all-time high of 69%.

The rate for black players, a longtime concern, jumped five points to 61% of those who entered school in 2003, the association's latest study showed.

But the annual academic scorecard was less flattering to many of the sport's top-tier programs. Seven of the top 10—including No. 1 Auburn, No. 2 Oregon and No. 3 Boise State—and 16 of the top 25 in the current Bowl Championship Series standings fell beneath the sport's four-year average. Oklahoma and Arizona graduated fewer than half of their players.

The numbers also were low in the other marquee college sport, men's basketball, where three of last season's eight NCAA regional finalists and more than one in five programs overall had four-year rates beneath 50%.

Those multi-year averages count players who arrived on campus from 2000–2003, giving them six years to earn degrees.

Awards ceremony for the ladies moguls at the Vancouver 2010 Olympic Games.

SOLUTION

We would choose the year to be the independent variable and the number of medals won to be the dependent variable. Then the number of medals won is a function of the year. The function gives the most medals won by any country in a given year.

Note that in this example it would not be appropriate to label the date as the dependent variable because there were three different years in which 29 medals were won. Therefore, the year doesn't depend only on the number of medals won. Another way to say this is that the year is not a function of the number of medals won. Also note that using the country as the dependent variable isn't appropriate because typically, the function value represents quantitative information.

TRY IT YOURSELF 2.1

The NCAA Men's Division I Basketball Tournament is played in March of each year. The following table shows the most bids to this tournament by any athletic conference in the given year. The conferences having that number of bids are also listed.

Year	Most bids by a conference	Conferences having the most bids
2005	6	Big East, Big 12
2006	8	Big East
2007	7	ACC
2008	8	Big East
2009	7	ACC, Big East, Big 10
2010	8	Big East
2011	11	Big East

Choose independent and dependent variables for this table and explain what the corresponding function means.

The answer is provided at the end of this section.

Data tables are meant to convey information. But the tables themselves are sometimes dry and difficult to decipher. There are important tools that help us understand

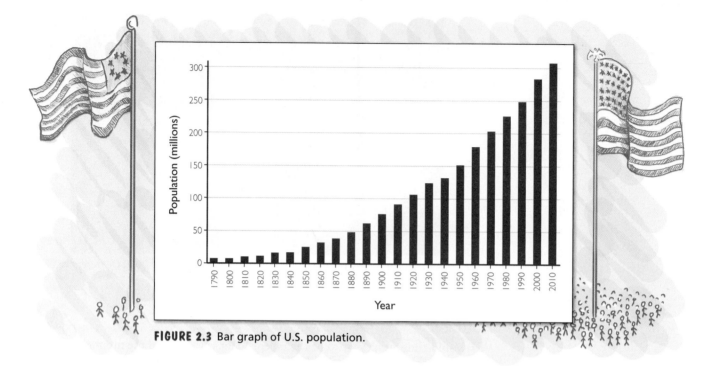

FIGURE 2.3 Bar graph of U.S. population.

information presented through a visual display. One of these is the *bar graph*. Bar graphs are useful in displaying data from tables of relatively modest size. For large data tables, other types of visual representation may be more appropriate. To make a bar graph of the census data in Table 2.1, we let the horizontal axis represent the independent variable (the date in this case), and we let the heights of the vertical bars show the value of the function (the U.S. population, in millions) for the corresponding date. The completed bar graph is in **Figure 2.3**. It shows how the population increased with time.

Percentage change

The apparent regularity of the graph in Figure 2.3 hides some important information. We can find out more about U.S. population growth by calculating the *percentage change*, that is, the *percentage increase* or *percentage decrease*. Population growth is often measured in this way, and in the article opening this section, the growth rate of Facebook was measured in terms of percentages.

Key Concept

The **percentage change** or **relative change** in a function is the percentage increase in the function from one value of the independent variable to another.

The formula for percentage change from one function value to another is

$$\text{Percentage change} = \frac{\text{Change in function}}{\text{Previous function value}} \times 100\%.$$

In the case of the census data, this change is the *percentage* increase in population from decade to decade. The following Quick Review is provided as a reminder of how to make percentage calculations.

Quick Review Percentage Change

Suppose the function value changes from 4 to 5. The *change*, or *absolute change*, from 4 to 5 is simply $5 - 4 = 1$. The *percentage change*, or *relative change*, from 4 to 5 is the change in the function divided by the previous function value, written as a percentage:

$$\text{Percentage change} = \frac{\text{Change in function}}{\text{Previous function value}} \times 100\% = \frac{1}{4} \times 100\% = 25\%.$$

This percentage change of 25% says that 5 is 25% more than 4.

If, on the other hand, the function value decreases from 5 to 4, then the change from 5 to 4 is $4 - 5 = -1$. The percentage change from 5 to 4 is the change in the function divided by the previous function value, written as a percentage:

$$\text{Percentage change} = \frac{\text{Change in function}}{\text{Previous function value}} \times 100\% = \frac{-1}{5} \times 100\% = -20\%.$$

This percentage change of -20% says that 4 is 20% less than 5.

EXAMPLE 2.2 Calculating percentage change: U.S. population

Calculate the percentage change in the U.S. population from 1790 to 1800. Round your answer in percentage form to the nearest whole number.

SOLUTION

To find the percentage change, we divide the change in the population from 1790 to 1800 by the population in 1790 and then multiply by 100%:

$$\text{Percentage change from 1790 to 1800} = \frac{\text{Change in function}}{\text{Previous function value}} \times 100\%$$

$$= \frac{\text{Change from 1790 to 1800}}{\text{Population in 1790}} \times 100\%$$

$$= \frac{5.31 \text{ million} - 3.93 \text{ million}}{3.93 \text{ million}} \times 100\%$$

$$= \frac{1.38}{3.93} \times 100\%.$$

This is about $0.35 \times 100\%$ or 35%. Between 1790 and 1800, the population increased by 35%.

TRY IT YOURSELF 2.2

Calculate the percentage change in the U.S. population from 1800 to 1810. Round your answer in percentage form to the nearest whole number.

The answer is provided at the end of this section.

The percentage change during each decade is shown in the accompanying table and as a bar graph in **Figure 2.4**.

Decade	Percentage change	Decade	Percentage change
1790–1800	35%	1900–1910	21%
1800–1810	36%	1910–1920	15%
1810–1820	33%	1920–1930	16%
1820–1830	34%	1930–1940	7%
1830–1840	33%	1940–1950	15%
1840–1850	36%	1950–1960	18%
1850–1860	36%	1960–1970	13%
1860–1870	23%	1970–1980	11%
1870–1880	30%	1980–1990	10%
1880–1890	25%	1990–2000	13%
1890–1900	21%	2000–2010	10%

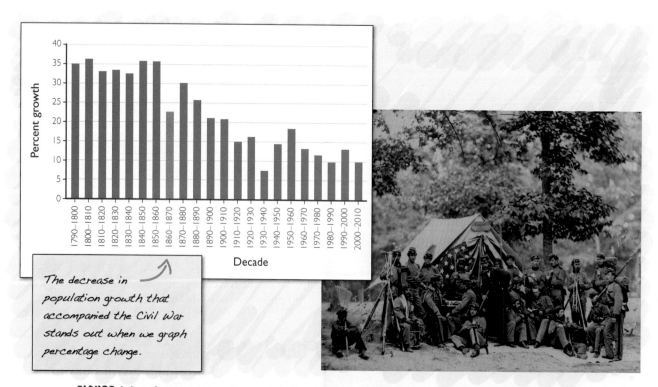

The decrease in population growth that accompanied the Civil War stands out when we graph percentage change.

FIGURE 2.4 Left: Percentage change in U.S. population. Right: Engineers of the 8th N.Y. State Militia, 1861.

As we look at each percentage change, some things stand out. Note that between 1790 and 1860, the percentage changes in each decade are approximately the same—between 33% and 36%. That is, the percentage increase in population was nearly constant over this period.

The decade of the Civil War (1860–1870) jumps out. Even if we didn't know the historical significance of this decade, the bar graph of percentage change would alert us that something important may have occurred in that decade. Observe that the significance of this decade is not apparent from the graph of population shown in Figure 2.3. It is interesting to note that percentage population growth never returned to its pre-Civil War levels. The percentage change generally declined to its current level of 10% to 13% per decade.

SUMMARY 2.1 **Tables and Percentage Change**

1. A data table shows a relationship between two quantities. We designate one quantity as the independent variable and think of the other quantity, the dependent variable, as a function of the independent variable.

2. A bar graph provides a visual display of a table that can help us understand what the information shows.

3. Percentage change can provide information that may not be readily apparent from the raw data. We calculate percentage change from one function value to another using

$$\text{Percentage change} = \frac{\text{Change in function}}{\text{Previous function value}} \times 100\%.$$

EXAMPLE 2.3 Creating and analyzing data: World population

The following table shows the world population (in billions) on the given date:

Date	1950	1960	1970	1980	1990	2000	2010
Population (billions)	2.56	3.04	3.71	4.45	5.29	6.09	6.85

a. Identify the independent variable and the function, and make a bar graph that displays the data.

b. Make a table and bar graph showing percentage changes between decades.

c. Comment on what your graphs tell you. Compare the second graph with the corresponding graph for U.S. population growth.

SOLUTION

a. The independent variable is the date, and the function gives the world population (in billions) on that date. To create a bar graph, we list dates across the horizontal axis and make bars representing population for each date. The finished product is in **Figure 2.5**.

b. We calculate the percentage change from one date to the next using

$$\text{Percentage change} = \frac{\text{Change in population}}{\text{Previous population}} \times 100\%.$$

So the percentage increase from 1950 to 1960 is

$$\text{Percentage change} = \frac{3.04 - 2.56}{2.56} \times 100\%$$

or about 19%.

The table is given below, and the bar graph is in **Figure 2.6**.

Decade	1950–1960	1960–1970	1970–1980	1980–1990	1990–2000	2000–2010
% change	19%	22%	20%	19%	15%	12%

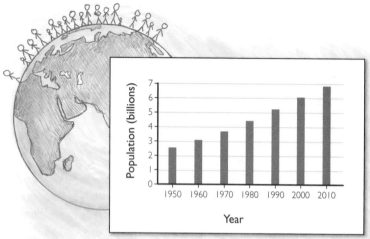

FIGURE 2.5 World population.

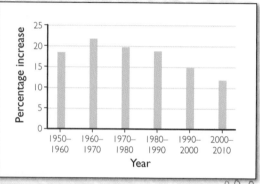

FIGURE 2.6 Percentage growth of world population.

c. The world population is steadily increasing, but from 1960 on the percentage change is declining. Starting from 1960, the percentage change drops from 22% to 16% per decade. The percentage increase for the world population is still larger than that for the United States.

Average growth rate

Percentage change focuses on the dependent variable. Note that when we calculated the percentage change in U.S. population from 1790 to 1800, only population figures were used. The actual calculation did not involve the dates at all. This way of measuring change makes it difficult to compare growth over time periods of different lengths. To tell at what rate the function changes with respect to the independent variable, we calculate the *average growth rate*.[2] It is the most useful measure of change that we will encounter.

The average growth rate is always calculated over a given interval, often a time period (such as 1790 to 1800). To calculate the average growth rate, we divide the change in the function over that interval by the change in the independent variable.

Key Concept

The average growth rate of a function over an interval is the change in the function divided by the change in the independent variable.

---

[2] This quantity is also known as the *average rate of change*.

We calculate the average growth rate using the formula

$$\text{Average growth rate} = \frac{\text{Change in function}}{\text{Change in independent variable}}.$$

EXAMPLE 2.4 Calculating average growth rate: U.S. population

Calculate the average growth rate in the population of the United States from 1940 to 1950. Explain its meaning.

SOLUTION

The change in time (the independent variable) is 10 years. According to Table 2.1 on page 74, the change in population (the function) over that interval of time is $151.33 - 132.16 = 19.17$ million people. So the average growth rate in population over this decade is

$$\frac{\text{Change in function}}{\text{Change in independent variable}} = \frac{19.17}{10} = 1.917 \text{ million people per year.}$$

This result tells us that from 1940 to 1950 the U.S. population grew, *on average*, by about 1.92 million people each year.

TRY IT YOURSELF 2.4

Show that the average growth rate in the population of the United States from 1950 to 1960 is about 2.80 million people per year. Explain the meaning of this rate.

The answer is provided at the end of this section.

Note that the average growth rate from 1940 to 1950 is not the same as that from 1950 to 1960. In general, average growth rates vary from decade to decade. In some cases, the average growth rate may be negative. This indicates a decline rather than an increase. The following example illustrates this possibility.

EXAMPLE 2.5 Calculating a negative growth rate: Population of Russia

The population of Russia declined from about 146 million in 2000 to about 143 million in 2007. Calculate the average growth rate over this period and explain its meaning.

SOLUTION

The change in population is negative, $143 - 146 = -3$ million. The change in time is 7 years. Thus,

$$\text{Average growth rate} = \frac{\text{Change in function}}{\text{Change in independent variable}} = \frac{-3}{7}$$

or about -0.429 million per year. This means that over this interval the population declined, on average, by about 429,000 people per year.

TRY IT YOURSELF 2.5

It is projected that the population of Russia will decline from 143 million in 2007 to about 111 million in 2050. Use this projection to predict the average growth rate from 2007 to 2050. Explain the meaning of this rate.

The answer is provided at the end of this section.

It is always important to pay attention to the units of measurement. This is particularly true for average growth rates. The units of the average growth rate are the units of the function divided by the units of the independent variable. See the example above, where the average growth rate is measured in millions of people per year.

In practical settings, the average growth rate almost always has a familiar meaning, and proper use of units can help determine that meaning. For example, if the function gives the distance driven in an automobile and the independent variable is time, the average growth rate is the average velocity, which typically is measured in miles per hour. Thus, if you make a 100-mile trip and the trip takes two hours, the average growth rate is 50 miles per hour. In other settings, the average growth rate may represent the average number of traffic accidents per week, the average number of births per thousand women, or the average number of miles per gallon.

EXAMPLE 2.6 Interpreting the average growth rate: Tuition cost

Assume that the independent variable is the year and the function gives the tuition cost, in dollars, at your university. Give the units of the average growth rate and explain in practical terms what that rate means.

SOLUTION

The change in the independent variable is the elapsed time measured in years, and the change in the function is the tuition increase measured in dollars. So the units of the average growth rate are dollars per year. The average growth rate tells how much we expect the tuition to increase each year. (A tuition decrease would result in a negative growth rate, but unfortunately, tuition usually goes up these days, not down.)

TRY IT YOURSELF 2.6

Assume that the independent variable is the day and the function is the cumulative number of flu cases reported to date. Give the units of the average growth rate and explain in practical terms what that rate means.

The answer is provided at the end of this section.

Interpolation

Once we know the average rate of change, we can obtain more information from the data. Suppose we want to estimate the U.S. population in the year 1945. According to Example 2.4, between 1940 and 1950, the U.S. population grew at an average rate of 1.917 million people per year. So it is reasonable to assume that over the 5-year period from 1940 to 1945 the population increased by approximately $5 \times 1.917 = 9.585$ million. We know that the population in 1940 was 132.16 million. We can then estimate that the population in 1945 was

Population in 1940 + 5 years of increase = $132.16 + 9.585 = 141.745$ million,

or about 141.75 million.

Using average growth rates to estimate values between data points is known as *interpolation*.

Key Concept

Interpolation is the process of estimating unknown values between known data points using the average growth rate.

EXAMPLE 2.7 Interpolating data: Opinions on legalizing marijuana

In the fall of 2005, 37.7% of college freshmen in the United States believed that marijuana should be legalized.[3] In the fall of 2008, that figure was 41.3%. Use these figures to estimate the percentage in the fall of 2007. The actual figure for 2007 was 38.2%. What does this say about how the growth rate in the percentage varied over time?

Medical marijuana dispensary in Los Angeles.

SOLUTION

We want to interpolate, so first we find the average growth rate. The change in the independent variable from 2005 to 2008 is three years. The change in the dependent variable over that period is $41.3 - 37.7 = 3.6$ percentage points. Hence, the average growth rate from the fall of 2005 to the fall of 2008 was

$$\frac{\text{Change in function}}{\text{Change in independent variable}} = \frac{3.6}{3} = 1.2 \text{ percentage points per year.}$$

This means that, on average, the figure increased by 1.2 percentage points each year over this period. The year 2007 represents two years of increase from 2005, so the percentage in 2007 was about

$$\text{Percentage in 2005} + 2 \text{ years of increase} = 37.7 + 2 \times 1.2 = 40.1 \text{ percent.}$$

Our estimate of 40.1% for the fall of 2007 is higher than the actual figure of 38.2%. This means that the growth rate of the percentage was lower from 2005 to 2007 than it was from 2005 to 2008—the figure grew more quickly from 2007 to 2008.

TRY IT YOURSELF 2.7

Figure 2.1, the graph accompanying the article at the beginning of this chapter, indicates that on January 1, 2007, there were approximately 13 million Facebook users. One claim is that usage is increasing by an average of 250,000, or 0.25 million, users

---

[3]Data from the annual CIRP Freshman Surveys, 2005–2008, by the Higher Education Research Institute at UCLA.

per day. Use this information to estimate the number of users on January 11, 2007, which is 10 days after January 1. Give your answer in millions to one decimal place.

The answer is provided at the end of this section.

Now we use these concepts to study the spread of a disease.

EXAMPLE 2.8 Interpolating data: SARS

Severe Acute Respiratory Syndrome (SARS) is a viral respiratory disease. There was a serious outbreak initially in China from November 2002 to July 2003. There were 8096 known cases and 774 deaths worldwide. The following table from the World Health Organization shows the cumulative number of SARS cases reported on certain dates in March and April 2003:

Date	March 26	March 31	April 5	April 10	April 15
Number of cases	1323	1622	2416	2781	3235

People in Beijing wearing masks as protection against the spread of SARS.

a. Calculate the average growth rate of cases from March 26 to March 31. Be sure to express your answer using proper units.

b. Use your answer from part a to estimate the cumulative number of SARS cases by March 28. The actual cumulative number on March 28 was 1485. What does this say about how the growth rate varied over time?

SOLUTION

a. The independent variable is the date, and the function is the cumulative number of SARS cases reported. The average growth rate from March 26 to March 31 is

$$\frac{\text{Change in reported cases}}{\text{Elapsed time}} = \frac{1622 - 1323}{5} = 59.8 \text{ new cases per day.}$$

b. There were 1323 cases on March 26, and we expect to see about 59.8 new cases on March 27 and again on March 28. Thus, we make our estimate as follows:

Estimated cases on March 28 = Cases on March 26 + 2 × Average new cases per day

$$= 1323 + 2 \times 59.8 = 1442.6,$$

or about 1443 cases. Our estimate using interpolation is relatively good but somewhat lower than the actual value of 1485. The fact that interpolation gives an underestimate indicates that the growth rate between March 26 and March 28 was higher than the growth rate between March 26 and March 31—the number of new cases per day was higher over the first part of the five-day period.

We should be aware that sometimes interpolation works well and sometimes it does not. We will return to this question in Chapter 3, but for now it suffices to say that interpolation works better when the values in the table are closer together.

Extrapolation

It is also possible to use average growth rates to make estimates beyond the limits of available data and thus to forecast trends. This process is called *extrapolation*. As with interpolation, extrapolation requires a little judgment and common sense.

Key Concept

Extrapolation is the process of estimating unknown values beyond known data points using the average growth rate.

As an example, let's extrapolate beyond the limits of the table of SARS cases to estimate the number of cases on April 16. We begin by calculating the average number of new cases per day from April 10 to April 15:

$$\text{Average number of new cases per day} = \frac{3235 - 2781}{5} = 90.8 \text{ cases per day.}$$

So on April 16th we expect 90.8 more cases than on April 15, for a total of $3235 + 90.8 = 3325.8$. Rounding gives an estimate of 3326 cases on April 16.

EXAMPLE 2.9 Extrapolating data: An epidemic

In a certain epidemic, there is an average of 12 new cases per day. If a total of 500 cases have occurred as of today, what cumulative number of cases do we expect three days from now?

SOLUTION

We expect 12 new cases for each of three days, for a total of 36 new cases. So the cumulative number of cases expected three days from now is $500 + 36 = 536$.

TRY IT YOURSELF 2.9

In a certain epidemic, there is an average of 13 new cases per day. If a total of 200 cases have occurred as of today, what cumulative number of cases do we expect four days from now?

The answer is provided at the end of this section.

Now let's see what happens if we try to extrapolate farther beyond the limits of the SARS data by predicting the number of cases on April 30. We start with 3235 cases on April 15 and use the rate of change of 90.8 new cases per day for 15 days to obtain

$$\text{Expected cases on April 30} = 3235 + 15 \times 90.8 = 4597 \text{ cases.}$$

The actual number of SARS cases on April 30, as reported by the World Health Organization, was 5663. That's 1066 (or about 23%) more cases than our estimate.

This example illustrates why one should be cautious in using extrapolation to make estimates well beyond the limits of the data.

SUMMARY 2.2   Interpolation and Extrapolation

1. The average growth rate of a function over an interval is calculated using

$$\text{Average growth rate} = \frac{\text{Change in function}}{\text{Change in independent variable}}.$$

2. To make an estimate by interpolation or extrapolation from a function value, multiply the average growth rate (e.g., cases per day, accidents per week, miles per hour) by the number of increments (e.g., days, weeks, hours) in the independent variable and add the result to the function value.

3. At best, interpolation and extrapolation give estimates. They are most reliable over short intervals.

EXAMPLE 2.10 Interpolating and extrapolating data: Age of first-time mothers

The following table shows the average age, in years, of first-time mothers in the given year:

Year	1970	1980	1990	2000
Average age	21.4	22.7	24.2	24.9

a. Estimate the average age of first-time mothers in 1997.

b. Predict the average age of first-time mothers in 2005.

c. Predict the average age of first-time mothers in the year 3000. Explain why the resulting figure is not to be trusted.

SOLUTION

a. We estimate the average age in 1997 by interpolating. For this we need the average growth rate between 1990 and 2000. The average age changed from 24.2 years to 24.9 years over a 10-year period, so

$$\text{Average growth rate} = \frac{24.9 - 24.2}{10} = 0.07 \text{ year per year.}$$

Thus, the average age of first-time mothers increased at a rate of 0.07 year per year over this decade. There are seven years from 1990 to 1997, and the average age increased, on average, by 0.07 year each year. We therefore expect the average age in 1997 to be $24.2 + 7 \times 0.07 = 24.69$ years, or about 24.7 years.

b. In this case, we estimate the average age by extrapolating. For this we again need the average growth rate between 1990 and 2000, which we found in part a to be 0.07 year per year. We therefore estimate an increase of 0.07 year over each of the five years from 2000 to 2005:

Estimated average age in 2005 $= 24.9 + 5 \times 0.07 = 25.25$ years,

or about 25.3 years.

c. This is exactly like part b. The average growth rate from 1990 to 2000 is 0.07 year per year. Because the year 3000 is 1000 years after the year 2000, this growth rate gives a prediction for the year 3000 of $24.9 + 1000 \times 0.07 = 94.9$ years. Our projection for the average age in the year 3000 of first-time mothers is 95 years. This number is absurd and clearly illustrates the danger of extrapolating too far beyond the limits of the given data.

Try It Yourself answers

Try It Yourself 2.1: Choosing variables: NCAA Basketball Tournament   We choose the year to be the independent variable and the greatest number of bids received by any conference to be the dependent variable. Then the most bids received is a function of the year: The function gives the most bids received by any conference in a given year.

Try It Yourself 2.2: Calculating percentage change: U.S. population   36%.

Try It Yourself 2.4: Calculating average growth rate: U.S. population   The average growth rate over this decade is

$$\frac{179.32 - 151.33}{10} = 2.799 \text{ million people per year},$$

or about 2.80 million people per year. This says that from 1950 to 1960, the U.S. population grew, on average, by about 2.80 million people each year.

Try It Yourself 2.5: Calculating a negative growth rate: Population of Russia -0.744 million per year. If the estimate is correct, the population will decline, on average, by 744,000 per year.

Try It Yourself 2.6: Interpreting the average growth rate: Flu cases   The units of the average growth rate are flu cases per day. The average growth rate measures the number of new flu cases reported each day.

Try It Yourself 2.7: Interpolating data: Facebook users   15.5 million users.

Try It Yourself 2.9: Extrapolating data: An epidemic   252 total cases.

Exercise Set 2.1

1. Home runs. The following table shows the total number of home runs scored by major league baseball players in the given year:

Year	Home runs
2001	5458
2002	5059
2003	5207
2004	5451
2005	5017
2006	5386
2007	4957
2008	4878
2009	5042
2010	4613

Choose independent and dependent variables for this table and explain what the corresponding function means.

2. More basketball. The following table shows the number of bids to the NCAA Men's Basketball Tournament for the Big 12 Conference in the given year:

Year	Number of bids by Big 12
2005	6
2006	4
2007	4
2008	6
2009	6
2010	7
2011	5

Choose independent and dependent variables for this table and explain what the corresponding function means.

Children in kindergarten   The following table shows the percentage of children in the United States between the ages of 3 and 5 who are enrolled in public and nonpublic nursery school and kindergarten programs:

Date	1970	1975	1980	1985	1990	1995	2000
Percentage	37.5	48.6	52.5	54.6	59.4	61.8	64.0

Exercises 3 and 4 refer to this table.

3. Use interpolation to estimate the percentage of children enrolled in 1998.

4. Use extrapolation to estimate the percentage of children enrolled in 2005.

Customer complaints The following table shows the number of customer complaints against airlines operating in the United States during the given year:

Year	2002	2004	2006	2008	2010
Number of complaints	9466	7452	8325	10,648	16,508

Exercises 5 through 7 refer to this table.

5. Use interpolation to estimate the number of complaints in 2007.

6. Complete the table below with the absolute numbers and the percentages by which the complaints changed over each two-year period. Round your final answers to the nearest whole number.

Range	2002–2004	2004–2006	2006–2008	2008–2010
Change in complaints				
Percentage change				

7. Use the second row of the table you completed in Exercise 6 to estimate the number of complaints in 2001. The actual number was 16,508. What events in 2001 might cause the actual number to be higher than your estimate? Explain.

Grades The bar graph in **Figure 2.7** shows the percentage of college freshmen in the United States, as of the fall of the given year, whose average grade in high school was between an A- and an A+. Exercises 8 through 11 refer to this bar graph.

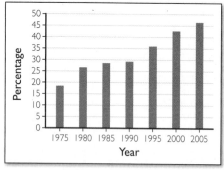

FIGURE 2.7 Freshmen with high school grades of "A".

8. Describe in general how the percentage changed over time.

9. Find the average yearly growth rate in the percentage from 2000 to 2005.

10. Use your answer to Exercise 9 to estimate the percentage in the year 2008.

11. The actual percentage in 2008 was 47.2%. Compare this with your answer to Exercise 10. What does this say about the rate of growth in the percentage after 2005, as compared with the rate of growth in the percentage during the first part of that decade?

College expenditures The table below shows total expenditures for colleges and universities in millions of dollars. All data are in current dollars.

Year	1970	1980	1990	2000	2010
Expenditures	21,043	56,914	134,656	236,784	461,000

Exercises 12 through 15 refer to this table.

12. What, in general, does the table tell you about spending on colleges and universities from 1970 through 2010?

13. Use interpolation to complete the following table:

Year	1970	1975	1980	1990	1991	2000	2010
Expenditures	21,043		56,914	134,656		236,784	461,000

14. Make a table that shows the absolute change in expenditures over each 10-year period.

15. Estimate expenditures in 2015. Explain your reasoning.

Attitudes on abortion The following table shows the percentage of U.S. college freshmen in the given year who believed abortion should be legal:

Year	1980	1985	1990	1995	2000	2005
Percentage	53.7	56.4	65.5	59.9	53.9	55.2

Exercises 16 through 18 refer to this table.

16. Calculate the average growth rate per year from 1995 to 2000 in the percentage of college freshmen who believed that abortion should be legal.

17. Use your calculation from Exercise 16 to estimate the percentage of college freshmen in 1997 who believed that abortion should be legal.

18. Use extrapolation to estimate the percentage of college freshmen in 2008 who believed that abortion should be legal. *Note:* The actual number was 58.2%.

Inflation Inflation is a measure of the buying power of your dollar. For example, if inflation this year is 5%, then a given item would be expected to cost 5% more at the end of the year than at the beginning of the year. **Figure 2.8** shows the rate of

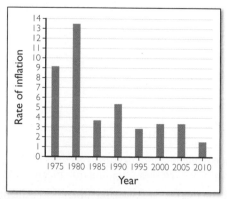

FIGURE 2.8 Rate of inflation.

inflation in the United States for the given years. This information is needed for Exercises 19 through 22.

19. Use interpolation to estimate the rate of inflation in the United States for each of the years 1996 through 1999. Round your answer to one decimal point as a percentage.

20. Assume that men's clothing followed the rate of inflation. If a man's suit cost $300 at the beginning of 1995, what did it cost at the end of 1995? Round your answer to the nearest dollar.

21. It is a fact that in 1949 the rate of inflation in the United States was negative, and the value was −2%. If a man's suit cost $100 at the beginning of 1949, what did it cost at the end of 1949?

22. Extreme inflation can have devastating effects. In 1992 inflation was 100% in Lebanon and Bulgaria.[4] If a man's suit cost $100 at the beginning of a year with that rate of inflation, what would it cost at the end of the year?

Exercises 23 through 27 are suitable for group work.

School costs The following table shows the average total cost, measured as dollars expended per pupil, of operating a high school with the given enrollment. This table is based on a study of economies of scale in high school operations from the 1960s.

Enrollment	170	250	450	650
Average total cost	532	481	427	413

Exercises 23 through 27 refer to this table.

23 Estimate the average cost per student for a high school with an enrollment of 350 students.

24. The actual cost per student with an enrollment of 350 is $446 per pupil. Use this information to compare how the average cost changes from an enrollment of 250 to 350 with how the average cost changes from an enrollment of 350 to 450.

25. What is the average growth rate of the average cost per additional student enrolled, when moving from an enrollment of 450 students to an enrollment of 650 students?

26. What would you estimate for the average cost per student at an enrollment of 750 students?

27. Some policy analysts use tables such as this to argue that small schools should be consolidated to make the operations more cost-effective. On the basis of your answers to previous parts, determine which is likely to save more money (in terms of average total cost): consolidation that results in increasing the enrollment by 100 starting at the low end of the scale (say, from 250 to 350) or such consolidation starting at a higher level (say, from 350 to 450). *Note*: These computations, of course, represent only one side of a complex problem. School consolidation can be devastating to small rural communities.

Historical events The table for percentage change in U.S. population on page 78 contains some anomalies that we did not discuss in the text. Exercises 28 through 30 refer to this table.

28. The percentage increase from 1930 to 1940 is markedly reduced. What historical event may account for the fact that this percentage is much smaller than expected? *Suggestion*: Use the Internet or the library to investigate what happened on "Black Thursday."

29. The population increased by a smaller percentage from 1910 to 1920 than the preceding or following decades. There were two historical events that may have accounted for this. One was a great "pandemic" of influenza. Look up the history of the great influenza pandemic and write a brief account of it. What do you think was the other great historical event that may have influenced the population growth in this decade?

30. Look at the table on page 78. What might account for the large percentage increase from 1950 to 1960?

The following exercises are designed to be solved using technology such as calculators or computer spreadsheets. For assistance, see the technology supplement.

Enrollment in French The table below shows U.S. higher education enrollment (in thousands) in French for the given year.

Year	1983	1986	1990	1995	1998
Enrollment	128.2	121.0	113.3	96.3	89.0

This information is needed for Exercises 31 through 33.

31. Explain in general terms what the data on French enrollment show.

32. For each period shown in the table (1983–1986, 1986–1990, etc.), calculate the average growth rate per year in enrollment in French.

33. To find the average decrease when the average growth rate is negative, just drop the minus sign. Make a bar graph displaying the average decrease per year in enrollment in French.

AP examinations The following table shows the number in thousands of AP foreign language and literature examinations taken by males and females for the given languages. Data are for the year 2005.

Subject	French	German	Latin	Spanish
Male	6.6	2.3	4.0	39.1
Female	15.4	2.3	3.9	72.9

This information is used in Exercises 34 and 35.

34. Make a bar graph showing the data. *Suggestion*: Use the different languages for categories and put a male bar and a female bar for each one.

35. For which language were the fewest examinations taken?

Buying power The following table shows how the buying power of a dollar is reduced with inflation. The second line in the table gives the percent reduction of a dollar's buying power

---

[4]See www.theodora.com/wfb/1992/rankings/inflation_rate_pct_1.html.

in one year due to the corresponding inflation rate. Exercises 36 and 37 refer to this table.

Annual inflation rate	10%	20%	30%	40%	50%
Buying power reduction	9.1%	16.7%	23.1%	28.6%	33.3%

Annual inflation rate	60%	70%	80%	90%	100%
Buying power reduction	37.5%	41.2%	44.4%	47.4%	50.0%

36. Make a bar graph for this table with the inflation rate on the horizontal axis.

37. If the buying power of a dollar declined by 25% during a year, what was the inflation rate during that year? *Suggestion*: First note that 25% is between 23.1% and 28.6%, so the answer must be an inflation rate between 30% and 40%. Interpolate to find the answer more accurately.

2.2 Graphs: Picturing growth

TAKE AWAY FROM THIS SECTION

Interpret growth rates displayed by graphs.

The following is an excerpt from an article at the Web site of the magazine *New Scientist*.

IN THE NEWS 2.3

Climate: The Great Hockey Stick Debate

FRED PEARCE

It is a persuasive image. Dubbed "the hockey stick" soon after it was first drawn, the graph in **Figure 2.9** shows the average temperature over the past 1000 years. For the first 900 or so years there is little variation, like the shaft of an ice-hockey stick. Then, in the 20th century, comes a sharp rise like a hockey stick's blade. The graph seems proof at a glance that we are drastically altering the climate of our planet.

So it is not surprising that the Intergovernmental Panel on Climate Change (IPCC) chose to put the graph in the summary for policymakers in its 2001 report. Some of the scientists must have hoped that the image would become an icon of climate change.

An icon it has certainly become, but not always for the reasons those scientists hoped. For the skeptics who dispute that global warming is real, or say it's nothing to worry about, the graph was like a red rag to a bull. They made it the focus of their attacks, hoping that by demolishing the hockey stick graph they would destroy the credibility of climate scientists and the notion of global warming as a phenomenon caused by human activity.

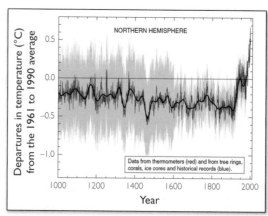

FIGURE 2.9 Global warming.

At the center of the public controversy over global warming is a graph. The graph is shown in Figure 2.9. In this graph red represents historical change in temperatures from the twentieth century, blue represents reconstructed change in temperatures from the eleventh through the nineteenth centuries, black represents an average of the change in temperatures, and grey indicates the variability of the information.

There is controversy regarding both the accuracy of the graph and its interpretation. But at the heart of the controversy is how the graph changes and what that change means. From 1900 on (the "blade" portion of the "hockey stick"), the character of the graph is different from the earlier part—the "hockey stick handle." It is this change in character that concerns scientists, politicians, and the general public.

Why are graphs so often used by both the scientific community and the popular media? A graph is a good example of the old saying that a picture is worth a thousand words. Graphs allow us to visualize information, which helps us see patterns that may not be readily apparent from a table. For example, a graph reveals visually where the function is increasing or decreasing and how rapid the change is. In this section, we study various types of graphs and learn how to interpret these patterns.

In the preceding section, we used bar graphs to present data from tables. In this section, we consider two other ways to present data: scatterplots and line graphs. These ways of presenting data are often seen in the media, and it is important to be familiar with them, to interpret them, to analyze them critically, and to be aware of their advantages and disadvantages.

Quick Review Making Graphs

Typically, we locate points on a graph using a table such as the one below, which shows the number of automobiles sold by a dealership on a given day.

Day (independent variable)	1	2	3	4
Cars sold (dependent variable)	3	6	5	2

When we create a graph of a function, the numbers on the horizontal axis correspond to the independent variable, and the numbers on the vertical axis correspond to the dependent variable. For this example, we put the day (the independent variable) on the horizontal axis, and we put the number of cars sold (the dependent variable) on the vertical axis.

The point corresponding to day 3 is denoted by (3, 5) and is represented by a single point on the graph. To locate the point (3, 5), we begin at the origin and move 3 units to the right and 5 units upward. This point is displayed on the graph as one of the four points of that graph. The aggregate of the plotted points is the graph of the function.

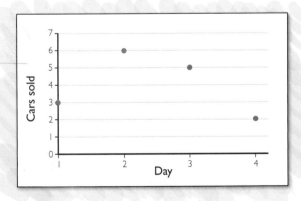

Scatterplots

Often we have a collection of isolated data points and want to make a visual display in order to search for patterns. When the values of the independent variable are evenly spaced, a bar graph may be used. When the values of the independent variable are not evenly spaced, a scatterplot is often used to represent the data. Scatterplots are also used to represent large numbers of data points that cannot be reasonably displayed using a bar graph.

Key Concept

A scatterplot is a graph consisting of isolated points, with each dot corresponding to a data point.

An interesting example comes from recording information on over 100 eruptions of the Old Faithful geyser at Yellowstone National Park. These data show the waiting time between eruptions in terms of the duration of the eruption. The graph is shown in **Figure 2.10**. Because each eruption is a separate data point, the scatterplot is simply a plot of the 100 or more separate points.

Certainly, it would be impractical and messy to try to represent these data using a bar graph. Moreover, the scatterplot neatly indicates that there are two types of eruptions: those with short duration and short waiting time, and those with long duration and long waiting time. Scatterplots are often used to discover patterns in the data. For example, the line segments that have been added in Figure 2.10 suggest such patterns.

Line graphs and smoothed line graphs

Many times connecting the data points on a scatterplot makes patterns easier to spot. The result of connecting the dots is called a *line graph*.

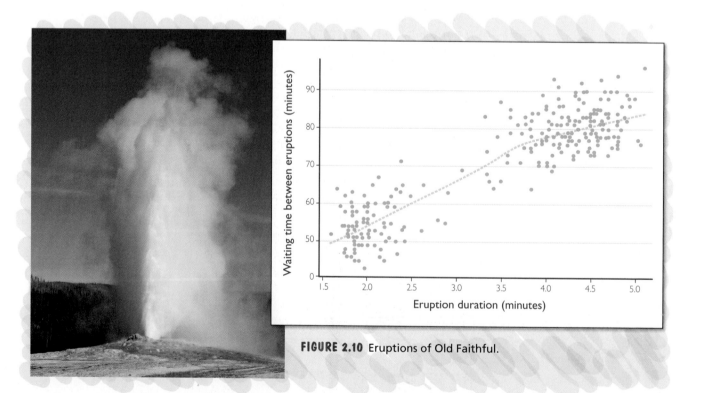

FIGURE 2.10 Eruptions of Old Faithful.

Key Concept

To make a **line graph**, we begin with a scatterplot and join the adjacent points with straight line segments.

An interesting example comes from the following table, which shows the running speed of various animals as a function of the length:[5]

Animal	Length (inches)	Speed (feet per second)
Deermouse	3.5	8.2
Chipmunk	6.3	15.7
Desert crested lizard	9.4	24.0
Grey squirrel	9.8	24.9
Red fox	24.0	65.6
Cheetah	47.0	95.1

We make the line graph in two steps. First we make the scatterplot shown in **Figure 2.11**. Then we join adjacent points with line segments to get the line graph shown in **Figure 2.12**.

One advantage of a line graph over a scatterplot is that it allows us to estimate the values between the points in the scatterplot. Suppose, for example, we want to know how fast a 15-inch-long animal can run. There is no data point in the table corresponding to 15 inches, but we can use the graph to estimate it. In **Figure 2.13** on the following page, we find the 15-inch point on the horizontal axis (halfway between 10 and 20) and then go up to the line graph. We cross the graph at a level corresponding to about 40 feet per second on the vertical axis. So, these segments allow us to fill gaps in data by estimation.

An interesting fact about filling in gaps using line segments is that the value of the function (running speed in this case) determined in this way is exactly the same as that obtained by interpolation, which we discussed in Section 2.1. Line graphs provide a graphical representation of the average growth rate from one point to the next.

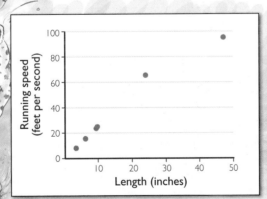

FIGURE 2.11 Scatterplot of running speed versus length.

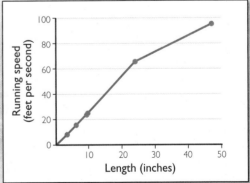

FIGURE 2.12 Line graph by adding segments.

---

[5]This table is adapted from J. T. Bonner, *Size and Cycle* (Princeton, NJ: Princeton University Press, 1965). We will encounter it again in the next chapter when we study trend lines.

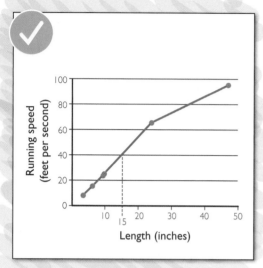

FIGURE 2.13 Information added by the line graph.

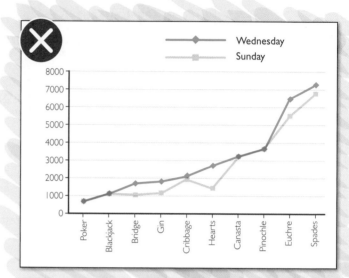

FIGURE 2.14 When a line graph is not appropriate.

One must take care not to use a line graph when there is nothing between categories. For example, **Figure 2.14** shows two line graphs. These graphs show the number of people playing various card games at the Yahoo! Web site on a certain Wednesday and a certain Sunday in the spring of 2001.[6] The line segment joining blackjack to bridge would indicate that there are card games in between them. A bar graph would be appropriate here.

EXAMPLE 2.11 Making a scatterplot and a line graph: Running speed and temperature

The running speed of ants varies with the ambient temperature. Here are data collected at various temperatures:

Temperature (degrees Celsius)	Speed (centimeters per second)
25.6	2.62
27.5	3.03
30.4	3.56
33.0	4.17

First make a scatterplot of the data showing the speed as the function and the temperature as the independent variable, then make a line graph using these data.

SOLUTION

To make a scatterplot, we plot the data points, each of which corresponds to a point on the graph: $(25.6, 2.62), \ldots, (33.0, 4.17)$. The scatterplot is the graph of these points, as shown in **Figure 2.15**. We join the points with line segments to get the line graph in **Figure 2.16**.

---

[6]This figure is taken from http://cnx.org/content/m10927/latest/, where it is used to illustrate the misuse of line graphs.

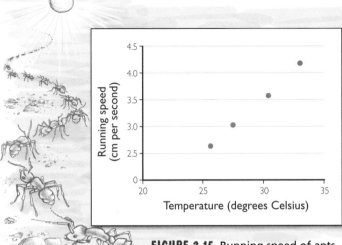

FIGURE 2.15 Running speed of ants versus temperature: scatterplot.

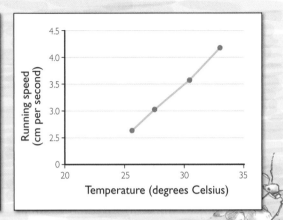

FIGURE 2.16 Running speed of ants versus temperature: line graph.

TRY IT YOURSELF 2.11

Use the following table to make a scatterplot of the height of a sunflower as a function of its age. Then make a line graph.

Age (days)	Height (centimeters)
5	17.43
35	100.88
41	128.68
61	209.84
70	231.23

The answer is provided at the end of this section.

Interpreting line graphs: Growth rates

A key feature of line graphs is that they display maximums and minimums as well as visual evidence of growth rates. Consider the following example, which is taken from the Web site of the National Cancer Institute. The table below shows the five-year cancer mortality rates per 100,000 person-years in the United States. The data are sorted by race and sex.

	Five-year cancer mortality rates			
Period	Black male	White male	Black female	White female
1950–54		174.15		145.68
1955–59		181.24		139.34
1960–64		185.92		133.99
1965–69		194.14		131.81
1970–74	252.6	201.37	148.64	130.71
1975–79	276.51	206.88	149.43	131.31
1980–84	299.42	210.92	158.11	135.13
1985–89	312.03	212.73	164.09	138.65
1990–94	315.51	211.41	169.48	140.25
1995–99	287.72	196.66	162.97	134.60
2000–04	254.19	181.95	152.77	128.50

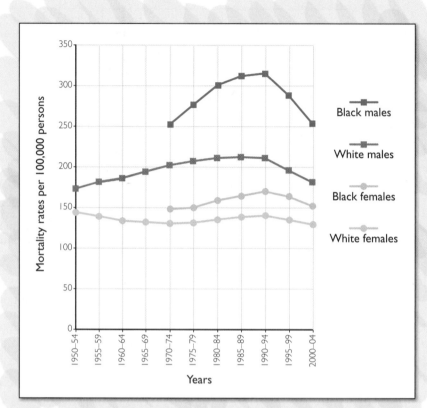

FIGURE 2.17 Five-year cancer mortality rates.

Figure 2.17 shows line graphs of these data. This figure reveals several trends that are not readily apparent from the table. For one, it is immediately clear that cancer mortality rates among black males are much higher than for any of the other categories presented. These rates were on the rise from 1970–74 to 1990–94 and then fell sharply through 2000–04, ending at a level about the same as in 1970–74.

The graph shows that cancer mortality rates among white women reached a minimum of about 130 per 100,000 at the point representing the period 1970–74. Those rates began increasing slightly after that time and then decreased to the lowest level in 2000–04. Also, we see that cancer mortality rates for white males reached their peak at the point representing the period 1985–89. It would take considerable effort to see the trends above using only the table.

It is important to note not only how high a graph is but also how steep it is: A steeper graph indicates a growth rate of greater magnitude. For example, we observe that the graph of mortality rates for black males begins to level off from 1980–84 to 1990–94. That is, the *rate* of increase appears to be slowing. Rates for black women appear to have increased at a fairly steady rate from 1975–79 to 1990–94, and then decreased at a steady rate from 1990–94 to 2000–04. The line graph makes apparent these changes in growth rates, far more so than the numerical data themselves do.

The sign of the growth rate for a graph can be seen by examining where the graph increases or decreases. A positive growth rate is indicated when the graph increases as we move from left to right along the horizontal axis, and a negative growth rate is indicated by a decreasing graph. For example, the mortality rate for white women is decreasing from 1950–54 to 1970–74 and also from 1990–94 to 2000–04, indicating negative growth rates during those periods. The increasing graph from 1970–74 to 1990–94 indicates a positive growth rate.

> ### SUMMARY 2.3   Growth Rates and Graphs
>
> **1.** The growth rate of data is reflected in the steepness of the graph. Steeper graphs indicate a growth rate of greater magnitude.
>
> **2.** An increasing graph indicates a positive growth rate, and a decreasing graph indicates a negative growth rate.

"This is where the person who makes this graph started taking an anti-depressant."

EXAMPLE 2.12   Interpreting line graphs: Income

The line graph in **Figure 2.18** shows the yearly gross income in thousands of dollars for a small business from 2002 through 2011. Explain what this graph says about the yearly income. Pay particular attention to the rate of growth of income.

SOLUTION

Income increased between the years 2002 and 2004. After that it remained constant until 2006, when it began to decline. Income reached a minimum in 2009 and increased after that.

The growth rate of income was positive from 2002 to 2004 and from 2009 to 2011. The fact that the graph is steeper from 2009 to 2011 than from 2002 to 2004 means that income was growing at a faster rate from 2009 to 2011.

Because income declined from 2006 to 2009, the growth rate was negative over that period. From 2004 to 2006, income did not change, meaning there was no growth. So the growth rate over that period was zero.

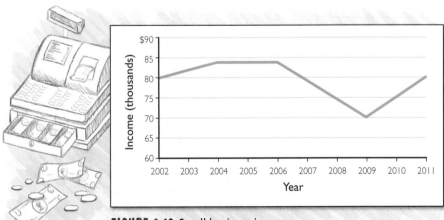

FIGURE 2.18 Small business income.

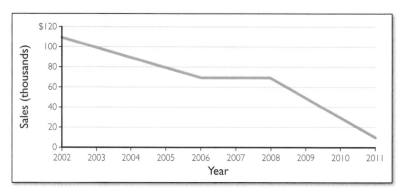

FIGURE 2.19 Sales.

TRY IT YOURSELF 2.12

The graph in **Figure 2.19** shows sales in thousands of dollars from 2002 to 2011. Describe the rate of growth in sales over this period.

The answer is provided at the end of this section.

Growth rates for smoothed line graphs

Sometimes it is more appropriate to join points with a curve rather than a straight line. The result is a *smoothed line graph*.

Key Concept

A smoothed line graph is made from a scatterplot by joining data points smoothly with curves instead of line segments.

A smoothed line graph may be appropriate when the growth rate is continuous. For example, **Figure 2.20** shows the weight of a baby over the first 12 months of life. Weight changes continuously, not in jerks. Hence, a smooth graph is appropriate.

Smoothed line graphs that are accurate contain more information about growth rates than do straight line graphs. For example, **Figure 2.21** shows the growth of money in a savings account, and **Figure 2.22** shows a man's height as a function of his age. Both graphs are increasing and so display a positive growth rate. But the

FIGURE 2.20 Smoothed graph showing a child's weight.

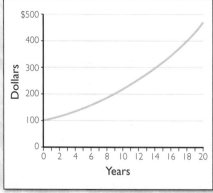

FIGURE 2.21 Money in a savings account.

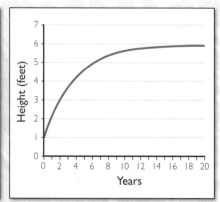

FIGURE 2.22 Height versus age.

shapes of the two curves are different. The graph in Figure 2.21 gets steeper as we move to the right, indicating an increasing growth rate. That is, the value of the account is increasing at an increasing rate. This is typical of how savings accounts grow, as we shall see in Chapter 4.

On the other hand, the graph in Figure 2.22 gets less steep as we move to the right, indicating a decreasing growth rate. That is, the graph is increasing at a decreasing rate. Practically speaking, this means that our growth in height slows as we age.

EXAMPLE 2.13 Making graphs with varying growth rates: Cleaning a waste site

In cleaning toxic waste sites, typically the amount of waste eliminated decreases over time. That is, the graph of the amount of toxic waste remaining as a function of time is decreasing at a decreasing rate. Sketch an appropriate graph for the amount of toxic waste remaining as a function of time.

SOLUTION

The graph should be decreasing, but it should become less steep as we move to the right. One possibility is shown in **Figure 2.23**.

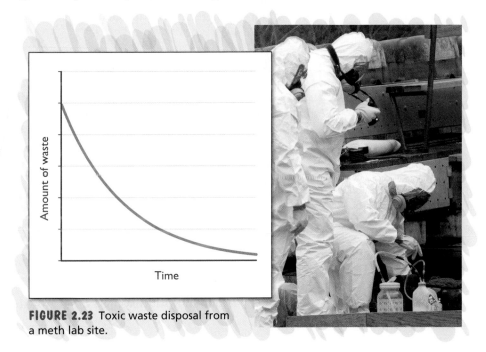

FIGURE 2.23 Toxic waste disposal from a meth lab site.

TRY IT YOURSELF 2.13

As an automobile ages, gas mileage typically decreases, and it decreases at an increasing rate. Sketch a possible graph showing gas mileage as a function of time.

The answer is provided at the end of this section.

Growth rates in practical settings

In practical settings, the growth rate has a familiar meaning. For example, the graph in **Figure 2.24** on the following page shows a population that increases for a time and then begins to decrease. The growth rate in this context is, as the name suggests, the rate of population growth. In this graph, the growth rate is positive in Year 0. As we move toward Year 3, it remains positive but becomes less and less so until we reach

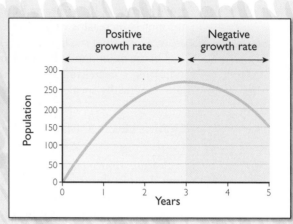

FIGURE 2.24 A population graph.

Year 3, where the growth rate is 0. Beyond Year 3 the growth rate gets progressively more and more negative through Year 5.

EXAMPLE 2.14 Describing growth rates using graphs: Deer population

Ecologists have closely studied the deer population on the George Reserve in Michigan.[7] A graph of the population over time is shown in **Figure 2.25**. Here, the horizontal axis represents the number of years since observations began, and the vertical axis represents the number of deer present.

  a. Give a brief general description of how the deer population changes over the 10-year period shown. Your description should include the growth rate of the population and how that rate changes. Explain what this means for the population in the long run.

  b. Estimate the time at which the population is growing at the fastest rate.

SOLUTION

  a. The population grew throughout the 10-year observation. Over about the first 4 years of the observation, the graph is increasing and getting steeper, so the population shows increasingly rapid growth over this period. This means that

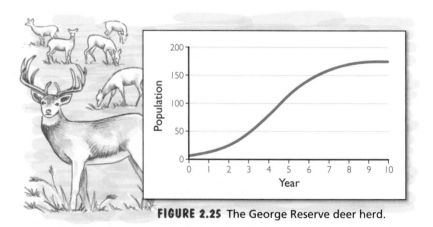

FIGURE 2.25 The George Reserve deer herd.

[7] One interesting source is Dale R. McCullough, *The George Reserve Deer Herd* (Ann Arbor: University of Michigan Press, 1979).

not only is the population increasing over this period, but also the *rate* of population growth is increasing. From about 4 years on, we see that the graph continues to increase but becomes less steep. Thus, the population continues to grow, but at a decreasing rate. The graph levels off as we approach Year 10. This flattening of the graph indicates that the growth rate is declining to near 0, and the population is stabilizing at about 175 deer.

b. The population is growing at the fastest rate where the graph is the steepest. We estimate that this occurs at about Year 4.

Try It Yourself answers

Try It Yourself 2.11: Making a scatterplot and a line graph: Sunflower growth

Scatterplot:

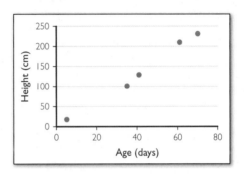

Line graph:

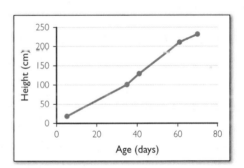

Try It Yourself 2.12: Interpreting line graphs: Sales growth   The rate of growth in sales is negative from 2002 to 2006 and from 2008 to 2011. The growth rate is more negative over the second period. From 2006 to 2008, the growth rate is 0.

Try It Yourself 2.13: Making graphs with varying growth rates: Gas mileage   The graph should be decreasing and become steeper as we move to the right. The accompanying figure gives one example.

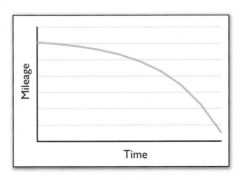

Exercise Set 2.2

1. Prescription drugs. Here is a table showing retail prescription drug sales in billions of dollars for different U.S. sales outlets.

Sales outlet	Traditional chain	Independent	Mass merchant	Super- markets	Mail order
Sales (billions of dollars)	73.4	36.7	17.4	21.8	33.5

Represent these data with a bar graph, a scatterplot, and a line graph. Which type of graph do you think best represents the data?

2. Graduation rates. According to the NCAA, 62% of female athletes at Division I schools who entered college in 1984 graduated within six years.[8] This increased to nearly 70% in 1988 and remained at that rate through 1992. By 2002 it had reached 72%. Of their male counterparts, only 46% of students who entered in 1984 graduated within six years. This number increased steadily to just over 54% in 1989 and steadily declined to 52% in 1992. By 2002 it had reached 57%. On one set of axes, sketch two line graphs displaying this information, one for females and another for males.

3. An investment. You invest $1000 in a risky fund. Over the first six months, your investment loses half its value. Over the next six months, the value increases until it reaches $1000. Make a smoothed line graph that shows your investment over the 12-month period.

4. Paper airplanes. The following table shows the gliding distance in meters of a certain type of paper airplane and its wingspan in centimeters:

Wingspan (centimeter)	5	10	15	20
Glide distance (meter)	5	8	9	13

Make a scatterplot of these data with wingspan on the horizontal axis.

5. A man's height. A child is 38 inches tall at age 3. He grows to 72 inches at age 17, at which time he has achieved his adult height. Make a line graph showing his height as a function of his age over the first 20 years.

6. A population of foxes. Sketch a graph of a population of foxes that reflects the following information: The population grew rapidly from 1990 to 2000. In 2000 a disease caused the rate of growth to slow until the population reached a maximum in 2010. From 2010 onward, the fox population declined very slowly.

7. A flu epidemic. During a certain outbreak of the flu, there were 20 new cases initially reported. The number of new cases increased each day for five days until a maximum of 35 new cases was reported. From Day 5 on, the number of new cases declined. Sketch a line graph of new cases of flu as a function of time in days.

8. Making a graph. Sketch a graph that starts from the vertical axis at 10 and has the following properties: The graph is initially decreasing at an increasing rate, but after a bit the graph decreases at a decreasing rate.

9. Making a graph. Sketch a graph that is increasing at a decreasing rate from 0 to 3, decreasing at an increasing rate from 3 to 6, and decreasing at a decreasing rate from 6 on.

10. A newspaper. Suppose a certain newspaper has recorded daily sales beginning in January 2012. On the first day of that month, 200 copies were sold. From January through May, sales increased each day until they hit a high of 800 copies on June 1. From there sales decreased each day to 400 copies on December 1. Sales declined most rapidly at some point in August. Sketch a graph reflecting this information. *Suggestion*: Locate the points you know (January 1, June 1, and December 1) first.

11. Bears. A population of bears is introduced into a game preserve. Over the first five years, the bear population shows a negative growth rate. The growth rate is positive over the next five years. Sketch a possible graph of the bear population as a function of time.

12. Wolves. A population of wolves is introduced into a game preserve. Over the first five years, the wolf population shows a positive growth rate. The growth rate is negative over the next five years. Sketch a possible graph of the wolf population as a function of time.

Blogosphere growth The number of blogs (or weblogs) grew rapidly in the opening years of the twenty-first century. **Figure 2.26** is taken from a report measuring that growth. The graph requires a bit of explanation. For example, for the date 11-May-03, the value of 40 on the vertical axis says that 40 days earlier there were half as many blogs as there were on this date. Hence, the graph shows the doubling time as a function of the date. Interpreting the graph can be a little tricky: A larger doubling time indicates a *smaller* percentage rate of growth in the number of blogs.

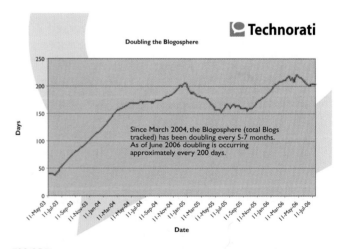

FIGURE 2.26 Doubling times for number of blogs.

---

[8]*NCAA Graduation Rates Summary*, 1999.

Exercises 13 through 15 refer to Figure 2.26.

13. The caption on the graph says that since March 2004, the number of blogs has been doubling every five to seven months. Explain how this can be determined from the graph.

14. Between 11-Jul-03 and 11-Mar-04, the graph is increasing. What does this say about the percentage rate of growth in the number of blogs?

15. Do you expect the number of blogs to grow without any limit? Use your answer to this question to sketch what the graph of doubling times should look like over a long period in the future.

16. Population growth. The graphs in **Figures 2.27 and 2.28** show the population (in thousands) of a certain animal in two different reserves, reserve A and reserve B. In each case, the horizontal axis represents the number of years since observation began.

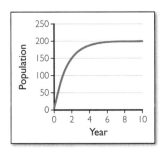

FIGURE 2.27 Population in reserve A.

FIGURE 2.28 Population in reserve B.

In **Figures 2.29 and 2.30** are two graphs of population growth rates. Let's call them growth rate I and growth rate II. One of the graphs of growth rate describes the population in reserve A, and the other describes the population in reserve B. Which is which? For each reserve, give a careful explanation of how the graph of population is related to the appropriate graph of growth rate.

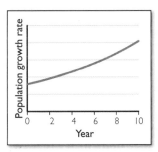
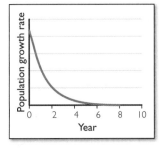

FIGURE 2.29 Population growth rate I.

FIGURE 2.30 Population growth rate II.

17. Magazine sales. An executive for a company that publishes a magazine is presented with two graphs, one showing sales (in thousands of dollars), and the other showing the rate of growth in sales (in thousands of dollars per month). Someone forgot to label the vertical axis for each graph, and the executive doesn't know which is which. Let's call the graphs mystery curve 1 and

mystery curve 2. They are shown in **Figures 2.31 and 2.32**. (In each case, the horizontal axis represents the number of months since the start of 2010.)

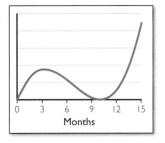
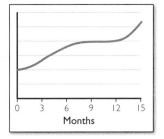

FIGURE 2.31 Mystery curve 1.   **FIGURE 2.32** Mystery curve 2.

Help the executive out by identifying which curve shows sales and which shows the rate of growth in sales. Give a careful explanation of how the graph of sales is related to the graph of growth in sales.

18. Location and speed. The graph in **Figure 2.33** shows the location of a vehicle moving east on a straight road. The location is measured as the distance east of a fixed observation point. The graph in **Figure 2.34** shows the speed of the same vehicle, as measured by speedometer readings. In each case, the horizontal axis represents the time in hours since measurements began.

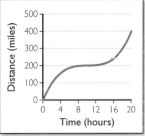
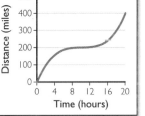
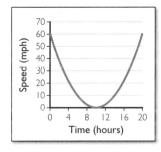

FIGURE 2.33 Location of vehicle.   **FIGURE 2.34** Speed of vehicle.

In this situation, the speedometer reading measures the growth rate of distance from your starting point. In Example 2.14 on page 100, we saw how the graph of population growth rate is related to the graph of population. Explain how the graph of speed in Figure 2.34 is related to the graph of location in Figure 2.33, and write a travel story that matches these graphs.

19. Historical figures: Descartes and Fermat. Graphs and charts have been used throughout recorded history. Their incorporation into mathematics should be credited to any number of early mathematicians. Certainly, two on any such list are René Descartes and Pierre de Fermat. Write a brief report on their development of graphs in mathematics.

20. Historical figures: Oresme. Descartes and Fermat are famous early mathematicians, but lesser known is Nicole Oresme, who preceded both Fermat and Descartes by 300 years. He employed graphs in mathematics long before either. Write a brief report on his use of graphs in mathematics.

2.3 Misleading graphs: Should I believe my eyes?

The following is an excerpt from an article appearing on the *New York Times* Web site.

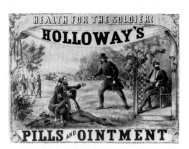

A vintage drug ad.

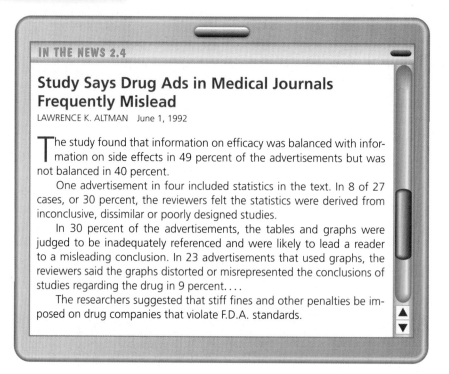

IN THE NEWS 2.4

Study Says Drug Ads in Medical Journals Frequently Mislead

LAWRENCE K. ALTMAN   June 1, 1992

The study found that information on efficacy was balanced with information on side effects in 49 percent of the advertisements but was not balanced in 40 percent.

One advertisement in four included statistics in the text. In 8 of 27 cases, or 30 percent, the reviewers felt the statistics were derived from inconclusive, dissimilar or poorly designed studies.

In 30 percent of the advertisements, the tables and graphs were judged to be inadequately referenced and were likely to lead a reader to a misleading conclusion. In 23 advertisements that used graphs, the reviewers said the graphs distorted or misrepresented the conclusions of studies regarding the drug in 9 percent. . . .

The researchers suggested that stiff fines and other penalties be imposed on drug companies that violate F.D.A. standards.

Graphs can be helpful visual tools, but sometimes they can be misleading (intentionally or otherwise), as the above article suggests. It is remarkable that a significant percentage of graphs associated with drug advertisements in medical journals were found to distort facts.

There are many ways that graphs can mislead us into drawing incorrect, inaccurate, or inappropriate conclusions about data.[9] We illustrate some of the most common types of misleading graphs in this section.

Misleading by choice of axis scale

Consider the bar graphs in **Figures 2.35 and 2.36**, which show federal defense spending in billions of dollars for the given year.

If we look carefully at the years represented and the spending reported, we see that the two graphs represent exactly the same data—yet they don't look the same. For example, Figure 2.36 gives the visual impression that defense spending doubled from 2000 to 2002, and Figure 2.35 suggests that defense spending did increase from 2000 to 2002, but not by all that much. What makes the graphs so different?

The key to understanding the dramatic difference between these two graphs is the range on the vertical axis. In Figure 2.35 the vertical scale goes from 0 to 400 billion dollars, and in Figure 2.36 that scale goes from 250 to 350 billion dollars. The shorter vertical range exaggerates the changes from year to year.

---

[9] A classic reference here is Darrell Huff's *How to Lie with Statistics* (New York: W.W. Norton, 1954).

"How close to the truth do you want to come, sir?"

One way to assess the actual change in defense spending is to find the percentage change. Let's consider the change from 2000 to 2002. In both graphs, we see that the amount spent in 2000 was about 300 billion dollars, and the amount spent in

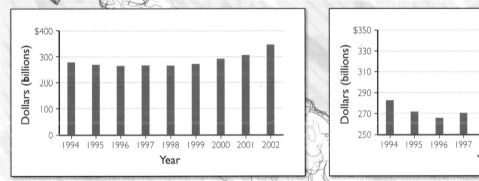

FIGURE 2.35 Defense spending.

FIGURE 2.36 Defense spending again.

2002 was about 350 billion dollars. First we find the absolute change:

$$\text{Change} = 350 - 300 = 50 \text{ billion dollars.}$$

Now to find the percentage change, we divide by the spending in 2000:

$$\text{Percentage change} = \frac{50}{300} \times 100\%.$$

This is about 17%. The percentage increase is fairly significant, but it is not nearly as large as that suggested by Figure 2.36, which gives the impression that spending has almost doubled—that is, increased by 100%.

It is striking to see how easily someone with a particular point of view can make visual displays to suit his or her own position. If we wanted others to believe that defense spending remained nearly stable over this time period, which graph would we use? On the other hand, which graph would we use if we wanted others to believe that defense spending showed a dramatic increase over this time period?

EXAMPLE 2.15 Analyzing a choice of scale: A misleading graph from CNN

An article from the Media Matters[10] Web site asserts: In presenting the results of a CNN/*USA Today*/Gallup poll, CNN.com used a visually distorted graph that falsely conveyed the impression that Democrats far outnumber Republicans and Independents in thinking the Florida state court was right to order Terri Schiavo's feeding tube removed.[11] CNN presented the graph shown in **Figure 2.37** to show the response to the following question: "Based on what you have heard or read about the case, do you agree with the court's decision to have the feeding tube removed?" CNN.com responded to criticisms by replacing Figure 2.37 with **Figure 2.38** at its Web site.

a. What percent of Republicans polled agreed with the court's decision? What percent of Democrats polled agreed with the court's decision?

b. Fill in the blank in this sentence: The fraction of Democrats agreeing with the court's decision was ___ percent more than the fraction of Republicans agreeing with the court's decision. (Round your answer to the nearest whole number.)

c. What impression does CNN's original graph give about how Democrats and Republicans compared on this question?

SOLUTION

a. The graph shows that 54% of Republicans polled agreed with the court's decision and 62% of Democrats polled agreed with the court's decision.

b. To find the answer, we divide the difference of eight percentage points by the percentage for Republicans:

$$\text{Percentage change} = \frac{\text{Difference in percentages}}{\text{Republican percentage}} \times 100\%$$

$$= \frac{8}{54} \times 100\%,$$

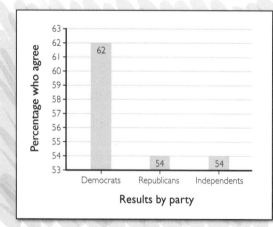

FIGURE 2.37 Graphic initially posted by CNN.com.

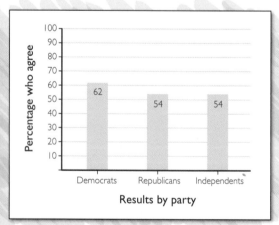

FIGURE 2.38 Replacement graphic posted by CNN.com.

---

[10]This article, and the graphic shown in Figure 2.38, appear at the Web site http://mediamatters.org/items/200503220005.

[11]Terri Schiavo was a young woman diagnosed as being in a persistent vegetative state. She had been kept on life-sustaining feeding tubes for several years. Her husband eventually got court permission to disconnect the life support, and she subsequently died.

or about 15%. We conclude that the fraction of Democrats agreeing was about 15% larger than the fraction of Republicans agreeing.

c. In the original graph, the bar representing Democrats is 9 units high, and the bar representing Republicans is 1 unit high. This gives the impression that the fraction of Democrats agreeing was nine times as large as the fraction of Republicans agreeing, which is 800% more instead of 15% more.

TRY IT YOURSELF 2.15

The graphs in Figures 2.37 and 2.38 represent exactly the same data. Explain what accounts for the distortion in the first graph, as compared with the second one.

The answer is provided at the end of this section.

Default ranges on graphs generated by calculators and computers

Graphing software for calculators and computers has default methods for deciding how to scale the axes for graphs. These defaults often give good results, but sometimes the graphs will be misleading unless they are adjusted. Here is an excerpt from an article that appeared in the *Daily Oklahoman* concerning scores on the ACT college entrance exam in Oklahoma and the nation. We note that ACT scores range from 1 to 36.

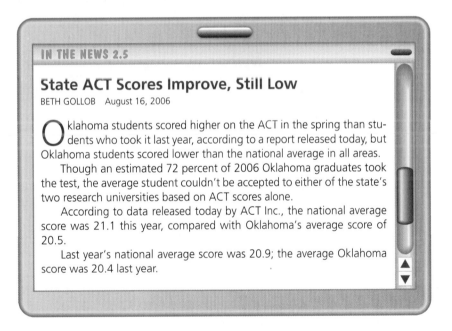

IN THE NEWS 2.5

State ACT Scores Improve, Still Low

BETH GOLLOB   August 16, 2006

Oklahoma students scored higher on the ACT in the spring than students who took it last year, according to a report released today, but Oklahoma students scored lower than the national average in all areas.

Though an estimated 72 percent of 2006 Oklahoma graduates took the test, the average student couldn't be accepted to either of the state's two research universities based on ACT scores alone.

According to data released today by ACT Inc., the national average score was 21.1 this year, compared with Oklahoma's average score of 20.5.

Last year's national average score was 20.9; the average Oklahoma score was 20.4 last year.

The article is accurate. It states, among other things, that the national average was 21.1 and the Oklahoma average was 20.5. Now consider **Figure 2.39**, which shows the graph that accompanied this article. This bar graph is based on data that compare the subject area test scores in Oklahoma to those in the nation as a whole. It certainly gives the impression that Oklahoma scores are far below national scores—especially in math.

Is this impression accurate? Are the Oklahoma scores really as far below the national scores as the bar graph makes it appear? We'll explore that question further in Exercises 10 through 12, where the scale on the vertical axis is examined more closely.

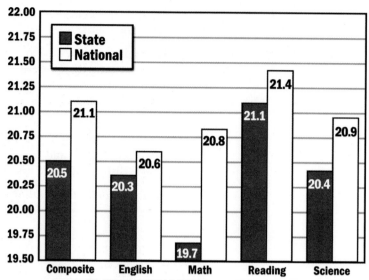

FIGURE 2.39 Bar graph of ACT scores.

Even though the graph may be a misleading representation of the data, the reporter might be trying to give a fair and accurate presentation. It is worth noting that the graph shown in Figure 2.39 is exactly the graph obtained by accepting the default ranges offered by one of the most popular graphing packages.

We emphasize that, although a graph may or may not be *intended* to deceive, the effect is the same. It is up to us to think critically so as not to allow ourselves to be misled.

CALCULATION TIP 2.1 Graphing on Calculators

A graphing calculator or computer software usually has default settings for the scales it uses on the two axes. When we plot a graph with a calculator or computer, we need to be alert to those scales. We might need to adjust them manually to obtain a graph that conveys an appropriate picture.

EXAMPLE 2.16 Choosing scales for graphs: Presenting a point of view on gasoline prices

The following table shows the average price of a gallon of regular gasoline in January of the given year:

Year	1997	1998	1999	2000	2001	2002	2003	2004	2005	2006	2007
Average price	$1.28	$1.10	$0.93	$1.27	$1.40	$1.09	$1.44	$1.51	$1.77	$2.24	$2.33

Your debate team must be prepared to present a case for the following proposition and against it.

Proposition: The average price of a gallon of regular gasoline showed a significant increase from 1997 through 2007.

Make a bar graph of the data that you would use for an argument in support of the proposition. Make a second bar graph that you would use for an argument against the proposition.

SOLUTION

To support the proposition, you want a graph that emphasizes the increase in gas prices. A narrow range on the vertical scale emphasizes the differences in data. In

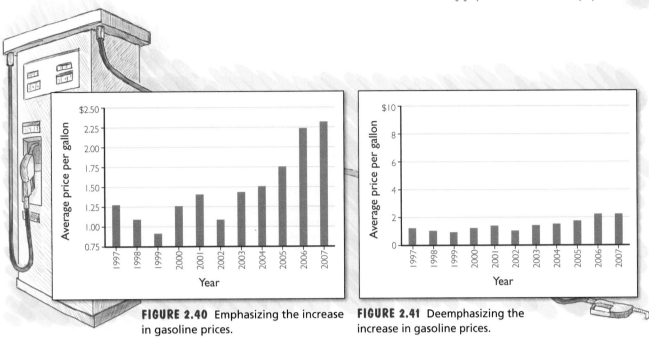

FIGURE 2.40 Emphasizing the increase in gasoline prices.

FIGURE 2.41 Deemphasizing the increase in gasoline prices.

Figure **2.40**, we have made a graph to support the proposition using a vertical range from $0.75 to $2.50. To argue against the proposition, we want a graph that deemphasizes the differences in data. A wider vertical range will accomplish this. In **Figure 2.41**, we have used a vertical range from $0 to $10.00. This graph would be used in an argument against the proposition.

TRY IT YOURSELF 2.16

The table below shows average ACT scores for 1990 through 2000, which, according to the ACT, "is the first decade ever in which the national average increased substantially."

Year	1990	1991	1992	1993	1994	1995	1996	1997	1998	1999	2000
ACT scores	20.6	20.6	20.6	20.7	20.8	20.8	20.9	21.0	21.0	21.0	21.0

Make a bar graph that you would use to argue that average ACT scores increased significantly during this period. Make a second bar graph that you would use to argue that average ACT scores did not increase significantly over this period.

The answer is provided at the end of this section.

Misleading by misrepresentation of data: Inflation

The scatterplot in **Figure 2.42** appeared in an article by Frederick Klein in the July 5, 1979, edition of the *Wall Street Journal*. The graph shows the total amount of currency (in billions of dollars) in circulation. The plot appeared with the caption, "Over a quarter-century the currency pile-up steepens." The article uses this graph to bolster the case that "the amount of currency in individual hands is soaring." On the surface, the graph makes a strong case for the author, and his conclusion is technically accurate—that is, the amount of currency in circulation was increasing. But this conclusion ignores the *value* of the currency.

The price of many commodities increases over time. For instance, the median price of an existing one-family house in 1968 was $20,100, the median price of a house in 1982 was \$67,800, and the median price of a house in 2009 was \$172,100.[12] This increase in home values is for the most part a reflection of increases in prices overall. The overall increase in prices is measured by the *inflation rate*.[13]

Another way of looking at this question is in terms of the value of a dollar. For example, from 1950 to 2000 prices, on average, increased by 596%. This means that a dollar would buy the same amount of goods in 1950 as the dollar plus \$5.96, or \$6.96, would buy in 2000. So a "1950 dollar" has the same buying power as 6.96 "2000 dollars." When we *adjust a graph for inflation*, we report all currency amounts in *constant dollars*—for example, we could state the buying power in terms of the year 2000.

In **Figure 2.43**, we have reproduced the scatterplot after adjusting for inflation. Using constant dollars means we are using the same measure of value in 1950 and 1980. Before adjusting for inflation, the vertical axis represents the amount of currency. After the adjustment, the vertical axis represents the buying power of the currency, which is a better measure of wealth. We see that, in terms of buying power, the amount of currency increased much less dramatically than the original graph suggested. It could be argued that the graph in Figure 2.42 presents a somewhat distorted view of the amount of currency because of the changing value of the dollar over time.

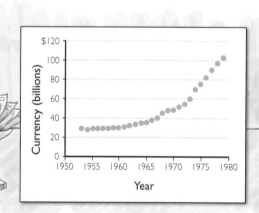

FIGURE 2.42 Amount of currency in circulation.

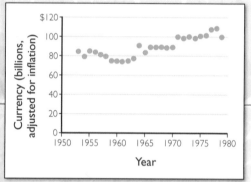

FIGURE 2.43 Adjusted for inflation in constant 1979 dollars.

Another example is shown in **Figure 2.44**. This figure shows the federal minimum wage by year. The blue (lower) graph shows the minimum wage (referred to as *nominal*), which increased from 1938 through 2010. The red graph shows the minimum wage in constant 2007 dollars. It is the blue graph adjusted to 2007 dollars. Because the red graph shows the minimum wage in terms of a fixed value of the dollar, it provides a better comparison of minimum wages earned from year to year than does the blue graph. It is interesting that the value of the minimum wage did not always increase (as one might be led to believe by the blue graph) but actually peaked in 1968.

Let's see how we find the graph showing constant 2007 dollars. For example, the nominal minimum wage in 1968 was \$1.60. According to the U.S. Bureau of Labor, the inflation rate from 1968 to 2007 was 490%, meaning that prices increased by 490%. Hence, one "1968 dollar" has the same buying power as the dollar plus 4.90, or 5.90 "2007 dollars." Therefore, we convert the value of \$1.60 for the minimum wage in 1968 to 2007 dollars as follows:

$$1.60 \text{ "1968 dollars"} = 1.60 \times 5.90 = 9.44 \text{ "2007 dollars."}$$

We note that this result agrees with the location of the peak of the red graph in Figure 2.44.

---

[12] Prices don't always go up. Housing prices fell quite sharply from 2007 to 2009.

[13] Inflation is calculated using the *consumer price index*. We will look more closely at inflation in Chapter 4.

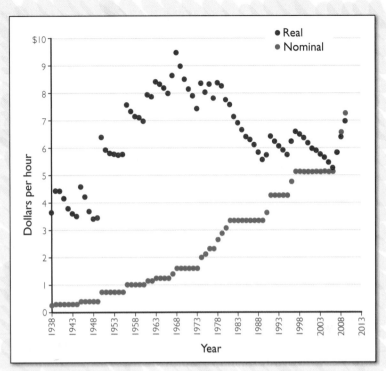

FIGURE 2.44 Minimum wage.

"I guess it's an OK job. Minimum wage plus all you can eat. I tried to quit once but couldn't get out the door."

SUMMARY 2.4   Adjusting for Inflation

- It is important to know whether graphs involving currency are adjusted for inflation.

- If inflation from Year 1 to Year 2 is r as a decimal, one Year-1 dollar has the same purchasing power as $1 + r$ Year-2 dollars.

- If inflation from Year 1 to Year 2 is r as a decimal, we express Year-1 dollars in constant Year-2 dollars using

$$D \text{ "Year-1 dollars"} = D(1 + r) \text{"Year-2 dollars."}$$

EXAMPLE 2.17 Adjusting for inflation: Gasoline prices

The following table shows the average cost of a gallon of regular gasoline in the given year. These data are plotted in **Figure 2.45**.

Year	1970	1980	1990	2000	2010
Price per gallon	$0.36	$1.25	$1.16	$1.51	$2.78

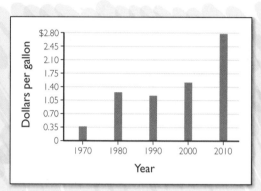

FIGURE 2.45 Gasoline prices.

Inflation rates are shown in the following table:

Time span	1970–2010	1980–2010	1990–2010	2000–2010
Inflation	451%	154%	64%	26%

a. Use the graph in Figure 2.45 to determine, without adjusting for inflation, in what year gasoline was most expensive.

b. Complete the following table showing gasoline prices in constant 2010 dollars:

Year	1970	1980	1990	2000	2010
Price per gallon	$0.36	$1.25	$1.16	$1.51	$2.78
Price in 2010 dollars					

c. Make a bar graph showing gasoline prices in constant 2010 dollars.

d. In what year were gasoline prices, adjusted for inflation, the highest?

SOLUTION

a. Gasoline was most expensive in 2010.

b. To find the price per gallon of 1970 gas in terms of "2010 dollars," we use the inflation rate of 451%. Expressed as a decimal, this number is $r = 4.51$. Now we use the constant-dollars formula in Summary 2.4:

$$D \text{ "1970 dollars"} = D(1 + r) \text{ "2010 dollars"}$$

$$0.36 \text{ "1970 dollars"} = 0.36 \times (1 + 4.51) \text{ "2010 dollars"}$$

$$= 1.98 \text{ "2010 dollars."}$$

So the price of gas in 1970 was $1.98 in "2010 dollars." This gives the first entry in the table. We find the remaining entries in a similar fashion. The table below shows the result.

Year	1970	1980	1990	2000	2010
Price per gallon	$0.36	$1.25	$1.16	$1.51	$2.78
Price in 2000 dollars	0.36×5.51 $= \$1.98$	1.25×2.54 $= \$3.18$	1.16×1.60 $= \$1.86$	$\$1.51 \times 1.26$ $= \$1.90$	$2.78

c. The bar graph giving prices in 2000 dollars is shown below.

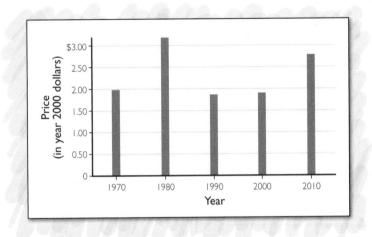

d. The price of gasoline, adjusted for inflation, was highest in 1980. Comparing this answer with the answer to part a shows the value of adjusting for inflation.

These examples lead us to a question about the example on defense spending we considered at the beginning of this section. Were the graphs in Figures 2.35 and 2.36 adjusted for inflation? The answer is that they were not. You will have the opportunity to look further at this question in Exercise 25.

Misleading by using insufficient data

A graph can give an accurate picture—provided sufficient data are used to produce it. If the data are insufficient, graphs can mislead us into drawing inaccurate conclusions. In **Figure 2.46**, we show a bar graph of U.S. passenger car production by year. The graph seems to show that automobile production declined steadily from 1960 to 2000. If we look closely at the labels on the horizontal axis, we see that we have recorded data points only once per decade. In **Figure 2.47**, we have added data

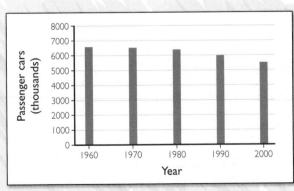

FIGURE 2.46 Car production.

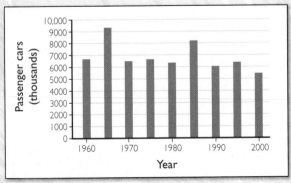

FIGURE 2.47 Car production revisited.

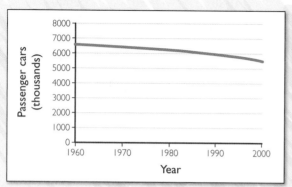
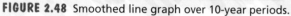

FIGURE 2.48 Smoothed line graph over 10-year periods.

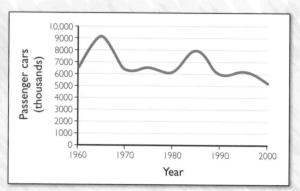

FIGURE 2.49 Smoothed line graph over 5-year periods.

points for each five-year interval. This graph does show an overall pattern of decline in car production, but it also shows that there were fluctuations in production. It is clear what makes the graphs different: We used more data points to make the graph in Figure 2.47.

In the case of bar graphs or scatterplots, we can tell how many data points have been plotted, but in the case of smoothed line graphs, there may be no indication of how many data points were used. To illustrate this point, we consider again the data on the production of passenger cars, but now we use smoothed line graphs. A smoothed line graph using data points once per decade is shown in **Figure 2.48**, and a graph using data from five-year intervals is shown in **Figure 2.49**. Unlike the preceding bar graphs, there is no way to tell how many data points were used to make the new graphs, and hence there is no way to determine which gives a more accurate picture of automobile production.

2009 Honda FCX Clarity vehicle assembly.

If we see a smoothed line graph in the media, we may not know how many data points were used in producing it. If there's a reason to question the accuracy, the wisest course of action is to look for the original data source.

Pictorial representations

Often data are presented using images designed to grab our attention. These images can be helpful if they are properly constructed. But they can be misleading if they are not properly presented. One common graphical device is the *pie chart*. Pie charts are used to display how constituent parts make up a whole. They are circular regions divided into slices like pieces of a pie. The pie represents the whole, and the slices represent individual parts of the whole. For example, if one item is 30% of the whole, then its corresponding slice should make up about 30% of the area of the pie.

EXAMPLE 2.18 Making a pie chart: Undergraduate enrollment at UC Davis

Make a pie chart showing the following enrollment data from the University of California, Davis:

UC Davis Undergraduates	23,499
Agricultural and Environmental Sciences	4819
Engineering	2950
Letters and Science	10,243
Biological Sciences	5361
Teaching Credential	126

SOLUTION

First we calculate the percentage of the whole represented by each category of student. For example, the 4819 students in Agricultural and Environmental Sciences represent

$$\frac{4819}{23,499} \times 100\%,$$

or about 20.5% of the whole undergraduate population of 23,499. The resulting percentages are shown in the table below.

UC Davis Undergraduates	23,499	100%
Agricultural and Environmental Sciences	4819	20.5%
Engineering	2950	12.6%
Letters and Science	10,243	43.6%
Biological Sciences	5361	22.8%
Teaching Credential	126	0.5%

Our pie must be divided in these proportions. The result is shown in **Figure 2.50**.

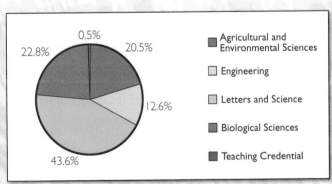

FIGURE 2.50 UC Davis undergraduate enrollment.

Pie charts such as the one in Figure 2.50 can be helpful in understanding data. As a contrast, consider the accompanying "handcuff pie chart." It is taken from a Web site and presumably tells us something about repeat offenders. Recall that the sections of a pie chart represent percentages of a whole. It is unclear what the sections in this chart are intended to represent.

Other types of graphical descriptions can also be misleading. For example, the currency representation on the left is supposed to show the value of the dollar under various presidents. But notice that the "Carter dollar" is labeled as being worth

1958—Eisenhower

1963—Kennedy

1968—Johnson

1973—Nixon

1978—Carter

1984—Reagan

1990—Bush

1994—Clinton

2006—Bush

A visual currency representation claiming to show the value of a dollar.

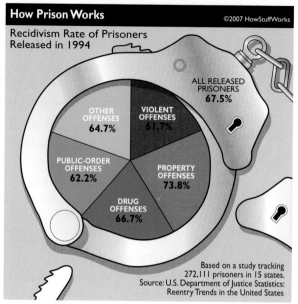

A strange pie chart.

44 cents, compared to $1.00 for the "Eisenhower dollar." Hence, the "Carter dollar" should be almost half the size of the "Eisenhower dollar." In fact, it is less than an eighth of the size in area of the "Eisenhower dollar." The picture itself gives a false impression of the relative values of currency during these two presidential administrations.

There is a similar difficulty in a fuel economy chart. Note that the length of the line segment corresponding to 18 miles per gallon is less than a quarter of the length of the line segment for 27.5 miles per gallon. The figure gives an inflated view of the progress in automobile efficiency.

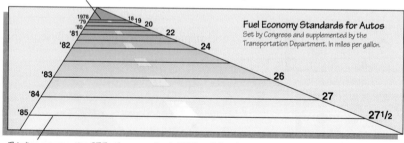

A visual representation claiming to show changes in gas mileage.

SUMMARY 2.5 **Cautionary Notes**

1. The range of the axes can distort graphs.

2. Be careful that data are accurately represented. For example, data that depend on the value of currency may need to be adjusted for inflation.

3. Any graph can be misleading if insufficient data are plotted. It may be particularly difficult to determine how many data points were used to produce a smoothed line graph.

4. Visual representations can be misleading if the relative sizes of the items pictured do not correspond to the relative sizes of the data.

5. If in doubt, we should always go to the original data source to determine the accuracy of a pictorial representation.

There are many other ways of presenting graphs that distort our perception. You will be asked to comment on some of these in the exercises.

Try It Yourself answers

Try It Yourself 2.15: Analyzing a choice of scale: A misleading graph from CNN
The scale on the vertical axis in Figure 2.37 is compressed.

Try It Yourself 2.16: Choosing scales for graphs: Presenting a point of view on ACT scores

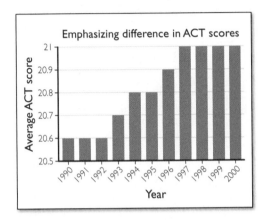 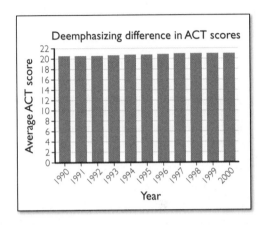

Exercise Set 2.3

1. Female graduation rates. According to the NCAA, 62% of female athletes at Division I schools who entered college in 1984 graduated within six years.[14] This increased to nearly 70% in 1988 and remained at that rate through 1992. By 2002, this had increased to 72%. Make a line graph that emphasizes the increase in graduation rates.

2. Male graduation rates. According to the NCAA, only 46% of male athletes who entered college in 1984 graduated within six years. This number increased steadily to just over 54% in 1989 and steadily declined to 52% in 1992. By 2002, it had reached 57%. Make a line graph that deemphasizes the variation in graduation rates.

3. Cancer mortality rates. Consider the graph of cancer mortality rates for white males shown in Figure 2.17 in Section 2.2. The original table of data appears on page 95. Make a bar graph of the data that emphasizes the variation in mortality rates from 1950 through 2004.

4. Female coaches. The following table is taken from data provided in the report *Women in Intercollegiate Sport: A Longitudinal, National Study; Twenty-Nine Year Update*, by L. J. Carpenter and R. V. Acosta. It shows the percentage of female head coaches for women's teams (all divisions, all sports) in the given year. Make a line graph of the data that deemphasizes the decline in the percentage of female head coaches from 1980 through 2010.

Year	1986	1992	1998	2004	2010
Percent	50.6	48.3	47.4	44.1	42.6

5. Informative graph? Figure 2.51 shows the distribution of education level among the staff positions at a certain university. Discuss the usefulness of this graph. *Suggestion*: Look at the first bar.

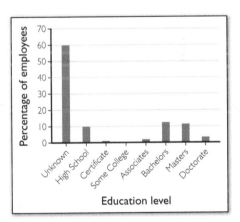

FIGURE 2.51 Total university education distribution.

6. Donations. An article in the Oklahoma State University newspaper *The Daily O'Collegian* looked at the seven largest gifts to the university. The graph in **Figure 2.52** accompanied the article. In a few sentences, explain what you think the graph is trying to convey. In particular, what information do the lines connecting the dots convey? Can you suggest a better way to present these data?

---

[14]*NCAA Graduation Rates Summary*, 1999.

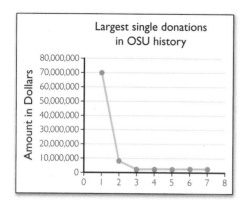

FIGURE 2.52 Donations to OSU.

7. Pets. The graph in **Figure 2.53** shows the number of pets of various types on my street. What information do the lines connecting the dots convey? Can you suggest a better way to present these data?

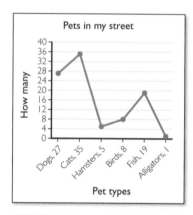

FIGURE 2.53 Pets.

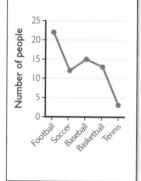

FIGURE 2.54 Sports.

8. Sports. The graph in **Figure 2.54** shows the favorite sports of people in my class. What information do the lines connecting the dots convey? Can you suggest a better way to present these data?

9. Drivers. Figure 2.55 shows the number (in tens) of fatally injured drivers involved in automobile accidents in the state of

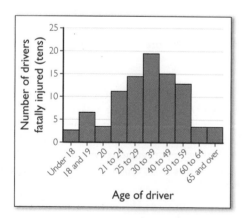

FIGURE 2.55 Fatal automobile accidents.

New York. Comment on the grouping of ages. The graph shows many more fatal accidents for drivers 30 to 39 than for drivers 65 and older. Do you suspect that there are more drivers in one age group than in another? What additional information would you need to make a judgment as to which age group represented the safest drivers?

Oklahoma ACT scores Exercises 10 through 12 refer to the In the News feature 2.5 on page 107 and the graph in Figure 2.39, which accompanied the original article.

10. Calculate, for the composite and each of the subject areas, how many percentage points lower than the national average the Oklahoma scores are.

11. Sketch a bar graph of the data in Figure 2.39 using a scale of 0 to 22 on the vertical axis.

12. Explain how someone might prefer the original bar graph or the bar graph you made in Exercise 11, depending on the impression to be conveyed. Which graph do you think more accurately reflects the numbers you found in Exercise 10?

13. Wedding costs. The graph in **Figure 2.56** was taken from the Web site of *The Wedding Report*. It shows average wedding costs in thousands of dollars for the given year.

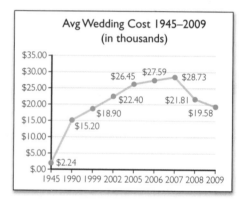

FIGURE 2.56 Average wedding costs.

a. Over which of the time periods indicated by the horizontal axis do wedding costs appear to be increasing most rapidly?

b. What is the average growth rate per year in wedding costs from 1945 to 1990? (First express your answer in thousands of dollars per year rounded to two decimal places, and then express it in dollars per year.)

c. What is the average growth rate per year in wedding costs from 2005 to 2006?

d. In view of your answers to parts b and c, do you think the graph in Figure 2.56 is misleading?

Web conferencing The graph in **Figure 2.57** shows the worldwide revenue in millions of dollars for the Web conferencing market from 2001 through 2007. It appeared in the December 2003/January 2004 edition of *Technology Review*. Exercises 14 and 15 refer to this graph.

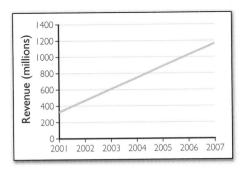

FIGURE 2.57 Revenue for Web conferencing.

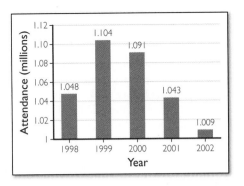

FIGURE 2.59 Tulsa State Fair attendance.

14. In light of the date of publication of the article, comment on your estimation of the reliability of the graph in Figure 2.57.

15. Can you think of a reason why the person who produced the graph in Figure 2.57 made it a straight line?

16. Stock charts. Consider the graph in **Figure 2.58**, which tracks the Dow Jones Industrial Average over a specific week. What do you notice about the scale on the vertical axis? It is common for graphs that track the Dow (and other stock market indices) over relatively short periods of time to use a similar vertical scale. What is the advantage of such a scale in this situation?

19. Sketch a bar graph using 0 to 1.2 million on the vertical axis. Does this give a more accurate display of the data? Why or why not?

20. Minorities employed. The bar graph in **Figure 2.60** shows how the percent of federal employees who are minorities has changed with time.[15] By adjusting the vertical range, make a bar graph that might accompany an article with the title "Federal employment of minorities shows little change."

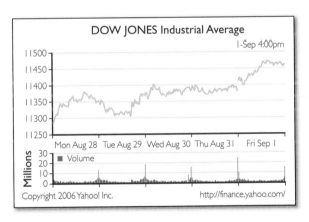

FIGURE 2.58 Dow Jones Industrial Average.

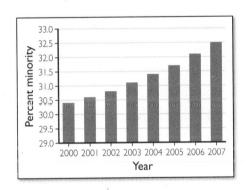

FIGURE 2.60 Minorities employed at federal level.

Aging employees   The following table shows the average age in years of federal employees at various times. It is used in Exercises 21 through 24.

Year	2000	2001	2002	2003	2004	2005	2006	2007
Average age	46.3	46.5	46.5	46.7	46.8	46.9	46.9	47.0

Tulsa State Fair   The bar graph in **Figure 2.59** appeared in a report of the Tulsa County Public Facilities Authority. The graph shows Tulsa State Fair attendance in millions from 1998 through 2002. Exercises 17 through 19 refer to this graph.

17. In a few sentences, explain what you think the graph in Figure 2.59 appears to show regarding State Fair attendance.

18. Make a table of percentage change in attendance for each year.

21. Make a bar graph that emphasizes the increase in average age.

22. Make a bar graph that gives an impression of relative stability in average age.

23. By what percentage did the average age increase from 2000 to 2007?

[15]Data, but not graph, from *Statistical Abstract of the United States, 2003.* Only federal civilian nonpostal employees included.

24. In light of the result of Exercise 23, which of the graphs you made in Exercises 21 and 22 is a better representation of the change in average age?

25. Adjusting defense spending for inflation. The amount in billions of constant 2000 dollars spent on defense in the given year is shown in the table below. Make a bar graph of the data and compare the information provided by this graph with the graphs in Figures 2.35 and 2.36.

Year	1994	1995	1996	1997	1998	1999	2000	2001	2002
2000 dollars	322.8	305.9	289.2	288.4	282.6	283.7	294.5	297.5	330.8

Adjusting for inflation Exercises 26 through 29 use the following tables, which show inflation over the given period:

Time span	1970–2010	1980–2010	1990–2010	2000–2010
Inflation	451%	154%	64%	26%

Time span	2001–07	2002–07	2003–07	2004–07	2005–07	2006–07
Inflation	17%	15%	13%	9%	5%	3%

26. Geology salaries. The following table shows the average starting salary for geologists for the given year:

Year	2001	2003	2005	2007
Salary	$57,900	$65,000	$67,800	$82,200

a. Make a bar graph of starting salaries for geologists.

b. Complete the following table showing starting salaries in constant 2007 dollars:

Year	2001	2003	2005	2007
Salary 2007 dollars	$57,900	$65,000	$67,800	$82,200

c. Make a bar graph of salaries in constant 2007 dollars.

d. Compare the graphs from parts a and c.

27. Gold prices. The following table shows the average price per ounce of gold in the given year:

Year	1970	1980	1990	2000	2010
Price per ounce	$35.94	$612.56	$383.51	$279.11	$1224.53

a. Make a bar graph of gold prices.

b. Complete the following table showing gold prices in constant 2010 dollars:

Year	1970	1980	1990	2000	2010
Price per ounce 2010 dollars	$35.94	$612.56	$383.51	$279.11	$1224.53

c. Make a bar graph of gold prices in constant 2010 dollars.

28. Crude oil prices. The following table shows the average price per barrel of crude oil in the given year:

Year	1970	1980	1990	2000	2010
Price per barrel	$3.39	$37.42	$23.19	$27.39	$71.21

a. Complete the following table showing oil prices in constant 2010 dollars:

Year	1970	1980	1990	2000	2010
Price per barrel 2010 dollars	$3.39	$37.42	$23.19	$27.39	$71.21

b. Make a bar graph of oil prices in constant 2010 dollars.

c. Was crude oil more expensive in 1990 or 2000? Explain your answer.

29. Super Bowl ticket prices. The following table shows the price of a ticket to the Super Bowl in the given year:

Year	1970	1980	1990	2000	2010
Ticket price	$15	$30	$125	$325	$500

a. Complete the following table showing ticket prices in constant 2010 dollars:

Year	1970	1980	1990	2000	2010
Ticket price 2010 dollars	$15	$30	$125	$325	$500

b. Make a bar graph of ticket prices in constant 2010 dollars.

c. Compare ticket prices in constant dollars for 1970 and 2010.

30. Federal receipts. The pie chart in **Figure 2.61** on the following page shows the sources of federal receipts in 2009–2010. The total amount of federal receipts was about 2.162 billion dollars. Complete the table below.

Source	Income tax	Insurance and retirement	Corporate tax	Excise tax	Other
Percent					
Amount					

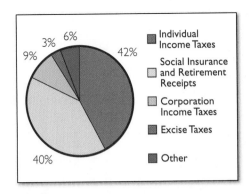

FIGURE 2.61 Federal receipts in 2009–2010.

31. A cake. The following table shows ingredients in a hypothetical recipe for a cake as a percent of the total by weight. (The authors cannot recommend actual use of this recipe.)

Ingredients	Flour	Milk	Sugar	Eggs
Percent	50	25	15	10

Represent the data in this table in two ways: by sketching a bar graph and by sketching a pie chart.

Exercise 32 is suitable for group work.

32. Health care. The graphical presentation in **Figure 2.62** is intended to compare among several countries health-care spending with life expectancy. The graph appeared in the January 2010 issue of *National Geographic* and generated a fair amount of discussion on various blogs.[16] This type of graphical presentation is called a *parallel coordinate plot*. Some suggested that a scatterplot would be more suitable, but others preferred the original plot. Summarize the reasons on both sides and give your own conclusions. Include a scatterplot of these data, either one of your own devising or one from the Web.

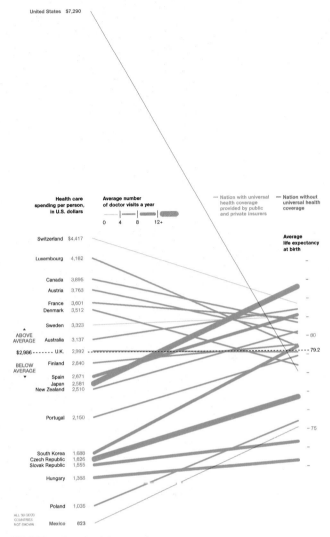

FIGURE 2.62 Health-care data.

CHAPTER SUMMARY

In our age, we are confronted with an overwhelming amount of information. Analyzing quantitative data is an important aspect of our lives. Often this analysis requires the ability to understand visual displays such as graphs, and to make sense of tables of data. A key point to keep in mind is that growth rates are reflected in graphs and tables.

Measurements of growth: How fast is it changing?
Data sets can be presented in many ways. Often the information is presented using a table. To analyze tabular information, we choose an independent variable. From this point of view, the table presents a function, which tells how one quantity (the dependent variable) depends on another (the independent variable).

A simple tool for visualizing data sets is the *bar graph*. See **Figure 2.63** for an example of a bar graph. A bar graph is often appropriate for small data sets when there is nothing to show between data points.

---

[16]http://blogs.ngm.com/blog_central/2010/01/the-other-health-care-debate-lines-vs-scatterplot.

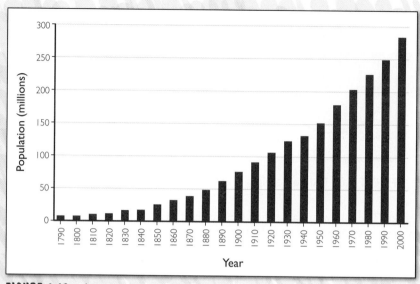

FIGURE 2.63 A bar graph.

Both absolute and percentage changes are easily calculated from data sets, and such calculations often show behavior not readily apparent from the original data. We calculate the percentage change using

$$\text{Percentage change} = \frac{\text{Change in function}}{\text{Previous function value}} \times 100\%.$$

One example of the value of calculating such changes is the study of world population in Example 2.3.

To find the average growth rate of a function over an interval, we calculate the change in the function divided by the change in the independent variable:

$$\text{Average growth rate} = \frac{\text{Change in function}}{\text{Change in independent variable}}.$$

Keeping track of the units helps us to interpret the average growth rate in a given context.

Data sets often have gaps. With no further information, there is no way to accurately fill in the gaps. But *interpolation* using the average growth rate is available to provide estimates. Such estimates should always be viewed critically.

Extending data sets by *extrapolation* can be helpful in predicting function values, but doing so is a risky business, especially if we go far beyond the limits of the data set. This is illustrated in Example 2.10, a study of the age of first-time mothers.

Graphs: Picturing growth

Graphical presentations provide a visual display that can significantly aid understanding, usually by conveying information about growth rates. There are various types of graphical displays, and each has its advantages as well as its limitations. The three most common types of graphical representations are *bar graphs* (mentioned above), *scatterplots*, and *line graphs*. A commonly used variation on the third of these is the *smoothed line graph*. Examples of the second and third type are shown in **Figures 2.64 and 2.65**. A scatterplot can be used to display and study the relationship between any two sets of quantitative data. A line graph, if properly used, enhances the picture given by a scatterplot by filling gaps in the data with estimates.

In general, these four types of graphical displays allow us to locate not only intervals of increase and decrease but also rates of increase and decrease.

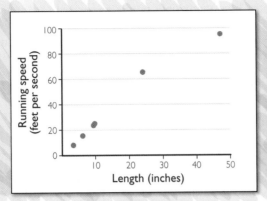

FIGURE 2.64 A scatterplot.

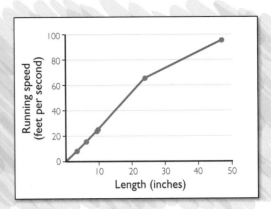

FIGURE 2.65 A line graph.

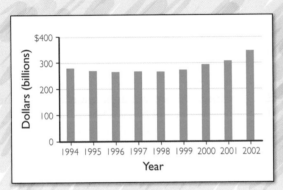

FIGURE 2.66 Defense spending.

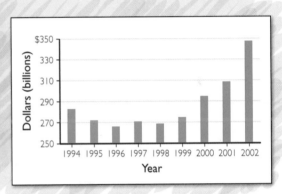

FIGURE 2.67 The same defense spending data.

Misleading graphs: Should I believe my eyes?

Graphical representations are sometimes misleading, whether or not this is intentional. A typical case is shown in the graphs of defense spending in **Figures 2.66 and 2.67**. The graphs represent identical data. The difference in appearance is due to the scale on the vertical axes.

When the data in a graph depend on the value of currency, it is important to know whether the currency values have been adjusted for inflation. For example, the graph in **Figure 2.68** shows the average price in dollars per barrel of crude oil in the given year. **Figure 2.69** shows that price in constant 2000 dollars. Using constant dollars gives a more accurate way of comparing the true cost of crude oil at different times.

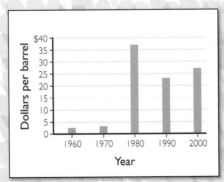

FIGURE 2.68 Price per barrel of oil.

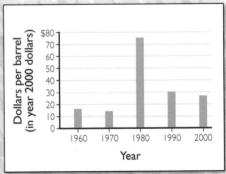

FIGURE 2.69 Crude price adjusted for inflation.

Any graph can be misleading if insufficient data are plotted. It may be particularly difficult to determine how many data points were used to produce a smoothed line graph.

Often data are presented using images designed to grab our attention. One common graphical representation is the *pie chart*, which is used to display how constituent parts make up a whole. In general, visual representations can be misleading if the relative sizes of the items pictured do not correspond to the relative sizes of the data.

KEY TERMS

independent variable, p. 73
dependent variable, p. 73
function, p. 73
percentage change, p. 76

relative change, p. 76
average growth rate, p. 80
interpolation, p. 82
extrapolation, p. 85

scatterplot, p. 92
line graph, p. 93
smoothed line graph, p. 98

CHAPTER QUIZ

1. The following table shows the population (in millions) of greater New York City in the given year:

Date	1900	1920	1940	1960	1980	2000
Population (millions)	3.44	5.62	7.46	7.78	7.07	8.01

Make a table and a bar graph showing the percentage change over each 20-year period. Round your answers in percentage form to the nearest whole number.

Answer

Date range	1900–20	1920–40	1940–60	1960–80	1980–2000
Percentage change	63%	33%	4%	−9%	13%

If you had difficulty with this problem, see Example 2.3.

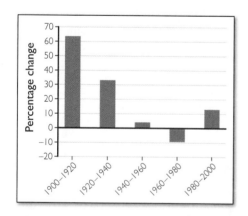

2. The cumulative number of disease cases from an epidemic are shown in the following table:

Day	1	5	8	12	15
Number of cases	22	35	81	91	96

a. Make a new table showing the average number of new cases per day over each period. Round your answers to two decimal places.

b. Use your answer from part a to estimate the cumulative number of cases by Day 11.

Answer

a.

Period	1–5	5–8	8–12	12–15
Average new cases per day	3.25	15.33	2.50	1.67

b. Interpolation gives 88.5, so the cumulative number of cases is about 89.

If you had difficulty with this problem, see Example 2.8.

3. The following table shows the barometric pressure in (millimeters of mercury) at various altitudes above sea level:

Altitude (thousand feet)	0	20	40	60	80	100
Pressure (mm mercury)	760	350	141	54	21	8

Make a scatterplot and a line graph of the data. Use the altitude for the independent variable and the pressure for the dependent variable.

Answer See **Figures 2.70** and **2.71**.

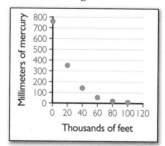

FIGURE 2.70 Pressure versus altitude: scatterplot.

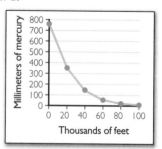

FIGURE 2.71 Pressure versus altitude: line graph.

If you had difficulty with this problem, see Example 2.11.

4. **Figure 2.72** shows a population of elk. Describe the growth rate in the elk population over the 10-year period. Approximately when was the population declining fastest?

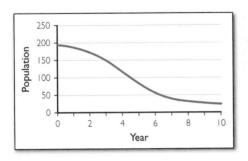

FIGURE 2.72 A population of elk.

Answer The population is declining throughout the 10-year period, so the growth rate is negative. The rate of decline increases until about Year 4, when the rate starts to level off. The time of fastest decline is around Year 4.

If you had difficulty with this problem, see Example 2.14.

5. The following table shows total sales in the given month:

Month	Jan.	Feb.	March	April	May
Sales	$5322	$5483	$6198	$6263	$6516

Make a bar graph that emphasizes the increase in sales over the five-month period, and make a second bar graph that de-emphasizes the increase.

Answer To emphasize the increase, we use a narrow range on the vertical axis. One possibility is shown in **Figure 2.73**. To deemphasize the increase, we use a wide range on the vertical axis. One possibility is shown in **Figure 2.74**.

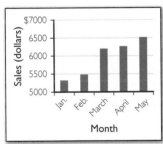

FIGURE 2.73 Emphasizing growth in sales.

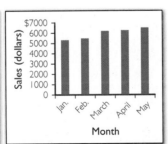

FIGURE 2.74 Deemphasizing growth in sales.

If you had difficulty with this problem, see Example 2.16.

6. The following table shows the cost of a certain item in the given year:

Year	1960	1970	1980	1990	2000
Cost	$23.00	$35.00	$65.00	$80.00	$116.00

Time span	1960–2000	1970–2000	1980–2000	1990–2000
Inflation	484%	337%	102%	30%

a. Complete the following table showing the cost in constant 2000 dollars:

Year	1960	1970	1980	1990	2000
Cost	$23.00	$35.00	$65.00	$80.00	$116.00
2000 dollars					

b. Make a bar graph showing the cost in constant 2000 dollars.

c. When was the item most expensive, taking inflation into account?

Answer

a.

Year	1960	1970	1980	1990	2000
Cost	$23.00	$35.00	$65.00	$80.00	$116.00
2000 dollars	$134.32	$152.95	$131.30	$104.00	$116.00

b.

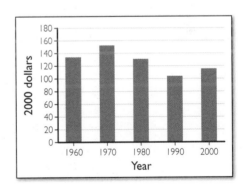

c. The maximum cost, adjusted for inflation, occurred in 1970.

If you had difficulty with this problem, see Example 2.17.

LINEAR AND EXPONENTIAL CHANGE: COMPARING GROWTH RATES

The following article is from the *Huffington Post*.

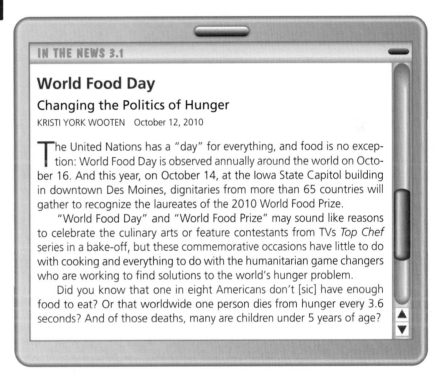

IN THE NEWS 3.1

World Food Day
Changing the Politics of Hunger
KRISTI YORK WOOTEN   October 12, 2010

The United Nations has a "day" for everything, and food is no exception: World Food Day is observed annually around the world on October 16. And this year, on October 14, at the Iowa State Capitol building in downtown Des Moines, dignitaries from more than 65 countries will gather to recognize the laureates of the 2010 World Food Prize.

"World Food Day" and "World Food Prize" may sound like reasons to celebrate the culinary arts or feature contestants from TVs *Top Chef* series in a bake-off, but these commemorative occasions have little to do with cooking and everything to do with the humanitarian game changers who are working to find solutions to the world's hunger problem.

Did you know that one in eight Americans don't [sic] have enough food to eat? Or that worldwide one person dies from hunger every 3.6 seconds? And of those deaths, many are children under 5 years of age?

Thomas Robert Malthus, 1766–1834.

Concern over world hunger is nothing new. The English demographer and political economist Thomas Robert Malthus (1766–1834) is best known for his influential views on population growth. In 1798 he wrote the following in *An Essay on the Principle of Population*.

I say that the power of population is indefinitely greater than the power in the earth to produce subsistence for man. Population, when unchecked, increases in a geometrical ratio whereas the food-supply grows at an arithmetic rate.

The power of population is so superior to the power of the earth to produce subsistence for man, that premature death must in some shape or other visit the human race. The vices of mankind are active and able ministers of depopulation. They are the precursors in the great army of destruction, and often finish the dreadful work themselves. But should they fail in this war of extermination, sickly seasons, epidemics, pestilence, and plague advance in terrific array, and sweep off their thousands and tens of thousands.

Note the distinction Malthus makes between a "geometrical ratio" and an "arithmetic rate." In contemporary terms we would say that population growth is *exponential* and food production growth is *linear*. Inherent properties of these two types of growth led Malthus to his dire predictions. His analysis is regarded today as an oversimplification of complex issues, but these specific types of growth remain crucial tools in analyzing demand versus availability of Earth's resources.

Change can follow various patterns, but linear and exponential patterns are of particular importance because they occur so often and in such significant applications. We will examine what they mean in this chapter. We will also consider the logarithm and its applications.

3.1 Lines and linear growth: What does a constant rate mean?

TAKE AWAY FROM THIS SECTION

Understand linear functions and consequences of a constant growth rate.

The following article is from the Baseball Analysts Web site.

IN THE NEWS 3.2

Was the 1990s Home Run Production Out of Line?

DAVID VINCENT   May 30, 2007

In the last five years, baseball fans have read and heard a lot of commentary from politicians and the media about what a travesty the home run totals have been since the mid-1990s....

[**Figure 3.1**] shows a graph of the home run production rate for all major league players each year since 1919. One can easily see a gradual increase from 1919 to the present. The numbers in the charts do not represent the total homers hit in the major leagues for any one season but rather the home run production rate (homers per 500 plate appearances)....

[The] trend line [in Figure 3.1] shows the steady increase in home run production from 1919 through 2006....

In 1994, the production rate reached 13.8 homers per 500 plate appearances, only the second time in history that the rate climbed above 13.0. From 1994 through the present, the production rate has been above the trend line with the exception of 2005. The highest point in the chart is 2000 when the production rate reached 15.0....

As a side note about the last 13 years, ... the home run production rate from 1994 through 2006... has held fairly steady through the period and, contrary to pronouncements by the commissioner, the production rate has not dropped in the years since Major League Baseball instituted its drug testing policy.... [T]he rate has held steady since 2001, slowly undulating around the 14.0 per 500 plate appearance [level].

This analysis makes use of a straight line to help us understand home runs in baseball. Straight lines and the relationships they represent frequently occur in the media, and every critical reader needs a basic understanding of them.

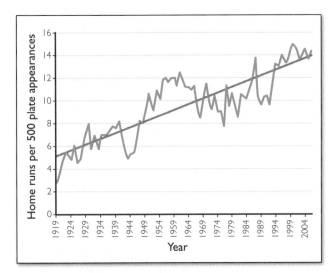

FIGURE 3.1 Home runs per 500 plate appearances.

Hank Aaron broke Babe Ruth's career home run record in 1974.

Recognizing linear functions

We have noted that in the media we often see graphics showing straight lines, such as the trend line in Figure 3.1. A straight line represents a special type of function, called a *linear* function.

A simple example will help us understand linear functions. Suppose we start with $100 in a cookie jar and each month we add $25. We can think of the number of months as the independent variable and the amount of money in the cookie jar as the dependent variable. In this view, the balance in the cookie jar is a function of the number of months since we began saving. In this case, the function grows by the same amount each month ($25). Functions with a constant growth rate are *linear functions*.

Key Concept

A function is called **linear** if it has a constant growth rate.

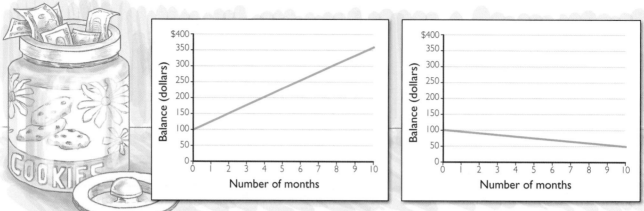

FIGURE 3.2 Adding money to a cookie jar: Positive growth rate corresponds to an increasing linear function.

FIGURE 3.3 Taking money from a cookie jar: Negative growth rate corresponds to a decreasing linear function.

We note that the growth rate may be a negative number, which indicates a decreasing function. For example, suppose we start with money in a cookie jar and withdraw $5 each month. Then the balance is changing by the same amount each month, so it is a linear function of the number of months since we started withdrawing money. In this case, the growth rate is −$5 (negative 5 dollars) per month.

Not all functions are linear. A person's height is not a linear function of her age. If it were, she would grow by the same amount each year of her life.

An important feature of linear functions is that their graphs are straight lines. This is illustrated in **Figures 3.2 and 3.3**, where we have graphed the functions in our preceding examples. For comparison, a graph of typical height versus age for females is shown in **Figure 3.4**. It is not a straight line, and this is further evidence that height as a function of age is not a linear function.

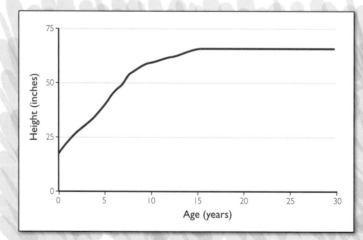

FIGURE 3.4 Height is not a linear function of age.

EXAMPLE 3.1 Determining what is linear and what is not: Total cost and salary

In each part below, a function is described. Find the growth rate of the function and give its practical meaning. Make a graph of the function. Is the function linear?

**Trust me, it's just an expression: I'm a crow and
I can tell you I don't always fly in a straight line...**

a. For my daughter's wedding reception, I pay $500 rent for the building plus $15 for each guest. This describes the total cost of the reception as a function of the number of guests.

b. My salary is initially $30,000, and I get a 10% salary raise each year for several years. This describes my salary as a function of time.

SOLUTION

a. The growth rate in this case is the extra cost incurred for each additional guest, which is $15. The additional cost is the same for each additional guest, so the growth rate is constant. Because the growth rate is constant, the total cost of the reception is a linear function of the number of guests. The graph of cost versus number of guests is shown in **Figure 3.5**. It is a straight line, as expected.

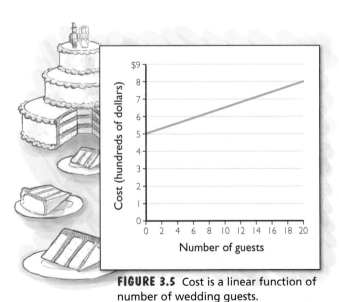

FIGURE 3.5 Cost is a linear function of number of wedding guests.

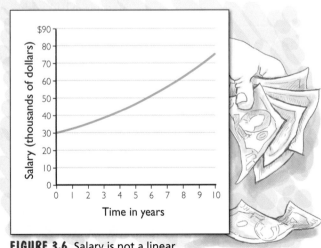

FIGURE 3.6 Salary is not a linear function of time.

b. The growth rate in this case is the increase per year in my salary. The first year my salary increased by 10% of $30,000, or $3000, so my new annual salary is $33,000. The second year my salary increased by 10% of $33,000, or $3300. Therefore, the growth rate is not the same each year, so my salary is not a linear function of time in years. A graph showing salary over several years with a 10% increase each year is illustrated in **Figure 3.6** (page 133). The graph is not a straight line, which is further verification that salary is not a linear function of time in years.

TRY IT YOURSELF 3.1

One of the following is a linear function, and one is not. In each case, determine the practical meaning of the growth rate, then determine whether or not the given function is linear.

a. One inch is the same as 2.54 centimeters. Consider distance in centimeters to be a function of distance in inches.

b. Consider the area of a square to be a function of the length of a side. **Figure 3.7** may be helpful here.

The answer is provided at the end of this section.

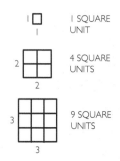

1   1 SQUARE UNIT

2   4 SQUARE UNITS

3   9 SQUARE UNITS

FIGURE 3.7 Areas of squares of different sizes.

Formulas for linear functions

Let's look again at the cookie jar example in which we start with $100 and add $25 each month. We use x to denote the number of months since we started saving and y to denote the balance in dollars after x months.

After $x = 1$ month, there is $y = 25 \times 1 + 100 = \$125$ in the cookie jar.

After $x = 2$ months, there is $y = 25 \times 2 + 100 = \$150$ in the cookie jar.

After $x = 3$ months, there is $y = 25 \times 3 + 100 = \$175$ in the cookie jar.

Following this pattern, we see that after x months, there will be

$$y = 25x + 100$$

dollars in the cookie jar.

In the preceding formula, the number $25 per month is the growth rate. The number $100 is the *initial value* because it is the amount with which we start. In other words, $100 is the function value (the value of y) when the independent variable x is 0. This way of combining the initial value with the growth rate is typical of how we obtain formulas for linear functions: The general formula for a linear function is

$$y = \text{Growth rate} \times x + \text{Initial value}.$$

It is customary to use m to denote the growth rate and b to denote the initial value. With these choices, the formula for a linear function is

$$y = mx + b.$$

Figure 3.8 shows how the growth rate m determines the steepness of a line. Moving one unit to the right corresponds to a rise of m units of the graph. That is why the growth rate m is often called the *slope*. For linear functions, we will use the terms *slope* and *growth rate* interchangeably.

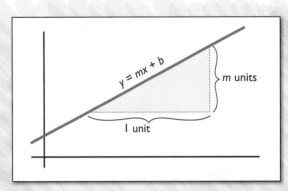

FIGURE 3.8 The growth rate m is the slope of the line.

SUMMARY 3.1   Linear Functions

- A linear function is a function with a constant growth rate.
- The graph of a linear function is a straight line.
- The formula for a linear function takes the form

$$y = \text{Growth rate} \times x + \text{Initial value}.$$

If m is the growth rate or slope and b is the initial value, the formula becomes

$$y = mx + b.$$

EXAMPLE 3.2 Interpreting formulas: Athletic records

Let L denote the length in meters of the winning long jump in the early years of the modern Olympic Games. We think of L as a function of the number n of Olympic Games since 1900. (Thus, because the Olympic Games were held every four years, $n = 0$ corresponds to the 1900 Olympics, $n = 1$ to the 1904 Olympics, and so on.) An approximate linear formula is $L = 0.14n + 7.20$. Identify the initial value and growth rate, and explain in practical terms their meaning.

SOLUTION

The initial value of the linear function $y = mx + b$ is b, and the growth rate or slope is m. Therefore, for $L = 0.14n + 7.20$, the initial value is 7.20 meters. It is the (approximate) length of the winning long jump in the 1900 Olympic Games. The growth rate is 0.14 meter per Olympic Game. This means that the length of the winning long jump increased by approximately 0.14 meter from one game to the next.

The Olympic symbol.

TRY IT YOURSELF 3.2

Let H denote the height in meters of the winning pole vault in the early years of the modern Olympic Games. We think of H as a function of the number n of Olympic Games since 1900. An approximate linear formula is $H = 0.20n + 3.3$. Identify the initial value and growth rate, and explain in practical terms the meaning of each.

The answer is provided at the end of this section.

Note that in Example 3.2, we used the variables n and L rather than x and y. There is nothing special about the names x and y for the variables, and it is common to use variable names that help us remember their practical meanings.

If we know the initial value and growth rate for a linear function, we can immediately write its formula. The method is shown in the following example.

The Saturn V carried the first men to the moon in 1969.

EXAMPLE 3.3 Finding linear formulas: Rocket

A rocket starting from an orbit 30,000 kilometers above the surface of Earth blasts off and flies at a constant speed of 1000 kilometers per hour away from Earth. Explain why the function giving the rocket's distance from Earth in terms of time is linear. Identify the initial value and growth rate, and find a linear formula for the distance.

SOLUTION

First we choose letters to represent the function and variable. Let d denote the distance in kilometers from Earth after t hours. The growth rate is the velocity, 1000 kilometers per hour. (The velocity is positive because the distance is increasing.) Because the growth rate is constant, d is a linear function of t. The initial value is 30,000 kilometers, which is the height above Earth at blastoff (that is, at time $t = 0$). We find the formula for d using

$$d = \text{Growth rate} \times t + \text{Initial value}$$
$$d = 1000t + 30{,}000.$$

TRY IT YOURSELF 3.3

Suppose that a stellar object is first detected at 1,000,000 kilometers from Earth and that it is traveling toward Earth at a speed of 2000 kilometers per hour. Explain why the function giving the object's distance from Earth in terms of time is linear. Identify the initial value and growth rate, and find a linear formula for the distance.

The answer is provided at the end of this section.

Finding and interpreting the slope

We have seen that it is easy to find the formula for a linear function if we know the initial value and the growth rate, or slope. But sometimes in linear relationships the growth rate m is not given. In such a situation, we calculate the average growth rate in the same way as we did in Chapter 2. We use the formula

$$\text{Slope} = \text{Growth rate} = \frac{\text{Change in function}}{\text{Change in independent variable}}.$$

When we write the equation of a linear function as $y = mx + b$, we can write the formula for the slope as

$$m = \text{Slope} = \text{Growth rate} = \frac{\text{Change in } y}{\text{Change in } x}.$$

A geometric interpretation of this formula is shown in **Figure 3.9**.

Recall from Section 2.1 that the units of the average growth rate are the units of the function divided by the units of the independent variable. In practical settings, the slope has a familiar meaning, and proper use of units can help us to determine that meaning.

Suppose, for example, that it is snowing at a steady rate, which means that the depth of snow on the ground is a linear function of the time since it started snowing. At some point during the snowfall, we find that the snow is 8 inches deep. Four hours

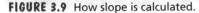

$$m = slope = \frac{change\ in\ y}{change\ in\ x}$$

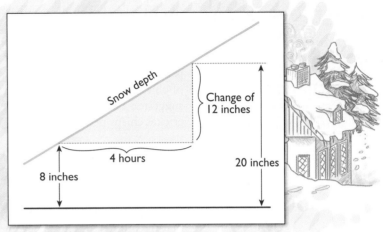

FIGURE 3.9 How slope is calculated.

FIGURE 3.10 Growth rate of snow accumulation.

later we find that the snow is 20 inches deep. Then the depth has increased by 12 inches over a 4-hour period. (See **Figure 3.10.**) That is a slope of

$$m = \frac{\text{Change in depth}}{\text{Change in time}} = \frac{12 \text{ inches}}{4 \text{ hours}} = 3 \text{inches per hour.}$$

The slope is measured in inches per hour, which highlights the fact that the slope represents the rate of snowfall.

EXAMPLE 3.4 Calculating the slope: Gas tank

Suppose your car's 20-gallon gas tank is full when you begin a road trip. Assume that you are using gas at a constant rate, so the amount of gas in your tank (in gallons) is a linear function of the time in hours you have been driving. After traveling for two hours, your fuel gauge reads three-quarters full. Find the slope of the linear function and explain in practical terms what it means.

SOLUTION

We need to know the change in the amount of gas in the tank. When the tank is three-quarters full, there are 15 gallons of gas left in the tank. The amount of gas in the tank has decreased by 5 gallons. This means that the change in gas is -5 gallons. This occurred over two hours of driving, so the slope is

$$
\begin{aligned}
m &= \frac{\text{Change in gas}}{\text{Change in time}} \\
&= \frac{-5 \text{ gallons}}{2 \text{ hours}} = -2.5 \text{ gallons per hour.}
\end{aligned}
$$

In practical terms, this means we are using 2.5 gallons of gas each hour.

An ice core from Antarctica.

TRY IT YOURSELF 3.4

Scientists believe that thousands of years ago the depth of ice in a glacier was increasing at a constant rate, so the depth in feet was a linear function of time in years. Using a core sample, they measured a depth of 25 feet at one time and a depth of 28 feet five years later. What is the slope of the linear function? Explain in practical terms the meaning of the slope.

The answer is provided at the end of this section.

The snowy tree cricket.

If an airplane is flying away from us along a straight line at a constant speed of 300 miles per hour, its distance from us increases by 300 miles for each hour that passes. This is a typical description of a linear function: If the function has the formula $y = mx + b$, a 1-unit increase in the independent variable x corresponds to a change in the function value y by an amount equal to the slope m. This interpretation is key to the way we use the slope in practical settings. For example, the frequency of chirping by crickets is related to the temperature. The relationship that has been suggested between the temperature T in degrees Fahrenheit and the number n of cricket chirps per minute is

$$T = 0.25n + 39.$$

Note that the slope of this linear function is 0.25. The slope is measured in degrees Fahrenheit per chirp per minute.

Suppose initially the temperature is 50 degrees Fahrenheit. An hour later we find that the number of chirps per minute has increased by 20. What is the new temperature? An increase of one chirp per minute corresponds to an increase of 0.25 degree Fahrenheit because 0.25 is the slope. Therefore, an increase by 20 chirps per minute corresponds to an increase in temperature of $20 \times 0.25 = 5$ degrees. The new temperature is $50 + 5 = 55$ degrees Fahrenheit.

SUMMARY 3.2   **Interpreting and Using the Slope**

- We calculate the slope of a linear function using

$$\text{Slope} = \text{Growth rate} = \frac{\text{Change in function}}{\text{Change in independent variable}}.$$

- When we write the equation of a linear function as $y = mx + b$, the formula for the slope becomes

$$m = \text{Slope} = \text{Growth rate} = \frac{\text{Change in } y}{\text{Change in } x}.$$

- For a linear function, a 1-unit increase in the independent variable corresponds to a change in the function value by an amount equal to the slope. For the linear function $y = mx + b$, each 1-unit increase in x corresponds to a change of m units in y.

EXAMPLE 3.5 Using the slope: Temperature and speed of sound

The speed of sound in air depends on the temperature. If T denotes the temperature in degrees Fahrenheit and S is the speed of sound in feet per second, the relation is given by the linear formula

$$S = 1.1T + 1052.3.$$

What is the speed of sound when the temperature is 0 degrees Fahrenheit? If the temperature increases by 1 degree Fahrenheit, what is the corresponding increase in the speed of sound?

SOLUTION

Note that the initial value of the linear function is 1052.3. Therefore, when the temperature is 0 degrees, the speed of sound is 1052.3 feet per second.

The slope of the linear function is 1.1 feet per second per degree Fahrenheit. So a 1-degree increase corresponds to a 1.1-foot-per-second increase in the speed of sound.

TRY IT YOURSELF 3.5

Consider again the linear formula in the example. Suppose a bright sun comes out one day and raises the temperature by 3 degrees. What is the corresponding increase in the speed of sound?

The answer is provided at the end of this section.

The next example illustrates the importance of correctly interpreting the slope of a linear function.

EXAMPLE 3.6 Interpreting the slope: Temperature conversion

Measuring temperature using the Fahrenheit scale is common in the United States, but use of the Celsius (sometimes referred to as *centigrade*) scale is more common in most other countries. The temperature in degrees Fahrenheit is a linear function of the temperature in degrees Celsius. On the Celsius scale, 0 degrees is the freezing temperature of water. This occurs at 32 degrees on the Fahrenheit scale. Also, 100 degrees Celsius is the boiling point (at sea level) of water. This occurs at 212 degrees Fahrenheit.

a. What is the slope of the linear function giving the temperature in degrees Fahrenheit in terms of the temperature in degrees Celsius? Use your answer to determine what increase in degrees Fahrenheit corresponds to a 1-degree increase on the Celsius scale.

b. Choose variable and function names, and find a linear formula that converts degrees Celsius to degrees Fahrenheit. Make a graph of the linear function.

c. A news story released by Reuters on March 19, 2002, said that the Antarctic peninsula had warmed by 36 degrees Fahrenheit over the past half-century. Such temperature increases would result in catastrophic climate changes worldwide, and it is surprising that an error of this magnitude could have slipped by the editorial staff of Reuters. We can't say for certain how the error occurred, but it is likely that the British writer saw a report that the temperature had increased by 2.2 degrees Celsius. Verify that a temperature of 2.2 degrees Celsius is about 36 degrees Fahrenheit.

d. The report described in part c did not say that the temperature was 2.2 degrees Celsius but that the temperature *increased* by 2.2 degrees Celsius. What increase in Fahrenheit temperature should the writer have reported?

SOLUTION

a. To find the slope, we need the change in degrees Fahrenheit for a given change in degrees Celsius. An increase on the Celsius scale from 0 to 100 degrees corresponds to an increase on the Fahrenheit scale from 32 to 212, a 180-degree increase. Therefore,

$$\text{Slope} = \frac{\text{Change in degrees Fahrenheit}}{\text{Change in degrees Celsius}}$$

$$= \frac{180}{100} = 1.8 \text{ degrees Fahrenheit per degree Celsius.}$$

Therefore, a 1-degree increase on the Celsius scale corresponds to a 1.8-degree increase on the Fahrenheit scale.

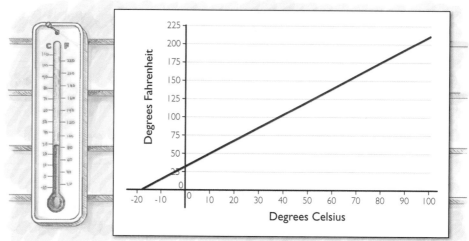

FIGURE 3.11 Fahrenheit temperature is a linear function of Celsius temperature.

b. We use F for the temperature in degrees Fahrenheit and C for the temperature in degrees Celsius. The linear relationship we seek is expressed by the formula $F = \text{Slope} \times C + \text{Initial value}$. We found in part a that the slope is 1.8. Now we need the initial value, which is the Fahrenheit temperature when the Celsius temperature is 0 degrees. That is 32 degrees Fahrenheit. Therefore, the formula is $F = 1.8C + 32$. We have graphed this function in **Figure 3.11**.

c. We put 2.2 degrees Celsius into the formula for converting Celsius to Fahrenheit:

$$F = 1.8C + 32$$
$$= 1.8 \times 2.2 + 32 = 35.96 \text{ degrees Fahrenheit.}$$

Rounding gives 36 degrees Fahrenheit.

d. The slope of the linear function that converts from Celsius to Fahrenheit is 1.8 degrees Fahrenheit per degree Celsius. As we noted in the solution to part a, this tells us that each 1-degree increase on the Celsius scale corresponds to a 1.8-degree increase on the Fahrenheit scale. So an increase of 2.2 degrees Celsius corresponds to an increase of $2.2 \times 1.8 = 3.96$ degrees Fahrenheit. The writer should have reported a warming of about 4 degrees Fahrenheit.

Linear data and trend lines

The following table is taken from a federal income tax table provided by the Internal Revenue Service.[1] It applies to a married couple filing jointly.

Taxable income	$41,000	$41,100	$41,200	$41,300	$41,400
Tax owed	$5424	$5439	$5454	$5469	$5484

In **Figure 3.12**, we have plotted these data points.

The data points appear to fall on a straight line, which indicates that we can use a linear function to describe taxable income, at least over the range given in the table. To verify this, we calculate the average growth rate over each of the income ranges.

---

[1] The actual table shows taxable income in increments of $50 rather than $100.

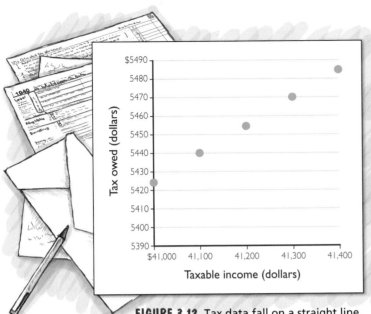
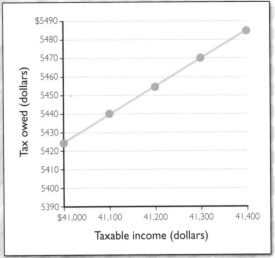

FIGURE 3.12 Tax data fall on a straight line.   **FIGURE 3.13** Line added to plot of tax data.

Recall from Chapter 2 that we find the average growth rate using

$$\text{Average growth rate} = \frac{\text{Change in function}}{\text{Change in independent variable}} = \frac{\text{Change in tax owed}}{\text{Change in taxable income}}.$$

These calculations are shown in the following table:

Interval	41,000 to 41,100	41,100 to 41,200	41,200 to 41,300	41,300 to 41,400
Change in income	$41,100 - 41,000$ $= 100$	$41,200 - 41,100$ $= 100$	$41,300 - 41,200$ $= 100$	$41,400 - 41,300$ $= 100$
Change in tax owed	$5439 - 5424$ $= 15$	$5454 - 5439$ $= 15$	$5469 - 5454$ $= 15$	$5484 - 5469$ $= 15$
Average growth rate	$\frac{15}{100} = 0.15$	$\frac{15}{100} = 0.15$	$\frac{15}{100} = 0.15$	$\frac{15}{100} = 0.15$

The fact that the average growth rate over each interval is the constant 0.15 tells us that the tax owed is a linear function of income with slope $0.15 per dollar. (If we had obtained different values for the average growth rate over different intervals, we would have concluded that the data are not related by a linear formula.)

This slope is known as the *marginal tax rate*. It tells us the additional tax owed on each additional dollar of income. For example, if you have $100 more in taxable income than someone else, your tax burden will be greater by $0.15 \times 100 = 15$ dollars. Let T be the tax owed in dollars, and let I be the taxable income (also in dollars) in excess of $41,000. The tax owed when $I = 0$ is $5424, and this is the initial value. Therefore,

$$T = \text{Slope} \times I + \text{Initial value}$$

$$T = 0.15I + 5424.$$

In **Figure 3.13**, we have added the graph of the line to the plot of the tax data. The graph passes through each of the data points, which verifies the validity of our formula.

"I'M WORTH MORE THAN $3,000 AS A FEDERAL INCOME TAX EXEMPTION, SO HOW ABOUT A RAISE IN MY ALLOWANCE?"

The linear function fits the tax data exactly. Many data sets are not perfectly linear like this, but they may be fairly well approximated by a linear function. And we may be able to get useful information from such an approximation. This is especially true when we have good reason to suspect that the relationship is approximately linear.

As a case in point, let's return to the example of running speed versus length discussed in Section 2.2. We recall the following table, which shows the running speed of various animals versus their length:[2]

Animal	Length (inches)	Speed (feet per second)
Deer mouse	3.5	8.2
Chipmunk	6.3	15.7
Desert crested lizard	9.4	24.0
Grey squirrel	9.8	24.9
Red fox	24.0	65.6
Cheetah	47.0	95.1

Figure 3.14 shows the scatterplot. We note that the points do not fall on a straight line, so the data in the table are not exactly linear. On the other hand, the points seem

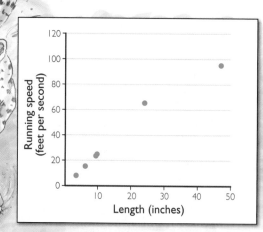

FIGURE 3.14 Scatterplot of running speed versus length.

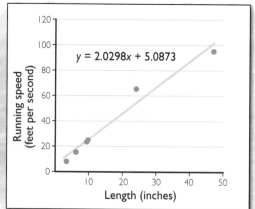

$$y = 2.0298x + 5.0873$$

FIGURE 3.15 Trend line added.

---

[2]This table is adapted from J. T. Bonner, *Size and Cycle* (Princeton, NJ: Princeton University Press, 1965).

to fall *nearly* on a straight line, so it may make sense to *approximate* the data using a linear function. Statisticians, scientists, and engineers frequently use a line called a *regression line* or *trend line* to make such approximations.

Key Concept

Given a set of data points, the **regression line** or **trend line** is a line that comes as close as possible (in a certain sense) to fitting those data.

The procedure for calculating a trend line for given data is called *linear regression*. An explanation of this procedure is outside the scope of this book, but linear regression is programmed into most calculators and commercial spreadsheets. All one has to do is enter the data.

In **Figure 3.15**, we have added the trend line produced by the spreadsheet program Excel©. We see that the points lie close to the line, so it appears to be a reasonable approximation to the data. Note that the spreadsheet gives both the graph and the linear formula.

The equation of the trend line is $y = 2.03x + 5.09$. (We have rounded the initial value and the slope to two decimal places.) This means that running speed S in feet per second can be closely estimated by

$$S = 2.03L + 5.09,$$

where L is the length measured in inches. We can use this formula to make estimates about animals not shown in the table. For example, we would expect a 20-inch-long animal to run

$$S = 2.03 \times 20 + 5.09 = 45.69 \text{ feet per second},$$

or about 45.7 feet per second. The most important bit of information we obtain from the trend line is its slope. In this case, that is 2.03 feet per second per inch. This value for the slope means that, in general, an animal that is 1 inch longer than another would be expected to run about 2.03 feet per second faster.

We can also gain information by studying the position of the data points relative to the trend line. For example, in Figure 3.15, the point for the red fox is above the trend line, so the red fox is faster than one might expect for an animal of its size.

EXAMPLE 3.7 Meaning of the equation for the trend line: Home runs

The article at the beginning of this section shows the trend line for the number of home runs (per 500 plate appearances) as a function of time. The formula for this trend line is $H = 0.1t + 5.1$, where H is the number of home runs per 500 plate appearances and t is the time in years since 1919. Explain in practical terms the meaning of the slope.

SOLUTION

The slope is 0.1, so each year the number of home runs per 500 plate appearances increases, on average, by 0.1.

TRY IT YOURSELF 3.7

Suppose the trend line for walks per 500 plate appearances is $W = 0.2t + 10.3$, where W is the number of walks per 500 plate appearances and t is the time in years since 1919. Explain in practical terms the meaning of the slope.

The answer is provided at the end of this section.

Try It Yourself answers

Try It Yourself 3.1: Determining what is linear and what is not: Distance and area

a. The growth rate for distance is the increase in distance measured in centimeters for each 1-inch increase in distance. That increase is 2.54 centimeters. The distance in centimeters is a linear function of the distance in inches.

b. The growth rate for the area is the increase in area for each unit increase in the length of a side. The area of a square is not a linear function of the length of a side.

Try It Yourself 3.2: Interpreting formulas: Athletic records   The initial value of 3.3 meters is the (approximate) height of the winning pole vault in the 1900 Olympic Games. The growth rate of 0.20 meter per Olympic game means that the height of the winning pole vault increased by approximately 0.20 meter from one game to the next.

Try It Yourself 3.3: Finding linear formulas: Distance from Earth   Let d denote the distance from Earth in kilometers after t hours. Then d is a linear function of t because the growth rate (the velocity) is constant. The initial value is 1,000,000 kilometers, and the growth rate is -2000 kilometers per hour (negative since the object is traveling toward Earth, so the distance is decreasing). The formula is $d = -2000t + 1{,}000{,}000$.

Try It Yourself 3.4: Calculating the slope: Ice depth   0.6 foot per year. The depth increased by 0.6 foot each year.

Try It Yourself 3.5: Using the slope: Temperature and speed of sound   3.3 feet per second

Try It Yourself 3.7: Meaning of the equation for the trend line: Walks   Each year the number of walks per 500 plate appearances increases, on average, by 0.2.

Exercise Set 3.1

Note: In some exercises, you are asked to find a formula for a linear function. If no letters are assigned to quantities in an exercise, you are expected to choose appropriate letters and also give appropriate units.

Linear or not?   In Exercises 1 through 12, you are asked to determine whether the relationship described represents a function that is linear. In each case, explain your reasoning. If the relationship can be represented by a linear function, find an appropriate linear formula, and explain in practical terms the meaning of the growth rate.

1. A staircase. Is the relationship between the height of a staircase and the number of steps linear?

2. A music club. Suppose that the cost of purchasing CDs from a music club is a flat membership fee of $30 plus $10 for each CD purchased. Is the amount of money you pay a linear function of the number of CDs you buy?

3. Another music club. Suppose that a music club charges $10 per CD but offers a 2% discount on orders larger than 15 CDs. Is the cost per order a linear function of the number of CDs you order?

4. Typing. Suppose that I wrote 150 words of my English paper yesterday and today I begin typing at a rate of 35 words per minute. Is the total number of words typed a linear function of the number of minutes since I began typing today?

5. A savings account. My savings account pays 10% per year, and the interest is compounded yearly. That is, each year accrued interest is added to my account balance. Is the amount of money in my savings account a linear function of the number of years since it was opened? *Suggestion:* Compare the interest earned the first year with the interest earned the second year.

6. Velocity of a falling rock. If we drop a rock near the surface of Earth, its velocity increases by 32 feet per second for each second the rock falls. Is the velocity of the rock a linear function of the time since it was dropped?

7. The distance a rock falls. If we drop a rock near the surface of the earth, physics tells us that the distance in feet that it falls is given by $16t^2$, where t is the number of seconds since the rock was dropped. Is the distance a rock falls a linear function of the time since it was dropped? *Suggestion:* Compare the distance fallen over the first second to that fallen over the next second.

8. Shipping costs. In order to ship items I bought on eBay, I pay a delivery company a flat fee of $10, plus $1.50 per pound. Is the shipping cost a linear function of the number of pounds shipped?

9. Gas mileage. My car gets 25 miles per gallon at 60 miles per hour. Is the number of miles I can drive at 60 miles per hour a linear function of the number of gallons of gas in the tank?

10. Bacteria. A certain type of bacterium reproduces by dividing each hour. If we start with just one bacterium, is the number of bacteria present a linear function of time?

11. Temperature. On a typical day, the temperature increases as the sun rises. It is hottest around midday, and then things cool down toward evening. Is the relationship between temperature and time of day linear?

12. Oklahoma Teachers' Retirement System. The method of calculating the compensation paid to retired teachers in the state of Oklahoma is as follows.[3] Yearly retirement income is 0.02 times the number of years of service times the average of the last three years of salary. Is retirement income for Oklahoma teachers a linear function of the number of years of service?

13. Income tax. One federal tax schedule states that if you are single and your taxable income is between $31,900 and $77,050, then you owe $4393 plus 25% of your taxable income over $31,900. Is the tax you owe a linear function of your taxable income if the taxable income is between $31,900 and $77,050? If so, find a formula giving tax owed T as a linear function of taxable income I over $31,900. Both T and I are measured in dollars.

14. Working on commission. A certain man works in sales and earns a base salary of $1000 per month, plus 5% of his total sales for the month. Is his monthly income a linear function of his monthly sales? If so, find a linear formula that gives monthly income in terms of sales.

15. Driving up a hill. Suppose that you are at the base of a hill and see a sign that reads "Elevation 3500 Feet." The road you are on goes straight up the hill to the top, which is 3 horizontal miles from the base. At the top, you see a sign that reads "Elevation 4100 Feet." What is the growth rate in your elevation with respect to horizontal distance as you drive up the road? Use E for elevation in feet and h for horizontal distance in miles, and find a formula that gives your elevation as a linear function of your horizontal distance from the base of the hill.

Price of Amazon's Kindle **Figure 3.16** shows how the price of Amazon's Kindle 2 e-book reader has decreased over time. (The data for 2011 are projected.) For example, the price was $349 when the Kindle was launched in February 2009. The price had dropped to $299 by July 2009. The graph is a straight line, and we will assume that the relationship is linear. This information is used in Exercises 16 through 19.

16. Use the information for February and July of 2009 to determine the decrease in price over each month.

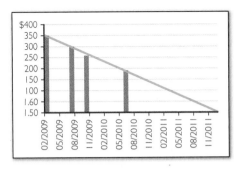

FIGURE 3.16 Price of the Kindle 2.

17. Use your answer to Exercise 16 to determine the slope of the linear function that gives the price in terms of the time in months since February 2009. Be careful about the sign of the slope.

18. Use your answer to Exercise 17 and the initial price to find a formula for the price, in dollars, as a linear function of the time in months since February 2009.

19. What price does the formula from Exercise 18 project for January of 2012 (35 months after the Kindle was launched)? *Note*: The data in the graph have been the basis for speculation that someday the Kindle would be free.

Speed of sound in the ocean Measuring the speed of sound in the ocean is an important part of marine research.[4] One application is the study of climate change. The speed of sound depends on the temperature, salinity, and depth below the surface. For a fixed temperature of 25 degrees Celsius and salinity of 35 parts per thousand, the speed of sound is a function of the depth. At the surface, the speed of sound is 1534 meters per second. For each increase in depth by 1 kilometer, the speed of sound increases by 17 meters per second. This information is used in Exercises 20 through 25.

20. Explain why the function expressing the speed of sound in terms of depth is linear.

21. Identify the slope and initial value of the linear function that gives the speed of sound in terms of depth. Explain in practical terms what each means.

22. What increase in the speed of sound is caused by a 2-kilometer increase in depth?

23. Use your answer to Exercise 22 to determine the speed of sound when the depth is 2 kilometers.

24. Use D for depth (in kilometers) and S for the speed of sound (in meters per second), and find a linear formula for S as a function of D.

25. Use your formula from Exercise 24 to calculate the speed of sound when the depth is 2 kilometers.

Speed on a curve On rural highways, the average speed S (in miles per hour) is related to the amount of curvature C (in degrees) of the road.[5] On a straight road ($C = 0$), the

---

[3] The actual calculation is somewhat more complicated.

[4] For further background, see www.dosits.org.

[5] A. Taragin, "Driver Performance on Horizontal Curves," *Proceedings of the Highway Research Board* **33** (Washington, DC: Highway Research Board, 1954), 446–466.

average speed is 46.26 miles per hour. This decreases by 0.746 mile per hour for each additional degree of curvature. This information is used in Exercises 26 through 29.

26. Explain why the function expressing S in terms of C is linear.

27. Find the slope of the linear function expressing S in terms of C. Explain in practical terms its meaning. *Suggestion*: Be careful about the sign of the slope.

28. Find a formula expressing S as a linear function of C.

29. What is the average speed if the degree of curvature is 15 degrees?

The IRS In 2003, the Internal Revenue Service collected about 1.952 trillion dollars. From 2003 through 2008, collections increased by (approximately) 178 billion (or 0.178 trillion) dollars per year. This information is used in Exercises 30 through 32.

30. Explain how the above information shows that the function expressing tax collections in terms of time from 2003 through 2008 is linear. Why do we have to limit the time interval to 2003 through 2008?

31. Find a formula for tax collections as a linear function of time from 2003 through 2008.

32. What would you estimate for the collections in 2006?

33. A real estate agency. Suppose that a real estate agency makes its money by taking a percentage commission of total sales. It also has fixed costs of $15,257 per month associated with rent, staff salaries, utilities, and supplies. In the month of August, total sales were $832,000 and net income (after paying fixed costs) was $9703. Since each dollar increase in sales results in the same increase (a certain percentage of that dollar) in net income, net income is a linear function of total sales. Find a linear formula that gives net income in terms of total sales. What percentage commission on sales is this company charging?

34. Meaning of marginal tax rate. Recall from the discussion on page 139 that the marginal tax rate is the additional tax you expect to pay for each additional dollar of taxable income. The marginal tax rate is usually expressed as a percentage. In the example, the marginal tax rate was 15%. Does that mean that your total tax is 15% of your taxable income? Specifically, if your taxable income is $41,300, is your total tax 15% of $41,300? Explain why or why not.

More on tax tables The following table shows for a single taxpayer the federal income tax owed for the given level of taxable income. Both are measured in dollars.

Taxable income	97,000	97,050	97,100	97,150	97,200	97,250	97,300
Tax owed	21,913	21,927	21,941	21,955	21,969	21,983	21,997

This information is used in Exercises 35 through 38.

35. Show that, over the range of taxable income shown in the table, the tax owed is a linear function of taxable income.

36. What tax do you owe if you have a taxable income of $97,000?

37. How much additional tax is due on each dollar of taxable income over $97,000?

38. Let A denote the amount (in dollars) of your taxable income in excess of $97,000 and T the tax (in dollars) you owe. Find a linear formula[6] that gives T in terms of A.

The flu The following table shows the total number of patients diagnosed with the flu in terms of the time in days since the outbreak started:

Time in days	0	5	10	15	20	25
Number of flu patients	35	41	47	53	59	65

This information is used in Exercises 39 through 42.

39. Show that the function giving the number of diagnosed flu cases in terms of time is linear.

40. Find the slope of the linear function from Exercise 39. Explain in practical terms the meaning of the slope.

41. Find a formula for the linear function in Exercise 39.

42. What would you expect to be the total number of diagnosed flu cases after 17 days?

43. High school graduates. The following table shows the number (in millions) graduating from high school in the United States in the given year:

Year	1994	1995	1996	1997	1998	1999	2000	2001
Graduating (millions)	2.52	2.60	2.66	2.77	2.81	2.90	2.76	2.55

The scatterplot of the data is shown in **Figure 3.17**. In **Figure 3.18**, the trend line has been added. Does the trend line appear to offer an appropriate way to analyze the data? Explain your reasoning.

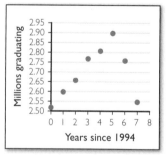

FIGURE 3.17 Scatterplot of high school graduates.

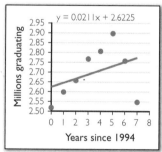

FIGURE 3.18 Trend line added.

Consumer price index The *Consumer Price Index* (CPI) is calculated by the U.S. Department of Labor. It is a measure of the relative price of a typical market basket of goods in the

---

[6]The formula does not apply for taxable incomes above $100,000.

given year. The percentage increase in the CPI from one year to the next gives one measure of the annual rate of inflation. The following table shows the consumer price index in December of the given year. Here, C represents the consumer price index, and t is the time in years since December 2005.

t	0	1	2	3	4
C	196.8	201.8	210.0	210.2	215.9

In **Figure 3.19**, we have plotted the data, and in **Figure 3.20**, we have added the trend line, which is given by $C = 4.66t + 197.62$. This information is used in Exercises 44 through 46.

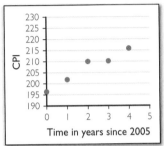

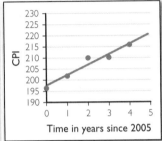

FIGURE 3.19 Plot of consumer price index.

FIGURE 3.20 Plot of consumer price index with trend line added.

44. Determine the slope of the trend line and explain in practical terms the meaning of the slope.

45. Which year shows a larger CPI than would be expected from the trend line? Interpret your answer.

46. What CPI does the trend line predict for December 2010? Round your answer to one decimal place. (The actual CPI in December 2010 was 219.2.)

Exercises 47 through 51 are suitable for group work.

A closer look at tax tables   The table below gives selected entries from the 2009 federal income tax table. It applies to a married couple filing jointly. Taxable income and tax owed are both in dollars.

Taxable income	$67,500	67,600	67,700	67,800	67,900	68,000
Tax owed	$9294	9309	9324	9339	9356	9381
Taxable income	$68,100	68,200	68,300	68,400	68,500	68,600
Tax owed	$9406	9431	9456	9481	9506	9531

This table is used in Exercises 47 through 49.

47. Make a table that shows the additional tax owed over each income span.

48. Your table in Exercise 47 should show that the data are not linear over the entire income range. But it should also show that the data are linear over each of two smaller ranges. Identify these ranges. *Note:* This is an example of a *piecewise-linear function.* The term just means that two linear functions are pasted together.

49. For each of the two ranges you identified in Exercise 48, calculate the marginal tax rate.

50. The table below gives selected entries from the 2010 federal income tax table. It applies to a married couple filing jointly. Taxable income and tax owed are both in dollars.

Taxable income	$67,500	67,600	67,700	67,800	67,900	68,000
Tax owed	$9291	9306	9321	9336	9351	9369
Taxable income	$68,100	68,200	68,300	68,400	68,500	68,600
Tax owed	$9394	9419	9444	9469	9494	9519

Repeat Exercises 47 through 49 for this table.

51. Compare your answer to Exercise 50 (for the tax year 2010) with your answers to Exercises 47 through 49 (for the tax year 2009). Consider especially the income level where the transition from one marginal tax rate to the next takes place. Examine tax tables from earlier years and investigate tax law to find out how the transition level is determined. As a start to your research, determine what the phrase *bracket creep* means.

Life expectancy   The following table shows the average life expectancy, in years, of a child born in the given year:

Year	2003	2004	2005	2006	2007
Life expectancy	77.1	77.5	77.4	77.7	77.9

If t denotes the time in years since 2003 and E the life expectancy in years, then it turns out that the trend line for these data is given by

$$E = 0.18t + 77.16.$$

The data and the trend line are shown in **Figure 3.21**. This information is used in Exercises 52 through 55.

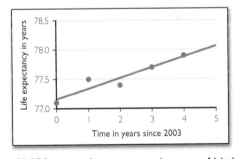

FIGURE 3.21 Life expectancy by year of birth.

52. What is the slope of the trend line, and what is its meaning in practical terms?

53. During which year was life expectancy clearly higher than would have been expected from the linear trend?

54. If the linear trend established by these data persisted through 2010, what would be the life expectancy of a child born in 2010? (Round your answer to one decimal point.) *Note:* Experts predict this life expectancy to be 78.3 years.

55. If the linear trend established by these data persisted through 2300, what would be the life expectancy of a child born in 2300? (Round your answer to 1 decimal point.) Does this age seem reasonable to you?

3.2 Exponential growth and decay: Constant percentage rates

The following article is from the Web site of CBS News.

IN THE NEWS 3.3

Yucca Mountain

Transporting Nuclear Waste May Put Millions at Risk
60 Minutes Report

July 25, 2004

(CBS) For nearly 50 years, the U.S. government and the nuclear regulatory industry have been trying to figure out what to do with massive quantities of deadly radioactive waste that has been piling up at nuclear power plants and munitions factories since the dawn of the atomic age.

Right now, it's sitting in temporary storage facilities, many of them near major metropolitan areas, vulnerable to accidents, environmental disasters and terrorism.

Every possible solution has been explored, from dumping it in the ocean to launching it towards the sun. Finally, President Bush, the Department of Energy, and the U.S. Congress decided that all of that nuclear waste should be moved to Nevada and buried under a mountain in the middle of the desert.

Needless to say, people in Nevada aren't crazy about this idea, and, as Correspondent Steve Kroft reported last fall, they believe most Americans will agree when they find out how the plan might affect them.

Yucca Mountain sits on federal land in Nevada, not far from Death Valley....

If the U.S. government has its way, this will be the final resting place for 70,000 tons of highly radioactive nuclear waste.[7]

Yucca Mountain, Nevada.

Dangerous radioactive materials, such as those referred to in the article, will eventually decay into relatively harmless matter. The problem is that this process takes an extremely long time because the way in which these materials decay is exponential, not linear. We will return to this question later in this section.

Recall that linear change occurs at a constant rate. Exponential change, on the other hand, is characterized by a constant *percentage* rate. The rates may be positive or negative. In certain idealized situations, a population of living organisms, such as bacteria, grows at a constant percentage rate. Radioactive substances provide an example of a negative rate because the amount decays rather than grows.

---

[7]As of this writing, the Obama administration has made moves to stop the development of the site, but some in Congress oppose these moves.

The nature of exponential growth

To illustrate the nature of exponential growth, we consider the simple example of bacteria that reproduce by division. Suppose there are initially 2000 bacteria in a petri dish and the population doubles every hour. We refer to 2000 as the *initial value* because it is the amount with which we start. After one hour the population doubles, so

$$\text{Population after 1 hour} = 2 \times \text{Initial population}$$
$$= 2 \times 2000 = 4000 \text{ bacteria.}$$

After one more hour the population doubles again, so

$$\text{Population after 2 hours} = 2 \times \text{Population after 1 hour}$$
$$= 2 \times 4000 = 8000 \text{ bacteria.}$$

In general, we find the population for the next hour by multiplying the current population by 2:

$$\text{Population next hour} = 2 \times \text{Current population.}$$

Let's find the percentage change in the population. Over the first hour the population grows from 2000 to 4000, an increase of 2000. The percentage change is then

$$\text{Percentage change} = \frac{\text{Change in population}}{\text{Previous population}} \times 100\% = \frac{2000}{2000} \times 100\% = 100\%.$$

Over the second hour the population grows from 4000 to 8000, another increase of 100%. In fact, doubling always corresponds to a 100% increase. Thus, the population shows a 100% increase each hour. This means that the population grows not by a constant number but by a constant *percentage*, namely 100%. Growth by a constant percentage characterizes exponential growth. In this example, the population size is an exponential function of time.

Key Concept

An exponential function is a function that changes at a constant percentage rate.

EXAMPLE 3.8 Determining constant percentage growth: Tripling and population size

If a population triples each hour, does this represent constant percentage growth? If so, what is the percentage increase each hour? Is the population size an exponential function of time?

SOLUTION

The population changes each hour according to the formula

$$\text{Population next hour} = 3 \times \text{Current population.}$$

To get a clear picture of what is happening, suppose we start with 100 individuals:

$$\text{Initial population} = 100$$
$$\text{Population after 1 hour} = 3 \times 100 = 300$$
$$\text{Population after 2 hours} = 3 \times 300 = 900.$$

Let's look at this in terms of growth:

$$\text{Growth over first hour} = 300 - 100 = 200 = 200\% \text{ increase over } 100$$
$$\text{Growth over second hour} = 900 - 300 = 600 = 200\% \text{ increase over } 300.$$

Multiplying by 3 results in a 200% increase. So the population is growing at a constant percentage rate, namely 200% each hour. Therefore, the population is an exponential function of time.

TRY IT YOURSELF 3.8

If a population quadruples each hour, does this represent constant percentage growth? If so, what is the percentage? Is the population size an exponential function of time?

The answer is provided at the end of this section.

Formula for exponential functions

Let's return to the example of an initial population of 2000 bacteria that doubles every hour. Some further calculations will show a simple but important pattern. If we let N denote the population t hours after we begin the experiment, then

When $t = 0$, $N = 2000 = 2000 \times 2^0$.
When $t = 1$, $N = 2000 \times 2 = 2000 \times 2^1$.
When $t = 2$, $N = 2000 \times 2 \times 2 = 2000 \times 2^2$.
When $t = 3$, $N = 2000 \times 2 \times 2 \times 2 = 2000 \times 2^3$.
When $t = 4$, $N = 2000 \times 2 \times 2 \times 2 \times 2 = 2000 \times 2^4$.

The pattern is evident. To calculate the population after t hours, we multiply the initial value 2000 by 2 raised to the power t:

$$N = 2000 \times 2^t.$$

In **Figure 3.22**, we have plotted the data points we calculated for the population. In **Figure 3.23**, we have added the graph of $N = 2000 \times 2^t$. This curve passes through each of the data points. This gives further evidence that the formula we proposed is correct.

This formula allows us to make calculations that would be difficult or impossible if we didn't have it. For example, we can find the population four and a half hours after the experiment begins. We just put 4.5 in place of t:

$$N = 2000 \times 2^{4.5}.$$

The result is 45,255 bacteria. (We rounded to the nearest whole number because we are counting bacteria, which don't occur as fractions.)

It is important to note the following two facts about this example:

• We find next hour's population by multiplying the current population by 2. The number 2 is called the *base* of this exponential function.

• The formula for the population is

$$N = \text{Initial value} \times \text{Base}^t.$$

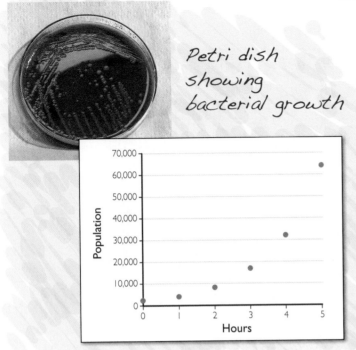

Petri dish showing bacterial growth

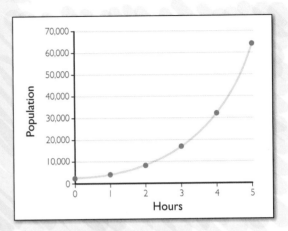

FIGURE 3.22 A population doubles each hour.

FIGURE 3.23 Adding the graph of the exponential function.

These observations are typical of exponential functions, as we summarize below.

SUMMARY 3.3   Exponential Formulas

The formula for an exponential function y of t is

$$y = \text{Initial value} \times \text{Base}^{t}.$$

An exponential function y of t is characterized by the following property: When t increases by 1, to find the new value of y, we multiply the current value by the base. In symbols,

$$y\text{-value for } t + 1 = \text{Base} \times y\text{-value for } t.$$

EXAMPLE 3.9 Finding an exponential formula: An investment

The value of a certain investment grows according to the rule

$$\text{Next year's balance} = 1.07 \times \text{Current balance}.$$

Find the percentage increase each year, and explain why the balance is an exponential function of time. Assume that the original investment is $800. Find an exponential formula that gives the balance in terms of time. What is the balance after 10 years?

SOLUTION

We multiply this year's balance by 1.07 to get next year's balance. So next year's balance is 107% of this year's balance. That is an increase of 7% per year. Because the balance grows by the same percentage each year, it is an exponential function of time.

"Ironically, despite our exponential growth, we have to downsize..."

Let B denote the balance in dollars after t years. We use the formula

$$B = \text{Initial value} \times \text{Base}^t.$$

The initial value is $800 because that is the original investment. The base is 1.07 because that is the multiplier we use to find next year's balance. This gives the formula

$$B = 800 \times 1.07^t.$$

To find the balance after 10 years, we substitute $t = 10$ into this formula:

$$\text{Balance after } t \text{ years} = 800 \times 1.07^t$$
$$\text{Balance after 10 years} = 800 \times 1.07^{10} = \$1573.72.$$

TRY IT YOURSELF 3.9

An investment grows according to the rule

$$\text{Next month's balance} = 1.03 \times \text{Current balance}.$$

Find the percentage increase each month. If the initial investment is $450, find an exponential formula for the balance as a function of time. What is the balance after two years?

The answer is provided at the end of this section.

Figure 3.24 shows the graph of the balance in Example 3.9.

The rapidity of exponential growth

Let's look again at the graph of the bacteria population we examined earlier. The graph of $N = 2000 \times 2^t$ is in **Figure 3.25**. The shape of this graph is characteristic of graphs of exponential growth: It is increasing at an *increasing* rate—unlike linear

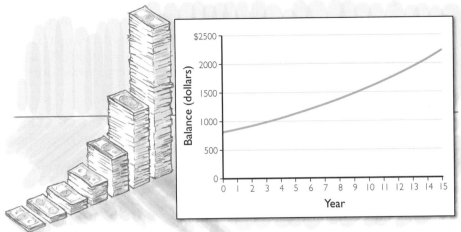

FIGURE 3.24 The balance of an investment.

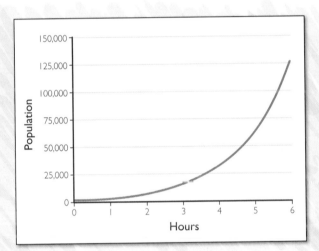

FIGURE 3.25 Graph of population growth.

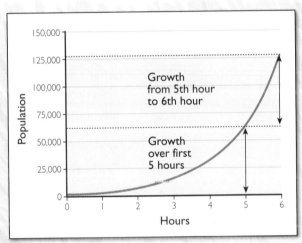

FIGURE 3.26 Comparing growth rates.

graphs, which have a *constant* growth rate. The rate of growth is rather slow to begin with, but as we move to the right, the curve becomes very steep, showing more rapid growth. Referring to **Figure 3.26**, we see that the growth from the fifth to the sixth hour is almost the same as the growth over the first five hours.

A useful illustration of the eventual high rate of exponential growth is provided by an ancient tale. A grateful (but naive) king grants a reward to a loyal subject by promising to give him two grains of wheat on the first square of a chess board, four grains on the second, eight on the third, and so on. In general, there will be 2^n grains of wheat on the nth square. On the last square of the chess board, the sixty-fourth, the king has promised to place

$$2^{64} = 18,446,744,073,709,551,616 \text{ grains of wheat.}$$

This number was calculated in the eleventh century by the Persian philosopher and mathematician al-Biruni.

There are about a million grains of wheat in a bushel, so the king's promise forces him to put over 18 trillion bushels of wheat on the 64th square. At \$3.50 per bushel, that amount of wheat would be worth about 65 trillion dollars—about five times the national debt of the United States on January 1, 2010. This illustrates how all exponential growth eventually reaches gargantuan proportions.

EXAMPLE 3.10 Calculating growth: An investment

Consider the investment from Example 3.9, where it was found that the balance B after t years is given by $B = 800 \times 1.07^t$ dollars. What is the growth of the balance over the first 10 years? Compare this with the growth from year 40 to year 50.

SOLUTION

The initial value of the investment is $800. In Example 3.9, we found that the balance after 10 years was $1573.72. This gives the growth over the first 10 years as

$$\text{Growth over first 10 years} = \$1573.72 - \$800 = \$773.72.$$

To calculate the growth from year 40 to year 50, we put $t = 40$ and $t = 50$ into the formula $B = 800 \times 1.07^t$:

$$\text{Balance after 40 years} = 800 \times 1.07^{40} = \$11{,}979.57$$
$$\text{Balance after 50 years} = 800 \times 1.07^{50} = \$23{,}565.62.$$

That is an increase of $23,565.62 - $11,979.57 = $11,586.05. This is almost 15 times the growth over the first 10 years.

TRY IT YOURSELF 3.10

After t months an investment has a balance of $B = 400 \times 1.05^t$ dollars. Compare the growth over the first 20 months with the growth from month 50 to month 70.

The answer is provided at the end of this section.

Relating percentage growth and base

Suppose that a certain job opportunity offers a starting salary of $50,000 and includes a 5% salary raise each year. In evaluating this opportunity, you may wish to know what your salary will be after 15 years. A common error would be to assume that a 5% raise each year for 15 years is the same as a $15 \times 5\% = 75\%$ raise. As we shall see, the actual increase is much more.

"No, we're not eliminating your position, Fischer. We're just eliminating your salary."

Our knowledge of exponential functions will allow us to find an appropriate formula and calculate the salary after 15 years. We reason as follows. Because the salary increases by the same percentage (namely 5%) every year, it is an exponential function of time in years. The initial value is \$50,000. Now we want to find the base. A 5% annual raise means that next year's salary is 105% of the current salary. To calculate 105% of a quantity, we multiply that quantity by 1.05. Therefore,

$$\text{Next year's salary} = 1.05 \times \text{Current salary.}$$

This says that the base is 1.05.

We now know that your salary is given by an exponential formula with initial value \$50,000 and base 1.05. Then your salary S in dollars after t years is given by

$$S = \text{Initial value} \times \text{Base}^t$$
$$S = 50,000 \times 1.05^t.$$

To find the salary after 15 years, we put $t = 15$ into this formula:

$$\text{Salary after 15 years} = 50,000 \times 1.05^{15} = \$103,946.41.$$

We note that this salary represents more than a 100% increase over the initial salary of \$50,000—much more than the 75% increase that some might expect.

The key observations in this example are that the salary is given by an exponential function and that the base corresponding to 5% is 1.05. If we write 5% as a decimal, we get 0.05. We can think of the base 1.05 as $1+0.05$. In general, suppose a quantity is growing at a constant percentage rate r in decimal form. Then we find the new value by multiplying the old value by $1+r$. Thus, the amount after t periods (years in this case) is an exponential function with base $1+r$. For exponential growth, the base is always greater than 1. The exponential formula for the amount is

$$\text{Amount} = \text{Initial value} \times (1+r)^t.$$

SUMMARY 3.4 **Exponential Growth**

1. A quantity grows exponentially when it increases by a constant percentage over a given period (day, month, year, etc.).

2. If r is the percentage growth per period, expressed as a decimal, then the base of the exponential function is $1+r$. For exponential growth, the base is always greater than 1. The formula for exponential growth is

$$\text{Amount} = \text{Initial value} \times (1+r)^t.$$

Here, t is the number of periods.

3. Typically, exponential growth starts slowly and then increases rapidly.

EXAMPLE 3.11 Finding formula from growth rate: Health care

U.S. health-care expenditures in 2010 reached 2.47 trillion dollars. In the near term this is expected to grow by 6.5% each year. Assuming that this growth rate continues, find a formula that gives health-care expenditures as a function of time. If this trend continues, what will health-care expenditures be in 2030?

SOLUTION

Because health-care expenditures are increasing at a constant percentage rate, they represent an exponential function of time. Let H denote the expenditures, in trillions of dollars, t years after 2010. We find the formula using

$$H = \text{Initial value} \times (1 + r)^t.$$

The initial value is the expenditures in 2010, namely 2.47 trillion dollars. Also, 6.5% as a decimal is 0.065, and this is the value we use for r:

$$1 + r = 1 + 0.065 = 1.065.$$

So the formula is

$$H = 2.47 \times 1.065^t \text{ trillion dollars.}$$

Now 2030 is 20 years after 2010. Thus to predict health-care expenditures in 2030, we use $t = 20$ in the formula for H:

$$\text{Expenditures in 2030} = 2.47 \times 1.065^{20} \text{ trillion dollars.}$$

The result is about 8.7 trillion dollars.

TRY IT YOURSELF 3.11

U.S. defense spending was about 719.2 billion dollars in 2010. Since that time it has grown by about 9.5% each year. Assume that this growth rate continues, and find a formula that gives defense spending as a function of time. What prediction does this formula give for defense spending in 2015?

The answer is provided at the end of this section.

In **Figure 3.27**, we have plotted the graph of health-care expenditures from Example 3.11.

Exponential decay

At the beginning of this section, we saw a news article about Yucca Mountain and the storage of nuclear waste. The decay of nuclear material is an example of exponential change that is decreasing instead of increasing.

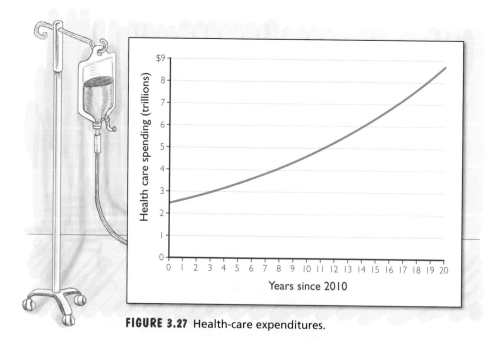

FIGURE 3.27 Health-care expenditures.

Let's examine how this works out in a simple example. Suppose we have a bacteria population that is dying off due to contaminants. Assume that initially the population size is 2000 and that the population is declining by 25% each hour. This means that 75% of the present population will still be there one hour from now. Because the change is by a constant percentage, this is an example of an exponential function. But because the population is declining, we call this process *exponential decay*.

Because the population after one hour is 75% of the current population, we find the population for the next hour by multiplying the current population by 0.75:

Population next hour $= 0.75 \times$ Current population.

This means that the population as an exponential function of time in hours has a base of 0.75. We obtain the formula for exponential decay just as we did for exponential growth. If N denotes the population size after t hours,

$$N = \text{Initial value} \times \text{Base}^t$$
$$N = 2000 \times 0.75^t.$$

We used the formula for this population to obtain the graph in **Figure 3.28**. Its shape is characteristic of graphs of exponential decay. Note that the graph is decreasing. It declines rapidly at first but then the graph levels off, showing a much smaller rate of decrease.

A key observation from the above example of population decline is that if the rate of decline is 25%, we find the base of 0.75 for the resulting exponential function using $1 - 0.25 = 0.75$. This procedure is typical of exponential decay.

Recall Summary 3.4, where the percentage increase per period was r as a decimal. In that case, the base was $1 + r$. If a quantity is declining at a constant percentage rate of r as a decimal, we use the same idea. But now the rate r is replaced by $-r$, which means that the base is $1 - r$. For exponential decay, the base is always less than 1.

In summary, we see that the formula for the amount left after t periods is

$$\text{Amount} = \text{Initial value} \times (1 - r)^t.$$

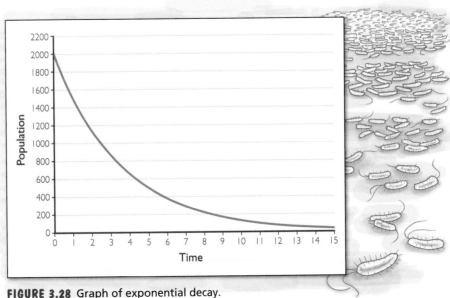

FIGURE 3.28 Graph of exponential decay.

> **SUMMARY 3.5**   **Exponential Decay**
>
> • A quantity decays exponentially when it decreases by a constant percentage over a given period (day, month, year, etc.).
>
> • Assume that a quantity decays exponentially, and let r denote the percentage decay rate per period, expressed as a decimal. Then the base of the exponential function is $1 - r$, and the formula for the function is
>
> $$\text{Amount} = \text{Initial value} \times (1 - r)^t.$$
>
> Here, t is the number of periods. For exponential decay, the base is always less than 1.
>
> • Typically, exponential decay is rapid at first but eventually slows.

EXAMPLE 3.12 Finding a formula for exponential decay: Antibiotics in the bloodstream

After antibiotics are administered, the concentration in the bloodstream declines over time. Amoxicillin is a common antibiotic used to treat certain types of infections. Suppose that 70 milligrams of amoxicillin are injected and that the amount of the drug in the bloodstream declines by 49% each hour. Find an exponential formula that gives the amount of amoxicillin in the bloodstream as a function of time since the injection. Another injection will be required when the level declines to 10 milligrams. Will another injection be required before five hours?

SOLUTION

Let A denote the amount (in milligrams) of amoxicillin in the bloodstream after t hours. When expressed as a decimal, 49% is 0.49. Hence, the base of the exponential function is

$$1 - r = 1 - 0.49 = 0.51.$$

Because the initial value is 70 milligrams, the formula is

$$A = \text{Initial value} \times (1 - r)^t$$
$$A = 70 \times 0.51^t.$$

To find the amount of amoxicillin in the blood after five hours, we put $t = 5$ into this formula:

$$\text{Amount remaining after 5 hours} = 70 \times 0.51^5 \text{ milligrams.}$$

The result is about 2.4 milligrams, which is less than the minimum of 10 milligrams. Therefore, another injection will be needed before five hours.

TRY IT YOURSELF 3.12

A certain population is initially 4000 and declines by 10% per year. Find an exponential formula that gives the population as a function of time. What will be the population after 10 years?

The answer is provided at the end of this section.

Figure 3.29 shows the amount of amoxicillin in the blood as a function of time for Example 3.12. The graph shows that the level declines to 10 milligrams in about three hours.

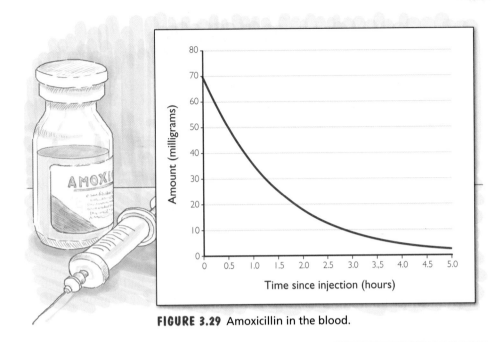

FIGURE 3.29 Amoxicillin in the blood.

EXAMPLE 3.13 Using the formula for exponential decay: The Beer-Lambert-Bouguer law

When light strikes the surface of water, its intensity decreases with depth. The *Beer-Lambert-Bouguer law* states that the percentage of decrease in intensity is the same for each additional unit of depth, so intensity is an exponential function of depth. In the waters near Cape Cod, Massachusetts, light intensity decreases by 25% for each additional meter of depth.

Light intensity decreases with depth.

a. Use I_0 for light intensity at the surface, and find an exponential formula for light intensity I at a depth of d meters in the waters off Cape Cod.

b. What is the percentage decrease in light intensity from the surface to a depth of 10 meters?

c. There is sufficient light for photosynthesis in marine phytoplankton to occur at ocean depths where 1% or more of surface light is available. Can photosynthesis of marine phytoplankton occur at a depth of 10 meters off Cape Cod? *Note:* During photosynthesis marine phytoplankton take in carbon dioxide and release oxygen. This process is crucial for maintaining life on Earth.

SOLUTION

a. Now 25% expressed as a decimal is $r = 0.25$. So the base of the exponential function is $1 - r = 1 - 0.25 = 0.75$. Using I_0 as initial intensity in the formula for percentage decrease gives

$$I = \text{Initial value} \times (1 - r)^d$$
$$I = I_0 \times 0.75^d.$$

b. We find the intensity at 10 meters using $d = 10$ in the formula from part a:

$$\text{Intensity at 10 meters} = I_0 \times 0.75^{10}.$$

This is about $0.06 I_0$. This result means that to find the intensity at a depth of 10 meters, we multiply surface intensity by 0.06. Therefore, 6% of surface intensity is left at a depth of 10 meters, so the intensity has decreased by 94%.

c. From part b we know that at a depth of 10 meters, the light intensity is 6% of light intensity at the surface. That is more than 1%, so the intensity is sufficient for photosynthesis.

Radioactive decay and half-life

The main reason for the concern expressed in the opening article about the storage of nuclear waste is that such waste is radioactive. Radioactive substances decay over time by giving off radiation or nuclear particles. This process is called *radioactive decay*. The rate of decay of a radioactive substance is normally measured in terms of its *half-life*.

Key Concept

The half-life of a radioactive substance is the time it takes for half of the substance to decay.

The long half-life of radioactive waste materials is one of the major concerns associated with the use of nuclear fuels for energy generation. Elements have different forms, called *isotopes*. The isotope Pu-239 of plutonium has a half-life of 24,000 years. Suppose that we start with 100 grams of plutonium-239. Then after 24,000 years there will be half of the original amount remaining, or 50 grams. After another 24,000 years half that amount will again decay, leaving 25 grams. This is summarized by the following formula:

$$\text{Amount left at end of a half-life} = \frac{1}{2} \times \text{Current amount}.$$

This formula means that the amount of plutonium-239 remaining after h half-lives is an exponential function with base $\frac{1}{2}$. We started with 100 grams, so after h half-lives the amount remaining is

$$\text{Amount remaining} = 100 \times \left(\frac{1}{2}\right)^{h} \text{ grams.}$$

We show the graph of remaining plutonium-239 in **Figure 3.30**.

Note that the horizontal axis represents time in half-lives, not years. The shape of the graph in Figure 3.30 epitomizes the concerns about nuclear waste. The leveling off of the curve means that dangerous amounts of the radioactive substance will be present for years to come. See also Exercise 20.

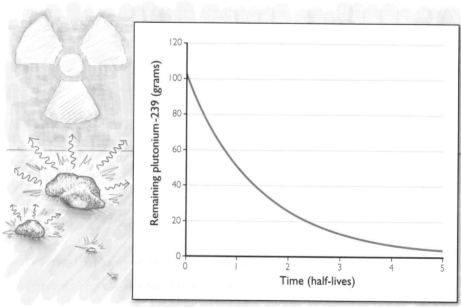

FIGURE 3.30 Plutonium-239 decay.

The formula

$$\text{Amount remaining} = 100 \times \left(\frac{1}{2}\right)^h \text{grams}$$

describes exponential decay with a percentage decrease (as a decimal) of $r = 1 - 1/2 = 0.5$ when time is measured in half-lives. In practice, we may want to know the amount remaining when time is measured in years, not in half-lives. For example, suppose we ask how much Pu–239 will remain after 36,000 years. The key here is to determine how many half-lives 36,000 years represents. Now one half-life is 24,000 years and $36,000/24,000 = 1.5$, so 36,000 years is 1.5 half-lives. Thus, to find the amount remaining after 36,000 years, we use $h = 1.5$ in the preceding formula:

$$\text{Amount remaining after 36,000 years} = 100 \times \left(\frac{1}{2}\right)^{1.5} \text{grams}.$$

This is about 35.4 grams.

Plutonium decays in a fashion typical of all radioactive substances. The half-life is different for each radioactive substance.

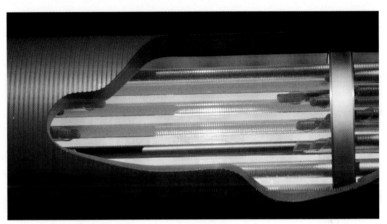

A cross-section display of nuclear fuel rods.

SUMMARY 3.6   Half-life

1. After h half-lives, the amount of a radioactive substance remaining is given by the exponential formula

$$\text{Amount remaining} = \text{Initial amount} \times \left(\frac{1}{2}\right)^h.$$

2. We can find the amount remaining after t years by first expressing t in terms of half-lives and then using the formula above.

Now we look at one important application of our study of radioactive decay. Measurements of a radioactive isotope of carbon can be used to estimate the age of organic remains. The procedure is called *radiocarbon dating*.

EXAMPLE 3.14 Applying radioactive decay: Radiocarbon dating

The isotope known as carbon-14 is radioactive and will decay into the stable form carbon-12. Assume that the percentage of carbon-14 in the air over the past 50,000 years has been about constant. As long as an organism is alive, it ingests air, and the level of carbon-14 in the organism remains the same. When it dies, it no longer absorbs carbon-14 from the air, and the carbon-14 in the organism decays, with a half-life of 5770 years.

Now, when a bit of charcoal or bone from a prehistoric site is found, the percent of carbon-14 remaining is measured. By comparing this with the percentage of carbon-14 in a living tree today, scientists can estimate the time of death of the tree from which the charcoal came.

Suppose a tree contained C_0 grams of carbon-14 when it was cut down. What percentage of the original amount of carbon-14 would we find if it was cut down 30,000 years ago?

SOLUTION

The amount C (in grams) remaining after h half-lives is given by

$$C = C_0 \times \left(\frac{1}{2}\right)^h.$$

Now 5770 years is one half-life, so 30,000 years is 30,000/5770 or about 5.20 half-lives. We use this value for h to calculate the amount remaining after 30,000 years:

$$\text{Amount after 30,000 years} = C_0 \times \left(\frac{1}{2}\right)^{5.2} \text{ grams.}$$

This is about $0.027 C_0$ grams. Thus, about 2.7% of the original amount of carbon-14 remains after 30,000 years.

TRY IT YOURSELF 3.14

A certain radioactive substance has a half-life of 450 years. If there are initially 20 grams, find a formula for the exponential function that gives the amount remaining after h half-lives. How much remains after 720 years? Round your answer to one decimal place.

The answer is provided at the end of this section.

In this example, we determined the percentage of carbon-14 from the age of the sample. In the next section, we return to the topic of radiocarbon dating and see how to estimate the age of the sample based on the percentage of carbon-14.

"Don't worry if you can't remember your age grandad, we'll get you carbon dated!"

Try It Yourself answers

Try It Yourself 3.8: Determining constant percentage growth: Quadrupling and population size Yes: Quadrupling represents 300% growth, so this is exponential.

Try It Yourself 3.9: Finding an exponential formula: An investment 3% per month. If B denotes the balance in dollars after t months, $B = 450 \times 1.03^t$. The balance after two years (24 months): $914.76.

Try It Yourself 3.10: Calculating growth: An investment   Growth over the first 20 months: $661.32. Growth from month 50 to month 70: $7583.61, about 11.5 times the earlier growth.

Try It Yourself 3.11: Finding formula from growth rate: Defense spending   If D denotes defense spending (in billions of dollars) t years after 2010, $D = 719.2 \times 1.095^t$. In 2015: 1132.2 billion dollars.

Try It Yourself 3.12: Finding a formula for exponential decay: Population decline   If N denotes the population after t years, $N = 4000 \times 0.90^t$. After 10 years, the population will be about 1395.

Try It Yourself 3.14: Applying radioactive decay: Half-life   If A is the amount (in grams) remaining, $A = 20 \times \left(\dfrac{1}{2}\right)^h$. After 720 years, there will be about 6.6 grams remaining.

Exercise Set 3.2

Note: In some exercises you are asked to find a formula for an exponential function. If no letters are assigned to quantities in an exercise, you are of course expected to choose them and give the appropriate units.

1. Exponential salary. Bob's salary grows by $3000 each year. Mary's salary grows by 3% each year. Which one has a salary that grows exponentially?

2. Linear or exponential. Water is pumped into a tank at a rate of 10 gallons per minute. Determine which type of function describes the volume of water in the tank: linear or exponential.

3. Heart health. One study measured the benefits of more sleep for healthier hearts.[8] Heart health was measured in terms of calcification of the coronary arteries. The main conclusion of the study was, "One hour more of sleep decreased the estimated odds of calcification by 33%." Rephrase this conclusion using the terminology of this section.

4. Large percent change. Chinese textile imports went up by 1000%. Were the imports

   a. 10 times the old figure?

   b. 11 times?

   c. 100 times?

   d. 110 times?

   e. none of these?

5. Balance. An investment grows according to the rule

   Next month's balance = 1.002 × Current balance.

Find the percentage increase each month, and explain why the balance is an exponential function of time.

Internet domain hosts   It is common with new technologies and new types of successful business ventures for growth to be approximately exponential. One example is the growth in the number of internet domain hosts between 1995 and 2005. One model is that the number of domain hosts grew according to

the rule

   Next year's number = 1.43 × Current number.

This information is used in Exercises 6 through 8.

6. Find the percentage increase each year, and explain why the number of hosts is an exponential function of time.

7. The number of domain hosts initially (in 1995) was 8.2 million. Find an exponential formula that gives the number of hosts in terms of time.

8. What number of Internet domains does your formula in Exercise 7 give for 2005? Round your answer to the nearest million.

9. German gold bonds. A certain gentleman acquired 54 gold bonds issued by the Weimar Republic in 1934. The value of each was 2500 troy ounces of gold. The bonds matured in 1954, at which time the value of each was 2500 troy ounces of gold at 1954 prices. The price of gold at that time was around $350 per troy ounce. The bonds could not be redeemed because in 1954 there was no single German country—it had been divided into two nations. But provisions of the bond stated that from 1954 onward the unredeemed bonds grew in value according to the exponential formula

   Value = Initial value × 1.00019^{365t},

where t is the number of years since 1954. What was the total value of this man's 54 bonds in 2012? If the bonds remain unredeemed in 2024, what will be their total value? Give your answers in billions of dollars rounded to two decimal places.

10. Headway. For traffic that flows on a highway, the *headway* is the average time between vehicles. On four-lane highways, the probability P (as a decimal) that the headway is at least t seconds when there are 500 vehicles per hour traveling one way is given by[9]

$$P = 0.87^t.$$

---

[8] C. R. King et al., "Short Sleep Duration and Incident Coronary Artery Calcification," *Journal of American Medical Association* **300** (2008), 2859–2866.

[9] See Institute of Traffic Engineers, *Transportation and Traffic Engineering Handbook*, ed. John E. Baerwald (Englewood Cliffs, NJ: Prentice-Hall 1976), p. 102.

Under these circumstances, what is the probability that the headway is at least 20 seconds? Round your answer to two decimal places.

Newton's law of heating   If we put a cool object into a preheated oven, Newton's law of heating tells us that the difference between the temperature of the oven and the temperature of the object decreases exponentially with time. The percentage rate of decrease depends on the material that is being heated. Suppose a potato initially has a temperature of 75 degrees and the oven is preheated to 375 degrees. Use the formula

$$D = 300 \times 0.98^t,$$

where D is the temperature difference between the oven and the potato, t is the time in minutes the potato has been in the oven, and all temperatures are measured in degrees Fahrenheit. Exercises 11 and 12 refer to this situation. Round your answers to the nearest degree.

11. What is the temperature difference after 30 minutes?

12. What is the temperature of the potato after 30 minutes?

Tsunami waves   Crescent City, California, is in an area that is subject to tsunamis. Its harbor was severely damaged by the tsunami caused by the earthquake off the coast of Japan on March 11, 2011. The probability P (as a decimal) that no tsunami with waves of 15 feet or higher will strike Crescent City over a period of t years is given by the formula[10]

$$P = 0.98^t.$$

This information is used in Exercises 13 and 14.

13. If you move to Crescent City and stay there for 20 years, what is the probability that you will witness no tsunami waves of 15 feet or higher? Round your answer to two decimal places.

14. What is the percentage decrease of the probability for each one-year increase in the time interval?

15. Population growth. Initially, a population is 500, and it grows by 2% each year. Explain why the population is an exponential function of time. Find a formula for the population at any time, and determine the population size after 5 years.

16. World population. The world population in 2010 was about 6.852 billion people. At that time, the population was increasing by about 1.11% per year. Assume that this percentage growth rate remains constant through 2025.[11] Explain why the population is an exponential function of time. What would you expect the world population to be in 2025?

17. United States population. According to the U.S. Census Bureau, the population of the United States in 2010 was 308.75 million people. The rate of growth in population was 0.57% per year. Assume that this rate of growth remains the same through 2025. Explain why the population is an exponential function of time. What would you predict the U.S. population to be in 2025?

18. Epidemics. In many epidemics, the cumulative number of cases grows exponentially, at least over a limited time. A recent example is the spread of the Influenza A (H1N1) virus, commonly called the swine flu, which affected Mexico and the United States especially. On April 24, 2009, there were seven cumulative cases of swine flu in the United States reported to the World Health Organization. Over the next two weeks, the cumulative number of cases in the United States increased by about 40% each day. If this growth rate had continued, what would have been the cumulative number of cases in the United States reported 30 days after April 24?

19. Library costs. The library at a certain university reported that journal prices had increased by 150% over a period of 10 years. The report concluded that this represented a price increase of 15% each year. If journal prices had indeed increased by 15% each year, what percentage increase would that give over 10 years? Round your answer as a percentage to the nearest whole number.

20. A nuclear waste site. Cesium-137 is a particularly dangerous by-product of nuclear reactors. It has a half-life of 30 years. It can be readily absorbed into the food chain and is one of the materials that would be stored in the proposed waste site at Yucca Mountain (see the article opening this section). Suppose we place 3000 grams of cesium-137 in a nuclear waste site.

   a. How much cesium-137 will be present after 30 years, or one half-life? After 60 years, or two half-lives?

   b. Find an exponential formula that gives the amount of cesium-137 remaining in the site after h half-lives.

   c. How many half-lives of cesium-137 is 100 years? Round your answer to two decimal places.

   d. Use your answer to part c to determine how much cesium-137 will be present after 100 years. Round your answer to the nearest whole number.

Folding paper   An ordinary sheet of paper (20-pound bond) is about a tenth of a millimeter thick. Each time you fold it in half, you double the thickness. Exercises 22 and 23 refer to folding this paper.

21. How would you estimate the thickness of a single sheet of paper?

22. Find an exponential formula that gives the thickness after f folds in:
   a. millimeters   **b.** kilometers

23. What would be the thickness in kilometers if you were physically able to fold it 50 times? *Note:* For comparison, the distance from Earth to the Sun is about 149,600,000 kilometers.

Inflation   The rate of inflation measures the percentage increase in the price of consumer goods. The rate of inflation in the year 2000 was 3% per year. To get a sense of what this rate would mean in the long run, let's suppose that it persists through 2020. Exercises 24 and 25 refer to this inflation.

---

[10]Taken from data provided on p. 294 of Robert L. Wiegel, ed., *Earthquake Engineering* (Englewood Cliffs, NJ: Prentice-Hall, 1970).

[11]In fact, for almost all of human history, the human population growth rate has been slowly increasing. However, in 1962 the rate reached a peak of 2.19% and declined steadily to its 2010 level of 1.11%.

24. What would be the cost in 2001 of an item that costs $100 in 2000?

25. What would be the cost in 2020 of an item that costs $100 in 2000?

Deflation From 1929 through the early 1930s, the prices of consumer goods actually decreased. Economists call this phenomenon *deflation*. The rate of deflation during this period was about 7% per year, meaning that prices decreased by 7% per year. To get a sense of what this rate would mean in the long run, let's suppose that this rate of deflation persisted over a period of 20 years. This is the situation for Exercises 26 and 27.

26. What would be the cost after one year of an item that costs $100 initially?

27. What would be the cost after 20 years of an item that costs $100 initially?

Blogosphere growth The number of blogs (or weblogs) grew rapidly for several years. According to one report in August 2006, since early in 2004 the number of blogs doubled in size every six months or so.[12] There were about 2 million blogs in March 2004. Exercises 28 through 32 use these facts.

28. Estimate the number of blogs in March 2006.

29. Assume that this trend continued, and estimate the number of blogs in March 2010. Is this number realistic?

30. Let N denote the number (in millions) of blogs and d the number of doubling periods since March 2004. Use a formula to express N as an exponential function of d.

31. Assume that this trend continued, and use your formula in Exercise 30 to estimate the number of blogs in December 2015. Round your answer to the nearest million.

32. In light of your answer to Exercise 29, decide whether your formula in Exercise 30 is realistic for all times. If not, suggest a limit on the number of doubling periods allowed.

33. Retirement options. This exercise illustrates just how fast exponential functions grow in the long term. Suppose you start work for a company at age 25. You are offered two rather unlikely retirement options:

   Retirement option 1: When you retire, you will receive $20,000 for each year of service.

   Retirement option 2: When you start work, the company deposits $2500 into a savings account that pays a monthly rate of 1.5%. When you retire, the account will be closed and the balance given to you.

How much will you have under the second plan at age 55? At age 65? Which retirement option is better if you plan to retire at age 55? Which if you plan to retire at age 65?

Cleaning dirty water Many physical processes are exponential in nature. A typical way of cleaning a tank of contaminated water is to run in clean water at a constant rate, stir, and let the mixture run out at the same rate. Suppose there are initially 100 pounds of a contaminant in a large tank of water. Assume that the cleaning method described above removes 10% of the remaining contaminant each hour. This information is used in Exercises 34 through 36. Round your answers to one decimal place.

34. Find an exponential formula that gives the number of pounds of contaminant left in the tank after t hours.

35. How much contaminant is *removed* during the first three hours?

36. How much contaminant is *removed* from the tenth ($t = 10$) to the thirteenth ($t = 13$) hour of the cleaning process?

Cleaning waste sites In many cases, removing dangerous chemicals from waste sites can be modeled using exponential decay. This is a key reason why such cleanups can be dramatically expensive. Suppose for a certain site there are initially 20 parts per million of a dangerous contaminant and that our cleaning process removes 5% of the remaining contaminant each day. This information is used in Exercises 37 and 38. Round your answers to two decimal places.

37. How much contaminant (in parts per million) is removed during the first two days?

38. How much contaminant is removed on days 10 and 11?

39. Ponzi schemes. Charles Ponzi became infamous in the early 1920s for running a classic "pyramid" investment scheme. Since that time such schemes have borne his name and are illegal. Even so, in 2009 a man named Bernard Madoff pleaded guilty to bilking investors out of billions of dollars in just such a scheme. Pyramid schemes work roughly like this. I get "rounds" of investors by promising a fat return on their money. I actually have no product or service to offer. I simply take money from second-round investors to pay off first-round investors while keeping a tidy sum for myself. Third-round investors pay off second-round investors, and so on. Suppose that there are 10 first-round investors, and for each investor in a round, I need 10 investors in the next round to pay them off. How many fifth-round investors do I need to pay off the fourth-round investors? After paying off the ninth-round investors, I take the money and run. How many investors get cheated out of their money?

Radiocarbon dating Radiocarbon dating was discussed in Example 3.14. Recall from that example that the half-life of carbon-14 is 5770 years. Exercises 40 and 41 use this information and refer to an organic sample that is 12,000 years old.

40. How many half-lives is 12,000 years? Round your answer to two decimal places.

41. Use your answer to Exercise 40 to determine what percentage of the original amount of carbon-14 remains after 12,000 years. Round your answer as a percentage to the nearest whole number.

42. Research project. The method of radiocarbon dating described in Example 3.14 is an example of a general technique known as *radiometric dating*. Investigate this general technique. What radioactive substances are used, and what are their half-lives? What are the limitations of this technique?

43. Twitter and Facebook. Consider the following news excerpt from *The Social*.

IN THE NEWS 3.4

Twitter's Growing Really, Really, Really, Really Fast

CAROLINE MCCARTHY   March 19, 2009

A small new survey from Nielsen about the five fastest growing "member community destinations" in the U.S. reveals what we all kind of knew already: Twitter is at the top. From February 2008 to February 2009, it clocked in at a whopping 1,382 percent growth rate. That's to be expected, considering the amount of press the still-without-a-business-model microblogging service has gotten in recent months.

In third place is Facebook, with 228 percent growth [from February 2008 to February 2009] according to Nielsen. That's not terribly surprising, as Facebook is still growing in the U.S. but not quite as exponentially as it once was.

What do you think the writer means by "not quite as exponentially?" Assume growth of 228% in 2008, and determine what growth rate in other years Facebook would need if it is to grow exponentially as we defined the term in this section.

The following exercises are designed to be solved using technology such as calculators or computer spreadsheets. For assistance, see the technology supplement.

Economic growth   The global rating agency Standard & Poor's predicted growth in Asian countries for 2011.[13] Exercises 44 and 45 refer to this report. In these exercises, the size of a country's economy is measured by the *gross domestic product* (GDP), which is the market value of the goods and services produced in that country in a given year.

44. In 2010 the GDP of India was about 1.4 trillion dollars, the fourth largest in the world. The report predicts that the GDP of India will grow by about 8.5% each year. Use these predictions and a spreadsheet to determine how long it will take for the economy of India to double in size from its level in 2010. (Round your answers to one decimal place.)

45. In 2010 the GDP of Japan was about 5.4 trillion dollars, the third largest in the world. The report predicts that the GDP of Japan will grow by about 1.8% each year. Use the predictions for India (as stated in Exercise 44) and Japan, along with a spreadsheet, to determine in what year the size of the economy of India will reach the size of the economy of Japan.

3.3 Logarithmic phenomena: Compressed scales

TAKE AWAY FROM THIS SECTION

Understand the use of logarithms in compressed scales and in solving exponential equations.

The article on the next page is a letter to the editor of the *New York Times*.

Caltech scientists Frank Press, Beno Gutenburg, Hugo Benioff, and Charles Richter.

---

[13]Asian Nations, India to Record High Growth in 2011: S&P, *The Economic Times*, March 16, 2011.

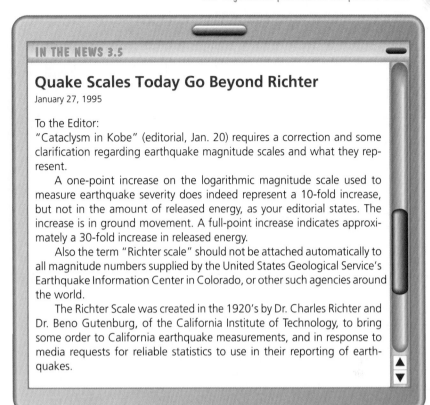

Quake Scales Today Go Beyond Richter

January 27, 1995

To the Editor:

"Cataclysm in Kobe" (editorial, Jan. 20) requires a correction and some clarification regarding earthquake magnitude scales and what they represent.

A one-point increase on the logarithmic magnitude scale used to measure earthquake severity does indeed represent a 10-fold increase, but not in the amount of released energy, as your editorial states. The increase is in ground movement. A full-point increase indicates approximately a 30-fold increase in released energy.

Also the term "Richter scale" should not be attached automatically to all magnitude numbers supplied by the United States Geological Service's Earthquake Information Center in Colorado, or other such agencies around the world.

The Richter Scale was created in the 1920's by Dr. Charles Richter and Dr. Beno Gutenburg, of the California Institute of Technology, to bring some order to California earthquake measurements, and in response to media requests for reliable statistics to use in their reporting of earthquakes.

The article above mentions the word "logarithm." The logarithm is a mathematical function that is used to define compressed scales for certain measurements, including magnitude of earthquakes, loudness of sound, acidity of substances, and brightness of stars. Logarithms are closely related to exponential functions, as we will see shortly.

In addition, the logarithm is the mathematical tool needed to solve certain exponential equations. Solving such equations is important for the applications that were discussed in the preceding section, such as the growth of investments and populations and finding the age of archaeological artifacts.

What is a logarithm?

We have noted that population growth is often modeled using an exponential function of time. But we could turn the model around and think of the time as a function of the population. This function would tell us, for example, when the population reaches 1000 and when it reaches 1,000,000. This kind of function, which reverses the effect of an exponential function, is called a *logarithmic function*. The corresponding graphs for a population of bacteria that doubles each hour are shown in **Figures 3.31 and 3.32**. As expected, the exponential graph shown in Figure 3.31 gets steeper as we move to the right. For example, over the first hour the population grows by 2 (from 2 to 4), but over the last hour shown the population grows by 8 (from 8 to 16). If we think of time as a function of population, this says that growth in population from 2 to 4 takes one hour, and growth in population from 8 to 16 also takes one hour. This observation explains the growth rate of the logarithmic function shown in Figure 3.32: Growth is rapid at first but slows as we move to the right.

We have said that the logarithm reverses the action of an exponential function. Here is a more precise definition: The *common logarithm*[14] of a number, often

---

[14]The common logarithm is also referred to as the "base 10 logarithm."

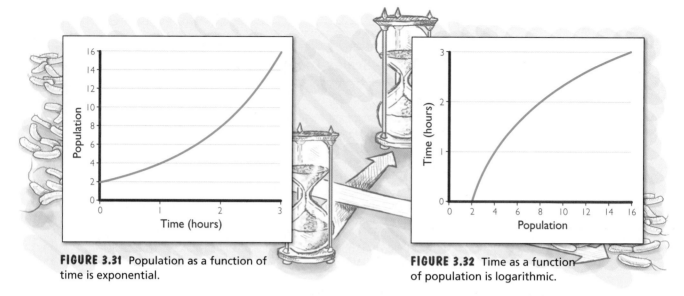

FIGURE 3.31 Population as a function of time is exponential.

FIGURE 3.32 Time as a function of population is logarithmic.

shortened to just "the log," is the exponent of 10 that gives that number. For example, the log of 100 is 2 because $100 = 10^2$. There are other kinds of logarithms besides the common logarithm, but in this section, we will use the term log or logarithm to mean the common logarithm.

Key Concept

The **common logarithm** of a positive number x, written log x, is the exponent of 10 that gives x. Formally,

$$\log x = t \text{ if and only if } 10^t = x.$$

Because the logarithmic equation $\log x = t$ means exactly the same thing as the exponential equation $x = 10^t$, we will often write both forms side by side to remind us of this fact. Thus, for example,

$$\log 10 = 1 \text{ because } 10^1 = 10; \ 1 \text{ is the exponent of 10 that gives 10}$$
$$\log 100 = 2 \text{ because } 10^2 = 100; \ 2 \text{ is the exponent of 10 that gives 100}$$
$$\log 1000 = 3 \text{ because } 10^3 = 1000; \ 3 \text{ is the exponent of 10 that gives 1000}$$
$$\log \frac{1}{10} = -1 \text{ because } 10^{-1} = \frac{1}{10}; \ -1 \text{ is the exponent of 10 that gives } \frac{1}{10}$$

For quantities that are not whole-number powers of 10, we use a scientific calculator or computer to find the logarithm.

EXAMPLE 3.15 Calculating logarithms: By hand and by calculator

What is the logarithm of

a. 1 million?

b. one thousandth?

c. 5?

SOLUTION

a. One million is 1 followed by 6 zeros, which is 10^6. Because 6 is the exponent of 10 that gives 1 million, the logarithm of 1 million is 6. In terms of symbols, we have $\log 1,000,000 = \log 10^6 = 6$.

b. One thousandth is $\dfrac{1}{1000} = 0.001 = 10^{-3}$. Because -3 is the exponent of 10 that gives one thousandth, we have $\log \dfrac{1}{1000} = -3$.

c. Because 5 is not a whole-number power of 10, we use a scientific calculator to calculate the logarithm. The result to three decimal places is 0.699. It's interesting to check this answer by using a calculator to find $10^{0.699}$. You will see that the result is very close to 5.

TRY IT YOURSELF 3.15

What is the logarithm of 1 billion?

The answer is provided at the end of this section.

The Richter scale

Now that we know what the logarithm is, let's see how it is used in real-life examples of compressed scales. A familiar example is the Richter scale used in measuring the *magnitude* of earthquakes. Charles Richter and Beno Gutenburg pioneered this method in the 1920s, as mentioned in the opening article.[15]

Direct measurement of earthquakes is done by *seismometers*, which record the *relative intensity* of ground movement. **Figure 3.33** shows a typical year of earthquakes measured by relative intensity. The graph gives the impression that there were only a few earthquakes over the period of a year, but in fact there were many thousands. The difficulty in graphing relative intensity is that the scale is very broad, stretching from 0 to 140,000,000, so only the very largest earthquakes appear above the horizontal axis. This phenomenon is why Richter and Gutenburg used the logarithm to change the scale

A seismometer being used in Haiti after the 2010 earthquake.

---

[15] That article implies there are other methods for measuring the magnitude of earthquakes. See Exercise 32 for one alternative method.

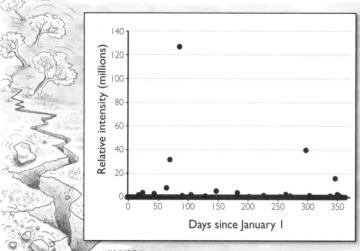

FIGURE 3.33 Earthquakes recorded by relative intensity.

FIGURE 3.34 Earthquakes recorded by magnitude.

Key Concept

The **relative intensity** of an earthquake is a measurement of ground movement. The **magnitude** of an earthquake is the logarithm of relative intensity:

$$\text{Magnitude} = \log(\text{Relative intensity}), \quad \text{Relative intensity} = 10^{\text{Magnitude}}.$$

If we measure earthquakes by magnitude rather than relative intensity, we obtain the graph in **Figure 3.34**, which shows that many earthquakes occurred.

These two graphs help us see how the "compressed scale" using the logarithm makes the differences among earthquakes more easily discernible.

An earthquake of magnitude 5.6 on the Richter scale would generally be considered moderate, and a 6.3-magnitude earthquake would be considered fairly strong. An earthquake with magnitude of 2.5 or less probably would go unnoticed by people in the area.

On March 11, 2011, an earthquake of magnitude 9.0 struck the Pacific Ocean off the coast of Japan. It caused a destructive tsunami; over 10,000 people were killed. The most powerful shock ever recorded was the Great Chilean Earthquake of 1960, which registered 9.5 on the Richter scale. On December 26, 2004, an earthquake of

Aftermath of the 2010 Haiti earthquake.

magnitude 8.9 struck the Indian Ocean near Indonesia; the resulting tsunami killed about 230,000 people. The earthquake that struck Haiti on January 12, 2010, was as devastating in terms of loss of life but much smaller in size (see Exercise 31).

The following table provides more information on interpreting the Richter scale in practical terms. Note that a small increase in the magnitude yields dramatic effects. That is one result of using a compressed scale.

Richter magnitude	Effects	Estimated number of earthquakes per year
2.5 or less	Generally not felt	900,000
2.6 to 5.4	Often felt, only minor damage	30,000
5.5 to 6.0	Slight damage to buildings	500
6.1 to 6.9	May cause lots of damage in populated areas	100
7.0 to 7.9	Major quake, serious damage	20
8.0 to 8.9	Can totally destroy communities	1
9.0 or higher	Rare, great quake. May cause major damage to areas as much as 1000 miles away	One every 20 years

EXAMPLE 3.16 Calculating magnitude and relative intensity: Earthquakes

If an earthquake has a relative intensity of 6700, what is its magnitude?

SOLUTION

The magnitude is the logarithm of relative intensity:

$$\text{Magnitude} = \log(\text{ Relative intensity})$$
$$= \log 6700.$$

Using a calculator, we find the magnitude to be 3.8. (The magnitude is usually rounded to one decimal place as we have done here.)

TRY IT YOURSELF 3.16

What is the relative intensity of an earthquake of magnitude 3.4? Round your answer to the nearest whole number.

The answer is provided at the end of this section.

Interpreting change on the Richter scale

To interpret the effect of a change in magnitude, we apply what we have learned about the growth of exponential functions. Recall Summary 3.3 from the preceding section: If y is an exponential function of t, increasing t by 1 unit causes y to be multiplied by the base. If we apply this fact to the exponential relationship

$$\text{Relative intensity} = 10^{\text{Magnitude}},$$

which has a base of 10, we see that a 1-point increase in magnitude corresponds to multiplying the relative intensity by 10. More generally, increasing the magnitude by t points multiplies relative intensity by 10^t. For example, a magnitude 6.3 quake is 10 times as intense as a magnitude 5.3 quake, and a magnitude 7.3 quake is $10^2 = 100$ times as intense as a magnitude 5.3 quake. This shows why, as noted in the table above, a relatively small increase on the Richter scale corresponds to a large increase in the relative intensity of a quake.

> **SUMMARY 3.7**   **Meaning of Magnitude Changes**
>
> • An increase of 1 unit on the Richter scale corresponds to increasing the relative intensity by a factor of 10.
>
> • An increase of t units in magnitude corresponds to increasing the relative intensity by a factor of 10^t.

This relationship lets us compare some historical earthquakes.

EXAMPLE 3.17 Interpreting magnitude changes: Comparing some important earthquakes

In 1994 an earthquake measuring 6.7 on the Richter scale occurred in Northridge, California. In 1958 an earthquake measuring 8.7 occurred in the Kuril Islands.[16] How did the intensity of the Northridge quake compare with that of the Kuril Islands quake?

SOLUTION

The Kuril Islands quake was $8.7 - 6.7 = 2$ points higher on the Richter scale. Increasing magnitude by 2 points multiplies relative intensity by 10^2. Thus, the Kuril Islands quake was $10^2 = 100$ times as intense as the Northridge quake.

TRY IT YOURSELF 3.17

In May 2008 an earthquake struck Sichuan, China. It had a magnitude of 8.0 and reportedly killed 70,000 people. The magnitude of 8.0 is a 1.3-point increase from the magnitude of the Northridge quake. How did the relative intensity of the Sichuan quake compare with that of the Northridge quake?

The answer is provided at the end of this section.

The decibel as a measure of sound

The human ear can hear sounds over a huge range of relative intensities—from rustling leaves to howling jet engines. Furthermore, the brain perceives large changes in sound intensity as smaller changes in loudness. Hence, it is useful to have a scale that measures sound like the human brain does. This is the *decibel* scale, abbreviated dB.

Key Concept

The **decibel** rating of a sound is 10 times the logarithm of its relative intensity:[17]

$$\text{Decibels} = 10 \log(\text{Relative intensity}).$$

The corresponding exponential equation is

$$\text{Relative intensity} = 10^{0.1 \times \text{Decibels}}.$$

---

[16]This is a string of islands stretching from Russia to Japan.

[17]The relative intensity of a sound is a measure of the sound power per unit area as compared to the power per unit area of a barely audible sound. Physicists use 10^{-12} watts per square meter as this base intensity.

This is used in the approximate form

$$\text{Relative intensity} = 1.26^{\text{Decibels}}.$$

Decibel readings are given as whole numbers.

If a sound doubles in intensity, we do not hear it as being twice as loud. As shown in the table below, a sound 100 times as intense as another sound is heard as 20 decibels, louder. For example, a vacuum cleaner is 20 decibels louder than normal conversation. A sound 1000 times as intense is heard as 30 decibels louder. For example, a large orchestra is 30 decibels louder than busy street traffic. Physicians warn that sustained exposure to noise levels over 85 decibels can lead to hearing impairment.

The following table lists the decibel reading and relative intensity of some familiar sounds.[18] It illustrates the compressed nature of the decibel scale.

Sound	Decibels	Relative intensity
Threshold of audibility	0	1
Rustling leaves	10	10
Whisper	20	100
Normal conversation	60	1,000,000
Busy street traffic	70	10,000,000
Vacuum cleaner	80	10^8
Large orchestra	100	10^{10}
Front row at rock concert	110	10^{11}
Pain threshold	130	10^{13}
Jet takeoff	140	10^{14}
Perforation of eardrum	160	10^{16}

The graph in **Figure 3.35** shows the decibel reading in terms of the relative intensity. This graph shows how our brains actually perceive loudness. This means that a unit increase in the relative intensity has more effect on our perception of soft sounds than it does on our perception of louder sounds. Our brains can distinguish small differences in soft sounds but find the same difference almost indistinguishable for

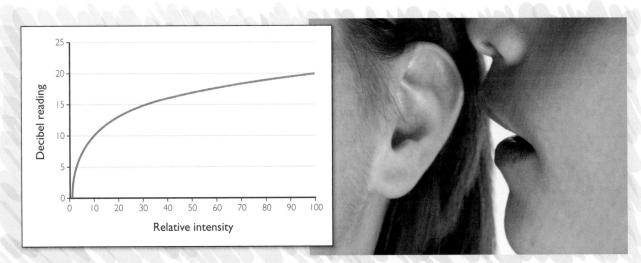

FIGURE 3.35 Left: Decibel reading in terms of relative intensity. Right: A whisper is about 20 decibels.

[18]The loudness of a sound decreases as we move away from the source. See Exercise 14 for a discussion.

louder sounds. That explains why doubling the relative intensity does not double our perception of a sound.

Because relative intensity is an exponential function of decibels, we can determine what happens to relative intensity when decibels increase. The base of the exponential equation

$$\text{Relative intensity} = 1.26^{\text{Decibel}}$$

is 1.26. Hence, an increase of one decibel multiplies relative intensity by 1.26, and an increase of t decibels multiplies relative intensity by 1.26^t.

SUMMARY 3.8   **Increasing Decibels**

- An increase of one decibel multiplies relative intensity by 1.26.
- An increase of t decibels multiplies relative intensity by 1.26^t.

EXAMPLE 3.18 Interpreting decibel changes: Vacuum cleaner to bulldozer

According to the preceding table, the sound from a vacuum cleaner is about 80 decibels. An idling bulldozer produces a sound that is about 85 decibels. How does the relative intensity of the idling bulldozer compare with that of the vacuum cleaner?

SOLUTION

According to Summary 3.8, an increase of t decibels multiplies relative intensity by 1.26^t. Therefore, increasing the number of decibels by five multiplies the intensity by 1.26^5 or about 3.2. Thus, the sound from the bulldozer is about 3.2 times as intense as that of the vacuum cleaner.

TRY IT YOURSELF 3.18

When a bulldozer is working, the noise can reach 93 decibels. How does the intensity of the sound of the working bulldozer compare with that of the vacuum cleaner? Round your answer to the nearest whole number.

The answer is provided at the end of this section.

Now we see in a practical example how doubling the intensity affects the loudness.

EXAMPLE 3.19 Calculating doubled intensity: Adding a speaker

Suppose we have a stereo speaker playing music at 60 decibels. What decibel reading would be expected if we add a second speaker?

SOLUTION

One might expect the music to be twice as loud, but we have doubled the *intensity* of the sound, not the loudness. The relative intensity of a 60-decibel speaker is

$$\text{Relative intensity} = 1.26^{\text{Decibels}}$$
$$= 1.26^{60}.$$

With the second speaker added, the new relative intensity is doubled:

$$\text{New relative intensity} = 2 \times 1.26^{60}.$$

Does adding a second speaker double the sound?

Therefore, the decibel reading of the pair of speakers is given by

$$\text{Decibels} = 10 \log(\text{Relative intensity})$$
$$= 10 \log(2 \times 1.26^{60}).$$

This is about 63 decibels.

The conclusion of this example may seem counter-intuitive: Adding a second 60-decibel speaker does not double the decibel level, which is the perceived loudness. Far from doubling loudness, adding another speaker only increases the decibel reading from 60 to 63. This interesting phenomenon is roughly true of all five human senses, including sight, and is known as *Fechner's law*.[19] This law states that psychological sensation is a logarithmic function of physical stimulus.

Other common logarithmic phenomena, such as apparent brightness of stars and acidity of solutions, will be introduced in the exercises.

Solving exponential equations

When we introduced the logarithm, we said that it reversed the effect of an exponential function. This means that logarithms allow us to find solutions to exponential equations—solutions that we could not otherwise find. In order to proceed, we first need to be aware of some basic rules of logarithms. We state them here without proof. Derivations can be found in most books on college algebra.

Properties of Logarithms

Logarithm rule 1: $\log(A^t) = t \log(A)$

Logarithm rule 2: $\log(AB) = \log(A) + \log(B)$

Logarithm rule 3: $\log\left(\dfrac{A}{B}\right) = \log(A) - \log(B)$

Now let's return to the problem of population growth in order to illustrate how logarithms are used to solve an equation.

Suppose we have a population that is initially 500 and grows at a rate of 0.5% per month. How long will it take for the population to reach 800?

---

[19] Gustav T. Fechner lived from 1801 to 1887.

Because the population is growing at a constant percentage rate, it is an exponential function. The monthly percentage growth rate as a decimal is $r = 0.005$. Hence, the base of the exponential function is $1 + r = 1.005$. Therefore, the population size N after t months is given by

$$N = \text{Initial value} \times (1 + r)^t$$
$$= 500 \times 1.005^t.$$

To find out when the population N reaches 800, we need to solve the equation

$$800 = 500 \times 1.005^t.$$

We divide both sides by 500 to simplify:

$$800 = 500 \times 1.005^t$$
$$\frac{800}{500} = \frac{500}{500} \times 1.005^t$$
$$1.6 = 1.005^t.$$

To complete the solution, we need to find the unknown exponent t. This is where logarithms play a crucial role. We apply the logarithm function to both sides:

$$\log 1.6 = \log(1.005^t).$$

Now, according to logarithm rule 1,

$$\log(1.005^t) = t \log 1.005.$$

Therefore,

$$\log 1.6 = t \log 1.005.$$

Finally, dividing by $\log 1.005$ gives

$$t = \frac{\log 1.6}{\log 1.005}.$$

To complete the calculation requires a scientific calculator, which gives about 94.2 months for t. Thus, the population reaches 800 in about seven years and 10 months.

This method of solution applies to any equation of the form $A = B^t$.

SUMMARY 3.9   **Solving Exponential Equations**

The solution for t of the exponential equation $A = B^t$ is
$$t = \frac{\log A}{\log B}.$$

EXAMPLE 3.20 Solving exponential equations: Growth of an investment

An investment is initially $5000 and grows by 10% each year. How long will it take the account balance to reach $20,000? Round your answer in years to one decimal place.

SOLUTION

The balance is growing at a constant percentage rate, so it is an exponential function. Because 10% as a decimal is $r = 0.1$, the base is $1 + r = 1.1$. The initial investment is $5000, so we find the balance B (in dollars) after t years using

$$B = \text{Initial value} \times (1 + r)^t$$
$$B = 5000 \times 1.1^t.$$

A stock certificate.

In order to find when the balance is $20,000, we need to solve the equation

$$20{,}000 = 5000 \times 1.1^t.$$

We first divide both sides by 5000 to get

$$\frac{20{,}000}{5000} = \frac{5000}{5000} \times 1.1^t$$
$$4 = 1.1^t.$$

This is an exponential equation of the form $A = B^t$ with $A = 4$ and $B = 1.1$. We use Summary 3.9 to find the solution:

$$t = \frac{\log A}{\log B} = \frac{\log 4}{\log 1.1}.$$

This is about 14.5 years.

TRY IT YOURSELF 3.20

An investment is initially $8000 and grows by 3% each month. How long does it take for the value of the investment to reach $10,000? Round your answer in months to one decimal place.

The answer is provided at the end of this section.

Doubling time and more

It is a fact that when a quantity grows exponentially, it will eventually double in size. (See Exercise 2 for more information.) Further, the doubling time depends only on the base—it does not depend on the initial amount. Doubling occurs over and over, and always over the same time span. Suppose we want to find how long it takes the exponential function

$$\text{Initial value} \times \text{Base}^t$$

to double. For this, we need to solve the equation

$$2 \times \text{Initial value} = \text{Initial value} \times \text{Base}^t.$$

Dividing by the initial value gives

$$2 = \text{Base}^t.$$

We solve this equation using Summary 3.9 with $A = 2$ and $B = \text{Base}$:

$$\text{Doubling time} = t = \frac{\log 2}{\log(\text{Base})}.$$

A similar formula applies for any multiple of the initial value.

SUMMARY 3.10   Doubling Time and More

Suppose a quantity grows as an exponential function with a given base. The time t required to multiply the initial value by K is

$$\text{Time required to multiply by } K \text{ is } t = \frac{\log K}{\log(\text{Base})}.$$

The special case $K = 2$ gives the doubling time:

$$\text{Doubling time} = \frac{\log 2}{\log(\text{Base})}.$$

EXAMPLE 3.21 Finding doubling time: An investment

Suppose an investment is growing by 7% each year. How long does it take the investment to double in value? Round your answer in years to one decimal place.

SOLUTION

Because there is constant percentage growth, the balance is an exponential function. The base is $1 + r = 1.07$. To find the doubling time, we use the formula in Summary 3.10 with Base = 1.07:

$$\text{Doubling time} = \frac{\log 2}{\log(\text{Base})}$$

$$= \frac{\log 2}{\log 1.07}.$$

This is about 10.2 years.

TRY IT YOURSELF 3.21

An investment is growing by 2% per month. How long does it take for the investment to double in value? Round your answer in months to one decimal place.

The answer is provided at the end of this section.

Investments may grow exponentially, but radioactive substances decay exponentially, as we saw in the last section. There, we discussed at some length the half-life of radioactive substances. The first formula in Summary 3.10 can be used to answer questions about radioactive decay as well as growth of investments. The next example does this.

EXAMPLE 3.22 Radiocarbon dating

In Example 3.14 of the preceding section, we discussed how carbon-14, with a half-life of 5770 years, is used to date objects. Recall that the amount of a radioactive material decays exponentially with time, and the base of that exponential function is 1/2 if time is measured in half-lives.

Living organisms absorb carbon-14 (radiocarbon) during their lifetimes.

Radiocarbon decays at a known rate. Paleontologists are able to determine the age of charcoal by measuring the amount of carbon-14 it contains.

Carbon-14 decays into nitrogen-14, emitting an electron. A radiation counter records the number of electrons emitted when a small piece of the charcoal is burned.

Suppose the charcoal from an ancient campfire is found to contain only one-third of the carbon-14 of a living tree. How long ago did the tree that was the source of the charcoal die? Give the answer first in half-lives and then in years rounded to the nearest hundred.

SOLUTION

The amount of carbon-14 present is an exponential function of time in half-lives, and the base is 1/2. Because we want to know when only one-third of the original amount is left, we use $K = 1/3$ in the formula from Summary 3.10:

$$\text{Time to multiply by 1/3 is } t = \frac{\log K}{\log(\text{Base})}$$
$$= \frac{\log(1/3)}{\log(1/2)}.$$

This is about 1.58, so the time required is 1.58 half-lives. Because each half-life is 5770 years, 1.58 half-lives is

$$1.58 \times 5770 = 9116.6 \text{ years.}$$

Thus, the tree died about 9100 years ago.

Try It Yourself answers

Try It Yourself 3.15: Calculating logarithms: By hand and by calculator   The logarithm of 1 billion is 9.

Try It Yourself 3.16: Calculating magnitude and relative intensity: Earthquakes $10^{3.4}$ is about 2512.

Try It Yourself 3.17: Interpreting magnitude changes: Comparing some important earthquakes   The Sichuan quake was about 20 times as intense as the Northridge quake.

Try It Yourself 3.18: Interpreting decibel changes: Vacuum cleaner to bulldozer The sound from the working bulldozer is about 20 times as intense as the sound of the vacuum cleaner.

Try It Yourself 3.20: Solving exponential equations: Growth of an investment 7.5 months.

Try It Yourself 3.21: Finding doubling time: An investment   35.0 months.

Exercise Set 3.3

1. Log of a negative number? Use your calculator to attempt to find the log of −1. What does your calculator say? Explain why, according to our definition, the logarithm of a negative number does not exist.

2. Not doubling. In connection with the discussion of doubling times, we stated that when a quantity grows exponentially, it will eventually double in size. This property does not hold for all increasing functions. Draw a graph that starts at the point $x = 1$, $y = 1$ and grows forever but for which y never doubles, that is, y never reaches 2.

3. Kansas and California. In 1812 a magnitude 7.1 earthquake struck Ventura, California. In 1867 a magnitude 5.1 quake struck Manhattan, Kansas. How many times as intense as the Kansas quake was the California quake?

4. Ohio earthquake. In 1986 an earthquake measuring 5.0 on the Richter scale hit northeast Ohio. How many times as intense as a magnitude 2.0 quake was it?

5. Alaska and California. In 1987 a magnitude 7.9 quake hit the Gulf of Alaska. In 1992 a magnitude 7.6 quake hit Landers,

California. How do the relative intensities of the two quakes compare?

6. New Madrid quake. On December 16, 1811, an earthquake occurred near New Madrid, Missouri, that temporarily reversed the course of the Mississippi River. The quake had a magnitude of 8.8. On October 17, 1989, a calamitous quake measuring 7.1 on the Richter scale occurred in San Francisco during a World Series baseball game on live TV. How many times as intense as the San Francisco quake was the New Madrid quake?

Energy of earthquakes The energy released by an earthquake is related to the magnitude by an exponential function. The formula is

$$\text{Energy} = 25{,}000 \times 31.6^{\text{Magnitude}}.$$

The unit of energy in the above equation is a *joule*. One joule is approximately the energy expended in lifting 3/4 of a pound 1 foot. Exercises 7 through 9 refer to this formula.

7. On April 19, 1906, San Francisco was devastated by a 7.9-magnitude earthquake. How much energy was released by the San Francisco earthquake?

8. Use the fact that energy is an exponential function of magnitude to determine how a 1-unit increase in magnitude affects the energy released by an earthquake. Compare your answer with the figure given in the article opening this section.

9. We noted earlier that on December 26, 2004, an earthquake of magnitude 8.9 struck the Indian Ocean near Indonesia. How did the energy released by the Indonesian quake compare with that of the San Francisco quake of Exercise 7?

10. Jet engines. A jet engine close up produces sound at 155 decibels. What is the decibel reading of a pair of nearby jet engines? In light of the table on page 171, what does your answer suggest about those who work on the tarmac to support jet aircraft?

11. Stereo speakers. A speaker is playing music at 80 decibels. A second speaker playing the same music at the same decibel reading is placed beside the first. What is the decibel reading of the pair of speakers?

12. More on stereo speakers. Consider the lone stereo speaker in Exercise 11. Someone has the none-too-clever idea of doubling the decibel level. Would 100 of these speakers do the job? How about 1000? You should see a pattern developing. If you don't see it yet, try 10,000 speakers. How many speakers are required to double the decibel level?

13. Adding two speakers. Suppose a speaker is playing music at 70 decibels. If two more identical speakers (playing the same music) are placed beside the first, what is the resulting decibel level of the sound?

14. Decibels and distance. The loudness of a sound decreases as the distance from the source increases. Doubling the distance from a sound source such as a vacuum cleaner multiplies the relative intensity of the sound by 1/4. If the sound from a vacuum cleaner has a reading of 80 decibels at a distance of 3 feet, what is the decibel reading at a distance of 6 feet?

15. Apparent brightness. The apparent brightness on Earth of a star is measured on the *magnitude* scale.[20] The apparent magnitude m of a star is defined by

$$m = 2.5 \log I,$$

where I is the relative intensity. The accompanying exponential formula is

$$I = 2.51^m.$$

To find the relative intensity, we divide the intensity of light from the star Vega by the intensity of light from the star we are studying, with both intensities measured on Earth. Thus, apparent magnitude is a logarithmic function of relative intensity. The star Fomalhaut has a relative intensity of 2.9, in the sense that on Earth light from Vega is 2.9 times as intense

as light from Fomalhaut. What is the apparent magnitude of Fomalhaut? Round your answer to two decimal places. *Note:* The magnitude scale is perhaps the reverse of what you might think. The higher the magnitude, the dimmer the star. Some very bright stars have a negative apparent magnitude.

16. Finding relative brightness. *This is a continuation of Exercise 15.* A star has an apparent magnitude of 4. Use the exponential formula in Exercise 15 to find its relative intensity. Round your answer to one decimal place.

17. Acid rain. Acid rain can do serious damage to the environment as well as human health. Acidity is measured according to the *pH scale*. (For more information on the pH scale, see Exercise 18.) Pure water has a pH of 7, which is considered neutral. A pH less than 7 indicates an *acidic* solution, and a pH higher than 7 indicates a *basic* solution. (Lye and baking soda solutions are very basic; vinegar and lemon juice are acidic.) An increase of 1 on the pH scale causes acidity to be multiplied by a factor of 1/10. Equivalently, a decrease of 1 on the pH scale causes acidity to be multiplied by a factor of 10. If rainfall has a pH of 4, how many times as acidic as pure water is it? (It is worth noting that most acid rain in the United States has a pH of about 4.3.)

18. More on pH. Acidity is measured on the pH scale, as mentioned in Exercise 17. Here, pH stands for "potential of Hydrogen." Here is how the scale works.

The acidity of a solution is determined by the concentration H of hydrogen ions.[21] The formula is

$$pH = -\log H.$$

The accompanying exponential formula is

$$H = 0.1^{pH}.$$

Lower pH values indicate a more acidic solution. Normal rainfall has a pH of 5.6. Rain in the eastern United States often has a pH level of 3.8. How many times as acidic as normal rain is this?

19. Saving for a computer. You have $200 and wish to buy a computer. You find an investment that increases by 0.6% each month, and you put your $200 into the account. When will the account enable you to purchase a computer costing $500?

Spent fuel rods   The half-life of cesium-137 is 30 years. Suppose that we start with 50 grams of cesium-137 in a storage pool. Exercises 20 and 21 refer to this cesium.

20. Find a formula that gives the amount C of cesium-137 remaining after h half-lives.

21. How many half-lives will it take for there to be 10 grams of cesium-137 in the storage pool? (Round your answer to two decimal places.) How many years is that?

22. Cobalt-60. Cobalt-60 is subject to radioactive decay, and each year the amount present is reduced by 12.3%.

---

[20]The notion of magnitude for stars goes back to the ancient Greeks, who grouped the stars into six magnitude classes, with the brightest stars being of the first magnitude. The use of the term "magnitude" for stars motivated Richter to use the same term in his scale for earthquakes.

[21]The concentration is measured in moles per liter of the solution.

a. The amount of cobalt-60 present is an exponential function of time in years. What is the base of this exponential function?

b. What is the half-life of cobalt-60? Round your answer to one decimal place.

Inflation Suppose that inflation is 2.5% per year. This means that the cost of an item increases by 2.5% each year. Suppose that a jacket cost $100 in the year 2011. Exercises 23 through 25 refer to this situation.

23. Find a formula that gives the cost C in dollars of the jacket after t years.

24. How long will it take for the jacket to cost $200? Round your answer to the nearest year.

25. How long will it take for the jacket to cost $400? *Suggestion*: Once you have completed Exercise 24, you do not need a calculator for this exercise, because you have calculated the doubling time for this exponential function.

26. World population. If the per capita growth rate of the world population continues to be what it was in the year 2010, the world population t years after July 1, 2010, will be

$$6.852 \times 1.0111^t \text{ billion.}$$

According to this formula, when will the world population reach 8 billion?

A poor investment Suppose you make an investment of $1000 that you are not allowed to cash in for 10 years. Unfortunately, the value of the investment decreases by 10% per year. Exercises 27 and 28 refer to this investment.

27. How much money will be left after the end of the 10-year term?

28. How long will it be before your investment decreases to half its original value? Round your answer to one decimal place.

Economic growth The global rating agency Standard & Poor's predicted growth in Asian countries for 2011.[22] Exercises 29 and 30 refer to this report. In these exercises, the size of a country's economy is measured by the *gross domestic product* (GDP), which is the market value of the goods and services produced in that country in a given year.

29. In 2010 the GDP of the United States was about 14.7 trillion dollars, the largest in the world. The report predicts that the GDP of the United States will grow by about 2.8% each year. What is the projected value of the GDP in 2030 (to the nearest trillion)?

30. In 2010 the GDP of China was about 5.9 trillion dollars, the second largest in the world. The report predicts that the GDP of China will grow by about 9.6% each year. Use this prediction and your answer to Exercise 29 to determine whether China will have a larger economy than the United States by the year 2030. If so, determine in what year the economy of China will reach the level projected in Exercise 29. (Round up to the next year.)

31. Earthquake in Haiti. The earthquake that devastated Haiti on January 12, 2010, had a magnitude of 7.0 and killed hundreds of thousands of people. Six weeks later a quake of magnitude 8.8 struck southern Chile and killed about 500 people. Investigate why the much smaller earthquake in Haiti was so much more devastating.

32. Beyond Richter. The article opening this section implies that there are other methods than the Richter scale for measuring the magnitude of earthquakes. One alternative is the *moment magnitude scale*. Investigate how this scale is defined. Does the definition involve a logarithm?

Exercises 33 through 36 are suitable for group work.

33. Yahoo stock graphs. If you go to Yahoo's finance site at http://finance.yahoo.com and look up a stock, you will see a chart, or graph, showing the price plotted over time. Among the chart settings is the option to set the chart scale to either logarithmic or linear. Selecting the logarithmic option yields a plot of the logarithm of the stock price (note the scale on the vertical axis). Try this with Starbucks' stock (symbol SBUX). Print a copy of the graphs of SBUX over a five-year period using the Linear option and using the Log option. Compare the graphs and discuss the differences. Pick out a couple of other stocks and do the same thing with them.

The logarithmic Dow For graphs that have sharp jumps, applying the logarithm can dampen the jumps and give a clearer visual presentation of the data. Yahoo's finance site does this (see Exercise 33). **Figure 3.36** shows the logarithmic Dow from 1897 through 2010. Exercises 34 through 36 refer to the logarithmic Dow.

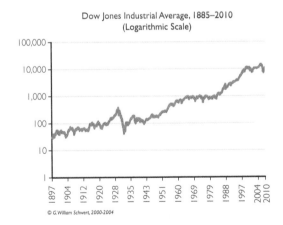

Dow Jones Industrial Average, 1885–2010
(Logarithmic Scale)

© G. William Schwert, 2000-2004

FIGURE 3.36 The logarithmic Dow.

34. In the late 1920s through the mid-1930s, there is a sharp drop. What historical event corresponds to this drop?

[22]Asian Nations, India to Record High Growth in 2011: S&P, *The Economic Times*, March 16, 2011.

35. Lay a straightedge on the picture and comment on the hypothesis that the logarithmic Dow is almost linear.

36. If we agree that the logarithmic Dow is almost linear, then it can be shown that the Dow itself is growing exponentially. Explain what this would mean in the long term for a diversified stock portfolio.

37. Historical figure: John Napier. John Napier is credited with inventing the idea of logarithms, but of a different kind than the common logs discussed here. Write a brief report on Napier's life and accomplishments.

38. Historical figure: Henry Briggs. Henry Briggs was an English mathematician notable for changing Napier's logarithm into what we have called the common logarithm. Write a brief report on Briggs's life and accomplishments.

39. Historical figure: Gustav T. Fechner. Gustav T. Fechner was mentioned in connection with Fechner's law. Write a brief report on Fechner's life and accomplishments.

40. Historical figure: Alexander J. Ellis. The name of Alexander J. Ellis is associated with the *cent*, a unit of measure for musical intervals that involves the logarithm. Write a brief report on Ellis and his work.

CHAPTER SUMMARY

This chapter discusses three basic types of functions: *linear*, *exponential*, and *logarithmic*. All three are important and occur naturally.

Lines and linear growth: What does a constant rate mean?

A linear function is one with a constant growth rate. The formula for a linear function is $y = mx + b$, where m is the growth rate or *slope* and b is the initial value. A 1-unit increase in x corresponds to a change of m units in y.

The growth rate, or slope, of a linear function can be calculated using

$$\text{Slope} = \text{Growth rate} = \frac{\text{Change in function}}{\text{Change in independent variable}}.$$

In practical settings, the slope always has an important meaning, and understanding that meaning is often the key to analyzing linear functions. When a linear function is given in a real-world context, proper use of the units of the slope (such as inches per hour) can help us determine the meaning of the slope.

The graph of a linear function is always a straight line. When data points almost fall on a straight line, it may be appropriate to approximate the data with a linear function. We use the terms *regression line* or *trend line* for linear approximations.

Exponential growth and decay: Constant percentage rates

Exponential functions are characterized by a constant *percentage* growth (or decay). For example, if a population grows by 5% each year, the population size is an exponential function of time: The percentage growth rate of this function is constant, 5% per year. The graph of an increasing exponential function has the characteristic shape shown in **Figure 3.37**. If a population is decreasing by 5% each year, the population shows exponential decay, which has the characteristic shape shown in **Figure 3.38**.

The formula for an exponential function is

$$\text{Amount} = \text{Initial value} \times \text{Base}^t.$$

An exponential function y of t is characterized by the following property: When t increases by 1, to find the new value of y, we multiply the current value by the base.

For constant percentage increase of r as a decimal, the base is $1 + r$. This gives the formula

$$\text{Amount} = \text{Initial value} \times (1 + r)^t.$$

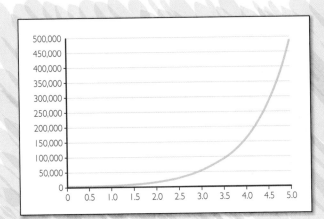

FIGURE 3.37 Increasing exponential function: exponential growth.

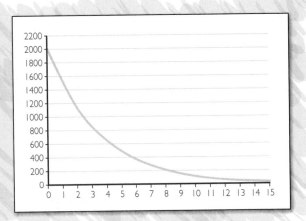

FIGURE 3.38 Decreasing exponential function: exponential decay.

For constant percentage decrease of r as a decimal, the base is $1 - r$. This gives the formula

$$\text{Amount} = \text{Initial value} \times (1 - r)^t.$$

The amount of a radioactive material decays over time as an exponential function. The base is $1/2$ if time is measured in half-lives.

Logarithmic phenomena: Compressed scales

The *common logarithm* of a positive number x is the exponent of 10 that gives x. The common logarithm of x is written $\log x$. Formally,

$$\log x = t \text{ if and only if } 10^t = x.$$

One of the most familiar occurrences of logarithmic functions is the Richter scale, which arises in the measurement of earthquakes. An increase of one point on the Richter scale corresponds to multiplying the relative intensity of the earthquake by 10. The logarithm provides the compression needed for the Richter scale.

The decibel scale is another logarithmic scale. It arises in the measurement of sound. In fact, according to *Fechner's law*, for all five of the human senses the perceived magnitude of a sensation can be described using the logarithm.

The graph in **Figure 3.39** shows the decibel reading in terms of the relative intensity. This figure shows the characteristic shape of the graph of a logarithmic function. It illustrates the decreasing growth rate that is expected from the applications to the Richter scale and human sensation.

Logarithms can be used to solve exponential equations. This solution yields exact formulas for the doubling time, tripling time, etc., of exponential growth.

KEY TERMS

linear, p. 129
regression line, p. 141
trend line, p. 141

exponential function, p. 147
half-life, p. 158
common logarithm, p. 166

relative intensity, p. 168
magnitude, p. 168
decibel, p. 170

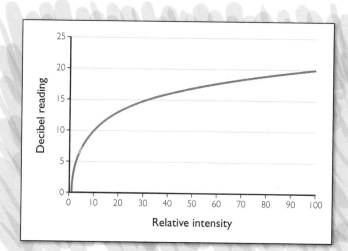

FIGURE 3.39 Decibel reading in terms of relative intensity.

CHAPTER QUIZ

1. For the following scenarios, explain the practical meaning of the growth rate, and determine which functions are linear.

   a. A book rental club membership charges $35 to join and then $4 for each book rented. Is the total amount spent on book club membership and rental a linear function of the number of books rented?

   b. Another book rental club charges $20 for membership, plus $5 for each of the first three books rented each month and only $2 for any additional books rented each month. Is the total amount spent on book club membership and rental a linear function of the number of books rented?

Answer In both cases, the growth rate is the extra rental cost for each additional book. The first scenario determines a linear function because the charge for each extra book is the same. The second scenario does not determine a linear function because the charge for each extra book varies depending on how many books you rent.

If you had difficulty with this problem, see Example 3.1.

2. It begins raining at 10:00 A.M. and continues raining at a constant rate all day. The rain gauge already contains 2 inches of water when the rain begins. Between 12:00 P.M. and 2:00 P.M., the level of the rain gauge increases by 1 inch. Find a linear formula that gives the height of water in the gauge as a function of the time since 10:00 A.M.

Answer If H denotes the height (in inches) of water in the gauge t hours after 10:00 A.M., $H = 2 + 0.5t$.

If you had difficulty with this problem, see Example 3.6.

3. Data for the yearly best 100-meter dash time for a small college from 2000 through 2005 were collected and fit with a trend line. The equation for the trend line is $D = 12 - 0.03t$, where t is the time in years since 2000 and D is the best time (in seconds) for that year. Explain in practical terms the meaning of the slope of the trend line.

Answer The slope is -0.03 second per year. It means that the best time was decreasing by about 0.03 second each year.

If you had difficulty with this problem, see Example 3.7.

4. My salary in 2000 was $48,000. I got a 3% raise each year for eight years. Find a formula that gives my salary as a function of the time since 2000. What was my salary in 2008?

Answer If we use S for my salary (in dollars) and t for the time in years since 2000, $S = 48,000 \times 1.03^t$. In 2008 my salary was $60,804.96.

If you had difficulty with this problem, see Example 3.11.

5. Uranium-235, which can be used to make atomic bombs, has a half-life of 713 million years. Suppose 2 grams of uranium-235 were placed in a "safe storage location" 150 million years ago. Find an exponential formula that gives the amount of uranium-235 remaining after h half-lives. How many half-lives is 150 million years? (Round your answer to two decimal places.) How much uranium-235 still remains in "safe storage" today?

Answer If A is the amount (in grams) remaining, $A = 2 \times (1/2)^h$. 150 million years is 0.21 half-life. The amount remaining is about 1.7 grams.

If you had difficulty with this problem, see Example 3.14.

6. One earthquake has a magnitude of 3.5, and another has a magnitude of 6.5. How do their relative intensities compare?

Answer   The larger earthquake is 1000 times as intense as the smaller quake.

If you had difficulty with this problem, see Example 3.17.

7. An idling bulldozer makes a sound of about 85 decibels. What would be the decibel reading of two idling bulldozers side-by-side?

Answer   88 decibels

If you had difficulty with this problem, see Example 3.19.

8. My salary is increasing by 4% each year. How long will it take for my salary to double?

Answer   About 17.7 years.

If you had difficulty with this problem, see Example 3.21.

PERSONAL FINANCE

T he following article from a Marlboro, Massachusetts, newspaper was written by a state representative.

IN THE NEWS 4.1

Let's Teach Financial Literacy

STEPHEN LEDUC   January 16, 2008

B oston—Last week legislation I filed requiring the Massachusetts Department of Education to include all aspects of personal finance as a major component in the existing math curriculum was favorably reported out by the Education Committee. . . .

Students need to learn personal finance math skills that will help them to succeed in life. This effort is intended to highlight the need for a comprehensive offering in the study of personal finance, with the understanding that the need for such knowledge cuts across all socio-economic segments of the population. . . .

Consider that saving levels in America are at their lowest since the Great Depression. Americans have accumulated $505 billion in credit card debt, and 81 percent of college freshmen have credit cards. Additionally, if we look at the current foreclosure crisis, we know that many of these problems could have been avoided if the public had a better understanding of credit and personal finance.

Our students are graduating from high school without knowing what debt can do to them or what compound interest can do for them. We need to teach our students the fundamentals of financial literacy. . . .

There are immediate, tangible benefits for students who are introduced to matters of personal finance as a component of their High School mathematics curriculum. It is very easy for students to pose the question, When am I ever going to use this? about derivatives and quadratic equations but, with regard to budgeting, debt, and personal investing, those practical applications are instantly apparent.

Debt is an issue for many college students.

This article makes clear that borrowing and saving money are a big part of our lives and that understanding these transactions can make a huge difference to us. In this chapter, we explore basic financial terminology and mechanisms. In Section 4.1, we examine the basics of compound interest and savings, and in Section 4.2, we look at borrowing. In Section 4.3, we consider long-term savings plans such as retirement funds. In Section 4.4, we focus on credit cards, and in Section 4.5, we discuss financial terms heard in the daily news.

4.1 Saving money: The power of compounding

The following article from the *New York Times* shows how some companies are encouraging employees to save for retirement.

IN THE NEWS 4.2

The Opt-Out Solution

TINA ROSENBERG   November 1, 2010

Americans don't save enough. In 2005, Americans' personal savings rate was negative for the first time since the Great Depression—instead of piling up savings, we are piling up debt. According to Financial Engines, an investment advisory firm that has surveyed the 401(k) retirement savings plans of 2.8 million people, only 28 percent of savers are on track to retire on 70 percent of our final salaries—and 70 percent may not be adequate to pay for health costs or travel. Worse, only one third of American workers participate in 401(k) savings plans at all.

Getting people to save more is a crucial challenge, and you might assume that it's complex. But some solutions are stunningly simple. The Minnesota-based Deluxe Corporation which provides services to small businesses, has long offered its 5,000 employees the opportunity to participate in 401(k) plans. Enrollment used to be 80 percent, with average savings between 5 and 6 percent of paychecks. Now it is about 90 percent, with savings rates above 7 percent of income. . . .

Before 2008, Deluxe's employees had to send in a form if they wanted to participate in the company's 401(k) plan. Starting that year, they had to send in a form to elect *not* to participate. In other words, the 401(k) plan went from opt-in, or voluntary enrollment, to opt-out, or automatic enrollment. . . .

Money management begins with saving, and (as the preceding article notes) Americans generally don't save enough. In this section, we see how to measure the growth of savings accounts.

Some readers may wish to take advantage of the following Quick Review of linear and exponential functions before proceeding.

Quick Review Linear Functions

Linear functions play an important role in the mathematics of finance. We recall their basic properties here. For additional information, see Section 1 of Chapter 3.

Linear functions: A linear function has a constant growth rate, and its graph is a straight line. The growth rate of the function is also referred to as the *slope*.

We find a formula for a linear function of t using

$$\text{Linear function} = \text{Growth rate} \times t + \text{Initial value}.$$

Example: If we initially have \$1000 in an account and add \$100 each year, the balance is a linear function because it is growing by a constant amount each year. After t years, the balance is

$$\text{Balance after } t \text{ years} = \$100t + \$1000.$$

The graph of this function is shown in **Figure 4.1**. Note that the graph is a straight line.

Quick Review Exponential Functions

Exponential functions play an important role in the mathematics of finance. We recall their basic properties here. For additional information, see Section 2 of Chapter 3.

Exponential functions: An increasing exponential function exhibits a constant percentage growth rate. If r is this percentage growth rate per period expressed as a decimal, the base of the exponential function is $1 + r$. We find a formula for an exponential function of the number of periods t using

$$\text{Exponential function} = \text{Initial value} \times (1 + r)^t.$$

Example: If we initially have \$1000 in an account that grows by 10% per year, the balance is exponential because it is growing at a constant percentage rate. Now 10% as a decimal is $r = 0.10$, so $1 + r = 1.10$. After t years, the balance is

$$\text{Balance after } t \text{ years} = \$1000 \times 1.10^t.$$

The graph of this function is shown in **Figure 4.2**. The increasing growth rate is typical of increasing exponential functions.

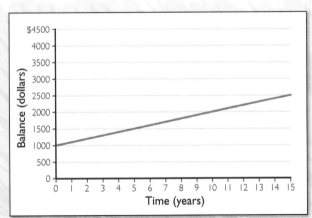

FIGURE 4.1 The graph of a linear function is a straight line.

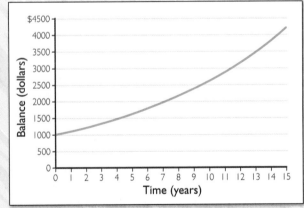

FIGURE 4.2 The graph of an exponential function (with a base greater than 1) gets steeper as we move to the right.

Investments such as savings accounts earn interest over time. Interest can be earned in different ways, and that affects the growth in value of an investment.

"This mattress fits up to $120,000 dollars."

Simple interest

The easiest type of interest to understand and calculate is *simple interest*.

Key Concept

The initial balance of an account is the **principal**. **Simple interest** is calculated by applying the interest rate to the principal only, not to interest earned.

Suppose, for example, that we invest $1000 in an account that earns simple interest at a rate of 10% per year. Then we earn 10% of $1000 or $100 in interest each year. If we hold the money for six years, we get $100 in interest each year for six years. That comes to $600 interest. If we hold it for only six months, we get a half-year's interest or $50.

Here is the formula for computing simple interest.

FORMULA 4.1 Simple Interest Formula

Simple interest earned $=$ Principal $\times$ Yearly interest rate (as a decimal) $\times$ Time in years.

EXAMPLE 4.1 Calculating simple interest: An account

We invest $2000 in an account that pays simple interest of 4% each year. Find the interest earned after five years.

SOLUTION

The interest rate of 4% written as a decimal is 0.04. The principal is $2000, and the time is five years. We find the interest by using these values in the simple interest formula (Formula 4.1):

Simple interest earned = Principal $\times$ Yearly interest rate $\times$ Time in years
$$= \$2000 \times 0.04/\text{year} \times 5 \text{ years} = \$400.$$

TRY IT YOURSELF 4.1

We invest $3000 in an account that pays simple interest of 3% each year. Find the interest earned after six years.

The answer is provided at the end of this section.

Compound interest

Situations involving simple interest are fairly rare and usually occur when money is loaned or borrowed for a short period of time. More often, interest payments are made in periodic installments during the life of the investment. The interest payments are credited to the account periodically, and future interest is earned not only on the original principal but also on the interest earned to date. This type of interest calculation is referred to as *compounding*.

Key Concept

Compound interest is paid on the principal and on the interest that the account has already earned. In short, compound interest includes *interest on the interest*.

To see how compound interest works, let's return to the $1000 investment earning 10% per year we looked at earlier, but this time let's assume the interest is compounded annually (at the end of each year). At the end of the first year, we earn 10% of $1000 or $100—the same as with simple interest. When interest is compounded, we add this amount to the balance, giving a new balance of $1100. At the end of the second year, we earn 10% interest on the $1100 balance:

Second year's interest $= 0.10 \times \$1100 = \$110$.

This amount is added to the balance, so after two years the balance is

Balance after 2 years $= \$1100 + \$110 = \$1210$.

An illustration of growth due to compound interest.

For comparison, we can use the simple interest formula to find out the simple interest earned after two years:

Simple interest after two years $=$ Principal $\times$ Yearly interest rate $\times$ Time in years

$= \$1000 \times 0.10/\text{year} \times 2 \text{ years} = \$200.$

The interest earned is $200, so the balance of the account is $1200.

After two years, the balance of the account earning simple interest is only $1200, but the balance of the account earning compound interest is $1210. Compound interest is always more than simple interest, and this observation suggests a rule of thumb for estimating the interest earned.

RULE OF THUMB 4.1 Estimating Interest

Interest earned on an account with compounding is always at least as much as that earned from simple interest. If the money is invested for a short time, simple interest can be used as a rough estimate.

The following table compares simple interest and annual compounding over various periods. It uses $1000 for the principal and 10% for the annual rate. This table shows why compounding is so important for long-term savings.

End of year	Simple interest			Yearly compounding		
	Interest	Balance	Growth	Interest	Balance	Growth
1	10% of $1000 = $100	$1100	$100	10% of $1000 = $100	$1100	$100
2	10% of $1000 = $100	$1200	$100	10% of $1100 = $110	$1210	$110
3	10% of $1000 = $100	$1300	$100	10% of $1210 = $121	$1331	$121
10	$100	$2000		$235.79	$2593.74	
50	$100	$6000		$10,671.90	$117,390.85	

To understand better the comparison between simple and compound interest, observe that for simple interest the balance is growing by the same *amount*, $100, each year. This means that the balance for simple interest is showing linear growth. For compound interest, the balance is growing by the same *percent*, 10%, each year. This means that the balance for compound interest is growing exponentially. The graphs of the account balances are shown in **Figure 4.3**. The widening gap between the two graphs shows the power of compounding.

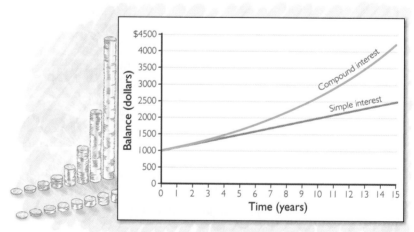

FIGURE 4.3 Balance for simple interest is linear, and balance for compound interest is exponential.

EXAMPLE 4.2 Calculating compound interest: Annual compounding

You invest $500 in an account that pays 6% compounded annually. What is the account balance after two years?

SOLUTION

Now 6% expressed as a decimal is 0.06. The first year's interest is 6% of $500:

$$\text{First year's interest} = 0.06 \times \$500 = \$30.00.$$

This interest is added to the principal to give an account balance at the end of the first year of $530.00. We use this figure to calculate the second year's interest:

$$\text{Second year's interest} = 0.06 \times \$530.00 = \$31.80.$$

We add this to the balance to find the balance at the end of two years:

$$\text{Balance after 2 years} = \$530.00 + \$31.80 = \$561.80.$$

TRY IT YOURSELF 4.2

Find the balance of this account after four years.

The answer is provided at the end of this section.

Other compounding periods and the APR

Interest may be compounded more frequently than once a year. For example, compounding may occur semi-annually, in which case the *compounding period* is half a year. Compounding may also be done quarterly, monthly, or even daily. To calculate the interest earned, we need to know the *period interest rate*.

Key Concept

The **period interest rate** is the interest rate for a given compounding period (for example, a month). Financial institutions report the **annual percentage rate** or **APR**. To calculate this, they multiply the period interest rate by the number of periods in a year.

The following formula shows how the APR is used in calculations to determine interest rates.

> **FORMULA 4.2** APR Formula
>
> $$\text{Period interest rate} = \frac{\text{APR}}{\text{Number of periods in a year}}.$$

Suppose, for example, that we invest $500 in a savings account that has an APR of 6% and compounds interest monthly. Then there are 12 compounding periods each year. We find the monthly interest rate using

$$\text{Monthly interest rate} = \frac{\text{APR}}{12} = \frac{6\%}{12} = 0.5\%.$$

"We can give you a 12% rate if you never withdraw it."

Each month we add 0.5% interest to the current balance. The following table shows how the account balance grows over the first few months:

End of month	Interest earned	New balance	Percent increase
1	0.5% of $500.00 = $2.50	$502.50	0.5%
2	0.5% of $502.50 = $2.51	$505.01	0.5%
3	0.5% of $505.01 = $2.53	$507.54	0.5%
4	0.5% of $507.54 = $2.54	$510.08	0.5%

Compound interest formula

So far we have calculated the interest by hand to see how the balance grows due to compounding. Now we simplify the process by giving a formula for the balance.

If r is the period interest rate expressed as a decimal, we find the balance after t periods using the following:

> **FORMULA 4.3   Compound Interest Formula**
>
> $$\text{Balance after } t \text{ periods} = \text{Principal} \times (1 + r)^t.$$

Here is an explanation for the formula: Over each compounding period, the balance grows by the same percentage, so the balance is an exponential function of t. That percentage growth is r as a decimal, and the initial value is the principal. Using these values in the standard exponential formula

$$\text{Exponential function} = \text{Initial value} \times (1 + r)^t$$

gives the compound interest formula.

Let's find the formula for the balance if $500 is invested in a savings account that pays an APR of 6% compounded monthly. The APR as a decimal is 0.06, so in decimal form the monthly rate is $r = 0.06/12 = 0.005$. Thus, $1 + r = 1.005$. By Formula 4.3,

$$\text{Balance after } t \text{ months} = \text{Principal} \times (1 + r)^t$$
$$= \$500 \times 1.005^t.$$

We can use this formula to find the balance of the account after five years. Five years is 60 months, so we use $t = 60$ in the formula:

$$\text{Balance after 60 months} = \$500 \times 1.005^{60} = \$674.43.$$

The APR by itself does not determine how much interest an account earns. The number of compounding periods also plays a role, as the next example illustrates.

EXAMPLE 4.3 Calculating values with varying compounding periods: Value of a CD

Suppose we invest $10,000 in a five-year certificate of deposit (CD) that pays an APR of 6%.

a. What is the value of the mature CD if interest is compounded annually? (Maturity refers to the end of the life of a CD. In this case, maturity occurs at five years.)

b. What is the value of the mature CD if interest is compounded quarterly?

c. What is the value of the mature CD if interest is compounded monthly?

d. What is the value of the mature CD if interest is compounded daily?

e. Compare your answers from parts a–d.

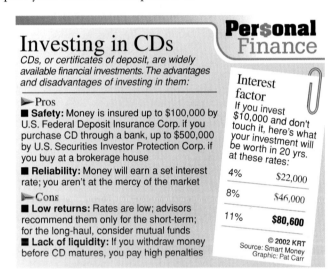

Investing in CDs

Per$onal Finance

CDs, or certificates of deposit, are widely available financial investments. The advantages and disadvantages of investing in them:

▶ Pros
■ **Safety:** Money is insured up to $100,000 by U.S. Federal Deposit Insurance Corp. if you purchase CD through a bank, up to $500,000 by U.S. Securities Investor Protection Corp. if you buy at a brokerage house
■ **Reliability:** Money will earn a set interest rate; you aren't at the mercy of the market

▶ Cons
■ **Low returns:** Rates are low; advisors recommend them only for the short-term; for the long-haul, consider mutual funds
■ **Lack of liquidity:** If you withdraw money before CD matures, you pay high penalties

Interest factor
If you invest $10,000 and don't touch it, here's what your investment will be worth in 20 yrs. at these rates:

4% $22,000

8% $46,000

11%   **$80,600**

© 2002 KRT
Source: Smart Money
Graphic: Pat Carr

SOLUTION

a. The annual compounding rate is the same as the APR. Now 6% as a decimal is $r = 0.06$. We use $1 + r = 1.06$ and $t = 5$ years in the compound interest formula (Formula 4.3):

$$\text{Balance after 5 years} = \text{Principal} \times (1 + r)^t$$
$$= \$10{,}000 \times 1.06^5$$
$$= \$13{,}382.26.$$

b. Again, we use the compound interest formula. To find the quarterly rate, we divide the APR by 4. The APR as a decimal is 0.06, so as a decimal, the quarterly rate is

$$r = \text{Quarterly rate} = \frac{\text{APR}}{4} = \frac{0.06}{4} = 0.015.$$

Thus, $1 + r = 1.015$. Also five years is 20 quarters, so we use $t = 20$ in the compound interest formula:

$$\text{Balance after 20 quarters} = \text{Principal} \times (1 + r)^t$$
$$= \$10{,}000 \times 1.015^{20}$$
$$= \$13{,}468.55.$$

c. This time we want the monthly rate, so we divide the APR by 12:

$$r = \text{Monthly rate} = \frac{\text{APR}}{12} = \frac{0.06}{12} = 0.005.$$

Also, five years is 60 months, so

$$\begin{aligned}
\text{Balance after 60 months} &= \text{Principal} \times (1 + r)^t \\
&= \$10,000 \times 1.005^{60} \\
&= \$13,488.50.
\end{aligned}$$

d. We assume that there are 365 days in each year, so as a decimal, the daily rate is

$$r = \text{Daily rate} = \frac{\text{APR}}{365} = \frac{0.06}{365}.$$

This is $r = 0.00016$ to five decimal places, but for better accuracy, we won't round this number. (See Calculation Tip 4.1 below.) Five years is $5 \times 365 = 1825$ days, so

$$\begin{aligned}
\text{Balance after 1825 days} &= \text{Principal} \times (1 + r)^t \\
&= \$10,000 \times \left(1 + \frac{0.06}{365}\right)^{1825} \\
&= \$13,498.26.
\end{aligned}$$

e. We summarize the results above in the following table:

Compounding period	Balance at maturity
Yearly	$13,382.26
Quarterly	$13,468.55
Monthly	$13,488.50
Daily	$13,498.26

This table shows that increasing the number of compounding periods increases the interest earned even though the APR and the number of years stay the same.

CALCULATION TIP 4.1 Rounding

Some financial calculations are very sensitive to rounding. In order to obtain accurate answers, when you use a calculator, it is better to keep all the decimal places rather than to enter parts of the formula that you have rounded. You can do this by either entering the complete formula or using the memory key on your calculator to store numbers with lots of decimal places.

For instance, in part d of the example above, we found the balance after 1825 days to be $13,498.26. But if we round the daily rate to 0.00016, we get $10,000 \times 1.00016^{1825} = \$13,390.72$. Rounding significantly affects the accuracy of the answer.

More information on rounding is given in Appendix 3.

> **SUMMARY 4.1** **Compound Interest**
>
> **1.** With compounding, interest is earned each period on both the principal and whatever interest has already accrued.
>
> **2.** Financial institutions report the annual percentage rate (APR).
>
> **3.** If interest is compounded n times per year, to find the period interest rate, we divide the APR by n:
>
> $$\text{Period rate} = \frac{\text{APR}}{n}.$$
>
> **4.** We can calculate the account balance after t periods using the compound interest formula (Formula 4.3):
>
> $$\text{Balance after } t \text{ periods} = \text{Principal} \times (1 + r)^t.$$
>
> Here, r is the period interest rate expressed as a decimal. Many financial formulas, including this one, are sensitive to round-off error, so it is best to do all the calculations and then round.

APR versus APY

Suppose we see one savings account offering an APR of 4.32% compounded quarterly and another one offering 4.27% compounded monthly. What we really want to know is which one pays us more money at the end of the year. This is what the *annual percentage yield* or APY tells us.

Key Concept

The **annual percentage yield** or **APY** is the actual percentage return earned in a year. Unlike the APR, the APY tells us the actual percentage growth per year, including returns on investment due to compounding.

A federal law passed in 1991 requires banks to disclose the APY.

To understand what the APY means, let's look at a simple example. Suppose we invest $100 in an account that pays 10% APR compounded semi-annually. We want

"A FIVE YEAR CD? WHO'S GOT TIME TO LISTEN TO THAT?!"

to see how much interest is earned in a year. The period interest rate is

$$\frac{\text{APR}}{2} = \frac{10\%}{2} = 5\%,$$

so we take $r = 0.05$. The number of periods in a year is $t = 2$. By the compound interest formula (Formula 4.3), the balance at the end of one year is

$$\text{Principal} \times (1 + r)^t = \$100 \times 1.05^2 = \$110.25.$$

We have earned a total of \$10.25 in interest. As a percentage of \$100, that is 10.25%. This number is the APY. It is the actual percent interest earned over the period of one year. The APY is always at least as large as the APR.

Here is a formula for the APY. In the formula, the APY and APR are in decimal form, and n is the number of compounding periods per year.

FORMULA 4.4   APY Formula

$$\text{APY} = \left(1 + \frac{\text{APR}}{n}\right)^n - 1.$$

An algebraic derivation of this formula is shown in Algebraic Spotlight 4.1 below. Let's apply this formula in the example above with 10% APR compounded semi-annually. Compounding is semi-annual so the number of periods is $n = 2$, and the APR as a decimal is 0.10:

$$\begin{aligned}
\text{APY} &= \left(1 + \frac{\text{APR}}{n}\right)^n - 1 \\
&= \left(1 + \frac{0.10}{2}\right)^2 - 1 \\
&= 0.1025.
\end{aligned}$$

As a percent, this is 10.25%—the same as we found above.

ALGEBRAIC SPOTLIGHT 4.1 Calculating APY

Suppose we invest money in an account that is compounded n times per year. Then the period interest rate is

$$\text{Period rate} = \frac{\text{APR}}{n}.$$

The first step is to find the balance after one year. Using the compound interest formula (Formula 4.3), we find the balance to be

$$\text{Balance after 1 year} = \text{Principal} \times \left(1 + \frac{\text{APR}}{n}\right)^n.$$

How much money did we earn? We earned

$$\text{Balance minus Principal} = \text{Principal} \times \left(1 + \frac{\text{APR}}{n}\right)^n - \text{Principal}.$$

If we divide these earnings by the amount we started with, namely, the principal, we get the percentage increase:

$$\text{APY as a decimal} = \left(1 + \frac{\text{APR}}{n}\right)^n - 1.$$

EXAMPLE 4.4 Calculating APY: An account with monthly compounding

We have an account that pays an APR of 10%. If interest is compounded monthly, find the APY. Round your answer as a percentage to two decimal places.

SOLUTION

We use the APY formula (Formula 4.4). As a decimal, 10% is 0.10, and there are $n = 12$ compounding periods in a year . Therefore,

$$\text{APY} = \left(1 + \frac{\text{APR}}{n}\right)^n - 1$$
$$= \left(1 + \frac{0.10}{12}\right)^{12} - 1.$$

To four decimal places, this is 0.1047. Thus, the APY is about 10.47%.

TRY IT YOURSELF 4.4

We have an account that pays an APR of 10%. If interest is compounded daily, find the APY. Round your answer as a percentage to two decimal places.

The answer is provided at the end of this section.

Using the APY

We can use the APY as an alternative to the APR for calculating compound interest. **Table 4.1** gives both the APR and the APY for CDs from First Command Bank.

TABLE 4.1 Rates from First Command Bank

CD rates		
30-Day	APR	APY
$1000 – $99,999.99	3.21%	3.25%
$100,000+	3.26%	3.30%
90-Day	APR	APY
$1000 – $9999.99	3.22%	3.25%
$10,000 – $99,999.99	3.26%	3.30%
$100,000+	3.35%	3.40%
1-Year	APR	APY
$1000 – $9999.99	3.31%	3.35%
$10,000 – $99,999.99	3.35%	3.40%
$100,000+	3.55%	3.60%
18-Month	APR	APY
$1000 – $9999.99	3.56%	3.60%
$10,000 – $99,999.99	3.59%	3.65%
$100,000+	3.74%	3.80%
2-Year	APR	APY
$1000 – $9999.99	3.80%	3.85%
$10,000 – $99,999.99	3.84%	3.90%
$100,000+	3.98%	4.05%

The APY can be used directly to calculate a year's worth of interest. For example, suppose we purchase a one-year $100,000 CD from First Command Bank. Table 4.1 gives the APY for this CD as 3.60%. So after one year the interest we will earn is

One year's interest = 3.60% of Principal = $0.036 \times \$100,000 = \$3600$.

EXAMPLE 4.5 Using APY to find value: CD

Suppose we purchase a one-year CD from First Command Bank for $25,000. What is the value of the CD at the end of the year?

SOLUTION

According to Table 4.1, the APY for this CD is 3.40%. This means that the CD earns 3.40% interest over the period of one year. Therefore,

One year's interest = 3.4% of Principal = $0.034 \times \$25,000 = \$850$.

We add this to the principal to find the balance:

Value after 1 year = $\$25,000 + \$850 = \$25,850$.

TRY IT YOURSELF 4.5

Suppose we purchase a one-year CD from First Command Bank for $125,000. What is the value of the CD at the end of the year?

The answer is provided at the end of this section.

We can also use the APY rather than the APR to calculate the balance over several years if we wish. The APY tells us the actual percentage growth per year, including interest we earned during the year due to periodic compounding.[1] Thus, we can think of compounding annually (regardless of the actual compounding period) using the APY as the annual interest rate. Once again, the balance is an exponential function. In this formula, the APY is in decimal form.

> ### FORMULA 4.5 APY Balance Formula
>
> Balance after t years = Principal $\times (1 + \text{APY})^t$.

EXAMPLE 4.6 Using APY balance formula: CD balance at maturity

Suppose we earn 3.6% APY on a 10-year $100,000 CD. Calculate the balance at maturity.

SOLUTION

The APY in decimal form is 0.036 and the CD matures after 10 years, so we use $t = 10$ in the APY balance formula (Formula 4.5):

$$\text{Balance after 10 years} = \text{Principal} \times (1 + \text{APY})^t$$
$$= \$100,000 \times 1.036^{10}$$
$$= \$142,428.71.$$

---

[1]The idea of an APY normally applies only to compound interest. It is not used for simple interest.

TRY IT YOURSELF 4.6

Suppose we earn 4.1% APY on a 20-year $50,000 CD. Calculate the balance at maturity.

The answer is provided at the end of this section.

SUMMARY 4.2   APY

1. The APY gives the true (effective) annual interest rate. It takes into account money earned due to compounding.

2. If n is the number of compounding periods per year,

$$\text{APY} = \left(1 + \frac{\text{APR}}{n}\right)^n - 1.$$

Here, the APR and APY are both in decimal form.

3. The APY is always at least as large as the APR. When interest is compounded annually, the APR and APY are equal. When compounding is more frequent, the APY is larger than the APR. The more frequent the compounding, the greater the difference.

4. The APY can be used to calculate the account balance after t years:

$$\text{Balance after } t \text{ years} = \text{Principal} \times (1 + \text{APY})^t.$$

Here, the APY is in decimal form.

Future and present value

Often we invest with a goal in mind, for example, to make a down payment on the purchase of a car. The amount we invest is called the *present value*. The amount the account is worth after a certain period of time is called the *future value* of the original investment. Sometimes we know one of these two and would like to calculate the other.

Key Concept

The **present value** of an investment is the amount we initially invest. The **future value** is the value of that investment at some specified time in the future.

If the account grows only by compounding each period at a constant interest rate after we make an initial investment, then the present value is the principal, and the future value is the balance given by the compound interest formula (Formula 4.3):

$$\text{Balance after } t \text{ periods} = \text{Principal} \times (1 + r)^t,$$

so

$$\text{Future value} = \text{Present value} \times (1 + r)^t.$$

We can rearrange this formula to obtain

$$\text{Present value} = \frac{\text{Future value}}{(1 + r)^t}.$$

In these formulas, t is the total number of compounding periods, and r is the period interest rate expressed as a decimal.

Which of these two formulas we should use depends on what question we are trying to answer. By the way, if we make regular deposits into the account, then the formulas are much more complicated. We will examine that situation in Section 4.3.

EXAMPLE 4.7 Calculating present and future value: Investing for a car

You would like to have $10,000 to buy a car three years from now. How much would you need to invest now in a savings account that pays an APR of 9% compounded monthly?

SOLUTION

In this problem, we know the future value ($10,000) and would like to know the present value. The monthly rate is $r = 0.09/12 = 0.0075$ as a decimal, and the number of compounding periods is $t = 36$ months. Thus,

$$\text{Present value} = \frac{\text{Future value}}{(1 + r)^t}$$
$$= \frac{\$10,000}{1.0075^{36}}$$
$$= \$7641.49.$$

Therefore, we should invest $7641.49 now.

TRY IT YOURSELF 4.7

Find the future value of an account after four years if the present value is $900, the APR is 8%, and interest is compounded quarterly.

The answer is provided at the end of this section.

Doubling time for investments

Exponential functions eventually get very large. This means that even a modest investment today in an account that pays compound interest will grow to be very large in the future. In fact, your money will eventually double and then double again.

In Section 3 of Chapter 3, we used logarithms to find the doubling time. Now we introduce a quick way of estimating the doubling time: The **Rule of 72** says that the doubling time in years is about 72/APR. Here, the APR is expressed as a percentage, not as a decimal. The estimate is fairly accurate if the APR is 15% or less.

SUMMARY 4.3   Doubling Time Revisited

- The exact doubling time is given by the formula

$$\text{Number of periods to double} = \frac{\log 2}{\log(\text{Base})} = \frac{\log 2}{\log(1 + r)}.$$

Here, r is the period interest rate as a decimal.

- We can approximate the doubling time using the *Rule of 72*:

$$\text{Estimate for doubling time} = \frac{72}{\text{APR}}.$$

Here, the APR is expressed as a percentage, not as a decimal, and time is measured in years. The estimate is fairly accurate if the APR is 15% or less, but we emphasize that this is only an approximation.

EXAMPLE 4.8 Computing doubling time: An account with quarterly compounding

Suppose an account earns an APR of 8% compounded quarterly. First estimate the doubling time using the Rule of 72. Then calculate the exact doubling time and compare the result with your estimate.

SOLUTION

The Rule of 72 gives the estimate

$$\text{Estimate for doubling time} = \frac{72}{\text{APR}} = \frac{72}{8} = 9 \text{ years.}$$

To find the exact doubling time, we need the period interest rate r. The period is a quarter, so $r = 0.08/4 = 0.02$. Putting this result into the doubling time formula, we find

$$\text{Number of periods to double} = \frac{\log 2}{\log(1 + r)}$$
$$= \frac{\log 2}{\log(1 + 0.02)}.$$

The result is about 35.0. Therefore, the actual doubling time is 35.0 quarters, or eight years and nine months. Our estimate of nine years was three months too high.

TRY IT YOURSELF 4.8

Suppose an account has an APR of 12% compounded monthly. First estimate the doubling time using the Rule of 72. Then calculate the exact doubling time in years and months.

The answer is provided at the end of this section.

Try It Yourself answers

Try It Yourself 4.1: Calculating simple interest: An account $540.

Try It Yourself 4.2: Calculating compound interest: Annual compounding $631.24.

Try It Yourself 4.4: Calculating APY: An account with daily compounding 10.52%.

Try It Yourself 4.5: Using APY to find value: CD $129,500.

Try It Yourself 4.6: Using APY balance formula: CD balance at maturity $111,682.36.

Try It Yourself 4.7: Calculating present and future value: An account with quarterly compounding $1235.51.

Try It Yourself 4.8: Computing doubling time: An account with monthly compounding The Rule of 72 gives an estimate of six years. The exact formula gives 69.7 months, or about 5 years and 10 months.

Exercise Set 4.1

In exercises for which you are asked to calculate the APR or APY as the final answer, round your answer as a percentage to two decimal places.

1. Simple interest. Assume a three-month CD purchased for $2000 pays simple interest at an annual rate of 10%. How much total interest does it earn? What is the balance at maturity?

2. More simple interest. Assume a 30-month CD purchased for $3000 pays simple interest at an annual rate of 5.5%. How much total interest does it earn? What is the balance at maturity?

3. Make a table. Suppose you put $3000 in a savings account at an APR of 8% compounded quarterly. Fill in the table below. (Calculate the interest and compound it by hand each quarter rather than using the compound interest formula.)

Quarter	Interest earned	Balance
		$3000.00
1	$	$
2	$	$
3	$	$
4	$	$

4. Make another table. Suppose you put $4000 in a savings account at an APR of 6% compounded monthly. Fill in the table below. (Calculate the interest and compound it by hand each month rather than using the compound interest formula.)

Month	Interest earned	Balance
		$4000.00
1	$	$
2	$	$
3	$	$

Compound interest calculated by hand Assume we invest $2000 for one year in a savings account that pays an APR of 10% compounded quarterly. Exercises 5 through 7 refer to this account.

5. Make a table to show how much is in the account at the end of each quarter. (Calculate the interest and compound it by hand each quarter rather than using the compound interest formula.)

6. Use your answer to Exercise 5 to determine how much total interest the account has earned after one year.

7. Compare the earnings to what simple interest or semi-annual compounding would yield.

8. Using the compound interest formula. *This is a continuation of Exercise 5.* In Exercise 5, we invested $2000 for one year in a savings account with an APR of 10% compounded quarterly. Apply the compound interest formula (Formula 4.3) to see whether it gives the answer for the final balance obtained in Exercise 5.

9. The difference between simple and compound interest. Suppose you invest $1000 in a savings account that pays an APR of 6%. If the account pays simple interest, what is the balance in the account after 20 years? If interest is compounded monthly, what is the balance in the account after 20 years?

Calculating interest Assume an investment of $7000 earns an APR of 6% compounded monthly for 18 months. Exercises 10 and 11 refer to this investment.

10. How much money is in your account after 18 months?

11. How much interest has been earned?

12. Getting rich. Assume an investment of $100 earns an APR of 5% compounded annually. Calculate the balance after 190 years. Would you feel cheated if you had paid to attend the seminar depicted in the accompanying cartoon?

13. **Retirement options.** At age 25, you start work for a company and are offered two retirement options.

> **Retirement option 1:** When you retire, you will receive a lump sum of $30,000 for each year of service.

> **Retirement option 2:** When you start to work, the company deposits $15,000 into an account that pays a monthly interest rate of 1%, and interest is compounded monthly. When you retire, you get the balance of the account.

Which option is better if you retire at age 65? Which is better if you retire at age 55?

14. **Compound interest.** Assume an 18-month CD purchased for $7000 pays an APR of 6% compounded monthly. What is the APY? Would the APY change if the investment were $11,000 for 30 months with the same APR and with monthly compounding?

15. **More compound interest.** Assume a 24-month CD purchased for $7000 pays an APY of 4.25% (and of course interest is compounded). How much do you have at maturity?

16. **Interest and APY.** Assume a one-year CD purchased for $2000 pays an APR of 8% that is compounded semi-annually. How much is in the account at the end of each compounding period? (Calculate the interest and compound it by hand each period rather than using the compound interest formula.) How much total interest does it earn? What's the APY?

17. **More interest and APY.** Assume a one-year CD purchased for $2000 pays an APR of 8% that is compounded quarterly. How much is in the account at the end of each compounding period? (Calculate the interest and compound it by hand each period rather than using the compound interest formula.) How much total interest does it earn? What's the APY?

A CD from First Command Bank Suppose you buy a two-year CD for $10,000 from First Command Bank. Exercises 18 through 20 refer to this CD. See Table 4.1 for the rates.

18. Use the APY from Table 4.1 to determine how much interest it earns for you at maturity.

19. Assume monthly compounding and use the APY formula in part 2 of Summary 4.2 to find the APY from the APR. Compare this to the APY in Table 4.1.

20. Assume monthly compounding. Use the APR in Table 4.1 and the compound interest formula to determine how much interest the CD earns for you at maturity.

21. **Some interest and APY calculations.** Parts b and c refer to the rates at First Command Bank shown in Table 4.1.

   a. Assume that a one-year CD for $5000 pays an APR of 8% that is compounded quarterly. How much total interest does it earn? What is the APY?

   b. If you purchase a one-year CD for $150,000 from First Command Bank, how much interest will you receive at maturity? Is compounding taking place? Explain.

   c. If you purchase a two-year CD for $150,000 from First Command Bank, the APY (4.05%) is greater than the APR (3.98%) because compounding is taking place. We are not told, however, what the compounding period is. Use the APR to calculate what the APY would be with monthly compounding. How does your answer compare to the APY in the table?

True or false In Exercises 22 through 27, answer "True" if the statement is always true and "False" otherwise. If it is false, explain why.

22. Principal is the amount you have in your account.

23. If the APY is greater than the APR, the cause must be compounding.

24. The APY can be less than the APR over a short period of time.

25. If the interest earned by a savings account is compounded semi-annually, equal amounts of money are deposited twice during the year.

26. If the interest earned by a savings account paying an APR of 8% is compounded quarterly, 2% of the current balance is deposited four times during the year.

27. If a savings account pays an APR of 8% compounded monthly, 8% of the balance is deposited each month during the year.

28. **Interest and APR.** Assume that a two-year CD for $4000 pays an APY of 8%. How much interest will it earn? Can you determine the APR?

29. Find the APR. Sue bought a six-month CD for $3000. She said that at maturity it paid $112.50 in interest. Assume this was simple interest, and determine the APR.

First Command Bank Exercises 30 and 32 refer to the rates at First Command Bank shown in Table 4.1.

30. Larry invests $99,999 and Sue invests $100,000, each for one year. How much more did Sue earn than Larry?

31. Compute the amount earned on $100,000 invested at 3.26% APR in one year if compounding is taking place daily.

32. Considering the results of Exercise 31 and the APY given in Table 4.1, how often do you think First Command Bank is compounding interest on 30-day $100,000 CDs?

33. Future value. What is the future value of a 10-year investment of $1000 at an APR of 9% compounded monthly? Explain what your answer means.

34. Present value. What is the present value of an investment that will be worth $2000 at the end of five years? Assume an APR of 6% compounded monthly. Explain what your answer means.

35. Doubling again. You have invested $2500 at an APR of 9%. Use the Rule of 72 to estimate how long it will be until your investment reaches $5000, and how long it will be until your investment reaches $10,000.

36. Getting APR from doubling rate. A friend tells you that her savings account doubled in 12 years. Use the Rule of 72 to estimate what the APR of her account was.

37. Find the doubling time. Consider an investment of $3000 at an APR of 6% compounded monthly. Use the formula that gives the exact doubling time to determine exactly how long it will take for the investment to double. (See the first part of Summary 4.3. Be sure to use the monthly rate for r.) Express

your answer in years and months. Compare this result with the estimate obtained from the Rule of 72.

The following exercises are designed to be solved using technology such as calculators or computer spreadsheets. For assistance, see the technology supplement.

38. Find the rate. According to a 2004 article by Doug Abrahms, the U.S. House of Representatives passed a bill to distribute funds to members of the Western Shoshone tribe. The article says:

> The Indian Claims Commission decided the Western Shoshone lost much of their land to gradual encroachment. The tribe was awarded $26 million in 1977. That has grown to about $145 million through compound interest, but the tribe never took the money.

Assume monthly compounding and determine the APR that would give this growth in the award over the 27 years from 1977 to 2004. *Note*: This can also be solved without technology using algebra.

39. Solve for the APR. Suppose a CD advertises an APY of 8.5%. Assuming the APY was a result of monthly compounding, solve the equation

$$0.085 = \left(1 + \frac{\text{APR}}{12}\right)^{12} - 1$$

to find the APR. *Note*: This can also be solved without technology using algebra.

40. Find the compounding period. Suppose a CD advertises an APR of 5.10% and an APY of 5.20%. Solve the equation

$$0.052 = \left(1 + \frac{0.051}{n}\right)^n - 1$$

for n to determine how frequently interest is compounded.

4.2 Borrowing: How much car can you afford?

TAKE AWAY FROM THIS SECTION

Be able to calculate the monthly payment on a loan.

The article on the next page appeared in *U.S. News and World Report*.

© Mike Baldwin / Cornered

STUDENT LOANS

"If you miss a payment, we show up and embarrass you in front of your friends."

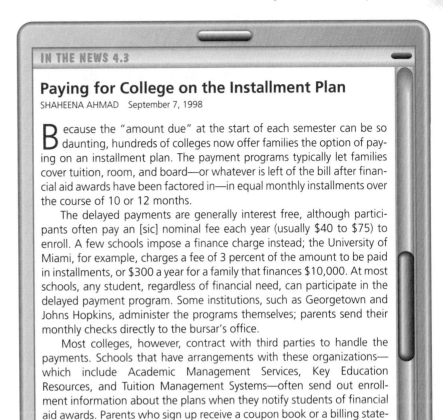

IN THE NEWS 4.3

Paying for College on the Installment Plan

SHAHEENA AHMAD   September 7, 1998

Because the "amount due" at the start of each semester can be so daunting, hundreds of colleges now offer families the option of paying on an installment plan. The payment programs typically let families cover tuition, room, and board—or whatever is left of the bill after financial aid awards have been factored in—in equal monthly installments over the course of 10 or 12 months.

The delayed payments are generally interest free, although participants often pay an [sic] nominal fee each year (usually $40 to $75) to enroll. A few schools impose a finance charge instead; the University of Miami, for example, charges a fee of 3 percent of the amount to be paid in installments, or $300 a year for a family that finances $10,000. At most schools, any student, regardless of financial need, can participate in the delayed payment program. Some institutions, such as Georgetown and Johns Hopkins, administer the programs themselves; parents send their monthly checks directly to the bursar's office.

Most colleges, however, contract with third parties to handle the payments. Schools that have arrangements with these organizations—which include Academic Management Services, Key Education Resources, and Tuition Management Systems—often send out enrollment information about the plans when they notify students of financial aid awards. Parents who sign up receive a coupon book or a billing statement each month and send their checks to the outside service—starting almost immediately after their son or daughter graduates from high school.

Installment payments are a part of virtually everyone's life. The goal of this section is to explain how such payments are calculated and to see the implications of installment plans.

Installment loans and the monthly payment

When we borrow money to buy a car or a house, the lending institution typically requires that we pay back the loan plus interest by making the same payment each month for a certain number of months.

Key Concept

With an **installment loan** you borrow money for a fixed period of time, called the **term of the loan**, and you make regular payments (usually monthly) to pay off the loan plus interest accumulated during that time.

To see how the monthly payment is calculated, we consider a simple example. Suppose you need $100 to buy a calculator but don't have the cash available. Your sister is willing, however, to lend you the money at a rate of 5% per month, provided you repay it with equal payments over the next two months. (By the way, this is an astronomical APR of 60%.)

How much must you pay each month? Because the loan is for only two months, you need to pay at least half of the $100, or $50, each of the two months. But that doesn't account for the interest. If we were to pay off the loan in a lump sum at the end of the two-month term, the interest on the account would be calculated in the same way as for a savings account that pays 5% per month. Using the compound interest formula (Formula 4.3 from the preceding section), we find

$$\text{Balance owed after 2 months} = \text{Principal} \times (1 + r)^t$$
$$= \$100 \times 1.05^2 = \$110.25$$

This would be two monthly payments of approximately $55.13. This amount over-estimates your monthly payment, however: In the second payment, you shouldn't have to pay interest on the amount you have already repaid.

We will see shortly that the correct payment is $53.78 for each of the two months. This is not an obvious amount, but it is reasonable—it lies between the two extremes of $50 and $55.13.

When we take out an installment loan, the amount of the payment depends on three things: the amount of money we borrow (sometimes called the *principal*), the interest rate (or APR), and the term of the loan.

Each payment reduces the balance owed, but at the same time interest is accruing on the outstanding balance. This makes the calculation of the monthly payment fairly complicated, as you might surmise from the following formula.[2]

FORMULA 4.6   Monthly Payment Formula

$$\text{Monthly payment} = \frac{\text{Amount borrowed} \times r(1 + r)^t}{((1 + r)^t - 1)}$$

Here, t is the term in months and $r = \text{APR}/12$ is the monthly interest rate as a decimal.

A derivation of this formula is presented at the end of this section in Algebraic Spotlight 4.2 and Algebraic Spotlight 4.3.

If you want to borrow money, the monthly payment formula allows you to determine in advance whether you can afford that car or home you want to buy. The formula also lets you check the accuracy of any figure that a potential lender quotes.

Let's return to the earlier example of a $100 loan to buy a calculator. We have a monthly rate of 5%. Expressed as a decimal, 5% is 0.05, so $r = 0.05$ and $1 + r = 1.05$. Because we pay off the loan in two months, we use $t = 2$ in the monthly payment formula (Formula 4.6):

$$\text{Monthly payment} = \frac{\text{Amount borrowed} \times r(1 + r)^t}{((1 + r)^t - 1)}$$
$$= \frac{\$100 \times 0.05 \times 1.05^2}{(1.05^2 - 1)}$$
$$= \$53.78.$$

---

[2] This formula is built in as a feature on some hand-held calculators and most computer spreadsheets. Loan calculators are also available on the Web.

EXAMPLE 4.9 Using the monthly payment formula: College loan

You need to borrow $5000 so you can attend college next fall. You get the loan at an APR of 6% to be paid off in monthly installments over three years. Calculate your monthly payment.

SOLUTION

The monthly rate as a decimal is

$$r = \text{Monthly rate} = \frac{\text{APR}}{12} = \frac{0.06}{12} = 0.005.$$

This gives $1 + r = 1.005$. We want to pay off the loan in three years or 36 months, so we use a term of $t = 36$ in the monthly payment formula:

$$\begin{aligned}\text{Monthly payment} &= \frac{\text{Amount borrowed} \times r(1+r)^t}{((1+r)^t - 1)} \\ &= \frac{\$5000 \times 0.005 \times 1.005^{36}}{(1.005^{36} - 1)} \\ &= \$152.11.\end{aligned}$$

TRY IT YOURSELF 4.9

You borrow $8000 at an APR of 9% to be paid off in monthly installments over four years. Calculate your monthly payment.

The answer is provided at the end of this section.

Suppose you can afford a certain monthly payment and you'd like to know how much you can borrow to stay within that budget. The monthly payment formula can be rearranged to answer that question.

FORMULA 4.7 Companion Monthly Payment Formula

$$\text{Amount borrowed} = \frac{\text{Monthly payment} \times ((1+r)^t - 1)}{(r \times (1+r)^t)}.$$

Before you shop for a car, know how much car you can afford.

EXAMPLE 4.10 Computing how much I can borrow: Buying a car

We can afford to make payments of $250 per month for three years. Our car dealer is offering us a loan at an APR of 5%. What price automobile should we be shopping for?

SOLUTION

The monthly rate as a decimal is

$$r = \text{Monthly rate} = \frac{0.05}{12}.$$

To four decimal places, this is 0.0042, but for better accuracy, we won't round this number. Now three years is 36 months, so we use $t = 36$ in the companion payment formula (Formula 4.7):

$$\text{Amount borrowed} = \frac{\text{Monthly payment} \times ((1+r)^t - 1)}{(r \times (1+r)^t)}$$

$$= \frac{\$250 \times ((1 + 0.05/12)^{36} - 1)}{((0.05/12) \times (1 + 0.05/12)^{36})}$$

$$= \$8341.43.$$

We should shop for cars that cost $8341.43 or less.

TRY IT YOURSELF 4.10

We can afford to make payments of $300 per month for four years. We can get a loan at an APR of 4%. How much money can we afford to borrow?

The answer is provided at the end of this section.

SUMMARY 4.4 **Monthly Payments**

In parts 1 and 2, the monthly rate r is the APR in decimal form divided by 12, and t is the term in months.

1. The monthly payment is

$$\text{Monthly payment} = \frac{\text{Amount borrowed} \times r(1+r)^t}{((1+r)^t - 1)}.$$

2. A companion formula gives the amount borrowed in terms of the monthly payment:

$$\text{Amount borrowed} = \frac{\text{Monthly payment} \times ((1+r)^t - 1)}{(r \times (1+r)^t)}.$$

3. These formulas are sensitive to round-off error, so it is best to do calculations all at once, keeping all the decimal places rather than doing parts of a computation and entering the rounded numbers.

EXAMPLE 4.11 Calculating monthly payment and amount borrowed: A new car

Suppose we need to borrow $15,000 at an APR of 9% to buy a new car.

a. What will the monthly payment be if we borrow the money for $3\frac{1}{2}$ years? How much interest will we have paid by the end of the loan?

b. We find that we cannot afford the $15,000 car because we can only afford a monthly payment of $300. What price car can we shop for if the dealer offers a loan at a 9% APR for a term of $3\frac{1}{2}$ years?

SOLUTION

a. The monthly rate as a decimal is

$$r = \text{Monthly rate} = \frac{\text{APR}}{12} = \frac{0.09}{12} = 0.0075.$$

We are paying off the loan in $3\frac{1}{2}$ years, so $t = 3.5 \times 12 = 42$ months. Therefore, by the monthly payment formula (Formula 4.6),

$$\text{Monthly payment} = \frac{\text{Amount borrowed} \times r(1+r)^t}{((1+r)^t - 1)}$$

$$= \frac{\$15{,}000 \times 0.0075 \times 1.0075^{42}}{(1.0075^{42} - 1)}$$

$$= \$417.67.$$

Now let's find the amount of interest paid. We will make 42 payments of $417.67 for a total of $42 \times \$417.67 = \$17{,}542.14$. Because the amount we borrowed is $15,000, this means that the total amount of interest paid is $\$17{,}542.14 - \$15{,}000 = \$2542.14$.

b. The monthly interest rate as a decimal is $r = 0.0075$, and there are still 42 payments. We know that the monthly payment we can afford is $300. We use the companion formula (Formula 4.7) to find the amount we can borrow on this budget:

$$\text{Amount borrowed} = \frac{\text{Monthly payment} \times ((1+r)^t - 1)}{(r \times (1+r)^t)}$$

$$= \frac{\$300 \times (1.0075^{42} - 1)}{(0.0075 \times 1.0075^{42})}$$

$$= \$10{,}774.11.$$

This means that we can afford to shop for a car that costs no more than $10,774.11.

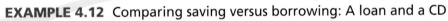

The next example shows how saving compares with borrowing.

EXAMPLE 4.12 Comparing saving versus borrowing: A loan and a CD

a. Suppose we borrow $5000 for one year at an APR of 7.5%. What will the monthly payment be? How much interest will we have paid by the end of the year?

b. Suppose we buy a one-year $5000 CD at an APR of 7.5% compounded monthly. How much interest will be paid at the end of the year?

c. In part a the financial institution loaned us $5000 for one year, but in part b we loaned the financial institution $5000 for one year. What is the difference in the amount of interest paid? Explain why the amounts are different.

SOLUTION

a. In this case, the principal is $5000, the monthly interest rate r as a decimal is $0.075/12 = 0.00625$, and the number t of payments is 12. We use the monthly payment formula (Formula 4.6):

$$\text{Monthly payment} = \frac{\text{Amount borrowed} \times r(1+r)^t}{((1+r)^t - 1)}$$

$$= \frac{\$5000 \times 0.00625 \times 1.00625^{12}}{(1.00625^{12} - 1)}$$

$$= \$433.79.$$

We will make 12 payments of $433.79 for a total of $12 \times \$433.79 = \$5205.48$. Because the amount we borrowed is $5000, the total amount of interest paid is $205.48.

b. To calculate the interest earned on the CD, we use the compound interest formula (Formula 4.3 from the preceding section). The monthly rate is the same as in part a. Therefore,

$$\text{Balance} = \text{Principal} \times (1 + r)^t$$
$$= \$5000 \times 1.00625^{12}$$
$$= \$5388.16.$$

This means that the total amount of interest we earned is \$388.16.

c. The interest earned on the \$5000 CD is \$182.68 more than the interest paid on the \$5000 loan in part a.

Here is the explanation for this difference: When we save money, the financial institution credits our account with an interest payment—in this case, every month. We continue to earn interest on the full \$5000 and on those interest payments for the entire year. When we borrow money, however, we repay the loan monthly, thus decreasing the balance owed. We are being charged interest only on the balance owed, not on the full \$5000 we borrowed.

One common piece of financial advice is to reduce the amount you borrow by making a *down payment*. Exercises 15 through 17 explore the effect of making a down payment on the monthly payment and the total interest paid.

Estimating payments for short-term loans

The formula for the monthly payment is complicated, and it's easy to make a mistake in the calculation. Is there a simple way to give an estimate for the monthly payment? There are, in fact, a couple of ways to do this. We give one of them here and another when we look at home mortgages.

One obvious estimate for a monthly payment is to divide the loan amount by the term (in months) of the loan. *This would be our monthly payment if no interest were charged.* For a loan with a relatively short term and not too large an interest rate, this gives a rough lower estimate for the monthly payment.[3]

> **RULE OF THUMB 4.2** Monthly Payments for Short-Term Loans
>
> For all loans, the monthly payment is *at least* the amount we would pay each month if no interest were charged, which is the amount of the loan divided by the term (in months) of the loan. This would be the payment if the APR were 0%. It's a rough estimate of the monthly payment for a short-term loan if the APR is not large.

EXAMPLE 4.13 Estimating monthly payment: Can we afford it?

The largest monthly payment we can afford is \$800. Can we afford to borrow a principal of \$20,000 with a term of 24 months?

SOLUTION

The rule of thumb says that the monthly payment is at least \$20,000/24 = \$833.33. This is more than \$800, so we can't afford this loan.

---

[3] If the term is at most five years and the APR is less than 7.5%, the actual payment is within 20% of the ratio. If the term is at most two years and the APR is less than 9%, the actual payment is within 10% of the ratio.

TRY IT YOURSELF 4.13

The largest monthly payment we can afford is $450. Can we afford to borrow a principal of $18,000 with a term of 36 months?

The answer is provided at the end of this section.

For the loan in the preceding example, our rule of thumb says that the monthly payment is *at least* $833.33. Remember that this amount does not include the interest payments. If the loan has an APR of 6.6%, for example, the actual payment is $891.83.

Once again, a rule of thumb gives an estimate—not the exact answer. It can at least tell us quickly whether we should be shopping on the BMW car lot.

Amortization tables and equity

When you make payments on an installment loan, part of each payment goes toward interest, and part goes toward reducing the balance owed. An *amortization table* is a running tally of payments made and the outstanding balance owed.

Key Concept

An **amortization table** or **amortization schedule** for an installment loan shows for each payment made the amount applied to interest, the amount applied to the balance owed, and the outstanding balance.

EXAMPLE 4.14 Making an amortization table: Buying a computer

Suppose we borrow $1000 at 12% APR to buy a computer. We pay off the loan in 12 monthly payments. Make an amortization table showing payments over the first six months.

SOLUTION

The monthly rate is 12%/12 = 1%. As a decimal, this is $r = 0.01$. The monthly payment formula with $t = 12$ gives

$$\text{Monthly payment} = \frac{\text{Amount borrowed} \times r(1+r)^t}{((1+r)^t - 1)}$$
$$= \frac{\$1000 \times 0.01 \times 1.01^{12}}{(1.01^{12} - 1)}$$
$$= \$88.85.$$

Because the monthly rate is 1%, each month we pay 1% of the outstanding balance in interest. When we make our first payment, the outstanding balance is $1000, so we pay 1% of $1000, or $10.00, in interest. Thus, $10 of our $88.85 goes toward interest, and the remainder, $78.85, goes toward the outstanding balance. So after the first payment, we owe:

Balance owed after 1 payment = $1000.00 − $78.85 = $921.15.

When we make a second payment, the outstanding balance is $921.15. We pay 1% of $921.15 or $9.21 in interest, so $88.85 − $9.21 = $79.64 goes toward the balance due. This gives the balance owed after the second payment:

Balance owed after 2 payments = $921.15 − $79.64 = $841.51.

If we continue in this way, we get the following table:

Payment number	Payment	Applied to interest	Applied to balance owed	Outstanding balance
				$1000.00
1	$88.85	1% of $1000.00 = $10.00	$78.85	$921.15
2	$88.85	1% of $921.15 = $9.21	$79.64	$841.51
3	$88.85	1% of $841.51 = $8.42	$80.43	$761.08
4	$88.85	1% of $761.08 = $7.61	$81.24	$679.84
5	$88.85	1% of $679.84 = $6.80	$82.05	$597.79
6	$88.85	1% of $597.79 = $5.98	$82.87	$514.92

TRY IT YOURSELF 4.14

Suppose we borrow $1000 at 30% APR and pay it off in 24 monthly payments. Make an amortization table showing payments over the first three months.

The answer is provided at the end of this section.

The loan in Example 4.14 is used to buy a computer. The amount you have paid toward the actual cost of the computer (the principal) at a given time is referred to as your *equity* in the computer. For example, the table above tells us that after four payments we still owe $679.84. This means we have paid a total of $1000.00 − $679.84 = $320.16 toward the principal. That is our equity in the computer.

Key Concept

If you borrow money to pay for an item, your **equity** in that item at a given time is the part of the principal you have paid.

EXAMPLE 4.15 Calculating equity: Buying land

You borrow $150,000 at an APR of 6% to purchase a plot of land. You pay off the loan in monthly payments over 10 years.

a. Find the monthly payment.

b. Complete the four-month amortization table below.

Payment number	Payment	Applied to interest	Applied to balance owed	Outstanding balance
				$150,000.00
1				
2				
3				
4				

c. What is your equity in the land after four payments?

SOLUTION

a. The monthly rate is APR/12 = 6%/12 = 0.5%. As a decimal, this is $r = 0.005$. We use the monthly payment formula with $t = 10 \times 12 = 120$ months:

$$\text{Monthly payment} = \frac{\text{Amount borrowed} \times r(1+r)^t}{((1+r)^t - 1)}$$

$$= \frac{\$150,000 \times 0.005 \times 1.005^{120}}{(1.005^{120} - 1)}$$

$$= \$1665.31.$$

b. For the first month, the interest we pay is 0.5% of the outstanding balance of $150,000:

$$\text{First month interest} = \$150,000 \times 0.005 = \$750.$$

Now $750 of the $1665.31 payment goes to interest and the remainder, $1665.31 − $750 = $915.31, goes toward reducing the principal. The balance owed after one month is

$$\text{Balance owed after 1 month} = \$150,000 - \$915.31 = \$149,084.69.$$

This gives the first row of the table. The completed table is shown below.

Payment number	Payment	Applied to interest	Applied to balance owed	Outstanding balance
				$150,000.00
1	$1665.31	0.5% of $150,000.00 = $750.00	$915.31	$149,084.69
2	$1665.31	0.5% of $149,084.69 = $745.42	$919.89	$148,164.80
3	$1665.31	0.5% of $148,164.80 = $740.82	$924.49	$147,240.31
4	$1665.31	0.5% of $147,240.31 = $736.20	$929.11	$146,311.20

c. The table from part b tells us that after four payments we still owe $146,311.20. So our equity is

$$\text{Equity after 4 months} = \$150,000 - \$146,311.20 = \$3688.80.$$

The graph in **Figure 4.4** shows the percentage of each payment from Example 4.15 that goes toward interest, and **Figure 4.5** shows how equity is built. Note that in the early months, a large percentage of the payment goes toward interest. For long-term loans, an even larger percentage of the payment goes toward interest early on. This means that equity is built slowly at first. The rate of growth of equity increases over the life of the loan. Note in Figure 4.5 that when you have made half of the payments (60 payments), you have built an equity of just over $60,000—much less than half of the purchase price.

Home mortgages

A home mortgage is a loan for the purchase of a home. It is very common for a mortgage to last as long as 30 years. The early mortgage payments go almost entirely

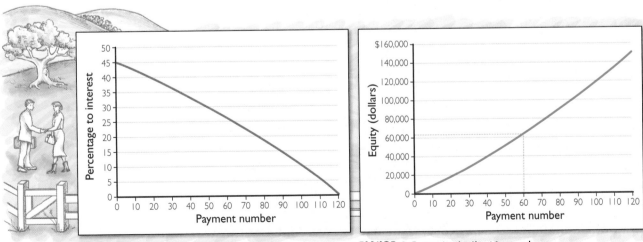

FIGURE 4.4 Percentage of payment that goes to interest: 10-year loan.

FIGURE 4.5 Equity built: 10-year loan.

toward interest, with a small part going to reduce the principal. As a consequence, your home equity grows very slowly. For a 30-year mortgage of $150,000 at an APR of 6%, **Figure 4.6** shows the percentage of each payment that goes to interest, and **Figure 4.7** shows the equity built. Your home equity is very important because it tells you how much money you can actually keep if you sell your house, and it can also be used as collateral to borrow money.

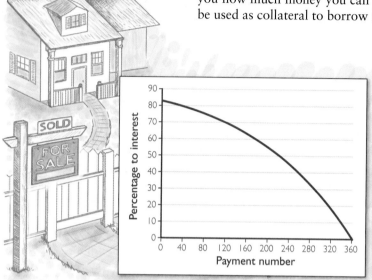

FIGURE 4.6 Percentage of payment that goes to interest: 30-year mortgage.

FIGURE 4.7 Equity built: 30-year mortgage.

EXAMPLE 4.16 Computing interest: 30-year mortgage

Your neighbor took out a 30-year mortgage for $300,000 at an APR of 9%. She says that she will wind up paying more in interest than for the home (that is, the principal). Is that true?

SOLUTION

We first need to find the monthly payment. The monthly rate as a decimal is

$$r = \text{Monthly rate} = \frac{\text{APR}}{12} = \frac{0.09}{12} = 0.0075.$$

" So what happens if we can't meet our monthly mortgage repayments?"

Because the loan is for 30 years, we use $t = 30 \times 12 = 360$ months in the monthly payment formula:

$$\text{Monthly payment} = \frac{\text{Amount borrowed} \times r(1+r)^t}{((1+r)^t - 1)}$$

$$= \frac{\$300{,}000 \times 0.0075 \times 1.0075^{360}}{(1.0075^{360} - 1)}$$

$$= \$2413.87.$$

She will make 360 payments of $2413.87, for a total of

$$\text{Total amount paid} = 360 \times \$2413.87 = \$868{,}993.20.$$

The interest paid is the excess over $300,000, or $568,993.20. Your neighbor paid almost twice as much in interest as she did for the home.

TRY IT YOURSELF 4.16

Find the interest paid on a 25-year mortgage of $450,000 at an APR of 7.2%.

The answer is provided at the end of this section.

Now we see the effect of varying the term of the mortgage on the monthly payment.

EXAMPLE 4.17 Determining monthly payment and term: Choices for mortgages

You need to secure a loan of $250,000 to purchase a home. Your lending institution offers you three options:

Option 1: A 30-year mortgage at 8.4% APR.

Option 2: A 20-year mortgage at 7.2% APR.

Option 3: A 30-year mortgage at 7.2% APR, including a fee of 4 *loan points*.
Note: "Points" are a fee you pay for the loan in return for a decrease in the interest rate. In this case, a fee of 4 points means you pay 4% of the loan, or $10,000. One way to do this is to borrow the fee from the bank by just adding the $10,000 to the amount you borrow. The bank keeps the $10,000 and the other $250,000 goes to buy the home.
Determine the monthly payment for each of these options.

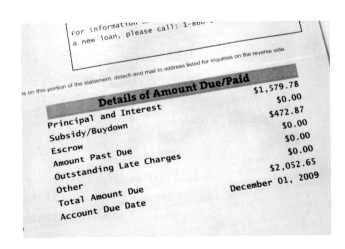

SOLUTION

Option 1: The monthly rate as a decimal is

$$r = \text{Monthly rate} = \frac{\text{APR}}{12} = \frac{0.084}{12} = 0.007.$$

We use the monthly payment formula with $t = 360$ months:

$$\begin{aligned} \text{Monthly payment} &= \frac{\text{Amount borrowed} \times r(1+r)^t}{((1+r)^t - 1)} \\ &= \frac{\$250{,}000 \times 0.007 \times 1.007^{360}}{(1.007^{360} - 1)} \\ &= \$1904.59. \end{aligned}$$

Option 2: The APR for a 20-year loan is lower. (It is common for loans with shorter terms to have lower interest rates.) An APR of 7.2% is a monthly rate of $r = 0.072/12 = 0.006$ as a decimal. We use the monthly payment formula with $t = 20 \times 12 = 240$ months:

$$\begin{aligned} \text{Monthly payment} &= \frac{\text{Amount borrowed} \times r(1+r)^t}{((1+r)^t - 1)} \\ &= \frac{\$250{,}000 \times 0.006 \times 1.006^{240}}{(1.006^{240} - 1)} \\ &= \$1968.37. \end{aligned}$$

The monthly payment is about $60 higher than for the 30-year mortgage, but you pay the loan off in 20 years rather than 30 years.

Option 3: Adding 4% to the amount we borrow gives a new loan amount of $260,000. As with Option 2, the APR of 7.2% gives a monthly rate of $r = 0.006$. The term of 30 years means that we put $t = 360$ months into the monthly payment formula:

$$\begin{aligned} \text{Monthly payment} &= \frac{\text{Amount borrowed} \times r(1+r)^t}{((1+r)^t - 1)} \\ &= \frac{\$260{,}000 \times 0.006 \times 1.006^{360}}{(1.006^{360} - 1)} \\ &= \$1764.85. \end{aligned}$$

Getting the lower interest rate makes a big difference in the monthly payment, even with the 4% added to the original loan balance. This is clearly a better choice than Option 1 if we consider only the monthly payment. Option 3 does require borrowing more money than Option 1 and could have negative consequences if you need to sell the home early. Further, comparing the amount of interest paid shows that Option 2 is the best choice from that point of view—if you can afford the monthly payment.

Adjustable-rate mortgages

In the summer of 2007, a credit crisis involving home mortgages had dramatic effects on many people and ultimately on the global economy. A significant factor in the crisis was the widespread use of *adjustable-rate mortgages* or ARMs.

Key Concept

A **fixed-rate mortgage** keeps the same interest rate over the life of the loan. In the case of an **adjustable-rate mortgage** or **ARM**, the interest rate may vary over the life of the loan. The rate is often tied to the **prime interest rate**, which is the rate banks must pay to borrow money.

One advantage of an ARM is that the initial rate is often lower than the rate for a comparable fixed-rate mortgage. One disadvantage of an ARM is that a rising prime interest rate may cause significant increases in the monthly payment.

EXAMPLE 4.18 Comparing monthly payments: Fixed-rate and adjustable-rate mortgages

We want to borrow $200,000 for a 30-year home mortgage. We have found an APR of 6.6% for a fixed-rate mortgage and an APR of 6% for an ARM. Compare the initial monthly payments for these loans.

SOLUTION

Both loans have a principal of $200,000 and a term of $t = 360$ months. For the fixed-rate mortgage, the monthly rate in decimal form is $r = 0.066/12 = 0.0055$. The monthly payment formula gives

$$\text{Monthly payment} = \frac{\text{Amount borrowed} \times r(1+r)^t}{((1+r)^t - 1)}$$
$$= \frac{\$200,000 \times 0.0055 \times 1.0055^{360}}{(1.0055^{360} - 1)}$$
$$= \$1277.32.$$

For the ARM, the initial monthly rate in decimal form is $r = 0.06/12 = 0.005$. The monthly payment formula gives

$$\text{Monthly payment} = \frac{\text{Amount borrowed} \times r(1+r)^t}{((1+r)^t - 1)}$$
$$= \frac{\$200,000 \times 0.005 \times 1.005^{360}}{(1.005^{360} - 1)}$$
$$= \$1199.10.$$

The initial monthly payment for the ARM is almost $80 less than the payment for the fixed-rate mortgage—but the payment for the ARM could change at any time.

TRY IT YOURSELF 4.18

We want to borrow $150,000 for a 30-year home mortgage. We have found an APR of 6% for a fixed-rate mortgage and an APR of 5.7% for an ARM. Compare the monthly payments for these loans.

The answer is provided at the end of this section.

Example 4.18 illustrates why an ARM may seem attractive. The following example illustrates one potential danger of an ARM. It is typical of what happened in 2007 to many families who took out loans in 2006 when interest rates fell to historic lows.

EXAMPLE 4.19 Using an ARM: Effect of increasing rates

Suppose a family has an annual income of $60,000. Assume that this family secures a 30-year ARM for $250,000 at an initial APR of 4.5%.

    a. Find the family's monthly payment and the percentage of its monthly income used for the mortgage payment.

b. Now suppose that after one year, the rate adjusts to 6%. Find the family's new monthly payment and the percentage of its monthly income used now for the mortgage payment.

SOLUTION

a. The monthly rate as a decimal is $0.045/12 = 0.00375$, so with $t = 360$ the monthly payment on the family's home is

$$\text{Monthly payment} = \frac{\text{Amount borrowed} \times r(1 + r)^t}{((1 + r)^t - 1)}$$

$$= \frac{\$250{,}000 \times 0.00375 \times 1.00375^{360}}{(1.00375^{360} - 1)}$$

$$= \$1266.71.$$

An annual income of \$60,000 is a monthly income of \$5000. To find the percentage of the family's monthly income used for the mortgage payment, we calculate $1266.71/5000$. The result is about 0.25, so the family is paying 25% of its monthly income for the mortgage payment.

b. The APR has increased to 6%, so $r = 0.06/12 = 0.005$. To find the new monthly payment, we use a loan period of 29 years or $29 \times 12 = 348$ months. For the principal, we use the balance owed the bank after one year of payments. We noted earlier in this section that home owners build negligible equity in the first year of payments, so we will get a very good estimate of the new monthly payment if we use \$250,000 as the balance owed the bank. The formula gives

$$\frac{\$250{,}000 \times 0.005 \times 1.005^{348}}{(1.005^{348} - 1)} = \$1517.51.$$

The family is now paying about $\$1517.51 - \$1266.71$ or about \$250 more per month, and the fraction of monthly income used for the mortgage payment is about $1517.51/5000$ or 0.30. The family is now paying 30% of its monthly income for the mortgage payment.

In the example, the extra burden on family finances caused by the increase in the interest rate could lead to serious problems. Further rate adjustments could easily result in the loss of the family home.

Estimating payments on long-term loans

The payment on a long-term loan is *at least* as large as the monthly interest on the original balance. This is a pretty good estimate for mortgages in times of moderate-to-high interest rates.[4]

RULE OF THUMB 4.3 Monthly Payments for Long-Term Loans

For all loans, the monthly payment is *at least* as large as the principal times the monthly interest rate as a decimal. This is a fairly good estimate for a long-term loan with a moderate or high interest rate.

---

[4]If the term is at least 30 years and the APR is at least 6%, then the actual payment is within 20% of the estimate. If the term is at least 30 years and the APR is greater than 8%, then the actual payment is within 10% of the estimate.

Let's see what this rule of thumb would estimate for the payment on a mortgage of $100,000 at an APR of 7.2%. The monthly rate is 7.2%/12 = 0.6%. The monthly interest on the original balance is 0.6% of $100,000:

$$\text{Monthly payment estimate} = \$100,000 \times 0.006 = \$600.$$

If this is a 30-year mortgage, the monthly payment formula gives the value $678.79 for the actual payment. This is about 13% higher than the estimate.

We should note that most home mortgage payments also include taxes and insurance. These vary from location to location but can be significant.[5] For many, a home mortgage is the most significant investment they will ever make. It is crucial to understand clearly both the benefits and the costs of such an investment.

ALGEBRAIC SPOTLIGHT 4.2 Derivation of the Monthly Payment Formula, Part I

First we derive a formula for the balance still owed on an installment loan after a given number of payments. Suppose we borrow B_0 dollars and make monthly payments of M dollars. Suppose further that interest accrues on the balance at a monthly rate of r as a decimal. Let B_n denote the account balance (in dollars) after n months. Our goal is to derive the formula

$$B_n = B_0(1+r)^n - M\frac{(1+r)^n - 1}{r}.$$

Each month we find the new balance B_{n+1} from the old balance B_n by first adding the interest accrued rB_n and then subtracting the payment M. As a formula, this is

$$B_{n+1} = B_n + rB_n - M = B_n(1+r) - M.$$

If we put $R = 1 + r$, this formula can be written more compactly as

$$B_{n+1} = B_n R - M.$$

Repeated application of this formula gives

$$B_n = B_0 R^n - M(1 + R + R^2 + \cdots + R^{n-1}).$$

To finish, we use the *geometric sum formula*, which tells us that

$$1 + R + R^2 + \cdots + R^{n-1} = \frac{R^n - 1}{R - 1}.$$

This gives

$$B_n = B_0 R^n - M\frac{R^n - 1}{R - 1}.$$

Finally, we recall that $R = 1 + r$ and obtain the desired formula

$$B_n = B_0(1+r)^n - M\frac{(1+r)^n - 1}{r}.$$

---

[5] See the group of exercises on affordability beginning with Exercise 39.

ALGEBRAIC SPOTLIGHT 4.3 Derivation of the Monthly Payment Formula, Part II

Now we use the account balance formula to derive the monthly payment formula. Suppose we borrow B_0 dollars at a monthly interest rate of r as a decimal and we want to pay off the loan in t monthly payments. That is, we want the balance to be 0 after t payments. Using the account balance formula we derived in Algebraic Spotlight 4.2, we want to find the monthly payment M that makes $B_t = 0$. That is, we need to solve the equation

$$0 = B_t = B_0(1+r)^t - M\frac{(1+r)^t - 1}{r}$$

for M. Now

$$0 = B_0(1+r)^t - M\frac{(1+r)^t - 1}{r}$$

$$M\frac{(1+r)^t - 1}{r} = B_0(1+r)^t$$

$$M = \frac{B_0 r(1+r)^t}{((1+r)^t - 1)}.$$

Because B_0 is the amount borrowed and t is the term, this is the monthly payment formula we stated earlier.

Try It Yourself answers

Try It Yourself 4.9: Using the monthly payment formula: College loan $199.08.

Try It Yourself 4.10: Computing how much I can borrow: Buying a car $13,286.65.

Try It Yourself 4.13: Estimating monthly payment: Can we afford it?   No: The monthly payment is at least $500.

Try It Yourself 4.14: Making an amortization table: Buying a computer

Payment number	Payment	Applied to interest	Applied to balance owed	Outstanding balance
				$1000.00
1	$55.91	2.5% of $1000.00 = $25.00	$30.91	$969.09
2	$55.91	2.5% of $969.09 = $24.23	$31.68	$937.41
3	$55.91	2.5% of $937.41 = $23.44	$32.47	$904.94

Try It Yourself 4.16: Computing how much interest: 25-year mortgage $521,445.00.

Try It Yourself 4.18: Comparing monthly payments: Fixed-rate and adjustable-rate mortgages   The monthly payment of $870.60 for the ARM is almost $30 less than the payment of $899.33 for the fixed-rate mortgage.

Exercise Set 4.2

Rounding in the calculation of monthly interest rates is discouraged. Such rounding can lead to answers different from those presented here. For long-term loans, the differences may be pronounced.

1. Car payment. To buy a car, you borrow $20,000 with a term of five years at an APR of 6%. What is your monthly payment? How much total interest is paid?

2. Truck payment. You borrow $18,000 with a term of four years at an APR of 5% to buy a truck. What is your monthly payment? How much total interest is paid?

3. Estimating the payment. You borrow $25,000 with a term of two years at an APR of 5%. Use Rule of Thumb 4.2 to estimate your monthly payment, and compare this estimate with what the monthly payment formula gives.

4. Estimating a mortgage. You have a home mortgage of $110,000 with a term of 30 years at an APR of 9%. Use Rule of Thumb 4.3 to estimate your monthly payment, and compare this estimate with what the monthly payment formula gives.

5. Affording a car. You can get a car loan with a term of three years at an APR of 5%. If you can afford a monthly payment of $450, how much can you borrow?

6. Affording a home. You find that the going rate for a home mortgage with a term of 30 years is 6.5% APR. The lending agency says that based on your income, your monthly payment can be at most $750. How much can you borrow?

7. No interest. A car dealer offers you a loan with no interest charged for a term of two years. If you need to borrow $18,000, what will your monthly payment be? Which rule of thumb is relevant here?

Interest paid For Exercises 8 through 11, assume you take out a $2000 loan for 30 months at 8.5% APR.

8. What is the monthly payment?

9. How much of the first month's payment is interest?

10. What percentage of the first month's payment is interest? (Round your answer to two decimal places as a percentage.)

11. How much total interest did you pay at the end of the 30 months?

More saving and borrowing In Exercises 12 through 14, we compare saving and borrowing as in Example 4.12.

12. Suppose you borrow $10,000 for two years at an APR of 8.75%. What will your monthly payment be? How much interest will you have paid by the end of the loan?

13. Suppose you buy a two-year $10,000 CD at an APR of 8.75% compounded monthly. How much interest will you be paid by the end of the period?

14. In Exercise 12, the financial institution loaned you $10,000 for two years, but in Exercise 13, you loaned the financial institution $10,000 for two years. What is the difference in the amount of interest paid?

Down payment In Exercises 15 through 17, we examine the benefits of making a down payment.

15. You want to buy a car. Suppose you borrow $15,000 for two years at an APR of 6%. What will your monthly payment be? How much interest will you have paid by the end of the loan?

16. Suppose that in the situation of Exercise 15, you make a down payment of $2000. This means that you borrow only $13,000. Assume that the term is still two years at an APR of 6%. What will your monthly payment be? How much interest will you have paid by the end of the loan?

17. Compare your answers to Exercises 15 and 16. What are the advantages of making a down payment? Why would a borrower not make a down payment? *Note:* One other factor to consider is the interest one would have earned on the down payment if it had been invested. Typically, the interest rate earned on investments is lower than that charged for loans.

18. Amortization table. Suppose we borrow $1500 at 4% APR and pay it off in 24 monthly payments. Make an amortization table showing payments over the first three months.

Payment number	Payment	Applied to interest	Applied to balance owed	Outstanding balance
				$1500.00
1				
2				
3				

19. Another amortization table. Suppose we borrow $100 at 5% APR and pay it off in 12 monthly payments. Make an amortization table showing payments over the first three months.

Payment number	Payment	Applied to interest	Applied to balance owed	Outstanding balance
				$100.00
1				
2				
3				

Term of mortgage There are two common choices for the term of a home mortgage: 15 years or 30 years. Suppose you need to borrow $90,000 at an APR of 6.75% to buy a home. Exercises 20 through 25 explore mortgage options for the term.

20. What will your monthly payment be if you opt for a 15-year mortgage?

21. What percentage of your first month's payment will be interest if you opt for a 15-year mortgage? (Round your answer to two decimal places as a percentage.)

22. How much interest will you have paid by the end of the 15-year loan?

23. What will your monthly payment be if you opt for a 30-year mortgage?

24. What percentage of your first month's payment will be interest if you opt for a 30-year mortgage? (Round your answer to two decimal places as a percentage.)

25. How much interest will you have paid by the end of the 30-year loan? Is it twice as much as for a 15-year mortgage?

Formula for equity Here is a formula for the equity built up after k monthly payments:

$$\text{Equity} = \frac{\text{Amount borrowed} \times ((1+r)^k - 1)}{((1+r)^t - 1)},$$

where r is the monthly interest rate as a decimal and t is the term in months. Exercises 26 and 27 use this formula for a mortgage of $100,000 at an APR of 7.2% with two different terms.

26. Assume that the term of the mortgage is 30 years. How much equity will you have halfway through the term of the loan? What percentage of the principal is this? (Round your answer to one decimal place as a percentage.)

27. Suppose now that instead of a 30-year mortgage, you have a 15-year mortgage. Find your equity halfway through the term of the loan, and find what percentage of the principal that is. (Round your answer to one decimal place as a percentage.) Compare this with the percentage you found in Exercise 26. Why should the percent for the 15-year term be larger than the percent for the 30-year term?

Car totaled In order to buy a new car, you finance $20,000 with no down payment for a term of five years at an APR of 6%. After you have the car for one year, you are in an accident. No one is injured, but the car is totaled. The insurance company says that before the accident, the value of the car had decreased by 25% over the time you owned it, and the company pays you that depreciated amount after subtracting your $500 deductible. Exercises 28 through 31 refer to this situation.

28. What is your monthly payment for this loan?

29. How much equity have you built up after one year? *Suggestion*: Use the formula for equity stated in connection with Exercises 26 and 27.

30. How much money does the insurance company pay you? (Don't forget to subtract the deductible.)

31. Can you pay off the loan using the insurance payment, or do you still need to make payments on a car you no longer have? If you still need to make payments, how much do you still owe? (Subtract the payment from the insurance company.)

Rebates   When interest rates are low, some automobile dealers offer loans at 0% APR, as the following excerpt from an article at AutoLoanDaily.com shows.

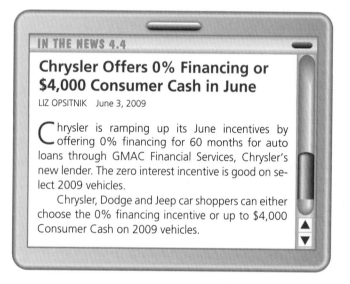

IN THE NEWS 4.4

Chrysler Offers 0% Financing or $4,000 Consumer Cash in June

LIZ OPSITNIK   June 3, 2009

Chrysler is ramping up its June incentives by offering 0% financing for 60 months for auto loans through GMAC Financial Services, Chrysler's new lender. The zero interest incentive is good on select 2009 vehicles.

Chrysler, Dodge and Jeep car shoppers can either choose the 0% financing incentive or up to $4,000 Consumer Cash on 2009 vehicles.

"Zero percent financing" means the obvious thing—that no interest is being charged on the loan. So if we borrow $1200 at 0% interest and pay it off over 12 months, our monthly payment will be $1200/12 = $100.

Suppose you are buying a new truck at a price of $20,000. You plan to finance your purchase with a loan you will repay over two years. The dealer offers two options: either dealer financing with 0% interest, or a $2000 rebate on the purchase price. If you take the rebate, you will have to go to the local bank for a loan (of $18,000) at an APR of 6.5%. Exercises 32 through 34 refer to these options.

32. What would your monthly payment be if you used dealer financing?

33. What would your monthly payment be if you took the rebate?

34. Should you take the dealer financing or the rebate? How much would you save over the life of the loan by taking the option you chose?

35. Too good to be true? A friend claims to have found a really great deal at a local loan agency not listed in the phone book: The agency claims that its rates are so low that you can borrow $10,000 with a term of three years for a monthly payment of $200. Is this too good to be true? Be sure to explain your answer.

36. Is this reasonable? A lending agency advertises in the paper an APR of 12% on a home mortgage with a term of 30 years. The ad claims that the monthly payment on a principal of $100,000 will be $10,290. Is this claim reasonable? What should the ad have said the payment would be (to the nearest dollar)? What do you think happened here?

37. Question for thought. Why would it be unwise (even if it were allowed) to charge the purchase of a home to a credit card? Go beyond the credit limit to consider factors such as interest rates, budgeting, etc.

38. Microloans and flat interest. The 2006 Nobel Peace Prize was awarded to Muhammad Yunus and the Grameen Bank he founded. The announcement of the award noted the development of *microloans* for encouraging the poor to become entrepreneurs. Microloans involve small amounts of money but fairly high interest rates. In contrast to the installment method discussed in this section, for microloans typically interest is paid at a flat rate: The amount of interest paid is not reduced as the principal is paid off. Investigate flat rate loans, the reasons why microloans are often structured this way, and any downsides to such an arrangement.

Exercises 39 through 44 are suitable for group work.

Affordability   Over the past 30 years, interest rates have varied widely. The rate for a 30-year mortgage reached a high of 14.75% in July 1984, and it reached a low of 4.64% in October 2010. A significant impact of lower interest rates on society is that they enable more people to afford the purchase of a home. In the following exercises, we consider the purchase of a home that sells for $125,000. Assume that we can make a down payment of $25,000, so we need to borrow $100,000. We assume that our annual income is $40,000 and that we have no other debt. In Exercises 39 through 44, we determine whether we can afford to buy the home at the high and low rates mentioned above.

39. What is our monthly income?

40. Lending agencies usually require that no more than 28% of the borrower's monthly income be spent on housing. How much does that represent in our case?

41. The amount we will spend on housing consists of our monthly mortgage payment plus property taxes and hazard insurance. Assume that property taxes plus insurance total $250 per month, and subtract this from the answer to Exercise 40 to determine what monthly payment we can afford.

42. Use your answer to Exercise 41 to determine how much we can borrow if the term is 30 years and the interest rate is the historic high of 14.75%. Can we afford the home?

43. Use your answer to Exercise 41 to determine how much we can borrow if the term is 30 years and the interest rate is the historic low of 4.64%. Can we afford the home now?

44. What is the difference in the amount we can borrow between the lowest and highest rates?

45. Some history. In the United States before the 1930s, home ownership was not standard—most people rented. In part, this was because home loans were structured differently: A large down payment was required, the term of the loan was five years or less, the regular payments went toward interest, and the principal was paid off in a lump sum at the end of the term. Home mortgages as we know them came into being through the influence of the Federal Housing Administration, established by Congress in the National Housing Act in 1934, which provided insurance to lending agencies. Find more details on the history of home mortgages, and discuss why the earlier structure of loans would discourage home ownership. Does the earlier structure have advantages?

The following exercises are designed to be solved using technology such as calculators or computer spreadsheets. For assistance, see the technology supplement.

46. Equity. You borrow $15,000 with a term of four years at an APR of 8%. Make an amortization table. How much equity have you built up halfway through the term?

47. Finding the term. You want to borrow $15,000 at an APR of 7% to buy a car, and you can afford a monthly payment of $500. To minimize the amount of interest paid, you want to take the shortest term you can. What is the shortest term you can afford? *Note:* Your answer should be a whole number of years. Rule of Thumb 4.2 should give you a rough idea of what the term will be.

4.3 Saving for the long term: Build that nest egg

TAKE AWAY FROM THIS SECTION
Prepare for the future by saving now.

The following article was found on the Bankrate.com Web site.

IN THE NEWS 4.5

Retirement Planning for 20-Somethings
LESLIE HAGGIN GEARY   April 23, 2007

It's easy to understand why retirement doesn't loom large on the horizon for 20-somethings. Young workers are more concerned with kick-starting careers, not ending them in the long-distant future.

But it's worth noting that the very fact that you're young gives you a huge edge if you want to be rich in retirement. . . .

Consider this scenario: If you begin saving for retirement at 25, putting away $2000 a year for just 40 years, you'll have around $560,000, assuming earnings grow at 8 percent annually. Now, let's say you wait until you're 35 to start saving. You put away the same $2000 a year, but for three decades instead, and earnings grow at 8 percent a year. When you're 65 you'll wind up with around $245,000—less than half the money.

Seems like a no-brainer, right? Save a little now and reap big rewards later.

Unfortunately, many of today's youngest workers pass on the opportunity to save for retirement early, when the beauty of compounding interest can work its magic and maximize savings. A recent study by human resources consultant Hewitt Associates found that just 31 percent of Generation Y workers (those born in 1978 or later, now in the thick of their 20s) who are eligible to put money into a 401(k) retirement savings plan to do so. That's less than half of the 63 percent of workers between ages 26 and 41 who do invest in employer-sponsored savings accounts. . . .

"These years of saving in your early 20s are your prime years. If you deny yourself the opportunity, it will just set you back with retirement planning in the long run," says [one author]. "You've got to have balance."

In Section 4.1, we discussed how an account grows if a lump sum is invested. Many long-term savings plans such as retirement accounts combine the growth power of compound interest with that of regular contributions. Such plans can show truly remarkable growth, the article above indicates. We look at such accounts in this section.

"... and help my parents to pick the right investments for my college education."

Saving regular amounts monthly

Let's look at a plan where you deposit $100 to your savings account at the end of each month, and let's suppose the account pays you a monthly rate of 1% on the balance in the account. (That is an APR of 12% compounded monthly.)

At the end of the first month, the balance is $100. At the end of the second month, the account is increased by two factors—interest earned and a second deposit. Interest earned is 1% of $100 or $1.00, and the deposit is $100. This gives the new balance as

$$\text{New balance} = \text{Previous balance} + \text{Interest} + \text{Deposit}$$
$$= \$100 + \$1 + \$100 = \$201.$$

Table 4.2 tracks the growth of this account through 10 months. Note that at the end of each month, the interest is calculated on the previous balance and then $100 is added to the balance.

TABLE 4.2 Regular Deposits into an Account

At end of month number	Interest paid on previous balance	Deposit	Balance
1	$0.00	$100	$100.00
2	1% of $100.00 = $1.00	$100	$201.00
3	1% of $201.00 = $2.01	$100	$303.01
4	1% of $303.01 = $3.03	$100	$406.04
5	1% of $406.04 = $4.06	$100	$510.10
6	1% of $510.10 = $5.10	$100	$615.20
7	1% of $615.20 = $6.15	$100	$721.35
8	1% of $721.35 = $7.21	$100	$828.56
9	1% of $828.56 = $8.29	$100	$936.85
10	1% of $936.85 = $9.37	$100	$1046.22

TRY IT YOURSELF 4.21

Suppose we have a savings account earning 4% APR. We deposit $40 to the account at the end of each month for 10 years. What is the future value for this savings arrangement?

The answer is provided at the end of this section.

Determining the savings needed

People approach savings in different ways—some are committed to depositing a certain amount of money each month into a savings plan, and others save with a specific purchase in mind. Suppose you are nearing the end of your sophomore year and plan to purchase a car when you graduate in two years. If the car you have your eye on will cost $20,000 when you graduate, you want to know how much you will have to save each month for the next two years to have $20,000 at the end.

We can rearrange the regular deposits balance formula to tell us how much we need to deposit regularly in order to achieve a goal (that is, a future value) such as this. In the following formula, r is the monthly interest rate APR/12 as a decimal, and t is the number of deposits you will make to reach your goal.

FORMULA 4.9   Deposit Needed Formula

$$\text{Needed deposit} = \frac{\text{Goal} \times r}{((1+r)^t - 1)}.$$

For example, if your goal is $20,000 to buy a car in two years, and if the APR is 6%, we can use this formula to find how much we need to deposit each month. The monthly rate as a decimal is

$$r = \frac{\text{APR}}{12} = \frac{0.06}{12} = 0.005.$$

We use $t = 2 \times 12 = 24$ deposits in the formula:

$$\begin{aligned}
\text{Needed deposit} &= \frac{\text{Goal} \times r}{((1+r)^t - 1)} \\
&= \frac{\$20,000 \times 0.005}{(1.005^{24} - 1)} \\
&= \$786.41.
\end{aligned}$$

You need to deposit $786.41 each month so you can buy that $20,000 car when you graduate.

EXAMPLE 4.22 Computing deposit needed: Saving for college

How much does your younger brother need to deposit each month into a savings account that pays 7.2% APR in order to have $10,000 when he starts college in five years?

SOLUTION

We want to achieve a goal of $10,000 in five years, so we use Formula 4.9. The monthly interest rate as a decimal is

$$r = \frac{\text{APR}}{12} = \frac{0.072}{12} = 0.006,$$

and the number of deposits is $t = 5 \times 12 = 60$:

$$\text{Needed deposit} = \frac{\text{Goal} \times r}{((1 + r)^t - 1)}$$
$$= \frac{\$10{,}000 \times 0.006}{(1.006^{60} - 1)}$$
$$= \$138.96.$$

He needs to deposit $138.96 each month.

TRY IT YOURSELF 4.22

How much do you need to deposit each month into a savings account that pays 9% APR in order to have $50,000 for your child to use for college in 18 years?

The answer is provided at the end of this section.

> **SUMMARY 4.5**   **Monthly Deposits**
>
> Suppose we deposit a certain amount of money at the end of each month into a savings account that pays a monthly interest rate of $r = \text{APR}/12$ as a decimal. The balance in the account after t months is given by the regular deposits balance formula (Formula 4.8):
>
> $$\text{Balance after } t \text{ deposits} = \frac{\text{Deposit} \times ((1 + r)^t - 1)}{r}.$$
>
> The ending balance is called the future value for this savings arrangement.
>
> A companion formula (Formula 4.9) gives the monthly deposit necessary to achieve a given balance:
>
> $$\text{Needed deposit} = \frac{\text{Goal} \times r}{((1 + r)^t - 1)}.$$

Saving for retirement

As the article at the beginning of this section points out, college students often don't think much about retirement, but early retirement planning is important.

EXAMPLE 4.23 Finding deposit needed: Retirement and varying rates

Suppose that you'd like to retire in 40 years and you want to have a future value of $500,000 in a savings account. (See the article at the beginning of this section.) Also suppose that your employer makes regular monthly deposits into your retirement account.

a. If you can expect an APR of 9% for your account, how much do you need your employer to deposit each month?

b. The formulas we have been using assume that the interest rate is constant over the period in question. Over a period of 40 years, though, interest rates can vary widely. To see what difference the interest rate can make, let's assume a constant APR of 6% for your retirement account. How much do you need your employer to deposit each month under this assumption?

SOLUTION

a. We have a goal of $500,000, so we use Formula 4.9. The monthly rate as a decimal is

$$r = \text{Monthly rate} = \frac{\text{APR}}{12} = \frac{0.09}{12} = 0.0075.$$

The number of deposits is $t = 40 \times 12 = 480$, so the needed deposit is

$$\text{Needed deposit} = \frac{\text{Goal} \times r}{((1 + r)^t - 1)}$$

$$= \frac{\$500,000 \times 0.0075}{(1.0075^{480} - 1)}$$

$$= \$106.81.$$

b. The computation is the same as in part a except that the new monthly rate as a decimal is

$$r = \frac{0.06}{12} = 0.005.$$

We have

$$\text{Needed deposit} = \frac{\text{Goal} \times r}{((1 + r)^t - 1)}$$

$$= \frac{\$500,000 \times 0.005}{(1.005^{480} - 1)}$$

$$= \$251.07.$$

Note that the decrease in the interest rate from 9% to 6% requires that the monthly deposit more than doubles if you are to reach the same goal. The effect of this possible variation in interest rates is one factor that makes financial planning for retirement complicated.

Retirement income: Annuities

How much income will you need in retirement? That's a personal matter, but we can analyze what a *nest egg* will provide.

Key Concept

Your nest egg is the balance of your retirement account at the time of retirement. The monthly yield is the amount you can withdraw from your retirement account each month.

Your nest egg provides for you in retirement.

Once you retire, there are several ways of using your retirement funds. One method is to withdraw each month only the interest accrued over that month; the principal remains the same. Under this arrangement, your nest egg will never be reduced; you'll be living off the interest alone. An arrangement like this is called a *perpetuity* because the constant income continues indefinitely. The reader is invited to explore this arrangement further in Exercises 29 through 36.

With a perpetuity, the original balance at retirement remains untouched, but if we are willing to reduce the principal each month, we won't need to start with as large a nest egg for our given monthly income. In this situation, we receive a constant monthly payment, part of which represents interest and part of which represents a reduction of principal. Such an arrangement is called a *fixed-term annuity* because, unlike a perpetuity, this arrangement will necessarily end after a fixed term (when we have spent the entire principal).

Key Concept

An annuity is an arrangement that withdraws both principal and interest from your nest egg. Payments end when the principal is exhausted.

An annuity works just like the installment loans we considered in Section 4.2, only someone is paying *us* rather than the other way around.[7] In fact, the formula for the monthly payment applies in this situation, too. We can think of the institution that holds our account at retirement as having borrowed our nest egg; it will pay us back in monthly installments over the term of the annuity. In the following formula, r is the monthly rate (as a decimal), and t is the term (the number of months the annuity will last).

FORMULA 4.10 Annuity Yield Formula

$$\text{Monthly annuity yield} = \frac{\text{Nest egg} \times r \times (1+r)^t}{((1+r)^t - 1)}.$$

"We could both avoid this daily
annoyance, sir, if you'd buy
me an *annuity!*"

EXAMPLE 4.24 Finding annuity yield: 20-year annuity

Suppose we have a nest egg of $800,000 with an APR of 6% compounded monthly. Find the monthly yield for a 20-year annuity.

SOLUTION

We use the annuity yield formula (Formula 4.10). With an APR of 6%, the monthly rate as a decimal is

$$r = \text{Monthly rate} = \frac{\text{APR}}{12} = \frac{0.06}{12} = 0.005.$$

The term is 20 years, so we take $t = 20 \times 12 = 240$ months:

$$\begin{aligned}
\text{Monthly annuity yield} &= \frac{\text{Nest egg} \times r \times (1+r)^t}{((1+r)^t - 1)} \\
&= \frac{\$800,000 \times 0.005 \times 1.005^{240}}{(1.005^{240} - 1)} \\
&= \$5731.45.
\end{aligned}$$

TRY IT YOURSELF 4.24

Suppose we have a nest egg of $1,000,000 with an APR of 6% compounded monthly. Find the monthly yield for a 25-year annuity.

The answer is provided at the end of this section.

---

[7] We assume that the payments are made at the end of the month.

How large a nest egg is needed to achieve a desired annuity yield? We can answer this question by rearranging the annuity yield formula (Formula 4.10).

> **FORMULA 4.11   Annuity Yield Goal**
>
> $$\text{Nest egg needed} = \frac{\text{Annuity yield goal} \times ((1+r)^t - 1)}{(r \times (1+r)^t)}.$$

Here, r is the monthly rate (as a decimal), and t is the term (in months) of the annuity.

The following example shows how we use this formula.

EXAMPLE 4.25   Finding nest egg needed for annuity: Retiring on a 20-year annuity

Suppose our retirement account pays 5% APR compounded monthly. What size nest egg do we need in order to retire with a 20-year annuity that yields $4000 per month?

SOLUTION

We want to achieve an annuity goal, so we use Formula 4.11. The monthly rate as a decimal is $r = 0.05/12$, and the term is $t = 20 \times 12 = 240$ months:

$$\text{Nest egg needed} = \frac{\text{Annuity yield goal} \times ((1+r)^t - 1)}{(r \times (1+r)^t)}$$
$$= \frac{\$4000 \times ((1 + 0.05/12)^{240} - 1)}{((0.05/12) \times (1 + 0.05/12)^{240})}$$
$$= \$606,101.25.$$

TRY IT YOURSELF 4.25

Suppose our retirement account pays 9% APR compounded monthly. What size nest egg do we need in order to retire with a 25-year annuity that yields $5000 per month?

The answer is provided at the end of this section.

The balance at retirement (the nest egg) is called the *present value* of the annuity. Future value and present value depend on perspective. When you started saving, the balance at retirement was the future value. When you actually retire, it becomes the present value.

The obvious question is, how many years should you plan for the annuity to last? If you set it up to last until you're 80 and then you live until you're 85, you're in trouble. What a retiree often wants is to have the monthly annuity payment continue for as long as he or she lives. Insurance companies offer such an arrangement, and it's called a *life annuity*. How does it work?

The insurance company makes a statistical estimate of the life expectancy of a customer, which is used in determining how much the company will probably have to pay out.[8] Some customers will live longer than the estimate (and the company may lose money on them) but some will not (and the company will make money on them). The monthly income for a given principal is determined from this estimate using the formula for the present value of a fixed-term annuity.

---

[8] Such calculations are made by professionals known as *actuaries*.

For a fixed-term annuity, the formulas for monthly payments apply. Let r be the monthly interest rate (as a decimal) and t the term (in months) of the annuity.

1. To find the monthly yield provided by a nest egg, we use

$$\text{Monthly annuity yield} = \frac{\text{Nest egg} \times r \times (1+r)^t}{((1+r)^t - 1)}.$$

2. To find the nest egg needed to provide a desired income, we use

$$\text{Nest egg needed} = \frac{\text{Annuity yield goal} \times ((1+r)^t - 1)}{(r \times (1+r)^t)}.$$

The balance at retirement (the nest egg) is called the present value of the annuity.

ALGEBRAIC SPOTLIGHT 4.4 Derivation of the Regular Deposits Balance Formula

Suppose that at the end of each month, we deposit money (the same amount each month) into an account that pays a monthly rate of r as a decimal. Our goal is to derive the formula

$$\text{Balance after } t \text{ deposits} = \frac{\text{Deposit} \times ((1+r)^t - 1)}{r}.$$

In Algebraic Spotlight 4.2 from Section 4.2, we considered the situation where we borrow B_0 dollars and make monthly payments of M dollars at a monthly interest rate of r as a decimal. We found that the balance after t payments is

$$\text{Balance after } t \text{ payments} = B_0(1+r)^t - M\frac{(1+r)^t - 1}{r}.$$

Making payments of M dollars per month subtracts money from the balance. A deposit adds money rather than subtracting it. The result is to add the term involving M instead of subtracting it:

$$\text{Balance after } t \text{ deposits} = B_0(1+r)^t + M\frac{(1+r)^t - 1}{r}.$$

Now the initial balance is zero, so $B_0 = 0$. Therefore,

$$\text{Balance after } t \text{ deposits} = M\frac{(1+r)^t - 1}{r}.$$

Because M is the amount we deposit, we can write the result as

$$\text{Balance after } t \text{ deposits} = \frac{\text{Deposit} \times ((1+r)^t - 1)}{r}.$$

Try It Yourself answers

Try It Yourself 4.20: Verifying a balance: Regular deposits into an account $\$303.01 + 0.01 \times \$303.01 + \$100 = \$406.04$.

Try It Yourself 4.21: Using the balance formula: Saving money regularly $\$5889.99$.

Try It Yourself 4.22: Computing deposit needed: Saving for college $\$93.22$.

Try It Yourself 4.24: Finding annuity yield: 25-year annuity $\$6443.01$.

Try It Yourself 4.25: Finding nest egg needed for annuity: Retiring on a 25-year annuity $\$595,808.11$.

Exercise Set 4.3

1. Saving for a car. You are saving to buy a car, and you deposit $200 at the end of each month for two years at an APR of 4.8% compounded monthly. What is the future value for this savings arrangement? That is, how much money will you have for the purchase of the car after two years?

2. Saving for a down payment. You want to save $20,000 for a down payment on a home by making regular monthly deposits over five years. Take the APR to be 6%. How much money do you need to deposit each month?

3. Planning for college. At your child's birth, you begin contributing monthly to a college fund. The fund pays an APR of 4.8% compounded monthly. You figure your child will need $40,000 at age 18 to begin college. What monthly deposit is required?

A table You have a savings account into which you invest $50 at the end of every month, and the account pays you an APR of 9% compounded monthly. Exercises 4 through 7 refer to this account.

4. Fill in the following table. (Don't use the regular deposits balance formula.)

At end of month number	Interest paid on prior balance	Deposit	Balance
1			
2			
3			
4			

5. Use the regular deposits balance formula (Formula 4.8) to determine the balance in the account at the end of four months. Compare this to the final balance in the table from Exercise 4.

6. Use the regular deposits balance formula to determine the balance in the account at the end of four years.

7. Use the regular deposits balance formula to determine the balance in the account at the end of 20 years.

Another table You have a retirement account into which your employer invests $75 at the end of every month, and the account pays an APR of 5.25% compounded monthly. Exercises 8 through 11 refer to this account.

8. Fill in the following table. (Don't use the regular deposits balance formula.)

At end of month number	Interest paid on prior balance	Deposit	Balance
1			
2			
3			
4			

9. Use the regular deposits balance formula (Formula 4.8) to determine the balance in the account at the end of four months. Compare this to the final balance in the table from Exercise 8.

10. Use the regular deposits balance formula to determine the balance in the account at the end of four years.

11. Use the regular deposits balance formula to determine the balance in the account at the end of 20 years.

Is this reasonable? Jerry calculates that if he makes a deposit of $5 each month at an APR of 4.8%, then at the end of two years he'll have $100. Benny says that the correct amount is $135. This information is used in Exercises 12 through 14. Rule of Thumb 4.4 should be helpful here.

12. What was the total amount deposited (ignoring interest earned)? Whose answer is ruled out by this calculation? Why?

13. Suppose the total amount deposited ($5 per month for two years) is instead put as a lump sum at the beginning of the two years as principal in an account earning an APR of 4.8%. Use the compound interest formula (with monthly compounding) from Section 4.1 to determine how much would be in the account after two years. Whose answer is ruled out by this calculation? Why?

14. Find the correct balance after two years.

Saving for a boat Suppose you want to save in order to purchase a new boat. In Exercises 15 and 16, take the APR to be 7.2%.

15. If you deposit $250 each month, how much will you have toward the purchase of a boat after three years?

16. You want to have $13,000 toward the purchase of a boat in three years. How much do you need to deposit each month?

17. Fixed-term annuity. You have a 20-year annuity with a present value (that is, nest egg) of $425,000. If the APR is 7%, what is the monthly yield?

18. Life annuity. You have set up a life annuity with a present value of $350,000. If your life expectancy at retirement is 21 years, what will your monthly income be? Take the APR to be 6%.

Just a bit more You begin working at age 25, and your employer deposits $300 per month into a retirement account that pays an APR of 6% compounded monthly. You expect to retire at age 65. Use this information in Exercises 19 and 20.

19. What will be the size of your nest egg at age 65?

20. Suppose you are allowed to contribute $100 each month in addition to your employer's contribution. What will be the size of your nest egg at age 65? Compare this with your answer to Exercise 19.

Just a bit longer You begin working at age 25, and your employer deposits $250 each month into a retirement account that pays an APR of 6% compounded monthly. You expect to retire at age 65. Use this information for Exercises 21 and 22.

21. What will be the size of your nest egg when you retire?

22. Suppose instead that you arranged to start the regular deposits two years earlier, at age 23. What will be the size of your nest egg when you retire? Compare this with your answer from Exercise 21.

Planning to retire on an annuity You plan to work for 40 years and then retire using a 25-year annuity. You want to arrange a retirement income of $4500 per month. You have access to an account that pays an APR of 7.2% compounded monthly. Use this information for Exercises 23 and 24.

23. What size nest egg do you need to achieve the desired monthly yield?

24. What monthly deposits are required to achieve the desired monthly yield at retirement?

Retiring You want to have a monthly income of $2000 from a fixed-term annuity when you retire. Take the term of the annuity to be 20 years, and assume an APR of 6% over the period of investment covered in Exercises 25 and 26.

25. How large will your nest egg have to be at retirement to guarantee the income described above?

26. You plan to make regular deposits for 40 years to build up your savings to the level you determined in Exercise 25. How large must your monthly deposit be?

Deposits at the beginning In this section, we considered the case of regular deposits at the end of each month. If deposits are made at the beginning of each month, then the formula is a bit different. The adjusted formula is

$$\text{Balance after } t \text{ deposits} = \text{Deposit} \times (1 + r) \times \frac{((1 + r)^t - 1)}{r}.$$

Here, r is the monthly interest rate (as a decimal), and t is the number of deposits. The extra factor of $1 + r$ accounts for the interest earned on the deposit over the first month after it's made.

Suppose you deposit $200 at the beginning of each month for five years. Take the APR to be 7.2% for Exercises 27 and 28.

27. What is the future value? In other words, what will your account balance be at the end of the period?

28. What would be the future value if we had made the deposits at the end of each month rather than at the beginning? Explain why it is reasonable that your answer here is smaller than that from Exercise 27.

29. Retirement income: perpetuities. If a retirement fund is set up as a *perpetuity*, one withdraws each month only the interest accrued over that month; the principal remains the same. For example, suppose you have accumulated $500,000 in an account with a monthly interest rate of 0.5%. Each month, you can withdraw $500,000 × 0.005 = $2500 in interest, and the nest egg will always remain at $500,000. That is, the $500,000 perpetuity has a monthly yield of $2500. In general, the monthly yield for a perpetuity is given by the formula

Monthly perpetuity yield = Nest egg × Monthly interest rate.

In this formula, the monthly interest rate is expressed as a decimal.

Suppose we have a perpetuity paying an APR of 6% compounded monthly. If the value of our nest egg (that is, the present value) is $800,000, find the amount we can withdraw each month. *Note:* First find the monthly interest rate.

30. Perpetuity yield. *Refer to Exercise 29 for background on perpetuities.* You have a perpetuity with a present value (that is, nest egg) of $650,000. If the APR is 5% compounded monthly, what is your monthly income?

31. Another perpetuity. *Refer to Exercise 29 for background*

on perpetuities. You have a perpetuity with a present value of $900,000. If the APR is 4% compounded monthly, what is your monthly income?

32. Comparing annuities and perpetuities. *Refer to Exercise 29 for background on perpetuities.* For 40 years, you invest $200 per month at an APR of 4.8% compounded monthly, then you retire and plan to live on your retirement nest egg.

    a. How much is in your account on retirement?

    b. Suppose you set up your account as a perpetuity on retirement. What will your monthly income be? (Assume that the APR remains at 4.8% compounded monthly.)

    c. Suppose now you use the balance in your account for a life annuity instead of a perpetuity. If your life expectancy is 21 years, what will your monthly income be? (Again, assume that the APR remains at 4.8% compounded monthly.)

    d. Compare the total amount you invested with your total return from part c. Assume that you live 21 years after retirement.

33. Perpetuity goal: How much do I need to retire? *Refer to Exercise 29 for background on perpetuities.* Here is the formula for the nest egg needed for a desired monthly yield on a perpetuity:

$$\text{Nest egg needed} = \frac{\text{Desired monthly yield}}{\text{Monthly interest rate}}.$$

In this formula, the monthly interest rate is expressed as a decimal.

If your retirement account pays 5% APR with monthly compounding, what present value (that is, nest egg) is required for you to retire on a perpetuity that pays $4000 per month?

34. Desired perpetuity. *Refer to the formula in Exercise 33.* You want a perpetuity with a monthly income of $3000. If the APR is 7%, what does the present value need to be?

Planning to retire on a perpetuity You plan to work for 40 years and then retire using a perpetuity. You want to arrange to have a retirement income of $4500 per month. You have access to an account that pays an APR of 7.2% compounded monthly. Use this information for Exercises 35 and 36.

35. *Refer to the formula in Exercise 33.* What size nest egg do you need to achieve the desired monthly yield?

36. What monthly deposits are required to achieve the desired monthly yield at retirement?

Exercises 37 through 40 are suitable for group work.

Starting early, starting late In Exercises 37 through 40, we consider the effects of starting early or late to save for retirement. Assume that each account considered has an APR of 6% compounded monthly.

37. At age 20, you realize that even a modest start on saving for retirement is important. You begin depositing $50 each month into an account. What will be the value of your nest egg when you retire at age 65?

38. Against expert advice, you begin your retirement program at age 40. You plan to retire at age 65. What monthly contributions do you need to make to match the nest egg from Exercise 37?

39. Compare your answer to Exercise 38 with the monthly deposit of $50 from Exercise 37. Also compare the total amount deposited in each case.

40. Let's return to the situation in Exercise 37: At age 20, you begin depositing $50 each month into an account. Now suppose that at age 40, you finally get a job where your employer puts $400 per month into an account. You continue your $50 deposits, so from age 40 on, you have two separate accounts working for you. What will be the total value of your nest egg when you retire at age 65?

41. Retiring without interest. Suppose we lived in a society without interest. At age 25, you begin putting $250 per month into a cookie jar until you retire at age 65. At age 65, you begin to withdraw $2500 per month from the cookie jar. How long will your retirement fund last?

42. History of annuities. The origins of annuities can be traced back to the ancient Romans. Look up the history of annuities and write a report on it. Include their use in Rome, in Europe during the seventeenth century, in colonial America, and in modern American society.

43. History of actuaries. Look up the history of the actuarial profession and write a report on it. Be sure to discuss what mathematics are required to become an actuary.

The following exercises are designed to be solved using technology such as calculators or computer spreadsheets. For assistance, see the technology supplement.

44. How much? You begin saving for retirement at age 25, and you plan to retire at age 65. You want to deposit a certain amount each month into an account that pays an APR of 6% compounded monthly. Make a table that shows the amount you must deposit each month in terms of the nest egg you desire to have when you retire. Include nest egg sizes from $100,000 to $1,000,000 in increments of $100,000.

45. How long? You begin working at age 25, and your employer deposits $350 each month into a retirement account that pays an APR of 6% compounded monthly. Make a table that shows the size of your nest egg in terms of the age at which you retire. Include retirement ages from 60 to 70.

4.4 Credit cards: Paying off consumer debt

TAKE AWAY FROM THIS SECTION

Be savvy in paying off credit cards.

The accompanying excerpt is from an article at the Web site of CBS News. The subject is credit card debt among students in the United States, which might sound familiar to you.

IN THE NEWS 4.6

Beware of Student Credit Cards

TATIANA MORALES   September 3, 2003

NEW YORK—The average undergraduate leaves school with a debt of $18,900. That's up 66 percent from five years ago, according to a new study by loan provider Nellie Mae. A large part of this is, of course, student loans, which more and more students report needing these days, thanks to ballooning tuition bills.

However, *The Early Show* financial advisor Ray Martin reports, a growing part of this debt is unnecessary; it's a result of four years of charging pizza and shoes and booze on new credit cards.

College students are prime targets for credit card companies, which set up tables on campus and entice students to sign up for new cards with promises of free T-shirts or other goodies. Unfortunately, many students eagerly apply for credit and use it unwisely.

Students **double** their credit card debt and **triple** the number of cards in their wallets between the time they arrive on campus and graduation, Nellie Mae found. Another scary finding: by the time college students reach their senior year, 31 percent carry a balance of $3000 to $7000.

Credit cards are convenient and useful. They allow us to travel without carrying large sums of cash, and they sometimes allow us to defer cash payments, even interest-free, for a short time. In fact, owning a credit card can be a necessity: Most hotels and rental car companies require customers to have a credit card. Some sources say that the average credit card debt per American is almost $8000.[9]

The convenience comes at a cost, however. For example, some credit cards carry an APR much higher than other kinds of consumer loans. The APR depends on various factors, including the customer's credit rating. According to the article above, credit cards intended for students may range from 10% to 19.8%. At the time of this writing, the site www.indexcreditcards.com/lowaprcreditcards.html reveals cards with APRs ranging from 7.24% to 22.9%.

In this section, we explore credit cards in some detail. In particular, we look at how payments are calculated, the terminology used by credit card companies, and the implications of making only the minimum required payments.

"I didn't have time to cut the lawn, so I used your credit card to have it carpeted. Do you like the cool color I picked out?"

Credit card basics

Different credit cards have different conditions. Some have annual fees, and the interest rates vary. Many of them will apply no finance charges if you pay the full amount owed each month. In deciding which credit card to use, it is very important that you read the fine print and get the card that is most favorable to you.

Here is a simplified description of how to calculate the finance charges: Start with the balance shown in the statement from the previous month, subtract payments you've made since the previous statement, and add any new purchases. That is the amount that is subject to finance charges.[10] Thus, the formula for finding the amount subject to finance charges is

Amount subject to finance charges = Previous balance − Payments + Purchases.

The finance charge is calculated by applying the monthly interest rate (the APR divided by 12) to this amount.

---

[9] Averages can be a bit misleading. We will say more about this in Chapter 6.

[10] The most common method of calculation actually uses the *average daily balance*. That is the average of the daily balances over the payment period (usually one month). This means that a pair of jeans purchased early in the month will incur more finance charges than the same pair purchased late in the month. Most experts consider this a fair way to do the calculation. See Exercises 29 and 30.

Let's look at a card that charges 21.6% APR. Suppose your previous statement showed a balance of $300, you made a payment of $100 in response to your previous statement, and you have new purchases of $50. The amount subject to finance charges is

Amount subject to finance charges = Previous balance − Payments + Purchases

$$= \$300 - \$100 + \$50 = \$250.$$

Because the APR is 21.6%, we find the monthly interest rate using

$$\text{Monthly interest rate} = \frac{\text{APR}}{12} = \frac{21.6\%}{12} = 1.8\%.$$

The finance charge is 1.8% of $250:

$$\text{Finance charge} = 0.018 \times \$250 = \$4.50.$$

This makes your new balance

New balance = Amount subject to finance charges + Finance charge

$$= \$250 + \$4.50 = \$254.50.$$

Your new balance is $254.50.

A credit card statement.

EXAMPLE 4.26 Calculating finance charges: Buying clothes

Suppose your Visa card calculates finance charges using an APR of 22.8%. Your previous statement showed a balance of $500, in response to which you made a payment of $200. You then bought $400 worth of clothes, which you charged to your Visa card. Complete the following table:

	Previous balance	Payments	Purchases	Finance charge	New balance
Month 1					

SOLUTION

We start by entering the information given.

	Previous balance	Payments	Purchases	Finance charge	New balance
Month 1	$500.00	$200.00	$400.00		

In the first month, the amount subject to finance charges is

Amount subject to finance charges = Previous balance − Payments + Purchases
$$= \$500 - \$200 + \$400 = \$700.$$

The APR is 22.8%, and we divide by 12 to get a monthly rate of 1.9%. In decimal form, this calculation is $0.228/12 = 0.019$. Therefore, the finance charge on $700 is

Finance charge $= 0.019 \times \$700 = \$13.30.$

That gives a new balance of

New balance = Amount subject to finance charges + Finance charge
$$= \$700.00 + \$13.30 = \$713.30.$$

Your new balance is $713.30. Here is the completed table.

	Previous balance	Payments	Purchases	Finance charge	New balance
Month 1	$500.00	$200.00	$400.00	1.9% of $700 = $13.30	$713.30

TRY IT YOURSELF 4.26

This is a continuation of Example 4.26. You make a payment of $300 to reduce the $713.30 balance and then charge a TV costing $700. Complete the following table:

	Previous balance	Payments	Purchases	Finance charge	New balance
Month 1	$500.00	$200.00	$400.00	$13.30	$713.30
Month 2					

The answer is provided at the end of this section.

SUMMARY 4.7   **Credit Card Basics**

The formula for finding the amount subject to finance charges is

Amount subject to finance charges = Previous balance − Payments + Purchases.

The finance charge is calculated by applying the monthly interest rate (the APR divided by 12) to this amount. The new balance is found by using the formula

New balance = Amount subject to finance charges + Finance charge.

Making only minimum payments

Most credit cards require a minimum monthly payment that is normally a fixed percentage of your balance. We will see that if you make only this minimum payment, your balance will decrease very slowly and will follow an exponential pattern.

Minimal payments reduce balances minimally.

EXAMPLE 4.27 Finding next month's minimum payment: One payment

We have a card with an APR of 24%. The minimum payment is 5% of the balance. Suppose we have a balance of $400 on the card. We decide to stop charging and to pay it off by making the minimum payment each month. Calculate the new balance after we have made our first minimum payment, and then calculate the minimum payment due for the next month.

SOLUTION

The first minimum payment is

$$\text{Minimum payment} = 5\% \text{ of balance} = 0.05 \times \$400 = \$20.$$

The amount subject to finance charges is

$$\text{Amount subject to finance charges} = \text{Previous balance} - \text{Payments} + \text{Purchases}$$
$$= \$400 - \$20 + \$0 = \$380.$$

The monthly interest rate is the APR divided by 12. In decimal form, this is $0.24/12 = 0.02$. Therefore, the finance charge on $380 is

$$\text{Finance charge} = 0.02 \times \$380 = \$7.60.$$

That makes a new balance of

$$\text{New balance} = \text{Amount subject to finance charges} + \text{Finance charge}$$
$$= \$380 + \$7.60 = \$387.60.$$

The next minimum payment will be 5% of this:

$$\text{Minimum payment} = 5\% \text{ of balance} = 0.05 \times \$387.60 = \$19.38.$$

TRY IT YOURSELF 4.27

This is a continuation of Example 4.27. Calculate the new balance after we have made our second minimum payment, and then calculate the minimum payment due for the next month.

The answer is provided at the end of this section.

Let's pursue the situation described in the previous example. The following table covers the first four payments if we continue making the minimum 5% payment each month:

Month	Previous balance	Minimum payment	Purchases	Finance charge	New balance
1	$400.00	5% of $400.00 = $20.00	$0.00	2% of $380.00 = $7.60	$387.60
2	$387.60	5% of $387.60 = $19.38	$0.00	2% of $368.22 = $7.36	$375.58
3	$375.58	5% of $375.58 = $18.78	$0.00	2% of $356.80 = $7.14	$363.94
4	$363.94	5% of $363.94 = $18.20	$0.00	2% of $345.74 = $6.91	$352.65

The table shows that the $400 is not being paid off very quickly. In fact, after four payments, the decrease in the balance is only about $47 (from $400 to $352.65). If we look closely at the table, we will see that the balances follow a pattern. To find the pattern, let's look at how the balance changes in terms of percentages:

Month 1: The new balance of $387.60   is 96.9% of   the initial balance of $400.
Month 2: The new balance of $375.58   is 96.9% of   the Month 1 new balance of $387.60.
Month 3: The new balance of $363.94   is 96.9% of   the Month 2 new balance of $375.58.
Month 4: The new balance of $352.65   is 96.9% of   the Month 3 new balance of $363.94.

The balance exhibits a constant percentage change. This makes sense: Each month, the balance is decreased by a constant percentage due to the minimum payment and increased by a constant percentage due to the finance charge. This pattern indicates that the balance is a decreasing exponential function. That conclusion is supported by the graph of the balance shown in **Figure 4.8**, which has the classic shape of exponential decay.

Because of this exponential pattern, we can find a formula for the balance on our credit card in the situation where we stop charging and pay off the balance by making the minimum payment each month.

FORMULA 4.12 Minimum Payment Balance Formula

Balance after t minimum payments = Initial balance $\times ((1 + r)(1 - m))^t$.

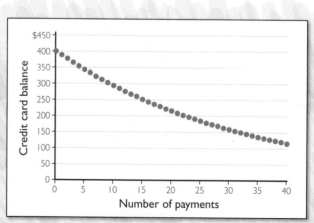

FIGURE 4.8 The balance from minimum payments is a decreasing exponential function.

In this formula, r is the monthly interest rate and m is the minimum monthly payment as a percent of the balance. Both r and m are in decimal form. You should not round off the product $(1+r)(1-m)$ when performing the calculation. We provide a derivation of this formula in Algebraic Spotlight 4.5 at the end of this section.

EXAMPLE 4.28 Using the minimum payment balance formula: Balance after two years

We have a card with an APR of 20% and a minimum payment that is 4% of the balance. We have a balance of $250 on the card, and we stop charging and pay off the balance by making the minimum payment each month. Find the balance after two years of payments.

SOLUTION

The APR in decimal form is 0.2, so the monthly interest rate in decimal form is $r = 0.2/12$. To avoid rounding, we leave r in this form. The minimum payment is 4% of the new balance, so we use $m = 0.04$. The initial balance is $250. The number of payments for two years is $t = 24$. Using the minimum payment balance formula (Formula 4.12), we find

$$\text{Balance after 24 minimum payments} = \text{Initial balance} \times ((1 + r)(1 - m))^t$$
$$= \$250 \times ((1 + 0.2/12)(1 - 0.04))^{24}$$
$$= \$250 \times ((1 + 0.2/12)(0.96))^{24}$$
$$= \$139.55.$$

TRY IT YOURSELF 4.28

We have a card with an APR of 22% and a minimum payment that is 5% of the balance. We have a balance of $750 on the card, and we stop charging and pay off the balance by making the minimum payment each month. Find the balance after three years of payments.

The answer is provided at the end of this section.

We already noted that the credit card balance is not paid off quickly when we make only the minimum payment each month. The reason for this is now clear: The balance is a decreasing exponential function, and such functions typically decrease very slowly in the long run. The next example illustrates the dangers of making only the minimum monthly payment.

EXAMPLE 4.29 Paying off your credit card balance: Long repayment

Suppose you have a balance of $10,000 on your Visa card, which has an APR of 24%. The card requires a minimum payment of 5% of the balance. You stop charging and begin making only the minimum payment until your balance is below $100.

    a. Find a formula that gives your balance after t monthly payments.

    b. Find your balance after five years of payments.

    c. Determine how long it will take to get your balance under $100.[11]

Wise credit card use avoids debt.

---

[11] It is worth noting that paying in this fashion will never actually get your balance to exactly 0. But doing so will eventually make the balance small enough that you could pay it off in a lump sum. Normally, credit cards take care of this with a requirement such as "The minimum payment is 5% or $20, whichever is larger."

d. Suppose that instead of the minimum payment, you want to make a fixed monthly payment so that your debt is clear in two years. How much do you pay each month?

SOLUTION

a. The minimum payment as a decimal is $m = 0.05$, and the monthly rate as a decimal is $r = 0.24/12 = 0.02$. The initial balance is $10,000. Using the minimum payment balance formula (Formula 4.12), we find

$$\text{Balance after } t \text{ minimum payments} = \text{Initial balance} \times ((1 + r)(1 - m))^t$$
$$= \$10,000 \times ((1 + 0.02)(1 - 0.05))^t$$
$$= \$10,000 \times 0.969^t.$$

b. Now five years is 60 months, so we put $t = 60$ into the formula from part a:

$$\text{Balance after 60 months} = 10,000 \times 0.969^{60} = \$1511.56.$$

After 5 years, we still owe over $1500.

c. We will show two ways to determine how long it takes to get the balance down to $100.

Method 1: Using a logarithm: We need to solve for t the equation

$$\$100 = \$10,000 \times 0.969^t.$$

The first step is to divide each side of the equation by $10,000:

$$\frac{100}{10,000} = \frac{10,000}{10,000} \times 0.969^t$$
$$0.01 = 0.969^t.$$

In Section 3.3, we learned how to solve exponential equations using logarithms. Summary 3.9 tells us that the solution for t of the equation $A = B^t$ is $t = \dfrac{\log A}{\log B}$. Using this formula with $A = 0.01$ and $B = 0.969$ gives

$$t = \frac{\log 0.01}{\log 0.969} \text{ months.}$$

This is about 146.2 months. Hence, the balance will be under $100 after 147 monthly payments, or more than 12 years of payments.

Method 2: Trial and error: If you want to avoid logarithms, you can solve this problem using trial and error with a calculator. The information in part b indicates that it will take some time for the balance to drop below $100. So we might try 10 years, or 120 months. Computation using the formula from part a shows that after 10 years, the balance is still over $200, so we should try a larger number of months. If we continue in this way, we find the same answer as that obtained from Method 1: The balance drops below $100 at payment 147, which represents over 12 years of payments. (Spreadsheets and many calculators will create tables of values that make problems of this sort easy to solve.)

d. Making fixed monthly payments to clear your debt is like considering your debt as an installment loan: Just find the monthly payment if you borrow $10,000 to buy (say) a car at an APR of 24% and pay the loan off over 24 months. We use the monthly payment formula from Section 4.2:

$$\text{Monthly payment} = \frac{\text{Amount borrowed} \times r(1 + r)^t}{((1 + r)^t - 1)}.$$

Recall that t is the number of months taken to pay off the loan, in this case 24, and that r is the monthly rate as a decimal, which in this case is 0.02. Hence,

$$\text{Monthly payment} = \frac{\$10{,}000 \times 0.02 \times 1.02^{24}}{(1.02^{24} - 1)} = \$528.71.$$

So a payment of \$528.71 each month will clear the debt in two years.

SUMMARY 4.8   *Making Minimum Payments*

Suppose we have a balance on our credit card and decide to stop charging and pay off the balance by making the minimum payment each month.

1. The balance is given by the exponential formula

Balance after t minimum payments $=$ Initial balance $\times ((1 + r)(1 - m))^t$.

In this formula, r is the monthly interest rate and m is the minimum monthly payment as a percent of the balance. Both r and m are in decimal form.

2. The product $(1 + r)(1 - m)$ should not be rounded when the calculation is performed.

3. Because the balance is a decreasing exponential function, the balance decreases very slowly in the long run.

Further complications of credit card usage

Situations involving credit cards can often be even more complicated than those discussed in previous examples in this section. For instance, in each of those examples, there was a single purchase and no further usage of the card. Of course, it's more common to make purchases every month. We also assumed that your payments were made on time, but there are substantial penalties for late or missed payments.

Use credit carefully.

Another complication is that credit card companies sometimes have "specials" or promotions in which you are allowed to skip a payment. And then there are *cash advances*, which are treated differently from purchases. Typically, cash advances incur finance charges immediately rather than after a month; that is, a cash advance is treated as carrying a balance immediately. Also, cash advances incur higher finance charges than purchases do.

Another complication occurs when your credit limit is reached (popularly known as "maxing out your card"). The credit limit is the maximum balance the credit card company allows you to carry. Usually, the limit is based on your credit history and your ability to pay. When you max out a credit card, the company will sometimes raise your credit limit, if you have always made required payments by its due dates.

In addition to all these complications, there are often devious hidden fees and charges, as one can see in the following excerpt from an editorial in the *St. Petersburg Times*. An amended version of the bill referred to in the editorial was signed into law by President Barack Obama as the Credit Card Accountability Responsibility and Disclosure (CARD) Act of 2009. Among other things, the law requires that credit card companies inform cardholders how long it will take to pay off the balance if they make only the minimum payment.

IN THE NEWS 4.7

Cancel Credit Card Abuses
May 11, 2009

The credit card industry has figured out all sorts of ways to trap consumers into paying high fees and interest charges. Some of those practices that one would think would be illegal, such as retroactively raising interest rates on current balances, are fairly common and raise billions of dollars in profits from unwitting cardholders. Finally, it appears Congress will put an end to the worst elements of the credit card business.

Last month in an overwhelming and bipartisan vote, the House approved the Credit Card Holders' Bill of Rights. It has the support of President Barack Obama, who met recently with credit card executives only to tell them that he would sign the bill. Now the Senate is expected to take up the issue this week, and it should not water down key consumer protections.

Americans owe more than $960 billion in credit card debt and should be held accountable for that spending. But unethical "gotcha" practices of the industry for late or incomplete payments exact too high a financial punishment.

For example, a common gimmick is for companies to avoid applying payments to the debt that carries the highest interest rate, a practice that the House bill would outlaw. It also would ban double-cycle billing, a particularly noxious practice in which interest is charged on prior balances that have been paid off if the consumer revolves a balance the next month. So if a consumer pays off a $600 balance one month but leaves $100 balance unpaid the next month, he is charged interest on $700.

ALGEBRAIC SPOTLIGHT 4.5 Derivation of the Minimum Payment Balance Formula

Suppose a credit card has an initial balance. Assume that we incur no further charges and make only the minimum payment each month. Suppose r is the monthly interest rate and m is the minimum monthly payment as a percentage of the new balance. Both r and m are in decimal form. Our goal is to derive the formula

$$\text{Balance after } t \text{ minimum payments} = \text{Initial balance} \times ((1+r)(1-m))^t.$$

Here is the derivation. Assume that we have made a series of payments, and let B denote the balance remaining. We need to calculate the new balance. First we find the minimum payment on the balance B. To do so, we multiply B by m:

$$\text{Minimum payment} = mB.$$

Next we find the amount subject to finance charges:

$$\begin{aligned}
\text{Amount subject to finance charges} &= \text{Previous balance} - \text{Payments} \\
&= B - mB \\
&= B(1 - m).
\end{aligned}$$

To calculate the finance charge, we apply the monthly rate r to this amount:

$$\text{Finance charge} = r \times B(1 - m).$$

Therefore, the new balance is

$$\begin{aligned}
\text{New balance} &= \text{Amount subject to finance charges} + \text{Finance charge} \\
&= B(1 - m) + r B(1 - m) \\
&= B \times ((1 - m) + r(1 - m)) \\
&= B \times ((1 + r)(1 - m)).
\end{aligned}$$

We can write this as

$$\text{New balance} = \text{Previous balance} \times ((1+r)(1-m))$$

Therefore, to find the new balance each month, we multiply the previous balance by $(1 + r)(1 - m)$. That makes the balance after t payments an exponential function of t with base $(1+r)(1-m)$. The initial value of this function is the initial balance, so we have the exponential formula

$$\text{Balance after } t \text{ minimum payments} = \text{Initial balance} \times ((1 + r)(1 - m))^t.$$

This is the minimum payment balance formula.

Try It Yourself answers

Try It Yourself 4.26: Calculating finance charges: Buying clothes

	Previous balance	Payments	Purchases	Finance charge	New balance
Month 1	$500.00	$200.00	$400.00	$13.30	$713.30
Month 2	$713.30	$300.00	$700.00	$21.15	$1134.45

Try It Yourself 4.27: Finding next month's minimum payment: One payment New balance: $375.58; minimum payment: $18.78.

Try It Yourself 4.28: Using the minimum payment balance formula: Balance after three years $227.59.

Exercise Set 4.4

1. Calculating balances. You have a credit card with an APR of 16%. You begin with a balance of $800, in response to which you make a payment of $400. The first month you make charges amounting to $300. You make a payment of $300 to reduce the new balance, and the second month you charge $600. Complete the following table:

	Previous balance	Payments	Purchases	Finance charge	New balance
Month 1					
Month 2					

2. Calculating balances. You have a credit card with an APR of 20%. You begin with a balance of $600, in response to which you make a payment of $400. The first month you make charges amounting to $200. You make a payment of $300 to reduce the new balance, and the second month you charge $100. Complete the following table:

	Previous balance	Payments	Purchases	Finance charge	New balance
Month 1					
Month 2					

3. Calculating balances. You have a credit card with an APR of 22.8%. You begin with a balance of $1000, in response to which you make a payment of $200. The first month you make charges amounting to $500. You make a payment of $200 to reduce the new balance, and the second month you charge $600. Complete the following table:

	Previous balance	Payments	Purchases	Finance charge	New balance
Month 1					
Month 2					

4. Calculating balances. You have a credit card with an APR of 12%. You begin with a balance of $200, in response to which you make a payment of $75. The first month you make charges amounting to $50. You make a payment of $75 to reduce the new balance, and the second month you charge $60. Complete the following table:

	Previous balance	Payments	Purchases	Finance charge	New balance
Month 1					
Month 2					

5. A balance statement. Assume you start with a balance of $4500 on your Visa credit card. During the first month, you charge $500, and during the second month, you charge $300. Assume that Visa has finance charges of 24% APR and that each month you make only the minimum payment of 2.5% of the balance. Complete the following table:

Month	Previous balance	Minimum payment	Purchases	Finance charge	New balance
1	$4500.00				
2					

6. A balance statement. Assume you start with a balance of $4500 on your Visa credit card. Assume that Visa has finance charges of 12% APR and that each month you make only the minimum payment of 3% of the balance. During the first month, you charge $300, and during the second month, you charge $600. Complete the following table:

Month	Previous balance	Minimum payment	Purchases	Finance charge	New balance
1	$4500.00				
2					

New balances Assume that you have a balance of $4500 on your Visa credit card and that you make no more charges. Assume that Visa charges 21% APR and that each month you make only the minimum payment of 2.5% of the balance. This is the setting for Exercises 7 through 9.

7. Find a formula for the balance after t monthly payments.

8. What will the balance be after 30 months?

9. What will the balance be after 10 years?

Paying tuition on your American Express card at the maximum interest rate You have a balance of $10,000 for your tuition on your American Express credit card. Assume that you make no more charges on the card. Also assume that American Express charges 24% APR and that each month you make only the minimum payment of 2% of the balance. This is the setting for Exercises 10 through 13.

10. Find a formula for the balance after t monthly payments.

11. How much will you owe after 10 years of payments?

12. How much would you owe if you made 100 years of payments?

13. Find when the balance would be less than $50.

Paying tuition on your American Express card You have a balance of $10,000 for your tuition on your American Express credit card. Assume that you make no more charges on the card. Also assume that American Express charges 12% APR and that each month you make only the minimum payment of 4% of the balance. This is the setting for Exercises 14 and 15.

14. Find a formula for the balance after t monthly payments.

15. How long will it take to get the balance below $50?

Paying off a Visa card You have a balance of $1000 on your Visa credit card. Assume that you make no more charges on the card and that the card charges 9.9% APR and requires a minimum payment of 3% of the balance. Assume also that you make only the minimum payments. This is the setting for Exercises 16 and 17.

16. Find a formula for the balance after t monthly payments.

17. Find how many months it takes to bring the balance below $50.

18. Balance below $200. You have a balance of $4000 on your credit card, and you make no more charges. Assume the card requires a minimum payment of 5% and carries an APR of 22.8%. Assume also that you make only the minimum payments. Determine when the balance drops below $200.

New balances   Assume that you have a balance of $3000 on your Visa credit card and that you make no more charges. Assume that Visa charges 12% APR and that each month you make only the minimum payment of 5% of the balance. This is the setting for Exercises 19 through 23.

19. Find a formula for the balance after t monthly payments.

20. What will the balance be after 30 months?

21. What will the balance be after 10 years?

22. At what balance do you begin making payments of $20 or less?

23. Find how many months it will take to bring the remaining balance down to the value from Exercise 22.

Paying off an American Express card   Assume that you have a balance of $3000 on your American Express credit card and that you make no more charges. Assume that American Express charges 21% APR and that each month you make only the minimum payment of 2% of the balance. This is the setting for Exercises 24 through 26.

24. Find a formula for the remaining balance after t monthly payments.

25. On what balance do you begin making payments of $50 or less?

26. Find how many months it will take to bring the remaining balance down to the value from Exercise 25.

27. Monthly payment. You have a balance of $400 on your credit card and make no more charges. Assume the card carries an APR of 18%. Suppose you wish to pay off the card in six months by making equal payments each month. What is your monthly payment?

28. What can you afford to charge? Suppose you have a new credit card with 0% APR for a limited period. The card requires a minimum payment of 5% of the balance. You feel you can afford to pay no more than $250 each month. How much can you afford to charge? How much could you afford to charge if the minimum payment were 2% instead of 5%?

Average daily balance   The most common way of calculating finance charges is not the simplified one we used in this section but rather the *average daily balance*. With this method, we calculate the account balance at the end of each day of the month and take the average. That average is the amount subject to finance charges. To simplify things, we assume that the billing period is one week rather than one month. Assume that the weekly rate is 1%. You begin with a balance of $500. On day 1, you charge $75. On day 3, you make a payment of $200 and charge $100. On day 6, you charge $200. Use this information for Exercises 29 and 30.

29. Assume that finance charges are calculated using the simplified method shown in this section. Find the account balance at the end of the week.

30. Assume that finance charges are calculated using the average daily balance. Find the account balance at the end of the week.

More on average daily balance   The method for calculating finance charges based on the average daily balance is explained in the setting for Exercises 29 and 30. As in those exercises, to simplify things, we assume that the billing period is one week rather than one month and that the weekly rate is 1%. You begin with a balance of $1000. On day 1, you charge $200. On day 3, you charge $500. On day 6, you make a payment of $400 and you charge $100. Use this information for Exercises 31 and 32.

31. Assume that finance charges are calculated using the simplified method shown in this section. Find the account balance at the end of the week.

32. Assume that finance charges are calculated using the average daily balance. Find the account balance at the end of the week.

33. Finance charges versus minimum payments. In all of the examples in this section, the monthly finance charge is always less than the minimum payment. In fact, this is always the case. Explain what would happen if the minimum payment were less than the monthly finance charge.

34. A bit of history. Credit cards are relatively new. Write a brief report on the introduction of credit cards into the U.S. economy.

The following exercises are designed to be solved using technology such as calculators or computer spreadsheets. For assistance, see the technology supplement.

35. Paying off a Visa card—in detail. Assume that you have a balance of $1000 on your Visa credit card and that you make no more charges on the card. Assume that Visa charges 12% APR and that the minimum payment is 5% of the balance each month. Assume also that you make only the minimum payments. Make a spreadsheet listing the items below for each month until the payment falls below $20.

Month	Previous balance	Minimum payment	Purchases	Finance charge	New balance
1	$1000.00				
2					
and so on					

36. Paying off an American Express card. Assume that you have a balance of $3000 on your American Express credit card and that you make no more charges on the card. Assume that American Express charges 20.5% APR and that the minimum payment is 2% of the balance each month. Assume also that you make only the minimum payments. Make a spreadsheet listing the items below for each month until the payment drops below $50.

Month	Previous balance	Minimum payment	Purchases	Finance charge	New balance
1	$3000.00				
2					
and so on					

4.5 Inflation, taxes, and stocks: Managing your money

The following excerpt is from an article at CNNMoney.com.

Ben Bernanke's High-Wire Act

Fed Chief, in First of Two Days of Testimony on Capitol Hill, Acknowledges Troubling Signs About Economic Growth But Also Raises Concerns About Inflation

DAVID ELLIS   February 28, 2008

WASHINGTON (CNNMoney.com)—For Federal Reserve Chairman Ben Bernanke, running the central bank has become an increasingly challenging high-wire balancing act.

All of Wall Street was watching the Fed chairman on Wednesday when he headed to Capitol Hill to outline the trio of challenges facing the Fed: an economy at risk of falling into a recession, topsy-turvy financial markets and the rising risk of inflation.

"We do face a difficult situation," Bernanke told members of the House Financial Services Committee, marking the first day of his two-day semi-annual hearing on the Fed's monetary policy. "The challenge for us is to balance those risks and decide at any given time which is more serious."

...Bernanke's comments were in line with the Fed's latest economic outlook and remarks he delivered alongside Treasury Secretary Henry Paulson before a Senate panel nearly two weeks ago.

At the time, the two policymakers warned of slower economic growth in the coming year but said they believed the U.S. economy would avoid tipping into a recession, helped in part by the $170 billion economic stimulus package signed by President Bush on Feb. 13 and the most recent interest rate cuts by the Federal Reserve.

Ben Bernanke.

This article refers to financial issues such as inflation, financial markets, and a stimulus package that involved tax rebates. We will explore such issues in this section.

CPI and the inflation rate

The above article states the concerns of the Federal Reserve Board about inflation. But what is inflation, and how is it measured?

Inflation is calculated using the *Consumer Price Index* (CPI), which is a measure of the price of a certain "market basket" of consumer goods and services relative to a predetermined benchmark. When the CPI goes up, we have inflation. When it goes down, we have *deflation*.

According to the U.S. Department of Labor, this "market basket" consists of commodities in the following categories:

- **Food and beverages** (breakfast cereal, milk, coffee, chicken, wine, service meals and snacks)

- **Housing** (rent of primary residence, owners' equivalent rent, fuel oil, bedroom furniture)

- **Apparel** (men's shirts and sweaters, women's dresses, jewelry)

- **Transportation** (new vehicles, airline fares, gasoline, motor vehicle insurance)

- **Medical care** (prescription drugs and medical supplies, physicians' services, eyeglasses and eye care, hospital services)

- **Recreation** (televisions, pets and pet products, sports equipment, admissions)

- **Education and communication** (college tuition, postage, telephone services, computer software and accessories)

- **Other goods and services** (tobacco and smoking products, haircuts and other personal services, funeral expenses)

Key Concept

The **Consumer Price Index (CPI)** is a measure of the average price paid by urban consumers for a "market basket" of consumer goods and services.

A "market basket" of goods and services measures inflation.

The *rate of inflation* is measured by the percentage change in the CPI.

Key Concept

An increase in prices is referred to as **inflation**. The **rate of inflation** is measured by the percentage change in the Consumer Price Index over time. When prices decrease, the percentage change is negative; this is referred to as **deflation**.

TABLE 4.3 Historical Inflation

December	CPI	Inflation rate	December	CPI	Inflation rate
1945	18.2	—	1978	67.7	9.0%
1946	21.5	18.1%	1979	76.7	13.3%
1947	23.4	8.8%	1980	86.3	12.5%
1948	24.1	3.0%	1981	94.0	8.9%
1949	23.6	−2.1%	1982	97.6	3.8%
1950	25.0	5.9%	1983	101.3	3.8%
1951	26.5	6.0%	1984	105.3	3.9%
1952	26.7	0.8%	1985	109.3	3.8%
1953	26.9	0.7%	1986	110.5	1.1%
1954	26.7	−0.7%	1987	115.4	4.4%
1955	26.8	0.4%	1988	120.5	4.4%
1956	27.6	3.0%	1989	126.1	4.6%
1957	28.4	2.9%	1990	133.8	6.1%
1958	28.9	1.8%	1991	137.9	3.1%
1959	29.4	1.7%	1992	141.9	2.9%
1960	29.8	1.4%	1993	145.8	2.7%
1961	30.0	0.7%	1994	149.7	2.7%
1962	30.4	1.3%	1995	153.5	2.5%
1963	30.9	1.6%	1996	158.6	3.3%
1964	31.2	1.0%	1997	161.3	1.7%
1965	31.8	1.9%	1998	163.9	1.6%
1966	32.9	3.5%	1999	168.3	2.7%
1967	33.9	3.0%	2000	174.0	3.4%
1968	35.5	4.7%	2001	176.7	1.6%
1969	37.7	6.2%	2002	180.9	2.4%
1970	39.8	5.6%	2003	184.3	1.9%
1971	41.1	3.3%	2004	190.3	3.3%
1972	42.5	3.4%	2005	196.8	3.4%
1973	46.2	8.7%	2006	201.8	2.5%
1974	51.9	12.3%	2007	210.0	4.1%
1975	55.5	6.9%	2008	210.2	0.1%
1976	58.2	4.9%	2009	215.9	2.7%
1977	62.1	6.7%	2010	219.2	1.5%

Inflation reflects a decline of the purchasing power of the consumer's dollar. Besides affecting how much we can afford to buy, the inflation rate has a big influence on certain government programs that impact the lives of many people.

For example, about 48 million Social Security beneficiaries, 20 million food stamp recipients, 4 million military and Federal Civil Service retirees and survivors, and 27 million children who eat subsidized lunches at school are affected by the CPI because these benefits are adjusted periodically to compensate for inflation.

Table 4.3 shows the annual change in prices in the United States over a 65-year period. For example, from December 1945 to December 1946, the CPI changed from 18.2 to 21.5, an increase of $21.5 - 18.2 = 3.3$. Now we find the percentage change:

$$\text{Percentage change} = \frac{\text{Change in CPI}}{\text{Previous CPI}} \times 100\% = \frac{3.3}{18.2} \times 100\%.$$

This is about 18.1%, and that is the inflation rate for this period shown in the table. Usually, we round the inflation rate as a percentage to one decimal place.

EXAMPLE 4.30 Calculating inflation: CPI increase to 205

Suppose the CPI increases this year from 200 to 205. What is the rate of inflation for this year?

SOLUTION

The change in the CPI is $205 - 200 = 5$. To find the percentage change, we divide the increase of 5 by the original value of 200 and convert to a percent:

$$\text{Percentage change} = \frac{\text{Change in CPI}}{\text{Previous CPI}} \times 100\% = \frac{5}{200} \times 100\% = 2.5\%.$$

Therefore, the rate of inflation is 2.5%.

TRY IT YOURSELF 4.30

Suppose the CPI increases this year from 215 to 225. What is the rate of inflation for this year?

The answer is provided at the end of this section.

"When you take out food, energy, taxes, insurance, housing, transportation, healthcare, and entertainment, inflation remained at a 20 year low."

Some countries have experienced very high rates of inflation, sometimes referred to as *hyperinflation*. Table 4.4 lists the five countries with the highest rates of inflation for 2008. Note that the inflation rate in Zimbabwe for 2008 is estimated to be 11.2 million percent.

If the rate of inflation is 10%, one may think that the buying power of a dollar has decreased by 10%, but that is not the case. Inflation is the percentage change in

TABLE 4.4 Examples of Hyperinflation in 2008

Country	Estimated inflation rate
Zimbabwe	11,200,000%
Ethiopia	41%
Venezuela	31%
Guinea	30%
Mongolia	28%

prices, and that is not the same as the percentage change in the value of a dollar. To see the difference, let's imagine a frightening inflation rate of 100% this year. With such a rate, an item that costs $200 this year will cost $400 next year. This means that my money can buy only half as much next year as it can this year. So the buying power of a dollar would decrease by 50%, not by 100%.

The following formula tells us how much the buying power of currency decreases for a given inflation rate.

FORMULA 4.13   Buying Power Formula

$$\text{Percent decrease in buying power} = \frac{100\,i}{100 + i}.$$

Here, i is the inflation rate expressed as a percent, not a decimal. Usually, we round the decrease in buying power as a percentage to one decimal place.

The buying power formula is derived in Algebraic Spotlight 4.6 at the end of this section.

EXAMPLE 4.31 Calculating decrease in buying power: 5% inflation

Suppose the rate of inflation this year is 5%. What is the percentage decrease in the buying power of a dollar?

SOLUTION

We use the buying power formula (Formula 4.13) with $i = 5\%$:

$$
\begin{aligned}
\text{Percent decrease in buying power} &= \frac{100\,i}{100 + i} \\
&= \frac{100 \times 5}{100 + 5}.
\end{aligned}
$$

This is about 4.8%.

TRY IT YOURSELF 4.31

According to Table 4.4, in 2008 the rate of inflation for Venezuela was 31%. What was the percentage decrease that year in the buying power of the *bolívar* (the currency of Venezuela)?

The answer is provided at the end of this section.

A companion formula to the buying power formula (Formula 4.13) gives the inflation rate in terms of the percent decrease in buying power of currency.

FORMULA 4.14   Inflation Formula

$$\text{Percent rate of inflation} = \frac{100\,B}{100 - B}.$$

In this formula, B is the decrease in buying power expressed as a percent, not as a decimal.

EXAMPLE 4.32 Calculating inflation: 2.5% decrease in buying power

Suppose the buying power of a dollar decreases by 2.5% this year. What is the rate of inflation this year?

SOLUTION

We use the inflation formula (Formula 4.14) with $B = 2.5\%$:

$$\text{Percent rate of inflation} = \frac{100B}{100 - B}$$
$$= \frac{100 \times 2.5}{100 - 2.5}.$$

This is about 2.6%.

TRY IT YOURSELF 4.32

Suppose the buying power of a dollar decreases by 5.2% this year. What is the rate of inflation this year?

The answer is provided at the end of this section.

The next example covers all of the concepts we have considered so far in this section.

EXAMPLE 4.33 Understanding inflation and buying power: Effects on goods

Parts a through c refer to Table 4.3.

a. Find the 10-year inflation rate in the United States from December 2000 to December 2010.

b. If a sofa cost $100 in December 1966 and the price changed in accordance with the inflation rate in the table, how much did the sofa cost in December 1967?

c. If a chair cost $50 in December 1948 and the price changed in accordance with the inflation rate in the table, how much did the chair cost in December 1949?

d. According to Table 4.4, in 2008 the rate of inflation for Ethiopia was 41%. How much did the buying power of the currency, the *birr*, decrease during the year?

e. The inflation rate in Ukraine for 2008 was 25.0%. In Kenya, the buying power of the currency decreased by 20.3% in 2008. Which of these two countries had the larger inflation rate for 2008?

SOLUTION

a. In 2000 the CPI was 174.0, and in 2010 it was 219.2. The increase was $219.2 - 174.0 = 45.2$, so

$$\text{Percentage change} = \frac{\text{Change in CPI}}{\text{Previous CPI}} \times 100\% = \frac{45.2}{174.0} \times 100\%,$$

or about 26.0%. The 10-year inflation rate was about 26.0%.

b. According to the table, the inflation rate was 3% in 1967, which tells us that the price of the sofa increased by 3% during that year. Therefore, it cost $100 + 0.03 \times \$100 = \$103$ in December 1967.

c. The inflation rate is −2.1%, which tells us that the price of the chair decreased by 2.1% during that year. Because 2.1% of $50.00 is $1.05, in December 1949 the chair cost $50.00 − $1.05 = $48.95.

d. We use the buying power formula (Formula 4.13) with $i = 41\%$:

$$\text{Percent decrease in buying power} = \frac{100\,i}{100 + i}$$
$$= \frac{100 \times 41}{100 + 41},$$

or about 29.1%. The buying power of the Ethiopian birr decreased by 29.1%.

e. We want to find the inflation rate for Kenya. We know that the reduction in buying power was 20.3%, so we use the inflation formula (Formula 4.14) with $B = 20.3\%$:

$$\text{Percent rate of inflation} = \frac{100\,B}{100 - B}$$
$$= \frac{100 \times 20.3}{100 - 20.3},$$

or about 25.5%. The inflation rate of 25.5% in Kenya was higher than the inflation rate of 25.0% in Ukraine.

> **SUMMARY 4.9**   **Inflation and Reduction of Currency Buying Power**
>
> If the inflation rate is i (expressed as a percent), the change in the buying power of currency can be calculated using
>
> $$\text{Percent decrease in buying power} = \frac{100\,i}{100 + i}.$$
>
> A companion formula gives the inflation rate in terms of the decrease B (expressed as a percent) in buying power:
>
> $$\text{Percent rate of inflation} = \frac{100\,B}{100 - B}.$$

Income taxes

Refer to the included tax tables for the year 2000 from the Internal Revenue Service. **Table 4.5** shows tax rates for single people, and **Table 4.6** shows tax rates for married couples filing jointly. Note that the tax rates are applied to *taxable income*. The percentages in the tables are called *marginal rates*, and they apply only to earnings in excess of a certain amount. With a marginal tax rate of 15%, for example, the tax

TABLE 4.5 2000 Tax Table for Singles

If Taxable Income		The Tax is		
Is over	But not over	This amount	Plus this %	Of the excess over
Schedule X—Use if your filing status is Single				
$0	$26,250	—	15%	$0
26,250	63,550	$3,937.50	28%	26,250
63,550	132,600	14,381.50	31%	63,550
132,600	288,350	35,787.00	36%	132,600
288,350	—	91,857.00	39.6%	288,350

TABLE 4.6 2000 Tax Table for Married Couples Filing Jointly

If Taxable Income		The Tax is		
Is over	But not over	This amount	Plus this %	Of the excess over
Schedule Y-1—Use if your filing status is Married filing jointly or Qualifying widow(er)				
$0	$43,850	—	15%	$0
43,850	105,950	$6,577.50	28%	43,850
105,950	161,450	23,965.50	31%	105,950
161,450	288,350	41,170.50	36%	161,450
288,350	—	86,854.50	39.6%	288,350

owed increases by $0.15 for every $1 increase in taxable income. This makes the tax owed a linear function of the taxable income within a given range of incomes. The slope is the marginal tax rate (as a decimal).

Note that these marginal rates increase as you earn more and so move from one *tax bracket* to another. A system of taxation in which the marginal tax rates increase for higher incomes is referred to as a *progressive tax*.[12]

EXAMPLE 4.34 Calculating the tax: A single person

In the year 2000, Alex was single and had a taxable income of $70,000. How much tax did she owe?

SOLUTION

According to Table 4.5, Alex owed $14,381.50 plus 31% of the excess taxable income over $63,550. The total tax is

$$\$14,381.50 + 0.31 \times (\$70,000 - \$63,550) = \$16,381.00.$$

TRY IT YOURSELF 4.34

In the year 2000, Bob was single and had a taxable income of $50,000. How much tax did he owe?

The answer is provided at the end of this section.

Income taxes are part of financial planning.

A person's taxable income is obtained by subtracting certain *deductions* from total income. Everyone is allowed a lump sum "personal exemption" deduction, but other deductions can be more complicated and may include things like state and local taxes, home mortgage interest, and charitable contributions. Some people do not have very many of these kinds of deductions. In this case, they may choose not to itemize them but rather to take what is called a "standard deduction" (in addition to their personal exemption).

EXAMPLE 4.35 Comparing taxes: The "marriage penalty"

a. In the year 2000, Ann was single and made a salary of $30,000 per year. She did not itemize deductions and instead took the standard deduction of $4400 plus her personal exemption of $2800. What was Ann's taxable income, and how much income tax did she owe?

---

[12]In 1996 and 2000, magazine publisher Steve Forbes ran for president on a platform that included a 17% "flat tax." In that system everyone would pay 17% of their taxable income no matter what that income is. In a speech he said, "When we're through with Washington, the initials of the IRS will be RIP." Forbes did not win the Republican nomination.

b. In the year 2000, Bill was single and made a salary of $30,000 per year. Bill and Ann got married, and they filed jointly. A married couple making $60,000 per year filing jointly in the year 2000 was given a personal exemption of $2800 each plus a standard deduction as a couple of $7350 if they did not itemize. What was Bill and Ann's taxable income, and how much income tax did they owe if they did not itemize?

c. Explain what you notice about the amount of tax paid by Ann and Bill as separate single people versus the amount they pay as a married couple filing jointly.

SOLUTION

a. Ann's deductions came to $2800 + $4400 = $7200. This gives her a total taxable income of $30,000 − $7200 = $22,800. According to the tax table for singles, Ann owed 15% of $22,800, which is $3420, in income tax.

b. Ann and Bill's deductions came to $2 \times \$2800 + \$7350 = \$12,950$. This gave them a total taxable income of $60,000 − $12,950 = $47,050. According to the tax table for married couples shown in Table 4.6, they owe $6577.50 plus 28% of the excess taxable income over $43,850. That is

$$\$6577.50 + 0.28 \times (\$47,050 - \$43,850) = \$7473.50.$$

c. Ann and Bill as a married couple paid $7473.50 in taxes, but if they filed as single people, according to part a, their combined tax liability would be $2 \times \$3420 = \$6840$. That is a difference of $633.50. This means that because they were married and filed jointly, they had to pay an extra $633.50 in taxes with absolutely no change in their incomes.

"We were going over some of your returns
from a past life and..."

The disparity in part c of Example 4.35 is referred to as the "marriage penalty." Many people believe the marriage penalty is unfair, but there are those who argue that it makes sense to tax a married couple more than two single people. Can you think of some arguments to support the two sides of this question? The impact of the marriage penalty was reduced in the years following 2000. See Exercises 24 through 26.

Claiming deductions lowers your tax by reducing your taxable income. Another way to lower your tax is to take a *tax credit*. Here is how to apply a tax credit: Calculate the tax owed using the tax tables (making sure first to subtract from the total income any deductions), and then subtract the tax credit from the tax determined by the tables. Because a tax credit is subtracted directly from the tax you owe, a tax credit of $1000 has a much bigger impact on lowering your taxes than a deduction of $1000. That is the point of the following example.

EXAMPLE 4.36 Comparing deductions and credits: Differing effects

In the year 2000, Betty and Carol were single, and each had a total income of $75,000. Betty took a deduction of $10,000 but had no tax credits. Carol took a deduction of $9000 and had an education tax credit of $1000. Compare the tax owed by Betty and Carol.

SOLUTION

The taxable income of Betty is $75,000 − $10,000 = $65,000. According to the tax table in Table 4.5, Betty owes $14,381.50 plus 31% of the excess taxable income over $63,550. That tax is

$$\$14,381.50 + 0.31 \times (\$65,000 - \$63,550) = \$14,831.00.$$

Betty has no tax credits, so the tax she owes is $14,831.00.

The taxable income of Carol is $75,000 − $9000 = $66,000. According to the tax table in Table 4.5, before applying tax credits, Carol owes $14,381.50 plus 31% of the excess taxable income over $63,550. That tax is

$$\$14,381.50 + 0.31 \times (\$66,000 - \$63,550) = \$15,141.00.$$

Carol has a tax credit of $1000, so the tax she owes is

$$\$15,141.00 - \$1000 = \$14,141.00.$$

Betty owes

$$\$14,831.00 - \$14,141.00 = \$690.00$$

more tax than Carol.

TRY IT YOURSELF 4.36

In the year 2000, Dave was single and had a total income of $65,000. He took a deduction of $8000 and had a tax credit of $1800. Calculate the tax owed by Dave.

The answer is provided at the end of this section.

In Example 4.36, the effect of replacing a $1000 deduction by a $1000 credit was to reduce the tax owed by $690. This is a significant reduction in taxes and highlights the benefits of tax credits.

The Dow

In the late nineteenth century, tips and gossip caused stock prices to move because solid information was hard to come by. This prompted Charles H. Dow to introduce the Dow Jones Industrial Average (DJIA) in May 1896 as a benchmark to gauge the

The New York Stock Exchange.

state of the market. The original DJIA was simply the average price of 12 stocks that Mr. Dow picked himself. Today the Dow, as it is often called, consists of 30 "blue-chip" U.S. stocks picked by the editors of the *Wall Street Journal*. For example, in June 2009 General Motors was removed from the list as it entered bankruptcy protection. Here is the list as of this writing.

The 30 Dow Companies

- 3M Company
- Alcoa Incorporated
- American Express Company
- AT&T Incorporated
- Bank of America Corporation
- Boeing Company
- Caterpillar Incorporated
- Chevron Corporation
- Cisco Systems Incorporated
- Coca-Cola Company
- E.I. DuPont de Nemours & Company
- Exxon Mobil Corporation
- General Electric Company
- Hewlett-Packard Company
- Home Depot Incorporated
- Intel Corporation
- International Business Machines Corporation
- Johnson & Johnson
- JPMorgan Chase & Company
- Kraft Foods Incorporated
- McDonald's Corporation
- Merck & Company Incorporated
- Microsoft Corporation
- Pfizer Incorporated
- Procter & Gamble Company
- Travelers Companies Incorporated
- United Technologies Corporation
- Verizon Communications Incorporated
- Wal-Mart Stores Incorporated
- Walt Disney Company

As we said earlier, the original DJIA was a true average—that is, you simply added up the stock prices of the 12 companies and divided by 12. In 1928 a divisor of 16.67 was used to adjust for mergers, takeovers, bankruptcies, stock splits, and company substitutions. Today, they add up the 30 stock prices and divide by 0.132319125 (the *divisor*), or equivalently, multiply by 1/0.132319125 or about 7.56. This means that for every $1 move in any Dow company's stock price, the average changes by about 7.56 points. (The DJIA is usually reported using two decimal places.)

EXAMPLE 4.37 Finding changes in the Dow: Disney goes up

Suppose the stock of Walt Disney increases in value by $3 per share. If all other Dow stock prices remain unchanged, how does this affect the DJIA?

SOLUTION

Each $1 increase causes the average to increase by about 7.56 points. So a $3 increase would cause an increase of about $3 \times 7.56 = 22.68$ points in the Dow.

TRY IT YOURSELF 4.37

Suppose the stock of Microsoft decreases in value by $4 per share. If all other Dow stock prices remain unchanged, how does this affect the DJIA?

The answer is provided at the end of this section.

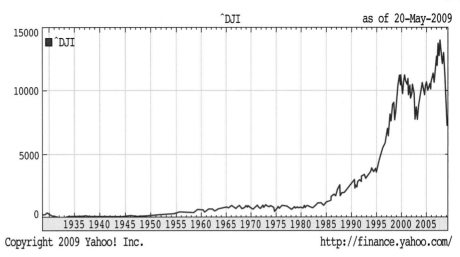

^DJI   as of 20-May-2009

Copyright 2009 Yahoo! Inc.   http://finance.yahoo.com/

FIGURE 4.9 This graph shows the movement of the Dow since the early 1900's.

The graph in **Figure 4.9** shows how the Dow has moved over the last several decades. In Exercises 32 through 35, we explore a few of the more common types of stock transactions.

ALGEBRAIC SPOTLIGHT 4.6 Derivation of the Buying Power Formula

Suppose the inflation rate is $i\%$ per year. We want to derive the buying power formula:

$$\text{Percent decrease in buying power} = \frac{100\,i}{100 + i}.$$

Suppose we could buy a commodity, say, one pound of flour, for one dollar a year ago. An inflation rate of $i\%$ tells us that today that same pound of flour would cost $1 + i/100$ dollars. To find the new buying power of the dollar, we need to know how much flour we could buy today for one dollar. Because one pound of flour costs $1 + i/100$ dollars, one dollar will buy

$$\frac{1}{1 + i/100} = \frac{100}{100 + i} \text{ pounds of flour.}$$

This quantity represents a decrease of

$$1 - \frac{100}{100 + i} = \frac{i}{100 + i} \text{ pounds}$$

from the one pound we could buy with one dollar a year ago. The percentage decrease in the amount of flour we can buy for one dollar is

$$\text{Percentage decrease} = \frac{\text{Decrease}}{\text{Previous value}} \times 100\% = \frac{\frac{i}{100+i}}{1} \times 100\% = \frac{100\,i}{100 + i}.$$

This is the desired formula for the percentage decrease in buying power.

Try It Yourself answers

Try It Yourself 4.30: Calculating inflation: CPI increase to 225   4.7%.

Try It Yourself 4.31: Calculating decrease in buying power: 31% inflation   23.7%.

Try It Yourself 4.32: Calculating inflation: 5.2% decrease in buying power   5.5%.

Try It Yourself 4.34: Calculating the tax: A single person $10,587.50.

Try It Yourself 4.36: Comparing deductions and credits: Differing effects $10,747.50.

Try It Yourself 4.37: Finding changes in the Dow: Microsoft goes down   The DJIA decreases by 30.24 points.

Exercise Set 4.5

Tables 4.7 and 4.8 show tax tables for the year 2010 that we will refer to in the exercises. There were some significant changes from 2000 to 2010.

1. Large inflation rate. The largest annual rate of inflation in the CPI table, Table 4.3 on page 250, was 18.1% for the year 1946. What year saw the next largest rate? Why do you think the inflation rate in the table for 1945 is blank?

Food inflation   The excerpt on the following page from an article at the Web site of the Press-Enterprise discusses inflation of food prices.

Exercises 2 through 4 refer to this article.

2. If your food bill was $3000 in 2005, what was it in 2006?

3. If your food bill was $3000 in 2006, what was it in 2007?

4. Assume that the rate of food inflation did double from 2007 to 2008. If your food bill was $3000 in 2007, what was it in 2008?

Inflation compounded   In Exercises 5 and 6, we see the cumulative effects of inflation. We refer to Table 4.3 on page 250.

5. Find the three-year inflation rate from December 1977 to December 1980.

6. Consider the one-year inflation rate for each of the three years from December 1977 to December 1980. Find the sum of these three numbers. Is the sum the same as your answer to Exercise 5?

TABLE 4.7 2010 Tax Table for Singles

If Taxable Income		The Tax is		
Is over	But not over	This amount	Plus this %	Of the excess over
Schedule X—Use if your filing status is Single				
$0	$8,375	—	10%	$0
8,375	34,000	$837.50	15%	8,375
34,000	82,400	4,681.25	25%	34,000
82,400	171,850	16,781.25	28%	82,400
171,850	373,650	41,827.25	33%	171,850
373,650	—	108,421.25	35%	373,650

TABLE 4.8 2010 Tax Table for Married Couples Filing Jointly

If Taxable Income		The Tax is		
Is over	But not over	This amount	Plus this %	Of the excess over
Schedule Y-1—Use if your filing status is Married filling jointly or Qualifying widow(er)				
$0	$16,750	—	10%	$0
16,750	68,000	$1,675.00	15%	16,750
68,000	137,300	9,362.50	25%	68,000
137,300	209,250	26,687.50	28%	137,300
209,250	373,650	46,833.50	33%	209,250
373,650	—	101,085.50	35%	373,650

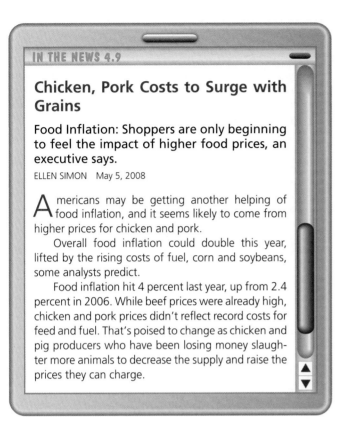

IN THE NEWS 4.9

Chicken, Pork Costs to Surge with Grains

Food Inflation: Shoppers are only beginning to feel the impact of higher food prices, an executive says.

ELLEN SIMON   May 5, 2008

Americans may be getting another helping of food inflation, and it seems likely to come from higher prices for chicken and pork.

Overall food inflation could double this year, lifted by the rising costs of fuel, corn and soybeans, some analysts predict.

Food inflation hit 4 percent last year, up from 2.4 percent in 2006. While beef prices were already high, chicken and pork prices didn't reflect record costs for feed and fuel. That's poised to change as chicken and pig producers who have been losing money slaughter more animals to decrease the supply and raise the prices they can charge.

More on compounding inflation   Here is a hypothetical CPI table.

Year	Hypothetical CPI	Inflation rate
1935	10	—
1936	20	%
1937	40	%
1938	80	%

Use this table for Exercises 7 through 10.

7. Fill in the missing inflation rates.

8. Find the three-year inflation rate from 1935 to 1938.

9. Find the sum of the three inflation rates during the three years from 1935 to 1938. Is your answer here the same as your answer to Exercise 8?

10. Use the idea of compounding from Section 4.1 to explain the observation in Exercise 9.

11. Guinean franc. Table 4.4 on page 251 shows that the inflation rate for Guinea in the year 2008 was 30%. By how much did the buying power of the Guinean franc (the currency in Guinea) decrease during 2008?

12. Find the inflation rate. Suppose the buying power of a dollar went down by 60% over a period of time. What was the inflation rate during that period?

13. Continuing inflation. Suppose that prices increase 3% each year for 10 years. How much will a jacket that costs $80 today cost in 10 years? *Suggestion:* The price of the jacket increases

by the same percentage each year, so the price is an exponential function of the time in years. You can think of the price as the balance in a savings account with an APY of 3% and an initial investment of $80.

14. More on continuing inflation. *This is a continuation of Exercise 13.* If prices increase 3% each year for 10 years, what is the percentage decrease in the buying power of currency over the 10-year period?

Flat tax   Steve Forbes ran for U.S. president in 1996 and 2000 on a platform proposing a 17% flat tax, that is, an income tax that would simply be 17% of each tax payer's taxable income. Suppose that Alice was single in the year 2010 with a taxable income of $30,000 and that Joe was single in the year 2010 with a taxable income of $300,000. Use this information for Exercises 15 through 19.

15. What was Alice's tax? Use the tax tables on page 260.

16. What was Joe's tax? Use the tax tables on page 260.

17. If the 17% flat tax proposed by Mr. Forbes had been in effect in 2010, what would Alice's tax have been?

18. What would Joe's tax be under the 17% flat tax?

19. When you compare Alice and Joe, what do you think about the fairness of the flat tax versus a progressive tax?

More on the flat tax   Let's return to Alice and Joe from Exercises 15 through 19. We learn that Joe actually made $600,000, but his taxable income was only $300,000 because of various deductions allowed by the system in 2010. Proponents of the flat tax say that many of these deductions should be eliminated, so the 17% flat tax should be applied to Joe's entire $600,000. Use this information for Exercises 20 through 23.

20. What would Joe's tax be under the 17% flat tax?

21. How much more tax would Joe pay than under the 2010 system?

22. How much income would Joe have to make for the 17% flat tax to equal the amount he pays in the year 2010 with a taxable income of $300,000?

23. When you compare Alice and Joe now, what do you think about the fairness of the flat tax versus a progressive tax?

2010 taxes   This exercise re-does Example 4.35 using the tax tables and deduction rates for the year 2010. In 2010 the personal exemption was $3650, the standard deduction for a single person was $5700, and the standard deduction for a couple filing jointly was $11,400. Use this information for Exercises 24 through 26.

24. In the year 2010, Ann was single and made a salary of $30,000 per year. She took the standard deduction (plus her personal exemption). What was Ann's taxable income, and how much income tax did she owe? Use the tax tables on page 260.

25. In the year 2010, Bill was single and made a salary of $30,000 per year also. Bill and Ann got married. They filed jointly and took the standard deduction (plus their personal exemptions). What was their taxable income, and how much income tax did they owe? Use the tax tables on page 260.

26. Explain what you notice about the amount of tax paid by Ann and Bill as separate single people versus the amount they pay as a married couple filing jointly.

27. Bracket creep. At the start of 2010, your taxable income was $34,000, and you received a cost-of-living raise because of inflation. In 2010, inflation was 1.5% and your raise resulted in a 1.5% increase in your taxable income. By how much, and by what percent, did your taxes go up over what they would have been without a raise? (Assume that you were single in 2010, and use the tax tables on page 260.) *Remark*: Note that your buying power remains the same, but you're paying higher taxes. Not only that, but you're paying at a higher marginal rate! This phenomenon is known as "bracket creep," and federal tax tables are adjusted each year to account for this.

28. Deduction and credit. In the year 2010, Ethan was single and had a total income of $55,000. He took a deduction of $9000 and had a tax credit of $1500. Calculate the tax owed by Ethan.

29. Moving DJIA. Suppose the stock of McDonald's increases in value by $2 per share. If all other Dow stock prices remain unchanged, how does this affect the DJIA?

30. Average price. What is the average price of a share of stock in the Dow list when the DJIA is 10,000?

31. Dow highlights. Use Figure 4.9 to determine in approximately what year the DJIA first reached 5000. About when did it first reach 10,000?

Exercises 32 through 35 are suitable for group work.

Stock market transactions   There are any number of ways to make (or lose) money with stock market transactions. In Exercises 32 through 35, we will explore a few of the more common types of transactions. (Fees for such transactions will be ignored.)

32. Market order. The simplest way to buy stock is the *market order*. Through your broker or online, you ask to buy 100 shares of stock X at market price. As soon as a seller is located, the transaction is completed at the prevailing price. That price will normally be very close to the latest quote, but the prevailing price may be different if the price fluctuates between the time you place the order and the time the transaction is completed. Suppose you place a market order for 100 shares of stock X and the transaction is completed at $44 per share. Two weeks later the stock value is $58 per share and you sell. What is your net profit?

33. Limit orders. If you want to insist on a fixed price for a transaction, you place a *limit order*. That is, you offer to buy (or sell) stock X at a certain price. If the stock can be purchased for that price, the transaction is completed. If not, no transaction occurs. Often limit orders have a certain expiration date. Suppose you place a limit order to buy 100 shares of stock X for $40 per share. When the stock purchase is completed, you plan to place immediately a limit order to sell stock X at $52 per share. The following table shows the value of stock X. On which days are these two transactions completed, and what is your profit?

Date	Day 1	Day 2	Day 3	Day 4	Day 5	Day 6
Market price	$44	$45	$40	$48	$52	$50

34. Stop loss and trailing stops. If you own a stock, a *stop loss* order protects you from large losses. For example, if you own 100 shares of stock X, bought at $45 per share, you might place a stop loss order for $40 per share. This order automatically sells your stock if the price drops to $40. No matter what happens, you can't lose more than $5 per share. A similar type of order that protects profits is the *trailing stop*. The trailing stop order sells your stock if the value goes below a certain percentage of the market price. If the market price remains the same or drops, the trailing stop doesn't change and acts like a stop loss order. If on the other hand, the market price goes up, the trailing stop follows it so that it protects profits. Suppose, for example, that you own 100 shares of stock X, that the market price is $40 per share, and that you place a trailing loss order of 5%. If the price drops by $2 (5% of $40), the stock is sold. If on the other hand, the market value increases to $44, then you will sell the stock when it declines by 5% of $44. That is, you sell when the market price drops to $41.80. Consider the table in Exercise 33, and suppose that you purchase 100 shares of the stock and place a 5% trailing stop order on day 1. On which day (if any) will your stock be sold?

35. Selling short. *Selling short* is the selling of stock you do not actually own but promise to deliver. Suppose you place an order to sell short 100 shares of stock X at $35. Eventually, your order must be *covered*. That is, you must sell 100 shares of stock X at a value of $35 per share. If when the order is covered, the value of the stock is less than $35 per share, then you make money; otherwise, you lose money. Suppose that on the day you must cover the sell short order, the price of stock X is $50 per share. How much money did you lose?

36. History: More on selling short. In 1992 George Soros "broke the Bank of England" by selling short the British pound. Write a brief report on his profit and exactly how he managed to make it.

37. History: The Knights Templar. The Knights Templar was a monastic order of knights founded in 1112 to protect pilgrims traveling to the Holy Land. Recent popular novels have revived interest in them. Some have characterized the Knights Templar as the first true international bankers. Report on the international aspects of their early banking activities.

38. History: The stock market. The Dow Jones Industrial Average normally fluctuates, but over the last half-century it has generally increased. Dramatic drops in the Dow (stock market crashes) can have serious effects on the economy. Report on some of the most famous of these. Be sure to include the crash of 1929.

39. History: The Federal Reserve Bank. The Federal Reserve Bank is an independent agency that regulates various aspects of American currency. Write a report on the Federal Reserve Bank. Your report should include the circumstances of its creation.

40. History: The SEC. The Securities and Exchange Commission regulates stock market trading in the United States. Write a report on the creation and function of the SEC.

The following exercises are designed to be solved using technology such as calculators or computer spreadsheets. For assistance, see the technology supplement.

Mortgage interest deduction Interest paid on a home mortgage is normally tax-deductible. That is, you can subtract the total mortgage interest paid over the year in determining your taxable income. This is one advantage of buying a home. Suppose you take out a 30-year home mortgage for $250,000 at an APR of 8% compounded monthly. Use this information for Exercises 41 through 44.

41. Determine your monthly payment using the monthly payment formula in Section 4.2.

42. Make a spreadsheet that shows for each month of the first year your payment, the amount that represents interest, the amount toward the principal, and the balance owed.

43. Use the results of Exercise 42 to find the total interest paid over the first year. (Round your answer to the nearest dollar.) That is what you get to deduct from your taxable income.

44. Suppose that your marginal tax rate is 27%. What is your actual tax savings due to mortgage payments? Does this make the $250,000 home seem a bit less expensive?

CHAPTER SUMMARY

This chapter is concerned with financial transactions of two basic kinds: saving and borrowing. We also consider important financial issues related to inflation, taxes, and the stock market.

Saving money: The power of compounding

The principal in a savings account typically grows by interest earned. Interest can be credited to a savings account in two ways: as *simple interest* or as *compound interest*. For simple interest, the formula for the interest earned is

Simple interest earned = Principal × Yearly interest rate (as a decimal) × Time in years.

Financial institutions normally compound interest and report the *annual percentage rate* or APR. The interest rate for a given compounding period is calculated using

$$\text{Period interest rate} = \frac{\text{APR}}{\text{Number of periods in a year}}.$$

We can calculate the account balance after t periods using the compound interest formula:

$$\text{Balance after } t \text{ periods} = \text{Principal} \times (1 + r)^t.$$

Here, r is the period interest rate expressed as a decimal, and it should not be rounded. In fact, it is best to do all the calculations and then round.

The *annual percentage yield* or APY is the actual percentage return in a year. It takes into account compounding of interest and is always at least as large as the APR. If n is the number of compounding periods per year,

$$\text{APY} = \left(1 + \frac{\text{APR}}{n}\right)^n - 1.$$

Here, both the APR and the APY are in decimal form. The APY can be used to calculate the account balance after t years:

$$\text{Balance after } t \text{ years} = \text{Principal} \times (1 + \text{APY})^t.$$

Here, the APY is in decimal form.

The *present value* of an investment is the amount we initially invest. The *future value* is the value of that investment at some specified time in the future. If the investment grows by compounding of interest, these two quantities are related by the compound interest formula. We can rearrange that formula to give the present value we need for a desired future value:

$$\text{Present value} = \frac{\text{Future value}}{(1 + r)^t}.$$

In this formula, t is the total number of compounding periods, and r is the period interest rate expressed as a decimal.

The *Rule of 72* can be used to estimate how long it will take for an account growing by compounding of interest to double in size. It says that the doubling time in years can be approximated by dividing 72 by the APR, where the APR is expressed as a percentage, not as a decimal. The exact doubling time can be found using the formula:

$$\text{Number of periods to double} = \frac{\log 2}{\log(1 + r)}.$$

Here, r is the period interest rate as a decimal.

Borrowing: How much car can you afford?

With an *installment loan*, you borrow money for a fixed period of time, called the *term* of the loan, and you make regular payments (usually monthly) to pay off the loan plus interest in that time. Loans for the purchase of a car or home are usually installment loans.

If you borrow an amount at a monthly interest rate r (as a decimal) with a term of t months, the monthly payment is

$$\text{Monthly payment} = \frac{\text{Amount borrowed} \times r(1 + r)^t}{((1 + r)^t - 1)}.$$

It is best to do all the calculations and then round. A companion formula tells how much you can borrow for a given monthly payment:

$$\text{Amount borrowed} = \frac{\text{Monthly payment} \times ((1 + r)^t - 1)}{(r \times (1 + r)^t)}.$$

For all loans, the monthly payment is *at least* the amount we would pay each month if no interest were charged, which is the amount of the loan divided by the term (in months) of the loan. This number can be used to estimate the monthly payment for a short-term loan if the APR is not large. For all loans, the monthly payment is *at least* as large as the principal times the monthly interest rate as a decimal. This number can be used to estimate the monthly payment for a long-term loan with a moderate or high interest rate.

A record of the repayment of a loan is kept in an *amortization table*. In the case of buying a home, an important thing for the borrower to know is how much *equity* he or she has in the home. The equity is the total amount that has been paid toward the principal, and an amortization table keeps track of this amount.

Some home loans are in the form of an *adjustable-rate mortgage* or ARM, where the interest rate may vary over the life of the loan. For an ARM, the initial rate is often lower than the rate for a comparable fixed-rate mortgage, but rising rates may cause significant increases in the monthly payment.

Saving for the long term: Build that nest egg

Another way to save is to deposit a certain amount into your savings account at the end of each month. If the monthly interest rate is r as a decimal, your balance is given by

$$\text{Balance after } t \text{ deposits} = \frac{\text{Deposit} \times ((1 + r)^t - 1)}{r}.$$

The ending balance is often called the *future value* for this savings arrangement. A companion formula gives the amount we need to deposit regularly in order to achieve a goal:

$$\text{Needed deposit} = \frac{\text{Goal} \times r}{((1+r)^t - 1)}.$$

Retirees typically draw money from their nest eggs in one of two ways: either as a *perpetuity* or as an *annuity*. An annuity reduces the principal over time, but a perpetuity does not. The principal (your nest egg) is often called the *present value*.

For an annuity with a term of t months, we have the formula

$$\text{Monthly annuity yield} = \frac{\text{Nest egg} \times r(1+r)^t}{((1+r)^t - 1)}.$$

In this formula, r is the monthly interest rate as a decimal. A companion formula gives the nest egg needed to achieve a desired annuity yield:

$$\text{Nest egg needed} = \frac{\text{Annuity yield goal} \times ((1+r)^t - 1)}{(r(1+r)^t)}.$$

Credit cards: Paying off consumer debt

Buying a car or a home usually involves a regular monthly payment that is computed as described earlier. But another way of borrowing is by credit card. If the balance is not paid off by the due date, the account is subject to finance charges. A simplified formula for the amount subject to finance charges is

$$\text{Amount subject to finance charges} = \text{Previous balance} - \text{Payments} + \text{Purchases}.$$

Suppose we have a balance on our credit card and decide to stop charging. If we make only the minimum payment, the balance is given by the exponential formula

$$\text{Balance after } t \text{ minimum payments} = \text{Initial balance} \times ((1+r)(1-m))^t.$$

In this formula, r is the monthly interest rate and m is the minimum monthly payment as a percent of the balance. Both r and m are in decimal form. The product $(1+r)(1-m)$ should not be rounded when the calculation is performed. Because the balance is a decreasing exponential function, the balance decreases very slowly in the long run.

Inflation, taxes, and stocks: Managing your money

The *Consumer Price Index* or CPI is a measure of the average price paid by urban consumers in the United States for a "market basket" of goods and services. The *rate of inflation* is measured by the percent change in the CPI over time.

Inflation reflects a decline of the buying power of the consumer's dollar. Here is a formula that tells how much the buying power of currency decreases for a given inflation rate i (expressed as a percent):

$$\text{Percent decrease in buying power} = \frac{100i}{100 + i}.$$

A key concept for understanding income taxes is the *marginal tax rate*. With a marginal tax rate of 30%, for example, the tax owed increases by $0.30 for every $1 increase in taxable income. Typically, those with a substantially higher taxable income have a higher marginal tax rate. To calculate our taxable income, we subtract any *deductions* from our total income. Then we can use the tax tables. To calculate the actual tax we owe, we subtract any *tax credits* from the tax determined by the tables.

The *Dow Jones Industrial Average* or DJIA is a measure of the value of leading stocks. It is found by adding the prices of 30 "blue-chip" stocks and dividing by a certain number, the *divisor*, to account for mergers, stock splits, and other factors. With the current divisor, for every $1 move in any Dow company's stock price, the average changes by about 7.56 points.

KEY TERMS

principal, p. 188
simple interest, p. 188
compound interest, p. 189
period interest rate, p. 191
annual percentage rate (APR), p. 191
annual percentage yield (APY), p. 195
present value, p. 199
future value, p. 199

Rule of 72, p. 201
installment loan, p. 205
term, p. 205
amortization table (or amortization schedule), p. 211
equity, p. 212
fixed-rate mortgage, p. 216
adjustable-rate mortgage (ARM), p. 216

prime interest rate, p. 216
nest egg, p. 229
monthly yield, p. 229
annuity, p. 229
Consumer Prime Index (CPI), p. 249
inflation, p. 249
rate of inflation, p. 249
deflation, p. 249

CHAPTER QUIZ

1. We invest $2400 in an account that pays simple interest of 8% each year. Find the interest earned after five years.

Answer $960

If you had difficulty with this problem, see Example 4.1.

2. Suppose we invest $8000 in a four-year CD that pays an APR of 5.5%.
a. What is the value of the mature CD if interest is compounded annually?
b. What is the value of the mature CD if interest is compounded monthly?

Answer a. $9910.60, b. $9963.60

If you had difficulty with this problem, see Example 4.3.

3. We have an account that pays an APR of 9.75%. If interest is compounded quarterly, find the APY. Round your answer as a percentage to two decimal places.

Answer 10.11%

If you had difficulty with this problem, see Example 4.4.

4. How much would you need to invest now in a savings account that pays an APR of 8% compounded monthly in order to have a future value of $6000 in a year and a half?

Answer $5323.64

If you had difficulty with this problem, see Example 4.7.

5. Suppose an account earns an APR of 5.5% compounded monthly. Estimate the doubling time using the Rule of 72, and calculate the exact doubling time. Round your answers to one decimal place.

Answer Rule of 72: 13.1 years; exact method: 151.6 months (about 12 years and 8 months)

If you had difficulty with this problem, see Example 4.8.

6. You need to borrow $6000 to buy a car. The dealer offers an APR of 9.25% to be paid off in monthly installments over $2\frac{1}{2}$ years.
a. What is your monthly payment?
b. How much total interest did you pay?

Answer a. $224.78, b. $743.40

If you had difficulty with this problem, see Example 4.11.

7. We can afford to make payments of $125 per month for two years for a used motorcycle. We're offered a loan at an APR of 11%. What price bike should we be shopping for?

Answer $2681.95

If you had difficulty with this problem, see Example 4.10.

8. Suppose we have a savings account earning 6.25% APR. We deposit $15 into the account at the end of each month. What is the account balance after eight years?

Answer $1862.16

If you had difficulty with this problem, see Example 4.21.

9. Suppose we have a savings account earning 5.5% APR. We need to have $2000 at the end of seven years. How much should we deposit each month to attain this goal?

Answer $19.57

If you had difficulty with this problem, see Example 4.22.

10. Suppose we have a nest egg of $400,000 with an APR of 5% compounded monthly. Find the monthly yield for a 10-year annuity.

Answer $4242.62

If you had difficulty with this problem, see Example 4.24.

11. Suppose your MasterCard calculates finance charges using an APR of 16.5%. Your previous statement showed a balance of $400, toward which you made a payment of $100. You then bought $200 worth of clothes, which you charged to your card. Complete the following table:

	Previous balance	Payments	Purchases	Finance charge	New balance
Month 1					

Answer

	Previous balance	Payments	Purchases	Finance charge	New balance
Month 1	$400.00	$100.00	$200.00	$6.88	$506.88

If you had difficulty with this problem, see Example 4.26.

12. Suppose your MasterCard calculates finance charges using an APR of 16.5%. Your statement shows a balance of $900, and your minimum monthly payment is 6% of that month's balance.
a. What is your balance after a year and a half if you make no more charges and make only the minimum payment?
b. How long will it take to get your balance under $100?

Answer a. $377.83, b. 46 monthly payments

If you had difficulty with this problem, see Example 4.29.

13. Suppose the CPI increases this year from 210 to 218. What is the rate of inflation for this year?

Answer 3.8%

If you had difficulty with this problem, see Example 4.30.

14. Suppose the rate of inflation last year was 20%. What was the percentage decrease in the buying power of currency over that year?

Answer 16.7%

If you had difficulty with this problem, see Example 4.31.

2002 NT7

Mars

Mercury

Venus

Earth

. 2002 NT7

Earth Distance: 0.711 AU
Sun Distance : 1.657 AU

Jul 24, 2002

5

INTRODUCTION TO PROBABILITY

The NASA Web site offers the following interesting bit of information.

IN THE NEWS 5.1

Asteroid 2002 NT7 Low Probability of Earth Impact in 2019

DON YEOMANS   July 24, 2002

Asteroid 2002 NT7 currently heads the list on our IMPACT RISKS Page because of a low-probability Earth impact prediction for February 1, 2019. While this prediction is of scientific interest, the probability of impact is not large enough to warrant public concern.

Discovered on July 9, 2002 by the LINEAR team, asteroid 2002 NT7 is in an orbit, which is highly inclined with respect to the Earth's orbit about the sun and in fact nearly intersects the orbit of the Earth. While the orbits of Earth and 2002 NT7 are close to one another at one point in their respective orbits, that does not mean that the asteroid and Earth themselves will get close to one another. Just after an asteroid like 2002 NT7 is discovered, the limited number of observations available do not allow its trajectory to be tightly constrained and the object's very uncertain future motion often allows a very low probability of an Earth impact at some future date. Just such a low probability impact has been identified for February 1, 2019 and a few subsequent dates. As additional observations of the asteroid are made in the coming months, and perhaps pre-discovery archival observations of this object are identified, the asteroid's orbit will become more tightly constrained and the future motion of the asteroid will become better defined. By far the most likely scenario is that, with additional data, the possibility of an Earth impact will be eliminated.

This article from NASA both warns and reassures us. The first thing we gather from the article is that there is a possibility that an asteroid will strike Earth in 2019. But all of NASA's highly trained scientists, mathematicians, and engineers with access to NASA's state-of-the-art equipment cannot tell us for sure whether the event will occur. The best they can do is to estimate the *probability* of the event. What are we to make of such information? Probabilities of this sort apply not only to asteroids but also to hurricanes, rainstorms, traffic accidents, sports contests, and winning the lottery. In this chapter, we examine the idea of probability, what it means, and how probabilities are calculated.

5.1 Calculating probabilities: How likely is it?

Probabilities arise in many contexts. The following article illustrates one of these contexts.

IN THE NEWS 5.2

2008 Bay Area Earthquake Probabilities

UNITED STATES GEOLOGICAL SURVEY (USGS)   April, 2008

I n April 2008, scientists and engineers released a new earthquake forecast for the State of California called the Uniform California Earthquake Rupture Forecast (UCERF). Compiled by USGS, Southern California Earthquake Center (SCEC), and the California Geological Survey (CGS), with support from the California Earthquake Authority, it updates the earthquake forecast made for the greater San Francisco Bay Area by the 2002 Working Group for California Earthquake Probabilities.

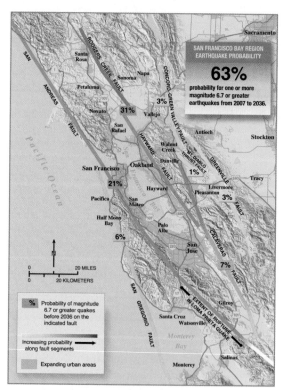

The accompanying figure shows the updated probabilities for earthquakes of magnitude 6.7 or greater in the next 30 years. The overall probability of a magnitude 6.7 or greater earthquake in the Greater Bay Area is 63%, about 2 out of 3, which is very close to the probability of 62% obtained by the 2002 Working Group.

In order to appreciate the significance of the article above, we need to understand the meaning of "probability." We use numbers to measure all kinds of things: our 5-foot-long dining table, our 6-pound cat, the 3% annual rate of inflation in prices, etc. Probability is just another kind of numerical measure. It expresses the likelihood that a specific event will occur.

People use the term *probability* in various ways—often casually or informally. We will begin by explaining some of these uses. Consider the following statements:

1. Your chances of being a TV star are zero.

2. If I toss a coin, the probability of getting a "head" is 1/2.

3. A recent poll tells us that if you bump into an American at random, there is a 40% chance you will be meeting a Democrat.

All of these statements have one thing in common: They assign a numerical value to the likelihood of a certain event. But they are quite different in nature. Let's take them one at a time.

Your chances of being a TV star are zero: This statement is expressing the idea that there's no way you're going to be a TV star. That's what we usually mean when we say your chances are zero. Similarly, if we say your chances of being a TV star are 100%, we are saying that you are absolutely certain to become a TV star. These statements don't give probabilities in the mathematical sense because they can't be measured with any degree of accuracy. Rather, they merely express opinions.

If I toss a coin, the probability of getting a "head" is 1/2: The probability one-half is obtained from the knowledge that if we toss a typical coin, there are only two possible outcomes (heads or tails), and they are equally likely to occur. To say the probability is 1/2 means that if we toss a coin a number of times, we expect to obtain a head about half the time. This fact is also expressed by saying that the probability is 50%. This number is a *theoretical probability*—the type we will be most concerned with in this chapter.

A recent Harris Poll tells us that if you bump into an American at random, there is a 40% chance you will be meeting a Democrat: This numerical value is known as an *empirical probability* because it comes from an actual measurement, as opposed to a theoretical calculation. It is obtained by polling a relatively small sample of the U.S. population and extrapolating to estimate a percentage of the whole population. (As we will see in Chapter 6, if the sample is properly chosen, this method can be quite accurate.)

Theoretical probability

To discuss theoretical probability, we need to make clear what we mean by an event. An *event* is a collection of specified outcomes of an experiment.[1] For example, suppose we roll a standard die (with faces numbered 1 through 6). One event is that the number showing is even. This event consists of three outcomes: a 2 shows, a 4 shows, or a 6 shows.

To define the probability of an event, it is common to describe the outcomes corresponding to that event as *favorable* and all other outcomes as *unfavorable*. If the event is that an even number shows when we toss a die, there are three favorable outcomes (2, 4, or 6) and three unfavorable outcomes (1, 3, or 5).

---

[1]The situations we will consider all have a finite number of outcomes, so we are considering only *discrete* probabilities.

Key Concept

If each outcome of an experiment is *equally likely*, the **probability** of an event is the fraction of favorable outcomes. We can express this definition using a formula:

$$\text{Probability of an event} = \frac{\text{Number of favorable outcomes}}{\text{Total number of possible equally likely outcomes}}.$$

We will often shorten this to

$$\text{Probability} = \frac{\text{Favorable outcomes}}{\text{Total outcomes}}.$$

It is important to recall that in the above definition, each outcome is assumed to be equally likely.

Suppose again that we roll a standard six-sided die. Let's calculate the probability that we will get a 1 or a 6. There are six possible outcomes—namely, the numbers 1 through 6—and each number is equally likely to occur. Two of these outcomes are favorable: 1 and 6. So the probability is

$$\frac{\text{Favorable outcomes}}{\text{Total outcomes}} = \frac{2}{6} = \frac{1}{3}.$$

What this means is that if we roll the die many times, we expect about one-third of the rolls to come up 1 or 6. Now 1/3 is about 0.33 or 33%, so we can say that the probability of a 1 or a 6 is 1/3 or 0.33 or 33%. All of these numbers are equivalent ways of expressing the same ratio (although 0.33 is only approximately equal to 1/3). It is common in this setting to indicate the probability of a 1 or 6 as $P(1 \text{ or } 6)$. With this notation, we write

$$P(1 \text{ or } 6) = \frac{1}{3}.$$

EXAMPLE 5.1 Calculating simple probabilities: An ace

If I draw a card at random from a standard deck of 52 cards, what is the probability I draw an ace?

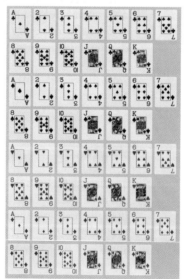

SOLUTION

There are 52 different cards that I might draw, and these outcomes are equally likely. There are four aces, so four outcomes are favorable. Thus,

$$P(\text{Ace}) = \frac{\text{Favorable outcomes}}{\text{Total outcomes}} = \frac{4}{52} = \frac{1}{13}.$$

Thus, the chance of drawing an ace is 1 in 13 or about 8%.

TRY IT YOURSELF 5.1

In a standard deck, the cards are divided into four suits, called spades, hearts, diamonds, and clubs, with 13 in each suit. If I draw a card at random from a standard deck, what is the probability I draw a diamond?

The answer is provided at the end of this section.

The probability of an event is always between 0 and 1 inclusive. (Exercise 10 asks you to provide an explanation of this.) An event has probability 0 if and only if it can never occur.[2] For example, if I draw five cards from a standard deck of cards, the probability that I get five aces is 0. At the other extreme, an event has probability 1 if and only if it will always occur. For example, if I flip a coin, the probability that I get either heads or tails is 1. Exercises 11 and 12 ask you to provide explanations of these statements.

SUMMARY 5.1   Probability

An event is a collection of specified outcomes of an experiment, and we describe each outcome corresponding to that event as favorable. Assume that there are a finite number of outcomes and that each outcome is equally likely.

1. The probability of an event equals the number of favorable outcomes, divided by the total number of possible outcomes:

$$\text{Probability} = \frac{\text{Favorable outcomes}}{\text{Total outcomes}}.$$

2. A probability may be expressed as a fraction, a decimal, or a percent.

3. A probability must be between 0 and 1 inclusive.

4. The probability of an event is 0 if and only if the event can never occur.

5. The probability of an event is 1 (or 100%) if and only if the event will always occur.

Distinguishing outcomes

One needs to be careful in determining what constitutes an outcome. To illustrate this point, consider a committee consisting of three members. They need to send one member as a representative to a meeting. They decide to choose the representative by writing their three names on slips of paper and drawing a slip at random from a basket.

Suppose first that the names of the members are Alan, Benny, and Jerry. What is the probability that Alan is chosen? There are three equally likely outcomes: Alan, Benny, and Jerry. Only one of these outcomes is favorable (Alan). Thus,

$$P(\text{Alan}) = \frac{\text{Favorable outcomes}}{\text{Total outcomes}} = \frac{1}{3}.$$

Now suppose that Jerry is replaced by another member whose name is also Alan. Thus, two of the three committee members are now named Alan, and one is named Benny. What is the probability that someone named Alan is chosen in this case?

To understand this situation, we imagine that members put their surnames on the slips, say, Alan Jones and Alan Smith. Now it is easy to see that there are three

---

[2] Remember that we are speaking only of *discrete* probabilities.

equally likely outcomes: Alan Jones, Alan Smith, and Benny. Two of these outcomes are favorable, so the probability of selecting a member named Alan is 2/3.

This example warns of a common error in calculating probabilities: If I choose a name from the basket, I will get either Alan or Benny. The temptation is to say that the probability of getting Alan is 1/2. The error here is that although one can think of having two outcomes, Alan and Benny, *they are not equally likely.* In fact, because there are two Alans in the hat and only one Benny, we are twice as likely to draw Alan as Benny. Often this sort of error is easy to spot, but we must be careful to avoid it. We can often avoid difficulties by distinguishing outcomes just as we used surnames, Jones and Smith, to distinguish the two Alans.

EXAMPLE 5.2 Distinguishing outcomes: Coins

Suppose I flip two identical coins. What is the probability that I get two heads?

SOLUTION

We stated that the two coins are identical, but as with the "two Alans" discussed above, it helps to imagine that the coins are labeled to make them distinguishable. So let's imagine that one of the coins is a nickel and the other is a dime.

First we must figure out how many equally likely outcomes there are. Either coin can come up as a head (H) or a tail (T). All possibilities are listed in the table below.

Nickel	Dime
H	H
H	T
T	H
T	T

There are four equally likely outcomes: HH, HT, TH, TT. Only one of them gives two heads, so the probability of getting two heads is

$$P(\text{HH}) = \frac{\text{Favorable outcomes}}{\text{Total outcomes}} = \frac{1}{4}.$$

TRY IT YOURSELF 5.2

Suppose I flip two identical coins. What is the probability that I get two tails?

The answer is provided at the end of this section.

These examples show how distinguishing things (Alan Jones and Alan Smith, or nickel and dime) that appear to be the same can ensure that all outcomes are equally likely.

EXAMPLE 5.3 Distinguishing outcomes: A traffic light

Suppose I have a 50-50 chance of getting through a certain traffic light without having to stop. I go through this light on my way to work and again on my way home.

a. What is the probability of having to stop at this light at least once on a workday?

b. What is the probability of not having to stop at all?

SOLUTION

a. A 50-50 chance means that the probability of stopping when I come to the light is 1/2 and the probability of not stopping is 1/2. For a trip from home to work and back again, the possibilities are listed in the table on next page.

To work	To home
Stop	Stop
Stop	Don't stop
Don't stop	Stop
Don't stop	Don't stop

This means that there are four equally likely outcomes: Stop–Stop, Stop–Don't stop, Don't stop–Stop, Don't stop–Don't stop. Three of these yield the result of stopping at least once, so the desired probability is

$$\frac{\text{Favorable outcomes}}{\text{Total outcomes}} = \frac{3}{4}.$$

b. One of the four possible outcomes (Don't stop–Don't stop) corresponds to not having to stop at all, so the desired probability is

$$\frac{\text{Favorable outcomes}}{\text{Total outcomes}} = \frac{1}{4}.$$

CALCULATION TIP 5.1 Equally Likely Outcomes

When calculating probabilities, take care that each of the outcomes considered is equally likely to occur. In cases involving identical items, such as coins or dice, it may help to label the items to distinguish them.

"How do you want it—the crystal mumbo-jumbo or statistical probability?"

Probability of non-occurrence

The calculations in Example 5.3 lead us to something important. There we found

$$P(\text{Stopping at least once}) = \frac{3}{4}$$

$$P(\text{Not stopping at all}) = \frac{1}{4}.$$

That is, the probability of not stopping at all is 1 minus the probability of stopping at least once. This is no accident, as the following observations show: Recall that

$$\text{Probability of event occurring} = \frac{\text{Favorable outcomes}}{\text{Total outcomes}}.$$

Then

$$\text{Probability of event not occurring} = \frac{\text{Unfavorable outcomes}}{\text{Total outcomes}}.$$

If we add the favorable and unfavorable outcomes, we get all outcomes. Hence,

$$\text{Probability of event occurring} + \text{Probability of event not occurring} = \frac{\text{Total outcomes}}{\text{Total outcomes}} = 1.$$

Rearranging this equation yields

$$\text{Probability of event not occurring} = 1 - \text{Probability of event occurring}.$$

SUMMARY 5.2   Probability of an Event Not Occurring

The probability that an event will not occur is 1 minus the probability that it will occur, or if expressed as a percent, 100% minus the probability that it will occur. As an equation, this is

$$\text{Probability of event not occurring} = 1 - \text{Probability of event occurring}.$$

EXAMPLE 5.4 Calculating probability of non-occurrence: English teachers

There are several sections of English offered. There are some English teachers I like and some I don't. I enroll in a section of English without knowing the teacher. A friend of mine has calculated that the probability that I get a teacher I like is

$$P(\text{Teacher I like}) = \frac{7}{17}.$$

What is the probability that I will get a teacher that I don't like?

SOLUTION

Our formula for non-occurrence applies:

$$\text{Probability of event not occurring} = 1 - \text{Probability of event occurring}.$$

Hence,

$$P(\text{Teacher I don't like}) = 1 - P(\text{Teacher I like})$$
$$= 1 - \frac{7}{17}$$
$$= \frac{10}{17}.$$

TRY IT YOURSELF 5.4

I am expecting letters. Some of these are bills and the rest are letters from friends. There is a letter in my mailbox. A friend has calculated the probability that it is a bill is

$$P(\text{The letter is a bill}) = \frac{15}{34}.$$

What is the probability that it is a letter from a friend?

The answer is provided at the end of this section.

Now we use the idea of distinguishing items in tandem with the formula for non-occurrence.

EXAMPLE 5.5 Distinguishing outcomes and non-occurence: Rolling dice

Suppose we toss a pair of standard six-sided dice.

a. What is the probability that a we get a 7 (i.e., the sum of the two faces is 7)?

b. What is the probability that we get any sum but 7?

SOLUTION

a. As we discussed previously, it is helpful to distinguish the dice. Let's imagine that one of them is red and the other is green. There are six possible outcomes for the red die (numbers 1 through 6). For each number on the red die, there are six possible numbers for the green die. That is a total of $6 \times 6 = 36$ possible outcomes: a 1 on the first die and a 1 on the second, a 1 on the first die and a 2 on the second, and so on.[3]

Now we need to know how many outcomes yield a 7 (what we would call a favorable outcome). We list the possibilities in a table.

Red die	Green die
1	6
2	5
3	4
4	3
5	2
6	1

Thus, there are six possible ways to get a 7. Therefore,

$$\text{Probability of a 7} = \frac{\text{Favorable outcomes}}{\text{Total outcomes}} = \frac{6}{36} = \frac{1}{6}.$$

This is about 0.17 or 17%.

b. We use Summary 5.2:

Probability of event not occurring = 1 − Probability of event occurring.

We find

$$\text{Probability of not getting a 7} = 1 - \frac{1}{6} = \frac{5}{6}.$$

This is about 0.83 or 83%.

Note that if you tried to compute this probability directly, you would have to count all the ways of getting a sum different from 7, which would be very tedious.

Probability of disjunction

Our first calculation in this section concerned the probability that when we roll a six-sided die, we will get a 1 or a 6. This is an example of finding the probability that either one event or another occurs. In accordance with the terminology of formal logic (see Section 1.3), we call this the probability of the *disjunction* of the two events. It turns out that there is a convenient formula for finding this probability.

---

[3]In Section 5.3, we will explain in more detail how to find the total number of outcomes in such situations.

Key Concept

Suppose A and B are two events. Their **disjunction** is the event that either A or B occurs. The probability of this disjunction is given by the formula

$$P(A \text{ or } B) = P(A) + P(B) - P(A \text{ and } B).$$

The rationale for this formula comes from the fact that if you add the probability of event A and the probability of event B, you are counting the event "A and B" twice. (Recall that in mathematics the disjunction allows for the possibility that both events occur.) Therefore, subtracting $P(A \text{ and } B)$ from the sum gives the correct answer.

EXAMPLE 5.6 Calculating probability of a disjunction: Choosing books

Suppose a librarian has a cart with 10 paperback algebra books, 15 paperback biology books, 21 hardbound algebra books, and 39 hardbound biology books. What is the probability that a book selected at random from this cart is an algebra book or a paperback book?

SOLUTION

Let A denote the event of choosing an algebra book and B the event of choosing a paperback book. We want to find $P(A \text{ or } B)$. In order to use the formula, we need to know three probabilities: $P(A)$, $P(B)$, and $P(A \text{ and } B)$. Altogether, there are $10 + 15 + 21 + 39 = 85$ books. Because there are a total of $10 + 21 = 31$ algebra books out of these 85 books, we have

$$P(A) = \frac{31}{85}.$$

Likewise, there are 25 paperback books out of 85 books, so we have

$$P(B) = \frac{25}{85}.$$

Finally, "A and B" is the event of choosing a book that is both algebra and paperback. There are 10 such books, so we have

$$P(A \text{ and } B) = \frac{10}{85}.$$

Now the formula gives

$$P(A \text{ or } B) = P(A) + P(B) - P(A \text{ and } B) = \frac{31}{85} + \frac{25}{85} - \frac{10}{85} = \frac{46}{85}.$$

Thus, the probability that a book selected at random is an algebra book or a paperback book is 46/85, which is about 0.54 or 54%.

TRY IT YOURSELF 5.6

In the situation of this example, find the probability that a book selected at random is a biology book or a paperback book.

The answer is provided at the end of this section.

Probability with area

Let's look at a different situation in which a theoretical probability can be determined. We opened this section with an article on estimating the probability of asteroids striking Earth. Such estimates relate probability to area. To explain the connection, we examine the probability related to a man-made object striking Earth.

In 1979, Skylab, the first space station of the United States, reentered Earth's atmosphere. At the time there was widespread concern that the debris would fall in a heavily populated area, but in the end the area most affected was a sparsely populated region in western Australia. The area was roughly rectangular in shape. If pieces of Skylab fell at random over this rectangle, what is the probability that a given region would be affected? To see how to answer this question, we first consider the simpler situation in which the rectangle is divided into three regions, designated by color as follows: 1/4 of the area is red, 1/4 is blue, and 1/2 is white. See **Figure 5.1**. If a given small piece of Skylab fell at random onto this rectangle, what is the probability that it would land on the color blue?

The Skylab space laboratory.

There are three possible outcomes: The piece lands on red, blue, or white. These outcomes are *not* equally likely, however. Because a lot more of the target is white, one would expect that there is a greater likelihood of the piece landing on white than on either of the other two colors. One way of looking at this problem is to cut the white part into two sections, White 1 and White 2, as shown in **Figure 5.2**.

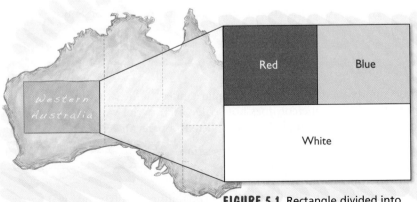

FIGURE 5.1 Rectangle divided into three regions.

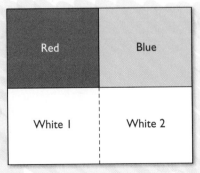

FIGURE 5.2 Cutting white into two sections.

Now because each of the four sections is the same size, it should be equally likely that the piece lands in any one of them. Because there are now four equally likely outcomes and only one of them is blue, we see that the probability of landing on blue is $1/4 = 0.25 = 25\%$. Observe that this number equals the percentage of the area that was labeled blue.

This observation holds in general. The probability of landing on a given color is the fraction of the area covered by that color. In this case, that means the probability of landing on white is $1/2$, the probability of landing on red is $1/4$, and the probability of landing on blue is $1/4$. As a practical matter, this means that an object falling to Earth at random is likely to strike an unpopulated region: Water covers about 70% of Earth's surface.

EXAMPLE 5.7 Finding probability using area: Meteor

The surface area of Earth is approximately 197 million square miles. North America covers approximately 9.37 million square miles, and South America covers approximately 6.88 million square miles. A meteor falls from the sky and strikes Earth. What is the probability that it strikes North or South America? Write your answer in decimal form rounded to three decimal places and also as a percentage.

SOLUTION

The total area covered by North and South America is $9.37 + 6.88 = 16.25$ million square miles. As a fraction of the surface area of Earth, that is $16.25/197$ or about 0.082. Hence, the probability that the meteor strikes North or South America is 0.082 or 8.2%.

TRY IT YOURSELF 5.7

Asia covers approximately 17.2 million square miles. In the situation above, what is the probability that the meteor does not strike Asia? Write your answer in decimal form rounded to three decimal places and also as a percentage.

The answer is provided at the end of this section.

Rain forecasts are actually estimates of probabilities involving areas. Exercise 46 explains this connection.

Defining probability empirically

Suppose I have a box containing 20 red jelly beans and 10 green ones. I select a jelly bean from the box without looking. There is a total of 30 jelly beans in the box, and 20 of them are red. Therefore, the probability of getting a red jelly bean is

$$\frac{20}{30} = \frac{2}{3}$$

or about 67%. This is a theoretical probability based on the known percentage of red jelly beans.

Now suppose we pick a jelly bean at random from the box, record its color, and toss it back into the box. If we repeat this experiment 3000 times, it is reasonable to expect that approximately 2/3 of the jelly beans chosen would be red. Thus, we would expect that about

$$3000 \times \frac{2}{3} = 2000$$

jelly beans would be red. Similarly, we would expect that approximately 1/3 of the jelly beans chosen, or 1000, would be green.

It would not be at all surprising if the numbers were something like 2005 red and 995 green, but we would definitely be surprised if the numbers were 130 red and 2870 green. In other words, we expect the actual *empirical* probabilities obtained by experiment to be fairly close to the theoretical probabilities.

This suggests that we can go backward. That is, we can do experiments to estimate the theoretical probability. Let's explore this idea.

An important feature of any manufacturing process is *quality control*. This term means just what the name implies: ensuring that high-quality products are produced. One aspect of quality control involves testing of finished products. Suppose, for example, that we make many thousands of lightbulbs each day. It is not practical to test each bulb we produce, but we can test a certain percentage of them.

Suppose that on a given day, we randomly choose 1000 bulbs for testing. We find that 64 of them are defective and 936 of them work perfectly.[4] On this basis, it is reasonable to estimate that about

$$\frac{64}{1000} = 0.064$$

or 6.4% of the bulbs we produce are, in fact, defective. That is, we estimate that a customer has about a 6.4% chance of purchasing a defective bulb with our name brand on it.

This is an example of calculation of an *empirical probability*. We don't know the actual probability but estimate it from a properly chosen sample.

Key Concept

The empirical probability of an event is a probability obtained by experimental evidence. It is the ratio of the number of favorable outcomes to the total number of outcomes in the experiment.

The idea seems simple enough, but the important question about an empirical probability is, how meaningful is it? For example, if I select three jelly beans from a jar and two of them are red while one is not, I surely can't deduce from this single experiment that exactly 2/3 of the jelly beans in the jar are red. However, suppose we perform this experiment numerous times and in 95% of them about 2/3 of the

---

[4]These numbers are far from acceptable for any reputable producer, but we use them for illustration.

jelly beans we selected were red. Now, we might start believing that about 2/3 of jelly beans in the jar really are red.

So one question is, how many samples should we take before we have some confidence in our empirical probability? And is there some way to measure a level of confidence? As it turns out, these questions are at the heart of what professional polling organizations do. We will return to these questions in Section 3 of the next chapter.

EXAMPLE 5.8 Finding empirical probability: Running a red light

Suppose a city wants to know the probability that an automobile going through the intersection of 5th and Main will run a red light. Suppose the city posted workers at the intersection, and over a five-week period it counted 16,652 vehicles passing through the intersection, of which 1432 ran a red light. Use these numbers to calculate an empirical probability that cars passing through the intersection will run a red light. Round your answer in decimal form to two places.

SOLUTION
Now 1432 out of 16,652 ran the light. So an empirical probability is

$$\frac{1432}{16,652},$$

which is about 0.09 or 9%.

TRY IT YOURSELF 5.8

At another intersection over the same period the city found that 19,221 cars passed through and that 2144 of them ran a red light. Using these numbers, calculate an empirical probability that cars passing through the intersection will run a red light. Round your answer in decimal form to two places.

The answer is provided at the end of this section.

Try It Yourself answers

Try It Yourself 5.1: Calculating simple probabilities: A diamond   13/52 = 1/4.

Try It Yourself 5.2: Distinguishing outcomes: Coins   1/4.

Try It Yourself 5.4: Calculating probability of non-occurrence: Mailbox   19/34.

Try It Yourself 5.6: Calculating probability of a disjunction: Choosing books 64/85 or about 0.75.

Try It Yourself 5.7: Finding probability using area: Meteor   0.913 or 91.3%.

Try It Yourself 5.8: Finding empirical probability: Running a red light 0.11 or 11%.

Exercise Set 5.1

In your answers to these exercises, leave each probability as a fraction unless you are instructed to do otherwise.

1. Three ways: Express each of the following probabilities as a fraction, a decimal, and a percent:

   a. 50%
             d. 65%

   b. 1/4
              e. 7/8

   c. 0.05
           f. 0.37

Probabilities and opinions   For Exercises 2 through 7, determine whether the statement gives a theoretical mathematical probability, an empirical mathematical probability, or just represents someone's opinion. Explain your answers.

2. The probability of two heads on a toss of two coins is 0.25.

3. The probability of humans going to Mars is 10% at most.

4. Your chances of dating him are zero.

5. My chances of getting a call from a telemarketer during dinner time are 95%.

6. The probability of a smoker getting heart disease is 80%.

7. There's a 50-50 chance of a terrorist attack in the United States sometime this year.

8. **More probabilities and opinions.** During the Cold War, someone from the government asked a famous mathematics professor if he could help them figure out the probability of a nuclear war between the United States and the Soviet Union. The professor said, "That's an absurd question." Can you explain why he said that? (By the way, this is a true story.)

9. **Rain.** Someone says that the probability of rain is always 50% because there are two outcomes—either it rains or it doesn't. Decide whether you agree or disagree with this, and explain your reasons.

10. **Probability between 0 and 1.** Explain why the probability of an event is always a number between 0 and 1 inclusive.

11. **Zero probability.** Explain why the probability of an event is 0 if and only if the event can never occur.

12. **100% probability.** Explain why the probability of an event is 1 (or 100%) if and only if the event is certain to occur.

13. **Red lights.** A city finds empirically that an automobile going through the intersection of 5th and Main will run a red light sometime during a given day with probability 9.2%. What is the probability that an automobile will *not* run a red light at that intersection on a given day? Write your answer as a percentage.

14. **An experiment.** Try an experiment like this one: Fill a box with 100 objects indistinguishable by feel, such as slips of paper. Mark 30 of them with an "X" and leave the other 70 unmarked. Now, draw a piece of paper from the box (without looking) and record whether it had an "X" or not. Return it to the box and shake up the box. Repeat this experiment 50 times.

Compare the theoretical probability of drawing a slip with an "X" to the empirical probability from your 50 trials. Discuss what you expected to happen versus what you observed. Try it again with 100 trials.

15. **Another experiment.** Do a survey of a number of students to find out what they are majoring in, and determine an empirical probability that a given student is a psychology major. Find out from your university administration the percentage of students who are psychology majors. Compare your empirical probability with the actual percentage of psychology majors. Discuss factors that may affect the results.

Boys and girls Alice and Bill are planning to have three children. Exercises 16 through 19 refer to this. (Assume it is equally likely for a boy or a girl to be born.)

16. What is the probability that all three of their children will be girls?

17. What is the probability that at least one will be a girl?

18. What is the probability that all three will be of the same gender?

19. What is the probability that not all three will be of the same gender?

Jelly beans Suppose there are five jelly beans in a box—two red and three green. This information is used in Exercises 20 and 21.

20. If a jelly bean is selected at random, what is the probability that it is red?

21. A friend claims that the answer to Exercise 20 is 1/2 because there are two outcomes, red and green. Explain what's wrong with that argument.

More jelly beans Suppose there are 15 jelly beans in a box—2 red, 3 blue, 4 white, and 6 green. A jelly bean is selected at random. Exercises 22 through 26 ask about possible outcomes of this event.

22. What is the probability that the jelly bean is red?

23. What is the probability that the jelly bean is not red?

24. What is the probability that the jelly bean is blue?

25. What is the probability that the jelly bean is red or blue?

26. What is the probability that the jelly bean is neither red nor blue?

Still more jelly beans Suppose I pick a jelly bean at random from a box containing five red and seven blue ones. I record the color and put the jelly bean back in the box. Exercises 27 through 30 refer to this experiment.

27. What is the probability of getting a red jelly bean both times if I do this twice?

28. What is the probability of getting a blue jelly bean both times if I do this twice?

29. If I do this three times, what is the probability of getting a red jelly bean each time?

30. If I do this three times, what is the probability of getting a blue jelly bean each time?

Tossing three coins Exercises 31 through 35 refer to the experiment of tossing three coins. The answers to some of them can be easily obtained from answers to earlier ones.

31. What is the total possible number of outcomes?

32. What is the probability of getting three heads?

33. What is the probability of getting one head and two tails?

34. What is the probability of getting anything but one head and two tails?

35. What is the probability of getting at least one tail?

36. **Tossing three dice.** You toss three dice and add up the faces that show. What is the probability that the sum will be 3, 4, or 5?

37. **Passing.** The probability of passing the math class of Professor Jones is 62%, the probability of passing Professor Smith's physics class 34%, and the probability of passing both is 30%. What is the probability of passing one or the other?

38. **The team.** The probability of making the basketball team is 32%, the probability of making the softball team is 34%, and the probability of making one or the other is 50%. What is the probability of making both teams?

A box Suppose the inside bottom of a box is painted with three colors: 1/3 of the bottom area is red, 1/6 is blue, and 1/2 is white. You toss a tiny pebble into the box without aiming. Exercises 39 and 40 refer to this.

39. What is the probability that the pebble lands on the color white?

40. What is the probability that the pebble does not land on the color blue?

41. Heart disease. In one study, researchers at the Cleveland Clinic looked at more than 120,000 patients enrolled in 14 international studies in the past 10 years. Among the patients with heart problems, researchers found at least one risk factor in 84.6% of the women and 80.6% of the men.

If a woman is selected at random from the women with heart problems, what is the probability she has none of the risk factors? If a man is selected at random from the men with heart problems, what is the probability he has none of the risk factors? Write both of your answers as percentages.

Health Exercises 42 through 44 refer to the accompanying table. The percentages are probabilities of 15-year survival without coronary heart disease, stroke, or diabetes in men aged 50 years with selected combinations of risk factors. Here "BMI" stands for *body mass index*, an index of weight adjusted for height. The World Health Organization gives the following classifications: "normal" 18.5 to 24.9; "grade 1 overweight" 25.0 to 29.9; "grade 2 overweight" 30.0 to 39.9; "grade 3 overweight" 40.0 or higher.

	BMI 20–24 normal weight		BMI 25–30 somewhat overweight		BMI 30+ very overweight	
	Active	Inactive	Active	Inactive	Active	Inactive
Never smoked	89%	83%	85%	77%	78%	67%
Former smokers	86%	78%	81%	71%	72%	59%
Current smokers	77%	67%	70%	56%	58%	42%

42. What is the probability that a 50-year-old man will not survive to age 65 without coronary heart disease, stroke, or diabetes if he smokes, is somewhat overweight, but is active? Write your answer as a percentage.

43. A 50-year-old man who has never smoked, has normal weight, and is active has a much higher probability of being healthy than a man who smokes, is very overweight, and is inactive. Is the probability twice as high, less than twice as high, or more than twice as high?

44. A man who used to smoke and who is very overweight and inactive is considering either starting an exercise program or dieting to lose enough weight to be somewhat overweight—but not both. According to the table, which choice would be better?

45. Skiing. Records kept by the National Ski Patrol and the National Safety Council show that, over a period of 14 years, skiers made 52.25 million visits to the slopes annually, with an average of 34 deaths each year. Assuming these figures are current, what is the probability that a skier will die in a skiing accident in a given year? Is your answer empirical or theoretical?

46. Weather. What does a meteorologist mean when he or she says there's a 30% chance of rain tomorrow? It means that the meteorologist is predicting that about 30% of the land area covered by the forecast (say, northeast Texas) will receive some rain during the time period of the forecast (which is usually 12 hours). If the forecast area is 5000 square miles and the forecast is for a 30% chance of rain, how much area should receive rain?

47. More on the weather. One of the weathermen on Channel 2 news in Reno, NV says he does not use percentages when forecasting precipitation. After reading the following article by meteorologist Jeff Haby, explain why you think that weatherman on Channel 2 news would say this.

IN THE NEWS 5.3

Percentages in Forecasts

JEFF HABY

Percentages in forecasts can give the public a general idea of how likely precipitation is, but percentages lack specifics and description. It is up to the meteorologist to describe the character of the precipitation. Percentages are associated too much with gambling and guess work to many people (example: There will be a 50% chance of rain *flips a coin*).

When convective activity is possible, description words can be used instead of chance probabilities. These description words include numerous (replaces 70%+ chance), scattered (replaces 40 to 60% chance), widely scattered (replaces 20 to 30% chance), and isolated (replaces less than 20% chance). Examples, "thunderstorms will be numerous across the forecast region tomorrow, most of you will be getting wet", "thunderstorms will be isolated tomorrow, most of you will miss the storms that develop, but a few of you may be in the right place at the right time to see a thunderboomer." The term scattered is a good term to use instead of (40, 50 or 60% chance of rain). Scattered implies that there will be rain, but there will also be areas that do not receive rain.

The character of the precipitation is also important to mention: severe or non-severe, heavy or light, long duration or short duration, wintry (type of winter precipitation) or non-wintry, convective or stratiform, slow moving or fast moving. Whatever precipitation character information you can sneak into your graphics or say when describing the graphics within the short amount of time will help the viewer or client prepare for the precipitation event.

48. Terrorism. The accompanying excerpt is from an article by Alex Fryer. Based on it, would you say that real probabilities are being used or just educated opinions? Explain your reasoning.

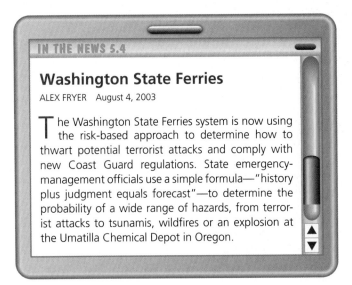

IN THE NEWS 5.4

Washington State Ferries

ALEX FRYER   August 4, 2003

The Washington State Ferries system is now using the risk-based approach to determine how to thwart potential terrorist attacks and comply with new Coast Guard regulations. State emergency-management officials use a simple formula—"history plus judgment equals forecast"—to determine the probability of a wide range of hazards, from terrorist attacks to tsunamis, wildfires or an explosion at the Umatilla Chemical Depot in Oregon.

49. Socks. You have four socks in your drawer, two blue and two brown. You get up early in the morning while it's dark, reach into your drawer, and grab two socks without looking. What is the probability that the socks are the same color? (*Hint*: If you take one sock first, what's the probability the second sock matches it?) If instead there are 10 blue and 10 brown socks, what is the probability that the socks you choose are the same color?

Odds   Another way of expressing probabilities is by the use of *odds*. Suppose I roll a single die and call getting a 4 a favorable outcome. The probability of this happening is of course 1/6, which means that we expect to see a 4 once out of every six rolls. But this can also be expressed in terms of odds. Here's how it works: There are six possible outcomes; one of these is favorable (getting a 4) and five are unfavorable. One then says that the odds *against* getting a 4 are 5 to 1. This could also be expressed by saying that the odds *in favor* of getting a 4 are 1 to 5.

In general, if we list all the possible equally likely outcomes, the *odds in favor of* a favorable outcome are expressed as the number of favorable outcomes *to* the number of unfavorable outcomes. To find the *odds against* it, we reverse the order of these two numbers.

This information is needed for Exercises 50 through 54

50. If we draw a card at random from a standard deck, the probability of getting an ace is 1/13. What are the odds in favor of drawing an ace? What are the odds against drawing an ace?

51. If we flip a pair of fair coins, what is the probability of getting one head and one tail? What are the odds in favor of getting one head and one tail?

52. Alaska comprises about 1/5 of the total surface area of the United States. If a meteor lands at some random point in the United States, what is the probability that it will land in Alaska? What are the odds in favor of it landing in Alaska?

53. The weatherman says that there is a 30% chance of rain in your area. What are the odds against your getting rain?

54. If an event has a probability of 10%, what are the odds in favor of it, and what are the odds against it?

More odds   We can calculate odds if we know the probability. But we can also go the other way. Suppose the odds in favor of an event are 3 to 2. We can interpret this as saying that there are three (equally likely) favorable outcomes and two unfavorable ones. That is a total of five outcomes. So the probability of the event occurring is 3/5. In general, if the odds in favor are p to q, then the probability is $p/(p + q)$. This information is used in Exercises 55 through 57.

55. Suppose the odds in favor of an event are 5 to 2. What is the probability that the event will occur? What is the probability that the event will *not* occur?

56. Suppose the odds against an event are 2 to 1. What is the probability that the event will occur? What is the probability that the event will *not* occur?

57. Odds in a horse race are, in fact, a reflection of money bet on the various horses rather than a statement of probability. But suppose for the moment that the money bet does indeed indicate the true mathematical odds. If *Whirlaway* is a 6 to 5 favorite to win, what is the probability that Whirlaway actually wins the race?

Exercises 58 through 60 involve hands-on experiments and are suitable for group work.

58. The Monty Hall problem. Monty Hall was the host of the popular game show *Let's Make a Deal*. At the end of each show, some lucky contestant was given the opportunity to choose one of three doors. Behind one door was a nice prize, and behind the other two were lesser prizes or even worthless items. The contestant had no information about which door might hide the prize, so his or her probability of getting the nice prize was one-third.

An interesting variant of this scenario gained notoriety when columnist Marilyn vos Savant posed a related question in her *Parade Magazine* article of September 9, 1990. Here is the question as stated by Mueser and Granberg in 1999.

> A thoroughly honest game-show host has placed a car behind one of three doors. There is a goat behind each of the other two doors. You have no prior knowledge that allows you to distinguish among the doors. "First you point toward a door," the host says, "then I'll open one of the other doors to reveal a goat. After I've shown you the goat, you may decide whether to stick with your initial choice of doors, or switch to the other remaining door. You win whatever is behind the door [you choose]."

The question is, should you switch doors or stay with your original choice? Or does it matter? The fun thing about this problem is that a number of Ph.D. mathematicians got the wrong answer!

We will present two analyses below. One is right and the other is wrong. Which is right? Explain why you chose the answer you did. If you can't decide which is right and which is wrong, you should try the experiment suggested in the next exercise.

Analysis number 1: It doesn't matter whether you switch The probability that I guessed right to begin with is 1/3. Either of the two remaining doors is equally likely to be a winning door or a losing door. One of them is certain to be a loser, and by telling me that one of the doors is a loser, the host has given me no new information on which to base my decision. Therefore, it makes no difference whether I change or not. In either case, my probability of winning is 1/3.

Analysis number 2: It does matter. You should switch doors If I have initially chosen the right door, then changing will cause me to lose. This happens one time out of three.

Suppose I have chosen one of the losing doors and the host shows me the other losing door. Then if I change, I win. This happens two times out of three.

So the strategy of changing allows me to win two times out of three. My probability of winning is increased from 1/3 to 2/3.

59. Experimenting with the Monty Hall problem. *This exercise is a continuation of Exercise 58.* With the help of one other person, it is not difficult to simulate the Monty Hall problem described above. Let one person act as host and hide pictures of one car and two goats under sheets of paper numbered 1 through 3. You be the contestant. Make your guesses and record your wins and losses if you never change and if you always change. If you repeat the game 100 times, you should have fairly good empirical estimates of the probability of winning using either strategy. Do the results make you change your mind about your answer to Exercise 58?

60. An experiment–Buffon's needle problem. The French naturalist Buffon posed the following problem—which he subsequently solved, and which bears his name today. Suppose a floor is marked by parallel lines, each a distance D apart. (Think of a hardwood floor.) Suppose we toss a needle of length L to the floor. (We assume that the length L of the needle is smaller than the distance D between the parallel lines on the floor.) What is the probability that the needle will land touching one of the parallel lines? It turns out that the correct answer is

$$\text{Probability} = \frac{2L}{\pi D}.$$

One may be surprised that the number π turns up in the answer, but this fact leads us to an interesting idea that can (at least theoretically) give a novel method of approximating the decimal value of π. Get on a nice, level hardwood floor and toss a needle onto the floor 100 times. Record the number n of times the needle touches one of the parallel lines. Then $n/100$ is an empirical estimate of the probability that the needle will touch a line. This gives

$$\frac{n}{100} \approx \frac{2L}{\pi D}.$$

(Here, we use the symbol $\approx$ to indicate approximate equality.) Rearranging gives

$$\pi \approx \frac{200L}{nD}.$$

Perform this experiment to get an estimate for the value of π. (Do not be disappointed if your answer is not very close to the correct value. The method is correct, but in practice, it takes a lot of tosses to get anywhere close to π.)

61. History. Pierre de Fermat and Blaise Pascal were French mathematicians and philosophers who lived in the seventeenth century. They are credited with some of the early development of probability theory. Pascal was an interesting figure who was also very much involved with religion. Write a paper about Pascal's life and beliefs. (Exercise 63 in Section 5.3 describes a specific question about probability that attracted the interest of Pascal, and Exercise 46 in Section 5.5 is concerned with his use of a concept from probability in a religious context.)

62. History. The formulation of the modern abstract theory of probability is usually attributed to the Russian mathematician A. N. Kolmogorov (1903–1987). He lived during the reign of the brutal dictator Joseph Stalin and once wrote a paper contradicting the academician Lysenko, who was popular with Stalin. This was a very brave act indeed. Write a paper about the life and work of Kolmogorov.

The following exercises are designed to be solved using technology such as calculators or computer spreadsheets. For assistance, see the technology supplement.

63. Random numbers. Simulate flipping a coin in the following way: Use a random-number generator to obtain a random sequence of 0's and 1's, with, say, 100 numbers altogether. Count how many 0's there are altogether. Is the result close to 50%? Is it exactly 50%? Does the result give a theoretical or an empirical probability? *Note:* Computer programs do not actually produce random numbers, in spite of the name, but they are close enough for our purposes.

64. More random numbers. Think of the real numbers between 0 and 10 as divided into two colors: red for those at least 0 and at most 6, blue for those greater than 6 and at most 10. Use a random-number generator to obtain a random sequence of real numbers between 0 and 10, with, say, 50 numbers altogether. (Be sure to get a random sequence of real numbers, not just a sequence of whole numbers.) Count how many "red" numbers there are altogether. What percentage of red numbers would you expect to get? Is the result of the experiment close to this?

5.2 Medical testing and conditional probability: Ill or not?

The following excerpt is from DailySkiff.com. It presents two distinctly different points of view regarding the testing of Texas high school students for steroid use.

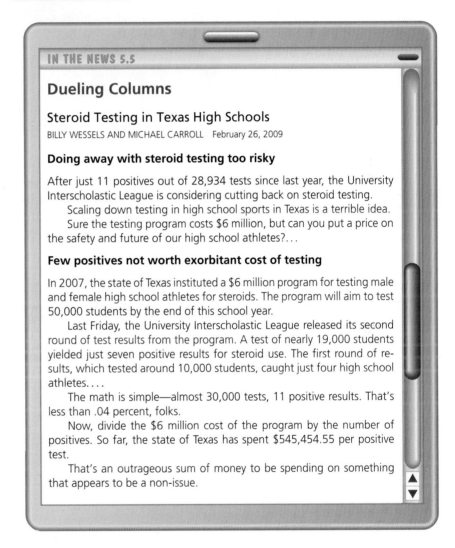

IN THE NEWS 5.5

Dueling Columns

Steroid Testing in Texas High Schools

BILLY WESSELS AND MICHAEL CARROLL   February 26, 2009

Doing away with steroid testing too risky

After just 11 positives out of 28,934 tests since last year, the University Interscholastic League is considering cutting back on steroid testing.

Scaling down testing in high school sports in Texas is a terrible idea.

Sure the testing program costs $6 million, but can you put a price on the safety and future of our high school athletes?...

Few positives not worth exorbitant cost of testing

In 2007, the state of Texas instituted a $6 million program for testing male and female high school athletes for steroids. The program will aim to test 50,000 students by the end of this school year.

Last Friday, the University Interscholastic League released its second round of test results from the program. A test of nearly 19,000 students yielded just seven positive results for steroid use. The first round of results, which tested around 10,000 students, caught just four high school athletes....

The math is simple—almost 30,000 tests, 11 positive results. That's less than .04 percent, folks.

Now, divide the $6 million cost of the program by the number of positives. So far, the state of Texas has spent $545,454.55 per positive test.

That's an outrageous sum of money to be spending on something that appears to be a non-issue.

Whether in the context of Texas high school athletes being tested for steroid use, Olympic athletes being tested for use of banned substances, or women being screened for breast cancer, medical tests are pervasive in modern society, and they have serious consequences. Some may be surprised to learn that medical tests are not as definitive as we might hope. In fact, they typically yield only probabilistic results. A positive result of a medical test for a disease does not tell us for certain that the patient actually has the disease—rather, it gives a probability that the disease is present. Similarly, a negative result does not guarantee that the disease is not present. In a famous case, Floyd Landis was stripped of his 2006 Tour de France cycling title because a test indicated that he had used banned substances. There was continuing controversy surrounding the case until 2010, when Landis admitted that he had used performance-enhancing drugs for most of his career.

Floyd Landis in the Tour de France.

What does it really mean to say that medical tests provide probabilistic results? A patient who has tested positive for a disease would certainly like to know what the test results really mean. That is the issue that we examine in this section.

Sensitivity and specificity

Accuracy is a crucial issue for medical tests. A medical test that gives unreliable results is useless in terms of medical practice and may unduly alarm patients. There are many ways of measuring accuracy, but certainly, an accurate test should detect a disease when it is present and give negative results when the disease is not present.[5]

The results of medical tests fall into four categories:

TRUE POSITIVES	FALSE POSITIVES
People who test positive and have the disease	*People who test positive but do not have the disease*
FALSE NEGATIVES	**TRUE NEGATIVES**
People who test negative but do have the disease	*People who test negative and do not have the disease*

Sensitivity and *specificity* are measures of a test's accuracy in these terms.

Key Concept

The **sensitivity** of a test is the probability that the test will detect the disease in a person who does have the disease.

---

[5] Although we refer to diseases, our discussion also applies to tests for banned substances, in which case having the disease means use of the substance.

"I have a positive attitude...
but it might be a false positive."

We calculate the sensitivity using the formula

$$\text{Sensitivity} = \frac{\text{True positives}}{\text{All who have the disease}} = \frac{\text{True positives}}{\text{True positives} + \text{False negatives}}.$$

The sensitivity measures a test's ability to detect correctly the presence of a disease. A test with high sensitivity has relatively few false negatives. If you actually have a disease, a high-sensitivity test is very likely to detect it.

Key Concept

The specificity of a test is the probability that the test will give a negative result for a person who does not have the disease.

We calculate the specificity using the formula

$$\text{Specificity} = \frac{\text{True negatives}}{\text{All who do not have the disease}} = \frac{\text{True negatives}}{\text{True negatives} + \text{False positives}}.$$

The specificity measures a test's ability to identify correctly the absence of a disease. A test with high specificity has relatively few false positives. If you are healthy, a high-specificity test is very likely to confirm that fact.

We normally write the sensitivity and specificity in percentage form and round the percentages to one decimal place when we calculate.

EXAMPLE 5.9 Estimating sensitivity and specificity: Test for TB

The accompanying table of data is adapted from a study of a test for tuberculosis (TB) among patients diagnosed with pulmonary TB (i.e., TB infection of the lungs).

	Has TB	Does not have TB
Test positive	571	8
Test negative	259	162

Use these data to estimate the sensitivity and specificity of this test.

SOLUTION

A total of $571 + 259 = 830$ people in this group have TB. Of these, 571 test positive, so there are 571 true positives. This indicates a sensitivity of

$$\text{Sensitivity} = \frac{\text{True positives}}{\text{All who have the disease}} = \frac{571}{830}.$$

This is about 0.688 or 68.8%. Hence, about 69% of the people who have TB will test positive, so about 31% of the people who have TB will test negative.

A total of $8 + 162 = 170$ do not have TB. Among these 162 test negative, there are 162 true negatives. This indicates a specificity of

$$\text{Specificity} = \frac{\text{True negatives}}{\text{All who do not have the disease}} = \frac{162}{170}.$$

This is about 0.953 or 95.3%. Thus, about 95% of the people who do not have TB will test negative, so about 5% of the people who do not have TB will test positive.

TRY IT YOURSELF 5.9

A new medical test is given to a population whose disease status is known. The results appear in the table below.

	Has disease	Does not have disease
Test positive	165	55
Test negative	29	355

Use these data to estimate the sensitivity and specificity of this test.

The answer is provided at the end of this section.

It is important to note that sensitivity and specificity depend only on the chemical and physical properties of a test and not on the population being tested. Hence, the calculations above give empirical estimates for theoretical probabilities. To explain this point, we consider lightbulbs of a certain brand. There is a theoretical probability that lightbulbs of this brand will burn out in the first month of use. To estimate this probability, we might observe the lights in a large building for a month to see how many burn out. That is, we can estimate the theoretical probability by calculating an empirical probability.

The next example shows how we can recover information by using the sensitivity and specificity.

EXAMPLE 5.10 Using sensitivity and specificity: Sensitivity of 93%

A certain test has a sensitivity of 93%. If 450 people who have the disease take this test, how many test positive?

SOLUTION

A sensitivity of 93% means that 93% of those who have the disease will test positive. Now 93% of 450 is about 419 people. Hence, we expect that about 419 of the 450 who have the disease will test positive.

TRY IT YOURSELF 5.10

The specificity of a test is 87%. If 300 people who do not have the disease take the test, how many will test negative?

The answer is provided at the end of this section.

The calculation of sensitivity and specificity for a medical test requires that we know whether a positive result (for example) represents a true positive or a false positive. This means that we need to be able to determine independently who in the test group is healthy and who is not. For this determination, we need an independent test that can be presumed to be accurate. Such a test is called a *gold standard*. A gold standard is very often expensive or difficult to administer. For example, in testing for HIV (a virus that causes AIDS), the gold standard would include a careful clinical examination. Researchers have developed less expensive tests for HIV, and the sensitivity and specificity provide a way to measure their accuracy in comparison with the gold standard.

Positive predictive value

A common misconception is that if one tests positive, and the test has a high sensitivity and specificity, then one should conclude that the disease is almost certainly present. But a more careful analysis shows that the situation is not so simple, and that *prevalence* plays a key role here.

Key Concept

The **prevalence** of a disease in a given population is the percentage of the population that has the disease.

Diabetes treatment may require monitoring of blood-glucose levels.

For example, the National Institutes of Health reports that 7.8% of the U.S. population have diabetes. That means the prevalence of diabetes in the United States is 7.8%. For comparison, the prevalence of epilepsy in the United States is less than 1%. The prevalence of a disease depends on the population under consideration. For example, the prevalence of HIV in Botswana in 2008 was almost 24%, but in the United States it was less than 1%.

A patient who has tested positive for a disease wants to know whether the disease is really present or not. That is, he or she wants to know the probability that the disease is actually present given that test results were positive. But this probability depends on the prevalence of the disease in the population being tested. A positive result for an HIV test administered among the residents of Botswana, where prevalence is high, is more likely to be accurate than is a positive result for the same test administered to residents of the United States, where prevalence is low.

We capture this notion of accuracy using the following terms: positive predictive value and negative predictive value.

Key Concept

For a population with a known prevalence, the **positive predictive value (PPV)** of a test is the probability that a person in the population who tests positive actually has the disease.

We calculate the positive predictive value for a given population using the formula

$$PPV = \frac{\text{True positives}}{\text{All positives}} = \frac{\text{True positives}}{\text{True positives} + \text{False positives}}.$$

The *negative predictive value* is defined in a similar fashion.

Key Concept

For a population with a known prevalence, the **negative predictive value (NPV)** of a test is the probability that a person in the population who tests negative, in fact, does not have the disease.

We calculate the negative predictive value for a given population using the formula

$$\text{NPV} = \frac{\text{True negatives}}{\text{All negatives}} = \frac{\text{True negatives}}{\text{True negatives} + \text{False negatives}}.$$

In our calculations, we normally round the PPV and NPV to one decimal place in percentage form.

EXAMPLE 5.11 Calculating PPV and NPV: Hepatitis C

Injection drug use is among the most important risk factors for hepatitis C. Two studies were done using a hepatitis C test with 90% sensitivity and specificity. One test involved a group that was representative of the overall U.S. population, which has a hepatitis C prevalence of about 2%. The other test was on a population of injection drug users where the prevalence is 57%. The results are shown in the table below.

	General population		Injection drug users	
	Has hep C	Does not have hep C	Has hep C	Does not have hep C
Test positive	9	49	257	21
Test negative	1	441	28	194

Calculate the PPV for each of the populations in the study.

SOLUTION

For the general population, there are 9 true positives and 49 false positives, for a total of $9 + 49 = 58$ positive results. Therefore,

$$\text{PPV} = \frac{\text{True positives}}{\text{All positives}}$$
$$= \frac{9}{58}.$$

This is about 0.155, so only 15.5% of those who test positive from the general population actually have hepatitis C. Applying this test to the general population would be of little value.

For the population of injection drug users, there are 257 true positives and 21 false positives, for a total of $257 + 21 = 278$ positive results. Therefore,

$$\text{PPV} = \frac{\text{True positives}}{\text{All positives}}$$
$$= \frac{257}{278}.$$

This is about 0.924, so 92.4% of those who test positive from the population of injection drug users do, in fact, have hepatitis C.

TRY IT YOURSELF 5.11

Calculate the NPV for each of the populations in this study.

The answer is provided at the end of this section.

> **SUMMARY 5.3**   **Prevalence, PPV, and NPV**
>
> **1.** The prevalence of a disease in a given population is the percentage of the population having the disease.
>
> **2.** Among a population of individuals, the positive predictive value (PPV) of a test is the probability that a person in the population who tests positive actually has the disease. The formula is
>
> $$PPV = \frac{\text{True positives}}{\text{All positives}} = \frac{\text{True positives}}{\text{True positives} + \text{False positives}}.$$
>
> **3.** Among a population of individuals, the negative predictive value (NPV) of a test is the probability that a person who tests negative actually does not have the disease. It is calculated using
>
> $$NPV = \frac{\text{True negatives}}{\text{All negatives}} = \frac{\text{True negatives}}{\text{True negatives} + \text{False negatives}}.$$
>
> **4.** The NPV and PPV depend on the prevalence of the disease in the population.

A case study: Crohn's disease

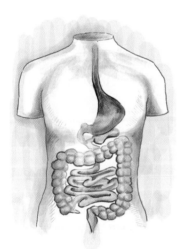

In this section, we want to clarify the ideas thus far developed by showing how they apply to a specific situation. Crohn's is an incurable inflammatory disease of the intestines. This serious disease afflicts many Americans. Here are some data from a study of a new screening test for Crohn's disease.[6]

	Has Crohn's	Does not have Crohn's
Test positive	36	25
Test negative	9	230

We can use these data to estimate the accuracy of this test in terms of sensitivity and specificity. The number of true positive tests in the study is 36, and the total number of patients tested who actually have the disease is 45 (36 true positives plus 9 false negatives). Therefore, we estimate that the sensitivity is $36/45 = 0.8$ or 80%. A sensitivity of 80% tells us that 80% of those people with Crohn's disease will test positive.

The number of true negative tests in the study is 230, and the total number of patients tested who do not have the disease is 255 (230 true negatives plus 25 false positives). Therefore, we estimate that the specificity is $230/255$ or about 90.2%. A specificity of 90.2% tells us that 90.2% of those people without Crohn's disease will test negative.

Consider a hypothetical patient who has tested positive for Crohn's disease. The patient wants to know whether the disease is actually present or not. That is, the patient wants to know the positive predictive value of the test. But, as we noted, the PPV depends on the population. To explore this point, we first consider applying the test to the entire population of the United States.

EXAMPLE 5.12 Using prevalence and calculating PPV: Crohn's disease nationwide

Crohn's is relatively rare in the United States, where the prevalence is about 0.17%. The population of the United States is about 300 million.

---

[6]These data are adapted from the illustration of sensitivity and specificity at www.fpnotebook.com/Prevent/Epi.

a. Approximately how many people in the United States suffer from Crohn's?

b. Use the sensitivity of 80% and specificity of 90.2% that we calculated above to complete the following table, which shows the results we would expect if everyone in the United States were tested for Crohn's. (Report your answers in millions and round to two decimal places.)

	Has disease	Does not have disease
Test positive (millions)		
Test negative (millions)		

c. Use the information from part b to calculate the PPV of the Crohn's test for the general U.S. population.

SOLUTION

a. Because the prevalence is 0.17%, to find the number of Americans suffering from Crohn's, we multiply the total population of 300 million by 0.0017:

$$\text{Number with Crohn's} = 0.0017 \times 300$$
$$= 0.51 \text{ million.}$$

Thus, over half a million people in the United States suffer from Crohn's.

b. Now 0.51 million people in the United States have Crohn's disease. Because the sensitivity of the test is 80%, to calculate the number of true positives, we multiply the number who have Crohn's by 0.8:

$$\text{True positives} = 0.8 \times 0.51$$

or about 0.41 million. This is the number of people who have Crohn's and test positive, so $0.51 - 0.41 = 0.10$ million Crohn's sufferers will test negative. This is the number of false negatives.

Because 0.51 million have Crohn's disease, $300 - 0.51 = 299.49$ million do not. Because the specificity of the test is 90.2%, to find the number of true negatives, we multiply 299.49 million by 0.902:

$$\text{True negatives} = 299.49 \times 0.902$$

or about 270.14 million. This is the number of people who do not have Crohn's and test negative, so $299.49 - 270.14 = 29.35$ million disease-free individuals will test positive. This is the number of false positives. The completed table is shown below.

	Has disease	Does not have disease
Test positive (millions)	0.41	29.35
Test negative (millions)	0.10	270.14

c. The total number of positive results is $0.41 + 29.35 = 29.76$ million. Because 0.41 million of these are true positives, we calculate the PPV as follows:

$$\text{PPV} = \frac{\text{True positives}}{\text{All positives}}$$
$$= \frac{0.41}{29.76}.$$

This is about 0.014, so the PPV for this population is 1.4%. That is, if everyone in the United States were tested for Crohn's, only 1.4% of people who test positive would actually have the disease.

The results of Example 5.12 seem to indicate that the Crohn's test is so ineffective that it is useless. This raises a big question: Why even bother with such a test if it is so ineffective in predicting the presence of the disease? Should a person who tests positive even be concerned?

The key is the population we regarded as the patient's peers—the entire population of the United States. Because the prevalence of Crohn's in the entire population is so low, we should not be surprised that even an accurate test has a hard time in picking out those with the disease. But because Crohn's disease is so rare, a doctor would never prescribe a test for it unless there was some reason to do so. Surely, it must be that the patient's medical profile (e.g., his or her symptoms, age, or family medical history) suggested the possible presence of the disease.

The point is that we should not think of the patient's peer group as everyone in the United States. Rather, we should restrict attention to the roughly 3 million Americans who have a medical profile suggesting the possibility of Crohn's disease. Surely among this group, a positive test will be a more reliable indicator of whether a person has Crohn's. In fact, among this group the screening test has a PPV of 63.1%. (See Exercises 12 and 13.) Thus, the probability that a patient in this group has Crohn's given that he or she tested positive is 63.1%. Comparing the PPV of 1.4% for the general population with the PPV of 63.1% for those with symptoms shows that this test is much more reliable when we choose an appropriate population of peers.

Conditional probability

Positive and negative predictive values (as well as the sensitivity and specificity) are examples of *conditional probabilities*.

Key Concept

A conditional probability is the probability that one event occurs given that another has occurred.

For example, suppose a card is drawn from a standard deck and laid face down on a table. What is the probability that it is a heart? There are four equally likely suits: clubs, diamonds, hearts, and spades. So the probability that the card is a heart is 1/4. Now let's change things just a little. Suppose that when the card was chosen we took a peek and noticed that it was a red card. Now, under these conditions, what is the probability that the card is a heart? What we're asking is the probability the card is a heart *given that* it is a red card. There are two equally likely red suits, hearts and diamonds. So with the additional information that the card is red, the probability that it is a heart is 1/2.

In the next example, a conditional probability is calculated in the context of medical testing.

EXAMPLE 5.13 Relating conditional probability and medical testing: TB

The accompanying table of data is adapted from a study of a test for TB among patients diagnosed with extrapulmonary TB (i.e., TB infection outside of the lungs).

	Has TB	Does not have TB
Test positive	446	15
Test negative	216	323

Calculate the conditional probability that a person tests positive given that the person has TB.

SOLUTION

There are $446 + 216 = 662$ people who have TB, and 446 of this group test positive. Hence, the probability that a person tests positive given that the person has TB is

$$P(\text{Positive test given TB is present}) = \frac{\text{True positives}}{\text{All who have TB}}$$
$$= \frac{446}{662}.$$

This is about 0.674 or 67.4%.

TRY IT YOURSELF 5.13

Use the table above to find the probability that a person tests negative given that the person does not have TB.

The answer is provided at the end of this section.

You may have noticed that the conditional probability in Example 5.13 is the same as the sensitivity: Both of them can be written as

$$\frac{\text{True positives}}{\text{All who have the disease}}.$$

This connection holds in general: The sensitivity is the conditional probability that a person tests positive given that the person has the disease. Similar reasoning shows that for the PPV the order is reversed: The PPV is the conditional probability that a person has the disease given that the person tests positive.

Try It Yourself answers

Try It Yourself 5.9: Estimating sensitivity and specificity: Test for disease Sensitivity: 85.1%; specificity: 86.6%.

Try It Yourself 5.10: Using sensitivity and specificity: Specificity of 87%   261.

Try It Yourself 5.11: Calculating PPV and NPV: Hepatitis C   General population: 99.8%; injection drug users: 87.4%.

Try It Yourself 5.13: Relating conditional probability and medical testing: TB 0.956 or 95.6%.

Exercise Set 5.2

In these exercises, round all answers in percentage form to one decimal place unless you are instructed otherwise.

1. 100% accurate? Your friend suggests the following as a way of doing medical testing: Just tell each person "tested" that he or she has the disease—the test result is always positive. What is the sensitivity of this test? What is the specificity?

Crohn's screening test   Exercises 2 through 4 refer to the table of data on p. 293 for the Crohn's screening test study.

2. What percentage of the individuals in the study were false negatives?

3. What percentage of the individuals in the study who had Crohn's tested negative?

4. For what percentage of the individuals in the study did the test return an incorrect result?

5. Lupus. The excerpt below is taken from the Medical University of South Carolina Web site.[7]

> Another problem we have when using diagnostic tests is that they are originally tested in populations that are homogeneously [i.e., entirely] positive or negative and/or designed to be used in patients with a high pretest probability of disease. When the tests are used in clinical practice, they sometimes fail to be helpful. For example, the Anti-Nuclear Antibody test is useful in confirming the diagnosis of SLE (Systemic Lupus Erythematosus) in patients with multiple clinical manifestations of the disease. The test, however, is often applied indiscriminately in patients where the physician is using a shotgun approach to diagnosis. Here, the test is diagnostically almost useless as the

---

[7]http://application.fnu.ac.fj/classshare/Medical_Science_Resources/MBBS/MBBS1-3/PBL/Sexually%20Transmitted%20Infections/HIV-AIDS/Sensitivity%20and%20Specificity%20of%20the%20HIV%20test.pdf.

number of false positive overwhelm the number of true positive results but it does keep the consulting Rheumatologist busy saying "no, the patient does not have SLE."

Explain the above in terms of sensitivity, specificity, and positive predictive value.

6. Sensitivity and specificity. The accompanying table gives the results of a screening test for a disease. Estimate the sensitivity and specificity of the test.

	Has disease	Does not have disease
Test positive	285	280
Test negative	15	420

7. Sensitivity and specificity. The accompanying table gives the results of a screening test for a disease. Estimate the sensitivity and specificity of the test.

	Has disease	Does not have disease
Test positive	15	12
Test negative	5	68

8. Extreme populations. Suppose *everyone* in Population 1 has a certain disease and *no one* in Population 2 has the disease. What is the NPV for each of these populations?

9. Negative predictive value. Would you expect the NPV of a given test to be higher in a population where prevalence is high or in a population where prevalence is low?

A disease Suppose a test for a disease has a sensitivity of 95% and a specificity of 60%. Further suppose that in a certain country 30% of the population has the disease. Exercises 10 and 11 refer to this situation.

10. Assume that the population is 10,000, and fill in the accompanying table.

	Has disease	Does not have disease	Totals
Test positive			
Test negative			
Totals			

11. *This exercise uses the results of Exercise 10.* What are the PPV and NPV of this test for this country?

More on screening test Recall the screening test for Crohn's that has a sensitivity of 80% and a specificity of 90.2% (see the discussion beginning on p. 293). In Exercises 12 and 13, we calculate the PPV of this test within a population of individuals who show symptoms of the disease.

12. Assume that the test is applied to the group of approximately 3 million Americans who suffer from symptoms that are very similar to those of Crohn's. Suppose also that all of the 0.51 million Crohn's sufferers belong to this group. Fill in the following table based on this group of 3 million. (Write your answers in millions and round to two decimal places.)

	Has disease	Does not have disease	Totals
Test positive (millions)			
Test negative (millions)			
Totals			

13. Use the table in Exercise 12 to show that the PPV of the test within this population of individuals showing symptoms of the disease is about 63.1%.

A disease Suppose a test for a disease has a sensitivity of 75% and a specificity of 85%. Further suppose that in a certain country, 20% of the population has the disease. Exercises 14 and 15 refer to this situation.

14. Assume that the population is 10,000, and fill in the accompanying table.

	Has disease	Does not have disease	Totals
Test positive			
Test negative			
Totals			

15. *This exercise uses the results of Exercise 14.* What are the PPV and NPV of this test for this country?

Testing for drug abuse Here is a scenario that came up in the nationally syndicated column called "Ask Marilyn" in *Parade Magazine*.[8] Suppose a certain drug test is 95% accurate, meaning that if a person is a user, the result is positive 95% of the time, and if she or he isn't a user, it's negative 95% of the time. Assume that 5% of all people are drug users. Exercises 16 through 22 refer to this drug test.

16. What is the sensitivity of this test?

17. What is the specificity of this test?

18. Assume that the population is 10,000, and fill in the accompanying table.

	Drug user	Not a drug user	Totals
Test positive			
Test negative			
Totals			

19. *This exercise uses the results of Exercise 18.* Suppose a randomly chosen person from the population in question tests positive. How likely is the individual to be a drug user? That is, what is the PPV of this test for this population?

20. *This exercise uses the results of Exercise 19.* If you found out a student was expelled because she tested positive for drug use with this test, would you feel that this action was justified? Explain.

21. *This exercise uses the results of Exercise 18.* Suppose a randomly chosen person from the population in question tests negative. How likely is the individual not to be a drug user? That is, what is the NPV of this test for this population?

---

[8]This column is available in detail at the Chance Web site: www.dartmouth.edu/~chance/course/Syllabi/mpls/handouts/section3_8.html.

22. *This exercise uses the results of Exercise 21.* If you found out that a student who had been accused of using drugs was exonerated because she tested negative for drug use with this test, would you feel that this action was justified? Explain.

23. Testing for Drug Abuse II. Repeat Exercises 18 through 22, this time assuming that the individual being tested is chosen from a population of which 20% are drug users. (This might be the case if the test is being applied to a prison population that has a known history of drug use.)

24. Conclusions. In Exercises 18 through 22, it was assumed that a drug test was applied to a population with 5% users. In Exercise 23, it was assumed that this drug test was applied to a population with 20% users. What do the results suggest about how the drug test could be used so as to be most effective?

Genetic testing   A certain genetic condition affects 5% of the population in a city of 10,000. Suppose there is a test for the condition that has an error rate of 1% (i.e., 1% false negatives and 1% false positives). Exercises 25 through 27 refer to this genetic condition.

25. Fill in the table below.

	Has condition	Does not have condition	Totals
Test positive			
Test negative			
Totals			

26. *This exercise uses the results of Exercise 25.* What is the probability (as a percentage) that a person has the condition if he or she tests positive?

27. *This exercise uses the results of Exercise 25.* What is the probability (as a percentage) that a person does not have the condition if he or she tests negative?

PSA test   The following is paraphrased from a Web site that discusses the sensitivity and specificity of the PSA (Prostate Specific Antigen) test for prostate cancer.[9] In a larger study (243 subjects), Ward and colleagues used a cutoff ratio of 0.15 to demonstrate 78% sensitivity and 69% specificity in the diagnosis of prostate cancer. In contrast, tPSA measurements alone yielded equivalent sensitivity but only 33% specificity. These reports are encouraging, but larger clinical studies are required for determination of the optimal fPSA:tPSA cutoff ratio. Exercises 28 through 31 refer to this PSA test.

28. What does the figure 78% sensitivity in the above study mean?

29. What does the figure 69% specificity in the above study mean?

30. Assume that 10% of the male population over the age of 50 has prostate cancer. What are the PPV and NPV of this test for this population? *Suggestion*: First make a table as in Exercise 10. Because the PPV and NPV don't depend on the size of the population, you may assume any size you wish. A choice of 10,000 for the population size works well.

31. Assume that 40% of the male population over the age of 70 has prostate cancer. What are the PPV and NPV of this test for this population?

HIV   Suppose a certain HIV test has both a sensitivity and specificity of 99.9%. This test is applied to a population of 1,000,000 people. Exercises 32 through 35 refer to this information.

32. Suppose 1% of the population is actually infected with HIV.

   a. Calculate the PPV. *Suggestion*: First make a table as in Exercise 10.

   b. Calculate the NPV.

   c. How many people will test positive who are, in fact, disease-free?

33. Suppose the population comes from the blood donor pool that has already been screened. In this population, 0.1% of the population is actually infected with HIV.

   a. Calculate the PPV.

   b. Calculate the NPV.

   c. How many people will test positive who are, in fact, disease-free?

34. Consider a population of drug rehabilitation patients for which 10% of the population is actually infected with HIV.

   a. Calculate the PPV.

   b. Calculate the NPV.

   c. How many people will test positive who are, in fact, disease-free?

35. Compare the values for the PPV you found in part a of Exercises 32 through 34, and explain what this says about testing for HIV.

Conditional probability and medical testing   Exercises 36 and 37 refer to the table in Example 5.13 on page 295.

36. Find the probability that a person tests positive given that the person does not have the disease.

37. Find the probability that a person does not have the disease given that the person tests negative. Which of the four basic quantities (sensitivity, specificity, PPV, NPV) does this number represent?

38. Conditional probability. You roll a fair six-sided die and don't look at it. What is the probability that it is a 5 given that your friend looks and tells you that it is greater than 2? Leave your answer as a fraction.

39. More conditional probability. In a standard deck of cards, the jack, queen, and king are "face cards." You draw a card from a standard deck. Your friend peeks and lets you know that your card is a face card. What is the probability that it is a queen given that it is a face card? Leave your answer as a fraction.

40. Shampoo. Your wife has asked you to pick up a bottle of her favorite shampoo. When you get to the store, you can't

---

[9] www.ivdtechnology.com/article/equimolar-psa-assays.

remember which brand she uses. There are nine shampoo brands on the shelf. If you select one, what is the probability that you get the right brand of shampoo? Of the nine bottles, four are blue, three are white, and two are red. You remember that your wife's shampoo comes in a blue bottle. What is the probability that you get the right bottle with this additional information? Leave your answers as fractions.

41. Conditional probability. The total area of the United States is 3.79 million square miles, and Texas covers 0.27 million square miles. A meteor falls from the sky and strikes Earth. What is the probability that it strikes Texas given that it strikes the United States?

42. This one is tricky. A family that you haven't yet met moves in next door. You know the couple has two children, and you discover that one of them is a girl. What is the probability that the other child is a boy? That is, what is the probability that one child is male given that the other child is female? Leave your answer as a fraction.

43. Formula for conditional probability. We express the probability that event A occurs given that event B occurs as

$$P(A \text{ given } B).$$

[In many books, the notation used is $P(A|B)$.] It turns out that the following basic formula for conditional probabilities holds:

$$P(A \text{ and } B) = P(A \text{ given } B) \times P(B).$$

In this exercise, we will verify this formula in the case that A is the event that a card drawn from a standard deck is a heart and B is the event that a card drawn from a standard deck is an ace. Leave your answers as fractions.

  a. Calculate $P(A \text{ and } B)$. (This is the probability that the card is the ace of hearts.)

  b. Calculate $P(A \text{ given } B)$. (This is the probability that the card is a heart given that it is an ace.)

  c. Calculate $P(B)$. (This is the probability that the card is an ace.)

  d. Use parts a, b, and c to verify the formula above in this case.

Bayes's theorem   Bayes's theorem on conditional probabilities states that, if $P(B)$ is not 0, then

$$P(A \text{ given } B) = \frac{P(B \text{ given } A) \times P(A)}{P(B)}.$$

Exercises 44, 46, and 47 can be solved using this formula. Leave your answers as fractions.

44. Cookies Mary bakes 10 chocolate chip cookies and 10 peanut butter cookies. Bill bakes 5 chocolate chip cookies and 10 peanut better cookies. The 35 cookies are put together and offered on a single plate. I pick a cookie at random from the plate.

  a. What is the probability that the cookie is chocolate chip?

  b. What is the probability that Mary baked the cookie?

  c. What is the probability that the cookie is chocolate chip given that Mary baked it?

  d. Use Bayes's theorem to calculate the probability that the cookie is baked by Mary given that it is chocolate chip. In this situation, Bayes's theorem tells us that

$$P(\text{Mary given Chocolate})$$
$$= \frac{P(\text{Chocolate given Mary}) \times P(\text{Mary})}{P(\text{Chocolate})}.$$

  e. Calculate without using Bayes's theorem the probability that the cookie is baked by Mary given that it is chocolate chip. *Suggestion:* How many of the chocolate chip cookies were baked by Mary?

45. *This is a warm-up exercise for Exercises 46 and 47. It does not require Bayes's theorem.* Alan plans to buy a car from one of two used car lots. For simplicity, we assume that only Toyotas and Chevrolets are available from these lots.

Here are two schemes that Alan might use to select a car for purchase:

  Scheme 1: Alan picks one of the two lots at random and then chooses a car at random from that lot.

  Scheme 2: Alan combines all of the cars from both lots into one list and picks a car at random from that list.

Consider the probability that Alan buys a Toyota. Is the probability the same regardless of which of the two schemes Alan uses? *Suggestion:* As a test case, think about the situation in which one lot has 99 Toyotas and no Chevrolets, and the other lot has no Toyotas and 1 Chevrolet. You don't have to compute the actual probabilities—just decide whether they are the same.

46. Yesterday car lot Alpha had two Toyotas and one Chevrolet for sale. Car lot Beta had three Toyotas and five Chevrolets for sale. This morning Alan bought a car, choosing one of the two lots at random and then choosing a car at random from that lot.

  a. What is the probability that Alan bought from Alpha?

  b. If Alan bought from Alpha, what is the probability he bought a Toyota?

  c. If Alan bought from Beta, what is the probability he bought a Toyota?

  d. It can be shown that the probability that Alan bought a Toyota is $25/48$. What is the probability Alan bought from Alpha given that he bought a Toyota? Give your answer both as a fraction and in decimal form rounded to two places. *Note:* In this situation, Bayes's theorem says

$$P(\text{Alpha given Toyota})$$
$$= \frac{P(\text{Toyota given Alpha}) \times P(\text{Alpha})}{P(\text{Toyota})}.$$

47. There are two bowls of cookies. Bowl Alpha has 5 chocolate chip cookies and 10 peanut butter cookies. Bowl Beta has 10 chocolate chip cookies and 10 peanut butter cookies. Jerry selects a bowl at random and chooses a cookie at random from that bowl.

  a. What is the probability that he chooses bowl Alpha?

  b. What is the probability that he gets a chocolate chip cookie given that he chooses from bowl Alpha?

  c. It can be shown that the probability that he gets a chocolate chip cookie is $5/12$. What is the probability that he chose from bowl Alpha given that he got a chocolate chip cookie?

Note: In this situation, Bayes's theorem says

$$P(\text{Alpha given Chocolate})$$
$$= \frac{P(\text{Chocolate given Alpha}) \times P(\text{Alpha})}{P(\text{Chocolate})}.$$

48. History. The tables used in this section are examples of *contingency tables*. These tables came into use in the twentieth century, but the concepts in probability behind them are often traced to Thomas Bayes, who lived in Britain in the eighteenth century. Research the life and work of Bayes and the ideas in probability and statistics currently associated with his name.

The following exercise is designed to be solved using technology such as a calculator or computer spreadsheet. For assistance, see the technology supplement.

49. Spreadsheet. Devise a spreadsheet that will allow the user to input the prevalence of a disease in a population along with the sensitivity and specificity of a test for the disease. The spreadsheet should then complete a table similar to that in Example 5.9 on p. 289 and report the PPV and NPV.

5.3 Counting and theoretical probabilities: How many?

TAKE AWAY FROM THIS SECTION

Learn how to count outcomes without listing all of them.

The following excerpt is from the *New York Times Magazine*.

IN THE NEWS 5.6

Sentencing by the Numbers

EMILY BAZELON   January 2, 2005

For decades, the science of predicting future criminality has been junk science—the guesswork of psychologists who were wrong twice as often as they were right. But today, the detailed collection of crime statistics is beginning to make it possible to determine which bad guys really will commit new offenses. In 2002, the Commonwealth of Virginia began putting such data to use: the state encourages its judges to sentence nonviolent offenders the way insurance agents write policies, based on a short list of factors with a proven relationship to future risk. If a young, jobless man is convicted of shoplifting, the state is more likely to recommend prison time than when a middle-aged, employed woman commits the same crime. . . .

Using these factors and a few others, including a defendant's adult and juvenile criminal records, Kern designed a simple 71-point scale of risk assessment as an aid for judges. If he scores 35 points or less, a defendant who would have otherwise gone to prison under Virginia sentencing guidelines is recommended for an alternative sanction like probation or house arrest. Anything above 35 means a recommendation of jail time. "Judges make risk assessments every day," Kern said. "Prosecutors do, too. Our model brings more equity to the process and ties the judgments being made to science."

Kern tested his model on prisoners released five years earlier and found that his ratings correctly predicted who would be reconvicted in three out of four cases. Of the felons who scored at or below the 35-point cutoff, 12 percent committed new crimes, compared with more than 38 percent for those who scored higher. After calculating that only a slight increase in recidivism would result, the state raised the 35-point cutoff to 38 points last July. Meanwhile, the growth of the state's prison population—which used to be more than twice the national average—has slowed nearly to a halt.

This article describes the use of probability calculations to aid a judge in sentencing decisions. In order to calculate the necessary probabilities, we need to be able to count the number of ways in which an event can happen. In this section, we will examine some fundamental counting methods.

Simple counting

Counting the number of outcomes in the case of flipping two coins as we did in Section 5.1 is not difficult. We distinguish the two coins (by thinking of a nickel and a dime for example) and then simply list the four possible outcomes: HH, HT, TH, TT.

If we are flipping three coins, the number of possible outcomes is much larger. Distinguishing the coins as a penny, nickel, and dime, we find the possibilities listed below.

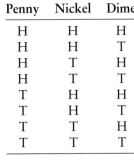

Penny	Nickel	Dime
H	H	H
H	H	T
H	T	H
H	T	T
T	H	H
T	H	T
T	T	H
T	T	T

That is a total of eight possible outcomes.

If there are many coins, it is impractical to make a list of all possible outcomes. For example, if you try to make such a list for 10 coins, you are unlikely to finish it. How can we compute the number of outcomes without writing down all such sequences? We do it by a systematic counting scheme. There are two possibilities for the first coin: a head or a tail. For each of these possibilities, there are also two possibilities for the second coin. To obtain the number of possibilities for two coins, we multiply these together to get $2 \times 2 = 2^2 = 4$. (As we noted earlier, these are HH, HT, TH, TT.) If we were flipping three coins, we would continue this counting process: For each of the four possibilities for the first two coins, there are two possibilities for the third coin. Thus, the total number of outcomes for flipping three coins is $4 \times 2 = 2^3 = 8$. (This agrees with the list HHH, HHT, HTH, HTT, THH, THT, TTH, TTT we made above.) If we continue in this way, we see that the number of possible outcomes in tossing 10 coins is $2^{10} = 1024$.

Armed with this information, we can calculate the probability of getting all heads in the toss of 10 coins. There is only one favorable outcome—namely, all heads. Therefore,

$$P(10 \text{ heads}) = \frac{\text{Favorable outcomes}}{\text{Total outcomes}}$$
$$= \frac{1}{1024}.$$

This is 0.00098 or 0.098%.

EXAMPLE 5.14 Counting outcomes: Tossing coins

How many possible outcomes are there if we toss 12 coins?

SOLUTION

Reasoning as above, we see that each extra coin doubles the number of outcomes. That gives a total of $2^{12} = 4096$ possible outcomes.

TRY IT YOURSELF 5.14

How many possible outcomes are there if we toss 15 coins?

The answer is provided at the end of this section.

The Counting Principle

The method we used to count outcomes of coin tosses works in a more general setting. Its formal statement is known as the *Counting Principle*. Suppose we perform two experiments in succession and that for each outcome of the first experiment there is the same number of outcomes for the second. For example, think of a menu at a diner where there are five entrées, and for each choice of entrée there are three side dishes available for that choice. The Counting Principle asserts that the number of possible outcomes for the two experiments is the number of possible outcomes for the first experiment times the number of possible outcomes for the second experiment. This extends to any number of such experiments. In the case of the diner, there are $5 \times 3 = 15$ possible combinations of an entrée with a side dish.

Key Concept

The **Counting Principle** tells us how to calculate the results of two experiments performed in succession. Suppose there are *N* outcomes for the first experiment. If for each outcome of the first experiment, there are *M* outcomes of the second experiment, then the number of possible outcomes for the two experiments is $N \times M$.

This principle extends to any number of such experiments.

Another way to think about this principle is as follows: Suppose there are two blanks

— —

that we wish to fill, and we have *N* choices for the first blank and *M* choices for the second blank:

$$\underline{N} \ \underline{M}$$

Then the total number of ways we can fill the two blanks is $N \times M$. This extends to any number of blanks.

As a simple application of the Counting Principle, let's look at product labeling. Suppose a company labels its products with a letter of the alphabet followed by a numeral (0 through 9), for example, model A4 or model D9. How many different labels can the company make? Let's think of this as filling in two blanks:

— —

A letter goes in the first blank and a numeral goes in the second blank. There are 26 letters that can go in the first blank and 10 numerals that can go in the second blank:

<u>26</u> <u>10</u>

According to the Counting Principle, the total number of labels is the number of outcomes for the first blank times the number of outcomes for the second blank. This is $26 \times 10 = 260$.

CALCULATION TIP 5.2 Filling in Blanks

When we want to know how many ways a set of items can be arranged, it is often helpful to think of filling a list of blank spaces with the items, one at a time.

EXAMPLE 5.15 Counting by filling in blanks: Codes

How many three-letter codes can we make if the first letter is a vowel (A, E, I, O, or U)? One such code is OXJ.

SOLUTION

We think of filling in three blanks:

— — —

There are 5 letters that can go in the first blank and 26 in each of the other two:

<u>5</u> <u>26</u> <u>26</u>

Applying the Counting Principle, we find a total of $5 \times 26 \times 26 = 3380$ codes.

TRY IT YOURSELF 5.15

How many three-letter codes can we make if the first letter is a consonant rather than a vowel? One such code is KXJ.

The answer is provided at the end of this section.

The Counting Principle allows us to calculate the total number of outcomes for fairly complex events. Suppose, for example, that we toss five dice and want to know the total number of outcomes. We think of tossing the dice one at a time. Applying the Counting Principle, we see that the total number of outcomes is

Outcomes for die #1 $\times$ Outcomes for die #2 $\times \cdots \times$ Outcomes for die #5.

Because each die has six possible outcomes, we are multiplying five 6's together, which gives the total number of outcomes as

$$6 \times 6 \times 6 \times 6 \times 6 = 6^5 = 7776.$$

We can also think of this in terms of five blanks, one for each die. Each blank has six possible outcomes:

<u>6</u> <u>6</u> <u>6</u> <u>6</u> <u>6</u>

We find $6 \times 6 \times 6 \times 6 \times 6 = 6^5$ outcomes, just as before.

EXAMPLE 5.16 Counting for complex events: License plates

Automobile license plates in the state of Nevada typically consist of three numerals followed by three letters of the alphabet, such as 072 ZXE.

a. How many such license plates are possible?

b. How many such plates are possible if we insist that on each plate no numeral can appear more than once and no letter can appear more than once?

SOLUTION

a. Let's think of a blank license tag with three slots for numbers and three slots for letters:

$$\underline{}\,\underline{}\,\underline{}\quad\underline{}\,\underline{}\,\underline{}$$
$$\text{numbers}\qquad\text{letters}$$

We need to figure out how many ways we can fill in the slots. For each number slot, there are 10 possible numerals (0 through 9) to use: 10 for the first blank, 10 for the second, and 10 for the third:

$$\underline{10}\,\underline{10}\,\underline{10}\quad\underline{}\,\underline{}\,\underline{}$$
$$\text{numbers}\qquad\text{letters}$$

For each letter slot, there are 26 possible letters of the alphabet: 26 for the first blank, 26 for the second, and 26 for the third:

$$\underline{10}\,\underline{10}\,\underline{10}\quad\underline{26}\,\underline{26}\,\underline{26}$$
$$\text{numbers}\qquad\text{letters}$$

Applying the Counting Principle gives $10 \times 10 \times 10 \times 26 \times 26 \times 26 = 17{,}576{,}000$ ways to fill in the six slots.

By the way, the population of Nevada is approximately 2,600,000. There should be enough license plate numbers for a while!

b. As in part a, we think of a blank plate with three slots for numbers and three slots for letters. We need to count the number of ways we can fill in these slots. But this time we are not allowed to repeat letters or numbers.

For the first number slot, we have 10 choices. For the second slot, we can't repeat the number in the first blank, so we have only nine choices. For the third slot, we can't repeat the numbers used in the first two slots, so we have a choice of only eight numbers:

$$\underline{10}\,\underline{9}\,\underline{8}\quad\underline{}\,\underline{}\,\underline{}$$
$$\text{numbers}\qquad\text{letters}$$

Now we fill in the blanks for the letters. There are 26 ways to fill in the first blank. We can't use that letter again, so there are only 25 ways to fill in the second blank. We can't use the first 2 letters, so we have a choice of only 24 letters for the third blank:

$$\underline{10}\,\underline{9}\,\underline{8}\quad\underline{26}\,\underline{25}\,\underline{24}$$
$$\text{numbers}\qquad\text{letters}$$

Now the Counting Principle gives

$$\text{Number of plates} = 10 \times 9 \times 8 \times 26 \times 25 \times 24 = 11{,}232{,}000.$$

Applying counting to probabilities

Let's see how to apply the Counting Principle to calculate probabilities when drawing cards. Recall that there are 52 playing cards in a standard deck. They are divided into four *suits*, called spades, hearts, diamonds, and clubs. Twelve of the cards (four kings, four queens, and four jacks) are called *face cards*.

Let's use what we have learned about counting to determine in how many ways we can draw two cards from a deck in order without replacement. That is, we draw a first card from a deck and, without putting it back, draw a second card. In this situation, drawing the 2 of spades for the first card and the 3 of spades for the second is different from drawing the 3 of spades for the first card and the 2 of spades for the second. We can think of filling in two blanks. There are 52 possibilities for the first, but only 51 for the second:

$$\underline{52}\ \underline{51}$$

The counting principle gives $52 \times 51 = 2652$ possible outcomes. These kinds of calculations will be helpful in calculating probabilities for cards.

EXAMPLE 5.17 Counting and calculating probabilities: Draw two cards

Suppose you draw a card, put it back in the deck, and draw another. What is the probability that the first card is an ace and the second one is a jack? (Assume that the deck is shuffled after the first card is returned to the deck.)

SOLUTION

In order to find the probability, we have two numbers to calculate: The number of ways we can draw two cards (the total number of possible outcomes), and the number of ways we can draw an ace followed by a jack (the number of favorable outcomes).

First we count the total number of ways to draw two cards in order from a deck. Think of filling two blanks:

$$\underline{\quad}\ \underline{\quad}$$

The first blank is filled by a card drawn from the full deck, and the second blank is also filled by a card drawn from a full deck (because we put the first card back in the deck). There are 52 possible cards to place in the first blank, and for each of these, there are 52 possible cards to place in the second blank:

$$\underline{52}\ \underline{52}$$

Using the Counting Principle, we find $52 \times 52 = 2704$ ways to fill the two blanks with two cards from the deck. This is the total number of possible outcomes.

Now we ask how many ways there are to fill these two blanks with an ace followed by a jack. There are four aces in the deck, so there are four choices to fill the first blank:

$$\underline{4}\ \underline{\quad}$$

After this has been done there are four jacks with which to fill the second blank:

$$\underline{4}\ \underline{4}$$

Again, using the Counting Principle, we find $4 \times 4 = 16$ ways to fill the two blanks with an ace followed by a jack. This is the number of favorable outcomes. So the probability of drawing an ace followed by a jack is

$$P(\text{Ace followed by jack}) = \frac{\text{Favorable outcomes}}{\text{Total outcomes}}$$
$$= \frac{16}{2704}.$$

This is about 0.006 or 0.6%.

TRY IT YOURSELF 5.17

Suppose you draw a card and, without putting it back, draw another. What is the probability that the first card is an ace and the second one is a jack?

The answer is provided at the end of this section.

The Counting Principle applies to elections just as it does to cards.

EXAMPLE 5.18 Counting: Election outcomes

Suppose there are 10 people willing to serve as an officer for a club. It's decided just to put the 10 names in a hat and draw three of them out in succession. The first name drawn is declared president, the second name vice president, and the third name treasurer.

a. How many possible election outcomes are there?

b. John is a candidate. What is the probability that he will be vice president?

c. Mary and Jim are also candidates along with John. What is the probability that all three will be selected to office?

d. What is the probability that none of the three (Mary, Jim, and John) will be selected to office?

e. What is the probability that at least one of the three (Mary, Jim, and John) will not be selected? *Suggestion:* Recall from Section 5.1 that if p is the probability of an event occurring, then the probability that the event will not occur is $1 - p$.

SOLUTION

a. Imagine that we have three blanks corresponding to president, vice president, and treasurer:

$$\overline{} \quad \overline{} \quad \overline{}$$
$$\text{P} \qquad \text{VP} \qquad \text{T}$$

There are 10 possibilities for the president blank. When that is filled, there are nine names left, so there are nine possibilities for the vice president blank. When that is filled, there are eight possibilities for the third blank:

$$\frac{10}{\text{P}} \quad \frac{9}{\text{VP}} \quad \frac{8}{\text{T}}$$

The Counting Principle gives

$$\text{Number of outcomes} = 10 \times 9 \times 8 = 720.$$

b. We must count the number of ways John can be vice president. In this case, John is not president, so there are nine possible names for the president blank. There is just one possibility, John, for the vice president blank, and that leaves eight possibilities for the treasurer blank:

$$\frac{9}{P} \quad \frac{1}{VP} \quad \frac{8}{T}$$

Hence,

Number of ways John is vice president $= 9 \times 1 \times 8 = 72.$

This is the number of favorable outcomes. Using these numbers, we find the probability we want:

$$\text{Probability John is vice president} = \frac{\text{Favorable outcomes}}{\text{Total outcomes}}$$

$$= \frac{72}{720}$$

$$= 0.1.$$

Thus, there is a 10% chance of John being selected as vice president.

Alternatively, we can think along the following lines. All we are interested in knowing is whose name goes in the vice president blank. There are 10 equally likely possibilities. Exactly one of them is John. Thus, the probability that he is vice president is $1/10 = 0.1$.

c. We want to arrange the three names John, Jim, and Mary in the three office-holder blanks. There are three possible names for the first, two for the second, and one for the third:

$$\frac{3}{P} \quad \frac{2}{VP} \quad \frac{1}{T}$$

Thus, the total number of ways these three candidates can all win office is $3 \times 2 \times 1 = 6$. These are the favorable outcomes. We calculated in part a that there are 720 possible outcomes. Using these values, we calculate

$$\text{Probability all three selected} = \frac{\text{Favorable outcomes}}{\text{Total outcomes}}$$

$$= \frac{6}{720}.$$

This is about 0.008 and represents a 0.8% chance that all three are selected.

d. If these three names don't appear in the officer blanks, then there are seven possibilities for the first, six for the second, and five for the third:

$$\frac{7}{P} \quad \frac{6}{VP} \quad \frac{5}{T}$$

That is a total of $7 \times 6 \times 5 = 210$ favorable outcomes. Thus, the probability that none will be selected is

$$\text{Probability none selected} = \frac{\text{Favorable outcomes}}{\text{Total outcomes}}$$

$$= \frac{210}{720}.$$

This is about 0.292, and it represents a 29.2% chance that none will be selected.

e. We found in part c that the probability that all three are selected is 6/720. The probability that at least one will not be selected is the probability that the event of all three being selected does not occur. This is 1 minus the probability that all three are selected, so the probability that at least one will not be selected is

$$1 - \frac{6}{720}$$

or about 0.992. Hence, there is a 99.2% chance that at least one will not be selected.

Independent events

Suppose we pick a person at random and consider the probability that the person is female and the probability that the person's last name is Smith. Knowing that the person is female does not affect the probability that the person's last name is Smith. These two events (picking a female and picking someone whose last name is Smith) are independent.

On the other hand, suppose we pick a person at random and look at the probability that the person is female and the probability that the person's first name is Mary. If we know that the person is female, the probability that the name will be Mary may be quite small. But if we don't know whether the person is female, the probability that the name will be Mary is much smaller. These two events (picking a female and picking someone named Mary) are not independent.

Key Concept

Suppose we perform an experiment. Two events are **independent** if knowing that one event occurs has no effect on the probability of the occurrence of the other.

In many practical situations, it is important to determine whether or not events are independent. For example, the events of being a smoker and having heart disease are not independent. The establishment of the link between smoking and heart disease was a milestone in public health policy.

There is an important formula involving probabilities of independent events. We look at an example to illustrate it. One of my favorite stores is having a sale. When I buy something, I get a ticket at the checkout stand that gives me either 10% off,

20% off, 30% off, 40% off, or 50% off. (We suppose that each of these tickets is equally likely.) I buy a pair of jeans in the morning and a shirt in the afternoon. What is the probability that I get at least 30% off in the morning and at least 40% off in the afternoon? These are independent events because winning one discount does not give any information about the other.

First we count the total number of possible outcomes. Because there are five possibilities for the morning and five for the afternoon, the Counting Principle tells us that there are $5 \times 5 = 25$ possible outcomes. Now let's count favorable outcomes. For the morning, 30%, 40%, and 50% are favorable, and for the afternoon, only 40% and 50% are favorable. With three favorable outcomes for the morning and two for the afternoon, the Counting Principle gives $3 \times 2 = 6$ favorable outcomes. So the probability we seek is

$$P(\text{At least } 30\% \text{ and At least } 40\%) = \frac{\text{Favorable outcomes}}{\text{Total outcomes}}$$
$$= \frac{6}{25}.$$

Now we calculate the two probabilities individually. For the morning, there are three favorable outcomes out of a total of five. So the probability of getting 30% or greater in the morning is 3/5. For the afternoon, there are two favorable outcomes out of a total of five. Thus, the probability of getting 40% or more in the afternoon is 2/5.

If we multiply the individual probabilities, we find the probability of 30% or more in the morning and 40% or more in the afternoon:

$$\frac{3}{5} \times \frac{2}{5} = \frac{6}{25},$$

so

$$P(\text{At least } 30\%) \times P(\text{At least } 40\%) = P(\text{At least } 30\% \text{ and At least } 40\%).$$

This result is no accident. Given two independent events, the probability that both occur is the product of the separate probabilities for the two events.[10] Here, each probability must be written as a decimal or a fraction, not as a percentage.

SUMMARY 5.4   Product Formula for Independent Events

Events A and B are independent exactly when the probability of Event A *and* Event B occurring is the probability of Event A times the probability of Event B. So A and B are independent exactly when

$$P(A \text{ and } B) = P(A) \times P(B).$$

Here, each probability must be written as a decimal or a fraction, not as a percentage.

The formula characterizes the situation when knowing that one event occurs has no effect on the probability of the occurrence of the other.

---

[10]In fact, our initial description of independent events is useful to build intuition, but the mathematical definition of independent events is that the product formula applies.

Using Frank Drake's famous equation, Betty calculates the
probability of finding intelligent life on a Saturday night.

EXAMPLE 5.19 Using the product formula: Defective products

Suppose that 1 in 500 digital cameras is defective and 3 in 1000 printers are defective.
On a shopping trip, I purchase a digital camera and a printer. What is the probability
that both the camera and the printer are defective?

SOLUTION

The probability of getting a defective camera is 1/500, and the probability of getting
a defective printer is 3/1000. The fact that one item is defective gives no information
about the other, so the two events are independent. We use the product formula:

$$P(\text{Defective camera and Defective printer}) = \frac{1}{500} \times \frac{3}{1000} = \frac{3}{500,000}.$$

This represents six times in a million.

TRY IT YOURSELF 5.19

Suppose that 20% of T-shirts in a stack are red and 10% of shorts in another stack
are red. You pick a T-shirt and pair of shorts at random. What is the probability both
are red?

The answer is provided at the end of this section.

Let's look further at the example above. Suppose I got home and found that my
digital camera was defective. I might reason as follows: "The probability of both the
camera and the printer being defective is so small that perhaps I can be assured that
the printer is fine." This is a bogus argument. The probability of both being defective
is indeed quite small. Nonetheless, the two events are independent, so knowing that
the camera is defective does not change the probability that the printer is defective.

Try It Yourself answers

Try It Yourself 5.14: Counting outcomes: Tossing coins $2^{15} = 32,768$.

Try It Yourself 5.15: Counting by filling in blanks: Codes 14,196.

Try It Yourself 5.17: Counting and calculating probabilities: Draw two cards
$\frac{16}{52 \times 51}$ or about 0.60%.

Try It Yourself 5.19: Using the product formula: Clothing color 0.02 or 2%.

Exercise Set 5.3

In your answers to these exercises, leave each probability as a fraction unless you are instructed to do otherwise.

1. Why use letters? Automobile license plates often use a mix of numerals and letters of the alphabet. Why do you suppose this is done? Why not use only numerals?

2. Bicycle speeds. A customer in a bicycle shop is looking at a bicycle with three gears on the front (technically, three chain-rings) and seven gears on the rear (technically, seven sprockets). He decides to look at a different bike because he wants more than 10 speeds. Has he counted correctly?

3. Shampoo. A company wants to market different types of shampoo using color-coded bottles, lids, and label print. It has four different colors of lids, five different colors of bottles, and two different colors of print for the label available. How many different types of shampoo can the company package?

Phone numbers   Local phone numbers consist of seven numerals, the first three of which are common to many users. Exercises 4 through 6 refer to phone numbers in various settings.

4. A small town's phone numbers all start with 337, 352, or 363. How many phone numbers are available?

5. Until 1996 all toll-free numbers started with the area code 800. Assume that all digits for the rest of a number were possible, and determine how many toll-free 800 numbers there were.

6. Today there are a total of five toll-free area codes: 800, 855, 866, 877, and 888. Assume that all digits for the rest of a number are possible, and determine how many toll-free numbers there are now.

7. I won the lottery. I just won the Powerball lottery and wanted to keep it a secret. But I told two close friends, who each told five other different people, and those in turn each told four other different people. How many people (other than me) heard the story of my lottery win? Be careful how you count.

8. Voting. I am voting in a state election. There are three candidates for governor, four candidates for lieutenant governor, three candidates for the state house of representatives, and four candidates for the senate. In how many different ways can I fill out a ballot if I vote once for each office?

9. My bicycle lock. My combination bicycle lock has four wheels, each of which has the numbers 0 through 20. Exactly one combination of numbers (in order) on the wheels (2-17-0-12, e.g.) will open the lock.

a. How many possible combinations of numbers are there on my bicycle lock?

b. Suppose that I have forgotten my combination. At random I try 1-2-3-4. What is the probability that the lock will open with this combination?

c. Out of frustration I try at random 50 different combinations. What is the probability that 1 of these 50 efforts will open the lock?

Counting faces   A classic children's toy consists of 11 different blocks that combine to form a rectangle showing a man's face.[11] Each block has four sides with alternatives for that part of the face; for example, the block for the right eye might have a closed eye on one side, a partially open eye on another, etc. A child can turn any block and so change that part of the face, and in this way many different faces can be formed. The next two exercises refer to this.

Changeable Charlie: A face formed by blocks.

10. How many different faces can be formed by the children's toy?

11. *This exercise uses the results of Exercise 10.* If a child makes one change each minute and spends 48 hours a week changing faces, how many years will it take to see every possible face?

Independent events   In each of Exercises 12 through 17, determine whether you think the given pair of events is independent. Explain your answer.

12. You collect Social Security benefits. You are over age 65.

13. You are in good physical condition. You work out regularly.

14. You suffer from diabetes. You are a registered Democrat.

15. You are over 6 feet tall. You prefer light-colored attire.

---

[11]One such toy was called Changeable Charlie.

16. You frequent casinos. You have lost money gambling.

17. You are male. You are a smoker.

18. **An old joke.** An old joke goes like this: A traveler reads in the newspaper that the probability of a passenger having a gun on an airplane is only one one-thousandth (0.001). He asked his math major friend what the probability would be of *two* passengers having guns on an airplane, and she replied that it would be the product of the probabilities or one one-millionth (0.001 × 0.001 = 0.000001). The traveler says, "Great, then I'm going to carry a gun next time I fly—that'll reduce the odds that someone else will have a gun!" Critique the traveler's logic.

19. **Coins.** We toss a coin twice. Use the product formula for independent events to calculate the probability that we get a tail followed by a head.

20. **A coin and a die.** We toss a coin and then roll a die. Use the product formula for independent events to calculate the probability that we get a tail followed by a 6.

Jelly beans Assume a box contains four red jelly beans and two green ones. We consider the event that a red bean is drawn. This information is used in Exercises 21 and 22.

21. Suppose I pick a jelly bean from the box without looking. I record the color and put the bean back in the box. Then I choose a bean again. Are the two events independent? What is the probability of getting a red bean both times?

22. Now suppose I pick a jelly bean from the same box without looking, but this time I do not put the bean back in the box. Then I choose a bean again. Are the two events independent? What is the probability of getting two red beans?

Two dice Exercises 23 through 26 refer to the toss of a pair of dice.

23. What is the probability that the total number of dots appearing on top is 5?

24. What is the probability that the total number of dots appearing on top is 7?

25. What is the probability that the total number of dots appearing on top is *not* 5?

26. What is the probability that the total number of dots appearing on top is *not* 7?

Marbles Suppose an opaque jar contains 5 red marbles and 10 green marbles. Exercises 27 and 28 refer to the experiment of picking two marbles from the jar without replacing the first one.

27. What is the probability of getting a green marble first and a red marble second?

28. What is the probability of getting a green marble and a red marble? (How is this exercise different from Exercise 27?)

More jelly beans Suppose I pick (without looking) a jelly bean from a box containing five red beans and seven blue beans. I record the color and put the bean back in the box. Then I choose a bean again. Exercises 29 and 30 refer to probabilities for this situation.

29. What is the probability of getting a red bean both times?

30. What is the probability of getting a blue bean both times?

Two cards Exercises 31 through 36 refer to choosing two cards from a thoroughly shuffled deck. Assume that the deck is shuffled after a card is returned to the deck.

31. If you put the first card back in the deck before you draw the next, what is the probability that the first card is a 10 and the second card is a jack?

32. If you do not put the first card back in the deck before you draw the next, what is the probability that the first card is a 10 and the second card is a jack?

33. If you put the first card back in the deck before you draw the next, what is the probability that the first card is a club and the second card is a diamond?

34. If you do not put the first card back in the deck before you draw the next, what is the probability that the first card is a club and the second card is a diamond?

35. If you do not put the first card back in the deck before you draw the next, what is the probability that both cards are diamonds?

36. If you do not put the first card back in the deck before you draw the next, what is the probability that the first card is a club and the second one is black?

Traffic lights Suppose a certain traffic light shows a red light for 45 seconds, a green light for 45 seconds, and a yellow light for 5 seconds. The next two exercises refer to this.

37. If you look at this light at a randomly chosen instant, what is the probability that you see a red or yellow light?

38. Each day you look at this light at a randomly chosen instant. What is the probability that you see a red light five days in a row?

Red light The city finds empirically that the probability that an automobile going through the intersection of 5th and Main will run a red light sometime during a given day is 9.2%, and the probability that an automobile going through the intersection of 7th and Polk will run a red light sometime during a given day is 11.7%. Suppose a car is picked at random at 5th and Main on a given day and another at random at 7th and Polk. Exercises 39 through 41 refer to this situation. Write each probability as a percentage rounded to one decimal place.

39. What is the probability that the car will not run a red light at 5th and Main?

40. What is the probability that the car will not run a red light at 7th and Polk?

41. What is the probability that neither car runs a red light?

A box Suppose that the inside bottom of a box is painted with three colors: 1/3 of the bottom area is red, 1/6 is blue, and 1/2 is white. You toss a tiny pebble into the box without aiming and note the color on which the pebble lands. Then you toss another tiny pebble into the box without aiming and note the color on which that pebble lands. Exercises 42 through 45 refer to these two trials.

42. Are the two trials independent?

43. What is the probability that both pebbles land on the color blue?

44. What is the probability that the first pebble lands on the color blue and the second pebble lands on red?

45. What is the probability that one of the pebbles lands on the color blue and the other pebble lands on red? (How is this exercise different from Exercise 44?)

Terrorism The next four exercises refer to the following two excerpts from an article by Alex Fryer.

46. Explain what Joe Myers means in Excerpt 1 when he talks about "multiplying probabilities." Does this have anything to do with independent events? Explain. Discuss his statements. Do you believe him?

47. In Excerpt 2, do you think Pi and Pn are subjective probabilities? How would you estimate them?

48. Propst's equation, Pe = Pi $\times$ Pn, in Excerpt 2 multiplies two probabilities. Does this have anything to do with independent events? Explain.

49. What do you think about the equation Pe = Pi $\times$ Pn in Excerpt 2? Does it make sense as it is explained in the article? Is it useful? Explain why or why not.

Lightbulbs I A room has three lightbulbs. Each one has a 10% probability of burning out within the month. The next three exercises refer to this. Write each probability as a percentage rounded to one decimal place.

50. What is the probability that all three will burn out within the month?

51. What is the probability that all three will be lit at the end of the month?

52. What is the probability that at the end of the month at least one of the bulbs will be lit?

53. Lightbulbs II. *This exercise is a bit more challenging.* It is important that a room have light during the next month. You want to ensure that the chances of the room going dark are less than 1 in a million. How many lightbulbs should you install in the room to guarantee this if each one has a 95% probability of staying lit during the month?

54. Murphy's law. "If something can go wrong, it will go wrong." This funny saying is called Murphy's law. Let's interpret this to mean "If something can go wrong, there is a very high probability that it will *eventually* go wrong."

Suppose we look at the event of having an automobile accident at some time during a day's commute. Let's assume that the probability of having an accident on a given day is 1 in a thousand or 0.001. That is, in your town, one of every thousand cars on a given day is involved in an accident (including little fender-benders). We also assume that having (or not having) an accident on a given day is independent of having (or not having) an accident on any other given day. Suppose you commute 44 weeks per year, 5 days a week, for a total of 220 days each year. In the following parts, write each probability in decimal form rounded to three places.

a. What is the probability that you have no accident over a year's time?

b. What is the probability that you will have at least one accident over a one-year period?

c. Repeat parts a and b for a 10-year period and for a 20-year period.

d. Does your work support the idea that there is a mathematical basis for Murphy's law as we interpreted it?

Party preference The following is excerpted from a report by Michael Barone.

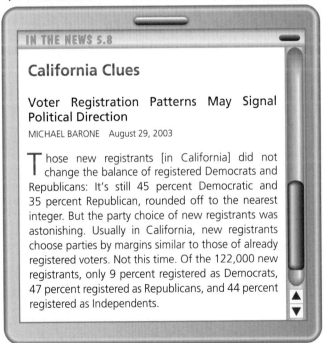

This information is used in Exercises 55 through 57. Write each probability as a percentage rounded to the nearest whole number.

55. Suppose that on 8/29/03 you chose a registered voter at random. What is the probability that you have chosen either a Democrat or a Republican?

56. Suppose you choose two people at random from among the new registrants. What is the probability that you get two Independents?

57. Challenge problem. Suppose you choose three people at random from among the new registrants. What is the probability that you get one Republican, one Democrat, and one Independent? *Suggestion:* First assume that you pick three people one at a time, and find the probability that you get a Republican, a Democrat, and an Independent, in that order. How should you modify this number to get the requested probability?

Exercises 58 through 61 are suitable for group work.

NCAA basketball tournament The NCAA basketball tournament in 2006 had several surprises, including the appearance in the Final Four of George Mason University. (The other teams making up the Final Four were Florida, LSU, and UCLA.) Assume that the tournament starts with four brackets having 16 teams each, and the winners of the brackets make it to the Final Four. ESPN sponsored a contest to predict, after the brackets were announced, the Final Four. There were 3 million entries in the contest. Exercises 58 through 61 use this information.

58. Choosing George Mason. What is the probability of correctly predicting that George Mason will reach the Final Four by choosing one team at random from each bracket? Use your answer to determine how many of the entries in ESPN's contest would have correctly chosen George Mason to reach the Final Four if all predictions had been made by random choice. *Note:* Of all the entries, 1853 correctly chose George Mason to reach the Final Four.

59. Choosing all four. What is the probability of correctly predicting the Final Four by choosing one team at random from each bracket? Use your answer to determine how many of the entries in the contest would have correctly predicted the Final Four if all predictions had been made by random choice. *Note:* Of all the entries, only four correctly predicted the Final Four.

60. Missing all four. Assume that you choose one team at random from each bracket. What is the probability that none of the teams you choose will reach the Final Four? What percentage (to the nearest whole number) of the entries in the contest would have missed each of the Final Four if all predictions had been made by random choice? *Note:* About 70% of the entries in the contest missed each team of the Final Four.

61. Picking the favorites. It was widely expected that Connecticut and Duke, which were in different brackets, would reach the Final Four. Assume that you choose one team at random from each bracket. What is the probability that you will choose either Duke or Connecticut (or both) to reach the Final Four? What percentage of the entries in the contest would have chosen either Duke or Connecticut (or both) for the Final Four if all predictions had been made by random choice? *Note:* About

87% of the entries in the contest picked Connecticut or Duke for the Final Four.

62. History. The Counting Principle has surely been known for a long time. In fact, it may be involved in a curious problem found in one of the oldest surviving mathematical texts, the Rhind papyrus from ancient Egypt. The papyrus itself dates from about 1650 B.C., but it may be a copy of a source 150 years older than that. The curious problem from the Rhind papyrus has the following data (where a *hekat* is a measure of volume):

Houses	7
Cats	49
Mice	343
Heads of wheat	2401
Hekat measures	16,807

The papyrus correctly gives 19,607 for the total of the numerical column. One historian has suggested that the context of the table is as follows: In an estate there are seven houses, each house had seven cats, each cat could eat seven mice, each mouse could eat seven heads of wheat, and each head of wheat could produce seven hekat measures of grain.[12] Use the Counting Principle to explain how this interpretation is consistent with the numbers in the table.

63. Double sixes. Chevalier de Méré was a gambler who lived in the seventeenth century. He asked whether or not one should bet even money on the occurrence of at least one "double six" during 24 throws of a pair of dice. This question was a factor that motivated Fermat and Pascal to study probability (see Exercise 61 in Section 5.1). Answer de Méré's question: What is the probability of at least one "double six" in 24 consecutive rolls of a pair of dice? Write your answer in decimal form rounded to two places. *Hint:* First find the probability of 24 consecutive rolls of a pair of dice *without* a "double six."

The following exercises are designed to be solved using technology such as calculators or computer spreadsheets. For assistance, see the technology supplement.

64. Factorials. Suppose you want to arrange five people in order in a waiting line. There are five choices for the person who goes first, four for the person who goes second, and so on. Thus, the number of ways to arrange five people is $5 \times 4 \times 3 \times 2 \times 1 = 120$. This sort of product arises often enough that it has a special notation:[13] We write $n!$ (read "n factorial") for $n \times (n-1) \times \cdots \times 1$. For example, $5! = 120$ by the above. Use technology to determine how many ways there are to arrange 12 people in order in a waiting line.

65. More on factorials. *This is a continuation of Exercise 64.* Use technology to determine how many ways there are to arrange 50 people in order in a waiting line. Can you find the answer for 200 people?

66. Listing cases. Devise a computer program that will list all possible outcomes of tossing n coins. *Warning:* If you actually run your program for even moderate values of n, you will get a very long output.

---

[12] See Howard Eves, *An Introduction to the History of Mathematics,* 6th ed. (Philadelphia: Saunders, 1990).

[13] The idea of factorials will be discussed more fully in the next section.

5.4 More ways of counting: Permuting and combining

This following article provides a basic introduction to the way genetic information is stored in DNA.

IN THE NEWS 5.9

SciWhys: What Is DNA and What Does It Do?

JONATHAN CROWE   February 28, 2011

DNA is like a long, thin chain—a chain that is constructed from a series of building blocks joined end-to-end...

There are only four different building blocks; these are represented by the letters A, C, G and T ... A single DNA molecule is composed of a mixture of these four building blocks, joined together one by one to form a long chain—and it is the order in which the four building blocks are joined together along the DNA chain that lies at the heart of DNA's information-storing capability...

Imagine we had a DNA molecule just two blocks long. Even with this tiny molecule, being able to draw upon four different building blocks at each of the two positions along the two-block chain makes 16 different molecules possible: AA AC AT AG CA CC CT CG TA TC TT TG GA GC GT GG...

When we consider that the DNA in a human cell is made up of around 3 billion of the four building blocks joined in sequence (that's 3,000,000,000), we can begin to imagine just how much variety is actually possible—and all from just four starting ingredients.

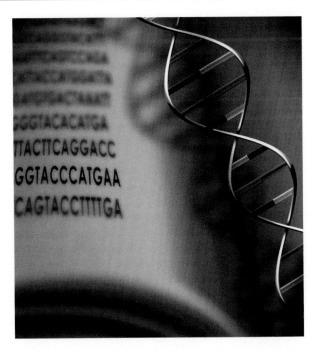

One major goal of understanding the sequence in a DNA molecule is the identification of genes that give rise to cancer. Understanding the genetics of cancer comes down to determining how four letters, A, C, G, T, go together in long strings. The number of possibilities is staggering, but there are mathematical tools to help in the enumeration. These tools, which ultimately rely on the Counting Principle, are considered in this section.

Permutations

Suppose we arrange a group of people into a line. The first in line gets first prize, the second in line gets second prize, and so on. Your place in line determines the prize you get. So the order in which people are arranged in the line is important. An arrangement where order is important is called a *permutation*.

Key Concept

A permutation of items is an arrangement of the items in a certain order. Each item can be used only once in the sequence.

Suppose, for example, that we use the letters A, B, and C to make product codes where each letter is used exactly once. How many such codes can we make? That is, how many permutations are there of the letters A, B, and C? We can list them all to get the answer. The permutations of the letters A, B, C, are ABC, ACB, BAC, BCA, CAB, CBA—six in all. When the number of items is large, it is not efficient to list all of the permutations. Instead, we need a faster way of calculating how many there are.

Think of making a permutation (or arrangement) by filling three blanks with the three letters without repetition, that is, using each letter once:

$$\underline{\quad}\ \underline{\quad}\ \underline{\quad}$$

There are three choices for the first blank. When that blank is filled, we have only two letters left, so there are two choices for the next blank. Now there is only one letter left, so we have just one choice for the last blank:

$$\underline{3}\ \underline{2}\ \underline{1}$$

The Counting Principle says that the total number of ways to fill all the blanks is $3 \times 2 \times 1 = 6$. This agrees with the number we found by making a list.

This kind of calculation occurs so often that it is given a special name and symbol. For a positive whole number n, the product of all the whole numbers from n down through 1 is denoted by n with an exclamation point: $n!$. This is pronounced "n factorial."

Although it may seem strange, zero factorial $(0!)$ is always defined to be the number 1. One way to justify this is by asserting that there's just one way to arrange no items. We'll see another reason why later. Here are some factorials:

$$0! = 1$$
$$1! = 1$$
$$2! = 2 \times 1 = 2$$
$$3! = 3 \times 2 \times 1 = 6$$
$$4! = 4 \times 3 \times 2 \times 1 = 24.$$

The method we used to find the number of arrangements of the letters A, B, C works in general. For example, suppose there are 10 people waiting to buy tickets to some event. In how many ways can we arrange the 10 people in a line at the ticket booth?

Thinking again of filling in slots, we see that there are 10 people who could be at the head of the line. That leaves nine who might be second, eight in third place, etc. So the total number of ways to arrange the 10 people is 10!, which equals 3,628,800. That's a pretty big number.

> **SUMMARY 5.5**   **Permutations of *n* Items**
>
> The number of permutations of *n* items is the number of ways to arrange *n* distinct items without repetition. The Counting Principle shows that this is *n* factorial, which is defined as
>
> $$n! = n \times (n-1) \times \cdots \times 2 \times 1.$$

EXAMPLE 5.20 Using factorials: Arranging four letters

In how many ways can I arrange the four letters A, B, C, D using each letter only once?

SOLUTION

The number of permutations of four letters is $4! = 4 \times 3 \times 2 \times 1 = 24$.

TRY IT YOURSELF 5.20

In how many ways can I arrange the five letters A, B, C, D, E using each letter only once?

The answer is provided at the end of this section.

Factorials get big fast—as we just saw, 10! = 3,628,800. To find large factorials on a calculator using the definition, you would have to do a lot of multiplications, which could be quite tedious. The accompanying tip should help.

> **CALCULATION TIP 5.3** Factorials with a Calculator or Computer
>
> • On the TI-83 and TI-84 graphing calculators, there's a factorial command $\boxed{!}$ located in the math probability menu. To find 10!, enter 10 MATH PRB $\boxed{4}$.
>
> • In the spreadsheet program Excel©, the command for n! is FACT(n). For example, to find 10!, enter FACT(10).
>
> • You can also find factorials on the calculator that comes with Microsoft Windows©. Choose the View menu in the calculator and click on Scientific to get the full size. You can see the $\boxed{n!}$ button in the fourth column, just above the $\boxed{1/x}$ button.

More on arrangements

Let's revisit the idea of putting people in order and awarding prizes. But this time, let's do it under the more realistic assumption that not everyone gets a prize. We think of selecting 3 lucky winners from a group of 10 people. The first person selected gets first prize, the second gets second prize, and the third gets third prize. Such an arrangement is called *a permutation of 10 people taken 3 at a time*.

Key Concept

The number of ways to select *k* items from *n* distinct items and arrange them in order is called the number of **permutations of *n* items taken *k* at a time.**

The Counting Principle allows us to calculate the number of permutations of this type.

EXAMPLE 5.21 Counting permutations: Three prizes

In how many ways can we select 3 people from a group of 10 and award them first, second, and third prizes? This is the number of permutations of 10 items taken 3 at a time.

SOLUTION

We think of filling 3 slots with names selected from the 10 people:

— — —

There are 10 names available for the first slot. But because nobody gets more than one prize, there are only 9 available for the second and then 8 available for the third:

$$\underline{10\ 9\ 8}$$

The Counting Principle shows that the number of ways to select the three winners is $10 \times 9 \times 8 = 720$.

TRY IT YOURSELF 5.21

In how many ways can we select four letters from the alphabet without repetition and then arrange them in a sequence?

The answer is provided at the end of this section.

Now we use our ideas about counting permutations to assign poll watchers.

EXAMPLE 5.22 Permutations: Poll watchers

A political organization wants to observe the voting procedures at five polling places. The organization has five poll watchers available.

a. In how many different ways can the poll watchers be assigned?

b. For each of the polling places, one of the available watchers lives in the precinct for that polling place. If the watchers are assigned at random, what is the probability that each one will be assigned to the polling place for the precinct where he or she lives?

c. Suppose now that the organization wants to observe only two of the polling places. In how many ways can the organization assign these places to two of the available watchers? This is the number of permutations of five items taken two at a time.

d. Assume now that there are 20 polling places and 20 poll watchers available. How many permutations are there of the 20 watchers?

SOLUTION

a. As we noted above, the number of ways to arrange five items is

$$5! = 5 \times 4 \times 3 \times 2 \times 1 = 120.$$

In terms of assigning poll watchers, there are five choices for the first polling place, followed by four choices for the next place, etc.

b. From part a there are 120 outcomes, and only one of them has each of the watchers assigned to his or her home precinct. So the probability that each one will be assigned to the polling place for the precinct where he or she lives is 1/120 or about 0.008.

c. There are five choices to watch the first polling place and four choices to watch the second polling place. That makes $5 \times 4 = 20$ different arrangements.

d. As we have noted, the number of arrangements of 20 items is 20!, which is (get ready!) 2,432,902,008,176,640,000. Yikes! This number is too big even for most calculators, which will report at best an estimate. Generally speaking, specialized mathematical software is needed to handle numbers of this size, and hand calculation is out of the question.

Calculations using factorials

We want to find a simpler way to write the number of permutations when the number of items is large. Let's look at what happens when we have lots of books to arrange on a shelf. Suppose, for example, that we have 20 different books and we want to arrange 9 of them (in order) on a shelf. That is, we want to calculate the number of permutations of 20 items taken 9 at a time. Thinking of filling in blanks, we see that there are 20 books for the first position, 19 for the second, down to 12 for the last position:

$$\underline{20}\ \underline{19}\ \underline{18} \ldots \underline{12}.$$

Thus, we find

$$20 \times 19 \times 18 \times 17 \times 16 \times 15 \times 14 \times 13 \times 12 \text{ arrangements.}$$

It would be very tedious to multiply all nine of these numbers, even with a calculator. But there's an easy way to find the answer on a calculator using the factorial command. Here's the trick: We write the expression

$$20 \times 19 \times 18 \times 17 \times 16 \times 15 \times 14 \times 13 \times 12$$

as

$$\frac{20 \times 19 \times 18 \times 17 \times 16 \times 15 \times 14 \times 13 \times 12 \times (11!)}{11!}.$$

Now we expand the factorial in the numerator to find

$$\frac{20 \times 19 \times 18 \times 17 \times 16 \times 15 \times 14 \times 13 \times 12 \times (11 \times 10 \times 9 \times 8 \times 7 \times 6 \times 5 \times 4 \times 3 \times 2 \times 1)}{11!} = \frac{20!}{11!}.$$

The result is that the number of permutations of 20 items taken 9 at a time is

$$\frac{20!}{11!} = \frac{20!}{(20 - 9)!}.$$

So we can divide two factorials on our calculator to find the number of permutations: 60,949,324,800.

This method is often useful, and we call attention to it in the accompanying summary.

SUMMARY 5.6   **Permutations of *n* Items Taken *k* at a Time**

The number of ways to select *k* items from *n* items and arrange them in order is the number of permutations of *n* items taken *k* at a time. We can calculate this number using the *permutations formula*

$$\text{Permutations of } n \text{ items taken } k \text{ at a time} = \frac{n!}{(n-k)!}.$$

EXAMPLE 5.23 Using the permutations formula: Talent show

The producer of a talent show has seven slots to fill. Twenty-five acts have requested a slot. Use the permutations formula to find the number of ways to select seven acts and arrange them in a sequence. Leave your answer in terms of factorials.

SOLUTION

We are selecting 7 of 25 acts, so we use the permutations formula for 25 items taken 7 at a time:

$$\text{Permutations of 25 items taken 7 at a time} = \frac{25!}{(25 - 7)!}$$
$$= \frac{25!}{18!}.$$

TRY IT YOURSELF 5.23

Use the permutations formula to find the number of ways for a producer to select 10 acts from 25 and arrange them in a sequence. Leave your answer in terms of factorials.

The answer is provided at the end of this section.

Sometimes we can simplify fractions involving factorials to avoid long calculations. Let's look, for example, at 55!/53!. We can save a lot of time in calculating this by hand if we notice that lots of terms cancel:

$$\frac{55!}{53!} = \frac{55 \times 54 \times 53 \times 52 \times \cdots \times 3 \times 2 \times 1}{53 \times 52 \times \cdots \times 3 \times 2 \times 1}$$

$$= \frac{55 \times 54 \times \cancel{53} \times \cancel{52} \times \cdots \times \cancel{3} \times \cancel{2} \times \cancel{1}}{\cancel{53} \times \cancel{52} \times \cdots \times \cancel{3} \times \cancel{2} \times \cancel{1}}$$

$$= 55 \times 54$$

$$= 2970.$$

> ### CALCULATION TIP 5.4 Permutations
>
> The TI-83 and TI-84 graphing calculators have a permutation command $\boxed{nPr}$ located in the math probability menu along with the factorial command. To find the number of permutations of n items taken r at a time, enter the number of items n, first, then the permutation key, then the number of items to select at a time, r. For example, to find the number of permutations of 10 items taken 4 at a time (10P4), enter 10 MATH PRB $\boxed{2}$ 4.
>
> In the spreadsheet Excel©, the command for the number of permutations of n items taken r at a time is PERMUT(n,r). For example, to find the number of permutations of 10 items taken 4 at a time, enter PERMUT(10,4).

Combinations

A permutation of a set of items is an arrangement of those items in a certain order. By contrast, a *combination* of a set of items is a collection that does not take into account the order in which they happen to appear.

Key Concept

A combination of a group of items is a selection from that group in which order is not taken into account. No item can be used more than once in a combination.

For example, if we consider the number of ways we can pick 5 people from a group of 10 and give the first a prize of $10, the second a prize of $9, and so on, it matters who comes first. Thus, we are interested in counting permutations. On the other hand, if we select 5 people from a group of 10 and give each of them a $10 prize, order makes no difference. Everybody gets the same prize. Here, we want to find the number of combinations, not the number of permutations.

To illustrate the point further, note that CAB and BAC are two different permutations of the letters A, B, and C. However, they would be considered the same *combination*.

How many combinations of the letters A, B, C are there taken two at a time? In other words, how many ways can we select two of the three letters if it makes no difference in which order the letters are considered ? (It may help to think of selecting a two-person committee from a group of three people named A, B, and C.) Simply by listing them, we can see that there are three possible *combinations*: AB, AC, and BC.

For comparison, we note that there are six *permutations* of these three letters taken two at a time: AB, BA, AC, CA, BC, and CB. That is, there are six ways of selecting two letters from a list of three if the order in which they are chosen matters. Note that in the list of six permutations above each combination appears twice—for example, AB and BA. The number of permutations, 6, is twice the number of combinations, 3. Thus, to find the number of combinations from the number of permutations, we divide 6 by 2. This observation is the key to finding an easy way to count combinations. We find the number of permutations, then look to see how many times each different combination is listed, and divide. Expressed as a formula, this is

$$\text{Combinations} = \frac{\text{Permutations}}{\text{Repeated listings}}.$$

In the case of permutations of k items, the number of times each combination is listed is $k!$. (Why?) This observation yields the formula in the following summary.

SUMMARY 5.7   **Combinations**

The number of ways to select k items from n items when order makes no difference is the number of **combinations of n items taken k at a time**. We can calculate this number using the *combinations formula*:

$$\text{Combinations of } n \text{ items taken } k \text{ at a time} = \frac{n!}{k!(n-k)!}.$$

EXAMPLE 5.24 Using the combinations formula: Three-person committee

Use the combinations formula to express the number of three-person committees I can select from a group of six people. Leave your answer in terms of factorials.

SOLUTION

This is the number of combinations of six people taken three at a time. We use the combinations formula $\dfrac{n!}{k!(n-k)!}$ with $n = 6$ and $k = 3$:

$$\text{Combinations of 6 people taken 3 at a time} = \frac{6!}{3!(6-3)!}$$
$$= \frac{6!}{3!3!}.$$

TRY IT YOURSELF 5.24

Use the combinations formula to express the number of four-person committees I can select from a group of nine people. Leave your answer in terms of factorials.

The answer is provided at the end of this section.

CALCULATION TIP 5.5 Combinations

If you enter the formula for combinations on a calculator, be sure to put parentheses around the denominator. For example, enter $\dfrac{8!}{5!3!}$ as 8!/(5!3!).

The TI-83 and TI-84 graphing calculators have a combination key $\boxed{nCr}$ located in the math probability menu. To find the number of combinations of n items taken r at a time, enter the number of items n first, then the combination key, then the number of items to select at one time, r. For example, to find the number of combinations of 10 items taken 4 at a time (10C4), enter 10 MATH PRB $\boxed{3}$ 4.

In the spreadsheet Excel©, the command to find the number of combinations of n items taken r at a time is COMBIN(n,r). For example, to find the number of combinations of 10 items taken 4 at a time, enter COMBIN(10,4).

© 1999 Randy Glasbergen. www.glasbergen.com

GLASBERGEN

"I couldn't do my homework because my computer has a virus and so do all my pencils and pens."

Hand calculation of combinations

Let's look more carefully at pencil-and-paper calculations of combinations. For example, consider the number of combinations of 10 items taken 6 at a time:

$$\text{Combinations of 10 items taken 6 at a time} = \frac{10!}{6!(10-6)!}$$
$$= \frac{10!}{6!4!}.$$

If we write this out, we find

$$\frac{10!}{6!4!} = \frac{10 \times 9 \times 8 \times 7 \times 6 \times 5 \times 4 \times 3 \times 2 \times 1}{(6 \times 5 \times 4 \times 3 \times 2 \times 1) \times (4 \times 3 \times 2 \times 1)}.$$

Look at the larger factorial on the bottom (6! in this case). It will always cancel with part of the top:

$$\frac{10 \times 9 \times 8 \times 7 \times \cancel{6} \times \cancel{5} \times \cancel{4} \times \cancel{3} \times \cancel{2} \times \cancel{1}}{(\cancel{6} \times \cancel{5} \times \cancel{4} \times \cancel{3} \times \cancel{2} \times \cancel{1}) \times (4 \times 3 \times 2 \times 1)} = \frac{10 \times 9 \times 8 \times 7}{4 \times 3 \times 2 \times 1}.$$

We will always get a whole number for the final answer. (Why?) Thus, the rest of the denominator will cancel as well. Note that the 4 × 2 in the denominator cancels with the 8 in the numerator, and the 3 divides into the 9:

$$\frac{10 \times \overset{3}{\cancel{9}} \times \cancel{8} \times 7}{\cancel{4} \times \cancel{3} \times \cancel{2} \times 1} = 10 \times 3 \times 7 = 210.$$

Following this procedure can significantly shorten manual computations of this sort.

CALCULATION TIP 5.6 Combinations

If you calculate a permutation or combination with pencil and paper, you can always cancel all the terms in the denominator before doing any multiplication. Calculating combinations involving large numbers is best done with a computer or calculator.

EXAMPLE 5.25 Calculating combinations by hand: Nine items taken five at a time

Calculate by hand the number of combinations of nine items taken five at a time.

SOLUTION

The number of combinations of nine items taken five at a time is

$$\text{Combinations of 9 items taken 5 at a time} = \frac{9!}{5!(9-5)!} = \frac{9!}{5!4!}.$$

We compute

$$\frac{9!}{5!4!} = \frac{9 \times 8 \times 7 \times 6 \times 5 \times 4 \times 3 \times 2 \times 1}{5 \times 4 \times 3 \times 2 \times 1 \times 4 \times 3 \times 2 \times 1}$$

$$= \frac{9 \times 8 \times 7 \times 6 \times \cancel{5} \times \cancel{4} \times \cancel{3} \times \cancel{2} \times \cancel{1}}{\cancel{5} \times \cancel{4} \times \cancel{3} \times \cancel{2} \times \cancel{1} \times 4 \times 3 \times 2 \times 1}$$

$$= \frac{9 \times 8 \times 7 \times 6}{4 \times 3 \times 2 \times 1}$$

$$= \frac{9 \times \cancel{8} \times 7 \times 6}{\cancel{4} \times 3 \times \cancel{2} \times 1}$$

$$= \frac{\overset{3}{\cancel{9}} \times 7 \times 6}{\cancel{3} \times 1}$$

$$= 3 \times 7 \times 6$$

$$= 126.$$

TRY IT YOURSELF 5.25

Calculate by hand the number of combinations of 10 items taken 7 at a time.

The answer is provided at the end of this section.

Here is an example of using combinations to count the number of three-judge panels used in the U.S. Ninth Circuit Court of Appeals, which covers several western states, including California.

EXAMPLE 5.26 Computing combinations: Three-judge panels

The *New York Times* ran an article on June 30, 2002, with the title, "Court That Ruled on Pledge Often Runs Afoul of Justices." The court in question is the Ninth Circuit Court, which ruled in 2002 that the Pledge of Allegiance to the flag is unconstitutional because it includes the phrase "under God." The article discusses the effect of having a large number of judges, and it states, "The judges have chambers throughout the circuit and meet only rarely. Assuming there are 28 judges, there are more than 3000 possible combinations of three-judge panels."

a. Is the article correct in stating that there are more than 3000 possible combinations consisting of three-judge panels? Exactly how many three-judge panels can be formed from the 28-judge court?

b. The article says that the judges meet rarely. Assume that there are 28 judges. How many 28-judge panels are there?

SOLUTION

a. We want to find the number of ways to choose 3 items from 28 items. We use the combinations formula with $n = 28$ and $k = 3$:

$$\text{Combinations of 28 items taken 3 at a time} = \frac{28!}{3!(28-3)!}$$

$$= \frac{28!}{3!25!}.$$

This fraction can be simplified by cancellation:

$$\frac{28!}{3!25!} = \frac{28 \times 27 \times 26 \times \cancel{25} \times \cancel{24} \times \cancel{23} \times \cdots \times \cancel{1}}{(3 \times 2 \times 1) \times (\cancel{25} \times \cancel{24} \times \cancel{23} \times \cdots \times \cancel{1})}$$

$$= \frac{28 \times 27 \times 26}{3 \times 2 \times 1}$$

$$= \frac{28 \times \overset{9}{\cancel{27}} \times \overset{13}{\cancel{26}}}{\cancel{3} \times \cancel{2} \times 1}$$

$$= 3276.$$

The article is correct in stating that there are more than 3000 possible three-judge panels. There are, in fact, 3276 such combinations.

b. We are really being asked, "In how many ways can we choose 28 judges from 28 judges to form a panel?" There is obviously only one way: All 28 judges go on the panel.

Let's see what happens if we apply the combinations formula to this simple case. We use the combinations formula with $n = 28$ and $k = 28$:

$$\text{Combinations of 28 items taken 28 at a time} = \frac{28!}{28!(28-28)!}$$

$$= \frac{28!}{28!0!}$$

$$= 1.$$

This is the same as the answer we found before. Remember that 0! is taken to be 1, and this formula is one reason why.

Probabilities with permutations or combinations

Now that we have more sophisticated ways of counting, we can calculate certain probabilities more easily.

EXAMPLE 5.27 Finding the probability: Selecting a committee

In a group of six men and four women, I select a committee of three at random. What is the probability that all three committee members are women?

SOLUTION
Recall that

$$\text{Probability} = \frac{\text{Favorable outcomes}}{\text{Total outcomes}}.$$

We first find the number of favorable outcomes, which is the number of three-person committees consisting of women. There are four women, so the number of ways to select a three-woman committee is the number of combinations of four items taken three at a time:

$$\text{Combinations of 4 items taken 3 at a time} = \frac{4!}{3!(4-3)!}$$

$$= \frac{4!}{3!1!}$$

$$= 4.$$

Next we find the number of ways to select a three-person committee from the 10 people (the total number of outcomes). This is the number of combinations of 10 people

taken 3 at a time:

$$\text{Combinations of 10 items taken 3 at a time} = \frac{10!}{3!(10-3)!}$$
$$= \frac{10!}{3!7!}$$
$$= 120.$$

So the probability of an all-female committee is

$$\frac{\text{Favorable outcomes}}{\text{Total outcomes}} = \frac{4}{120} = \frac{1}{30}.$$

This is about 0.03.

TRY IT YOURSELF 5.27

In a group of five men and five women, I select a committee of three at random. What is the probability that all three committee members are women?

The answer is provided at the end of this section.

The following article notes the popularity of poker among young people.

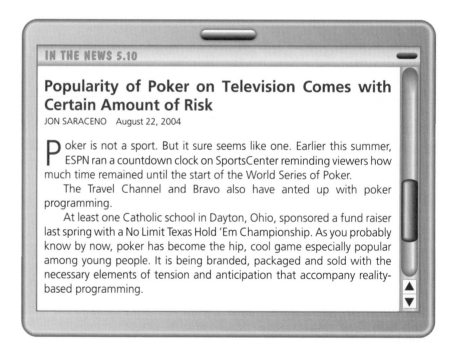

IN THE NEWS 5.10

Popularity of Poker on Television Comes with Certain Amount of Risk

JON SARACENO   August 22, 2004

Poker is not a sport. But it sure seems like one. Earlier this summer, ESPN ran a countdown clock on SportsCenter reminding viewers how much time remained until the start of the World Series of Poker.

The Travel Channel and Bravo also have anted up with poker programming.

At least one Catholic school in Dayton, Ohio, sponsored a fund raiser last spring with a No Limit Texas Hold 'Em Championship. As you probably know by now, poker has become the hip, cool game especially popular among young people. It is being branded, packaged and sold with the necessary elements of tension and anticipation that accompany reality-based programming.

EXAMPLE 5.28 Finding the probability: Poker hands

There are many different kinds of poker games, but in most of them the winner is ultimately decided by the best five-card hand. Suppose you draw five cards from a full deck.

   a. How many five-card hands are possible?

   b. What is the probability that exactly two of the cards are kings?

   c. What is the probability that all of them are hearts?

Various poker hands.

SOLUTION

a. The total number of hands is the number of combinations of 52 items taken 5 at a time:

$$\text{Combinations of 52 items taken 5 at a time} = \frac{52!}{5!(52-5)!}$$
$$= \frac{52!}{5!47!}$$
$$= 2,598,960.$$

(Thus, there are over 2.5 million different five-card poker hands.) This is the number we will use in the denominator of the remaining calculations.

b. We must count how many ways we can choose five cards from the deck so that exactly two are kings. To do this, we must choose two kings and three others. There are four kings, from which we choose two. The combinations formula gives

$$\text{Combinations of 4 items taken 2 at a time} = \frac{4!}{2!(4-2)!}$$
$$= \frac{4!}{2!2!}$$
$$= 6.$$

There are 48 cards that are not kings, and we must choose 3 from these. That is, we need the number of combinations of 48 items taken 3 at a time:

$$\text{Combinations of 48 items taken 3 at a time} = \frac{48!}{3!(48-3)!}$$
$$= \frac{48!}{3!45!}$$
$$= 17,296.$$

The Counting Principle says that there are $6 \times 17,296 = 103,776$ ways to do both. Therefore, the probability we seek is $103,776/2,598,960$. This is about 0.040 or 4.0%.

c. We need to count how many ways we can choose five cards from the deck so that all of them are hearts. There are 13 hearts, and the number of ways to choose 5 of them is the number of combinations of 13 items taken 5 at a time:

$$\text{Combinations of 13 items taken 5 at a time} = \frac{13!}{5!(13-5)!}$$
$$= \frac{13!}{5!8!}$$
$$= 1287.$$

So the probability is 1287/2,598,960. This is about 0.0005 or 0.05%.

Try It Yourself answers

Try It Yourself 5.20: Using factorials: Arranging five letters $5! = 120$.

Try It Yourself 5.21: Counting permutations: Four letters $26 \times 25 \times 24 \times 23 = 358,800$.

Try It Yourself 5.23: Using the permutations formula: Talent show $25!/15!$.

Try It Yourself 5.24: Using the combinations formula: Four-person committee $9!/(4!5!)$.

Try It Yourself 5.25: Calculating combinations by hand: Ten items taken seven at a time   120.

Try It Yourself 5.27: Finding the probability: Selecting a committee   1/12 or about 0.08.

Exercise Set 5.4

1. Candidates. There are eight candidates running for the presidential and vice-presidential nomination. How many ways are there of selecting presidential and vice-presidential candidates from the field?

Order or not   In Exercises 2 through 9 determine whether order makes a difference in the given situations.

2. Three people are selected at random from the class and are permitted to skip the final exam.

3. We put letters and numbers on a license plate.

4. I need some change, so I pick 5 coins from a dish containing 10 coins.

5. People are lined up at the polls to vote. Some are Democrats, and some are Republicans. We are interested in the outcome of the election.

6. We select three people for a committee. The first chosen will hold the office of president, the second vice president, and the third secretary.

7. We go to the deli and buy three kinds of cheeses for next week's lunches.

8. We select four men and six women to serve on a committee.

9. We put five people on a basketball team. They get the positions of point guard, shooting guard, small forward, power forward, and center.

10. Signals. The king has 10 (different) colored flags that he uses to send coded messages to his general in the field. For example, red-blue-green might mean "attack at dawn" and blue-green-red may mean "retreat." He sends the message by arranging three of the flags atop the castle wall. How many coded messages can the king send?

11. The Enigma machine. The *Enigma machine* was used by Germany in World War II to send coded messages. It has gained fame because it was an excellent coding device for its day and

Enigma machine.

because of the ultimately successful efforts of the British (with considerable aid from the Poles) to crack the Enigma code. The breaking of the code involved, among other things, some very good mathematics developed by Alan Turing and others. One part of the machine consisted of three rotors, each containing the letters A through Z. In order to read an encrypted message, it was necessary to determine the initial settings of the three rotors (e.g., PDX or JJN). How many different initial settings of the three rotors are there? The naval version of the Enigma machine had four rotors rather than three. How many initial settings were made possible by the rotors on the naval Enigma machine?

This is only the beginning of the problem of deciphering the Enigma code. Other parts of the machine allowed for many more initial settings.

12. A new Enigma machine. *This is a continuation of Exercise 11.* Suppose we invent a new Enigma machine with 10 rotors. Four of the rotors must be initially set to A, but the remaining rotors can be set to any letter from B to Z. How many initial settings are there for this new Enigma machine? *Suggestion:* First select four rotors that will be set to A. (In how many ways can you do that?) Now determine how many settings there are for the remaining six rotors. Finally, use the Counting Principle to put the two together.

13. Dave Barry. Consider the following news excerpt.

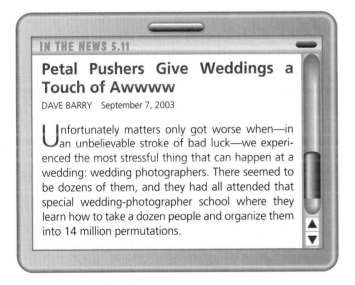

IN THE NEWS 5.11

Petal Pushers Give Weddings a Touch of Awwww

DAVE BARRY   September 7, 2003

Unfortunately matters only got worse when—in an unbelievable stroke of bad luck—we experienced the most stressful thing that can happen at a wedding: wedding photographers. There seemed to be dozens of them, and they had all attended that special wedding-photographer school where they learn how to take a dozen people and organize them into 14 million permutations.

Dave was trying to be funny, but how many permutations are there of 12 people? Was his figure of 14 million close?

Military draft   On December 1, 1969 (in the midst of the Vietnam War), the U.S. Selective Service System conducted a lottery to determine the order in which young men (ages 18 to 26) would be drafted into military service. The days of the year, including February 29, were assigned numbers from 1 to 366. Each such number was written on a slip of paper and placed in a plastic capsule. The 366 capsules were placed in a large bowl. Then capsules were drawn, one at a time, from the bowl. The first number drawn was 257, corresponding to September 14, and young men born on that day were the first group to be called up. This information is used in Exercises 14 through 16.

Military draft lottery in 1969.

14. Assume that capsules were drawn at random. (This point has been the subject of much discussion.) What is the probability that a given young man would be in the first group to be called up?

15. How many different outcomes are possible for such a drawing? Leave your answer in terms of factorials.

16. Young men born on the same day of the year were distinguished according to the first letter of their last, first, and middle names. This was done by randomly selecting the 26 letters of the alphabet. How many permutations of those 26 letters are there? Leave your answer in terms of factorials.

17. Candy. On Halloween, a man presents a child with a bowl containing eight different pieces of candy. He tells her that she may have three pieces. How many choices does she have?

18. More candy. On Halloween, a man presents a child with a bowl containing eight different pieces of candy and six different pieces of gum. He tells her that she may have four of each. How many choices does she have?

Committees   In Exercises 19 through 22, suppose there is a group of seven people from which we will make a committee.

19. In how many ways can we pick a three-person committee?

20. In how many ways can we pick a four-person committee? Compare your answer with what you found in Exercise 19.

21. In how many ways can we pick a five-person committee?

22. In how many ways can we pick a two-person committee? Compare your answer with what you found in Exercise 21.

Committees of men and women   We want to select a committee of five members from a group of eight women and six men. The order of selection is irrelevant. Exercises 23 through 26 refer to this information.

23. What is the total number of committees we can make?

24. How many committees can we make consisting of two women and three men?

25. How many committees can we make with Jack as committee chair? (Assume that only one member has that name!)

26. How many committees can we make with fewer men than women? *Suggestion:* There could be three women and two men, four women and one man, or five women.

27. Choosing 200 from 500. Is it true that the number of ways of choosing 200 items (not considering order) from 500 items is the same as the number of ways of choosing 300 items from 500 items? Explain how you arrived at your answer.

28. Mom's Country Kitchen. A dinner order at Mom's Country Kitchen consists of a salad with dressing, an entrée, and three sides. The menu offers a salad with your choice of ranch, bleu cheese, French, or Italian dressing; a choice of entrée from chicken-fried steak, meat loaf, fried chicken, catfish, roast beef, or pork chops; and, as sides, mashed potatoes, French fries, fried okra, black-eyed peas, green beans, corn, rice, pinto beans, or squash. How many possible dinner orders are available at Mom's?

Marbles In Exercises 29 through 32, we have a box with eight green marbles, five red marbles, and seven blue marbles. We choose three marbles from the box at random without looking. Write each probability in decimal form rounded to three places.

29. What is the probability they will all be green?

30. What is the probability they will all be red?

31. What is the probability they will all be blue?

32. What is the probability they will all be the same color?

33. Money. You have the following bills: a one, a two, a five, a ten, a twenty, and a hundred. In how many ways can you distribute three bills to one person and two bills to a second person?

34. Poll watchers. A certain political organization has 50 poll watchers. Of these, 15 are to be sent to Ohio, 15 to Pennsylvania, 15 to Florida, and the remaining 5 to Iowa. In how many ways can we divide the poll watchers into groups to be sent to the different states? First write your answer in terms of factorials, and then give an estimate such as "about 10^{20}." *Suggestion:* Choose 15 to go to Ohio, then choose 15 from the remaining 35 to go to Pennsylvania, and so on.

Cards In Exercises 35 through 37, we draw three cards from a full deck. Write each probability in decimal form rounded to four places.

35. What is the probability that they are all red?

36. What is the probability that they are all hearts?

37. What is the probability that they are all aces?

More cards In Exercises 38 through 40, we draw five cards from a full deck. Write each probability in decimal form rounded to four places.

38. What is the probability that you have at least one ace? *Suggestion:* First find the probability that you get no aces.

39. What is the probability that they are all face cards (jacks, queens, or kings)?

40. What is the probability that exactly three of them are jacks? *Suggestion:* Think of first picking three jacks. Then pick the remaining two cards at random.

The Ninth Circuit Court Exercises 41 and 42 refer to the article on the Ninth Circuit Court summarized in Example 5.26.

41. Suppose a three-judge panel has a designated chief justice and that we count two panels with the same members as being different if they have different chief justices. That is, we count the panel Adams, Jones, Smith, with Smith as chief, as a different panel than Adams, Jones, Smith, with Jones as chief. How many such panels can be formed from the 28-judge court?

42. Can you think of a reason why a large number of three-judge panels could have an adverse effect on the court's performance?

43. Birthdays. Consider the following problem: *Assume that Jack and Jane were both born in August and that they were born on different days. Find the probability that both were born in the first week of August.*

Here is a purported solution: The number of ways of choosing 2 days from the first 7 is $\frac{7!}{5!2!} = 21$. Next we get the number of ways of picking 2 birthdays out of 31 (because August has 31 days). There are 31 days for the first birthday and 30 for the second. That is $30 \times 31 = 930$ ways to choose 2 birthdays from 31. So the probability of both birthdays occurring in the first week is $\frac{21}{930}$.

What is wrong with this solution? What is the correct answer? *Suggestion:* Remember that counting using permutations shouldn't be mixed with counting using combinations.

Exercises 44 through 48 are more challenging and are suitable for group work. See also Exercise 51.

Changing locks The following article is concerned with the continuing plight of evacuees from Hurricanes Katrina and Rita in 2005.

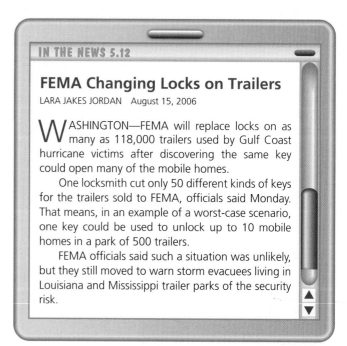

Exercises 44 through 48 refer to this article. In Exercises 45, 46, and 48, you do not need to simplify your answers; just leave your answers in terms of factorials. In one of the technology exercises, you will be asked to simplify those answers (see Exercise 51).

44. The article gives an example of a worst-case scenario. Is it really a worst-case scenario? Can you think of one that might be worse than this?

45. Suppose that a trailer park has 50 sets of 10 trailers, each keyed the same. (This is what the news article seems to suggest would be a "worst case.") If you have one of the 50 keys and try unlocking 10 trailers at random from the 500, what is the probability that you can open at least one of them? *Suggestion:* First calculate the probability that you can open none of them.

46. Suppose that a trailer park has 50 sets of 10 trailers, each keyed the same. If you have one of the 50 keys and you try unlocking 34 trailers at random from the 500, what is the probability that you can open at least one of them? *Suggestion:* First calculate the probability that you can open none of them.

47. Now let's assume that each of the 50 keys fits the same number of trailers and that there are 118,000 trailers. Then each key would fit $118{,}000/50 = 2360$ trailers. If we select two trailers at random from the 118,000 trailers, what is the probability that the same key will fit them both? That is, what is the probability that the two trailers are keyed the same? Write the answer as a simple fraction and in decimal form rounded to two places.

48. Let's make the same assumptions as in Exercise 47. If we select 500 trailers at random from the 118,000, what is the probability that we get 50 sets (one for each key) of 10 trailers each that are keyed the same? (This is what the news article seems to suggest would be a "worst case.")

49. History. The earliest detailed work on combinations was apparently done by Rabbi Abraham ben Meir ibn Ezra in the twelveth century in a text on astrology. Research the problems on combinations of interest to ibn Ezra, and determine some of his other scholarly pursuits.

50. History. Exercise 11 provides a very brief introduction to the German Enigma machine used in World War II. Investigate the Enigma code, including its eventual cracking by the code breakers at Bletchley Park, England, and report on your finding.

The following exercise is designed to be solved using technology such as a calculator or computer spreadsheet. For assistance, see the technology supplement.

51. More on changing locks. Simplify the answers to Exercises 45, 46, and 48. Write the probability in decimal form rounded to two places. *Warning:* For Exercise 48, this may require more sophisticated technology.

5.5 Expected value and the law of large numbers: Don't bet on it

> **TAKE AWAY FROM THIS SECTION**
>
> Mathematics provides tools that aid in risk assessment.

The following excerpt is from an article at theday.com. It refers to Foxwoods Resort Casino and Mohegan Sun, which are casinos in Connecticut.

> **IN THE NEWS 5.13**
>
> ## Another Soft Month for Slot Machine Revenues
>
> HEATHER ALLEN   September 16, 2008
>
> While Foxwoods Resort Casino saw a slight increase in slot revenues in August and Mohegan Sun experienced another loss, both of the tribally owned casinos seemed content with the latest numbers....
>
> The hold, or the average percent of total wagers that is kept by the casino, dropped at Foxwoods from 8.53 percent in August 2007 to 8.42 percent in August 2008. At Mohegan Sun, the drop in the hold percentage was more dramatic going from 8.91 percent last year to 8.18 percent in August.
>
> While the hold percentage decreased, both casinos saw their handles, or how much money patrons spend at the slot machines increase from the previous year.
>
> In August 2008, patrons spent $865 million in slots at Foxwoods compared to $843 million in August 2007. At Mohegan Sun that figure jumped to $948 million from $843 million in 2007.

This article illustrates the large amount of money wagered at casinos. Note that the percentage of wagers kept by these casinos is between 8% and 9%, representing a nice return for the owners. This percentage tells the customers that if they gamble at one of these casinos, they should expect to lose, on average over time, 8% or 9% of all that they wager.

Some form of legal gambling has spread to all but two states.[14] Examples include casinos, state lotteries, and parimutuel wagering on horse or dog racing. With gambling now so prevalent, an educated person should understand how it works. Gambling establishments have one thing in common: The odds are stacked against the player. In fact, a good title for our discussion could be, "Why you shouldn't gamble."

The mathematical concept of *expected value* is key to understanding games of chance. Theoretical expected values are connected to empirical observations through the *law of large numbers*. In this section, we will discuss these concepts and give applications to gambling, the insurance industry, and other areas. To introduce the concept of expected value, we take a quick look at lotteries.

Lotteries and expected value

Forty-three states now have a lottery, which can be a significant source of revenue for them. According to the state of Iowa, its lottery had sales of $3.05 billion over

© Mike Baldwin / Cornered

"And just how are we going to win if every time I buy a ticket, you eat it?"

---

[14]Hawaii and Utah are the exceptions.

a nine-year period. The lottery awarded $1.66 billion in prizes, spent $577.7 million on operations, and transferred $827.8 million to the state as profits (including $136.0 million in sales tax collected on the purchase of tickets). This means that only about 1.66 billion out of 3.05 billion, or about 54%, of the money collected was paid back to players in the form of prizes. The end result is that in the Iowa State Lottery players lose, on average, 46 cents of every dollar they spend on lottery tickets and, on average, the state takes in 46 cents of every dollar spent on lottery tickets.

The Iowa State Lottery illustrates the notion of *expected value*. It's not whether you win or how many times you win, it's the overall net payback you can expect that's important. Expected value has a precise mathematical description that we will introduce shortly. But for the moment, an informal notion of expected value will suffice.

Key Concept

Informally, the **expected value** for you of a game is a measure of the average amount you can expect to win (or lose) per play *in the long run*.

In the Iowa State Lottery, the expected value for the state would appear to be about 46 cents for each dollar played, on the basis of the empirical information above: For each dollar played, the state expects to take in 46 cents. Often the expected value is expressed as a percentage. For the Iowa State Lottery, the state apparently has an expected value of 46%. We can also look at the expected value from the point of view of the player. The expected value for the player is *negative* 46% or −46%, which means that, *on average over time*, the players expect to lose 46 cents for each dollar wagered.

In many lotteries, only a few people ever win anything. Some win millions and most win nothing. Therefore, when we speak of the expected value for a player in the Iowa State Lottery being −46%, we don't mean that all players really lose 46% of their wager. We mean that is the *average* loss per player. Another thing to keep in mind is that expected values should be applied over the long run.

EXAMPLE 5.29 Meaning of expected value: Lottery

Suppose you buy $20 worth of Iowa State Lottery tickets each week. How much should you expect to win or lose, on average, over the period of one year based on the lottery's expected value?

SOLUTION

Over the long run you would expect to lose about 46 cents of each dollar you spend on lottery tickets. Since there are 52 weeks in a year, you are betting $20 \times 52 = \$1040$. Therefore, you should expect to lose about $\$1040 \times 0.46 = \$478.40$.

TRY IT YOURSELF 5.29

Suppose the lottery in your state has an expected value for the player of -36%. If you spend $100 per month on lottery tickets, how much money do you expect to win or lose in an average year?

The answer is provided at the end of this section.

Expected value and the law of large numbers

Our informal definition of the expected value is that it tells the average amount you can expect to win or lose per play in the long run. As a practical matter, the expected value means very little for one or two plays of a game. But there is a *theoretical* expected value that has meaning for a single play of a game. Later we will give a formula for calculating the theoretical expected value, but first we explain the idea using slot machines.

Slot machines are designed so that each outcome has a certain probability and a certain payout. According to the manufacturers, the machines are designed so that every time you pull the handle, the result is independent of whatever has happened before. The overall result is that each machine has a theoretical expected value that can be set by the manufacturer.

According to the Nevada Gaming Control Board, the payback on all slot machines in a recent year was 94.93%, which means that casinos kept, on average, about 5% of all money put into them. This suggests that the theoretical expected value for the casino on slot machines is about 5% (and therefore the expected value for the player is -5%). A return of 5% might not seem enough to support a lavish casino, but it can because of the huge amount of money involved.

The connection between the theoretical expected value and the effect of that expected value in the long run is expressed by the *law of large numbers*, which applies to games for which repeated trials are independent. (Independent events were discussed in Section 3 of this chapter.)

CASINO

I'M NOT SURE, BUT I THINK ILLEGAL GAMBLING IS WHEN YOU WIN.

Key Concept

If you play a game many, many times then according to the **law of large numbers,** you will almost certainly win or lose approximately what the theoretical expected value of the game says you will win or lose.

The law of large numbers is actually a mathematical theorem and not just based on empirical observation.

The law of large numbers is the mathematical principle that assures lotteries and casinos of making money even though the players are going to win sometimes. Consider again the slot machines in Nevada. Even though the expected value for the player says you would expect to lose 5 cents for each dollar you put in the machine, the occasional win is no surprise. But now suppose you play a slot machine 1000 times. The expected value tells us that you would expect to be down about 5% of $1000, or $50. It would not be very surprising if you were down a few dollars more than $50 or a few dollars less, but it's very likely that you will have lost money. In the long term, the casino will come out ahead.

The fact that individual plays on a slot machine are independent means that you cannot predict what a slot machine will do based on what has happened before. Some gamblers do not behave as if they accept this. They seem to believe that if they play a slot machine many times without a payoff, the machine must be "due." This means that *surely* the machine must be ready to pay off after so many losses. This belief is the *gambler's fallacy*.

Key Concept

The gambler's fallacy is the belief that a string of losses in the past will be compensated by wins in the future.

To see why this belief is a fallacy, consider tossing a fair coin. If we toss 50 heads in a row, the probability of a tail on the 51st toss is still 1/2. In the same way, the next play of a slot machine has the same probability of a payout as any other play.

It may seem that the law of large numbers supports the gambler's fallacy. But the law of large numbers refers to the average payoff in the long run—it does not rule out streaks of losses or wins. The law of large numbers refers to future payoff, not payoff in the past. It does not mean that past losses will necessarily be made up by future wins or that past wins will be offset by future losses. It says that if you play many times from this point onward, your wins or losses will be very close to the prediction given by the expected value.

Calculating expected value

The theoretical expected value is calculated by multiplying the probability of each outcome by the amount you win (or by minus the amount if you lose) and adding these numbers. For a game with a payoff or a loss like the Iowa State Lottery, our expected value is given by the following formula:

$$\text{Expected value} = P(\text{Win}) \times \text{Profit} - P(\text{Lose}) \times \text{Loss}.$$

To illustrate the calculation, we look at a simple game. A fair die is rolled. If the result is a 1 or 2, I win \$9. If any other number comes up, I lose \$3. To calculate the expected value for me, we must first find the probability for a win and for a loss. For a win, there are two favorable outcomes (1 and 2) out of six total outcomes. Hence, the probability of a win is

$$P(\text{Win}) = \frac{\text{Favorable outcomes}}{\text{Total outcomes}}$$
$$= \frac{2}{6} = \frac{1}{3}.$$

We find the probability of a loss in a similar fashion: There are 4 losing numbers out of 6, so the probability of a loss is $4/6 = 2/3$. We use these probabilities to calculate the expected value for me:

$$\text{Expected value} = P(\text{Win}) \times \text{Profit} - P(\text{Lose}) \times \text{Loss}$$
$$= \frac{1}{3} \times \$9 - \frac{2}{3} \times \$3$$
$$= \$1.$$

My expected value for this game is \$1 per play. So if I play this game many times, I can expect to win an average of \$1 for each time I play. This is a very attractive game. Typical casino games are not as lucrative for the player, as is illustrated by the following example.

A standard roulette wheel has numbers 1 through 36 alternately colored red and black. It also has a green 0 and a green 00 called "double zero."[15] The wheel is spun with a small white ball inside. When the wheel stops, the ball falls into a numbered slot, which determines the winners. Successive spins of a wheel yield independent results.

There are many wagers you can place at the roulette table. For example, you may bet on the color (red or black), or on whether the number is odd or even, or on a single number. Note that, even though 0 is in fact an even number, in roulette both 0 and 00 count as neither odd nor even. These green numbers give the edge to the house.

EXAMPLE 5.30 Calculating expected value: A roulette bet

You are playing roulette, and you bet a dollar on the number 10. If 10 comes up, you win $35 (profit). If anything else comes up, you lose your dollar. What is your expected value for this bet?

SOLUTION

Because there are a total of 38 numbers on the roulette wheel (remember the 0 and 00), the probability of a 10 is 1/38, and this outcome wins $35. The probability of not getting a 10 is 37/38, and this outcome loses $1. We use these probabilities to calculate your expected value:

$$\text{Expected value} = P(\text{Win}) \times \text{Profit} - P(\text{Lose}) \times \text{Loss}$$

$$= \frac{1}{38} \times \$35 - \frac{37}{38} \times \$1$$

$$= -\frac{2}{38} \text{ dollar}$$

$$= -\frac{1}{19} \text{ dollar.}$$

This is about −5.3 cents, so your expected value for this bet is −5.3%. This means that if you make this roulette bet many times, you can expect to lose about 5.3% of all the money you wager.

TRY IT YOURSELF 5.30

If you bet $1 on an even number, you win $1 (profit) if any of the even numbers 2 through 36 come up, and you lose your dollar if 0, 00, or any odd number between 1 and 35 comes up. What is your expected value for this wager?

The answer is provided at the end of this section.

In addition to slot machines and roulette, casinos offer a variety of games, among which are blackjack, keno, craps, and various kinds of poker. But they all have one thing in common: The theoretical expected value for the player is negative. And no system of placing wagers can change this fact.[16]

Many people enjoy buying lottery tickets or going to casinos even though they assume they will probably lose some money. Some view it as buying entertainment, just as they might pay to go to a movie. A few will get lucky once in a while and win money, but the law of large numbers ensures that the vast majority will eventually lose and that casinos and state lotteries will be profitable for the proprietors.

---

[15] European roulette wheels have no double zero.

[16] "Card counting" at blackjack can produce a very small positive expected value for the player. But most casinos do not allow the practice.

Fair games

In probability theory, a game is called *fair* if its expected value is zero. Intuitively then, a fair game is one in which no participant has an advantage over others. Clearly, we would not call the game of roulette fair because it gives the edge to the casino.

Key Concept

A game is said to be fair if the expected value is zero.

For example, suppose Sue and Alice toss a coin. Sue wins $2 from Alice if the coin comes up heads and loses $2 to Alice if it comes up tails. Let's determine whether this is a fair game.

There are two outcomes: heads (Sue wins) and tails (Sue loses). The probability that Sue wins is 1/2, in which case she wins $2. The probability that Sue loses is also 1/2, and then she loses $2. Sue's expected value is

$$\text{Expected value} = P(\text{Win}) \times \text{Profit} - P(\text{Lose}) \times \text{Loss}$$
$$= \frac{1}{2} \times \$2 - \frac{1}{2} \times \$2$$
$$= \$0.$$

Because the expected value is 0, this is a fair game. If Sue plays this game many times, she can expect to lose about as much as she wins. This does not mean that she is *certain* to come out even in the end, but it would be unusual for her either to win or to lose a large amount of money over a long stretch.

EXAMPLE 5.31 Making a game fair: Roll a die

In a gambling game, I roll a single die. If it comes up a 6, I win. I lose $1 otherwise. How much money should I collect on a win to make this a fair game?

SOLUTION

The probability of a win is 1/6, and the probability of a loss is 5/6. Thus, we find that my expected value is

$$\text{Expected value} = P(\text{Win}) \times \text{Profit} - P(\text{Lose}) \times \text{Loss}$$
$$= \frac{1}{6} \times \text{Profit} - \frac{5}{6} \times \$1.$$

To make the game fair, the expected value should be 0. So the profit for a win should be $5.

TRY IT YOURSELF 5.31

In Example 5.30, if I place a bet on the number 10 on the roulette wheel, the casino has the advantage. How much profit should I get from a $1 bet on number 10 to make the game fair?

The answer is provided at the end of this section.

Now we give an application to guessing strategies for taking tests.

EXAMPLE 5.32 Understanding fair game: Guessing on the SAT

The SAT is widely used in college admissions. On multiple-choice questions with five possible answers, 1 point is added to the raw score for the correct answer, and

1/4 point is subtracted from the raw score for each incorrect answer. (Supplying no answer to a problem doesn't change the raw score.) If students randomly guess answers, what average raw score for a problem would we expect to see in the long run? In other words, what is your expected value of the raw score for a problem if the answers are filled in at random?

SOLUTION

We calculate the expected value using the formula

$$\text{Expected value} = P(\text{Win}) \times \text{Profit} - P(\text{Lose}) \times \text{Loss}.$$

When you take the SAT, winning corresponds to giving a correct answer, with a profit of 1 point, and losing corresponds to giving an incorrect answer, with a loss of 1/4 point.

Now we find the probabilities. There are five possible answers one can fill in, and one of them is correct. The probability of getting a correct answer by random guessing is thus 1/5, so the probability of an incorrect answer is 4/5.

Thus, the expected value of the score is

$$\text{Expected value} = P(\text{Win}) \times \text{Profit} - P(\text{Lose}) \times \text{Loss}$$
$$= \frac{1}{5} \times 1 \text{ point} - \frac{4}{5} \times \frac{1}{4} \text{ point}$$
$$= 0 \text{ points.}$$

This means that random guessing gives no advantage in the long run—which is the purpose of subtracting 1/4 point for incorrect answers. Because the expected value is 0, the awarding of points on the SAT exam is fair in the sense that we have been discussing.

TRY IT YOURSELF 5.32

Suppose that on SAT questions, you can always recognize one possible answer as incorrect. What is your expected value of the raw score for a problem if you guess at random among the remaining four choices? (Leave your answer as a fraction.) Is this "game" fair?

The answer is provided at the end of this section.

> **SUMMARY 5.8** **Expected Value**
>
> The theoretical expected value for you, the player in a game, is calculated as follows: If an outcome results in a win for you, multiply its probability by the profit you win. If an outcome results in a loss for you, multiply the probability by the *negative* of the amount you lose. Your expected value is the sum of all these positive and negative numbers. A simplified formula for this is
>
> $$\text{Expected value} = P(\text{Win}) \times \text{Profit} - P(\text{Lose}) \times \text{Loss}.$$
>
> The expected value is a measure of the average amount you can expect to win (or lose) per play *in the long run*.
> If the expected value is zero, the game is *fair*.

Risk assessment

So far we have discussed the concept of expected value primarily in the context of games of chance. But this concept is relevant whenever we weigh risks and benefits by considering the likelihood of various outcomes. Applications of expected values are found in various contexts, including manufacturing, finance, economics, and the insurance industry. Blaise Pascal, a famous mathematician and religious philosopher of the seventeenth century, applied the concept of expected value to belief in the existence of God. See Exercise 46.

The concept of expected value plays a key role in the insurance industry. Companies that insure property, for example, have to pay for catastrophic damages occasionally, such as those caused by a hurricane. The companies must set the costs of premiums to compensate for such losses but also provide a profit (i.e., a positive expected value) in the long run. The experts whom they employ to figure this out are called *actuaries*.

Storm damage.

EXAMPLE 5.33 Calculating expected profit: Insurance

Suppose a company charges a premium of $150 per year for an insurance policy for storm damage to roofs. Actuarial studies show that in case of a storm, the insurance company will pay out an average of $8000 for damage to a composition shingle roof and an average of $12,000 for damage to a shake roof. They also determine that out of every 10,000 policies, there are 7 claims per year made on composition shingle roofs and 11 claims per year made on shake roofs. What is the company's expected value (i.e., expected profit) per year of a storm insurance policy? What annual profit can the company expect if it issues 1000 such policies?

SOLUTION

This is just like a game where the insurance company accepts a bet of $150 from the customer and makes payouts to meet claims. To find the expected value, we multiply the probability of each outcome by the amount the company takes in (or by minus the amount if the company pays out) and add these numbers.

The probability of a composition shingle roof claim is $7/10,000 = 0.0007$, and the probability of a shake roof claim is $11/10,000 = 0.0011$. For every 10,000 policies, there are $7 + 11 = 18$ total claims. Hence, 9982 policies make no claim. So the probability of no claim is $9982/10,000 = 0.9982$.

Each composition shingle roof claim costs the company $7850 (which is $8000 minus the $150 premium), and each shake roof claim costs $11,850 (which is $12,000 minus the $150 premium). But each policy that makes no claim results in $150 profit for the company. Thus, the expected value for the company is

$$\text{Expected value} = P(\text{No claim}) \times \text{Profit} - P(\text{Composition claim})$$
$$\times \text{Cost} - P(\text{Shake claim}) \times \text{Cost}$$
$$= 0.9982 \times \$150 - 0.0007 \times \$7850 - 0.0011 \times \$11,850$$
$$= \$131.20.$$

This means that, on average, the insurance company earns a profit of $131.20 on each policy it writes. If the insurance company issues 1000 such policies, it can expect an annual profit of $131.20 × 1000 = $131,200.

TRY IT YOURSELF 5.33

Suppose a company charges an annual premium of $300 for a flood insurance policy. In case of a flood claim, the company will pay out an average of $200,000 per policy. Based on actuarial studies, it determines that in an average year there is one flood claim for every 5000 policies. What is the company's expected annual profit for a flood insurance policy? What annual profit can the company expect if it issues 1000 policies?

The answer is provided at the end of this section.

Here is an application of the concept of expected value to manufacturing.

EXAMPLE 5.34 Calculating expected cost: An assembly line

A company manufactures small electric motors at a cost of $50 each. Based on experience, quality control experts have come up with the following probabilities of defective motors in a given production run:

Probability of no defective motors: 30%.

Probability that 1% of motors are defective: 40%.

Probability that 2% of motors are defective: 20%.

Probability that 3% of motors are defective: 10%.

Note that the sum of the probabilities is 100%. When a defective motor is detected, it must be removed from the assembly line and replaced. This process adds an extra $15 to the cost of the replaced motor. What is the expected cost to the company of a batch of 1000 motors?

SOLUTION

You can think of a production run as a game where the company "wins" $50 for each good motor and "wins" $65 for each defective motor. The expected cost to the company is the expected value for the company of 1000 motors with the given probabilities. We proceed as follows.

If there were no defective motors in the production run, the cost to the company would be $50 × 1000 = $50,000. The probability of this happening is 30%, so we multiply $50,000 by 0.3 to get $15,000.

Now suppose that 1% of the motors in the production run are defective. Then there are 10 defective motors, which costs the company $15 × 10 = $150 extra. Therefore, the cost to the company is $50,150. The probability of this happening is 40%, so we multiply $50,150 by 0.4 to get $20,060.

Now suppose that 2% of the motors in the production run are defective. Then there are 20 defective motors, which costs the company $15 × 20 = $300 extra. Therefore, the cost to the company is $50,300. The probability of this happening is 20%, so we multiply $50,300 by 0.2 to get $10,060.

Finally, suppose that 3% of the motors in the production run are defective. Then there are 30 defective motors, which costs the company $15 × 30 = $450 extra. Therefore, the cost to the company is $50,450. The probability of this happening is 10%, so we multiply $50,450 by 0.1 to get $5045.

To get the expected value, we add up the results of the previous four calculations. Thus, we find

Expected value $= \$15,000 + \$20,060 + \$10,060 + \$5045 = \$50,165.$

The expected cost to the company of a batch of 1000 motors is \$50,165.

TRY IT YOURSELF 5.34

Suppose the probability of no defective motors is 40%, the probability of 1% defective motors is 30%, the probability of 2% defective motors is 15%, and the probability of 3% defective motors is 15%. Assume that the motors cost \$50 each and replacement adds \$15 to the cost. Find the expected cost to the company of a run of 2000 motors.

The answer is provided at the end of this section.

Try It Yourself answers

Try It Yourself 5.29: Meaning of expected value: Lottery   Lose \$432.

Try It Yourself 5.30: Calculating expected value: A roulette bet -5.3%.

Try It Yourself 5.31: Making a game fair: Another roulette bet   \$37.

Try It Yourself 5.32: Understanding fair game: Guessing on the SAT   1/16; no.

Try It Yourself 5.33: Calculating expected profit: Insurance   \$260; \$260,000.

Try It Yourself 5.34: Calculating expected cost: An assembly line   \$100,315.

Exercise Set 5.5

In these exercises, express the expected value as a percentage, rounded to one decimal place, unless you are instructed otherwise.

1. Fair games I. Explain why casinos do not offer fair games.

2. Fair games II. I told my friend Bob that I learned in math class that no casino games are fair. Bob said, "That's not true. My Mom works in a casino and I know they don't cheat." What should I say to explain myself to Bob?

3. Amount bet. Nevada casinos statewide won \$9.56 billion during a recent year. Assume that overall the casinos take in 3.75% of the money wagered. How much money was bet? Round your answer to the nearest billion dollars.

4. The handle. A casino's gross revenue—all the money wagered by players—is referred to in the business as the *handle*. The *hold* is the profit the casino retains after players are paid. (See the article opening this section.) If a casino's hold is \$1,500,000 in a month and it takes in 4.5% of the money wagered, how much is the handle? Round your answer in millions of dollars to two decimal places.

Expected value   Bill and Larry toss two coins. If both coins come up heads, Bill pays Larry \$4. Larry pays Bill \$1 otherwise. Exercises 5 through 9 refer to this game.

5. What is the expected value of this game for Bill?

6. What is the expected value of this game for Larry?

7. Bill and Larry play the game 100 times. Who should come out ahead and by how much?

8. Bill and Larry play the game 100 times, and Larry comes out \$23 ahead. Use only this empirical information to estimate the expected value of this game for Bill and the expected value for Larry.

9. Explain the difference, in general, between the theoretical expected value and an estimate of the expected value obtained using empirical information. (You may use the game above as an example.)

10. Double or nothing. You are playing a game with a friend in which you toss a fair coin. Your friend pays you a nickel if it's heads, and you pay your friend a nickel if it's tails. You win 10 times in a row and want to stop, but your friend says, "Let's play one more time, double or nothing. I'm due to win." What's the expected value of this game? What is your response to your friend?

11. Shooting free-throws. A basketball player who typically makes 80% of his free-throws has missed his first two attempts in a game. You hear a sports commentator say that the player is therefore "due" to make his next attempt. Analyze this statement. Consider whether the gambler's fallacy applies here. Are successive free-throws independent events?

12. Gambler's fallacy. One sports fan in describing how his mediocre team had gone from a five-game winning streak to a five-game losing streak had this summary: "The law of averages was bound to catch up with a team like this." Is this statement an application of the law of large numbers, an example of the gambler's fallacy, or neither? Explain your answer.

13. Martingale method. Suppose a game pays even money. That is, if you win, you get whatever you bet, and if you lose, you lose that same amount. Here's a strategy called the *Martingale method*.

If you win, bet the same amount again. But every time you lose, you play again and double your previous bet. For example, suppose you bet a dollar in such a game. If you win, you bet $1 again. If you lose, then you play again but this time you bet $2. If you win, you're back to even and you bet $1 next time. If you lose, you play again and bet $4. If you win, you're back to even, and so forth. This strategy seems appealing because no matter how many times you lose, eventually you're bound to win one time, and that puts you back to even. If you win twice in a row, you have a profit. So in theory you can't lose in the end. But this strategy is flawed. Can you find the flaw?

14. A coin game. In a game you bet $1 and toss two coins. If two heads come up, you win a profit of 75 cents. If two tails come up, no one wins anything. Otherwise, you lose your dollar. What is your expected value for this game? Don't round your answer.

15. How much will we win? We decide to open a casino where we offer only one game. We win a dollar 60% of the time, and our customer wins a dollar 40% of the time. How much profit would we expect to have if 5000 games are played?

16. Make it fair. You play a gambling game with your friend in which you win 30% of the time and lose 70% of the time. When you lose, you lose $1. What profit should you earn when you win in order for the game to be fair?

17. Voting. A poll taken in September 2008 showed Obama leading McCain by 54% to 46% in the U.S. presidential race. Let's assume that these numbers represent probabilities. I wished to make a bet with my friend on who would win the presidency, and we agreed to base our bet on that poll. I predicted a victory for Obama and would win a profit of $1 if I were right. Assuming we wished this to be a fair game, how much should I have agreed to pay my friend if he was right and McCain was elected?

18. Card game. A card game goes like this: You draw a card from a 52-card deck. If it is a face card (jack, queen or king), you win $5; otherwise, you lose $2. What is your expected value for this game?

19. Modified card game. Exercise 18 described a card game in which you draw a card from a 52-card deck. Suppose that this game is modified so that if you draw an ace, then no money changes hands. What is your expected value for this game?

20. Expected value I. What is your expected value for a game where your probability of winning is 1/6 and your profit is four times your wager?

21. Expected value II. Consider the following game: The probability of winning a profit of twice your wager is 1/7, the probability of winning a profit equal to your wager is 2/7, and all

other probabilities of winning are 0. What is your expected value for this game?

22. Expected value III. What is your expected value for a game whose probabilities and outcomes are given in the following table?

Probability	Profit
0.2	You win 1.5 times your wager.
0.4	You win your wager.
0.2	You lose 2 times your wager.
0.2	You win nothing.

23. Expected value IV. What is your expected value for a game whose probabilities and outcomes are given in the following table?

Probability	Profit
0.1	You win your wager.
0.4	You win half your wager.
0.2	You lose 3 times your wager.
0.3	You win nothing.

24. Exam score I. Each question on a multiple-choice exam has four choices. One of the choices is the correct answer, worth 5 points, another choice is wrong but still carries partial credit of 1 point, and the other two choices are worth 0 points. If a student picks answers at random, what is the expected value of his or her score for a problem? If the exam has 30 questions, what is his or her expected score?

25. Exam score II. Each question on a multiple-choice exam has five choices. One of the choices is the correct answer, worth 5 points. There are two other choices that are wrong but still carry partial credit of 2 points each. The other two choices are worth 0 points. If a student picks answers at random, what is the expected value of his or her score for a problem? If the exam has 40 questions, what is his or her expected score?

26. Insurance I. Suppose a company charges an annual premium of $450 for a fire insurance policy. In case of a fire claim, the company will pay out an average of $100,000. Based on actuarial studies, it determines that the probability of a fire claim in a year is 0.004. What is the expected annual profit of a fire insurance policy for the company? What annual profit can the company expect if it issues 1000 policies?

27. Insurance II. Suppose a company charges an annual premium of $100 for an insurance policy for minor injuries. Actuarial studies show that in case of an injury claim, the company will pay out an average of $900 for outpatient care and an average of $3000 for an overnight stay in the hospital. They also determine that, on average, each year there are five claims made that result in outpatient care for every 1000 policies and three claims made that result in an overnight stay out of every 1000 policies. What is the expected annual profit of an insurance policy for the company?

28. Tires. A company manufactures tires at a cost of $60 each. The following are probabilities of defective tires in a given production run: probability of no defective tires: 10%; probability of 1% defective tires: 30%; probability of 2% defective tires:

40%; probability of 3% defective tires: 20%. When a defective tire is detected, it must be removed from the assembly line and replaced. This process adds an extra $5 to the cost of the replaced tire. What is the expected cost to the company of a batch of 3000 tires?

29. Chairs. A company manufactures chairs at a cost of $30 each. The following are probabilities of defective chairs in a given production run: probability of 1% defective chairs: 40%; probability of 2% defective chairs: 35%; probability of 3% defective chairs: 25%. When a defective chair is detected, it must be removed from the assembly line and replaced. This process adds an extra $10 to the cost of the replaced chair. What is the expected cost to the company of a batch of 500 chairs?

Expected rate of return In finance the notion of expected value is used to analyze investments for which the investor has an estimate of the chances associated with various returns (and losses). For example, suppose you have the following information about one of your investments: With a probability of 0.9, the investment will return 20 cents for every dollar you invest, and with a probability of 0.1, the investment will lose 50 cents for every dollar you invest. The expected rate of return for this investment is calculated the way we calculate the expected value of a game: Multiply the probability of each outcome by the amount you earn (or by minus the amount if you lose) and add up these numbers. Exercises 30 and 31 refer to this concept.

30. Calculate the expected rate of return for the investment described above.

31. Here is information about another investment: With a probability of 0.5, it will return 20 cents for every dollar you invest; with a probability of 0.3, it will return 10 cents for every dollar you invest; and otherwise, it will lose 90 cents for every dollar you invest. Calculate the expected rate of return for this investment. Is this a good investment?

32. Betting on 11. In a dice game, you win if the two dice come up 11. Otherwise, you lose $1. What should be the profit for winning to make this game fair?

33. A fair game. You roll a pair of dice. You bet $1 that you will get a 2, 3, or 12. If you win, you are paid $9 (a profit of $8). Show that this is a fair game.

34. The real game. In Exercise 33, you rolled a pair of dice, and you bet $1 that you will get a 2, 3, or 12. In the casino game called *craps*, the casino pays you $8 (a profit of $7) if these numbers come up on your "opening roll." What is your expected value for this bet expressed as a percent?

35. Betting on red. Suppose you bet $1 on red at the roulette wheel. There are 18 red numbers and 20 nonred numbers (including the green 0 and 00). You win $1 if a red number comes up, and you lose $1 if any other number comes up. What is your expected value for this wager?

36. Even or odd. In roulette, you may bet whether the number that comes up is odd (or even). (Remember that 0 does not count as either odd or even.) The wager pays even money, that is, if you win, you get whatever you bet, and if you lose, you lose that same amount. What is your expected value for this bet?

Roulette and gambler's fallacy Roulette wheels in Monte Carlo have no double zero, so there are 37 numbers on a wheel. Exercises 37 through 40 use this fact.

37. What is the probability of getting an even number on one spin of a roulette wheel in Monte Carlo? Express your answer as a fraction and as a percentage rounded to one decimal place.

38. You spin a roulette wheel in Monte Carlo 26 times. What is the probability of getting an even number on all 26 spins? Express your answer as a fraction, then express the chances using a phrase such as "1 in 5 billion."

39. In 1913 at one fair roulette table in Monte Carlo, an even number came up on 26 spins in a row. What was the probability of getting an even number on the 27th spin?

40. Explain how the historical event described in Exercise 39 is consistent with the law of large numbers.

Powerball Powerball is a lottery in which most U.S. states participate. To play, you pay $1 and choose five white numbers between 1 and 59. Then you choose a single red ball, the Powerball, from a second set of numbers that range from 1 to 39. To win the Powerball jackpot, you must match all five white numbers, plus the Powerball number. Like other lottery games, varying amounts of money can also be won by matching fewer numbers, allowing for nine different ways to win a Powerball prize. Here they are:

- 5 white + Powerball = Powerball jackpot prize, $20 million minimum (Probability of winning: 1 in 195 million)
- 5 white = $200,000 prize (Probability of winning: 1 in 5.1 million)
- 4 white + Powerball = $10,000 prize (Probability of winning: 1 in 700,000)
- 4 white = $100 prize (Probability of winning: 1 in 19,000)
- 3 white + Powerball = $100 prize (Probability of winning: 1 in 14,000)
- 3 white = $7 prize (Probability of winning: 1 in 359)
- 2 white + Powerball = $7 prize (Probability of winning: 1 in 787)
- 1 white + Powerball = $4 prize (Probability of winning: 1 in 123)
- Powerball only = $3 prize (Probability of winning: 1 in 62)

The Powerball jackpot starts at $20 million and increases after each drawing with no Powerball jackpot hit. The largest Powerball jackpot on record (as of 2011) is $365 million, won on February 18, 2006, by a ticket sold in Nebraska. Exercises 41 through 44 refer to Powerball as described above.

41. Powerball players may choose to have the numbers generated randomly by the computer—called a Quick Pick—rather than choosing the numbers themselves. Do you think it matters? Why?

42. How many possible combinations of five numbers chosen from 1 through 59 are there?

43. How many possible combinations of six Powerball numbers are there, where five white numbers are chosen from

1 through 59 and one red Powerball number is chosen from 1 through 39?

44. Suppose several people chose the same winning numbers for the jackpot. Wouldn't that cause the state to lose money? (This may require some investigation.)

45. History. Historically, many of the ideas from probability were developed to understand games of chance. See Exercise 63 in Section 5.3 about "double sixes." Write a paper about this historical connection.

46. History. Blaise Pascal was an eminent scientist, mathematician, and religious philosopher. His work led to the develop-

ment of probability theory and the notion of expected value. In a section of his writings, he discusses belief in the existence of God in terms of expected value. Write a paper about *Pascal's wager*.

47. History. Choose one of the games discussed in this section and write a paper about its history and evolution.

48. History. If your state or a neighboring state has a lottery, write a paper about its history, the revenue it generates, and how it operates.

CHAPTER SUMMARY

The mathematical study of probability is an attempt to model and understand uncertainty. We examine how a probability is obtained, what it means, and where it is useful in the real world.

Calculating probabilities: How likely is it?

A probability is a numerical measure of the likelihood of an event. It can be theoretical, empirical, or subjective (based merely on opinion). An example of a statement using a theoretical probability is, "The probability of a fair coin coming up heads is 50%." This represents a theoretical probability because there are two equally likely events that can occur. Therefore, in theory one of them should happen half of the time. An example of a statement using an empirical probability is, "Based on a sample of some of the lightbulbs on a production line, I estimate that the probability of a lightbulb being defective is 1%." This represents an empirical probability because it is based on experimental evidence. An example of a statement using a subjective probability is, "My probability of becoming a rock star is one in a million." This probability is based on opinion: The number is plucked out of thin air.

To calculate a theoretical probability, we designate the outcomes in which we are interested as *favorable* and all other outcomes as *unfavorable*. In the experiments we consider, we make sure that each outcome is equally likely. We have the formula

$$\text{Probability of an event} = \frac{\text{Number of favorable outcomes}}{\text{Total number of possible outcomes.}}$$

For example, suppose we want to find the theoretical probability that I roll a 3 with two dice. We need to count the number of possible outcomes for the two dice, which is $6 \times 6 = 36$. The number of favorable outcomes, that is, those that give a 3, is 2 (either one and two or two and one). Then the probability is $2/36 = 1/18$. In cases involving identical items, such as coins or dice, it may help to label the items to distinguish them.

Here are two useful formulas:

$$\text{Probability of event not occurring} = 1 - \text{Probability of event occurring}$$

and

$$P(A \text{ or } B) = P(A) + P(B) - P(A \text{ and } B).$$

Medical testing and conditional probability: Ill or not?

One important application of mathematical probability involves medical testing. The terminology used in such tests includes: true positives, false positives, true negatives, false negatives, sensitivity, specificity, prevalence, positive predictive value (PPV), and negative predictive value (NPV). Two key formulas are

$$PPV = \frac{\text{True positives}}{\text{All positives}} = \frac{\text{True positives}}{\text{True positives} + \text{False positives}}$$

and

$$NPV = \frac{\text{True negatives}}{\text{All negatives}} = \frac{\text{True negatives}}{\text{True negatives} + \text{False negatives}}.$$

A *conditional probability* is the probability that one event occurs given that another has occurred. Positive and negative predictive values, as well as sensitivity and specificity, are examples of conditional probabilities.

Counting and theoretical probabilities: How many?

To calculate a theoretical probability, one often needs to calculate the number of ways something can happen. The most basic way of counting for this purpose is the *Counting Principle*. It says that if we perform two experiments in succession, with N possible outcomes for the first and M possible outcomes for the second, the number of possible outcomes for the two experiments is $N \times M$.

Suppose we perform an experiment. Two events are *independent* if knowing that one event occurs has no effect on the probability of the occurrence of the other. Events A and B are independent exactly when the probability of Event A *and* Event B occurring is the probability of Event A times the probability of Event B, that is,

$$P(A \text{ and } B) = P(A) \times P(B).$$

More ways of counting: Permuting and combining

More sophisticated ways of counting that can be deduced from the Counting Principle involve permutations and combinations. A *permutation* of items is an arrangement of them in a certain order. The number of permutations of n distinct objects is $n \times (n-1) \times \cdots \times 2 \times 1$. This expression is denoted by $n!$. The number of ways to select k items from n items and arrange them in order is the number of permutations of n objects taken k at a time. It is given by the formula

$$\frac{n!}{(n-k)!}.$$

The number of ways to select k items from n items when order makes no difference is the number of combinations of n objects taken k at a time. The formula for calculating this is

$$\frac{n!}{k!(n-k)!}.$$

Expected value and the law of large numbers: Don't bet on it

The *expected value* for you of a game is a measure of the average amount you can expect to win (or lose) per play in the long run. The theoretical expected value is calculated by multiplying the probability of each outcome by the amount you win (or by minus the amount if you lose) and adding up these numbers. The expected value can be estimated from empirical information (such as the percentage of money paid out as prizes in a lottery). The concept of expected value is applied not only to games of chance but also in such areas as manufacturing, finance, economics, and insurance.

The *law of large numbers* says that if you play a game many times, then you will almost certainly win or lose approximately what the theoretical expected value of the game says you will win or lose. Because the expected value for the player for games of chance in a casino is always negative, the law of large numbers ensures that casinos will make a profit and that in the long term gamblers will lose money.

KEY TERMS

probability, p. 272
disjunction, p. 278
empirical probability, p. 281
sensitivity, p. 288
specificity, p. 289
prevalence, p. 291
positive predictive value (PPV),
    p. 291

negative predictive value (NPV),
    p. 291
conditional probability, p. 295
Counting Principle, the, p. 302
independent, p. 308
permutation, p. 316
permutations of *n* items taken *k* at a
    time, p. 318

combination, p. 321
combination of *n* items taken *k* at a
    time, p. 322
expected value, p. 333
law of large numbers, p. 334
gambler's fallacy, p. 335
fair, p. 337

CHAPTER QUIZ

1. If I draw a card at random from a standard deck of 52 cards, what is the probability I draw a red queen? Write your answer as a fraction.

Answer $2/52 = 1/26$

If you had difficulty with this problem, see Example 5.1.

2. Suppose I flip two identical coins. What is the probability I get a head and a tail? Write your answer as a fraction.

Answer $1/2$

If you had difficulty with this problem, see Example 5.2.

3. Suppose a softball team has 25 players. Assume that 13 bat over .300, 15 are infielders, and 5 are both. What is the probability that a player selected at random is either batting over .300 or an infielder? Write your answer as a fraction.

Answer $23/25$

If you had difficulty with this problem, see Example 5.6.

4. A new medical test is given to a population whose disease status is known. The results are in the table below.

	Has disease	Does not have disease
Test positive	300	100
Test negative	10	200

Use these data to estimate the sensitivity and specificity of this test. Write your answers in percentage form rounded to one decimal place.

Answer Sensitivity 96.8%; specificity 66.7%

If you had difficulty with this problem, see Example 5.9.

5. The table below shows test results for a certain population.

	Has disease	Does not have disease
Test positive	10	30
Test negative	5	40

Find the positive predictive value and negative predictive value for this test for this population. Write your answers in percentage form rounded to one decimal place.

Answer PPV 25.0%; NPV 88.9%

If you had difficulty with this problem, see Example 5.11.

6. How many four-letter codes can we make if the first letter is A, B, or C and the remaining letters are D or E? Letters may be repeated. One such code is ADDE.

Answer 24

If you had difficulty with this problem, see Example 5.15.

7. It has been found that only 1 in 100 bags of M&M's have five more than the average number of M&Ms. Only 1 in 20 boxes of Cracker Jacks has a really good prize. I buy a box of Cracker Jacks and a bag of M&Ms. What is the probability that I get a really good prize and five more than the average number of M&Ms? Write your answer as a fraction.

Answer $1/2000$

If you had difficulty with this problem, see Example 5.19.

8. In how many ways can I choose 4 people from a group of 10 and make them president, vice president, secretary, and treasurer? Write your answer in terms of factorials.

Answer $10!/6!$

If you had difficulty with this problem, see Example 5.21.

9. In how many ways can I select a 4-person committee from a group of 10 people? Leave your answer in terms of factorials.

Answer $10!/(4!6!)$

If you had difficulty with this problem, see Example 5.24.

10. The expected value for the player of a certain lottery is -20%. If you spend \$5 each week on lottery tickets, how much money do you expect to win or lose over a 10-week period?

Answer Lose \$10

If you had difficulty with this problem, see Example 5.29.

11. You roll a single die. If you roll 1 or 2, you win \$10. Otherwise, you lose \$4. What is your expected value for this game?

Answer 67 cents

If you had difficulty with this problem, see Example 5.30.

12. A company manufactures coats at a cost of $20 each. The following are probabilities of defective coats in a given production run: probability of no defective coats: 5%; probability of 1% defective coats: 75%; probability of 2% defective coats: 20%. When a defective coat is detected, it must be removed from the assembly line and replaced. This process adds an extra $10 to the cost of the replaced coat. What is the expected cost to the company of a batch of 4000 coats?

Answer $80,460

If you had difficulty with this problem, see Example 5.34.

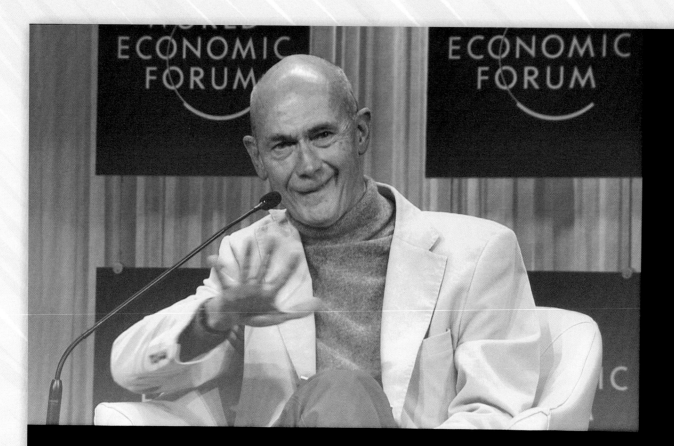

STATISTICS

The following article is a Reuters news release.

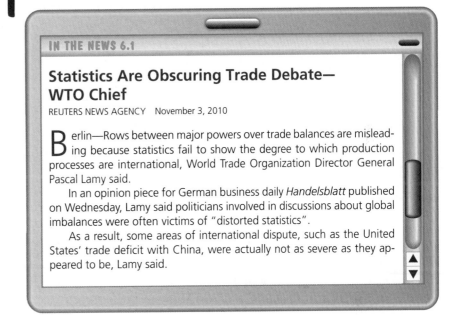

IN THE NEWS 6.1

Statistics Are Obscuring Trade Debate— WTO Chief

REUTERS NEWS AGENCY   November 3, 2010

Berlin—Rows between major powers over trade balances are misleading because statistics fail to show the degree to which production processes are international, World Trade Organization Director General Pascal Lamy said.

In an opinion piece for German business daily *Handelsblatt* published on Wednesday, Lamy said politicians involved in discussions about global imbalances were often victims of "distorted statistics".

As a result, some areas of international dispute, such as the United States' trade deficit with China, were actually not as severe as they appeared to be, Lamy said.

This article refers to "statistics" and makes the case that serious economic issues may be affected when the results of statistical studies are distorted. Statistics is the science of gathering and analyzing data with the objective of either describing the data or drawing conclusions from them. *Descriptive statistics* refers to organizing data in a form that can be more easily understood. *Inferential statistics* refers to the process of drawing conclusions from the data, along with some estimate of how accurate those conclusions are. In this chapter, we will examine both descriptive and inferential statistics in various contexts, including polling and medical research.

6.1 Data summary and presentation: Boiling down the numbers

TAKE AWAY FROM THIS SECTION

Know the statistical terms used to summarize data.

Usually, it is very difficult to make sense of large amounts of raw data unless they are organized and summarized in some coherent way. This summary is often given in terms of a few key numbers.

The following article is excerpted from a progress report of the San Diego Achievement Forum, which calls itself "a network of researchers and higher education leaders with a strong interest in K–12 student performance." It presents test scores in terms of *quartiles*, which provide a way of summarizing data.

IN THE NEWS 6.2

Achievement in San Diego
A Progress Report
THE SAN DIEGO ACHIEVEMENT FORUM   2002

The progress being realized by the [San Diego City School] District is also evident in the multi-year movement of its students from lower to higher quartiles of student performance on the Stanford-9 [test]. For example, many students have moved out of the lowest quartile of performance over the past four years in the subject area of Reading. Between 1998 and 2002 the District has reduced the percentage of students scoring in the bottom quartile in this subject area by nine percentile points. . . . Over the past four years the percentage of students scoring in the lowest quartile in mathematics has declined by 11 percentile points, while the percentage of students in the top quartile has risen from 24 percent to 31 percent.

Mean, median, and mode

The *mean* (also called the *average*) and the *median*, the middle data point, are the most common ways of summarizing a data set using a single number. The *mode*, the most frequently occurring data point, is sometimes used as well.

These numbers give us a "representative" sense of the data. They are defined as follows.

Key Concept

The **mean** (also called the **average**) of a list of numbers is the sum of the numbers divided by the number of entries in the list.

The **median** of a list of numbers is the middle number. It is obtained by first ordering the data from smallest to largest. If there is an odd number of data points, pick the middle value. If there is an even number of data points, take the average of the middle two numbers.

If a number in a data set occurs more frequently than any other number, it is called the **mode** of the data. If there are two such numbers, the data set is called **bimodal**. If there are more than two such numbers, the data set is **multimodal**.

The mean is the usual average you expect your teacher to calculate from test scores: Add the scores and divide by the number of exams. For example, if your test scores are

$$60, 90, 70, 60, 80,$$

the mean is

$$\frac{60 + 90 + 70 + 60 + 80}{5} = \frac{360}{5} = 72.$$

To find the median, we order the test scores from smallest to largest and pick out the middle score:

$$60, 60, \mathbf{70}, 80, 90.$$

Thus, the median score is 70. Suppose the list included one more score, an 86. Then there would have been two middle scores:

$$60, 60, \mathbf{70}, \mathbf{80}, 86, 90.$$

In this case, we find the median by averaging the middle two scores:

$$\text{Median} = \frac{70 + 80}{2} = 75.$$

The mode is 60 because that score occurs more than any of the others.

Often data are presented using a *frequency table*, which shows how often each data point occurs. Frequency tables can be used to calculate the mean, median, and mode, as the next example illustrates.

EXAMPLE 6.1 Calculating mean, median, and mode: Chelsea FC

The Chelsea Football Club (FC) is a British soccer team. The accompanying table shows the goals scored (by either team) in the games played by Chelsea FC between September 2007 and May 2008. The data are arranged according to the total number of goals scored in each game.

Goals scored by either team	0	1	2	3	4	5	6	7	8
Number of games	7	14	20	11	3	2	1	2	2

Find the mean, median, and mode for the number of goals scored per game. Round the mean to one decimal place.

Chelsea Football Club.

SOLUTION

To find the mean, we add the data values (the total number of goals scored) and divide by the number of data points. To find the total number of goals scored, for each entry we multiply the goals scored by the corresponding number of games. Then we add. For example, there were 20 games in which exactly 2 goals were scored, so these games contribute $20 \times 2 = 40$ goals. The total number of goals scored is

$$7 \times 0 + 14 \times 1 + 20 \times 2 + 11 \times 3 + 3 \times 4 + 2 \times 5 + 1 \times 6 + 2 \times 7 + 2 \times 8 = 145.$$

Now we find the number of data points, which is the total number of games:

$$7 + 14 + 20 + 11 + 3 + 2 + 1 + 2 + 2 = 62.$$

The mean is the total number of goals scored divided by the number of games played, so

$$\text{Mean} = \frac{145}{62}$$

or about 2.3.

Now we find the median. The number of games is 62, which is even, so we count from the bottom to find the 31st and 32nd lowest total goal scores. These are both 2, so

$$\text{Median} = 2.$$

Because 2 occurs most frequently as the number of goals (20 times), we have

$$\text{Mode} = 2.$$

Thus on average, the teams combined to score 2.3 goals per game. Half of the games had goals totaling 2 or more, and the most common number of goals scored in a Chelsea FC game was 2.

TRY IT YOURSELF 6.1

Find the mean, median, and mode of the data for the Chelsea FC, counting only games with no more than 2 goals scored. Round the mean to one decimal place.

The answer is provided at the end of this section.

One type of representative number may be preferred over another depending on the context. For example, the median is often used for home prices, and the mean is used for batting averages. The following example helps to explain this practice.

EXAMPLE 6.2 Choosing between mean and median: Home prices

The following list gives home prices (in thousands of dollars) in a small town:

$$80, 120, 125, 140, 180, 190, 820.$$

The list includes the price of one luxury home. Calculate the mean and median of this data set. (Round the mean to one decimal place.) Which of the two is more appropriate for describing the housing market?

SOLUTION

To calculate the mean we add the data values and divide by the number of data points:

$$\text{Mean} = \frac{80 + 120 + 125 + 140 + 180 + 190 + 820}{7} = \frac{1655}{7},$$

or about 236.4 thousand dollars. This is the average price of a home.

The list of seven prices is arranged in order, so the median is the fourth value, 140 thousand dollars.

Note that the mean is higher than the cost of every home on the market except for one—the luxury home. The median of 140 thousand dollars is more representative of the market.

TRY IT YOURSELF 6.2

To calculate the batting average for a baseball player, we record a 1 when a batter gets a hit and a 0 when he does not. The batting average is the average of this list of 1's and 0's. (This is the number of hits divided by the number of at-bats.) The "batting median" for a player would be the median of the list of 1's and 0's recorded for a batter. All major league baseball players get hits less than half the time they go to bat. Show that this median for a major league baseball player is 0, and explain why the batting median would not be a good way to represent the batting record.

The answer is provided at the end of this section.

An *outlier* is a data point that is significantly different from most of the others. For example, the luxury home in Example 6.2 represents an outlier.

Key Concept

An outlier is a data point that is significantly different from most of the data.

The skewing of average home prices by outliers is illustrated in the following article from the Web site of the *New York Sun*.

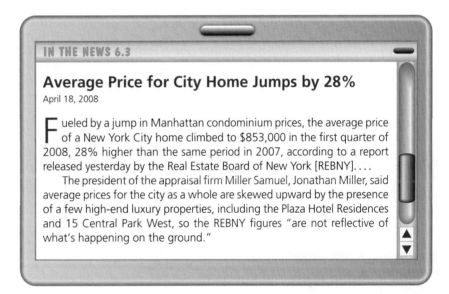

IN THE NEWS 6.3

Average Price for City Home Jumps by 28%

April 18, 2008

Fueled by a jump in Manhattan condominium prices, the average price of a New York City home climbed to $853,000 in the first quarter of 2008, 28% higher than the same period in 2007, according to a report released yesterday by the Real Estate Board of New York [REBNY]. . . .

The president of the appraisal firm Miller Samuel, Jonathan Miller, said average prices for the city as a whole are skewed upward by the presence of a few high-end luxury properties, including the Plaza Hotel Residences and 15 Central Park West, so the REBNY figures "are not reflective of what's happening on the ground."

The five-number summary

We have seen that the mean, median, and mode can often provide helpful summary information about data. But a single value is limited in what it can convey and can even be misleading. For instance, in Example 6.2 the median gives us some information about housing prices. But that number does not tell us anything about the spread of the data. Are most of the house prices clustered close to the median, or are they spread out over a wide range of prices?

A device called the *five-number summary* is helpful in answering such questions. In order to explain the five-number summary, we first define what is meant by a *quartile*.

Key Concept

- The **first quartile** of a list of numbers is the median of the lower half of the numbers in the list.

- The **second quartile** is the same as the median of the list.

- The **third quartile** is the median of the upper half of the numbers in the list.

If the list has an even number of entries, it is clear what we mean by the "lower half" and "upper half" of the list. If the list has an odd number of entries, eliminate the median from the list. This new list now has an even number of entries. The "lower half" and "upper half" refer to this new list.

We note that in everyday usage we may speak of a number that is *in the first quartile*. We mean that it is less than the first quartile value, which means that it is in the lowest one-quarter of the data. We say that a number is *in the second quartile* if it is between the first quartile value and the median—in the next-to-lowest one-quarter of the data. Similarly, being *in the third quartile* means being in the next-to-highest one-quarter of the data, and being *in the fourth quartile* means being in the highest one-quarter of the data. In the News 6.2, which opened this section, used these terms.

We use quartiles to make the *five-number summary* for data.

Key Concept

The **five-number summary** of a list of numbers consists of the minimum, the first quartile, the median, the third quartile, and the maximum.

The five-number summary does a much better job of describing data than any single number could. If the five numbers are relatively close together, the summary tells us that the data are fairly closely bunched about the median. If, on the other hand, the five numbers are widely separated, we know that the data are spread out as well.

EXAMPLE 6.3 Calculating the five-number summary: Income of celebrities

Each year *Forbes* magazine publishes a list it calls the Celebrity 100, which "includes film and television actors, models, chefs, athletes, authors and musicians, [and] is a measure of entertainment-related earnings and media visibility (exposure in print, television, radio and online)." The accompanying table shows the top nine names on the list for 2009, ordered according to the ranking of *Forbes*. The table also gives the incomes of the celebrities between June 2008 and June 2009.

Oprah Winfrey.

Celebrity	Income (millions of dollars)
Angelina Jolie	27
Oprah Winfrey	275
Madonna	110
Beyonce Knowles	87
Tiger Woods	110
Bruce Springsteen	70
Steven Spielberg	150
Jennifer Aniston	25
Brad Pitt	28

Calculate the five-number summary for this list of incomes.

SOLUTION

First we arrange the incomes in order:

$$25 \quad 27 \quad 28 \quad 70 \quad 87 \quad 110 \quad 110 \quad 150 \quad 275.$$

The minimum is $25 million (Jennifer Aniston), the median (the fifth in order of the nine incomes) is $87 million (Beyonce Knowles), and the maximum is $275 million (Oprah Winfrey). The "lower half" of the list consists of the four numbers less than the median, which are

$$25 \quad 27 \quad 28 \quad 70.$$

The median of this lower half is 27.5, so the first quartile of incomes is $27.5 million. The "upper half" of the list consists of the four numbers greater than the median, which are

$$110 \quad 110 \quad 150 \quad 275.$$

The median of this upper half is 130, so the third quartile of incomes is $130 million. Thus, the five-number summary is

$$\text{Minimum} = \$25 \text{ million}$$
$$\text{First quartile} = \$27.5 \text{ million}$$
$$\text{Median} = \$87 \text{ million}$$
$$\text{Third quartile} = \$130 \text{ million}$$
$$\text{Maximum} = \$275 \text{ million.}$$

TRY IT YOURSELF 6.3

Calculate the five-number summary for the incomes in Example 6.3 if the celebrity ranked number 10, Kobe Bryant, is added to the list. His income was $45 million.

The answer is provided at the end of this section.

What does the five-number summary in Example 6.3 show? All of the incomes are between $25 million and $275 million. The median income is $87 million, so as many of the incomes are above that level as are below. The lowest one-quarter of all incomes are between $25 million and $27.5 million, the next one-quarter are between $27.5 million and the median of $87 million, the next one-quarter are between the median of $87 million and $130 million, and the highest one-quarter are between $130 million and $275 million. The summary provided by this breakdown gives a general sense of how the incomes are spread out.

In our example, the number of data points was small, but if we were working with 100,000 data points, for example, the five-number summary would be a big help.

Boxplots

There is a commonly used pictorial display of the five-number summary known as a *boxplot* (also called a *box and whisker diagram*). **Figure 6.1** shows the basic

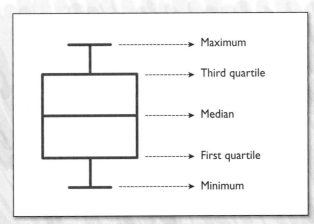

FIGURE 6.1 The basic boxplot diagram.

geometric figure used in a boxplot. Note that the bottom of the figure is the minimum and the top is the maximum. The box area represents the region from the first to the third quartile. (The width of the box has no significance.) The median is marked with a horizontal line inside the box. A greater length between the top and bottom "whiskers" indicates a greater spread from the maximum to the minimum, and a longer box indicates a greater spread from the first to third quartiles. The location of the median within the box gives information about how the data between the first and third quartile are distributed about the median.

EXAMPLE 6.4 Making and interpreting a boxplot: Gas mileage

A report on the greenercars.org Web site shows 2011-model cars that score well in terms of environmental impact. Here are the data, with mileage measured in miles per gallon (mpg) listed in order of the Web site's "green score."

Model	City mileage (mpg)	Highway mileage (mpg)
Toyota Prius	51	48
Honda Civic Hybrid	40	43
Honda CR-Z	35	39
Toyota Yaris	29	35
Audi A3	30	42
Hyundai Sonata	22	35
Hyundai Tucson	23	31
Chevrolet Equinox	22	32
Kia Rondo	20	27
Chevrolet Colorado/GMC Canyon	18	25

a. Find the five-number summary for city mileage.

b. Present a boxplot of city mileage.

c. Comment on how the data are distributed about the median. *Note:* The corresponding calculations for highway mileage are left as an exercise. See Exercise 8.

SOLUTION

a. The list for city mileage, in order from lowest to highest, is

$$18, 20, \mathbf{22}, 22, \underline{23, 29}, 30, \mathbf{35}, 40, 51.$$

The maximum is 51 mpg, and the minimum is 18 mpg. To find the median, we average the two numbers in the middle:

$$\text{Median} = \frac{23 + 29}{2} = 26 \text{ mpg}.$$

The lower half of the list is 18, 20, 22, 22, 23, and the median of this half is 22. Thus, the first quartile is 22 mpg. The upper half of the list is 29, 20, 35, 40, 51, and the median of this half is 35. Thus, the third quartile is 35 mpg.

b. The corresponding boxplot appears in **Figure 6.2**. The vertical axis is the mileage measured in miles per gallon.

c. Referring to the boxplot, we note that the first quartile is not far above the minimum, and the median is not far above the first quartile. The third quartile is well above the median, and the maximum is well above the third quartile. This

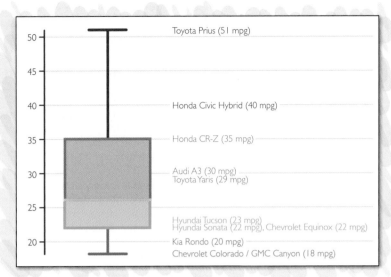

FIGURE 6.2 Boxplot for city mileage.

emphasizes the dramatic difference between the high-mileage cars (the hybrids) and ordinary cars.

Standard deviation

As we have noted, data sets can be clustered together (such as 10, 11, 12, 12, 13) or spread out (such as 1, 10, 20, 100, 300). One measure of how much data are spread out is the five-number summary. Another way to measure the spread of data is the *standard deviation*. This number is commonly denoted by the lowercase Greek letter σ (pronounced "sigma").

Key Concept

The standard deviation is a measure of how much the data are spread out from the mean. The smaller the standard deviation, the more closely the data are clustered about the mean.

For example, a data set of the yearly rainfall amounts in the Gobi Desert would show that values deviate very little from the mean because there is not a significant variation in rainfall from one year to the next. On the other hand, a data set of prices for new automobiles would have a wide range because high-end sports cars cost a great deal more than the average family automobile. For this data set, we would have a larger standard deviation. The standard deviation is used in many settings, and, like the mean, it is sensitive to outliers. The unit of measurement for the standard deviation is the same as the unit for the original data. For example, if the data show car prices in dollars, then the standard deviation is measured in dollars.

Deviation from the mean refers to the difference between data and the mean of the data. For example, let's say the mean price for a new automobile is \$22,000. If the price for a Mercedes is \$40,000, we say that the price deviates by \$18,000 from the mean because \$40,000 − \$22,000 = \$18,000. If the price for a Hyundai Elantra is \$15,000, we say that the price deviates by −\$7000 from the mean because \$15,000 − \$22,000 = −\$7000.

The following article from the Morningstar news site gives insight into the importance of the standard deviation.

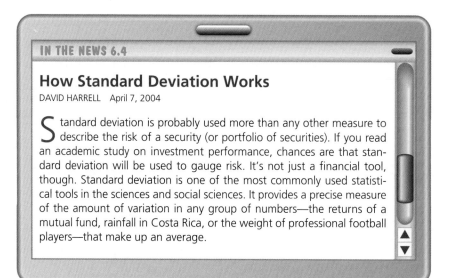

How Standard Deviation Works

DAVID HARRELL   April 7, 2004

Standard deviation is probably used more than any other measure to describe the risk of a security (or portfolio of securities). If you read an academic study on investment performance, chances are that standard deviation will be used to gauge risk. It's not just a financial tool, though. Standard deviation is one of the most commonly used statistical tools in the sciences and social sciences. It provides a precise measure of the amount of variation in any group of numbers—the returns of a mutual fund, rainfall in Costa Rica, or the weight of professional football players—that make up an average.

The formula for standard deviation looks intimidating. But, as we shall see, the actual calculation is not very difficult. Suppose the data points are

$$x_1, x_2, x_3, \ldots, x_n.$$

It is common practice to use the lowercase Greek letter μ (pronounced "mew") to denote the mean. The formula for the standard deviation is[1]

$$\sigma = \sqrt{\frac{(x_1 - \mu)^2 + (x_2 - \mu)^2 + \cdots + (x_n - \mu)^2}{n}}.$$

The template provided in the following Calculation Tip makes the process of calculating the standard deviation manageable.

CALCULATION TIP 6.1 Calculating Standard Deviation

To find the standard deviation of n data points, we first calculate the mean μ. The next step is to complete the following calculation template:

Data	Deviation	Square of deviation
$\vdots$	$\vdots$	$\vdots$
x_i	$x_i - \mu$	Square of second column
$\vdots$	$\vdots$	$\vdots$
		Sum of third column
		Divide the above sum by n
		and take the square root.

---

[1] There are actually two common definitions of the standard deviation of a data set. One is the **population standard deviation**, denoted by σ and defined as we have it here. The other is the **sample standard deviation**, denoted by s and defined in the same way except that the denominator n is replaced by $n - 1$. The sample standard deviation s is used when the data consist of a sample from a larger population. The population standard deviation σ is used when the data are not a sample but consist of the entire "population" of data. For large data sets, their ratio is very close to 1 (so they are nearly the same). The reason statisticians have different ways of defining these is due to a technical point about using a sample standard deviation as an "estimator" for the population standard deviation.

"It's the new keyboard for the statistics lab. Once you learn how
to use it, it will make computation of the standard deviation easier."

The procedure for calculating standard deviation is illustrated in the next example.

EXAMPLE 6.5 Calculating standard deviation: Baseball pitchers

Two leading pitchers in Major League Baseball for 2010 were Roy Halladay of the Philadelphia Phillies (Toronto Blue Jays before 2010) and Felix Hernandez of the Seattle Mariners. Their ERA (Earned Run Average—the lower the number, the better) histories are given in the table below.

Pitcher	ERA 2006	ERA 2007	ERA 2008	ERA 2009	ERA 2010
R. Halladay	3.19	3.71	2.78	2.79	2.44
F. Hernandez	4.52	3.92	3.45	2.49	2.27

Calculate the mean and the standard deviation for Hallady's ERA history. It turns out that the mean and standard deviation for Hernandez's ERA history are $\mu = 3.33$ and $\sigma = 0.85$. What comparisons between Halladay and Hernandez can you make based on these numbers?

SOLUTION

The mean for Halladay is

$$\mu = \frac{3.19 + 3.71 + 2.78 + 2.79 + 2.44}{5}.$$

This is about 2.98.

ERA x_i	Deviation $x_i - 2.98$	Square of deviation $(x_i - 2.98)^2$
3.19	$3.19 - 2.98 = 0.21$	$(0.21)^2 = 0.044$
3.71	$3.71 - 2.98 = 0.73$	$(0.73)^2 = 0.533$
2.78	$2.78 - 2.98 = -0.20$	$(-0.20)^2 = 0.040$
2.79	$2.79 - 2.98 = -0.19$	$(-0.19)^2 = 0.036$
2.44	$2.44 - 2.98 = -0.54$	$(-0.54)^2 = 0.292$
Sum of third column		0.945
Sum divided by $n = 5$, square root		$\sigma = \sqrt{0.945/5} = 0.43$

We make the calculation of the standard deviation using Calculation Tip 6.1. The first column shows the data points. We find the second column by subtracting the mean $\mu = 2.98$ to find the deviations. We calculate the third column by squaring the second column to find the squares of the deviations, which we round to three decimal places.

We conclude that the mean and the standard deviation for Halladay's ERA history are $\mu = 2.98$ and $\sigma = 0.43$.

Because Halladay's mean is smaller than Hernandez's mean of $\mu = 3.33$, over this period Halladay had a better pitching record. Also, Halladay's ERA had a smaller standard deviation than that of Hernandez (who had $\sigma = 0.85$), so Halladay was more consistent—his numbers are not spread as far from the mean.

TRY IT YOURSELF 6.5

Verify that the mean and the standard deviation for Hernandez's ERA history are $\mu = 3.33$ and $\sigma = 0.85$.

The answer is provided at the end of this section.

In the next example, we use the concepts from this section to compare two data sets.

EXAMPLE 6.6 Using standard deviation: Free-throw percentages

Below is a table showing the Eastern Conference NBA team free-throw percentages at home and away for the 2007–2008 season. At the bottom of the table, we have displayed the mean and standard deviation for each data set. You will be asked to verify these values in Exercise 14.

Team	Free-throw percentage at home	Free-throw percentage away
Toronto	81.2	77.6
Washington	78.2	75.4
Atlanta	77.2	75.2
Boston	77.1	74.3
Indiana	76.8	75.7
Detroit	76.7	74.4
Chicago	75.6	76.6
New Jersey	73.6	76.8
Milwaukee	73.3	76.6
Miami	72.7	75.5
New York	72.7	73.9
Orlando	72.1	75.4
Cleveland	71.7	74.8
Charlotte	71.4	74.7
Philadelphia	70.6	77.2
Mean	74.73	75.61
Standard deviation	2.95	1.09

What do these values for the mean and standard deviation tell us about free-throws shot at home compared with free-throws shot away from home? Does comparison of the minimum and maximum of each of the data sets support your conclusions?

SOLUTION

The means for free-throw percentages are 74.73 at home and 75.61 away, so on average the teams do somewhat better on the road than at home. This result is perhaps surprising. The standard deviation for home is 2.95 percentage points, which is considerably larger than the standard deviation of 1.09 percentage points away from home. This means that the free-throw percentages at home vary from the mean much more than the free-throw percentages away.

The difference between the maximum and minimum percentages shows the same thing: The free-throw percentages at home range from 70.6 to 81.2%, and the free-throw percentages away range from 73.9% to 77.6%. The plots of the data in **Figures 6.3 and 6.4** provide a visual verification that the data for home free-throws are more broadly dispersed than the data for away free-throws.

Histograms

In addition to numerical measures of distributions of data, such as the five-number summary and the standard deviation, it is often very useful to visualize a distribution of data, as we did with the boxplot. In cases where we are dealing with frequencies, that is, how many times an item occurs, bar graphs as described in Section 2.1 are useful representations.

Consider again the data in Example 6.1 showing goals scored in the games played by Chelsea FC between September 2007 and May 2008:

Goals scored by either team	0	1	2	3	4	5	6	7	8
Number of games	7	14	20	11	3	2	1	2	2

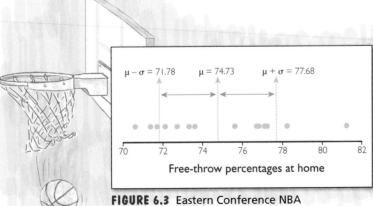

FIGURE 6.3 Eastern Conference NBA free-throw percentages at home.

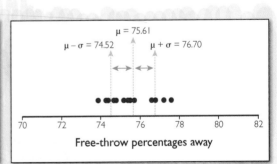

FIGURE 6.4 Eastern Conference NBA free-throw percentages away.

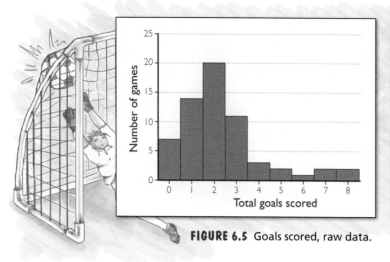

FIGURE 6.5 Goals scored, raw data.

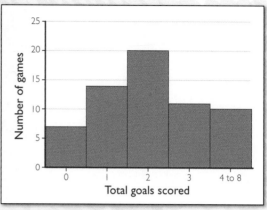

FIGURE 6.6 Goals scored, grouped data.

The two bar graphs in **Figures 6.5 and 6.6** show two ways to represent the data on frequencies of goals scored. The first graph uses the raw data. For the second graph, because there are relatively few games with a high number of goals scored, we group the highest score totals. In the category of 4 to 8 goals scored, there are $3 + 2 + 1 + 2 + 2 = 10$ games altogether.

Bar graphs with grouped data are especially useful in visualizing distributions of large data sets. These graphs are called *histograms*.

Key Concept

A histogram is a bar graph that shows the frequencies with which certain data occur.

Often we use histograms to reduce data to manageable chunks. Suppose, for example, we toss 1000 coins and write down the number of heads we got. We do this experiment a total of 1000 times. The accompanying table shows one part of the results from doing these experiments using a computer simulation.

Number of heads	451	457	458	459	461	462	463	464	465	467
Number of tosses (out of 1000)	2	2	1	3	3	2	1	3	1	1

The first entry shows that twice we got 451 heads, twice we got 457 heads, once we got 458 heads, and so on. Note that the entire table would include all possibilities from 0 through 1000. The raw data are a bit hard to digest because there are so many data points. The five-number summary provides one way to analyze the data. An alternative way to get a grasp of the data is to arrange them in groups—rather than to look at all the individual data points—and then draw a histogram.

For example, it turns out that the number of tosses yielding fewer than 470 heads is 23. Because 470 out of 1000 is 47%, this means that 23 tosses yielded less than 47% heads. We find the accompanying table by dividing the data into groups this way. (Here, "47% to 48%" means 470 through 479 heads, "48% to 49%" means 480 through 489 heads, and so on.)

Percent heads	Less than 47%	47% to 48%	48% to 49%	49% to 50%
Number of tosses	23	75	140	234

Percent heads	50% to 51%	51% to 52%	52% to 53%	At least 53%
Number of tosses	250	157	94	27

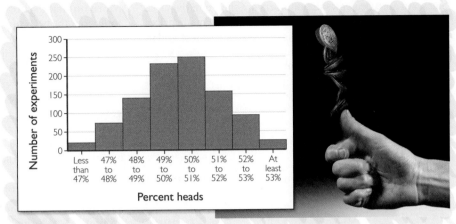

FIGURE 6.7 A histogram of coin tosses.

Figure 6.7 shows a histogram for this grouping of the data. This visual presentation makes the situation even clearer. We can clearly see that the vast majority of the tosses were between 47% and 53% heads. Note also how the number of heads increases near the 50% mark. This is what we would expect from tosses of fair coins.

Try It Yourself answers

Try It Yourself 6.1: Calculating mean, median, and mode: Chelsea FC   Mean = 1.3; median = 1; mode = 2.

Try It Yourself 6.2: Choosing between mean and median: Batting average   The median is 0 because there are more 0's than 1's in the list. This median doesn't distinguish between players.

Try It Yourself 6.3: Calculating the five-number summary: Income of celebrities

- Minimum = $25 million

- First quartile = $28 million

- Median = $78.5 million

- Third quartile = $110 million

- Maximum = $275 million

Try It Yourself 6.5: Calculating standard deviation: Baseball pitchers   Answers will vary.

Exercise Set 6.1

1. Homework scores. Your first eight homework scores are 92, 86, 78, 85, 95, 81, 88, and 90. Find the mean, median, and mode of these scores. Round the mean to one decimal place.

2. More on homework scores. *This is a continuation of Exercise 1.* Your first eight homework scores are listed in Exercise 1. Your ninth score is 15. (It was a bad week.) Find the mean and median of these nine scores. (Again, round the mean to one decimal place.) How would dropping the lowest score affect the mean and the median?

3. Exam scores. Suppose that one student's four exam scores for the semester are 75, 75, 75, and 75. Suppose that another student recorded scores of 100, 0, 100, 100. What are the mean and median scores for these two students? What grade does

each one earn if the cutoffs are A-90%, B-80%, C-70%, and D-60%, as determined by the mean?

It is debatable whether these students deserve the same grade. Give a good argument for why the second student deserves a "C" and a good argument for why the second student deserves a higher grade.

4. Home prices. Suppose we gathered the following list of home prices:

$80,000 $120,000 $120,000 $120,000 $122,000
$135,000 $135,000 $150,000 $160,000

Find the mean, median, and mode of these home prices (to the nearest thousand).

5. Sales. An auto dealer's sales numbers are shown in the table below. Find for each month the mean, median, and mode prices of the cars she sold. Round your answers to the nearest dollar.

	Number sold		
Price	May	June	July
$20,000	22	25	24
$15,000	49	24	24
$12,500	25	49	49

6. Land areas

a. Calculate the five-number summary of the land areas of the states in the U.S. Midwest.

State	Area (sq. miles)	State	Area (sq. miles)
Illinois	55,584	Missouri	68,886
Indiana	35,867	Nebraska	76,872
Iowa	55,869	North Dakota	68,976
Kansas	81,815	Ohio	40,948
Michigan	56,804	South Dakota	75,885
Minnesota	79,610	Wisconsin	54,310

b. Explain what the five-number summary in part a tells us about the land areas of the states in the Midwest.

c. Calculate the five-number summary of the land areas of the states in the U.S. Northeast.

State	Area (sq. miles)	State	Area (sq. miles)
Connecticut	4845	New York	47,214
Maine	30,862	Pennsylvania	44,817
Massachusetts	7840	Rhode Island	1045
New Hampshire	8968	Vermont	9250
New Jersey	7417		

d. Explain what the five-number summary in part c tells us about the land areas of the states in the Northeast.

e. Contrast the results from parts b and d.

7. Boxplot. A boxplot for a data set is shown in **Figure 6.8**. Estimate the minimum, first quartile, median, third quartile, and maximum of the data set.

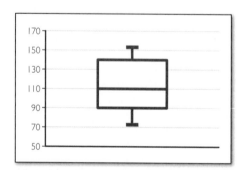

FIGURE 6.8 Boxplot.

8. Highway mileage. Find the five-number summary for the data on highway mileage in Example 6.4, and draw the corresponding boxplot.

More on math scores In the News 6.2 said, "Over the past four years the percentage of students scoring in the lowest quartile in mathematics has declined by 11 percentile points, while the percentage of students in the top quartile has risen from 24 percent to 31 percent." (Presumably, this refers to quartiles in a larger group of test-takers and not just the San Diego schools.) This sentence seems to be suggesting that things are going well. The next two exercises refer to this suggestion. In these exercises, we write T1 for the percentage of students in the lowest quartile (i.e., in the first quartile), T2 for the percentage of students in the second quartile, T3 for the percentage of students in the third quartile, and T4 for the percentage of students in the top quartile (i.e., in the fourth quartile).

9. Suppose that the original numbers were T1 = 25%, T2 = 25%, T3 = 26%, and T4 = 24%, and that four years later the numbers were T1 = 14%, T2 = 5%, T3 = 50%, and T4 = 31%. Is this an improvement? Is there a possible downside to the new numbers in comparison to the original ones?

10. Suppose that the original numbers were T1 = 25%, T2 = 25%, T3 = 26%, and T4 = 24%, and that four years later the numbers were T1 = 14%, T2 = 50%, T3 = 5%, and T4 = 31%. Is this an improvement? Is there a possible downside to the new numbers in comparison to the original ones?

11. Mathematics salaries. The first three columns of the following table are salary figures taken from an American Mathematical Society survey of mathematics faculty from Ph.D.-granting public institutions. The last two columns in the table (marked with an asterisk) were not reported by the AMS, so for the purposes of this exercise, we have just made up the (plausible) numbers in these columns.

Academic rank	First quartile	Median	Third quartile	Minimum*	Maximum*
Assistant professor	$57,400	$60,160	$63,810	$50,000	$65,000
Associate professor	60,650	67,140	74,010	55,000	80,000
Professor	83,770	98,490	120,510	75,000	130,000

a. Make a boxplot for each of the three categories.

b. Use the boxplots you made in part a to compare the dispersal of salaries among the three categories.

12. Employee absence. A certain company recorded the number of employee absences each week over a period of 10 weeks. The result is the data list 3, 5, 1, 2, 2, 4, 7, 4, 5, 5. Find the mean and standard deviation of the number of absences per week. Round the standard deviation to two decimal places.

13. Calculation. Find the mean and standard deviation of the following list of quiz scores: 75, 88, 65, 90. Round the standard deviation to two decimal places.

14. Free-throws. In the situation of Example 6.6, verify that the means for free-throws at home and away are 74.73% and 75.61%, respectively. Verify that the standard deviations for free-throws at home and away are 2.95 percentage points and 1.09 percentage points, respectively.

15. No calculation. Suppose every student in your class got a score of 85% on an exam. Find the mean and standard deviation of those exam scores by inspection without doing any calculations.

16. A challenge problem. A class has an equal number of boys and girls. The boys all got 78% on a test and the girls all got 84%. What are the mean and standard deviation of the test scores for the entire class?

17. Another challenge problem. A class has 9 boys and 21 girls. The class as a whole has a GPA (grade point average) of 2.96, and the boys have a GPA of 2.40. What is the GPA of the girls?

18. Tossing a die

   a. Perform the following experiment: Roll a fair die six times, and record a 1 when the die comes up 6 and a 0 otherwise. Fill in the blanks in the following table with your results:

Roll	1	2	3	4	5	6
Data						

   b. Calculate the mean and standard deviation of the data list you made in part a. Leave the mean as a fraction, and round the standard deviation in decimal form to two decimal places.

Spread   We have seen that the standard deviation σ measures the spread of a data set about the mean μ. *Chebyshev's inequality* gives an estimate of how well the standard deviation measures that spread. One consequence of this inequality is that for every data set at least 75% of the data points lie within two standard deviations of the mean, that is, between $\mu - 2\sigma$ and $\mu + 2\sigma$ (inclusive). For example, if $\mu = 20$ and $\sigma = 5$, then at least 75% of the data are at least $20 - 2 \times 5 = 10$ and at most $20 + 2 \times 5 = 30$. Exercises 19 through 21 illustrate this result.[2]

19. Consider the following data: 5, 10, 10, 10, 10, 10, 10, 15. Find the mean and the standard deviation. How many data points does the Chebyshev inequality promise will lie within two standard deviations of the mean?

20. Our statement of this result says "at least 75%," not "exactly 75%." What percent of the data points in Exercise 19 actually lie within two standard deviations of the mean?

21. An application. We have 1200 lightbulbs in our building. Over a 10-month period, we record the number of bulbs that burn out each month. The result is the data list 23, 25, 21, 33, 17, 39, 26, 24, 31, 22.

   a. What is the average number of bulbs that burn out each month?

   b. What is the standard deviation of these data? Round the standard deviation to one decimal place.

   c. Use Chebyshev's inequality and your answer to parts a and b to estimate how many replacement bulbs you should keep on hand so that for at least 75% of the months you don't have to acquire additional replacement bulbs.

22. Histogram. Consider an experiment in which we toss 20 coins. We perform this experiment 100 times and record

the number of experiments in which we got a given number of heads. The results are given in the following table:

Number of heads	5	6	7	8	9	10	11	12	13	14	15
Number of experiments	3	5	8	11	16	18	16	13	6	2	2

Arrange the data in the following groups: less than 35% heads, at least 35% but less than 45% heads, at least 45% but less than 55% heads, at least 55% but less than 65% heads, at least 65% heads. (Here, of course, 35% refers to 35% of the 20 coins, etc.) Make a histogram showing the grouped data.

Exercises 23 and 24 are suitable for group work.

23. Your class. Collect the heights of each of your classmates. Find the mean and standard deviation of these data and make a five-number summary and box plot of it.

24. Experiment. Do the following experiment: Toss 20 coins. Perform this experiment 25 times and record in the table below the number of experiments in which you got a given number of heads.

Number of heads	0	1	2	3	4	5	6	7	8	9	10
Number of experiments											

Number of heads	11	12	13	14	15	16	17	18	19	20
Number of experiments										

Arrange the data in the following groups: less than 35% heads, at least 35% but less than 45% heads, at least 45% but less than 55% heads, at least 55% but less than 65% heads, at least 65% heads. (Here, of course, 35% refers to 35% of the 20 coins, etc.) Make a histogram showing the grouped data.

25. History. The term *standard deviation* was coined by Karl Pearson (1857–1936), who was the founder of the world's first university statistics department. Write a brief biography of this colorful and controversial figure.

26. History. Pafnuty Lvovich Chebyshev (1821–1894) was a Russian mathematician. Write a brief biography, and investigate *Chebyshev's inequality*, sometimes called *Chebyshev's theorem*, which gives a more precise statement of how well the standard deviation measures the spread of a data set. (See Exercises 19 through 21 above.)

The following exercises are designed to be solved using technology such as calculators or computer spreadsheets. For assistance, see the technology supplement.

27. Standard deviations in Excel©. In this section, we discussed the fact that there are two versions of the standard deviation, one referred to as the "sample standard deviation" and the other as the "population standard deviation." In the Excel© spreadsheet, locate the formulas for these and use them to find both versions of the standard deviations for this list of home prices: $80,000 $120,000 $120,000 $120,000 $122,000 $135,000 $135,000 $150,000 $160,000.

---

[2]In the next section, we will see similar, but more precise, statements for special data sets, namely those that are *normally distributed*.

28. Computer simulation. Consider an experiment in which we toss 1000 coins. We perform the experiment 1000 times and record the percentage of heads for each experiment. Use a computer simulation to carry out this procedure. (This was done to get the data on coin tosses given on p. 362 in connection with the topic of histograms.) Report your results as a histogram. How does your histogram compare with that shown in Figure 6.7?

29. Five-number summary. *This is a continuation of Exercise 28.* Consider the data on coin tosses you found in Exercise

28. Use technology to find the five-number summary for the percentage of heads obtained for each experiment.

30. Another computer simulation. Consider an experiment in which we roll 1000 standard dice. We perform the experiment 1000 times and record for each experiment the percentage of dice coming up 6. Use a computer simulation to carry out this procedure. Report your results as a histogram. How does the shape of your histogram compare with that shown in Figure 6.7?

6.2 The normal distribution: Why the bell curve?

TAKE AWAY FROM THIS SECTION

Understand why the normal distribution is so important.

The following article appears at the Medscape Today Web site.

IN THE NEWS 6.5

Why Does Chutes and Ladders Explain Hemoglobin Levels?

Some Thoughts on the Normal Distribution

ANDREW J. VICKERS   September 7, 2006

Chutes and Ladders is a bit like a coin flip, in that there is exactly a 50:50 chance that I'll win a game. If you tell me that we are going to play a certain number of games, I can tell you the probability of each possible combination of wins and losses. As an easy example, if I play 2 games of Chutes and Ladders with my son, there is a 25% chance that I'll lose both, a 50% chance that we'll each win one, and a 25% chance that he'll throw a hissy fit. I can show this as a graph: The y-axis gives the probability that I'll win each particular number of games shown on the x-axis [**Figure 6.9**]. The math is a bit more complicated for 4 games, but as it turns out, there is a 37.5% chance that we split it with 2 games each, and a 6.25% chance of a total meltdown [**Figure 6.10**].

Something that you may notice here is that this second graph is starting to look a little bit like the bell-shaped curve that is usually described as the "normal" distribution. Now let's imagine a really wet weekend in which I play 100 games of Chutes and Ladders [**Figure 6.11**].

We now have something that really looks like a normal distribution. We also have something that looks very much like many natural biological phenomena. As an example, this is the distribution of hemoglobin in a cohort of Swedish men aged 40–50 taking part in a heart study [**Figure 6.12**].

If you concluded that the blood of middle-aged Swedes depended on games of Chutes and Ladders, you wouldn't be far wrong: Like the outcome of a dice-throwing game, a man's hemoglobin level is the result of numerous chance events—genes, environment, diet, lifestyle, and medical history—all added together. When you add up a lot of chance events, what you get is a normal distribution.

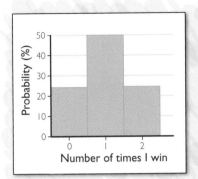

FIGURE 6.9 Graph for 2 games. The y-axis gives the probability that the author will win each particular number of games shown on the x-axis.

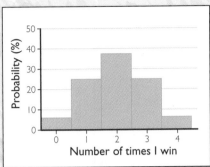

FIGURE 6.10 Graph for 4 games indicates a 37.5% chance that author and son split wins with 2 games each, and a 6.25% chance of a meltdown by author's son.

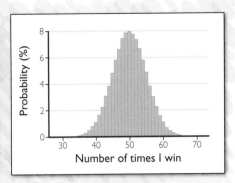

FIGURE 6.11 Graph for 100 games approaches normal distribution.

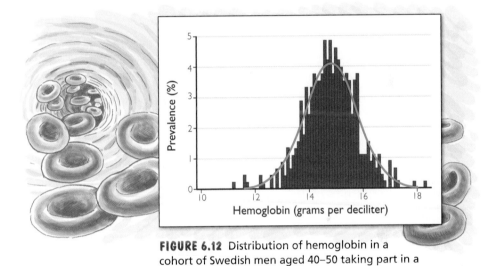

FIGURE 6.12 Distribution of hemoglobin in a cohort of Swedish men aged 40–50 taking part in a heart study.

The *distribution* of a set of data measures the frequency with which each data point occurs. In the preceding section, we introduced the histogram as a way of visualizing the way data are distributed in terms of frequency. Note that each of the graphs in Figure 6.9 through Figure 6.12 is a histogram.

In the News 6.5 refers to a very important distribution, the *normal distribution*. Results of many games of chance as illustrated in Figure 6.11 and hemoglobin levels in Swedish men as illustrated in Figure 6.12 provide examples of (approximately) normally distributed data, but, as the article suggests, many other common data sets are normally distributed as well. In this section, we see the importance of the normal distribution for both descriptive and inferential statistics, as defined at the beginning of this chapter: The normal distribution provides a very efficient way of organizing many types of data, and the Central Limit Theorem about the normal distribution is a powerful tool for drawing conclusions from data based on sampling.

The bell-shaped curve

Figure 6.13 shows the distribution of heights of adult males in the United States. A graph shaped like this one resembles a bell—thus, its common name, the *bell curve*. This bell-shaped graph is typical of normally distributed data.

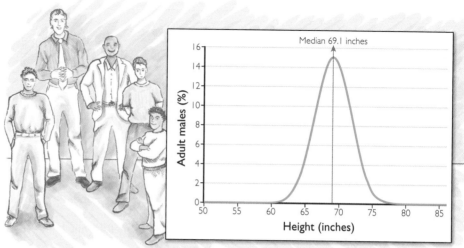

FIGURE 6.13 Heights of adult males are normally distributed.

Following are some important features of normally distributed data. These features are expressed in terms of the *mean* and *median*, which we studied in the preceding section.

The mean and median are the same: For normally distributed data, the mean and median are the same. Figure 6.13 indicates that the median height of adult males is 69.1 inches, or about 5 feet 9 inches. That means about half of males are taller than 69.1 inches and about half are shorter. Because the mean and median are the same, the average height of adult males is also 69.1 inches.

Most data are clustered about the mean: This is reflected in the fact that the vast majority of adult males are within a few inches of the mean. In fact, about 95% of adult males are within 5 inches of the mean—between 5 feet 4 inches and 6 feet 2 inches. This is shown in **Figure 6.14.** Only about 5% are taller or shorter than this. This is illustrated in **Figure 6.15.**

The bell curve is symmetric about the mean: This is reflected in the fact that the curve to the left of the mean is a mirror image of the curve to the right of the mean. In terms of heights, it means that there are about the same number of men 2 inches taller than the mean as there are men 2 inches shorter than the mean, and there are about the same number of men 6 feet 4 inches in height (7 inches above the mean) as men 5 feet 2 inches in height (7 inches below the mean). This is illustrated in **Figure 6.16.**

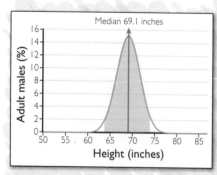

FIGURE 6.14 95% of adult males are within 5 inches of the median.

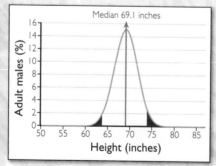

FIGURE 6.15 Relatively few men are very tall or very short.

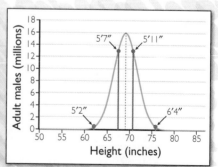

FIGURE 6.16 The bell curve is symmetric about the mean.

Key Concept

If data are **normally distributed:**

1. Their graph is a bell-shaped curve.

2. The mean and median are the same.

3. Most of the data tend to be clustered relatively near the mean.

4. The data are symmetrically distributed above and below the mean.

"There are lies, damn lies, and statistics. We're
looking for someone who can make all three
of these work for us."

Now we consider examples illustrating these properties.

EXAMPLE 6.7 Determining normality from a graph: IQ and income

Figure 6.17 shows the distribution of IQ scores, and **Figure 6.18** shows the percentage of American families and level of income. Which of these data sets appear to be normally distributed, and why?

SOLUTION

The IQ scores appear to be normally distributed because they are symmetric about the median score of 100, and most of the data are relatively close to this value. Family incomes do not appear to be normally distributed because they are not symmetric.

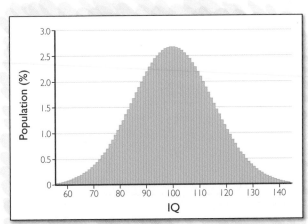

FIGURE 6.17 Percentage of population with given
IQ score.

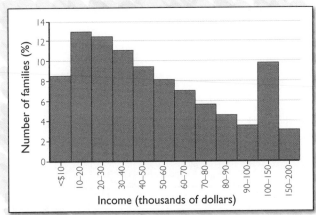

FIGURE 6.18 Percentage of families with given
income.

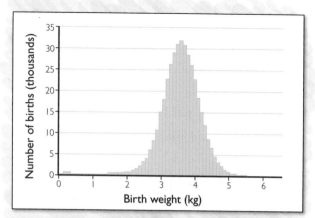

FIGURE 6.19 Number of births of a given weight in Norway.

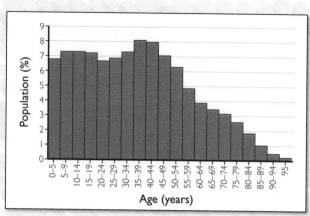

FIGURE 6.20 Percentage of Americans of a given age.

They are skewed toward the lower end of the scale, meaning there are many more families with low incomes than with high incomes.

TRY IT YOURSELF 6.7

Figure 6.19 shows birthweights of Norwegian children, and **Figure 6.20** shows the distribution of ages in America in 2000. Which of these data sets appears to be normally distributed, and why?

The answer is provided at the end of this section.

Mean and standard deviation for the normal distribution

Recall that in the preceding section we introduced the *standard deviation* for a set of data. We noted there that the standard deviation is a measure of the spread of the data about the mean. It turns out that for a normal distribution the mean and standard deviation completely determine the bell shape for the graph of the data. Let's see why.

The mean determines the middle of the bell curve—where it peaks—and the standard deviation determines how steep the curve is. A large standard deviation results in a very wide bell, and a small standard deviation results in a thin, steep bell. **Figures 6.21 and 6.22** both show a bell curve with mean 500. In Figure 6.21, the

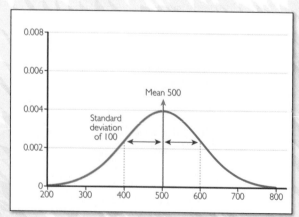

FIGURE 6.21 Normal curve with mean 500 and standard deviation 100.

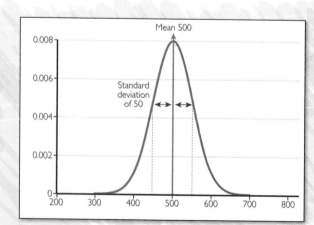

FIGURE 6.22 Normal curve with mean 500 and standard deviation 50.

standard deviation is 100, and the data are spread over a relatively wide range. In Figure 6.22, the standard deviation is 50. This smaller standard deviation reflects the fact that the data are bunched more tightly about the mean.

The following rule of thumb gives more information on the role of the standard deviation in the normal distribution.

RULE OF THUMB 6.1 Normal Data: 68-95-99.7% Rule

If a set of data is normally distributed:

• About 68% of the data lie within one standard deviation of the mean (34% within one standard deviation above the mean and 34% within one standard deviation below the mean). See **Figure 6.23**.

• About 95% of the data lie within two standard deviations of the mean (47.5% within two standard deviations above the mean and 47.5% within two standard deviations below the mean). See **Figure 6.24**.

• About 99.7% of the data lie within three standard deviations of the mean (49.85% within three standard deviations above the mean and 49.85% within three standard deviations below the mean). See **Figure 6.25**.

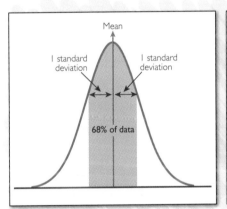

FIGURE 6.23 68% of data lie within one standard deviation of the mean.

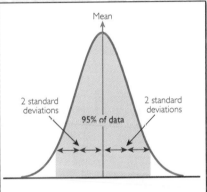

FIGURE 6.24 95% of data lie within two standard deviations of the mean.

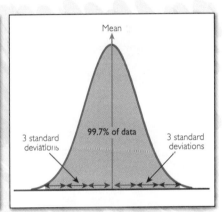

FIGURE 6.25 99.7% of the data lie within three standard deviations of the mean.

To illustrate the use of this rule of thumb, we consider scores on standardized tests. Most students are quite familiar with the ACT and SAT tests, which are widely used in college admissions. The data set in **Table 6.1** on the verbal section of the SAT Reasoning Test was obtained from the College Board Web site.

In **Figure 6.26**, we see a histogram of these data. The bell curve associated with the data is shown in **Figure 6.27**. These figures suggest that SAT verbal scores are normally distributed. This is indeed the case, and according to Table 6.1, the mean verbal score on the SAT is 507, and the standard deviation is 111. Applying the 68-95-99.7% rule to SAT verbal scores gives the following information:

• 68% of SAT scores are within one standard deviation of the mean. This means that 68% of scores are between $507 - 111 = 396$ and $507 + 111 = 618$. Because of the symmetry of the bell curve, about half of these, 34% of the total, are above the mean (between 507 and 618), and about 34% of the scores are below the mean (between 396 and 507).

• 95% of SAT scores are within two standard deviations of the mean. This means that 95% of scores are between $507 - 2 \times 111 = 285$ and $507 + 2 \times 111 = 729$.

TABLE 6.1 SAT Verbal Scores

SAT Verbal	Number of students	Percent of students
750–800	25,114	1.8%
700–749	41,283	2.9%
650–699	88,799	6.3%
600–649	152,518	10.8%
550–599	204,601	14.5%
500–549	243,120	17.3%
450–499	245,615	17.5%
400–449	190,406	13.5%
350–399	116,808	8.3%
300–349	58,372	4.2%
250–299	25,332	1.8%
200–249	14,356	1%
Total	1,406,324	100%

Mean 507
Standard deviation 111

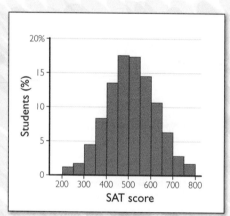

FIGURE 6.26 SAT histogram. **FIGURE 6.27** SAT bell curve plot.

- 99.7% of SAT scores are within three standard deviations of the mean. This means that 99.7% of scores are between $507 - 3 \times 111 = 174$ and $507 + 3 \times 111 = 840$.[3]

EXAMPLE 6.8 Interpreting the 68-95-99.7% rule: Heights of males

We noted earlier that adult male heights in the United States are normally distributed, with a mean of 69.1 inches. The standard deviation is 2.65 inches. What does the 68-95-99.7% rule tell us about the heights of adult males?

SOLUTION

Because 68% of adult males are within one standard deviation of the mean, 68% of adult males are between $69.1 - 2.65 = 66.45$ inches (5 feet 6.45 inches) and

---

[3] Because the number of scores above 800 is so small, the College Board lumps together all scores 800 or higher and reports them as 800. Similarly, the lowest reported score is 200.

69.1 + 2.65 = 71.75 inches (5 feet 11.75 inches) tall. Also, 95% of adult males are within two standard deviations of the mean. This means that 95% are between 69.1 − 2 × 2.65 = 63.8 inches and 69.1 + 2 × 2.65 = 74.4 inches tall. Furthermore, 99.7% of males are within three standard deviations of the mean. This means that 99.7% are between 69.1 − 3 × 2.65 = 61.15 inches and 69.1 + 3 × 2.65 = 77.05 inches tall.

TRY IT YOURSELF 6.8

The heights of adult American women are normally distributed, with a mean 64.1 inches and standard deviation of 2.5 inches. What does the 68-95-99.7% rule tell us about the heights of adult females?

The answer is provided at the end of this section.

The next example illustrates how to apply the 68-95-99.7% rule to the problem of counting the number of data points within a given range.

EXAMPLE 6.9 Counting data using the 68-95-99.7% rule: Weights of apples

The weights of apples in the fall harvest are normally distributed, with a mean weight of 200 grams and standard deviation of 12 grams. **Figure 6.28** shows the weight distribution of 2000 apples. In a supply of 2000 apples, how many will weigh between 176 and 224 grams?

SOLUTION

Apples weighing 176 grams are 200 − 176 = 24 grams below the mean, and apples weighing 224 grams are 224 − 200 = 24 grams above the mean. Now 24 grams represents 24/12 = 2 standard deviations. So the weight range of 176 grams to 224 grams is within two standard deviations of the mean. Therefore, about 95% of data points will lie in this range. This means that about 95% of 2000, or 1900 apples, weigh between 176 and 224 grams. This conclusion is illustrated in **Figure 6.29**.

TRY IT YOURSELF 6.9

The weights of oranges in one year's harvest are normally distributed, with a mean weight of 220 grams and standard deviation of 35 grams. In a supply of 4000 oranges, how many will we expect to weigh between 185 and 255 grams?

The answer is provided at the end of this section.

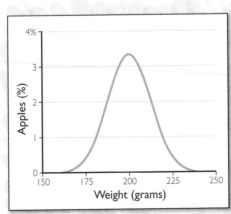

FIGURE 6.28 Apples with mean 200 grams and standard deviation 12 grams.

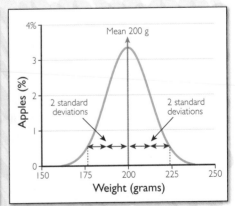

FIGURE 6.29 Apples between 176 and 224 grams.

z-scores

The standard deviation is such a useful tool when dealing with normal distributions that it is often used as a unit of measurement. Instead of saying that a SAT score of 618 is 111 points above the mean of 507, often one says that the score is one standard deviation above the mean. A score of 452 is 55 points or approximately one-half of a standard deviation below the mean. When used as a unit of measurement in this way, the standard deviation is often referred to as a *z-score*.

Key Concept

In a normal distribution, the **z-score** or **standard score** for a data point is the number of standard deviations that point lies above or below the mean. For data points above the mean the **z**-score is positive, and for data points below the mean the **z**-score is negative.

When we calculate the *z*-score, we normally round it to one decimal place.

Let's calculate the *z*-score for SAT verbal scores of 650 and 320. Recall that the mean score is 507 and the standard deviation is 111. A SAT score of 650 is 650 − 507 = 143 points above the mean score of 507. We need to express this difference in terms of standard deviations. One standard deviation is 111 points, so 143 points is

$$z\text{-score for } 650 = \frac{143}{111} \text{ standard deviations above the mean.}$$

This is about 1.3 standard deviations above the mean. Similarly, a SAT score of 320 is 507 − 320 = 187 points below the mean. This is 187/111 or about 1.7 standard deviations below the mean. Because the score is below the mean, the *z*-score is negative:

$$z\text{-score for } 320 = -1.7.$$

"You're kidding! You count S.A.T.s?"

EXAMPLE 6.10 Calculating z-scores: Weights of newborns

The weights of newborns in the United States are approximately normally distributed. The mean birthweight (for single births) is about 3332 grams (7 pounds 5 ounces). The standard deviation is about 530 grams. Calculate the *z*-score for a newborn weighing 3700 grams. (That is just over 8 pounds 2 ounces.)

SOLUTION

A 3700-gram newborn is 3700 − 3332 = 368 grams above the mean weight of 3332 grams. We divide by the number of grams in one standard deviation to find the z-score:

$$\text{z-score for 3700 grams} = \frac{368}{530}$$

or about 0.7.

TRY IT YOURSELF 6.10

In the situation of the example, calculate the z-score for a newborn weighing 3000 grams.

The answer is provided at the end of this section.

The following formulas may be useful:

$$\text{z-score} = (\text{Data point} - \text{Mean})/\text{Standard deviation}$$

or equivalently,

$$\text{Data point} = \text{Mean} + \text{z-score} \times \text{Standard deviation}.$$

Percentile scores

Recall that in Section 6.1 we discussed *quartiles*. There is a more general term, *percentile*, that tells how a data point is positioned in relation to other data points in a normal distribution.

Key Concept

The **percentile** for a number relative to a list of data is the percentage of data points that are less than or equal to that number.

For example, the median is always the 50th percentile because half the data are below the median. Similarly, the first quartile is always the 25th percentile because 25% of the data are below it, and the third quartile is always the 75th percentile because 75% of the data are below it.

It turns out that a SAT verbal score of 660 has a percentile score of 91.6%, which means that 91.6% of SAT verbal scores were 660 or lower. These percentiles are shown in **Figures 6.30 and 6.31**.

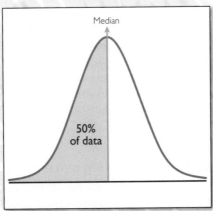

FIGURE 6.30 The percentile for the median is 50%.

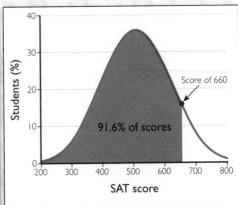

FIGURE 6.31 91.6% of SAT verbal scores were 660 or lower.

TABLE 6.2 Percentile from z-Score

z-score	Percentile	z-score	Percentile
−2.7	0.35	0.0	50.00
−2.6	0.47	0.1	53.98
−2.5	0.62	0.2	57.93
−2.4	0.82	0.3	61.79
−2.3	1.07	0.4	65.54
−2.2	1.39	0.5	69.15
−2.1	1.79	0.6	72.57
−2.0	2.28	0.7	75.80
−1.9	2.87	0.8	78.81
−1.8	3.59	0.9	81.59
−1.7	4.46	1.0	84.13
−1.6	5.48	1.1	86.43
−1.5	6.68	1.2	88.49
−1.4	8.08	1.3	90.32
−1.3	9.68	1.4	91.92
−1.2	11.51	1.5	93.32
−1.1	13.57	1.6	94.52
−1.0	15.87	1.7	95.54
−0.9	18.41	1.8	96.41
−0.8	21.19	1.9	97.13
−0.7	24.20	2.0	97.73
−0.6	27.43	2.1	98.21
−0.5	30.85	2.2	98.61
−0.4	34.46	2.3	98.93
−0.3	38.21	2.4	99.18
−0.2	42.07	2.5	99.38
−0.1	46.02	2.6	99.53
0.0	50.00	2.7	99.65

Table 6.2 gives the information needed to calculate the percentile from the z-score.

For example, a z-score of 0.5 corresponds to a percentile score of 69.15. A percentile score of 93.32 corresponds to a z-score of 1.5.

Let's see how this table helps in understanding SAT verbal scores. Suppose you scored 650 on the verbal section of the SAT. How does your test score compare with scores of other students? We calculated earlier that the z-score for a SAT score of 650 is 1.3. Table 6.2 shows a percentile of 90.32 for a z-score of 1.3. We conclude that about 90.3% of all SAT verbal scores were 650 or less.

We also found that the z-score for a SAT score of 320 is −1.7. The table shows a percentile of 4.46 for a z-score of −1.7. Thus, about 4.5% of all SAT verbal scores were 320 or less. Expressed differently, this says that about 95.5% of all SAT verbal scores were higher than 320.

EXAMPLE 6.11 Calculating percentiles: Length of illness

The average length of illness for flu patients in a season is normally distributed, with a mean of 8 days and standard deviation of 0.9 day. What percentage of flu patients will be ill for more than 10 days?

SOLUTION

Ten days is 2 days above the mean of 8 days. This gives a z-score of 2/0.9 or about 2.2. Table 6.2 gives a percentile of about 98.6% for this z-score. This means that about 98.6% of patients will recover in 10 days or less. Thus, only about 100% − 98.6% = 1.4% will be ill for more than 10 days.

TRY IT YOURSELF 6.11

In the preceding situation, what percentage of flu patients will recover in seven days or less?

The answer is provided at the end of this section.

Next we revisit the data on weights of newborns.

EXAMPLE 6.12 Using the normal distribution: Birthweight

Recall from Example 6.10 that the weights of newborns in the United States are approximately normally distributed. The mean birthweight (for single births) is about 3332 grams (7 pounds 5 ounces). The standard deviation is about 530 grams.

a. What percentage of newborns weigh more than 8 pounds (3636.4 grams)?

b. Low birthweight is a medical concern. The American Medical Association defines low birthweight to be 2500 grams (5 pounds 8 ounces) or less. What percentage of newborns are classified as low-birthweight babies?

SOLUTION

a. We first calculate the z-score. Now 3636.4 grams is 304.4 grams above the mean of 3332 grams because 3636.4 − 3332 = 304.4. One standard deviation is 530 grams, so 304.4 grams has a z-score of

$$z\text{-score} = \frac{304.4}{530}$$

or about 0.6. Consulting Table 6.2, we find that this represents a percentile of about 72.6%. This means that about 72.6% of newborns weigh 8 pounds or less. So about 100% − 72.6% = 27.4% of newborns weigh more than 8 pounds.

b. Again, we calculate the z-score: 2500 grams is 3332 − 2500 = 832 grams below the mean of 3332 grams. This is 832/530 or about 1.6 standard deviations below the mean. That is a z-score of −1.6. Table 6.2 shows a percentile of about 5.5% for a z-score of −1.6. Hence, about 5.5% of newborns are classified as low-birthweight babies.

The Central Limit Theorem

Recall the last paragraph of In the News 6.5, which opened this section: "If you concluded that the blood of middle-aged Swedes depended on games of Chutes and Ladders, you wouldn't be far wrong: Like the outcome of a dice-throwing game, a man's hemoglobin level is the result of numerous chance events—genes, environment, diet, lifestyle, and medical history—all added together. When you add up a lot of chance events, what you get is a normal distribution."

This is a pretty nice layman's explanation of one part of an important mathematical result known as the *Central Limit Theorem*. The theorem suggests a reason why many common measurements are normally distributed.

Key Concept

The Central Limit Theorem

According to the **Central Limit Theorem**, percentages obtained by taking many samples of the same size from a population are approximately normally distributed.

- The mean $p\%$ of the normal distribution is the mean of the whole population.

- If the sample size is n, the standard deviation of the normal distribution is

$$\text{Standard deviation} = \sigma = \sqrt{\frac{p(100 - p)}{n}} \text{ percentage points.}$$

Here, p is a percentage, not a decimal.

To see an application of the Central Limit Theorem, suppose that 60% of the registered voters in the state of Oklahoma in September 2008 intended to vote for John McCain for president. Suppose we took a poll of 100 (randomly selected) registered Oklahoma voters and calculated the percentage of those polled who intended to vote for McCain. The Central Limit Theorem tells us that the results of such a poll are (approximately) normally distributed with mean $p = 60\%$ because that is the mean for this population.[4] The sample size is $n = 100$, so the standard deviation is

$$\sigma = \sqrt{\frac{p(100 - p)}{n}}$$
$$= \sqrt{\frac{60(100 - 60)}{100}}.$$

This is about 4.9 percentage points.

How often would such a poll lead us to believe that Oklahoma would not vote for McCain? That is, what percentage of such polls would show that 50% or fewer voters intend to vote for McCain? To answer this, we first find the z-score for a sample that yields 50% for McCain. Now 50% is 10 percentage points below the mean of 60%. Expressed in terms of standard deviations, this number is 10/4.9 or about 2.0 standard deviations below the mean. That is a z-score of -2.0. Table 6.2 gives a percentile of about 2.3% for this z-score. So only 2.3% of such polls would report 50% or less support for John McCain.

We will look further into polling in the next section.

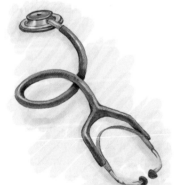

EXAMPLE 6.13 Using the Central Limit Theorem: Untreated patients

For a certain disease, 30% of untreated patients can be expected to improve within a week. We observe a population of 50 patients and record the percentage who improve within a week. According to the Central Limit Theorem, the results of such a study will be approximately normally distributed.

a. Find the mean and standard deviation for this normal distribution. Round the standard deviation to one decimal place.

b. Find the percentage of test groups of 50 patients in which more than 40% improve within a week.

---

[4] Strictly speaking, we should say that if many such polls, all of the same size, were taken from the same population, then the results would be approximately normally distributed.

SOLUTION

a. The mean is the percentage of untreated patients who will improve within a week, which is $p = 30\%$. The sample size is $n = 50$. This gives a standard deviation of

$$\sigma = \sqrt{\frac{p(100 - p)}{n}}$$
$$= \sqrt{\frac{30 \times 70}{50}}$$

or about 6.5 percentage points.

b. We first calculate the z-score for 40%, which is 10 percentage points above the mean of 30%. One standard deviation is 6.5 percentage points, so

$$z\text{-score } = \frac{10}{6.5}$$

or about 1.5. Table 6.2 gives a percentile of about 93.3%. This means that in 93.3% of test groups, we expect that 40% or fewer will improve within a week. Only $100\% - 93.3\% = 6.7\%$ of test groups will show more than 40% improving within a week.

TRY IT YOURSELF 6.13

On your campus, 20% of the students are education majors. Suppose you choose a random sample of 75 students and record the percentage of education majors. The results of such a survey are approximately normally distributed.

a. Find the mean and standard deviation for this normal distribution. Round the standard deviation to one decimal place.

b. What percentage of randomly selected groups of 75 students on your campus will include 10% or fewer education majors?

The answer is provided at the end of this section.

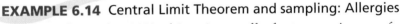

Significance of apparently small deviations

One consequence of the Central Limit Theorem is the observation that apparently small deviations in surveys can be significant. The following example illustrates this point.

EXAMPLE 6.14 Central Limit Theorem and sampling: Allergies

Assume we know that 20% of Americans suffer from a certain type of allergy. Suppose we take a random sample of 100,000 Americans and record the percentage who suffer from this allergy.

a. The Central Limit Theorem says that percentages from such surveys will be normally distributed. What is the mean of this distribution?

b. What is the standard deviation of the normal distribution in part a? Round your answer to two decimal places.

c. Suppose we find that in a town of 100,000 people, 21% suffer from this allergy. Is this an unusual sample? What does the answer to such a question tell us about this town?

SOLUTION

a. The mean is the percentage of people from the general population who suffer from the allergy. Therefore, the mean is $p = 20\%$.

b. For a sample size of 100,000, we find

$$\sigma = \sqrt{\frac{p(100 - p)}{n}}$$
$$= \sqrt{\frac{20 \times 80}{100,000}}$$

or about 0.13 percentage point.

c. Our sample of 21% is one percentage point larger than the mean of 20%. That gives

$$z\text{-score} = \frac{1}{0.13}$$

or about 7.7. This score is far larger than any z-score in Table 6.2. The chance of getting a random sample 7.7 standard deviations above the mean is virtually zero. There is almost no chance that in a randomly chosen sample of this size, 21% will suffer from this allergy. Thus, this is a truly anomalous sample: This town is not representative of the total population of Americans. Its allergy rate is highly unusual.

To summarize: It may seem counterintuitive, but even though 21% seems to be only a little larger than 20%, such a percentage is, in fact, a highly improbable result for a truly random sample of 100,000 people. Note that it is the large size of the sample that makes the standard deviation small and thus makes such a small percentage variation unlikely.

Health issues and the Fallon leukemia cluster

In Example 6.14, we found that when the sample size is large, even small deviations from the mean are highly unlikely. In such a situation, the sample almost certainly did not happen by pure chance.

A good real-world example of this kind of phenomenon is the leukemia cluster discovered around the town of Fallon in Churchill County, Nevada.[5] (Fallon happens to be the home of the Navy's Top Gun school at the Fallon Naval Air Station.) In the summer of 2000, it was learned that five cases of leukemia in children had been diagnosed in Churchill County within a few months of each other. In a four-year period, 16 children had been diagnosed, all of whom had lived in Churchill County for varying lengths of time prior to diagnosis. The office of the Nevada State Epidemiologist noted that the average rate is about three childhood cases per 100,000 children. This fact suggests, based on the size of the population of Churchill County, that one case would be expected about every five years in that county.

This statistical fact alerted the authorities that it was extremely unlikely that the cases were happening by pure chance—there was probably something extraordinary causing them.

In subsequent studies, the metal tungsten has emerged as a prime suspect. University of Arizona scientists said recent tests show that Fallon has up to 13 times as much tungsten in its dust as other Nevada cities. They said tests also have found elevated levels of tungsten in tree rings in Fallon and three other towns with leukemia clusters.

---

[5] http://www.cdc.gov/nceh/clusters/fallon/.

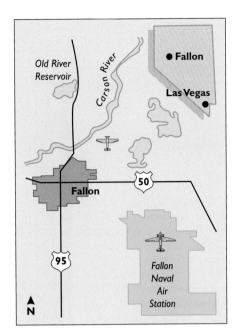

Examples like this illustrate how vital mathematics and statistics are to our well-being.

Try It Yourself answers

Try It Yourself 6.7: Determining normality from a graph: Birthweights and ages Birthweights of Norwegian children are normally distributed, but ages of Americans are not.

Try It Yourself 6.8: Interpreting the 68-95-99.7% rule: Heights of females   68% of adult American women are between 61.6 and 66.6 inches tall, 95% are between 59.1 and 69.1 inches, and 99.7% are between 56.6 and 71.6 inches.

Try It Yourself 6.9: Counting data using the 68-95-99.7% rule: Weights of oranges 2720 oranges.

Try It Yourself 6.10: Calculating *z*-scores: Weights of newborns -0.6.

Try It Yourself 6.11: Calculating percentiles: Length of illness   About 13.6%.

Try It Yourself 6.13: Using the Central Limit Theorem: Education majors
   a. The mean is 20%. The standard deviation is 4.6 percentage points.

   b. 1.4%.

Exercise Set 6.2

1. Which is normal? Consider the data distributions in Figures 6.32, 6.33, 6.34, and 6.35. Which of the distributions appear to be normal? If a distribution does not appear to be normal, explain why.

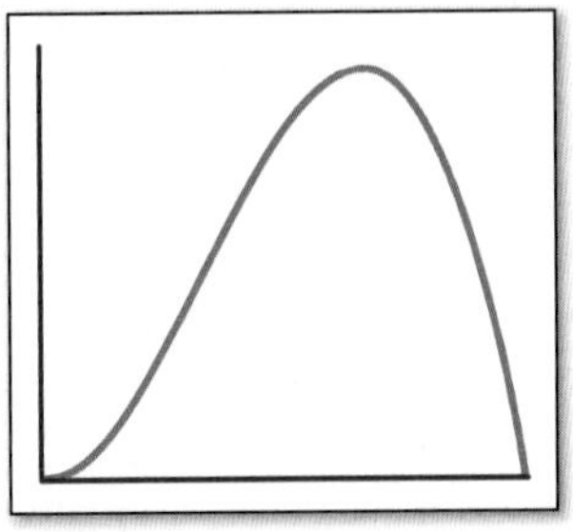

FIGURE 6.32 Distribution 1.

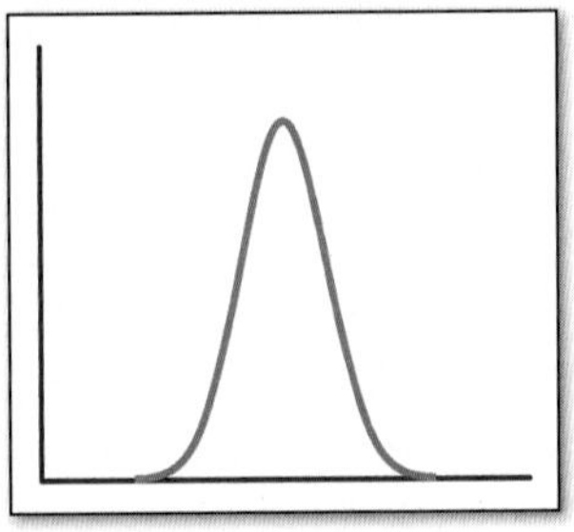

FIGURE 6.33 Distribution 2.

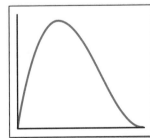

FIGURE 6.34 Distribution 3.

FIGURE 6.35 Distribution 4.

2. Getting heavier. The following are excerpts from an article that appeared in the *New York Times*.

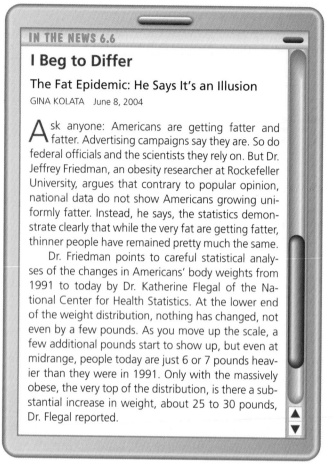

IN THE NEWS 6.6

I Beg to Differ

The Fat Epidemic: He Says It's an Illusion

GINA KOLATA   June 8, 2004

Ask anyone: Americans are getting fatter and fatter. Advertising campaigns say they are. So do federal officials and the scientists they rely on. But Dr. Jeffrey Friedman, an obesity researcher at Rockefeller University, argues that contrary to popular opinion, national data do not show Americans growing uniformly fatter. Instead, he says, the statistics demonstrate clearly that while the very fat are getting fatter, thinner people have remained pretty much the same.

Dr. Friedman points to careful statistical analyses of the changes in Americans' body weights from 1991 to today by Dr. Katherine Flegal of the National Center for Health Statistics. At the lower end of the weight distribution, nothing has changed, not even by a few pounds. As you move up the scale, a few additional pounds start to show up, but even at midrange, people today are just 6 or 7 pounds heavier than they were in 1991. Only with the massively obese, the very top of the distribution, is there a substantial increase in weight, about 25 to 30 pounds, Dr. Flegal reported.

It may be reasonable to assume that body weights of Americans are normally distributed. In view of the report above, is it reasonable to think that the distribution of weights was normal both before and after the increase in weight of already overweight people? Explain.

It's cold in Madison, Wisconsin   Figure 6.36 is a histogram of average yearly temperature in Madison, Wisconsin, over a

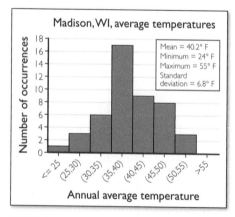

FIGURE 6.36 Average temperatures in Madison.

47-year period. It shows a mean of 40.2 degrees Fahrenheit and a standard deviation of 6.8 degrees. This information is used in Exercises 3 and 4.

3. Judging simply from the appearance of the histogram, do you think average temperatures in Madison are approximately normally distributed? Explain your reasoning.

4. Challenge problem. If the data were normally distributed, then about 68% of the data would be within one standard deviation of the mean. Use the histogram to estimate the percentage of the data that are within one standard deviation of the mean. (Give your answer as a percentage to the nearest whole number.) Does your answer support the proposition that the data are approximately normally distributed?

5. Fingerprints. Ridge counts on fingerprints are approximately normally distributed, with a mean of about 140 and standard deviation of 50. What does the 68-95-99.7% rule tell us about ridge counts on fingerprints?

Diastolic blood pressure   In a study of 500 men, diastolic blood pressure was found to be approximately normally distributed, with a mean of 82 millimeters (mm) of mercury and standard deviation of 10 millimeters. This information is used in Exercises 6 and 7.

6. Use the 68-95-99.7% rule to determine what percentage of the test group had a diastolic pressure between 72 millimeters and 92 millimeters of mercury.

7. *This is a continuation of Exercise* 6. Use your answer to Exercise 6 to determine how many men in the test group had a diastolic pressure between 72 millimeters and 92 millimeters of mercury.

8. Calculating the z-score. It is found that the number of raisins in a box of a popular cereal is normally distributed, with a mean of 133 raisins per box and standard deviation of 10 raisins. My cereal box has 157 raisins. What is the z-score for this box of cereal? Round your answer to one decimal place.

9. More on z-scores. The number of nuts in a can of mixed nuts is found to be normally distributed, with a mean of 500 nuts and standard deviation of 20 nuts. My can of mixed nuts has only 475 nuts. What is the z-score for this can of nuts? Round your answer to one decimal place.

Standardized exam scores   Exercises 10 through 13 refer to scores on standardized exams with results that are normally distributed. Round your answers to one decimal place.

10. Suppose you have a score that puts you 1.5 standard deviations below the mean. What is your percentile score?

11. Suppose you have a score that puts you in the 60th percentile. What is your z-score?

12. Suppose you have a score of 360 on the verbal section of the SAT Reasoning Test. What is your z-score? (See Table 6.1 on p. 372.) What is your percentile score?

13. Suppose you have a score of 720 on the verbal section of the SAT Reasoning Test. What is your z-score? (See Table 6.1 on p. 372.) What is your percentile score?

IQ A person's IQ (Intelligence Quotient) is supposed to be a measure of his or her "intelligence." The IQ test scores are normally distributed and are scaled to a mean of 100 and a standard deviation of 15. This information is used in Exercises 14 and 15.

14. If your IQ is 115, what is your percentile score? Round your answer to one decimal place.

15. It is often said that a genius is someone with an IQ of 140 or more. What is the z-score for a 140 IQ? (Round your answer to one decimal place.) What percentage of the population scores as a genius? Round your answer as a percentage to one decimal place. *Note*: The actual number is 0.4%, but you may get a slightly different value due to rounding.

SAT scores In Exercises 16 through 18, we suppose you had a score of 670 on the verbal section of the SAT Reasoning Test (see Table 6.1 on p. 372). Round your answers to one decimal place.

16. What is the z-score for this SAT score?

17. What is your percentile score?

18. Explain in practical terms the meaning of your percentile score.

Birthweights in Norway The National Institute of Environmental Health Sciences reports that over a six-year period birthweights (in grams) of newborns in Norway were distributed normally (see **Figure 6.37**), with a mean of

3668 grams and standard deviation of 511 grams. Exercises 19 through 22 refer to these data.

19. What is the z-score for a newborn weighing 4000 grams? Round your answer to one decimal place.

20. What is the z-score for a newborn weighing 3000 grams? Round your answer to one decimal place.

21. Newborns weighing less than 2500 grams are classified as having a low birthweight. What percentage of newborns suffered from low birthweight? Round your answer as a percentage to the nearest whole number.

22. Would you expect the distribution to remain normal if birthweights were reported in pounds?

Campinas, Brazil The average maximum monthly temperature in Campinas, Brazil, is 29.9 degrees Celsius. The standard deviation in maximum monthly temperature is 2.31 degrees. Assume that maximum monthly temperatures in Campinas are normally distributed. This information is used in Exercises 23 and 24.

23. Use Rule of Thumb 6.1 to fill in the following two blanks: "95% of the time the maximum monthly temperature is between _____ and _____." Round your answers to one decimal place.

24. What percentage of months would have a maximum temperature of 35 degrees or higher? Round your answer as a percentage to one decimal place.

Male and female height in America The heights of adult men in America are normally distributed, with a mean of 69.1 inches and standard deviation of 2.65 inches. Exercises 25 through 28 use this information. Round your answers as percentages to the nearest whole number.

25. Tall. What percentage of adult males in America are over 6 feet tall?

26. Short. What percentage of adult males in America are under 5 feet 7 inches tall?

27. Heights of women. The heights of American women between the ages of 18 and 24 are approximately normally distributed. The mean is 64.1 inches, and the standard deviation is 2.5 inches. What percentage of such women are over 5 feet 9 inches tall?

28. Dancing. It is sometimes said that women don't like to dance with shorter men. To get a dancing partner, a 6-foot-tall woman draws from a hat the name of an adult American male. What is the probability that he will be as tall as she? Give your answer as a percentage rounded to the nearest whole number.

29. Curving grades. When exam scores are low, students often ask the teacher whether he or she is going to "curve" the grades. The hope is that by curving a low score on the exam, the students will wind up getting a higher letter grade than might otherwise be expected. The term *curving* grades, or *grading on a curve*, comes from the bell curve of the normal distribution. If we assume that scores for a large number of students are distributed normally (as with SAT scores) and we also assume that the class average should be a "C," then a teacher might award grades as listed in **Table 6.3**.

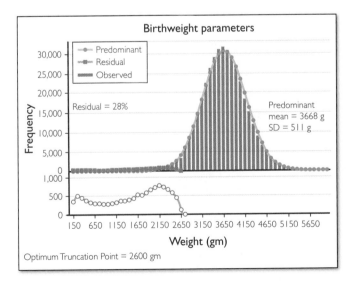

FIGURE 6.37 Birthweights in Norway.

TABLE 6.3 Curving Grades

A	1.5 standard deviations above the mean or higher
B	0.5 to 1.5 standard deviations above the mean
C	within 0.5 standard deviation of the mean
D	0.5 to 1.5 standard deviations below the mean
F	1.5 standard deviations below the mean or lower

Suppose a teacher curved grades using the bell curve as in Table 6.3 and that the grades were indeed normally distributed. What percent of students would get a grade of "A"? What percent of students would get a grade of "B"? Round each answer as a percentage to one decimal place. *Suggestion*: To find the percentage of students getting a grade of "B," subtract the percentage of students 0.5 standard deviation or less above the mean from the percentage of students 1.5 standard deviations or less above the mean.

30. More on curving grades. Suppose an exam had an average (mean) score of 55% and a standard deviation of 15%. If the teacher curved grades using the bell curve as in Table 6.3 from Exercise 29, what score would be necessary to receive an "A"? How about a "B"? A "C"? Round each answer as a percentage to one decimal place.

31. A medical test. It is known that in the absence of treatment, 70% of the patients with a certain illness will improve. The Central Limit Theorem tells us that the percentages of patients in groups of 400 that improve in the absence of treatment are approximately normally distributed.

    a. Find the mean and standard deviation of the normal distribution given by the Central Limit Theorem. Round the standard deviation to one decimal place.

    b. In what percentage of such groups (of 400) would 275 or fewer improve? Round your answer to one decimal place as a percentage.

Lots of coins Suppose we toss 100 fair coins and record the percentage of heads. Exercises 32 through 35 refer to this situation.

32. According to the Central Limit Theorem, the percentages of heads resulting from such experiments are approximately normally distributed. Find the mean and standard deviation of this normal distribution.

33. What is the z-score for 55 heads? How about 60 heads?

34. What percentage of tosses of 100 fair coins would show more than 55 heads? Round your answer to one decimal place as a percentage.

35. What percentage of tosses of 100 fair coins would show more than 60 heads? Round your answer to one decimal place as a percentage.

Contracting diseases It is known that under ordinary conditions, 15% of people will not contract a certain disease. Consider the situation where test groups of 500 were selected and the percentages of those who did not contract the disease were recorded. Exercises 36 through 37 use this context.

36. According the Central Limit Theorem, percentages of groups of 500 who don't contract the disease are approximately normally distributed. Find the mean and standard deviation of this normal distribution. Round the standard deviation to one decimal place.

37. For what percentage of such groups (of 500) will 80 or fewer contract the disease? Round your answer to one decimal place as a percentage.

38. Marbles in a jar. A large jar is filled with marbles. Of these marbles, 25% are red and the rest are blue. We randomly draw 100 marbles.

    a. According to the Central Limit theorem, the percentages of red marbles in such samples are approximately normally distributed. Find the mean and standard deviation of this normal distribution. Round the standard deviation to one decimal place.

    b. Our friend pulls 100 marbles from the jar, supposedly at random, and we find that 20 of them are red. Do we have reason to suspect our friend of cheating? Explain why you answered as you did.

39. Voting for president. In a certain presidential election, 38% of voting-age Americans actually voted. In our town of 200,000, 40% of voting-age citizens voted. What should we conclude regarding our town?

40. Overweight Americans. About 32% of Americans over age 20 are obese.

    a. If we take random samples of 500 Americans over 20 and record the percent of people in the sample who are obese, we expect to get data that are approximately normally distributed. Find the mean and standard deviation of this normal distribution. Round the standard deviation to one decimal place.

    b. Would it be unusual to find a random sample of 500 Americans over 20 in which 35% were obese?

41. Allergies. About 20% of Americans suffer from some form of allergies.

    a. The Central Limit Theorem tells us that percentages of allergy sufferers in random samples of 1000 people are approximately normally distributed. Find the mean and standard deviation of this normal distribution. Round the standard deviation to one decimal place.

    b. How unusual would it be to find a random sample of 1000 people in which 25% suffered from allergies?

42. Diabetes. About 6.3% of Americans suffer from diabetes. This exercise examines the percentage of people in samples of 2000 who have diabetes.

    a. The Central Limit Theorem tells us that these percentages are approximately normally distributed. Find the mean and standard deviation. Round the standard deviation to two decimal places.

    b. How unusual would it be for such a sample to show 7% suffering from diabetes?

43. Tall people. Take the average height for adult males in America to be 5 feet 9 inches. Suppose for the moment that half are taller and half are shorter. Let's take samples of 1000 men and record the percentage of tall (over 5 feet 9 inches) men. Would it be unusual to find such a sample with 45% short people? Would it be unusual to find that the percentage of tall people in a sample of 1000 college basketball players is over 80%?

Exercise 44 is suitable for group work.

44. Heights in your class—an experiment. Record the heights of students in your class and make a histogram of the data. Does it appear that these data are normally distributed? Explain why you gave the answer you did.

45. History. Write a paper on the history of the bell curve and the normal distribution. Some names to consider are de Moivre, Gauss, Jouffret, and Laplace.

46. IQ and the bell curve. There are controversies associated with the connection between the bell curve and the measurement of intelligence. The book *The Bell Curve: Intelligence and Class Structure in American Life* caused quite a stir when it was published.[6] Summarize and discuss these controversies.

47. History. Proofs of the Central Limit Theorem were given early in the twentieth century. The ideas involved are much older because certain other distributions have been studied for a long time. Summarize the results of de Moivre and Laplace

about the binomial distribution and its relationship to the normal distribution.

48. History. Write a paper about the men who are credited with proving the Central Limit Theorem: Lyapunov and Lindeberg.

The following exercises are designed to be solved using technology such as calculators or computer spreadsheets. For assistance, see the technology supplement.

49. A computer simulation. Consider an experiment in which we toss 500 coins. We repeat the experiment 500 times and record the percentage of heads for each experiment. Use a computer simulation to carry out this procedure, and record your results as a histogram. Does it appear that your data are approximately normally distributed? Explain why you gave the answer you did.

50. Formula for a bell. Most people know about the number π, which is approximately 3.14159. But there is another very important number in mathematics, denoted by the letter e, which is approximately 2.71828. Plot the graph of $\dfrac{1}{\sqrt{2\pi}} e^{-x^2/2}$ for x between -2 and 2. Do you see a connection with the bell curve? If so, what is the mean?

6.3 The statistics of polling: Can we believe the polls?

TAKE AWAY FROM THIS SECTION

Understand margins of error and confidence levels in polls.

We see polls everywhere, especially in politics. Polling is one of the most visible applications of statistics. The article on the following page describes public opinion on a number of social and political issues.

Articles like this may seem straightforward, but in fact they involve some very subtle and deep mathematical ideas. Nevertheless, it is possible to gain enough understanding of the principles to be informed users of such information. That is our goal in this section.

[6]By Richard Herrnstein and Charles Murray (New York: The Free Press, 1994).

Poll: Public Opposes Increased Presidential Power

MARK SHERMAN   September 15, 2008

Washington—Americans strongly oppose giving the president more power at the expense of Congress or the courts, even to enhance national security or the economy, according to a new poll.

The Associated Press-National Constitution Center poll of views on the Constitution found people wary of governmental authority after years of controversy over the Bush administration's expansion of executive power, and especially skeptical of increasing the president's powers. . . .

Two-thirds of Americans oppose altering the balance of power among the three branches of government to strengthen the presidency, even when they thought that doing so would improve the economy or national security. . . .

In 2005, the Supreme Court ruled 5-4 that governments may seize people's homes and businesses—even against their will—for private economic development when there is a corresponding public purpose of bringing more jobs and tax revenue.

In the new poll of people's views on the Constitution, 75 percent disagreed. Opposition to the government power known as eminent domain was as strong among liberals as conservatives. . . .

The poll also found a split on whether governments should recognize gay marriage. But a majority said same-sex couples should be entitled to the same benefits as married, heterosexual couples. The answers to these questions revealed a sharp generational split. More than two-thirds of people under 35 favor recognition of gay marriage, compared with less than 40 percent of those 35 and older. . . .

The AP-National Constitution Center poll involved telephone interviews with 1000 adults nationwide. The survey was conducted Aug. 22–29 by Abt SRBI Inc. and had a margin of sampling error of plus or minus 3.1 percentage points.

Election polling

Professional polling organizations have a good record of predicting the outcomes of elections. Of course, even these professionals sometimes get things wrong. In the 2000 U.S. presidential election, early projections by some news organizations led them to claim that Al Gore had won Florida, which would have secured his election. Early exit polls in November 2004 indicated more voter support for John Kerry than ultimately proved to be the case. Nevertheless, polls are remarkably good at getting it right. How are they able to do this?

There are two important aspects of polling, one of which involves mathematical and statistical considerations. The preceding article told us how many people were interviewed and also mentioned something they called a "margin of error." How were the numbers in the article determined, what do they really mean, and how much faith should we put in them?

To answer these questions, we need to understand the basics of statistical inference. In the preceding section, we saw how the Central Limit Theorem about the normal distribution allows us to draw inferences about a large population from samples of the population. That discussion is the background for our study of polls in this section. Here, we highlight the problem of estimating the accuracy of those inferences.

A second aspect of polling involves "methodology," that is, the way the polls are conducted, which includes how the sample is chosen. This is at least as important as the mathematical aspects.

Basic terms: Margin of error, confidence interval, and confidence level

To understand how real polls work, we look first at an idealized poll. Suppose we have an opaque jar containing a large number of red marbles and blue marbles, and we want to know what percentage of the marbles are red. The only way to know this for certain is actually to count all the marbles to determine how many are red and how many are blue. But this may be impractical if there are hundreds or millions of them. Is there a way we might make an educated guess and, if so, can we tell how good that guess might be?

One way is to count the number of red marbles in a small sample. For example, suppose we shake the jar and then select a few marbles at random from the jar and find that 20% of them are red. Of course, we cannot conclude that exactly 20% of the marbles in the jar are red, but based on the sample we have chosen, that would certainly be a reasonable guess. The big question is, how confident can we be in the accuracy of this guess?

There are two issues to resolve:

1. How large a sample size should we use?

2. How do we quantify the level of confidence we have in our answer?

Let's suppose that, in fact, 25% of the marbles in the jar are red, but let's also assume we don't know that. If we draw 100 marbles from the jar, we can use the ideas developed in Section 5.4 to show that the probability there will be exactly 25 red ones is very small, only about 0.09. But the same ideas show that the probability a sample of 100 will contain between 23% and 27% (inclusive) is 0.44 or 44%. This is the key.

The probability that a sample *exactly* reflects the population may be quite small, but the probability that it *approximately* reflects the population is much greater. We can predict with some confidence that the number of red marbles will be in a certain range, say, 23% to 27%, even though we can't be very confident of the exact number.

Now suppose we chose a marble sample of a different size than 100. What effect would that have on the probability? The table below gives the approximate probabilities that between 23% and 27% of the marbles in our sample are red for various sample sizes, denoted by n. Remember that we're assuming the jar has a vast number of marbles and that 25% of the marbles in the jar are red.

Number n selected	100	500	1000	1500	2000	2500
Probability 23% to 27%	0.44	0.72	0.87	0.93	0.96	0.98

Note that the larger the sample size, the greater the probability that our results fall within the 23% to 27% range. It makes sense that a larger sample is more representative of the whole. We saw the same phenomenon in our discussion of the Central Limit Theorem in the preceding section, where we noted that large sample sizes make even small deviations from the mean unlikely.

Let's consider a sample size of 2000. In that case, the probability is 0.96 or 96% that we will get between 23% and 27% red marbles. This means that if we were to do many such drawings, 96% of the time we would expect them to be between 23% and 27% red. That is, for a sample size of 2000, we can expect that 96% of the time we would get a result within two percentage points of the correct percentage of red marbles, 25%.

The number 2% in this example is the *margin of error*. This number is usually reported as "plus or minus 2%" or as ±2%. We will simply say 2%, with the "plus or minus" understood.

What does a margin of error of 2% mean in terms of polls? Suppose that a polling organization samples 2000 marbles and finds that 24% are red. Their report might be: "One can say with 96% confidence that the percentage of red marbles is 24%, with a margin of error of two percentage points." In other words, with 96% confidence the true percentage of red marbles is between 22% and 26%. The interval 22% to 26% is the *confidence interval*, and the number 96% is the *confidence level*. The meaning of the confidence level is that about 96 out of 100 such samples will produce a confidence interval that includes the true percentage (25%) of red marbles.

"No, Kevin — there isn't any
margin of error on spelling tests."

Key Concept

The **margin of error** of a poll expresses how close to the true result (the result for the whole population) the result of the poll can be expected to lie. To find the **confidence interval**, adjust the result of the poll by adding and subtracting the margin of error. The **confidence level** of a poll tells the percentage of such polls in which the confidence interval includes the true result.

It is very important to note that it is *possible* for the polling organization to get an "unlucky" or unrepresentative sample, say, a sample with 8% red marbles or a sample with 80% red marbles. But this is quite unlikely: For a sample size of 2000, only about 4% of the times we take our sample do we expect it to be outside the 23% to 27% range (because 100% − 96% = 4%).

Back to the polls: Margin of error, confidence interval, and confidence level

Now let's apply what we have learned to a hypothetical idealized opinion poll. In this case, let's assume we are dealing with people instead of marbles and, instead of differentiating on the basis of marble color, we ask people what opinion they hold on a certain question.

As an example, suppose a poll reports with 95% confidence that 45% of the American people support foreign aid and that the margin of error is two percentage points. (Note that 95% is a common confidence level for polls conducted by Gallup, Harris, Zogby, and other professional polling organizations.) Let's recall what this means: If we conducted this poll thousands of times, and each time we wrote down the percentage of the people in the sample who support foreign aid, we would expect about 95% of those results to be within two percentage points of the true value.

Television networks conduct exit polls of local elections in Seoul, South Korea.

In brief, we can be fairly confident that the support of the entire population for foreign aid lies in the confidence interval between 43% and 47%.

SUMMARY 6.1 **Polls and Margin of Error**

Suppose that, based on random sampling, a poll reports the percentage of the population having a certain property (e.g., planning to vote for a certain candidate) with a margin of error m. Assuming that this margin is based on a 95% confidence level, we can say that if we conducted this poll 100 times, then we expect about 95 of those sample results to be within m percentage points of the true percentage having that property.

We reiterate that the result from a single poll does not determine for certain the truth about the population, not even within the range given by the margin of error. In the poll about foreign aid, we cannot be certain that between 43% and 47% of the entire population really do support aid. In fact, the theory says that we should not be surprised if about 5% of the time our sample will not be within two percentage points of the true value. Indeed, it is entirely possible that 70% support aid, although the chances are very remote.

Also, opinion polls are often taken periodically over time and usually by different organizations, so you don't have to trust just one poll. Your confidence should increase if several polls report similar results.

EXAMPLE 6.15 Interpreting polls: Approval of Congress

Explain the meaning of a poll that says 33% of Americans approve of what Congress is doing, with a margin of error of 4% and confidence level of 90%.

SOLUTION

In 90% of such polls, the reported approval of Congress will be within four percentage points of the true approval level. Thus, we can be 90% confident that the true level lies in the confidence interval between 29% and 37%.

TRY IT YOURSELF 6.15

Explain the meaning of a poll that says 35% of Americans approve of what the president is doing, with a margin of error of 3% and confidence level of 93%.

The answer is provided at the end of this section.

How big should the sample be?

The next question is, how big should the sample be to ensure a certain level of confidence? Polling organizations such as the Harris Poll use rather sophisticated methods to decide how many people should be questioned (the sample size) and how they should be chosen. But there is a simple rule of thumb that gives a reasonably good estimate for a confidence level of 95%: For a sample size of n, the margin of error is approximately $100/\sqrt{n}\%$.

> **RULE OF THUMB 6.2** Margin of Error
>
> For a 95% level of confidence, we can estimate the margin of error when we poll n people using
>
> $$\text{Margin of error} \approx \frac{100}{\sqrt{n}}\%.$$
>
> Here, the symbol $\approx$ means "is approximately equal to."

Normally, when we use this estimate, we round the margin of error to one decimal place. We note that this estimate for the margin of error is a consequence of the discussion in the preceding section. The key points are that about 95% of normal data lie within two standard deviations of the mean, and the Central Limit Theorem gives a formula for the standard deviation that depends on the sample size n.

To show how to use the formula, suppose we choose 1100 Americans at random and obtain their opinion on a certain issue. For a confidence level of 95% we get a margin of error of approximately $100/\sqrt{1100}\%$, or about 3.0%. It is perhaps surprising that we can, with 95% confidence, know the opinions of over 280 million people within three percentage points by sampling only 1100 people.

EXAMPLE 6.16 Estimating margin of error: Fashion survey

A recent Oricon fashion survey[7] asked 900 people, "Which Japanese male celebrity looks best in sneakers?" The winner was Kimura Takuya. What is the approximate margin of error for a 95% confidence level?

SOLUTION

We use the formula

$$\text{Margin of error} \approx \frac{100}{\sqrt{n}}\%$$

with $n = 900$:

$$\text{Margin of error} \approx \frac{100}{\sqrt{900}}\%.$$

Kimura Takuya.

[7] http://www.alafista.com/2008/04/06/oricon-poll-male-celeb-who-look-good-in-sneakers/.

Our margin of error is about 3.3%. We can be 95% confident that our poll result is within 3.3 percentage points of the true value.

TRY IT YOURSELF 6.16

If we conduct a poll of 1600 people, what is the approximate margin of error for a 95% confidence level?

The answer is provided at the end of this section.

Now we interpret the margin of error in terms of a range of estimates.

EXAMPLE 6.17 Finding confidence interval: Balance of power

In the News 6.7 at the beginning of this section (p. 386) gives the sample size for the poll as 1000 and the margin of error as 3.1%. Assume that the confidence level is 95%. (Because $100/\sqrt{1000}$ is about 3.16, this assumption is reasonable.) The poll reported that two-thirds of Americans oppose altering the balance of power in the branches of government. What is the confidence interval for this poll? Interpret your answer.

SOLUTION

The percentage of Americans opposed to such an alteration was found to be two-thirds or about 66.7% in the poll. Because the margin of error is 3.1%, we expect the actual percentage to be between $66.7 - 3.1 = 63.6\%$ and $66.7 + 3.1 = 69.8\%$. Therefore, the confidence interval is 63.6% to 69.8%.

Because the level of confidence is 95%, if we conducted this poll 100 times, then for about 95 of the results, the true value would lie in the confidence interval corresponding to the result. Therefore, we are 95% confident that the actual percentage lies between 63.6% and 69.8%.

TRY IT YOURSELF 6.17

In the same poll, 75% of those polled indicated that they disagreed with the Supreme Court's ruling on eminent domain. What is the confidence interval for this poll? Interpret your answer. (Assume that the confidence level is 95%.)

The answer is provided at the end of this section.

In view of what we have just learned, let's return to In the News 6.7 at the beginning of this section (p. 386). It says that "more than two-thirds of people under 35 favor recognition of gay marriage, compared with less than 40% of those 35 and older." Why do you suppose that more precise percentages were not given here? One explanation may be that this is just the writing style of this journalist. Another explanation may be that the margin of error is not the advertised 3.1% for this part of the poll. We would need a sample of about 1000 people for a margin of error of about 3.1%. Some of the people polled are over 35 and some are not, so if the sample size for each group is less than 1000, then the margins of error are larger.

EXAMPLE 6.18 Sample size: Katrina aftermath

The Kaiser Family Foundation polled 1294 residents of Orleans Parish in New Orleans in 2008 and found that 41% of the residents who had lived through Hurricane Katrina in 2005 report that their lives are still disrupted.[8]

---

[8] http://www.kff.org/kaiserpolls/pomr081008nr.cfm.

Aftermath of Hurricane Katrina.

a. The poll surveyed 1294 people. What is the approximate margin of error for a 95% confidence interval?

b. The poll of 1294 people found that 41% of respondents still had disrupted lives. Can we conclude with certainty that no more than 45% of residents' lives are still disrupted?

c. Suppose instead that the poll of 1294 people had found that 52% still had disrupted lives. Explain what we could conclude from this result. Could we assert with confidence that a majority of residents' lives are still disrupted by Katrina?

d. Suppose we wish to have a margin of error of two percentage points. Approximately how many people should we interview?

SOLUTION

a. To find the margin of error, we use the formula

$$\text{Margin of error} \approx \frac{100}{\sqrt{n}}\%$$

with $n = 1294$:

$$\text{Margin of error} \approx \frac{100}{\sqrt{1294}}\%$$

or about 2.8%.

b. Our answer to part a tells us that we can be 95% confident that the poll number of 41% is within 2.8 percentage points of the true percentage of all residents whose lives are still disrupted from Katrina. This means that if we conducted this poll 100 times, then about 95 of the results would be within 2.8 percentage points of the true value. Thus, it is very likely that the true value is between $41 - 2.8 = 38.2\%$ and $41 + 2.8 = 43.8\%$. Because the whole interval is below 45%, we can be quite confident (at a 95% level) that no more than 45% of residents' lives are still disrupted. On the other hand, we cannot make this conclusion with absolute certainty.

c. We can be 95% confident that the poll number of 52% is within 2.8 percentage points of the true percentage of all residents whose lives are still disrupted by Katrina. This means that if we conducted this poll 100 times, about 95 of the results would be within 2.8 percentage points of the true value. Thus, it is very likely that the true value is between $52 - 2.8 = 49.2\%$ and

$52 + 2.8 = 54.8\%$. Most of this interval falls above 50%, so we continue to have good reason to think that a majority of residents' lives are still disrupted. But, because a portion of the interval falls below 50%, we should be more cautious in drawing conclusions.

d. To calculate how large our sample should be, we use the formula

$$\text{Margin of error} \approx \frac{100}{\sqrt{n}}\%$$

and substitute 2% for the margin of error. Thus, we want to find n so that

$$2 = \frac{100}{\sqrt{n}}.$$

Rearranging the equation to solve for n gives

$$\sqrt{n} = \frac{100}{2},$$

so

$$\sqrt{n} = 50.$$

Hence,

$$n = 50^2 = 2500.$$

We should interview about 2500 people. Note that one Harris Poll with a 95% confidence level and a margin of error of 2% surveyed 2415 people—very close to the 2500 given by the formula.

The procedure in part d of Example 6.18 can be used to estimate the sample size needed to get a given margin of error: To get a margin of error of m (as a percentage), the sample size needs to be about $(100/m)^2$. This assumes a 95% level of confidence. In part d of the example, we had $m = 2\%$, and this formula gives us

$$\left(\frac{100}{2}\right)^2 = 50^2 = 2500.$$

RULE OF THUMB 6.3 Sample Size

For a 95% level of confidence, the sample size needed to get a margin of error of m percentage points can be approximated using

$$\text{Sample size} \approx \left(\frac{100}{m}\right)^2.$$

EXAMPLE 6.19 Estimating sample size: 4% margin of error

What sample size is needed to give a margin of error of 4% with a 95% confidence level?

SOLUTION

We use the approximate formula

$$\text{Sample size} \approx \left(\frac{100}{m}\right)^2$$

and put in 4 for m:

$$\text{Sample size} \approx \left(\frac{100}{4}\right)^2 = 625.$$

TRY IT YOURSELF 6.19

What sample size is needed to give a margin of error of 5% with a 95% confidence level?

The answer is provided at the end of this section.

The "statistical dead heat"

When polls show that a political race is tight, sometimes the media call the result a "statistical dead heat." The accompanying article from cnn.com is an example.

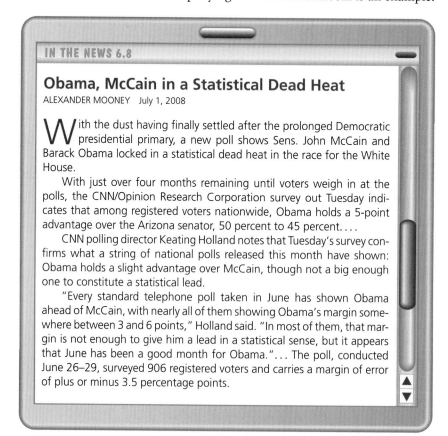

IN THE NEWS 6.8

Obama, McCain in a Statistical Dead Heat

ALEXANDER MOONEY   July 1, 2008

With the dust having finally settled after the prolonged Democratic presidential primary, a new poll shows Sens. John McCain and Barack Obama locked in a statistical dead heat in the race for the White House.

With just over four months remaining until voters weigh in at the polls, the CNN/Opinion Research Corporation survey out Tuesday indicates that among registered voters nationwide, Obama holds a 5-point advantage over the Arizona senator, 50 percent to 45 percent....

CNN polling director Keating Holland notes that Tuesday's survey confirms what a string of national polls released this month have shown: Obama holds a slight advantage over McCain, though not a big enough one to constitute a statistical lead.

"Every standard telephone poll taken in June has shown Obama ahead of McCain, with nearly all of them showing Obama's margin somewhere between 3 and 6 points," Holland said. "In most of them, that margin is not enough to give him a lead in a statistical sense, but it appears that June has been a good month for Obama."... The poll, conducted June 26–29, surveyed 906 registered voters and carries a margin of error of plus or minus 3.5 percentage points.

What does this article mean when it says that the race is a "statistical dead heat"? First of all, "dead heat" is a synonym for "tie." So this article seems to be saying that the race is, statistically speaking, a tie. But the poll has Obama ahead 50% to 45%, so why does the article say that the race is a statistical tie? What is going on here?

Assume a 95% confidence level. The ±3.5% margin of error in this poll means that we can say with 95% confidence that Obama's poll number is within 3.5 percentage points of his actual percentage—so it could be as low as $50 - 3.5 = 46.5\%$. We can say with 95% confidence that McCain's poll number is within 3.5 percentage points of his actual percentage, so it could be as high as $45 + 3.5 = 48.5\%$. Because the confidence intervals overlap, McCain could actually be ahead of Obama. In other words, the difference of five percentage points between them is "within the margin of error." This fact is no doubt what the reporter means by a "statistical dead heat." The fact that the difference of five percentage points between them is within the margin of error does not, however, mean that they are statistically tied.

The article also says, "Obama holds a slight advantage over McCain, though not a big enough one to constitute a statistical lead." This refers to the same idea as above—namely, that the difference of five percentage points between them is within the margin of error. But does this mean that we can say nothing at all about Obama having a lead over McCain? The answer is no. Probably Obama does have a lead, but the level of confidence in that statement is smaller than the 95% level we have been considering.

"YOU DON'T HAVE TO VOTE, SAM. THE NETWORKS PREDICTED THE WINNER THREE HOURS AGO."

A good explanation may be found in the following excerpt from an earlier online article written by Humphrey Taylor, chairman of the Harris Poll at the time.[9] It, in discussing the 2000 presidential election, clarifies very well the idea of a statistical dead heat.

> If one well-conducted poll shows a candidate with a two-point lead, there is a significant possibility that there is no such lead. However, even a two-point lead should never be described as a **"statistical dead heat"** as the probability is better than 50% that the lead is real. For this reason, **the words "statistical dead heat" should only be used when the candidates are actually tied.** Better still, avoid them altogether.
>
> When, as they do now, one poll after another shows the vice president with a lead, even a small one, the chances statistically that this is due to sampling error are so small as to be virtually non-existent.

Methodology of polling

In our initial discussion of polling, we assumed that we could obtain information from a random sample of the American public in the same way we could determine the color of randomly selected marbles. In fact, it is much harder to "randomly select" an American than it is to randomly select a marble, and it is far easier to determine a marble's color than it is to determine an individual's opinion on a given issue. In an idealized poll, all that is needed to get a more accurate poll is to increase the sample size. In practice, things are not so easy. Here is another quote from the article by Humphrey Taylor cited before: "The idea that the sample size is the main source of error, and that the bigger the sample the smaller the error, is dangerously misleading."

The professional polling organizations themselves are good sources of information on the difficulties of accurate polling. The following excerpt is from the Web site of the Harris Poll. It is the methodology section of a polling report.

---

[9] "Contrary to what much of the media have been saying, Gore has a statistically significant, if not huge, lead over Bush," September 18, 2000, http://www.harrisinteractive.com/vault/Harris-Interactive-Poll-Research-CONTRARY-TO-WHAT-MUCH-OF-THE-MEDIA-HAVE-BEEN-SAYIN-2000-09.pdf.

The American Public Strongly Supports Social Security Reform

January 3, 2007

The Harris Poll was conducted online within the United States between December 12 and 18, 2006 among 2,309 adults (aged 18 and over). Figures for age, sex, race, education, region and household income were weighted when necessary to bring them into line with their actual proportions in the population. Propensity score weighting was also used to adjust for respondents' propensity to be online....

With pure probability samples, with 100 percent response rate, it is possible to calculate the probability that the sampling error (but not other sources of error) is not greater than some number. With a pure probability sample of 2,309 adults one could say with a 95 percent probability that the overall results would have a sampling error of $+/-$ 2 percentage points. However that does not take other sources of error into account. This online survey is not based on a probability sample and therefore no theoretical sampling error can be calculated.

CNN host Wolf Blitzer.

The Harris Poll methodology statement above says, "This online sample is not based on a probability sample." A *probability sample* is one in which the participants are selected randomly, which is crucial to the mathematical basis of any poll. This particular Harris poll was conducted online, so the participants were not randomly selected. They were "self-selected," which means that they are people who happened to find the Web site and chose to respond. By contrast, one Zogby America poll states that the poll "involved 1013 likely voters selected randomly from throughout the 48 contiguous states using listed residential telephone numbers." This is arguably a probability sample, although it misses people without landline phones, for example.

Newspapers and TV shows often invite their audiences to respond to questions. For example, there is currently a CNN show hosted by Wolf Blitzer during which commentator Jack Cafferty asks a question and seeks a response from viewers. It is pointed out that this is not a "scientific poll," which means, among other things, this is not a probability sample—the respondents are not randomly chosen and may not represent a true cross section of the whole country. For example, the audience of such a show could be predominantly conservative or predominantly liberal, which would skew the poll. Because such a poll includes only those people who watch this particular show and want to be heard, it should be viewed skeptically.

Randomness of the sample is not the only methodology problem. In the News 6.9 notes several sources of error. These include issues such as question structure and order, which may have more effect on the results than the sample size.

Psychologists often play a role in composing the questions used in polls. The questions must be phrased in a way that elicits the desired information without causing an unwanted bias in the response. Here's an example to illustrate how the phrasing might easily affect the response. Consider how your own response might be affected by the different ways of wording the questions.

Question version 1: Do you support efforts to prevent fires in national forests?

Question version 2: Do you support efforts to prevent fires in national forests by thinning trees to maintain healthy forests?

Question version 3: Do you support efforts to prevent fires in national forests by increased road construction to allow commercial loggers to thin trees?

Question version 4: Do you support efforts to prevent fires in national forests even though fire is an essential part of maintaining a healthy ecosystem?

There are valid reasons to trust (perhaps with a grain of salt) reports from professional polling organizations such as Gallup, Harris, or Zogby that use random sampling and employ specialists from a number of areas to ensure that questions are properly composed and answers properly reported. But there is every reason to be skeptical of polls that do not use representative samples or are conducted by someone who has a point of view or vested interest to support.

We close this section with the following excerpt from the *New York Times*. It provides a striking example of why we should read polls with caution.

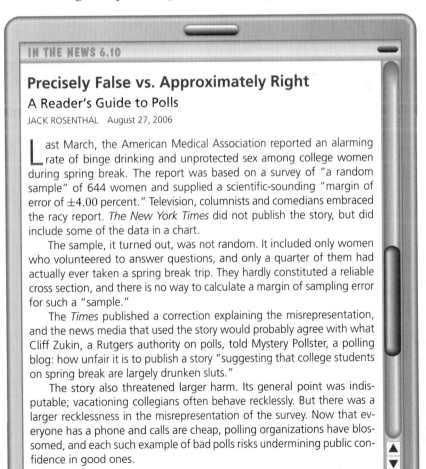

IN THE NEWS 6.10

Precisely False vs. Approximately Right
A Reader's Guide to Polls
JACK ROSENTHAL   August 27, 2006

Last March, the American Medical Association reported an alarming rate of binge drinking and unprotected sex among college women during spring break. The report was based on a survey of "a random sample" of 644 women and supplied a scientific-sounding "margin of error of ±4.00 percent." Television, columnists and comedians embraced the racy report. *The New York Times* did not publish the story, but did include some of the data in a chart.

The sample, it turned out, was not random. It included only women who volunteered to answer questions, and only a quarter of them had actually ever taken a spring break trip. They hardly constituted a reliable cross section, and there is no way to calculate a margin of sampling error for such a "sample."

The *Times* published a correction explaining the misrepresentation, and the news media that used the story would probably agree with what Cliff Zukin, a Rutgers authority on polls, told Mystery Pollster, a polling blog: how unfair it is to publish a story "suggesting that college students on spring break are largely drunken sluts."

The story also threatened larger harm. Its general point was indisputable; vacationing collegians often behave recklessly. But there was a larger recklessness in the misrepresentation of the survey. Now that everyone has a phone and calls are cheap, polling organizations have blossomed, and each such example of bad polls risks undermining public confidence in good ones.

Additional excerpts from this article appear in Exercise 30. The entire article is a good resource for anyone who wishes to understand the nature of polls, the difficulties involved in conducting them, and the concerns the public should have regarding them.

Try It Yourself answers

Try It Yourself 6.15: Interpreting polls: Presidential approval   In 93% of such polls, the reported approval rating of the president will be within three percentage points of the true approval rating. Thus, we can be 93% confident that the true approval rating is between 32% and 38%.

Try It Yourself 6.16: Estimating margin of error: Poll of 1600   2.5%.

Try It Yourself 6.17: Finding confidence interval: Eminent domain   71.9% to 78.1%. If we conducted this poll 100 times, for about 95 of the results, the true value would lie in the confidence interval corresponding to the result. Therefore, we are 95% confident that the actual percentage lies between 71.9% and 78.1%.

Try It Yourself 6.19: Estimating sample size: 5% margin of error   400.

Exercise Set 6.3

1. Estimating margin of error. A polling organization conducts a poll by making a random survey of 1500 people. Estimate the margin of error at a confidence level of 95%. Round your answer as a percentage to one decimal place.

2. Estimating sample size. A polling organization conducts a poll by making a random survey and is willing to accept a margin of error of 5% at a confidence level of 95%. What should the sample size be?

3. CNN poll. The last paragraph in In the News 6.8 (page 394) refers to a poll that "surveyed 906 registered voters and carries a margin of error of plus or minus 3.5 percentage points." Assume a 95% confidence level, and calculate the margin of error. Is your answer 3.5%? If not, how might you explain the difference between your answer and the one given in this news article?

The Patriot Act   A poll of 602 likely voters in Northern Nevada found a majority—60%—supported the Patriot Act as a tool to fight terrorism. The poll reported a margin of error of 4%. Exercises 4 through 6 refer to this poll.

4. Use the rule of thumb for the margin of error to estimate the margin of error for this poll, assuming a 95% confidence level. (Round your answer as a percentage to one decimal place.) How does your answer compare to the margin of error reported by the poll?

5. Explain in plain English what we can conclude from this poll. Can you assert with confidence that a majority of likely voters in northern Nevada support the Patriot Act?

6. Why would the poll settle for a 4% margin of error? Isn't this a bit large? Shouldn't they have polled more people to reduce the margin of error?

7. Kentucky poll. The following is a news excerpt. Compare the margin of error to what you obtain from the rule of thumb

in this section. (Assume that the level of confidence is 95%, and round your estimate for the margin of error as a percentage to one decimal place.)

IN THE NEWS 6.11

Beshear Leads Williams in Ky. Governor's Race by 21 Points after Ad Blitz, cn|2 Poll Shows

RYAN ALESSI   June 09, 2011

After airing ads for the last three weeks, Democratic Gov. Steve Beshear posted a 21-point lead over the Republican challenger David Williams, according to the latest cn|2 Poll.

Beshear and his running mate, Jerry Abramson, received 51% of support compared to just under 30% for Williams and his running mate, Agriculture Commissioner Richie Farmer. The independent ticket of Gatewood Galbraith and Dea Riley polled at below 6%, and nearly 14% said they were undecided....

The cn|2 Poll was conducted June 6-8 by live interviewers from Braun Research, Inc. The 802 respondents contacted for the survey voted in the 2007 or 2010 elections or both with about 88% of those surveyed having voted in both. The poll has a margin of error of 3.5 points....

An imaginary poll   Suppose we conduct a poll to determine American public opinion concerning our current relations with

Europe. We will give the respondents three options: approve, disapprove, or no opinion. Here are some hypothetical data.

Approve	36%
Disapprove	31%
No opinion	33%

Use these data for Exercises 8 and 9.

8. Suppose only 100 people were sampled. Estimate the margin of error for a 95% confidence level.

9. Suppose this poll has a 95% confidence level with a margin of error of 3.5%. Explain in practical terms what this means, and estimate how many people were sampled.

War in Afghanistan   Exercises 10 through 12 refer to the following excerpt from an article in the *Washington Post*.

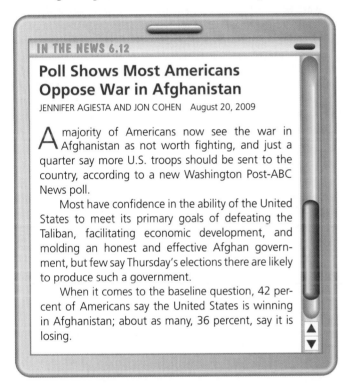

IN THE NEWS 6.12

Poll Shows Most Americans Oppose War in Afghanistan

JENNIFER AGIESTA AND JON COHEN   August 20, 2009

A majority of Americans now see the war in Afghanistan as not worth fighting, and just a quarter say more U.S. troops should be sent to the country, according to a new Washington Post-ABC News poll.

Most have confidence in the ability of the United States to meet its primary goals of defeating the Taliban, facilitating economic development, and molding an honest and effective Afghan government, but few say Thursday's elections there are likely to produce such a government.

When it comes to the baseline question, 42 percent of Americans say the United States is winning in Afghanistan; about as many, 36 percent, say it is losing.

10. Can you glean any information from this article regarding confidence level or margin of error?

11. In light of your answer to Exercise 10, what can you really conclude from this article?

12. Suppose we learn that this poll had sampled 1300 people and had a confidence level of 95%. Estimate the margin of error. Round your answer as a percentage to one decimal place.

13. Increasing sample size. If you double the sample size in a poll, what effect does that have on the margin of error? By what factor should the sample size in a poll be increased in order to cut the margin of error in half?

14. Proportionality. One quantity is *proportional* to another if it is a nonzero, constant multiple of the other. Thus, the statement "y is proportional to x" can be expressed using the formula $y = cx$, where c is a nonzero constant.

Determine which of the following is correct, and explain your answer. For a sample size of n, the margin of error is approximately proportional to:

a. n c. $\sqrt{n}$

b. $1/n$ d. $1/\sqrt{n}$.

90% confidence levels   There are estimates of the margin of error for confidence levels other than 95%. For a 90% confidence level and a sample size of n, the margin of error is approximately $82/\sqrt{n}\%$.

15. If we survey 1000 people, what is the approximate margin of error for a 90% confidence level? Round your answer as a percentage to one decimal place.

16. Suppose we conduct a poll using a representative sample of 1000 people to find whether Americans like salad. We find that 53% say they do. For a 90% confidence level, explain in plain English what we can conclude from this survey. Can you assert with confidence that a majority of Americans like salad?

17. Suppose our salad survey of 1000 people found that only 48% like salad. For a 90% confidence level, explain in plain English what we can conclude from this survey. Can you assert with confidence that less than a majority of Americans like salad?

18. Suppose we wish to have a margin of error of 2% with a 90% confidence level. Approximately how many people should we interview?

19. Sample size and margin of error. The margin of error for a given sample size presented in Rule of Thumb 6.2 is a good estimate. But it is not exactly right. To get the exact number, we need to make a complicated calculation using the methods developed in Section 5.4. Suppose 25% of Americans actually believe Congress is doing a good job, but we don't know this. So we select n Americans at random and ask their opinion about Congress. The probability that our poll will show between 23% and 27% (for a margin of error of 2%) depends on the sample size n and is given in the table below. (This table is the same as the one on p. 387.)

Number n selected	100	500	1000	1500	2000	2500
Probability 23% to 27%	0.44	0.72	0.87	0.93	0.96	0.98

a. For a 2% margin of error, what confidence level will a sample size of 1000 give?

b. We wish to have a margin of error of 2%. What sample size should we choose for a 72% confidence level?

20. Biased questions? Suppose you conduct a poll asking about foreign policy. You conceive of two possible ways to word the question you wish to ask:

First question: Do you approve of current U.S. policies toward Iran?

Second question: Do you approve of the president's policies toward Iran?

Do you think you would get the same result using either question? Explain why or why not.

21. The war in Afghanistan. Suppose you supported going to war in Afghanistan. Make up a question asking whether people supported going to war in Afghanistan that will bias the answer your way. Now do this supposing you did not support going to war in Afghanistan.

22. Tax cuts. Suppose you are in favor of a tax cut. Make up a question asking whether people support a tax cut that will bias the answer your way. Now do this supposing you are against a tax cut.

23. Polling your campus. If you poll students on your campus regarding how they intend to vote in an upcoming state election, do you believe the results would give a valid picture of what might happen in the statewide election? Explain why or why not.

Simulating a poll   Consider the following simulation of repeated polling. We toss a million or so fair coins on the floor and think of them as the voting population. Coins that show a head intend to vote Democratic. Coins that show a tail intend to vote Republican. We want to predict the upcoming election, so we do what the Harris Poll would do. We draw a random sample of 1000 coins from the floor and tabulate how they intend to vote.

But let's do more. Let's repeat this poll simulation 1000 times and tabulate the results. With the aid of computer simulation, we did exactly that. The number of polls showing the given percentage range of people intending to vote Democratic is shown in the accompanying table. (Here, "47% to 48%" means 470 through 479 people intending to vote Democratic, and so on.)

Percent voting Democratic	Less than 47%	47% to 48%	48% to 49%	49% to 50%
Number of polls	23	75	140	234

Percent voting Democratic	50% to 51%	51% to 52%	52% to 53%	At least 53%
Number of polls	250	157	94	27

Exercises 24 through 29 refer to this context.

24. Recall that these are fair coins. What is an accurate description of the voting population (in terms of percentage voting Democratic)? (For this question, the table with the polling data is irrelevant.)

25. Use Rule of Thumb 6.2 to estimate the margin of error for each of these polls for a confidence level of 95%. Round your answer as a percentage to the nearest whole number.

26. What percentage of polls gave predictions that within three percentage points accurately describe the voting population, as given in the answer to Exercise 24? Round your answer as a percentage to the nearest whole number.

27. Explain how your answers in Exercises 24, 25, and 26 are consistent with each other.

28. What percentage of polls gave predictions that within two percentage points accurately describe the voting population, as given in the answer to Exercise 24? Round your answer as a percentage to the nearest whole number.

29. Make a histogram summarizing the data in the table. (See Section 6.1.)

30. Be careful about polls. The following gives further excerpts from In the News 6.10 (page 397). In this article, there are five sections: False Precision, Sampling Error, Questions, Answers, and Intensity. In a few sentences, summarize the main points being made in each of these sections.

ADDITIONAL EXCERPTS FROM: Precisely False vs. Approximately Right: A Reader's Guide to Polls by Jack Rosenthal, The Public Editor, August 27, 2006.

False Precision

Beware of decimal places. When a polling story presents data down to tenths of a percentage point, what the pollster almost always demonstrates is not precision but pretension. A recent Zogby Interactive poll, for instance, showed that the candidates for the Senate in Missouri were separated by 3.8 percentage points. Yet the stated margin of sampling error meant the difference between the candidates could be seven points. The survey would have to interview unimaginably many thousands for that zero point eight to be useful.

Sampling Error

The *Times* and other media accompany poll reports with a box explaining how the random sample was selected and stating the sampling error. Error is actually a misnomer. What this figure actually describes is a range of approximation.

There's also a formula for calculating the error in comparing one survey with another. For instance, last May, a *Times/CBS News* survey found that 31 percent of the public approved of President Bush's performance; in the survey published last Wednesday, the number was 36 percent. Is that a real change? Yes. After adjustment for comparative error, the approval rating has gained by at least one point.

For a typical election sample of 1,000, the error rate is plus or minus three percentage points for each candidate, meaning that a 50-50 race could actually differ by 53 to 47. But the three-point figure applies only to the entire sample. How many of those are likely voters? In the recent Connecticut primary, 40 percent of eligible Democrats voted. Even if a poll identified the likely voters perfectly, there still would be just 400 of them, and the error rate for that number would be plus or minus five points. So to win confidence, a finding would have to exceed 55 to 45.

Questions

How questions are phrased can mean wide shifts, even with wholly neutral words. Men respond poorly, for instance, to questions asking if they are "worried" about something, so careful pollsters will ask if they are "concerned."

The classic "double negative" example came in July 1992, when a Roper poll asked, "Does it seem possible or does it seem impossible to you that the Nazi extermination of the Jews never happened?" The finding: one of every five Americans seemed to doubt that there was a Holocaust. How much did that startling finding result from the confusing question? In a follow-up survey, Roper asked a clearer question, and the number of doubters plunged from the original 22 percent to 1 percent....

The order of questions is another source of potential error. That's illustrated by questions asked by the Pew Research Center. Andrew Kohut, its president, says: "If you first ask people what they think about gay marriage, they are opposed. They vent. And if you then ask what they think about civil unions, a majority support that."

Answers

People never wish to look uninformed and will often answer questions despite ignorance of the subject. Some 40 years into the cold war, many respondents were still saying yes, Russia is a member of NATO. That's why, says Rob Daves, head of the American Association of Public Opinion Researchers, skillful pollsters will first ask, for new or sophisticated subjects, a scaling question like, How much do you know about this issue: a great deal, some, not at all?

Respondents also want to appear to be good citizens. When the *Times/CBS News* poll asks voters if they voted in the 2004 presidential election, 73 percent say yes. Shortly after the election, however, the Census Bureau reported that only 64 percent of the eligible voters actually voted.

Jon Krosnick, an authority on polling and politics at Stanford, uses the term "satisficing" to describe behavior when a pollster calls. If people find the subject compelling, they become engaged. If not, they answer impatiently. Either way, says Kathy Frankovich, director of surveys for *CBS News*, "people grab the first thing that comes to mind."

Intensity

How strongly people feel about an issue may be the most important source of poll misunderstanding. In survey after survey, half the respondents favor stronger gun controls—but don't care nearly as much as the 10 percent who want them relaxed.

Intensity can be measured by asking a scaled question: Is the issue of abortion so important that you will cast your vote because of a candidate's position? One of several important issues? Not important? Each added question increases the interview length, testing the respondent's patience and the pollster's budget. Nevertheless, on divisive issues, responsible pollsters will ask four, five, even a dozen questions, probing for true feelings.

31. History. Two famous examples of polls predicting the wrong outcome of an election are given by the American presidential elections in 1936 and 1948. Do research to discover who the candidates were, what happened in the elections, and what the polling organizations did wrong.

The following exercises are designed to be solved using technology such as calculators or computer spreadsheets. For assistance, see the technology supplement.

32. Margin of error. Make a table showing the approximate margin of error for sample sizes ranging from 1000 to 2000 in steps of 100. Assume that the level of confidence is 95%, and round each percentage to two decimal places.

33. Sample size. Make a table showing the sample size needed to get a given margin of error ranging from 2% to 3% in steps of 0.1 percentage point. Assume that the level of confidence is 95%, and round to the nearest whole number.

6.4 Statistical inference and clinical trials: Effective drugs?

> **TAKE AWAY FROM THIS SECTION**
>
> Understand statistical significance and *p*-values.

We have already examined the role of statistical inference in polling. Another important use of statistics is in medical research, where statistical inference is used to answer questions about the efficacy of medical treatments and drugs.

The article on the following page is an announcement of a clinical trial for a treatment of anemia induced by chemotherapy in cancer patients.

Experimental design

There are many factors involved in the setup of an experiment or clinical trial. In this case, the purpose of the trial is to determine the efficacy (or lack of efficacy) of a drug.

In a typical trial, one group of subjects is given the actual drug or treatment and another group is given a *placebo*. A placebo is a benign substance that contains no medication and that is outwardly indistinguishable from the real drug. One kind of

Celgene and Acceleron Initiate Phase 2/3 Study of ACE-011 (sotatercept) to Treat Chemotherapy-Induced Anemia in Patients with Lung Cancer

BENZINGA STAFF   June 2, 2011

Acceleron Pharma, Inc., a biopharmaceutical company developing protein therapeutics for cancer and orphan diseases, and Celgene Corporation (NASDAQ: CELG) today announced the initiation of the first part of a Phase 2/3 clinical study of ACE-011 (sotatercept) for the treatment of chemotherapy-induced anemia (CIA) in patients with metastatic non-small cell lung cancer (NSCLC)....

"We are excited to see the initiation of this late-stage clinical trial of sotatercept as there is enormous unmet medical need for a safe and effective alternative to ESAs for the treatment of chemotherapy-induced anemia," said Matthew Sherman, MD, Chief Medical Officer of Acceleron. "This study, along with an ongoing Phase 2 clinical trial in patients with end-stage renal disease on hemodialysis, will provide insight into the therapeutic potential of sotatercept as a first-in-class treatment for anemia."

The clinical trial is designed as a randomized, dose ranging study of sotatercept in patients with metastatic cancer followed by a double-blind, randomized, placebo-controlled study of sotatercept for chemotherapy-induced anemia (CIA) in patients with metastatic non-small cell lung cancer (NSCLC) treated with first-line platinum-based chemotherapeutic regimens.

placebo might be a simple sugar pill. Of course, the subjects do not know which they are getting.[10]

"The test results are in for our new drug. 9 out of 10 doctors recommended the placebo."

---

[10]It's interesting that taking a placebo can create an expectation of improved health and thus result in positive medical benefits. This is called the "placebo effect."

In some studies, the second group will be given a standard treatment rather than a placebo to compare the effect of a known drug with a proposed new drug. In these cases, the known drug will be disguised to look like the proposed drug.

Here are the typical key ingredients when planning a clinical trial:

1. Researchers first decide how much of a difference between treatment groups is medically important.

2. Then they decide how many subjects should be enrolled in the trial. The sample should include enough participants so that the researchers will have confidence in the results.

3. Next, part of the group is given the drug or treatment to be tested, and the other subjects are given the placebo. The group given the placebo is usually called the *control group*.

4. Finally, the researchers compare the subjects in the two groups to see whether there is a significant difference between their conditions after the trial. The key word here is "significant." That's where statistics comes in.

So far so good, but just as in polling, there are many possible pitfalls in such a study. One such pitfall is that the observer or the subject can inadvertently affect the results. For example, suppose a researcher has worked so hard to develop a drug that he or she has come to believe in it, and just *really* wants it to work. If the researcher knows which subjects have been given the drug and which the placebo, the chance exists that he or she may inadvertently interpret the condition of the subjects in favor of the drug.

That is why clinical trials are usually *double blind*, a term used in In the News 6.13 at the beginning of this section. In a double-blind experiment, neither the subjects nor the researchers know who is getting the drug and who is getting the placebo. A third party is the only one who knows which patient is receiving what.

Statistical significance and *p*-values

It happens by chance that some sick people just get better with no treatment at all and some get worse. Therefore, it would be extremely unusual for every subject in the group who received the drug being tested to get better and everyone in the control group to get worse. Usually, there will be some of each in both groups.

The trick is to figure out whether the difference (if any) between the results for the treated group and the control group is "significant" or could be happening simply by chance. Statistics provides a way to compare the results of the study with what would happen due to chance and thus to determine whether the results are *statistically significant*.

Key Concept

The results of a clinical trial are considered to be **statistically significant** if they are unlikely to have occurred by chance alone.

In this Key Concept, we have not said what we mean by "unlikely." We will explain this point shortly. First we present examples related to statistical significance.

EXAMPLE 6.20 Determining statistical significance: Coins and trucks

For each of these situations, determine whether the results described appear to be statistically significant.

    a. You flip a coin 40 times and get heads each time.

    b. You are driving and are passed by two pickup trucks in a row.

SOLUTION

a. By the Counting Principle from Section 5.3, the probability of getting 40 heads in a row by flipping a fair coin is $1/2^{40}$. This number is so unbelievably small that this result is very unlikely to have occurred by chance alone. Therefore, this result is statistically significant.

b. Pickup trucks are quite common in many areas, so being passed by two in a row does not seem terribly unlikely. This result does not seem statistically significant.

TRY IT YOURSELF 6.20

For each of these situations, determine whether the results described appear to be statistically significant.

a. You flip a coin twice and get heads both times.

b. You are driving and you pass three 1951 Studebaker cars in a row.

The answer is provided at the end of this section.

As the examples above indicate, statistical significance corresponds to our common-sense notion of being highly unusual. Although these examples are fairly straightforward, the statistical significance of results of clinical trials can be more subtle. For example, consider the following table showing the results of a hypothetical trial:

	Treated group	Control group
Improved	53	50
Got worse	47	50

Certainly, the treated group improved a little more than the control group, but is this difference statistically significant? The answer is not obvious. Statisticians involved with clinical trials determine the answer by calculating something called the *p-value*. The *p*-value measures how likely it is that the difference between the two groups in a clinical trial would occur by chance alone if the treatment had no effect. The *p*-value is a way to quantify the word "unlikely" that appeared in our explanation of the term *statistically significant*.

Key Concept

The *p*-value measures the probability that the outcome of a clinical trial would occur by chance alone if the treatment had no effect. A small *p*-value is usually interpreted as evidence that the event in question is unlikely to be due to chance alone. The results are usually accepted as statistically significant if the *p*-value is 0.05 = 5% or smaller.

Here is a very important real example involving clinical trials.

EXAMPLE 6.21 Determining statistical significance: Vioxx and *p*-values

The drug Vioxx was once used to treat arthritis. One published study compared the incidence of heart disease in patients treated with Vioxx versus patients treated with Aleve.[11] (Here, the group treated with Aleve is considered to be the control group.)

---

[11] Mukherjee, D., Nissen, S.E., and Topol, E.J., "Risk of Cardiovascular Events Associated With Selective COX-2 Inhibitors," *Journal of American Medical Association* (2001), 954–959.

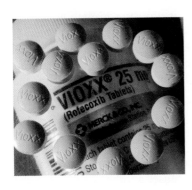

The study showed a much greater incidence of heart disease for the Vioxx patients, with a p-value of 0.002. Explain the meaning of the p-value in this test. Would the result of this study normally be accepted as statistically significant?

SOLUTION

If the treatment had no effect, the probability that the difference in results between the Vioxx and the Aleve patients would occur by chance alone is 0.002 or 2/10th of 1%. Because this probability is much less than the accepted threshold of 0.05, this result is certainly statistically significant.

TRY IT YOURSELF 6.21

The same study compared the effect of Vioxx with the effect of a placebo and also found a greater incidence of heart disease for the Vioxx patients, with a p-value of 0.04. Explain the meaning of the p-value in this case. Would the result of this study normally be accepted as statistically significant?

The answer is provided at the end of this section.

As the example indicates, in practice, one computes a p-value that measures the probability that the difference between the subjects in the control and test groups would occur by pure chance if the treatment had no effect. To explain in detail how this is done would take us too far afield, so we will use a simplified example to give the basic idea.

Idealized trials

Suppose we test a drug that we believe may prevent colds. There is no control group in this case, but suppose we know that, on average, 50% of people will get colds this winter. Suppose we treat 100 patients with this drug and find that 55 did not get a cold. Would we conclude that the drug is effective in preventing colds? Or did the fact that 55 patients did not get a cold just happen by chance?

Here's the idea: Without treatment, 50% of people will get colds, and from this fact it follows that the probability of 55 or more out of 100 subjects not getting a cold is 0.184 or 18.4%.[12] The number 0.184 is the p-value for this experiment. It says that in 18.4% of observations of 100 untreated subjects, we expect that 55 or more of the subjects will not get a cold. Therefore, having 55 or more stay healthy is not a highly unlikely event—we expect it to occur almost 20% of the time even without treatment. The benefits of the test drug would be in doubt if only 55 out of 100 patients we actually treated did not get a cold.

Suppose, on the other hand, that 65 of the patients taking the test drug did not get a cold. The p-value for this outcome turns out to be 0.002 = 0.2%. This is the probability that 65 or more patients will not get a cold if the drug had no effect. We'd be very surprised to see that many patients do not get a cold merely by chance. If 65 of the patients taking the test drug did not get a cold, it is reasonable to believe that something other than chance is at work.

Having a small p-value associated with a study suggests that the outcome is due to something other than chance alone, but it doesn't necessarily say what that something is. For this reason, it's important for the researchers to take care to avoid factors that may cause the two groups to differ other than the drug or procedure being tested. If this is done, then having a small p-value suggests that the drug or procedure was probably responsible.

---

[12]This number can be found using the ideas developed in Section 5.4.

In the end, no matter how good the study, there is always a probability, small though it may be, that the conclusion suggested by the results is incorrect. That is one reason why independent researchers might repeat a clinical trial. If a small *p*-value is obtained in several different trials, then we have much greater confidence in the result. This is important—after all, these results decide which drugs get on the market and which do not.

Correlation and causation

An important issue in medicine is deciding the cause of a certain disease or condition. How do we tell whether one condition actually causes another, or whether the two conditions are simply associated? Mathematically speaking, we would like to describe whether a change in one variable is related to changes in other variables. This is the context for the term *correlation*.

Key Concept

We say that two numerical variables are **positively correlated** if an increase in one of them accompanies an increase in the other. Similarly, two numerical variables are said to be **negatively correlated** if an increase in one of them accompanies a decrease in the other. If neither of these is true, the variables are called **uncorrelated**.

For example, the time required for a trip and the speed you drive are negatively correlated, because an increase in travel time accompanies a decrease in speed. By contrast, the price of a pizza and its diameter are positively correlated because an increase in price usually accompanies an increase in diameter.

In these examples, the correlations are the result of real relationships between the variables, but some correlations are the result of chance or coincidence, and we have to be careful not to draw unwarranted conclusions. Furthermore, even when two variables are connected in some way, we should not necessarily assume that one causes the other. For example, there is a positive correlation between ice cream sales and the number of shark attacks on swimmers. Both of these tend to rise during warm weather when more people swim and buy ice cream, but certainly neither causes the other.

EXAMPLE 6.22 Understanding correlation as an idea: Balloon

Determine whether the diameter of a balloon and its volume are positively correlated, negatively correlated, or uncorrelated.

SOLUTION

As diameter increases, so does the volume. Hence, these variables are positively correlated.

TRY IT YOURSELF 6.22

Determine whether the outdoor temperature and your home heating bill are positively correlated, negatively correlated, or uncorrelated.

The answer is provided at the end of this section.

There are statistical methods that quantify the extent to which two variables are, in fact, correlated. In particular, the method called *linear regression* quantifies the extent to which two variables are *linearly correlated*. We have already introduced linear regression in the context of trend lines in Section 3.1.

The following data show weekly expenditures in British pounds of households in the given region:[13]

Region	Alcohol	Tobacco
North	6.47	4.03
Yorkshire	6.13	3.76
Northeast	6.19	3.77
East Midlands	4.89	3.34
West Midlands	5.63	3.47
East Anglia	4.52	2.92
Southeast	5.89	3.20
Southwest	4.79	2.71
Wales	5.27	3.53
Scotland	6.08	4.51

We want to determine whether there is a linear correlation between the two variables, spending on alcohol and spending on tobacco. First we make a scatterplot of the data, as shown in **Figure 6.38**. The plot suggests a positive correlation between the two variables: For the most part, an increase in one accompanies an increase in the other. To determine whether this is a *linear* correlation, we add the trend (or regression) line, as seen in **Figure 6.39**. Note that the points seem roughly to fall along this line. This fact leads us to suspect that there is a linear relation between the two variables. In summary, we have evidence that a positive linear correlation exists between spending on alcohol in a region and spending on tobacco in that region.

Figure 6.39 also shows the equation of the trend line. We round the coefficients to get $y = 0.612x + 0.108$. Here, x represents spending on alcohol and y represents spending on tobacco, both in British pounds. This equation lets us do more than just establish a link between the variables. It quantifies, approximately, how the two variables are related. For example, if in a given region, households spend an average of 4 pounds per week on alcohol, we would expect $0.612 \times 4 + 0.108 = 2.556$, or about 2.56 pounds per week to be spent on tobacco. On the other hand, the data do not tell us that the consumption of alcohol causes tobacco usage, or vice versa.

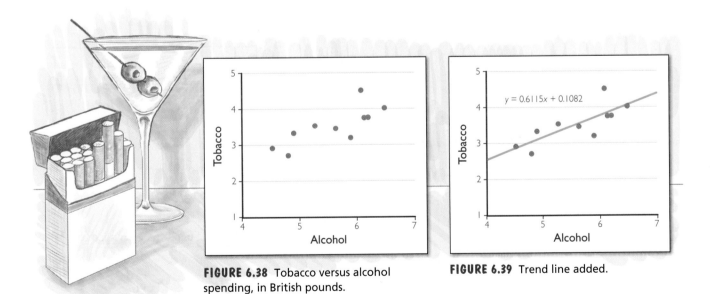

FIGURE 6.38 Tobacco versus alcohol spending, in British pounds.

FIGURE 6.39 Trend line added.

[13] Taken from *Family Expenditure Survey*, Department of Employment (British Official Statistics). We have omitted the data for Northern Ireland.

EXAMPLE 6.23 Determining correlation from data: Dental care

Consider the following data table comparing by state the percentages of people who visited the dentist the past year and the percentages of people over age 65 who have had all of their teeth extracted.

State	Dental visit	All teeth extracted
Oklahoma	58.0%	28.3%
Nevada	66.2%	18.4%
Alaska	66.9%	23.6%
Florida	68.7%	17.4%
New York	71.8%	17.5%
Connecticut	80.5%	12.8%

Figure 6.40 shows a plot of the data along with the trend line. Do these data show a correlation between people visiting the dentist within the past year and the elderly having all their teeth extracted? If so, what type of correlation exists?

SOLUTION

In general, as the percent visiting the dentist increases, the percent having had all their teeth extracted decreases. Therefore, the two variables are correlated, and the correlation is negative. The graph suggests that the correlation is linear.

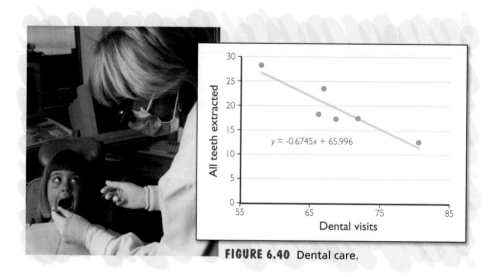

$y = -0.6745x + 65.996$

FIGURE 6.40 Dental care.

TRY IT YOURSELF 6.23

The following table shows the running speed of various animals versus their length. We discussed these data in Section 2.1 and in Section 3.1.

Animal	Length (inches)	Speed (feet per second)
Deermouse	3.5	8.2
Chipmunk	6.3	15.7
Desert crested lizard	9.4	24.0
Grey squirrel	9.8	24.9
Red fox	24.0	65.6
Cheetah	47.0	95.1

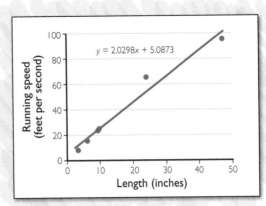

FIGURE 6.41 Length and speed.

Figure 6.41 shows a plot of the data along with the trend line. Are length and speed correlated? If so, what type of correlation is there?

The answer is provided at the end of this section.

Exactly how strong is the evidence of a linear correlation between alcohol and tobacco expenditures or between seeing the dentist regularly and having had all teeth extracted? For alcohol and tobacco purchases, consider again Figure 6.39. If the points were to lie exactly on the line, anyone would be convinced of a strong linear relationship. In fact, the points roughly follow the trend line, but most of them do not fall on the line. How do we decide how strong the correlation is?

Statisticians attempt to quantify the degree of linear correlation by using a number known as the *correlation coefficient*. The formula for calculating the correlation coefficient is complicated, and the job is best left to your calculator or computer. In the case of Figure 6.39, the correlation coefficient is 0.78. What does this number mean?

Key Concept

The correlation coefficient always lies between −1 and 1 inclusive. The closer a correlation coefficient is to 1, the greater the degree of positive linear correlation, with 1 indicating perfect positive linear correlation (the points lie exactly on the increasing trend line). The closer a correlation coefficient is to −1, the greater the degree of negative linear correlation, with −1 indicating perfect negative linear correlation (the points lie exactly on the decreasing trend line). A correlation coefficient near 0 indicates little if any *linear* correlation.

Note that when the coefficient is near 0, we can't conclude that there is no relationship between the variables. We can conclude only that no *linear* relationship exists.

The sign of the correlation coefficient matches the sign of the slope of the trend line. A positive correlation is indicated by a positive slope and an increasing line, and a negative correlation is indicated by a negative slope and a decreasing line. The graphs in **Figures 6.42 through 6.45** may help.

Interpreting correlation coefficients partly requires judgment and experience, just as it does with *p*-values and (as we saw in the preceding section) confidence levels. After all, the 95% confidence level is rather arbitrary. One rule of thumb is that a correlation coefficient whose magnitude is 0.8 or more indicates strong linear correlation, a coefficient whose magnitude is between 0.5 and 0.8 indicates moderate

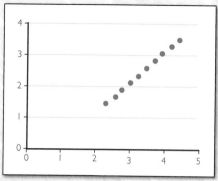

FIGURE 6.42 Correlation coefficient of 1 shows a strong linear relation with positive slope.

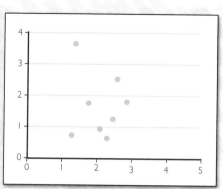

FIGURE 6.43 Correlation coefficient of 0.02 shows little if any linear relation.

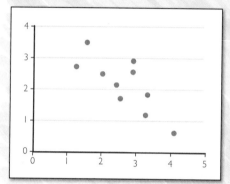

FIGURE 6.44 Correlation coefficient of −0.77 shows a moderately strong linear relation with negative slope.

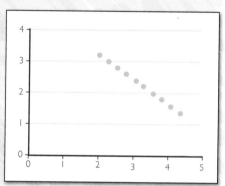

FIGURE 6.45 Correlation coefficient of −1 shows a strong linear relation with negative slope.

linear correlation, and a coefficient whose magnitude is less than 0.5 indicates weak linear correlation at best. By this measure, the correlation coefficient of 0.78 in the example concerning alcohol and tobacco spending is fairly good evidence of linear correlation.

RULE OF THUMB 6.4 Correlation Coefficient

A correlation coefficient whose magnitude is 0.8 or more indicates strong linear correlation, a coefficient whose magnitude is between 0.5 and 0.8 indicates moderate linear correlation, and a coefficient whose magnitude is less than 0.5 indicates weak linear correlation at best.

EXAMPLE 6.24 Interpreting correlation: Diversifying index funds

An *index fund* is a type of mutual fund that invests in a large number of diversified stocks. An index fund that invests in U.S. stocks is commonly considered to be a fairly safe investment. But if U.S. stock prices fall overall, it's possible that you would have been better off if you had also invested in a fund that is not strongly correlated with U.S. stocks.

One calculation shows that the correlation coefficient between the prices of U.S. stocks in the S&P 500 Index and stocks in the German market called the DAX averaged 0.528. Does this correlation coefficient indicate that there is a great advantage in adding a German stock index fund to your portfolio?

DAX traders.

SOLUTION

A correlation coefficient of 0.528 indicates moderate positive linear correlation. This says that a German index fund's prices will rise and fall, more or less, in a similar way to a U.S. index fund, so there is no great advantage in adding this investment.

TRY IT YOURSELF 6.24

Another calculation shows that the correlation coefficient between the prices of U.S. stocks and Japanese stocks (the S&P 500 Index and the Nikkei 225) averaged 0.138. Does this correlation coefficient indicate that there is an advantage in adding a Japanese stock index fund to your portfolio?

The answer is provided at the end of this section.

Causation

Now we return to the subject of causation and its relation to correlation. Consider the following excerpt from a news article.

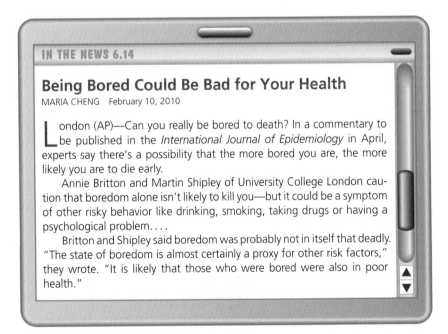

IN THE NEWS 6.14

Being Bored Could Be Bad for Your Health

MARIA CHENG   February 10, 2010

London (AP)—Can you really be bored to death? In a commentary to be published in the *International Journal of Epidemiology* in April, experts say there's a possibility that the more bored you are, the more likely you are to die early.

Annie Britton and Martin Shipley of University College London caution that boredom alone isn't likely to kill you—but it could be a symptom of other risky behavior like drinking, smoking, taking drugs or having a psychological problem. . . .

Britton and Shipley said boredom was probably not in itself that deadly. "The state of boredom is almost certainly a proxy for other risk factors," they wrote. "It is likely that those who were bored were also in poor health."

This article suggests that there is a link (a positive correlation) between boredom and early death. By noting that boredom is probably only a symptom associated with other risk factors, the article warns us against concluding that boredom is a *cause* of early death.

We have already noted that determining whether one event causes another event is much more difficult than determining mere correlation. Two phenomena can be correlated without one of them causing the other. When two quantities are correlated, there are basically three possibilities: One causes the other, both are caused by something else, or the correlation may simply be coincidence—just pure chance.

Here is an example: Suppose a mother notices that from time to time her baby gets a runny nose and a cough. The cough seems to come first most of the time, so she concludes that the cough causes the runny nose. In fact, the most likely explanation is that both are caused by a third factor: The baby has a cold. That they occur together, and thus are correlated, does not mean that one must cause the other.

For another example, consider again the data on tobacco and alcohol consumption in British households for various regions of Britain. Consumption of tobacco and consumption of alcohol in a region are certainly positively correlated because, in general, an increase in one corresponds to an increase in the other. (In fact, the correlation coefficient of 0.78 confirms the positive correlation.) It's possible that one of these somehow causes the other. But it's also possible that people who have a tendency to use one of these products also have a tendency to use the other.

A classic example of an argument about correlation versus causation is the connection between smoking and lung cancer. Many studies have shown that smoking and lung cancer are positively correlated. Although it is indisputable that smokers have a much higher rate of lung cancer than non-smokers, the tobacco industry has argued in the past that this does not necessarily tell us that smoking *causes* lung cancer. They have suggested that it may be like the baby with a cough and runny nose. The conditions may occur together, but how can we say one causes the other? Maybe, they argue, there is a gene or some unknown behavior that increases the propensity for both smoking and lung cancer.

Showing causation requires much more evidence than showing correlation, but enough research has been done to convince the medical community that smoking does, in fact, cause lung cancer.

Try It Yourself answers

Try It Yourself 6.20: Determining statistical significance: Coins and cars
a. The probability of getting heads twice in a row is 1/4, so this result is not statistically significant.

b. Certainly, 1951 Studebaker cars are few and far between, so it would be unusual enough to pass even one. Passing three such cars in a row is surely statistically significant.

Try It Yourself 6.21: Determining statistical significance: Vioxx and *p*-values   If the treatment had no effect, the probability that the difference in results between the Vioxx and the placebo patients occurred by chance alone is 0.04 or 4%. Because this is less than the accepted threshold of 0.05, this result is statistically significant.

Try It Yourself 6.22: Understanding correlation as an idea: Heating bill   They are negatively correlated.

Try It Yourself 6.23: Determining correlation from data: Running speed   The two variables have a positive linear correlation.

Try It Yourself 6.24: Interpreting correlation: Diversifying index funds.   Yes.

Exercise Set 6.4

1. Ethics. Suppose we want to test the idea that unregulated diets for children lead to serious health problems. Comment on the ethics of the following experimental design. We select two groups of small children. We let one group eat whatever they want whenever they want it. The second group eats what nutritionists and physicians tell us is a healthy diet. At the end of a few years, we analyze the health of each group.

Can you think of a more ethical design?[14]

2. *p*-value. A study to test the efficacy of a drug for the treatment of allergic rhinitis (inflammation of the nasal mucous membranes) reported a *p*-value as 0.012. Interpret carefully what this means.

3. An interesting drug. Does anything appear to be wrong in the following news release? Explain. *A drug company announced the results of a clinical trial for a certain drug for hangnails. Patients showed statistically significant improvement (p-value > 0.05) in the treatment group as compared to the control group.*

4. What caused it? In a clinical trial for patients with hangnails, the test group soaked their hands in warm water and were given an experimental drug. They showed statistically significant improvement (*p*-value < 0.03) as compared to the control group, which did nothing to treat their hangnails. The president of the drug company sponsoring this trial asserted that this *p*-value shows that the new drug had a 97% probability of helping hangnails. Discuss the validity of the president's comment.

5. Sunburn. The CEO of a certain drug company announced that the results of a clinical trial for an experimental drug in patients with sunburn showed statistically significant improvement. His justification for this claim was the following fact: 55% of the subjects treated with the drug got better in two days, but it is well known that, on average, only 50% of sunburn victims get better in two days. Discuss the validity of the CEO's comment. *Suggestion*: To find the *p*-value, see the discussion on p. 405 of a test drug for preventing colds.

6. What's the cause? A certain drug company announced that the results of a clinical trial for a new drug in patients with sunburn showed statistically significant improvement (*p*-value = 0.04). The president of the company asserted that this shows that 96% of the subjects in the trial were helped by the drug. Discuss the validity of the president's comment.

The efficacy of hormone replacement therapy Exercises 7 through 9 refer to In the News 6.15.

7. Explain what this sentence means: "The researchers intended to follow the women for a median of five years."

Hormone Replacement Therapy and Breast Cancer Relapse

NATIONAL CANCER INSTITUTE   February 5, 2004

For many years, HRT [Hormone Replacement Therapy] (usually a combination of the hormones estrogen and progestin) was widely prescribed to women to relieve…menopausal symptoms. It was also thought that HRT might reduce the risk of breast cancer, heart disease, and other conditions.

However, in July 2002, a large randomized clinical trial of estrogen and progestin in healthy postmenopausal women (part of the Women's Health Initiative) was stopped early when researchers found that women who took the hormones had an increased risk of developing breast cancer and heart disease. . . .

The effects of HRT on women who had already had breast cancer had not been studied in a randomized controlled trial, considered the "gold standard" in medical research. Because more than half of breast cancers are fueled by estrogen, some researchers worried that use of the hormone could stimulate recurrence of the disease. However, studies that simply observed breast cancer survivors for several years concluded that the risk of cancer recurring in HRT users was low.

In 1977, Swedish researchers began a randomized trial to determine whether a two-year course of HRT for menopausal symptoms was safe for women who had been treated for breast cancer. A total of 434 study participants were randomly assigned to receive either HRT or non-hormonal treatment for their menopausal symptoms. . . .

The researchers intended to follow the women for a median of five years. However after a median follow-up of just over two years, they found that 26 women in the HRT group—but only seven in the non-HRT group—had had a recurrence of breast cancer. They terminated the study concluding that short-term use of HRT posed an "unacceptably high risk" of breast cancer recurrence.

---

[14]Many such totally ethical studies have been done, and the results surprise no one.

8. Assume that half of the women in the study group received HRT and half (the control group) received non-hormonal treatment. What percent of those receiving HRT had a recurrence of breast cancer when the study was stopped? What percent of the control group had a recurrence of breast cancer? Round each answer as a percentage to the nearest whole number.

9. No *p*-value is mentioned, but do you think the researchers considered the difference between the numbers in Exercise 8 statistically significant? Why or why not?

Smarter people   Exercises 10 through 13 refer to the following article from the *New York Times*.

IN THE NEWS 6.16

Breast-Feeding Is Tied to Brain Power

January 6, 1998

A new study suggests that children who were breast-fed as babies do better in school and score higher on standardized tests in mathematics and reading....

The authors, Professors David M. Fergusson and L. John Horwood of Christchurch School of Medicine, subscribe to the theory that fatty acids that are present in breast milk but not in formula promote lasting brain development. The breast-fed children in the study tended to have mothers who were older, better-educated and wealthier. Skeptics say those factors, rather than the breast milk itself, could explain the findings.

But the authors wrote that they had adjusted for those factors and had still concluded: "There were small but consistent tendencies for increasing duration of breast-feeding to be associated with increased I.Q., increased performance on standardized tests, higher teacher ratings of classroom performance and better high school achievement."

---

10. Does the last paragraph suggest that the authors found the results to be statistically significant?

11. Consider young mothers who are unable to breast-feed their infants. Does the study indicate that they should look closely into formulas supplemented with the fatty acids referred to in the article?

12. How do you think the group of babies who did not get breast-fed were selected?

13. The third paragraph indicates that some of the results of this study were called into question. Use the concepts developed in the current section to explain these concerns.

Calculating *p*-values   Suppose that for a certain illness, the probability is 45% that a given patient will improve without treatment. Then the probability that at least *n* out of 20 patients will improve without treatment is given in the following table:

n	10	11	12	13	14	15	16
Probability at least *n* improve	0.41	0.25	0.13	0.058	0.021	0.0064	0.0015

We give an experimental drug to 20 patients who have this illness. Exercises 14 through 16 use this information.

14. Suppose 11 patients show improvement. What is the *p*-value?

15. Suppose we count the test significant if the *p*-value is 0.05 or less. How many patients must show improvement in order to make the test statistically significant?

16. Suppose 16 of the 20 subjects showed improvement. In light of the design of the test, would it be appropriate to recommend this drug as a "cure" for this disease?

17. Correlation as an idea. For each pair of variables, determine whether they are positively correlated, negatively correlated, or uncorrelated.

    a. Brightness of an oncoming car's headlights and your distance from the oncoming car

    b. The number of home runs hit by Barry Bonds in a year and my water bill for that year

    c. Damage in a city at the epicenter of an earthquake and the Richter scale reading of the earthquake

    d. A golfer's tournament ranking and her score

Intensity of light   If one moves away from a light source, the intensity of light striking one's eye decreases. **Figure 6.46** shows a graph of the relationship between the intensity of light and distance from the source. (We have not included the units because they are not relevant here.)

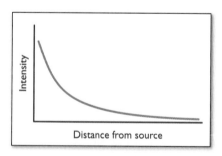

FIGURE 6.46 Intensity versus distance.

18. Is there a correlation between the intensity of light and distance from the source? If so, is the correlation positive or negative?

19. Does the relationship shown by the graph appear to be linear?

20. Correlation coefficient. Assume that **Figures 6.47 through 6.50** show data sets with correlation coefficients 0.99, 0.58, −0.10, and −0.63. Which figure shows data with which correlation coefficient?

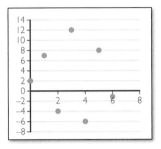

FIGURE 6.47

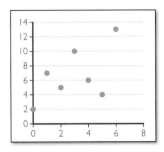

FIGURE 6.48

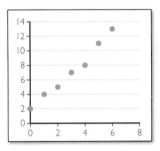

FIGURE 6.49

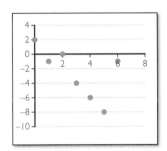

FIGURE 6.50

Advertising expense Figure 6.51 shows the effectiveness of spending on advertising by major American retail companies. Exercises 21 through 24 refer to this graph.

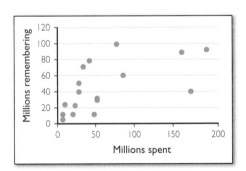

FIGURE 6.51 Spending and effectiveness.

21. Judging solely on the basis of the figure, does there appear to be a positive correlation between spending and effective advertising?

22. In **Figure 6.52** we have added the trend line. As the figure indicates, the equation for the trend line is $y = 0.3632x + 22.163$, where y is the number (in millions) remembering and x is the amount (in millions of dollars) spent on advertising. The next-to-last point on the graph (well below the line) corre-

sponds to Ford Motor Company. Does it appear that Ford was getting good value from its advertising dollars?

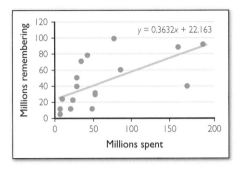

FIGURE 6.52 Regression line added.

23. Use the regression line in Figure 6.52 to estimate how many millions of people Ford should have expected to remember one of its ads. (See Exercise 22.) Round your answer to the nearest 10 million of those remembering.

24. The correlation coefficient was reported as 0.65. Comment on the hypothesis that there is a straight-line relation between advertising spending and effectiveness.

25. A study? Many times you hear testimonials regarding the health benefits of certain activities, foods, or herbal supplements. Suppose your trusted friend Jane tells you that in spite of a good diet and exercise and treatment from her physician, her blood pressure remained high. Then one year ago, she started a daily intake of 2 ounces of dried lemon peel. Her blood pressure returned to normal within 3 months and has remained so.

Discuss the pros and cons of Jane's testimonial regarding the curative properties of dried lemon peel in light of the concepts presented in this section.

26. Vegetarians. The following excerpt comes from an article published by the Vegetarian Society of the United Kingdom.

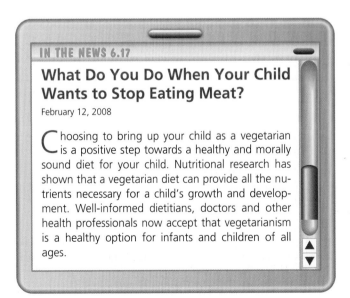

Your good friend Larry tells you that this article has convinced him to raise his child as a vegetarian. Discuss what you would say to Larry in light of the concepts presented in this section.

27. Potatoes contribute to athletics. Suppose a study showed that over 90% of gold medal winners from the 2010 Olympic Games in Vancouver eat potatoes as a regular part of their diet. In light of the concepts presented in this section, discuss whether it's proper to conclude that eating lots of potatoes will improve one's athletic abilities.

28. Coed housing. The following excerpt comes from an article that appeared in the Oklahoma State University newspaper, *The Daily O'Collegian*.

IN THE NEWS 6.18

Coed Housing Has Higher Rates of Sex, Binge Drinking

SEAN HARKIN   December 2, 2009

A recent study found students who live in coed housing are more likely to engage in binge drinking and sexual activities than those living in single-gender housing....

"Now, that doesn't mean coed housing causes it," said Matthew Brown, director of housing and residential life at OSU....

Brown said the study might have missed looking into the pre-existing attitudes or experiences of students prior to college.

"My guess is the students who choose to live in coed housing probably are also making choices to engage in alcohol use and sexual activity before they even got to college."

Discuss the relationship between correlation and causation suggested by this article.

29. Flu. Suppose a certain medical study shows that over 90% of patients with mild to moderate flu recover within 10 to 12 days if they are exposed to fresh, clean air. What can we conclude from this information about the relationship between breathing clean air and curing the flu? Explain the reasons for your answer.

30. The danger of pickles. Write a critique of the lighthearted article in In the News 6.19, which uses the notion of statistical correlation discussed in this chapter.

Exercises 31 and 32 are suitable for group work.

31. A research project. DPT (diphtheria, pertussis, tetanus) injections for infants are strongly recommended by the medical community. Sadly, many parents report cases of severe illness or even death following the injection. As a potential parent,

---

IN THE NEWS 6.19

Evils of Pickle Eating 101

PROFESSOR JAMESON   May 5, 2009

Pickles are associated with all the major diseases of the body. Eating them breeds war and Communism. They can be related to most airline tragedies. Auto accidents are caused by pickles. There exists a positive relationship between crime waves and consumption of this fruit of the cucurbit family. For example....

- Nearly all sick people have eaten pickles. The effects are obviously cumulative.

  99.9% of all people who die from cancer have eaten pickles.
  100% of all soldiers have eaten pickles.
  96.8% of all Communist sympathizers have eaten pickles.
  99.7% of the people involved in air and auto accidents ate pickles within 14 days preceding the accident.
  93.1% of juvenile delinquents come from homes where pickles are served frequently.

- Evidence points to the long term effects of pickle eating.

  Of the people born in 1839 who later dined on pickles, there has been a 100% mortality. All pickle eaters born between 1849 and 1859 have wrinkled skin, have lost most of their teeth, have brittle bones and failing eyesight if the ills of pickle eating have not already caused their death.
  Even more convincing is the report of a noted team of medical specialists: rats force fed with 20 pounds of pickles per day for 30 days developed bulging abdomens. Their appetites for WHOLESOME FOOD were destroyed.

In spite of all the evidence, pickle growers and packers continue to spread their evil. More than 120,000 acres of fertile U.S. soil are devoted to growing pickles. Our per capita consumption is nearly four pounds.

Eat orchid petal soup. Practically no one has as many problems from eating orchid petal soup as they do with eating pickles.

---

you may well be in a position to make this decision for your child. Use the Web and other available sources to research this issue carefully and write a brief report on your conclusions. Pay particular attention to the design of any studies you rely on.

32. A research project. Statisticians are quick to point out that *p*-values are sometimes misunderstood. For example, in a drug trial, the *p*-value is *not* the probability that the drug is ineffective. You will find many references to the incorrect use of *p*-values on the Web. Investigate these and write a report on your findings.

33. A research project. There has been a great deal of controversy concerning a possible link between autism and certain vaccinations. Use the Internet and other available sources to research this issue carefully and write a brief report on your conclusions.

34. History. Sir Ronald A. Fisher was a towering figure in the history of statistics. He helped develop statistics as a separate discipline, thought through the concepts involved in randomized experiments, and contributed in a fundamental way to the notions of inference we discuss in this chapter. Write a brief biography of Fisher highlighting his contributions to statistical inference.

The following exercises are designed to be solved using technology such as calculators or computer spreadsheets. For assistance, see the technology supplement.

35. Sales and price. The following table shows the number of units of a certain item sold by a store over a month in terms of the price in dollars charged by the store.

Price in dollars	1.00	1.10	1.20	1.30	1.40
Number sold	320	305	270	265	220

Use technology to find the trend line and correlation coefficient for the data. (Round the correlation coefficient to two decimal places.) Plot the data along with the trend line. What type of correlation is there between the price and the number sold? Interpret your answer.

36. Confirming example. Consider the table of data on p. 407 showing weekly expenditures on alcohol and tobacco. Use technology to find the trend line and correlation coefficient for the data. Round the coefficients for the trend line and the correlation coefficient to two decimal places.

37. Calculating *p*-values. If a fair coin is tossed 100 times, the probability that one gets at least 60 heads is given by

$$\frac{100!}{60!40!}0.5^{60}0.5^{40} + \frac{100!}{61!39!}0.5^{61}0.5^{39}$$

$$+ \frac{100!}{62!38!}0.5^{62}0.5^{38} + \cdots + \frac{100!}{100!0!}0.5^{100}0.5^{0}$$

$$= 0.5^{100}\left(\frac{100!}{60!40!} + \frac{100!}{61!39!} + \frac{100!}{62!38!} + \cdots + \frac{100!}{100!0!}\right).$$

This is the *p*-value if you toss 100 coins and get 60 heads. Use technology to calculate this number. Round your answer to three decimal places.

CHAPTER SUMMARY

Descriptive statistics refers to organizing data in a form that can be clearly understood and that describes what the data show. *Inferential statistics* refers to the attempt to draw valid conclusions from the data, along with some estimate of how accurate those conclusions are. We study both in this chapter.

Data summary and presentation: Boiling down the numbers

Four important measures in descriptive statistics are the *mean, median, mode,* and *standard deviation.* The mean (also called the average) of a list of numbers is their sum divided by the number of entries in the list. The median of a list of numbers is the "middle" of the list. There are as many numbers above it as there are below it. The mode is the most frequently occurring number in the list (if there is one). The standard deviation σ of a list of numbers measures how much the data are spread out from the mean. It is calculated as follows: Subtract the mean from each number in the list, square these differences, add all these squares, divide by the number of entries in the list, and then take the square root. As a formula, this is

$$\sigma = \sqrt{\frac{(x_1 - \mu)^2 + (x_2 - \mu)^2 + \ldots + (x_n - \mu)^2}{n}}$$

if the data points are $x_1, x_2, x_3, \ldots, x_n$ and μ is the mean.

A common method of describing data is the *five-number summary.* It consists of the minimum, the first quartile, the median, the third quartile, and the maximum. The first quartile is the median of the lower half of the list of numbers, and the third quartile is the median of the upper half. A *boxplot* is a pictorial display of a five-number summary.

A *histogram* is a bar graph that shows frequencies with which certain data appear. Histograms can be especially useful when we are dealing with large data sets.

The normal distribution: Why the bell curve?

The *normal distribution* arises naturally in analyzing data from standardized exam scores, heights of individuals, and many other situations. A plot of normally distributed data shows the classic *bell-shaped curve*. For normally distributed data, the mean and median are the same, and in fact the data are symmetrically distributed about the mean. Also, most of the data tend to be clustered relatively near the mean.

A normal distribution is completely determined by its mean and standard deviation. The size of the standard deviation tells us how closely the data are "bunched" about the mean. A larger standard deviation means a "fatter" curve, as shown in **Figure 6.53**, and a smaller standard deviation means a "thinner" curve, as shown in **Figure 6.54**, where the data are more closely bunched about the mean.

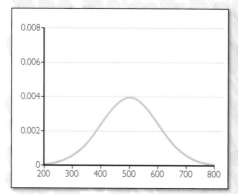

FIGURE 6.53 Normal distribution with mean 500 and standard deviation 100.

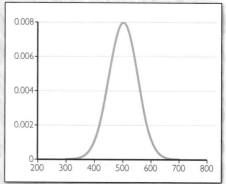

FIGURE 6.54 Normal distribution with mean 500 and standard deviation 50.

More precisely, for a normal distribution:

- About 68% of the data lie within one standard deviation of the mean.

- About 95% of the data lie within two standard deviations of the mean.

- About 99.7% of the data lie within three standard deviations of the mean.

For example, for the verbal section of the SAT Reasoning Test, scores were normally distributed with mean 507 and standard deviation 111. Thus, about 68% of scores fell between $507 - 111 = 396$ and $507 + 111 = 618$.

For a more precise tabulation of how data are distributed about the mean, we use *z-scores*. The *z*-score for a data point is the number of standard deviations that the data point lies above or below the mean. The *z*-score, together with a standard table of *z*-scores and percentiles, allows easy and complete analysis of normally distributed data.

One of the most striking facts about normal distributions is that, according to the *Central Limit Theorem*, data that consist of percentages of population samples (all of the same size) are approximately normally distributed. This allows us to use the normal distribution to analyze virtually any type of data from repeated samples.

The statistics of polling: Can we believe the polls?

Inferential statistics is used in attempting to draw conclusions about an entire population by collecting data from a small sample of the population. One of the most common applications in the media is to polling.

Polling involves a *margin of error*, a *confidence level*, and a *confidence interval*. To say that a poll has a margin of error of 3% with a confidence level of 95% means that if we conducted this poll 100 times, then we expect about 95 of those sample results to be within three percentage points of the true percentage having the given property. To find the confidence interval, adjust the result of the poll by adding and subtracting the margin of error.

For a 95% level of confidence, the margin of error when we poll n people is about $100/\sqrt{n}\%$, and the sample size needed to get a margin of error of m percentage points is about $(100/m)^2$.

There are significant and difficult questions of methodology involved in obtaining accurate results from polling.

Statistical inference and clinical trials: Effective drugs?

Another important use of inferential statistics is in clinical trials. Medical testing uses the notions of *statistical significance* and *p-values*. If the results of a clinical trial are unlikely to have occurred by chance alone, the results are considered to be statistically significant. Roughly speaking, the *p*-value measures the probability that the outcome of a clinical trial would occur by pure chance if the treatment had no effect. A small *p*-value is usually interpreted as evidence that it is unlikely the results are due to chance alone. The results are usually considered statistically significant if the *p*-value is 0.05 or smaller.

Two numerical variables are *positively correlated* if an increase in one of them accompanies an increase in the other. Two numerical variables are *negatively correlated* if an increase in one of them accompanies a decrease in the other. If neither of these is true, the variables are *uncorrelated*. Linear regression can quantify the extent to which two variables are *linearly* correlated.

Correlation can be the result of real connections between two variables, but some correlations are the result of chance or coincidence. Even when two variables are connected in some way, we should not assume causation.

KEY TERMS

mean (or average), p. 350
median, p. 350
mode, p. 350
bimodal, p. 350
multimodal, p. 350
outlier, p. 353
first quartile, p. 353
second quartile, p. 353
third quartile, p. 353

five-number summary, p. 354
standard deviation, p. 357
histogram, p. 362
normally distributed, p. 369
z-score (or standard score), p. 374
percentile, p. 375
Central Limit Theorem, p. 378
margin of error, p. 388
confidence interval, p. 388

confidence level, p. 388
statistically significant, p. 403
p-value, p. 404
positively correlated, p. 406
negatively correlated, p. 406
uncorrelated, p. 406
correlation coefficient, p. 409

CHAPTER QUIZ

1. The goals scored (by either team) in the games played by a soccer team are listed below according to the total number of goals scored in each game. Find the mean, median, and mode for these data. Round the mean to one decimal place.

Goals scored by either team	0	1	2	3	4	5	6	7	8
Number of games	7	14	11	12	3	2	1	2	2

Answer Mean: 2.4; median: 2; mode: 1.

If you had difficulty with this problem, see Example 6.1.

2. Calculate the five-number summary for this list of automobile prices:

$27,000 $19,500 $24,500 $25,600 $17,000
$32,700 $18,000 $27,800 $29,000 $30,200.

Answer The minimum is $17,000; the first quartile is $19,500; the median is $26,300; the third quartile is $29,000; the maximum is $32,700.

If you had difficulty with this problem, see Example 6.3.

3. Calculate the mean and standard deviation for this list of automobile prices:

$27,000 $19,500 $24,500 $25,600 $17,000
$32,700 $18,000 $27,800 $29,000 $30,200.

Answer Mean: $25,130; standard deviation: $5078.

If you had difficulty with this problem, see Example 6.5.

4. The weights of oranges in a harvest are normally distributed, with a mean weight of 150 grams and standard deviation of 10 grams. In a supply of 1000 oranges, how many will weigh between 140 and 160 grams? How many will weigh between 130 and 170 grams?

Answer 680 oranges between 140 and 160 grams; 950 oranges between 130 and 170 grams.

If you had difficulty with this problem, see Example 6.9.

5. The weights of oranges in a harvest are normally distributed, with a mean weight of 150 grams and standard deviation of 10 grams. Calculate the z-score for an orange weighing
a. 160 grams
b. 135 grams
Round your answers to one decimal place.

Answer
a. 1
b. −1.5

If you had difficulty with this problem, see Example 6.10.

6. For a certain disease, 40% of untreated patients can be expected to improve within a week. We observe a population of 80 untreated patients and record the percentage who improve within one week. According to the Central Limit Theorem, such percentages are approximately normally distributed. Find the mean and standard deviation of this normal distribution. Round the standard deviation to two decimal places.

Answer The mean is 40%. The standard deviation is $\sqrt{(40 \times 60)/80}$ or about 5.48 percentage points.

If you had difficulty with this problem, see Example 6.13.

7. Use the mean and standard deviation calculated in the preceding exercise to find the percentage of test groups of 80 untreated patients in which 50% or less improve within a week. Round your answer as a percentage to one decimal place.

Answer z-score $= 10/5.48$ or about 1.8 standard deviations above the mean. Table 6.2 (page 376) gives a corresponding percentage of about 96.4%.

If you had difficulty with this problem, see Example 6.13.

8. Explain the meaning of a poll that says 55% of Americans approve of the job the president is doing, with a margin of error of 3% and a confidence level of 95%.

Answer In 95% of such polls, the reported approval of the president will be within three percentage points of the true approval level. Thus, we can be 95% confident that the true level is between 52% and 58%.

If you had difficulty with this problem, see Example 6.15.

9. A poll asked 800 people their choice for mayor. What is the approximate margin of error for a 95% confidence level? Round your answer as a percentage to one decimal place.

Answer 3.5%.

If you had difficulty with this problem, see Example 6.16.

10. What sample size is needed to give a margin of error of 2.5% with a 95% confidence level?

Answer 1600.

If you had difficulty with this problem, see Example 6.19.

11. For each of these situations, determine whether the results described appear to be statistically significant.
a. You are walking down the street and encounter three people in a row with blond hair.
b. You cut a deck of playing cards 30 times and get an ace every time.

Answer
a. Not statistically significant
b. Statistically significant

If you had difficulty with this problem, see Example 6.20.

12. A clinical trial for an experimental drug finds that the drug is effective with a p-value of 0.08. Explain the meaning of the p-value in this case. Would this result normally be accepted as statistically significant?

Answer The probability that the outcome of the clinical trial would occur by chance alone if the drug had no effect is 0.08 (8%). This result would not normally be accepted as statistically significant because the p-value is larger than 0.05.

If you had difficulty with this problem, see Example 6.21.

13. Decide whether the following variables are positively correlated, negatively correlated, or uncorrelated.

a. Your grades and the number of hours per week you study
b. Your grades and the number of hours per week you party
c. Your grades and the number of cookies per week you eat
d. The daily average temperature and sales of cell phones
e. The daily average temperature and sales of overcoats
f. The daily average temperature and sales of shorts

Answer   Positively correlated: a and f; negatively correlated: b and e; uncorrelated: c and d.

If you had difficulty with this problem, see Example 6.22.

GRAPH THEORY

The following excerpt is from an article that appeared at the CBS News Web site.

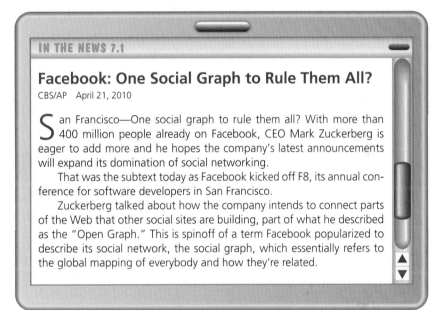

IN THE NEWS 7.1

Facebook: One Social Graph to Rule Them All?

CBS/AP   April 21, 2010

San Francisco—One social graph to rule them all? With more than 400 million people already on Facebook, CEO Mark Zuckerberg is eager to add more and he hopes the company's latest announcements will expand its domination of social networking.

That was the subtext today as Facebook kicked off F8, its annual conference for software developers in San Francisco.

Zuckerberg talked about how the company intends to connect parts of the Web that other social sites are building, part of what he described as the "Open Graph." This is spinoff of a term Facebook popularized to describe its social network, the social graph, which essentially refers to the global mapping of everybody and how they're related.

FIGURE 7.1 Mark Zuckerberg and the social graph.

This article shows one application of the mathematical concept of a *graph*, which in the simplest terms tells whether certain objects are somehow connected to each other.[1] For example, we can model Facebook using a graph if we say that two users are connected when they are friends. Graphs can be represented pictorially using dots and segments, as in the background of the image in **Figure 7.1** (on previous page). The dots are called *vertices*, and the segments connecting them are called *edges*.

Graph theory has important real-world applications, and in this chapter we will examine some of the basic ideas behind them. Serious applications of graph theory involve highly complex situations, such as fast routing of dynamic content on the World Wide Web, with huge numbers of vertices and edges that can be dealt with only by very powerful computers. We will focus on everyday applications of graphs but also will explain their importance to business, government, and industry.

7.1 Modeling with graphs and finding Euler circuits

TAKE AWAY FROM THIS SECTION

Know how to use graphs as models and how to determine efficient paths.

The following article appeared in the *New York Times*.

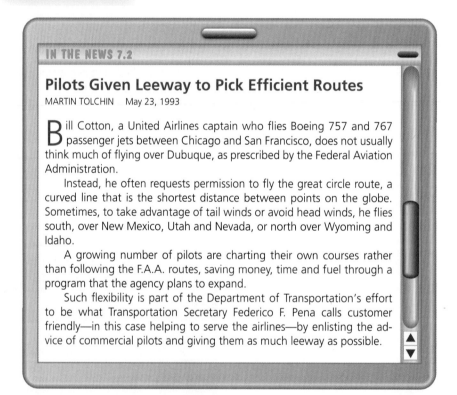

IN THE NEWS 7.2

Pilots Given Leeway to Pick Efficient Routes

MARTIN TOLCHIN   May 23, 1993

Bill Cotton, a United Airlines captain who flies Boeing 757 and 767 passenger jets between Chicago and San Francisco, does not usually think much of flying over Dubuque, as prescribed by the Federal Aviation Administration.

Instead, he often requests permission to fly the great circle route, a curved line that is the shortest distance between points on the globe. Sometimes, to take advantage of tail winds or avoid head winds, he flies south, over New Mexico, Utah and Nevada, or north over Wyoming and Idaho.

A growing number of pilots are charting their own courses rather than following the F.A.A. routes, saving money, time and fuel through a program that the agency plans to expand.

Such flexibility is part of the Department of Transportation's effort to be what Transportation Secretary Federico F. Pena calls customer friendly—in this case helping to serve the airlines—by enlisting the advice of commercial pilots and giving them as much leeway as possible.

Efficient routes for airplanes save money and get passengers to their destinations in a timely manner. Efficient routing is important for many business applications, from trucking to mail delivery. Typical options may consist of cities to pass through or highway routes to utilize. In this section, we study how to use graphs to find efficient routes.

---

[1]In earlier chapters, we talked about a graph as a pictorial representation of data or a formula. In this chapter, we are going to use the term in a completely different way.

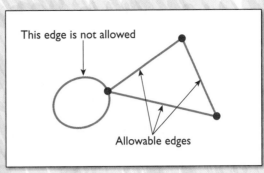

FIGURE 7.2 An edge is not allowed to start and end at the same vertex.

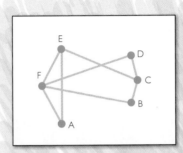

FIGURE 7.3 A connected graph.

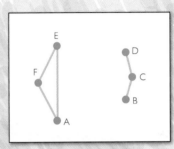

FIGURE 7.4 A disconnected graph.

Before we begin this study, let's agree that the graphs we will discuss have two properties: First, an edge cannot start and end at the same vertex. (See **Figure 7.2**.)

Second, the graphs are *connected*, that is, each pair of vertices can be joined by a sequential collection of edges, a *path*. The graph in **Figure 7.3** is connected, but the one in **Figure 7.4** is not connected—it comes in two pieces.

The first step in our study of efficient routes is learning how to use graphs to make models of real situations.

Modeling with graphs

Graphs can be used to represent many situations. For example, **Figure 7.5** shows a highway map where the vertices represent towns and the edges represent roads. In **Figure 7.6**, the vertices represent some users of Facebook, and the edges indicate that the users are friends on Facebook. The figure tells us, for example, that Lakesha and Juliana are friends because an edge connects them. Lakesha and Tom are not friends because no edge joins those two vertices.

FIGURE 7.5 A graph representing towns and roads.

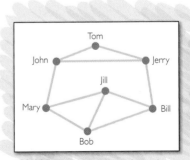

FIGURE 7.6 Facebook friends.

EXAMPLE 7.1 Making models: Little League

The Little League in our town has two divisions, the Red Division and the Blue Division. The Red Division consists of the Cubs, Lions, Bears, and Tigers. The Blue Division consists of the Red Sox, Yankees, and Giants. The inter-division schedule this season is:

Cubs play Red Sox, Yankees, and Giants.

Lions play Red Sox and Yankees.

Bears play Red Sox and Yankees.

Tigers play Red Sox and Giants.

Make a graphical representation of this schedule by letting the vertices represent the teams and connecting two vertices with an edge if they are scheduled to play this season.

SOLUTION

The appropriate graph is shown in **Figure 7.7**. Note, for example, that the edge joining the Cubs and Giants indicates they have a game scheduled. There is no edge joining the Tigers with the Yankees because they do not have a game scheduled.

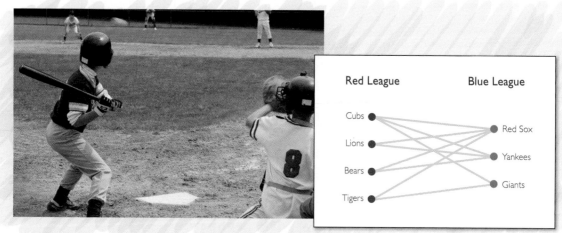

FIGURE 7.7 A Little League schedule.

TRY IT YOURSELF 7.1

Ferry routes connect Portsmouth and Poole, England, with Le Havre, Caen, Cherbourg, and St. Malo, France, and Bilbao, Spain (see the map on the next page).

Portsmouth has ferries to and from Le Havre, Caen, Cherbourg, St. Malo, and Bilbao.

Poole has ferries to and from Cherbourg and St. Malo.

Make a graph that shows these ferry routes.

The answer is provided at the end of this section.

Euler circuits

We noted at the beginning of this section that graphs can be used to find efficient routes. Let's consider the example of mail delivery. The edges in **Figure 7.8** represent streets along which mail must be delivered, and the vertices represent the post office

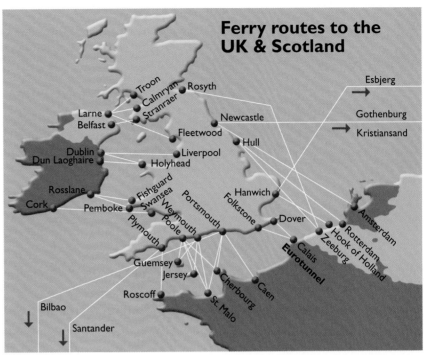

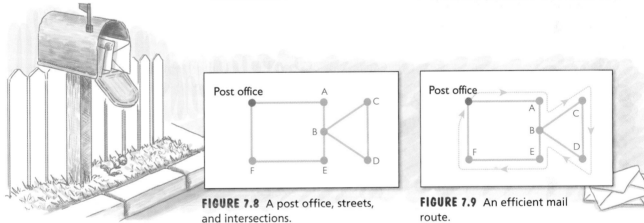

FIGURE 7.8 A post office, streets, and intersections.

FIGURE 7.9 An efficient mail route.

and street intersections. We have labeled one vertex as "post office" and the others as *A* through *F*. An efficient route for the mail carrier starts at the post office, travels along each street exactly once, and ends up back at the post office. Such a route is shown in **Figure 7.9**. We indicate the route by listing the vertices as we encounter them along the route:

Post office-*A*-*B*-*C*-*D*-*B*-*E*-*F*-Post office

The path for this postal route starts and ends at the same vertex and traverses each edge exactly once. Such a path is known as an *Euler circuit* or *Euler cycle* after the eighteenth-century Swiss mathematician Leonhard Euler (pronounced "oiler"). Euler is considered to be the father of modern graph theory. Euler circuits represent efficient paths for postal delivery, street sweepers, canvassers, and lots of other situations.

Key Concept

A **circuit** or **cycle** in a graph is a path that begins and ends at the same vertex. An **Euler circuit** or **Euler cycle** is a circuit that traverses each edge of the graph exactly once.

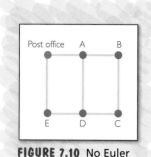

FIGURE 7.10 No Euler circuit.

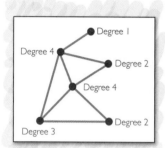

FIGURE 7.11 The degrees of vertices.

Such efficient routes are not always possible. For example, trial and error can quickly convince us that there is no Euler circuit for the graph in **Figure 7.10**. Any circuit that starts and ends at the post office must either omit a street or traverse some street more than once.

Degrees of vertices and Euler's theorem

We need to know how to determine whether a graph has an Euler circuit and, if it has one, how to find it. To do this, we need to look more closely at properties of graphs. The term *degree* is used to describe an important relationship between vertices and edges of a graph.

Key Concept

The **degree** of a vertex is the number of edges that touch that vertex. Some texts use **valence** instead of degree.

The numbers in **Figure 7.11** indicate the degree of each vertex.

EXAMPLE 7.2 Finding the degree of a vertex: An air route map

The graph in **Figure 7.12** shows a simplified air route map. It shows connections that are available between various cities. In the context of this graph, what is the meaning of the degree of a vertex? Find the degree of each of the vertices.

SOLUTION

Because each edge indicates a direct connection between cities, the degree of each vertex is the number of direct flights available from the given city. The degrees of the vertices are: Atlanta: degree 4; Chicago: degree 2; Los Angeles: degree 2; Miami: degree 2; New York: degree 4.

TRY IT YOURSELF 7.2

Figure 7.13 shows a simplified map of Amtrak rail service. Find the degree of each vertex.

The answer is provided at the end of this section.

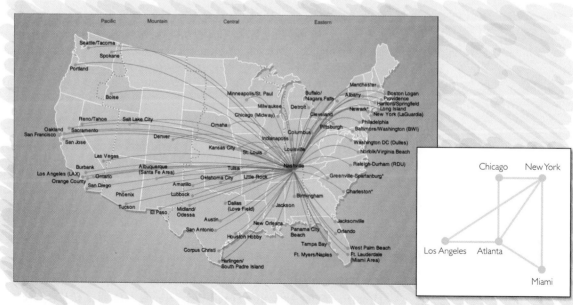

FIGURE 7.12 A simplified air route map.

FIGURE 7.13 A simplified Amtrak rail service map.

FIGURE 7.14 Edges of an Euler circuit come in pairs.

There is a crucial relationship between Euler circuits and the degrees of vertices. An Euler circuit goes into and then out of each vertex and uses each edge exactly once, as is illustrated in **Figure 7.14**. Because the edges meeting a vertex come in pairs (in-edges paired with out-edges), the number of edges at each vertex must be a multiple of 2. That is, the degree of each vertex is even.

In the eighteenth century, Leonhard Euler discovered that this degree condition completely characterizes when such circuits exist.

Theorem[2] (Euler, 1736) A connected graph has an Euler circuit precisely when every vertex has even degree.

The two mail routes shown in Figures 7.8 and 7.10 provide evidence for this theorem: In Figure 7.8, each vertex has even degree, and there is an Euler circuit. But in Figure 7.10, vertex A has degree 3, which is odd. Thus, there is no Euler circuit for Figure 7.10.

EXAMPLE 7.3 Applying Euler's theorem: Scenic trips

Does the air route map in Figure 7.12 allow for a scenic trip that starts and ends at New York and flies along each route exactly once? That is, is there an Euler circuit for the graph in Figure 7.12? If an Euler circuit exists, use trial and error to find one.

SOLUTION

Referring back to Example 7.2, we see that the degree of each vertex is even. So Euler's theorem guarantees the existence of an Euler circuit. Proceeding by trial and error, we find one such route:

New York - Miami - Atlanta - New York - Chicago - Atlanta - Los Angeles - New York

This route is shown in **Figure 7.15** on the following page. We note that this is not the only correct answer: There are many other Euler circuits for this graph and hence many scenic routes starting and ending at New York that use each airway exactly once.

---

[2] "Solutio Problematis ad Geometriam Situs Pertinentis (The solution of a problem relating to the geometry of position)," *Commentarii Academiae Scientiarum Imperialis Petropolitanae* 8 (1736; published 1741), 128–140.

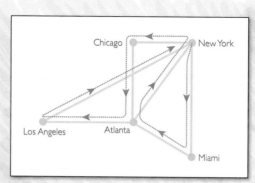

FIGURE 7.15 A scenic route that is an Euler circuit.

TRY IT YOURSELF 7.3

Does the Amtrak rail service map in Figure 7.13 allow for a scenic rail trip that begins and ends in New York and travels along each track route exactly once? In other words, is there an Euler circuit for this graph?

The answer is provided at the end of this section.

EXAMPLE 7.4 Applying Euler's theorem: The seven bridges of Königsberg

Königsberg is an old city that is now part of Russia and has been renamed Kaliningrad. It covers both banks of the Pregel River as well as two islands in the river. The banks and islands were connected by seven bridges, as shown in **Figure 7.16**. Tradition has it that the people of Königsberg enjoyed walking, and they wanted to figure out how to start at home, go for a walk that crossed each bridge exactly once, and arrive back home. Nobody could figure out how to do it. It was Leonhard Euler's solution of this problem that began the modern theory of graphs.

a. Make a graph that models Königsberg and its seven bridges.

b. Determine whether the desired walking path is possible.

FIGURE 7.16 Left: Königsberg (modern-day Kaliningrad). Right: The bridges of Königsberg.

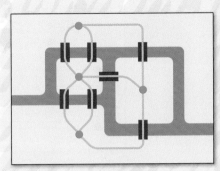

FIGURE 7.17 Preparing a graph model.

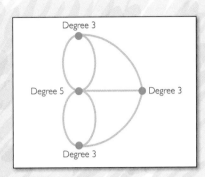

FIGURE 7.18 A graphical representation of the seven bridges.

SOLUTION

a. To make the model, we place a vertex on each bank of the river and on both islands. We connect these vertices by paths across the seven bridges. This is shown in **Figure 7.17**. We then obtain the graph shown in **Figure 7.18** by showing only the vertices and edges we have drawn. The degree of each vertex is noted in that figure.

b. The desired walking path would be an Euler circuit for the graph in Figure 7.18. But because this graph has a vertex of odd degree, it has no Euler circuit. The desired walking path in Königsberg does not exist.

> ### SUMMARY 7.1   Graphs and Euler Circuits
>
> **1.** A graph is a collection of vertices, some (or all) of which are joined by edges. Generally, we consider only connected graphs, and we do not allow edges to start and end at the same vertex.
>
> **2.** A circuit in a graph is a path (a sequential collection of edges) that begins and ends at the same vertex. An Euler circuit is a circuit that uses each edge exactly once.
>
> **3.** The degree of a vertex is the number of edges touching it.
>
> **4.** A connected graph has an Euler circuit precisely when each vertex has even degree.

Methods for making Euler circuits

We know that a graph has an Euler circuit if and only if each of its vertices has even degree. This degree condition is easy to check, but how can we actually find an Euler circuit? For "small" graphs this can be done by trial and error, but for a graph with hundreds or even thousands of vertices, trial and error is a hopeless method for finding Euler circuits.

Fortunately, there are several systematic methods for constructing Euler circuits. We will present two such methods here. The first works by combining smaller circuits to produce larger ones.

To illustrate the method, consider the graph in **Figure 7.19**. It has two obvious circuits, which we have marked as *cycle 1* and *cycle 2*. These two circuits meet at vertex *A*, but they share no common edge. We break the two circuits at this vertex

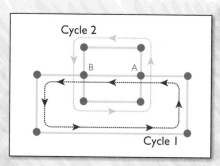

FIGURE 7.19 A graph with two circuits.

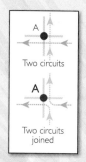

FIGURE 7.20 Breaking and rerouting at *A*.

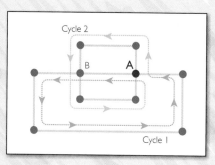

FIGURE 7.21 The resulting Euler circuit.

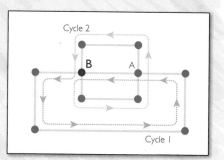

FIGURE 7.22 Breaking and rerouting at *B*.

and rejoin them, as shown in **Figure 7.20**. The result is that both circuits combine to make the longer circuit illustrated in **Figure 7.21**. This gives an Euler circuit. For a larger graph, we may have to put together a number of smaller circuits to obtain an Euler circuit.

We note that the two circuits in Figure 7.19 also meet at vertex *B*. We can find a different Euler circuit by breaking the circuits and rejoining at *B* rather than *A*. This is shown in **Figure 7.22**. When circuits meet at several vertices, we can break and reroute the circuits at any of these vertices.

A second method for making Euler circuits is known as *Fleury's algorithm*. For this method, we remove edges from the graph one at a time and add them to a path that will grow to become an Euler circuit. To lengthen this path, we add edges by applying repeatedly the following two-step procedure:

Step 1: Remove an edge from the graph and add it to the path so that the path is lengthened. But do not choose an edge whose removal will cut the graph into two pieces unless there is no other choice.

Step 2: Delete from the graph any isolated vertices resulting from the first step.

Repeat these two steps until an Euler circuit is constructed. Let's apply Fleury's algorithm to the graph in **Figure 7.23**. Starting at the vertex *A*, we have selected the three red edges to begin Fleury's algorithm. In **Figure 7.24**, we have used these edges to start an Euler circuit.

We have arrived at vertex *D*, and from here we might select any of the yellow edges in **Figure 7.25**. But removal of the edge *DC* would result in the disconnected graph shown in **Figure 7.26**.

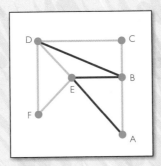

FIGURE 7.23 Selecting edges to start Fleury's algorithm.

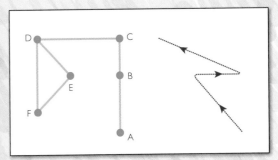

FIGURE 7.24 The beginning of an Euler circuit.

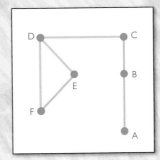

FIGURE 7.25 The three yellow edges are possibilities for continuing past vertex *D*.

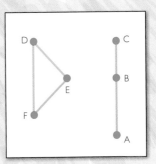

FIGURE 7.26 Removal of *DC* would result in a disconnected graph.

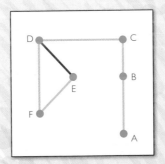

FIGURE 7.27 The edge *DE* is selected.

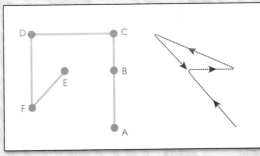

FIGURE 7.28 The edge *DE* is added to the path.

So we can choose either of the edges *DF* or *DE*, but not *DC*. In **Figure 7.27**, we have selected the edge *DE* and added that edge to the path in **Figure 7.28**.

We have arrived at the vertex *E*. Removal of the edge *EF*, as shown in **Figure 7.29**, will result in a disconnected graph, but there is no other choice, so we do so. We add this edge to the path, as shown in **Figure 7.30** on the following page. Note that this leaves the vertex *E* with no edges attached. We delete this isolated vertex and continue.

From this point on, we have no decisions to make, and the completed Euler circuit is shown in **Figure 7.31** on the following page.

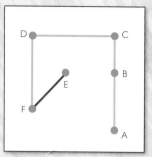

FIGURE 7.29 *EF* is the only choice even though its removal leaves an isolated vertex.

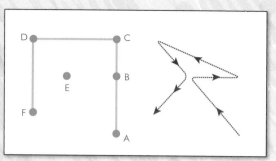

FIGURE 7.30 *EF* added to the path.

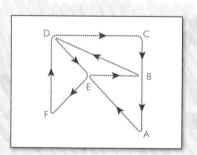

FIGURE 7.31 The completed Euler circuit.

EXAMPLE 7.5 Finding Euler circuits: A subway tour

Figure 7.32 shows a part of the New York City subway. Is it possible to take a tour of this part of the subway line that starts and ends at the 57 St & 7 Ave station and traverses each rail section exactly once? In other words, is there an Euler circuit for this graph? If an Euler circuit exists, find one.

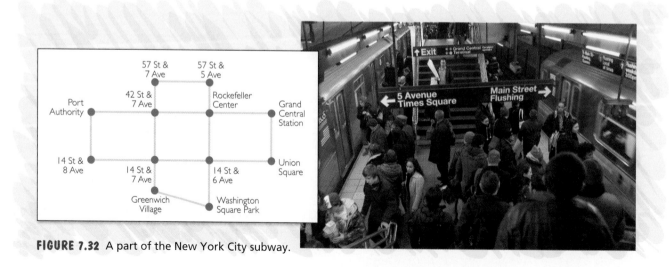

FIGURE 7.32 A part of the New York City subway.

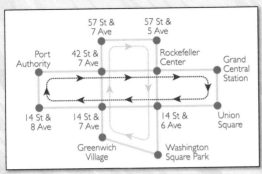

FIGURE 7.33 Two circuits.

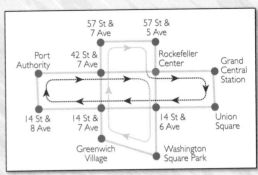

FIGURE 7.34 Patching together to make an Euler circuit.

SOLUTION

Each of the vertices has degree 2 or 4. Because each vertex has even degree, we are guaranteed an Euler circuit. One could find an Euler circuit by trial and error, but we apply the method of breaking and rerouting. In **Figure 7.33**, we have marked two circuits that share no common edges.

Circuit 1: 57 St & 7 Ave - 57 St & 5 Ave - Rockefeller Center - 14 St & 6 Ave - Washington Square Park - Greenwich Village - 14 St & 7 Ave - 42 St & 7 Ave - 57 St & 7 Ave

Circuit 2: Port Authority - 42 St & 7 Ave - Rockefeller Center - Grand Central Station - Union Square - 14 St & 6 Ave - 14 St & 7 Ave - 14 St & 8 Ave - Port Authority

In **Figure 7.34**, we have broken the circuits at Rockefeller Center and rerouted to combine the two circuits into one. The result is the required subway tour:
Euler circuit: 57 St & 7 Ave - 57 St & 5 Ave - Rockefeller Center - Grand Central Station - Union Square - 14 St & 6 Ave - 14 St & 7 Ave - 14 St & 8 Ave - Port Authority - 42 St & 7 Ave - Rockefeller Center - 14 St & 6 Ave - Washington Square Park - Greenwich Village - 14 St & 7 Ave - 42 St & 7 Ave - 57 St & 7 Ave

You should verify that Fleury's algorithm can also be used to produce an Euler circuit.

TRY IT YOURSELF 7.5

Find the Euler circuit obtained by breaking and rerouting circuit 1 and circuit 2 at 14 St & 7 Ave rather than at Rockefeller Center.

The answer is provided at the end of this section.

SUMMARY 7.2   Making Euler Circuits

Two circuits with a vertex in common (but no edge in common) can be put together to make a single longer circuit. Using this method repeatedly allows us to put smaller circuits together to form an Euler circuit when one exists.

Alternatively, we can make Euler circuits using Fleury's algorithm.

Eulerizing graphs

We have seen that in many practical settings, a most efficient route is an Euler circuit—if there is one. If there is no Euler circuit, any circuit traversing each edge must include some backtracking of edges, that is, going back along some paths we have

already taken. We still want a most efficient route possible in the sense that we have to backtrack as little as possible. We do this by *Eulerizing* graphs, temporarily adding duplicate edges.

We know that if all vertices have even degree, then we can find an Euler circuit. We will show how to find an efficient path when there are exactly two odd-degree vertices. The graph in **Figure 7.35** shows a part of the shuttle route in Yosemite National Park. Note that the vertices at Yosemite Creek and the Ansel Adams Gallery have degree 3 and all other vertices have even degree. Because there are odd-degree vertices, there is no Euler circuit: It is not possible to start and end at Yosemite Creek and travel each shuttle route without backtracking. We want to find a route that involves the least possible backtracking. The first step in our search for an efficient route is to pick a shortest[3] (in terms of the number of edges) path joining the odd-degree vertices, in this case Yosemite Creek and Ansel Adams Gallery. There is more than one choice for this, and any of them will serve. We have selected one shortest path from Yosemite Creek to the Ansel Adams Gallery and highlighted it in Figure 7.35.

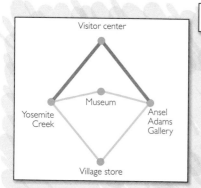

FIGURE 7.35 Step 1: Find shortest path between odd-degree vertices.

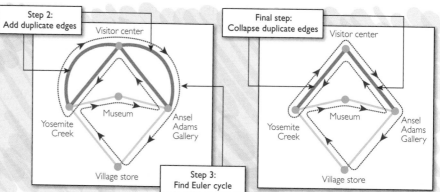

FIGURE 7.36 Step 2: Add duplicate edges. Step 3: Find an Euler circuit.

FIGURE 7.37 Final step: Collapse the duplicate edges.

The second step in our search is to add duplicate edges, as shown in **Figure 7.36**. Observe that after the addition of the duplicate edges, all vertices have even degree. Hence, there is an Euler circuit, as shown in Figure 7.36. We complete the process by collapsing the duplicate edges back to their originals. The result is a circuit starting and ending at Yosemite Creek that backtracks as little as possible (twice in this case), as shown in **Figure 7.37**.

Best route: Yosemite Creek - Visitor center - Ansel Adams Gallery - Village store - Yosemite Creek - Museum - Ansel Adams Gallery - Visitor center - Yosemite Creek

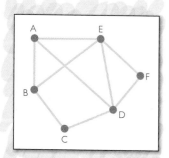

FIGURE 7.38 Street map for Example 7.6.

EXAMPLE 7.6 Eulerizing graphs: Streets in our town

The graph in **Figure 7.38** represents the streets and intersections in our town. We wish to start at vertex *A* and paint a centerline on each street. Find a route (starting and ending at *A*) that backtracks as little as possible.

SOLUTION

We note that every vertex has even degree except vertices *A* and *B*, which have degree 3. In **Figure 7.39**, we have added a duplicate edge between *A* and *B* and indicated an Euler circuit. Collapsing the duplicate edge, as shown in **Figure 7.40**, we find the efficient path *A-B-C-D-F-E-A-D-E-B-A* for the painters. It backtracks only once: from *A* to *B* and back.

---

[3]In large graphs, even the seemingly simple problem of finding a shortest path can be problematic.

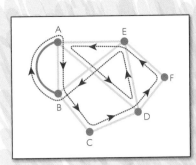

FIGURE 7.39 Adding a duplicate edge and indicating an Euler circuit.

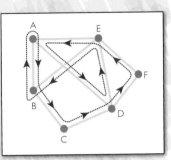

FIGURE 7.40 Collapsing the duplicate edge to get an efficient path.

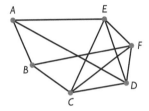

FIGURE 7.41 Street map for Try It Yourself 7.6.

TRY IT YOURSELF 7.6

The graph in **Figure 7.41** shows the streets and intersections in our town. We wish to start at vertex *A* and paint a centerline on each street. Find a route (starting and ending at *A*) that backtracks as little as possible.

The answer is provided at the end of this section.

Try It Yourself answers

Try It Yourself 7.1: Making models: Ferry routes   We use vertices for each of the towns. Edges indicate ferry routes.

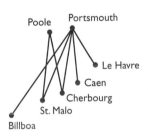

Try It Yourself 7.2: Finding the degree of a vertex: Rail service   Chicago: degree 7; Jacksonville: degree 2; Los Angeles: degree 3; New Orleans: degree 4; New York: degree 2; San Antonio: degree 3; San Francisco: degree 3; Seattle: degree 2; Washington: degree 4.

Try It Yourself 7.3: Applying Euler's theorem: A rail service map   Some of the vertices have odd degree, so it is not possible to take the desired trip.

Try It Yourself 7.5: Finding Euler circuits: A subway tour   The required Euler circuit is

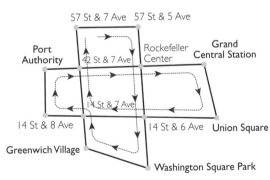

Try It Yourself 7.6: Eulerizing graphs: Streets in our town   An Eulerized graph is

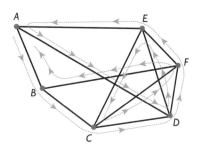

Exercise Set 7.1

1. Connectedness. All of the graphs we have considered are connected, in the sense that each pair of vertices can be joined by a path. One application of graphs that we gave at the beginning of this section involves efficient routes for airplanes. Explain what the property of connectedness means in that context. Also explain why it is important for graphs representing routes for airplanes to be connected.

2. Staying connected. For some applications, it is important that the associated graph have no edge whose removal disconnects the graph. In other words, each pair of vertices can be joined by a path even after any single edge is removed. Explain why this property is important in the case when the graph represents electric power transmission lines forming an electrical grid. (The edges represent power lines, and the vertices represent power plants, substations, consumers, and so forth.) How is this property of a graph related to the existence of redundant power lines between points?

3. Little League. The Little League in our town consists of the Red Division—the Cubs, Lions, Bears, and Tigers—and the Blue Division—the Red Sox, Yankees, and Giants. The inter-division schedule this season is:

Cubs play Red Sox, Yankees, and Giants.

Lions play Red Sox and Giants.

Bears play Red Sox and Yankees.

Tigers play Red Sox, Yankees, and Giants.

Make a graph to model this season's Little League schedule.

4. More Little League. The Little League in our town consists of the Red Division—the Cubs, Lions, Bears, and Tigers—and the Blue Division—the Red Sox, Yankees, and Giants. This season each team is to play exactly three teams from the other division. Can you make a graph to model this schedule? Is such a schedule possible?

5. A computer network. Each of the offices 501, 502, 503, 504, 505, 506, and 507 has a computer. Some of these are connected to others:

The computer in 501 connects to computers in 502, 503, 505, and 507.

The computer in 502 connects to the computer in 501, 503,

and 504.

The computer in 503 connects to all other computers.

Make a graph that shows the office computer network.

6. Bridges. Figure 7.42 shows islands in a river and bridges. Make a graph model of walking paths using vertices to indicate land masses and edges to indicate paths across bridges.

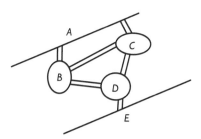

FIGURE 7.42 Islands and bridges for Exercise 6.

7. Birds. Here are characteristics of some birds:

The ruby-throated hummingbird is a hummingbird that lives in the eastern and central United States.

The calliope hummingbird is a hummingbird that lives in the northwestern United States.

The red-shafted flicker is a woodpecker that lives in the western United States.

The red-bellied woodpecker is a woodpecker that lives in the eastern United States.

The cinnamon teal is a duck that lives in the western United States.

The hooded merganser is a duck that lives in the eastern United States.

Make a graph with these birds as vertices. Use an edge to indicate that either the birds belong to the same family or their habitats overlap. For example, the ruby-throated and calliope are joined by an edge because they are both hummingbirds. The red-shafted flicker and calliope hummingbird are joined by an edge because their habitats overlap.

8. A computer network. The computers in a mall are connected.

> **Central security** connects to all computers.
>
> **The food court computer** is connected to central security and custodial services.
>
> **The custodial services computer** is connected to central security, food court, and Toys "R" Us.
>
> **The Toys "R" Us computer** is connected to central security and custodial services.

Make a graph modeling the mall computer network.

9. Degrees of vertices. Find the degrees of each of the vertices in **Figure 7.43**.

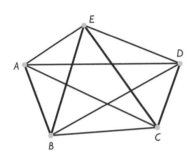

FIGURE 7.43

10. More degrees of vertices. Find the degrees of each of the vertices in **Figure 7.44**.

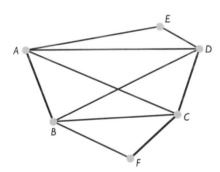

FIGURE 7.44

11. Interpreting Euler. The edges in a certain graph represent phone lines that must be maintained, and the vertices represent junctions. Explain why a worker maintaining the lines would like to find an Euler circuit for this graph.

Euler circuits   In Exercises 12 through 16, determine whether the given graph has an Euler circuit. If an Euler circuit exists, find one.

12. Refer to the graph in Figure 7.43.

13. Refer to the graph in Figure 7.44.

14. Refer to the graph in **Figure 7.45**.

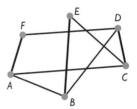

FIGURE 7.45

15. Refer to the graph in **Figure 7.46**.

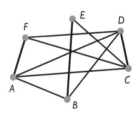

FIGURE 7.46

16. Refer to the graph in **Figure 7.47**.

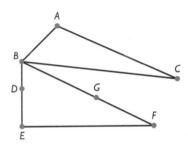

FIGURE 7.47

Euler circuits in practical settings   Exercises 17 through 27 involve Euler circuits in practical settings.

17. Mail routes. The edges in **Figure 7.48** represent streets along which there are mail boxes, and the vertices represent intersections. Either find a mail route that will not require any retracing by the mail carrier or explain why no such route exists. That is, either find an Euler circuit or explain why no such circuit exists.

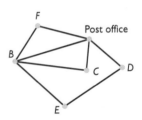

FIGURE 7.48

18. Snow plows. The edges in **Figure 7.49** on the following page represent streets that need to be cleared of snow, and the vertices represent intersections. Either find a route for the snow plow that will not require it to travel over any streets that have already been cleared of snow or explain why no such route exists. That is, either find an Euler circuit or explain why no such circuit exists.

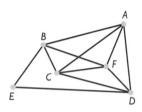

FIGURE 7.49

19. Paving streets. The edges in **Figure 7.50** represent unpaved roads, and the vertices represent intersections. Either find a route that will not require the paving machine to move along already paved streets or show that no such route exists. That is, either find an Euler circuit or explain why no such circuit exists.

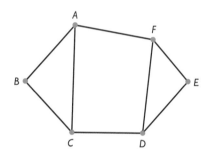

FIGURE 7.50

20. Utility ditches. The edges in **Figure 7.51** represent ditches that must be dug for utilities, and the vertices represent homes. Either find a route that will not require the ditch digger to travel along ditches that have already been dug or show that no such route exists. That is, either find an Euler circuit or explain why no such circuit exists.

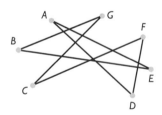

FIGURE 7.51

21. Street inspections. The edges in **Figure 7.52** represent streets that must be periodically inspected, and the vertices represent intersections. Either find a route that the inspectors can take to avoid traveling over previously inspected streets or show that no such route exists. That is, either find an Euler circuit or explain why no such circuit exists.

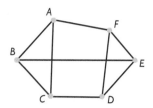

FIGURE 7.52

22. Repairing hiking paths. The edges in **Figure 7.53** represent hiking paths that must be repaired, and the vertices represent intersections. Either find a route that the crews can follow to avoid traveling over already repaired trails or show that no such route exists. That is, either find an Euler circuit or explain why no such circuit exists.

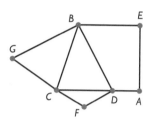

FIGURE 7.53

23. Package delivery. The edges in **Figure 7.54** represent streets along which packages must be delivered, and the vertices represent intersections. Either find a path that will allow the delivery truck to avoid streets where packages have already been delivered or show that no such path exists. That is, either find an Euler circuit or explain why no such circuit exists.

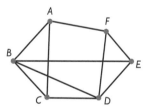

FIGURE 7.54

24. Police beats. The edges in **Figure 7.55** represent streets along which a policeman must walk his beat, and the vertices represent intersections. Either find a route that will allow the policeman to avoid retracing steps or show that no such route exists. That is, either find an Euler circuit or explain why no such circuit exists.

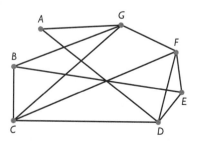

FIGURE 7.55

25. Fire inspections. The edges in **Figure 7.56** represent streets along which businesses must be inspected for fire safety, and the vertices represent intersections. Either find a route that allows the inspector to avoid retracing steps or show that no such route exists. That is, either find an Euler circuit or explain why no such circuit exists.

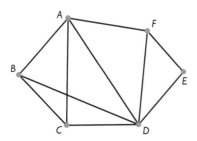

FIGURE 7.56

26. Phone line maintenance. The edges in **Figure 7.57** represent phone lines that must be periodically maintained, and the vertices represent junctions. Either find a maintenance path that does not require retracing or show that no such path exists. That is, either find an Euler circuit or explain why no such circuit exists.

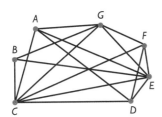

FIGURE 7.57

27. Inspecting for storm damage. The edges in **Figure 7.58** represent streets that must be inspected for storm damage, and the vertices represent intersections. Either find a route that will allow the inspector to avoid retracing or show that no such route exists. That is, either find an Euler circuit or explain why no such circuit exists.

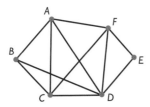

FIGURE 7.58

28. Making Euler circuits. The graph in **Figure 7.59** represents streets and intersections. We have marked three circuits, labeled cycle 1, cycle 2, and cycle 3. Patch these cycles together at vertices E and G to make an Euler circuit.

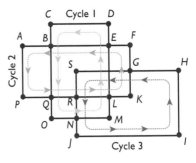

FIGURE 7.59

Efficient circuits The graphs in Exercises 29 through 32 each have exactly two vertices of odd degree, namely A and B. Eulerize these graphs to find a most efficient path that starts and ends at the same vertex and traverses each edge at least once.

29.

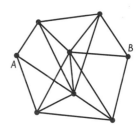

30.

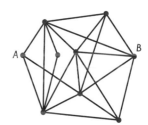

31.

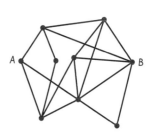

32.

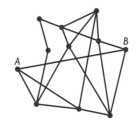

33. Euler paths. In some graphs where no Euler circuit exists, there is instead an *Euler path*. That is a path that starts at one vertex, traverses each edge exactly once, and ends at another vertex. Euler showed in 1736 that such a graph has exactly two vertices of odd degree (the beginning and ending points of the Euler path). It turns out that Fleury's algorithm produces an Euler path in this situation. In this exercise, we will show another way to find Euler paths when they exist. We will use the example of the graph in **Figure 7.60**.

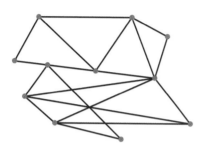

FIGURE 7.60 A graph with an Euler path.

a. In Figure 7.60, identify the two vertices of odd degree.

b. Make a new graph by adding an edge joining the two vertices of odd degree in part a.

c. Find an Euler circuit for the graph you made in part b.

d. Use the Euler circuit you found in part c to find an Euler path for the graph in Figure 7.60.

34. Coming up with applications. Write down some practical situations where finding Euler circuits may be important.

Exercises 35 through 38 are suitable for group work.

Talking to a computer Large graphs need to be handled by computers. Although computer graphics can be stunning, computers have difficulty extracting information from the pictures we have been using to represent graphs. So if we want to feed a graph to a computer, we must present it in some way other than a picture. One way of doing this is via an *incidence matrix*, which is an array of dashes, 1's, and 0's. We will explain what we mean with an example:

$$
\begin{array}{c@{\ }c}
 & \begin{array}{ccc} A & B & C \end{array} \\
\begin{array}{c} A \\ B \\ C \end{array} &
\left(\begin{array}{ccc}
- & 0 & 1 \\
0 & - & 1 \\
1 & 1 & -
\end{array} \right)
\end{array}
$$

In this matrix, the letters A, B, C represent the vertices of the graph. The dashes on the diagonal are just placeholders, and we ignore them. The "0" in the B column of the A row means there is no edge joining A to B. The "1" in the C column of the A row means there is an edge joining A to C. Similarly, the "1" in the C column of the B row means that there is an edge joining B to C. The completed graph represented by the

incidence matrix above is shown below. Exercises 35 through 38 use this idea.

35. Make the graph represented by the following incidence matrix:

$$
\begin{array}{c@{\ }c}
 & \begin{array}{cccccc} A & B & C & D & E & F \end{array} \\
\begin{array}{c} A \\ B \\ C \\ D \\ E \\ F \end{array} &
\left(\begin{array}{cccccc}
- & 0 & 1 & 1 & 0 & 1 \\
0 & - & 0 & 1 & 1 & 0 \\
1 & 0 & - & 1 & 1 & 1 \\
1 & 1 & 1 & - & 1 & 0 \\
0 & 1 & 1 & 1 & - & 1 \\
1 & 0 & 1 & 0 & 1 & -
\end{array} \right)
\end{array}
$$

36. Make the incidence matrix that describes the graph shown below.

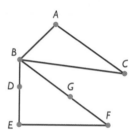

37. Can you make a graph with the following incidence matrix? Explain any difficulties you encounter.

$$
\begin{array}{c@{\ }c}
 & \begin{array}{cccccc} A & B & C & D & E & F \end{array} \\
\begin{array}{c} A \\ B \\ C \\ D \\ E \\ F \end{array} &
\left(\begin{array}{cccccc}
- & 0 & 1 & 1 & 0 & 1 \\
0 & - & 0 & 1 & 1 & 0 \\
1 & 0 & - & 1 & 1 & 1 \\
1 & 1 & 1 & - & 1 & 0 \\
0 & 1 & 1 & 1 & - & 1 \\
1 & 1 & 1 & 1 & 1 & -
\end{array} \right)
\end{array}
$$

38. Considering the results of the previous exercise, can you suggest a property that all incidence matrices of graphs must have?

39. History. Write a paper on the life and work of Leonhard Euler.

7.2 Hamilton circuits and traveling salesmen: Efficient routes

The following article appeared in the *New York Times*.

IN THE NEWS 7.3

Math Problem, Long Baffling, Slowly Yields

GINA KOLATA   March 12, 1991

A century-old math problem of notorious difficulty has started to crumble. Even though an exact solution still defies mathematicians, researchers can now obtain answers that are good enough for almost all practical applications.

The traveling salesman problem, as it is known, crops up in many practical applications, from the design of computer chips to the designation of work orders in factories.

Brute number-crunching by computers can now produce answers to most such problems, even though not an immaculate solution. "Everybody likes to point to the traveling salesman problem as a prototypically hard problem," said Dr. David Johnson of A.T.&T. Bell Laboratories in Murray Hill, N.J. But problems that a few years ago would have made scientists gasp in dismay are now being solved in a few hours of computer time.

The traveling salesman problem asks for the shortest tour around a group of cities. It sounds simple—just try a few tours out and see which one is shortest. But it turns out to be impossible to try all possible tours around even a small number of cities by enumerating them and looking for the shortest one. For example, if there are 100 cities, there are $100 \times 99 \times 98 \times 97$ and so on possible tours. This product is about equal to 10 to the 158th power, or 1 with 158 zeros after it.

In the preceding section, we studied Euler circuits for graphs. The path for the *traveling salesman problem* described in the article above is a different type of circuit, called a *Hamilton circuit*. With an Euler circuit, our goal is to traverse efficiently the *edges* by using each edge exactly once. For a Hamilton circuit, we want to traverse efficiently the *vertices*; that is, we want a path starting and ending at the same place that visits each vertex exactly once (though it may miss some edges altogether).

Key Concept

A Hamilton circuit in a graph is a circuit that visits each vertex exactly once.

To compare the two types of circuits, we consider a highway map. An Euler circuit would be appropriate if you wish to travel along each *road* exactly once, as shown in **Figure 7.61** on the following page. A Hamilton circuit would be appropriate if you wish to visit each *city* exactly once, as shown in **Figure 7.62** on the following page. Note that in a Hamilton circuit, some edges may not be used at all.

Hamilton circuits are named after William Rowan Hamilton.

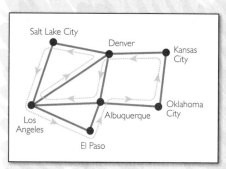

FIGURE 7.61 An Euler circuit.

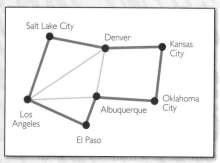

FIGURE 7.62 A Hamilton circuit shown in heavier edges.

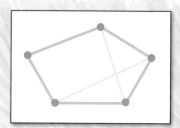

FIGURE 7.63 One Hamilton circuit.

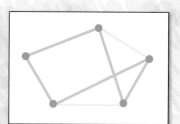

FIGURE 7.64 Another Hamilton circuit.

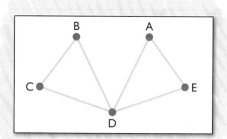

FIGURE 7.65 A graph with no Hamilton circuit.

A graph may have many Hamilton circuits, just as it may have many Euler circuits. For example, **Figures 7.63 and 7.64** show two different Hamilton circuits for the same graph. Thus, when you complete an exercise or follow one of our examples, you may well find a correct answer that is different from the circuit that we found.

Some graphs have no Hamilton circuit. A little trial and error will quickly convince you that the graph in **Figure 7.65** has no Hamilton circuit: Any circuit containing all the vertices must visit the middle vertex twice.

Hamilton circuits are very important in routing problems that arise with airlines, delivery services, Internet routing, etc. For example, suppose a delivery driver has dozens of packages to be delivered to various destinations in a day. She would want to travel to each destination exactly once and not go to (or pass) one destination many times. If the vertices are the destinations, along with headquarters, and the edges are the road routes between them, the desired efficient route is a Hamilton circuit from headquarters and back.

Making Hamilton circuits

We know that connected graphs have Euler circuits precisely when each vertex is of even degree. There is no such easy indicator for when a Hamilton circuit exists.

Furthermore, even when Hamilton circuits exist, they may be much more difficult to find than Euler circuits.

Although there is no nice recipe to tell us how to make Hamilton circuits, at least one observation is helpful when vertices of degree 2 are present. If we want to visit a city with only one road in and one road out, then any Hamilton circuit must use both roads. In general, if a graph has a vertex of degree 2, then each edge meeting that vertex must be part of any Hamilton circuit.

CALCULATION TIP 7.1 Vertices of Degree 2 and Hamilton Circuits

If a graph has a vertex of degree 2, then each edge meeting that vertex must be part of any Hamilton circuit.

Let's see how this observation can help us make Hamilton circuits. Suppose we need an efficient route that allows us to make deliveries from our terminal to the bank, the hardware store, Starbucks, Walmart, the Shoe Box, and Old Navy. That is, we want to find a Hamilton circuit for the graph in **Figure 7.66**. The vertices at the hardware store and at Walmart have degree 2. Hence, any Hamilton circuit must use the four roads highlighted in **Figure 7.67**.

It is now easy to see how to complete the Hamilton circuit, as shown in **Figure 7.68**.

It is helpful to keep in mind that as we try to extend a path to a Hamilton circuit, we cannot return to any vertex already used (except at the last step, when the path returns to its starting point). Therefore, no smaller circuit can be part of any Hamilton circuit.

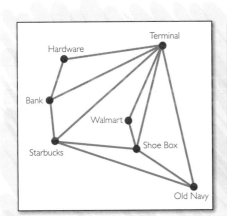

FIGURE 7.66 A delivery route problem.

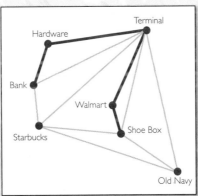

FIGURE 7.67 We must use edges that meet vertices of degree 2.

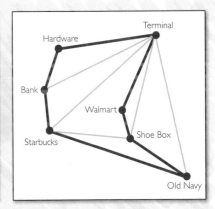

FIGURE 7.68 Completing the delivery route.

EXAMPLE 7.7 Finding a Hamilton circuit: Hiking trails

Figure 7.69 on the following page shows hiking trails and points of interest in Yellowstone National Park.

We want to find a path that starts and ends at Shoshone Lake and visits each point of interest exactly once. That is, we want to find a Hamilton circuit for the graph. Either find a Hamilton circuit or explain why no such circuit exists.

SOLUTION

Note that the vertices Grizzly Lake, Mammoth, Norris, and Lewis Lake all have degree 2. Thus in any Hamilton circuit, we must use the hiking paths that touch these vertices. The corresponding edges are highlighted in **Figure 7.70** on the following page. The result is that we already have a circuit at the top of Figure 7.70. Such a

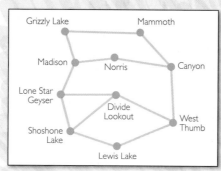

FIGURE 7.69 Hiking paths.

FIGURE 7.70 Edges meeting vertices of degree 2 must be used.

circuit cannot be part of any larger Hamilton circuit, and we conclude that it is not possible to take the desired walking path. We will have to settle for a path that visits at least one point of interest more than once.

TRY IT YOURSELF 7.7

Figure 7.71 shows a terminal and businesses from which items need to be picked up. Find a route, if one exists, that starts and ends at the terminal and visits each business exactly once. That is, either find a Hamilton circuit or explain why no such circuit exists.

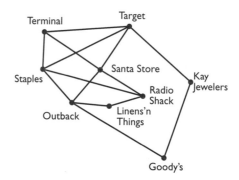

FIGURE 7.71 Terminal and businesses.

The answer is provided at the end of this section.

SUMMARY 7.3   **Hamilton Circuits**

1. A Hamilton circuit is a circuit that visits each vertex exactly once.

2. There is no set procedure for determining whether Hamilton circuits exist. Here is one helpful observation: If a graph has a vertex of degree 2, then each edge meeting that vertex must be part of any Hamilton circuit.

3. A Hamilton circuit cannot contain a smaller circuit.

FIGURE 7.72 A simple map.

The traveling salesman problem

Consider a trucking company that wants to send a truck from its home base in New York to Chicago, Miami, and New Orleans, and then home again. A simplified map with mileage is shown in **Figure 7.72**. The company could save a lot of money by using a shortest route. This is an example of the famous *traveling salesman problem*, which seeks to find a shortest route that visits each city once and then returns home. In other words, the problem is to find a "shortest" Hamilton circuit. As the article opening this section indicates, the traveling salesman problem has many important applications.

We can solve the traveling salesman problem in this case by listing the six possible routes and calculating the mileage for each:

- New York - Miami - Chicago - New Orleans - New York: Distance 5153 miles.
- New York - New Orleans - Chicago - Miami - New York: Distance 5153 miles. (*Note*: This is the reverse of the preceding trip.)
- New York - Miami - New Orleans - Chicago - New York: Distance 3938 miles.
- New York - Chicago - New Orleans - Miami - New York: Distance 3938 miles. (*Note*: This is the reverse of the preceding trip.)
- New York - Chicago - Miami - New Orleans - New York: Distance 4385 miles.
- New York - New Orleans - Miami - Chicago - New York: Distance 4385 miles. (*Note*: This is the reverse of the preceding trip.)

The shortest route is then New York - Miami - New Orleans - Chicago - New York or its reverse. It is worth noting that the mileage saved over alternative routes is significant.

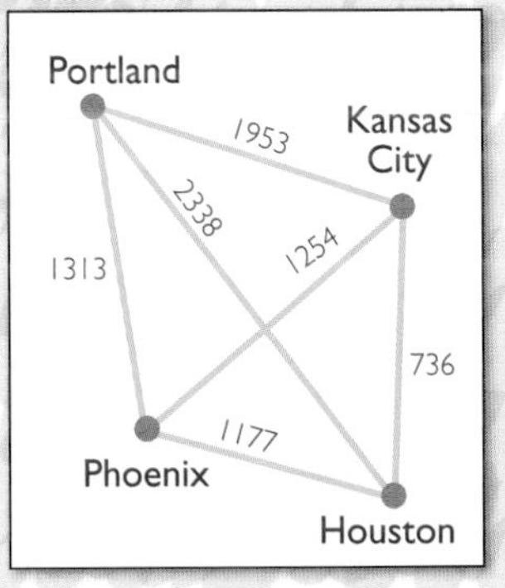

EXAMPLE 7.8 Finding by hand a shortest route: Delivery truck

The graph in **Figure 7.73** shows a delivery map for a trucking firm based in Kansas City. The firm needs a shortest route that will start and end in Kansas City and make stops in Houston, Phoenix, and Portland. That is, the trucking firm needs a solution of the traveling salesman problem for this map. Calculate the mileage for each possible route to find the solution.

SOLUTION

There are six possible routes altogether, but we need to list only three because we gain no new information by looking at the reverse of a route:

- Kansas City - Houston - Phoenix - Portland - Kansas City: Distance 5179 miles.
- Kansas City - Phoenix - Houston - Portland - Kansas City: Distance 6722 miles.
- Kansas City - Phoenix - Portland - Houston - Kansas City: Distance 5641 miles.

The shortest route of 5179 miles is the first one listed.

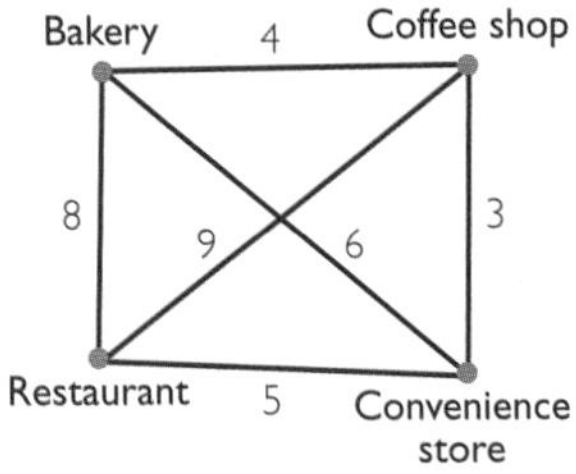

TRY IT YOURSELF 7.8

For the graph in **Figure 7.74**, find a shortest delivery route that starts and ends at the bakery and makes deliveries to the coffee shop, the restaurant, and the convenience store. That is, solve the traveling salesman problem for this graph.

The answer is provided at the end of this section.

FIGURE 7.73 Delivery map.

FIGURE 7.74 Delivery map for Try It Yourself 7.8.

The complexity of the traveling salesman problem

In each of the traveling salesman examples above, each destination is connected to every other by a direct route (an edge). In general, the traveling salesman problem applies to *complete graphs*, which are graphs where each vertex is connected to every other vertex by an edge. A complete graph on three vertices is a triangle. Complete graphs on four and five vertices are shown in **Figures 7.75 and 7.76**.

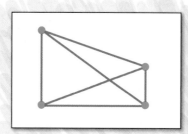

FIGURE 7.75 A complete graph on four vertices.

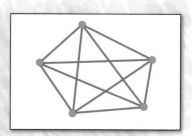

FIGURE 7.76 A complete graph on five vertices.

Key Concept

In a **complete graph**, each vertex is connected to every other vertex by an edge. The **traveling salesman problem** applies to complete graphs for which a distance (more generally, a value) is assigned to each edge. The problem is to find a shortest Hamilton circuit.

An example of the traveling salesman problem for five cities is shown in **Figure 7.77**. If we start at Seattle, it turns out that we need to check 12 separate routes (Hamilton circuits) to solve the problem for this map.

The number of Hamilton circuits for the traveling salesman problem starting at a given city, not counting reverse routes, is shown in the table on following page.[4]

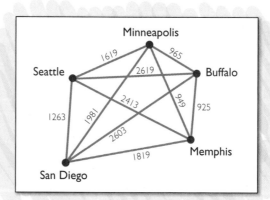

FIGURE 7.77 A traveling salesman problem for five cities.

---

[4]See Exercise 46.

Number of cities	Hamilton circuits
4	3
5	12
6	60
7	360
8	2520
9	20,160
10	181,440
11	1,814,400

Indeed, on a complete graph with n vertices, there are $(n-1)!/2$ Hamilton circuits starting at a given vertex, not counting reverse routes. (Recall that $k! = k \times (k-1) \times \ldots \times 1$. For example, $3! = 3 \times 2 \times 1 = 6$.)

This table means, for example, that if we deliver to 11 or more locations, we have to check over a million routes to find a shortest one. The article at the beginning of this section states that there are about 10^{158} possible tours of 100 cities. As a result, even myriads of the very fastest computers could not solve the traveling salesman problem for 100 cities within a reasonable time frame. The result is that solving the traveling salesman problem for a large number of cities by listing each possible route is totally impractical.[5]

In 1962, Procter & Gamble held a contest to find a shortest route between 33 U.S. cities. There are 131,565,418,466,846,765,083,609,006,080,000,000 possible routes for 33 cities, but amazingly, the contest did have some winners!

Nearest-neighbor algorithm

Solving the traveling salesman problem by listing each route is not practical if the number of cities is large. To get around this difficulty, researchers have developed a number of clever algorithms that find Hamilton circuits that are "nearly optimal." One of these is the *nearest-neighbor algorithm*, which starts at a vertex and makes a Hamilton circuit. At each step in the construction, we travel to the nearest vertex that has not already been used. If there are two or more vertices equally nearby, we just pick any one of those nearest vertices.

Key Concept

The **nearest-neighbor algorithm** constructs a Hamilton circuit in a complete graph by starting at a vertex. At each step, it travels to the nearest vertex not already visited (except at the final step, where it returns to the starting point). If there are two or more vertices equally nearby, any one of them may be selected.

Let's see how this algorithm would work starting from Seattle for the map in **Figure 7.78** on the following page. The nearest city from Seattle is San Diego, which is 1263 miles away. From there, the nearest city is Memphis, 1819 miles away. The nearest city (not already visited) from Memphis is Buffalo. From Buffalo, we have no choice but to travel to Minneapolis and then to Seattle. The resulting route is shown in **Figure 7.79** on the following page. The total mileage for this route is 6591 miles. This turns out to be the shortest Hamilton circuit, but as we shall see, the nearest-neighbor algorithm does not always produce a shortest route.

---

[5]There is a very good history of the traveling salesman problem at www.tsp.gatech.edu/index.html. A poster for the contest next described is shown at that site.

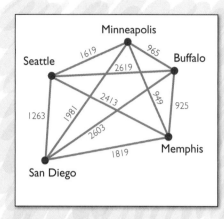

FIGURE 7.78 Traveling salesman problem for five cities.

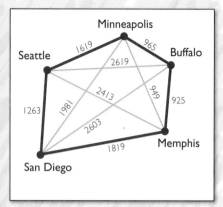

FIGURE 7.79 Result of the nearest-neighbor algorithm.

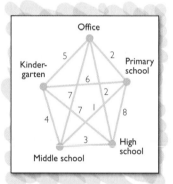

FIGURE 7.80 A map of schools.

EXAMPLE 7.9 Using the nearest-neighbor algorithm: Connecting schools

The local school board needs to visit the high school, middle school, primary school, and kindergarten each day. A map illustrating the office and schools, along with distances in miles, is shown in **Figure 7.80**.

a. Find a shortest path by listing all possible routes starting and ending at the office.

b. Use the nearest-neighbor algorithm starting at the office to approximate a shortest Hamilton circuit.

SOLUTION

a. There are 12 possible routes if we do not include reverse routes. We provide a visual display of each of these routes in **Figures 7.81** through **7.92**. The shortest Hamilton circuit, as shown in Figure 7.90, is the 16-mile route: Office - High school - Kindergarten - Middle school - Primary school - Office shown. We emphasize once more that there is no known way to get the shortest route without listing all the possibilities, and when only a moderate number of vertices are involved, the list becomes extraordinarily long. This is what makes the problem so difficult.

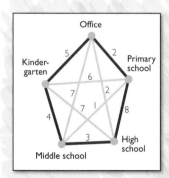

FIGURE 7.81 22 miles.

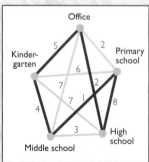

FIGURE 7.82 20 miles.

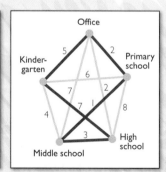

FIGURE 7.83 18 miles. (This is one of two routes that are found by the nearest-neighbor algorithm in part b.)

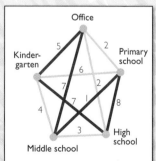

FIGURE 7.84 28 miles.

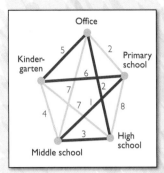

FIGURE 7.85 17 miles. (This is one of two routes that are found by the nearest-neighbor algorithm in part b.)

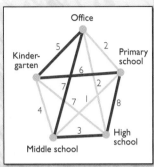

FIGURE 7.86 29 miles.

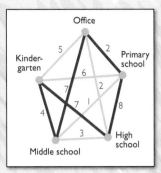

FIGURE 7.87 28 miles.

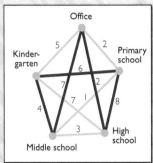

FIGURE 7.88 27 miles.

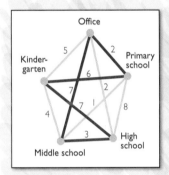

FIGURE 7.89 25 miles.

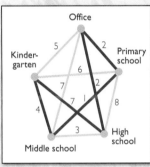

FIGURE 7.90 16 miles.

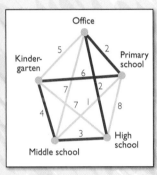

FIGURE 7.91 17 miles.

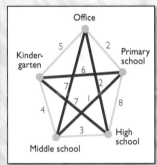

FIGURE 7.92 23 miles.

b. From the office, there are two nearest schools—both the primary school and high school are 2 miles away. We are free to choose either route. Suppose we choose to go first to the primary school. The closest school from the primary school is the middle school 1 mile away. The next nearest school (not already visited) is the high school. We complete the route by traveling to the kindergarten and then back to the office. The resulting route, which is shown in Figure 7.83, is Office - Primary school - Middle school - High school - Kindergarten - Office. This route is 18 miles long.

The reader can easily verify that if we had chosen to go first to the high school rather than the primary school, we would have gotten the route shown in Figure 7.85. That route is 17 miles long. We note that both of these routes are longer than the shortest route of 16 miles found in part a.

This example shows that the nearest-neighbor algorithm can fail to find a shortest route.

The cheapest link algorithm

Another method used to find approximate solutions of the traveling salesman problem is the *cheapest link algorithm*. This procedure begins by selecting the shortest edge in the graph and continues by selecting the shortest edge that has not already been chosen. (If there is more than one shortest edge, choose either.) But at each step, we must be careful not to violate the following rules:

Rule 1: Do not choose an edge that results in a circuit except at the final step, when a Hamilton circuit is constructed.

Rule 2: Do not choose an edge that would result in a vertex of degree 3.

The cheapest link algorithm is just as easy to implement as is the nearest-neighbor algorithm. Let's see how it works with the school route problem from Example 7.9. We first select the shortest edge in the graph, which is the 1-mile edge from the primary school to the middle school. This is shown in **Figure 7.93**. Next there are two edges of length 2, and we can choose both of them without violating either Rule 1 or Rule 2. The pieces thus far assembled are shown in **Figure 7.94**.

The next shortest edge is the 3-mile route joining the middle school to the high school. But we can't use this edge because it would make the circuit shown in **Figure 7.95**. Instead, we choose the 4-mile route from the middle school to the kindergarten, as shown in **Figure 7.96**.

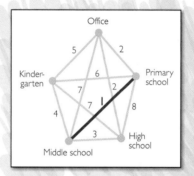

FIGURE 7.93 Selecting the shortest edge in the graph.

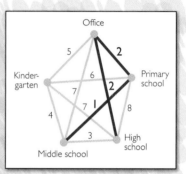

FIGURE 7.94 Adding two more edges.

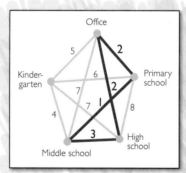

FIGURE 7.95 Choosing the 3-mile edge results in a circuit.

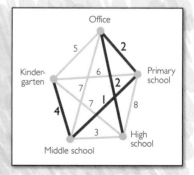

FIGURE 7.96 Use the 4-mile edge instead.

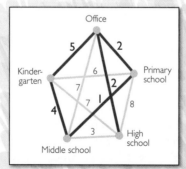

FIGURE 7.97 The edge from kindergarten to office violates both Rule 1 and Rule 2.

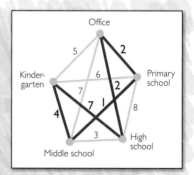

FIGURE 7.98 The completed Hamilton circuit.

The next shortest edge is the 5-mile route from the kindergarten to the office. But **Figure 7.97** shows that this choice violates both Rule 1 and Rule 2. The only other choice is the 7-mile route from the kindergarten to the high school, which results in the Hamilton circuit shown in **Figure 7.98**. We note that this is the shortest possible route, as we found in Example 7.9. But the cheapest link algorithm does not always produce the shortest path, as the next example shows.

EXAMPLE 7.10 Applying the cheapest link algorithm: A trucking company

We run a trucking company that makes deliveries in Seattle, Minneapolis, Buffalo, Memphis, and San Diego. The map is shown in **Figure 7.99**. Use the cheapest link algorithm to find an approximate solution of the traveling salesman problem.

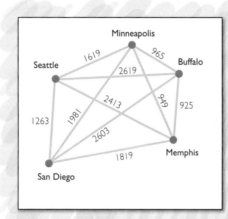

FIGURE 7.99 Five cities.

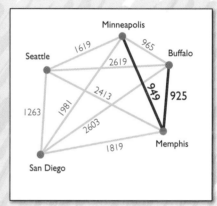

FIGURE 7.100 Choosing the first two segments of the route.

SOLUTION

We begin by choosing the two shortest edges in the graph, as shown in **Figure 7.100**.

The next shortest edge is the 965-mile route from Minneapolis to Buffalo. But we can't use this route because it would result in the circuit shown in **Figure 7.101**. We choose instead the next shortest edge, the 1263-mile route from San Diego to Seattle. The edges from Seattle to Minneapolis and San Diego to Buffalo complete the route, as shown in **Figure 7.102**.

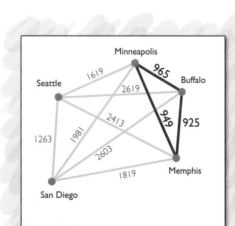

FIGURE 7.101 Adding the edge from Minneapolis to Buffalo makes a circuit.

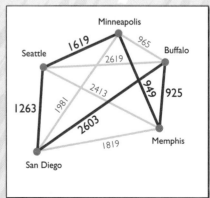

FIGURE 7.102 The completed route.

This route has a total length of $925 + 949 + 1263 + 1619 + 2603 = 7359$ miles. This is longer than the shortest route of 6591 miles that we found earlier using the nearest-neighbor algorithm (see Figure 7.79).

TRY IT YOURSELF 7.10

Figure 7.103 shows a map of the cities where our air freight company makes deliveries. Use the cheapest link algorithm to find an approximate solution of the traveling salesman problem.

The answer is provided at the end of this section.

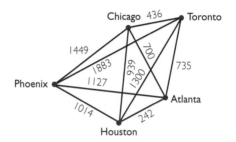

FIGURE 7.103 An air freight map.

SUMMARY 7.4   The Traveling Salesman Problem

1. A complete graph is one in which each pair of vertices is joined by an edge.

2. If a distance (more generally, a value) is assigned to each edge, the traveling salesman problem is to find a shortest Hamilton circuit.

3. Solving the traveling salesman problem is difficult when there are more than just a few vertices. However, there are algorithms that can give an approximate solution. One such is the nearest-neighbor algorithm. Another is the cheapest link algorithm.

The two algorithms that we have presented have the advantage of being simple, and they provide a flavor of how algorithms work. But, in fact, they are of questionable effectiveness in real situations. Much better (but more complex) algorithms are available that provide practical approximations to the optimal solution.

Try It Yourself answers

Try It Yourself 7.7: Finding a Hamilton circuit: Visiting stores   One Hamilton circuit is shown using highlighted edges.

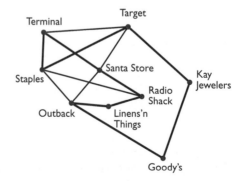

Try It Yourself 7.8: Finding by hand a shortest route: Delivery truck   The route Bakery - Coffee shop - Convenience store - Restaurant - Bakery (or its reverse) has the minimal length of 20 miles.

Try It Yourself 7.10: Applying the cheapest link algorithm: Air freight deliveries The route below has a length of 4275 miles.

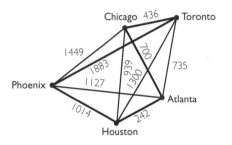

Exercise Set 7.2

1. Euler or Hamilton? The edges in a certain graph represent unpaved roads, and the vertices represent intersections. We want to pave all of the streets and not require the paving machine to move along streets already paved. Should we look for an Euler circuit or a Hamilton circuit?

2. Euler or Hamilton? The edges in a certain graph represent roads, and the vertices represent intersections. We want to inspect the traffic signs at each intersection and not visit any intersection twice. Should we look for an Euler circuit or a Hamilton circuit?

3. Euler but not Hamilton? Consider the graph in **Figure 7.104**. We noted on p. 444 that this graph (which is the same as the graph in Figure 7.65) has no Hamilton circuit. Does this graph have an Euler circuit?

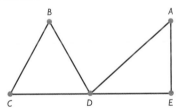

FIGURE 7.104 A graph with no Hamilton circuit.

4. Hamilton but not Euler? The graph in **Figure 7.105** has a Hamilton circuit, as is shown by the heavier edges. Does this graph have an Euler circuit?

FIGURE 7.105 A graph with a Hamilton circuit.

5. A park. **Figure 7.106** shows hiking trails and points of interest in a national park. Either find a route that begins at the park entrance, visits each point of interest once, and returns to

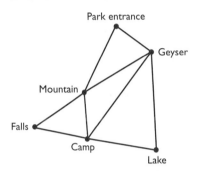

FIGURE 7.106 A park.

the park entrance, or show that no such route exists. That is, either find a Hamilton circuit or explain why no such circuit exists.

6. A bakery. Figure 7.107 shows a bakery and several sites where baked goods are to be delivered. Either find a route that begins and ends at the bakery and passes each store exactly once, or show that no such route exists. That is, either find a Hamilton circuit or explain why no such circuit exists.

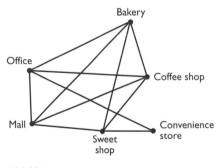

FIGURE 7.107 A bakery.

7. Computer network. Figure 7.108 on the following page shows a computer network. We want to route a message starting from computer *A*, passing to each other computer on the network exactly once, and returning to *A*. Either find such a path or show that no such path exists. That is, either find a Hamilton circuit or explain why no such circuit exists.

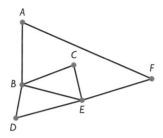

FIGURE 7.108 Computer network.

8. Visiting historical sites. Figure 7.109 shows some historical sites in Philadelphia. We want to take a walking tour beginning and ending at Franklin Square and visiting each site exactly once. Either find such a route or show that none exists. That is, either find a Hamilton circuit or explain why no such circuit exists.

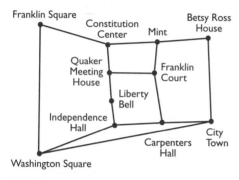

FIGURE 7.109 Visiting historical sites.

9. Internet connections. The edges in **Figure 7.110** represent fiber-optic lines. Vertex *A* is an internet service provider. The remaining vertices are homes that must be wired for an Internet connection. Either find a route that begins and ends at the provider location and visits each home exactly once, or show that no such route exists. That is, either find a Hamilton circuit or explain why no such circuit exists.

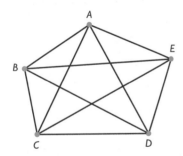

FIGURE 7.110 Internet connections.

10. Tour of western states. The vertices in **Figure 7.111** represent western states. Vertices are connected by an edge if the two states share a common border. We wish to take a scenic tour of the western United States. Either find a path that begins and ends at Washington and visits each state exactly once, or show that no such path exists. That is, either find a Hamilton circuit or explain why no such circuit exists.

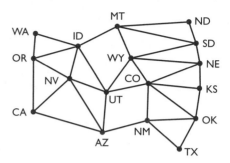

FIGURE 7.111 Tour of western states.

11. Tour of southern states. The vertices in **Figure 7.112** represent southern states. Vertices are connected by an edge if the two states share a common border. We wish to take a scenic tour of the southern United States. Either find a path that begins and ends at Missouri and visits each state exactly once, or show that no such path exists. That is, either find a Hamilton circuit or explain why no such circuit exists.

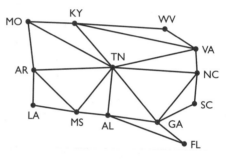

FIGURE 7.112 Tour of southern states.

12. Viewing holiday decorations. The edges in **Figure 7.113** represent streets. Vertex *A* is your home, and the remaining vertices are neighborhood homes that are displaying holiday decorations. Either find a path that begins and ends at your own home and passes each neighborhood home exactly once, or show that no such path exists. That is, either find a Hamilton circuit or explain why no such circuit exists.

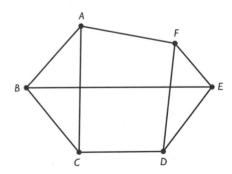

FIGURE 7.113 Viewing holiday decorations.

13. Holiday shopping. Figure 7.114 shows a shopping center. The vertices are stores, and the edges are sidewalks. We wish to visit each store for holiday shopping. Either find a route that begins and ends at T.J. Maxx and visits each store once or show that no such route exists. That is, either find a Hamilton circuit or explain why no such circuit exists.

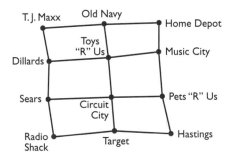

FIGURE 7.114 Holiday shopping.

14. **Checking in-home care patients.** The edges in **Figure 7.115** represent roads. The vertex marked "office" is the headquarters of an in-home care service. The remaining vertices are homes where patients must be visited each day. Either find a route beginning and ending at headquarters and visiting each patient home exactly once, or show that no such route exists. That is, either find a Hamilton circuit or explain why no such circuit exists.

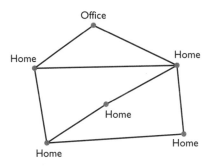

FIGURE 7.115 Checking in-home care patients.

15. **Meals on Wheels.** The edges in **Figure 7.116** represent roads. Meals on Wheels is an organization that provides meals for the elderly. The remaining vertices are homes of elderly people to which hot meals are to be delivered. Either find a path that begins and ends at the kitchen and passes each home exactly once, or show that no such path exists. That is, either find a Hamilton circuit or explain why no such circuit exists.

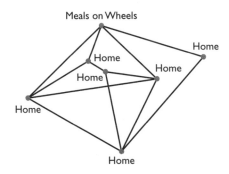

FIGURE 7.116 Meals on Wheels.

16. **Routing a message.** The vertices in **Figure 7.117** represent computers, and the edges represent connections. To check the network setup, we want to send a test message that starts at computer A, passes through each other computer exactly once,

and successfully returns to computer A. Either find a workable path or show that no such path exists. That is, either find a Hamilton circuit or explain why no such circuit exists.

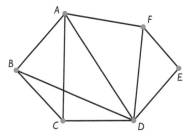

FIGURE 7.117 Routing a message.

Hamilton circuits In Exercises 17 through 20, either find a Hamilton circuit or show that none exists.

17. Either find a Hamilton circuit or show that none exists for the graph in **Figure 7.118**.

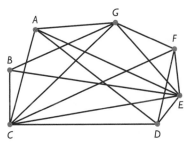

FIGURE 7.118

18. Either find a Hamilton circuit or show that none exists for the graph in **Figure 7.119**.

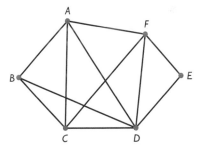

FIGURE 7.119

19. Either find a Hamilton circuit or show that none exists for the graph in **Figure 7.120**.

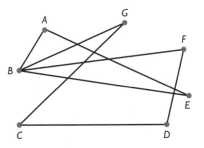

FIGURE 7.120

20. Either find a Hamilton circuit or show that none exists for the graph in **Figure 7.121**.

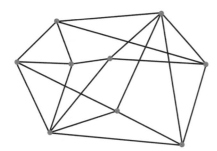

FIGURE 7.121

Traveling salesman problem   In Exercises 21 through 23, solve the traveling salesman problem for the given map.

21. Solve the traveling salesman problem for the map in **Figure 7.122** by calculating the mileage for each possible route. Use Ft. Wayne as a starting point.

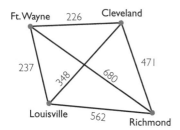

FIGURE 7.122

22. Solve the traveling salesman problem for the map in **Figure 7.123** by calculating the mileage for each possible route. Use New York as a starting point.

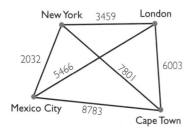

FIGURE 7.123

23. Solve the traveling salesman problem for the map in **Figure 7.124** by calculating the mileage for each possible route. Use Provo as a starting point.

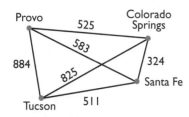

FIGURE 7.124

Nearest-neighbor algorithm   In Exercises 24 through 29, use the nearest-neighbor algorithm.

24. Use the nearest-neighbor algorithm starting at vertex A of the map in **Figure 7.125** to find an approximate solution of the traveling salesman problem.

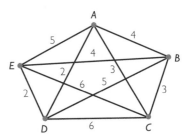

FIGURE 7.125

25. Use the nearest-neighbor algorithm starting at vertex A of the map in **Figure 7.126** to find an approximate solution of the traveling salesman problem.

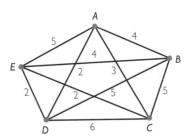

FIGURE 7.126

26. Use the nearest-neighbor algorithm starting at vertex A of the map in **Figure 7.127** to find an approximate solution of the traveling salesman problem.

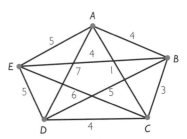

FIGURE 7.127

27. Use the nearest-neighbor algorithm starting at vertex A of the map in **Figure 7.128** to find an approximate solution of the traveling salesman problem.

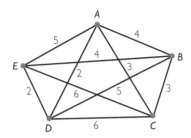

FIGURE 7.128

28. Use the nearest-neighbor algorithm starting at vertex *A* of the map in **Figure 7.129** to find an approximate solution of the traveling salesman problem.

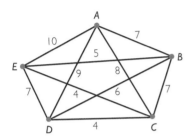

FIGURE 7.129

29. Use the nearest-neighbor algorithm starting at vertex *A* of the map in **Figure 7.130** to find an approximate solution of the traveling salesman problem.

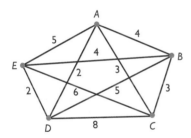

FIGURE 7.130

Cheapest link algorithm In Exercises 30 through 35, use the cheapest link algorithm.

30. Use the cheapest link algorithm to find an approximate solution of the traveling salesman problem for the map in Figure 7.125.

31. Use the cheapest link algorithm to find an approximate solution of the traveling salesman problem for the map in Figure 7.126.

32. Use the cheapest link algorithm to find an approximate solution of the traveling salesman problem for the map in Figure 7.127.

33. Use the cheapest link algorithm to find an approximate solution of the traveling salesman problem for the map in Figure 7.128.

34. Use the cheapest link algorithm to find an approximate solution of the traveling salesman problem for the map in Figure 7.129.

35. Use the cheapest link algorithm to find an approximate solution of the traveling salesman problem for the map in Figure 7.130.

36. Nearest-neighbor failure. This exercise provides an example like Example 7.9 where the nearest-neighbor algorithm fails to find a shortest route. Consider the map shown in **Figure 7.131**.

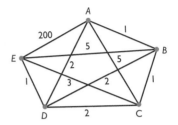

FIGURE 7.131

a. Use the nearest-neighbor algorithm starting at *A* to find an approximate shortest route. How long is the route you found?

b. By trial and error, find a shortest route. How does its length compare with that in part a?

37. Starting point for nearest-neighbor algorithm. This exercise refers to the map in Figure 7.130. In each part, use the nearest-neighbor algorithm starting at the indicated vertex to find an approximate solution of the traveling salesman problem.

a. Start at *A*. **b.** Start at *B*.

c. Start at *C*. **d.** Start at *D*.

e. Start at *E*.

f. Compare your answers to parts a through e. What does the result say about the choice of starting point for the nearest-neighbor algorithm?

38. Coming up with applications. Write down some practical situations where finding Hamilton circuits may be important.

Exercises 39 through 43 are suitable for group work.

An improved nearest-neighbor algorithm An improvement of the nearest-neighbor algorithm goes as follows. Apply the nearest-neighbor algorithm once for each vertex by starting the algorithm at that vertex. Take the best of these routes.

This information is needed for Exercises 39 through 43. In Exercises 39 through 42, apply the improved nearest-neighbor algorithm to the indicated graph.

39. Apply the improved nearest-neighbor algorithm to the graph in Figure 7.125 to find an approximate solution of the traveling salesman problem. Report the length of the shortest circuit you find.

40. Apply the improved nearest-neighbor algorithm to the graph in Figure 7.126 to find an approximate solution of the traveling salesman problem. Report the length of the shortest circuit you find.

41. Apply the improved nearest-neighbor algorithm to the graph in Figure 7.127 to find an approximate solution of the traveling salesman problem. Report the length of the shortest circuit you find.

42. Apply the improved nearest-neighbor algorithm to the graph in Figure 7.128 to find an approximate solution of the traveling salesman problem. Report the length of the shortest circuit you find.

43. An example. Find an example of a complete graph with distances for which the improved nearest-neighbor algorithm does not find the shortest path.

44. History. Hamilton circuits are named for Sir William Rowan Hamilton. Write a report on his scientific contributions.

45. Research problem. Locate four cities in the United States and find the distances between each using an atlas or the Internet. Solve the traveling salesman problem for these four cities.

The following exercise is designed to be solved using technology such as a calculator or computer spreadsheet. For assistance, see the technology supplement.

46. Counting routes. We noted that the number of Hamilton circuits to check for the traveling salesman problem is very large. Here is the formula for that number: If there are n cities, the number of Hamilton circuits starting at a given city, not counting reverse routes, is $(n - 1)!/2$. (Recall that $k! = k \times (k-1) \times \cdots \times 1$. For example, $3! = 3 \times 2 \times 1 = 6$.) How many distinct routes would need to be compared in solving by hand the traveling salesman problem for 50 cities? What about 70 cities? Round your answer to the nearest power of 10.

7.3 Trees: Why are spell checkers so fast?

TAKE AWAY FROM THIS SECTION

Understand how spell checkers use trees.

The following article is from the Web site of the National Science Foundation.

A part of the tree diagram used to analyze how e-mail petitions traveled to people's inboxes.

IN THE NEWS 7.4

How Did That Chain Letter Get to My Inbox?

May 16, 2008

Everyone who has an email account has probably received a forwarded chain letter promising good luck if the message is forwarded on to others—or terrible misfortune if it isn't. The sheer volume of forwarded messages such as chain letters, online petitions, jokes and other materials leads to a simple question—how do these messages reach so many people so quickly?

New research into these forwarded missives by Jon Kleinberg of Cornell University and David Liben-Nowell of Charleston College suggests a surprising explanation....

It had been assumed that the messages traveled to email users in much the same way that a disease spreads in an epidemic—people received the messages and passed them on to those they came in contact with, who, in turn, spread them to people they encountered, and so on. In recent years, some scientists, as well as marketers, have used the term, "viral," to describe this pattern.

Kleinberg and Liben-Nowell decided to study exactly how some selected messages were disseminated through the Internet....

Using this data, the researchers mapped out how these messages traveled from recipient to recipient on a tree diagram. A careful analysis of the diagram challenges some of the common assumptions about how messages spread, including the viral contagion theory. Rather than spreading like a virus, with each message producing many direct "descendents" in the tree diagram, the data suggest that people are selective in forwarding messages to others in their social networks. For example, the researchers discovered that 90 percent of the time, the messages produced only a single descendent.

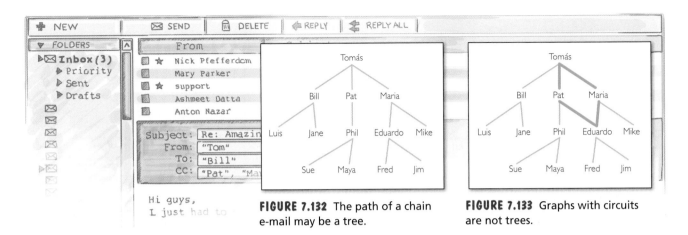

FIGURE 7.132 The path of a chain e-mail may be a tree.

FIGURE 7.133 Graphs with circuits are not trees.

The path typically taken by a chain letter or chain e-mail is a special kind of graph known as a *tree*. Suppose, for example, that you send an e-mail to some of your friends, who in turn forward it to some of their friends, and so on. The resulting path of the chain e-mail might look like the picture in **Figure 7.132**.

Informally, such a graph resembles an upside-down tree, in the sense that it has a starting vertex and the edges make branches that don't grow back together. To say that the branches don't grow back together is to say that the graph contains no circuits. If, for example, both Pat and Maria send the message to Eduardo, then the resulting graph is not a tree because it contains the circuit shown in heavy edges in **Figure 7.133**.

Key Concept

A tree is a graph that contains no circuits.

For the trees we consider, we always choose a starting vertex, typically drawn at the top.

The following terminology is a whimsical mixture of botanical and familial terms that are standard (we did not invent them). The vertex at the top of the tree is called the *root*. For example, Tomás is the root of the e-mail tree, as shown in **Figure 7.134** on the following page. There is a path leading from the root to any other vertex. The vertex in the path just before a given vertex is called its *parent*, and any vertex immediately after is its *child*. We see in Figure 7.134 that Maria is the parent of Eduardo and Mike, so Eduardo and Mike are children of Maria. Vertices that are not parents are *leaves*. These are typically at the bottom of the tree, as shown in Figure 7.134.

Every vertex of a tree is either a parent or a leaf, as shown in **Figure 7.135**. This gives us our first important formula for trees:

Number of vertices = Number of parents + Number of leaves.

A tree also has *levels*, which are shown in **Figure 7.136**. The root is level 0. The children of the root are level 1. The grandchildren are level 2, and so on. The largest level of the tree is the *height*. Thus, the height of the tree in Figure 7.136 is 3.

Binary trees

Perhaps one of the most common trees we see in everyday life, and certainly the one with the most hype, is the tree formed by the NCAA basketball tournament

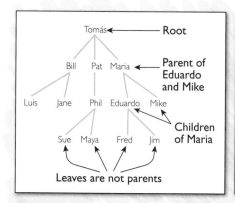

FIGURE 7.134 Root, parents, and children in trees.

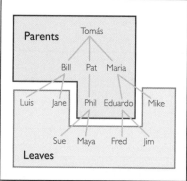

FIGURE 7.135 Vertices separated into parents and leaves.

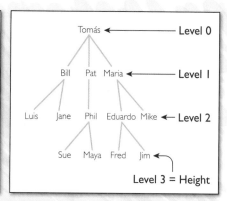

FIGURE 7.136 Levels of a tree.

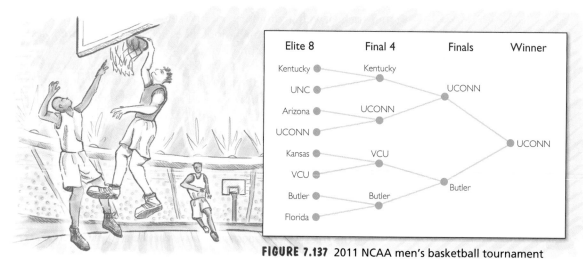

FIGURE 7.137 2011 NCAA men's basketball tournament is a binary tree.

brackets.[6] A piece of the 2011 tree is shown in **Figure 7.137**. This particular type of tree is known as a *binary tree* because each parent vertex has exactly two children.[7]

Key Concept

In a binary tree, each parent has exactly two children.

Because each parent in a binary tree has exactly two children, the number of children is twice the number of parents. But the only vertex in a tree that is not a child is the root. So, in a binary tree

$$\text{Number of vertices} = 2 \times \text{Number of parents} + 1.$$

---

[6] In our discussion of this tournament, we ignore the "play-in" games.

[7] In some texts, a binary tree is one where parents have *at most* two children. If each parent has *exactly* two children, those texts call the tree a *full binary tree*.

EXAMPLE 7.11 Counting in a binary tree: PTA and phone calls

There are 127 members of a PTA group. A school-closure alert plan requires one member to phone two others with the information. Each PTA member who receives a call phones two members who have not yet gotten the news. This continues until all members have the message. How many PTA members make phone calls?

SOLUTION

We represent the phone calls with a binary tree. Each vertex represents a PTA member, and edges indicate phone calls. A part of this tree is shown in **Figure 7.138**.

The parent vertices in the tree are the PTA members who make phone calls. The total number of vertices is the number of PTA members, 127. We can use this information to find the number of parent vertices:

$$\text{Number of vertices} = 2 \times \text{Number of parents} + 1$$
$$127 = 2 \times \text{Number of parents} + 1$$
$$126 = 2 \times \text{Number of parents}$$
$$63 = \text{Number of parents}.$$

Thus, 63 members make phone calls.

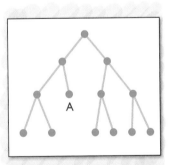

FIGURE 7.138 A phone-call binary tree.

TRY IT YOURSELF 7.11

How many members make no phone calls?

The answer is provided at the end of this section.

Note that in Figure 7.137, the elite 8 is at level 3, the final 4 is at level 2, finals are at level 1, and the winner is level 0. This bracket is an example of a *complete* binary tree, which has as many vertices as possible.

Key Concept

A complete binary tree has all the leaves at the highest level.

Examining the NCAA bracket in Figure 7.137 shows that it is an example of a complete binary tree. **Figure 7.139**, on the other hand, shows a tree that is not complete. The leaf A is not at the highest level. Intuitively, the tree has missing branches.

In a complete binary tree, each parent has exactly two children, and the only leaves are at the highest level. Hence, going from one level to the next doubles the number of vertices. At level 0 there is 1 vertex, the root. There are 2 vertices at level 1, 4 at level 2, 8 at level 3, and so on (see **Figure 7.140** on the following page). In general, there are 2^L vertices at level L. Note that in a tournament bracket like the one in **Figure 7.141** on the following page, the leaves are the entrants in the tournament, and the parent vertices represent games played.

Thinking of binary trees as tournaments is often useful. For example, there are 64 teams in the NCAA basketball tournament. Each game produces exactly one loser, and every team loses except for the tournament champion. Hence, there are 63 games played altogether. Since each game played represents a parent in the tournament tree, there are 63 parents in the tree.

This reasoning allows us to count vertices in any complete binary tree. In a complete binary tree of height H, there are 2^H vertices at the highest level. These are the leaves. We think of the leaves as tournament entrants and the parents as games

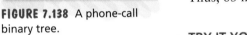

FIGURE 7.139 This binary tree is not complete.

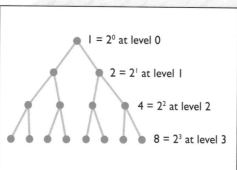

FIGURE 7.140 In a complete binary tree are 2^L vertices at level L.

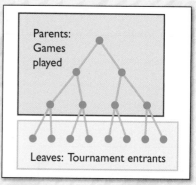

FIGURE 7.141 Entrants and games played in a tournament.

played. Each game produces one loser, and every entrant except one loses a game. Hence,

$$\text{Number of parents} = \text{Number of games played}$$
$$= \text{Number of losers}$$
$$= \text{Number of entrants} - 1.$$

Thus,

$$\text{Number of parents} = 2^H - 1.$$

Now we can find the total number of vertices in a complete binary tree of height H:

$$\text{Number of vertices} = 2 \times \text{Number of parents} + 1$$
$$= 2(2^H - 1) + 1$$
$$= 2 \times 2^H - 2 + 1.$$

Hence,

$$\text{Number of vertices} = 2^{H+1} - 1.$$

SUMMARY 7.5 **Formulas for Trees**

- In any tree,

  $$\text{Number of vertices} = \text{Number of parents} + \text{Number of leaves}.$$

- In a binary tree,

  $$\text{Number of vertices} = 2 \times \text{Number of parents} + 1.$$

- In a complete binary tree, there are 2^L vertices at level L.

- In a complete binary tree of height H:

  $$\text{Number of leaves} = 2^H$$
  $$\text{Number of parents} = 2^H - 1$$
  $$\text{Number of vertices} = 2^{H+1} - 1.$$

In 2009, Bernard Madoff pleaded guilty to operating a multi-billion-dollar Ponzi scheme. Many fortunes were lost when the scheme collapsed.

EXAMPLE 7.12 Applying formulas for trees: A Ponzi scheme

A shady broker offers a dubious investment opportunity. In week 0, he sells a $10,000 note promising to repay $11,000 in one week. In week 1, he sells two such notes and uses the proceeds to pay off the week 0 investor. In week 2, he sells 4 notes and pays off the two week 1 investors. Each week, he sells twice as many notes as the week before and uses the proceeds to pay off last week's investors. An investment scam of this sort is known as a Ponzi scheme.

a. How many notes did the broker sell in week 10?

b. What was the total number of notes sold by week 10?

c. How much money was collected by week 10?

d. How much money was paid out by week 10?

e. In week 10, the broker took his profits and left the country. How much money was lost by investors?

SOLUTION

a. We can model this scheme with a complete binary tree of height $H = 10$. Each vertex represents a sale. The number of sales in week 10 is the number of leaves, which is

$$\text{Number of leaves} = 2^H$$
$$= 2^{10}$$
$$= 1024.$$

Thus, 1024 notes were sold in week 10.

b. The number of notes sold is the total number of vertices, which is

$$\text{Number of vertices} = 2^{H+1} - 1$$
$$= 2^{11} - 1$$
$$= 2047.$$

Thus, 2047 notes were sold by week 10.

c. The broker sold 2047 notes at $10,000 each. Therefore, he collected a total of $2047 \times \$10,000 = \$20,470,000$.

d. The broker paid off the notes sold through week 9. The number of notes sold through week 9 is the number of vertices in a complete binary tree of height 9. Thus,

$$\text{Number sold through week } 9 = 2^{9+1} - 1$$
$$= 1023 \text{ notes.}$$

He paid $11,000 to each of these. That is a total of $1023 \times \$11,000 = \$11,253,000$.

e. He cheated investors out of $\$20,470,000 - \$11,253,000 = \$9,217,000$.

Spell checkers

Modern word processors come equipped with spell checkers that include a dictionary of as many as 100,000 or so words. When we type a word, the spell checker instantly compares that word with the words in its dictionary and highlights the word if there is no match. Did the spell checker instantly make 100,000 comparisons? No, even very fast computers aren't that fast. At the heart of the process is a binary tree. To illustrate the process, let's adopt a very simple dictionary containing only the following list of 13 one-letter "words":

$$A, C, E, G, I, K, \boxed{M}, O, Q, S, U, W, Y$$

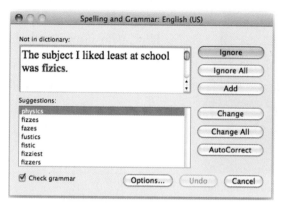

A typical spell checker in action.

We have put a box around M because it is the middle "word" in the list, and it will be the root of our tree—the place where we start our search. If we type in a letter, we will first compare it with M. If the letter we type happens to be M, we have a valid word. If it precedes M in the alphabet, we move to the left in our list; otherwise, we move to the right. These two possibilities correspond to two branches from the root.

Now we consider the two branches. Again, we go to the middle to make our comparison:

$$A, C, \underline{E}, G, I, K, \boxed{M}, O, Q, \underline{S}, U, W, Y$$

In the left branch A, C, E, G, I, K, there are two letters, E and G, that have equal claim to the middle position. We adopt the convention that if there are two middle

"Must have been an old can of alphabet soup. No spell checker.

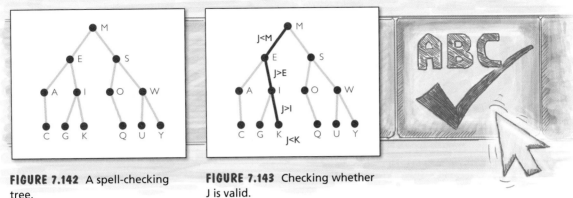

FIGURE 7.142 A spell-checking tree.

FIGURE 7.143 Checking whether J is valid.

words, we will take the one on the left. Thus, we chose E to make our spell-checking tree. In the right branch, there are also two letters that can claim the middle, and we chose the one on the left. Thus, we use S. We go to the middle of each remaining piece to make the next comparison.

The completed spell-checking tree is shown in **Figure 7.142**. At each stage, we move to the right if the letter we typed is "bigger" and to the left if it is "smaller," and we report a valid word if it is the same. Note that the tree is binary except at A and O. The path we would take if we typed in J, which is not in our list and so not a valid word, is highlighted in **Figure 7.143**. Note that we need to make a total of only four comparisons (not 13) to find out that J is not in our dictionary.

EXAMPLE 7.13 Making a spell checker

Make a spell-checking tree for the following dictionary:

> apple, boy, cat, dog, ear, fox, girl, hat, jet, kit, love, mom, net, out, pet

How many checks (or comparisons) are needed to test whether "cow" is a valid word? Show the path you take on the spell-checking tree to check whether "cow" is a valid word.

SOLUTION

We go to the middle of the list, which is "hat":

> apple, boy, cat, dog, ear, fox, girl, | hat |, jet, kit, love, mom, net, out, pet

Thus, "hat" goes at the top of our tree.

Next we go to the middle of the left and right sections:

> apple, boy, cat, <u>dog</u>, ear, fox, girl, | hat |, jet, kit, love, <u>mom</u>, net, out, pet

We get "dog" and "mom." These go on the first level of our tree. We continue to go to the middle of each piece to get the next comparison word. The completed tree is shown in **Figure 7.144** on the following page.

Four checks are needed to spell check "cow." The path of the check is shown in **Figure 7.145** on the following page.

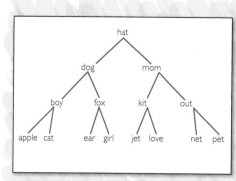

FIGURE 7.144 The spell-checking tree.

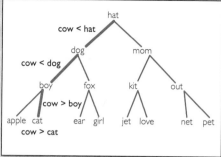

FIGURE 7.145 Checking "cow."

TRY IT YOURSELF 7.13

Make a spell-checking tree for the following dictionary:

box, cap, car, dip, eat, fur, him, hit, ink, joy, lid, man, now, pan, rat

Show the path you take on the spell-checking tree to check whether "it" is a valid word.

The answer is provided at the end of this section.

Efficiency of spell checkers

Now we return to the question of how a spell checker for a word processor is able to check words so quickly in a dictionary with, say, 100,000 words. In terms of the associated tree, the answer lies in the fact that the height is much smaller than the number of vertices.

We have seen that a complete binary tree with height H has $2^{H+1} - 1$ vertices. In a spell checking tree, we make checks at level 0, level 1, . . . , level H. That makes a total of $H + 1$ checks. We conclude that a spell-checking tree requiring a maximum of $c = H + 1$ checks can hold up to $2^c - 1$ words. Now we calculate that

$$2^{17} - 1 = 131,071.$$

This means that we can check a word against a 100,000-word dictionary easily—we need at most 17 checks. With only 3 more checks, we can accommodate a 1-million-word dictionary. That is why spell checkers work so fast. We note that this kind of binary branching is a formalized version of what we all do when we look up a word in the dictionary.

> **SUMMARY 7.6** Spell Checking
>
> A dictionary of $2^c - 1$ words can be accommodated using no more than c checks.

EXAMPLE 7.14 Analyzing a spell checker

a. You want to use a dictionary so that you have to check any word at most five times. How big a dictionary can you accommodate?

b. Estimate the number of checks needed if you have a 10,000-word dictionary.

SOLUTION

a. We know that with at most c checks allowed, a dictionary of $2^c - 1$ words can be accommodated. We want 5 checks, so we use $c = 5$ in this formula. We calculate that $2^5 - 1$ equals 31. This means that we can accommodate a dictionary with 31 words.

b. We want to choose the smallest c so that $2^c - 1$ is at least 10,000. We proceed by trial and error (with the help of a calculator) and find that $2^{14} - 1$ is greater than 10,000 but $2^{13} - 1$ is not. Thus at most, 14 checks are needed.

TRY IT YOURSELF 7.14

Estimate the number of checks needed for a spell-checking tree if we have a 1000-word dictionary.

The answer is provided at the end of this section.

Try It Yourself answers

Try It Yourself 7.11: Counting in a binary tree: PTA and phone calls   64 members do not make phone calls.

Try It Yourself 7.13: Making a spell checker   Here is the spell-checking tree.

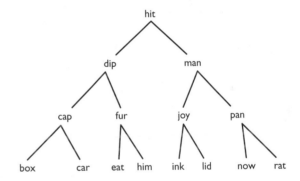

This is the path for checking "it."

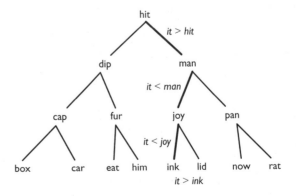

Try It Yourself 7.14: Analyzing a spell checker   10 checks.

Exercise Set 7.3

1. Making a tree. Make a tree with 7 vertices and exactly 4 parents.

2. Few leaves. Make a tree with 10 vertices and the fewest possible leaves.

3. Many leaves. Make a tree with 10 vertices and the most possible leaves.

4. A binary tree. Is it possible to make a binary tree with 18 vertices?

5. Another binary tree. Is it possible to make a binary tree using 256 vertices?

Nomenclature The tree in **Figure 7.146** is used in Exercises 6 through 8.

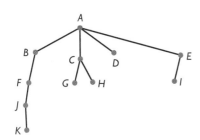

FIGURE 7.146

6. List all parents in Figure 7.146.

7. List all leaves in Figure 7.146.

8. List all children in Figure 7.146.

9. Making a spell-checking tree. Make a spell-checking tree for the dictionary:

and, bake, can, dig, even, fax, get, hive, in, is, it, jump, kite, loud, mate, never, open, quiet, rat, saw, top, under, vow, xray, yes, zebra

10. A spell checker. Make a spell-checking tree using the following dictionary:

air, bake, cow, din, eat, fire, game, home, it, jab, kick, let, moon, now, over, pal, quit, rain, sand

How many checks are needed to test whether "hope" is a valid word? How many checks are needed to test the word "cow"?

11. A bigger spell checker. We have a dictionary consisting of 100 words. If we use a binary tree as a spell checker, estimate how many checks might be required to test whether "name" is a valid word.

12. How big is the dictionary? List the sizes of dictionaries that can be accommodated by binary spell checkers if the number of checks required is between 3 and 10.

13. How many checks? Estimate how many checks are required if our dictionary contains 50,000 words.

14. A story. I tell a story to two of my friends. Some of them repeat the story to two others, and in turn some of these pass the story on. Each time the story is told, it is told to two new

people, and no one hears the story twice. Sometime later, 253 people know the story. How many people have related the story to others? *Suggestion:* Use the formula

$$\text{Number of vertices} = 2 \times \text{Number of parents} + 1.$$

15. A chain letter. For a chain letter, each mailer sends letters to 2 new people. Some letter recipients continue the chain, and some don't. No one gets the same letter twice. Sometime later, there are a total of 1000 letter recipients. (That means your graph model will have 1001 vertices because the initiator of the chain letter is not one of the 1000 recipients.) How many people received letters but did not mail anything? *Suggestion:* Use the formulas

$$\text{Number of vertices} = 2 \times \text{Number of parents} + 1$$

and

$$\text{Number of vertices} = \text{Number of parents} + \text{Number of leaves}.$$

16. Lucky nights. One night I put a silver dollar in a slot machine and got back two silver dollars. On the second night, I put these two dollars back in the machine one at a time, and each time two silver dollars came out. This pattern continued through the fifth night, after which I took my winnings and went home.

    a. How much money did I take home?

    b. How many silver dollars were fed into the machine?

17. Quality testing. *The second part of this exercise uses counting techniques from Section 5.4.* A certain computer chip is subjected to 10 separate tests before it can be sold. We represent the testing with a binary tree. The first vertex is test number 1. It meets two edges. The right-hand branch represents a "pass," and the left-hand branch represents a "fail." We then administer the next test and branch right or left depending on the performance of the chip. This continues until all 10 tests are administered. How many leaves are there in the completed tree? Suppose a chip must pass at least 7 of the tests before it can be marketed. How many leaves of the tree represent market-ready chips? *Suggestion:* Think of a chip with 10 blanks on it. We will stamp the blanks P or F. For a chip that passes exactly 7 tests, we want to choose 7 of the blanks on which to stamp a P.

18. An e-mail chain. An e-mail chain is initiated. On day 1, e-mails are sent to two people, and each is asked to forward the message to two new recipients the next day. Assume that each recipient complies with the request and that no one receives e-mails from two separate sources. How many people have been notified by day 6? Include the initiator of the e-mail chain.

19. The BCS. Each year the NCAA Division I football bowl games usher in a new debate regarding the Bowl Championship Series (BCS). One suggestion is that some bowl games be replaced by a tournament. Suppose eight teams are chosen for the championship tournament. How many games are played before a national champion is determined?

20. Division III. In NCAA Division III football, there is a year-end tournament involving 32 teams. How many teams play in the fourth round of the tournament?

Expanding the NCAA basketball tournament Suppose we decide to expand the NCAA basketball tournament to include 256 teams. Exercises 21 through 24 refer to this scenario.

21. How many games are played?

22. How many games are played in the third round?

23. How many teams lose in the first four rounds?

24. What round corresponds to the Elite Eight? (The Elite Eight are the final eight teams left in the tournament.)

25. Finishing your word. Some computer applications try to anticipate words you are typing and finish them automatically. For example, in your checking account application, when you type elec, your computer may immediately expand this to electric company. This trick is accomplished using a tree (which may be constructed by the software based on your previous typing). Suppose our computer knows that there are six words you are likely to type in the current field:

elephant, element, elfin, elvis, escalator, elevator

We make a *typing tree* associated with this list as shown in the figure below.

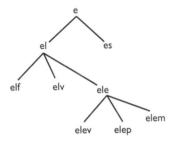

The vertices of the tree are made up of letter sequences that begin the test words. Note that the root has one letter, the level 1 vertices have two letters, and so on. When a leaf such as elf is reached, the computer knows that, if you intended to type one of the six likely words, then it must be *elfin*. Once the computer has constructed such trees, then the trick of finishing your words is easy.

Suppose your computer knows that you are likely to type one of the following words in the current field:

computer, container, comet, call, cask, carpet, chess, cheese, checker, chase, chin, chili

Make a typing tree for these words.

Exercise 26 is suitable for group work.

26. Testing for connectivity. A human being can quickly ascertain whether or not a graph is connected (i.e., comes in one piece) if she or he sees a picture of it. But computers deal with graphs that are not in picture form and that contain many thousands of vertices and edges. In such a case, it is not so easy to determine connectedness. Here, we will illustrate a method used by computers to solve this problem.

In the context of Exercises 35 through 38 in Section 7.1, we introduced the idea of an *incidence matrix* as a way to represent a graph that a computer can understand without a picture. Here is a review in terms of the following incidence matrix:

$$
\begin{array}{c c c c c c c}
 & A & B & C & D & E & F \\
A & - & 1 & 1 & 0 & 0 & 0 \\
B & 1 & - & 1 & 1 & 1 & 0 \\
C & 1 & 1 & - & 0 & 1 & 1 \\
D & 0 & 1 & 0 & - & 0 & 0 \\
E & 0 & 1 & 1 & 0 & - & 0 \\
F & 0 & 0 & 1 & 0 & 0 & - \\
\end{array}
$$

The letters A, B, C, D, E, F represent the vertices of the graph. The dashes on the diagonal are just placeholders, and we ignore them. The 1 in the B column of the A row means there is an edge joining A to B. The 1 in the C column of the A row means there is an edge joining A to C. The 0 in the D row of the A column means there is no edge joining A to D. Similarly, the 1 in the B column of the C row means there is an edge joining B to C. The completed graph represented by the incidence matrix above is shown below.

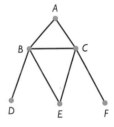

We want to determine whether a graph represented by an incidence matrix is connected, but we don't want to reproduce the entire graph. The method we present is known as a *tree search* because it employs the construction of a tree. We will show how it works using the example above. Begin with any vertex you like, say, A, and start the construction of a tree by listing the vertices adjacent to A. We get the list B, C. The partial tree associated with this list is shown below.

Note that we do not include the edge joining B to C. We are interested only in connecting to new vertices, not what connections may or may not exist among vertices already listed. Next we go to the first vertex of this list. We look only for new (other than A, B, or C) vertices adjacent to B. We find the new vertices D and E and thus obtain the following partial tree:

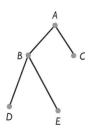

	A	B	C	D	E	F	G	H	I	J
A	–	1	1	0	0	0	0	0	0	0
B	1	–	1	1	1	1	0	0	0	0
C	1	1	–	1	0	0	1	0	0	0
D	0	1	1	–	0	1	0	0	0	0
E	0	1	0	0	–	0	0	0	0	0
F	0	1	0	1	0	–	1	0	0	0
G	0	0	1	0	0	1	–	0	0	0
H	0	0	0	0	0	0	0	–	1	1
I	0	0	0	0	0	0	0	1	–	1
J	0	0	0	0	0	0	0	1	1	–

Finally, we look for new vertices adjacent to C, and we find only F. Our final tree is shown below.

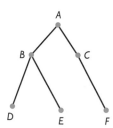

This search tree will either produce a tree including all the vertices (as it did here) or terminate without including all the vertices. In the first case, the graph is connected, and in the second case, it is not.

For the following incidence matrix, make a search tree starting at the vertex A to determine whether the graph represented is connected.

27. History. Minimal-cost spanning trees are very important in many applications. A number of efficient algorithms have been developed to find them. Report on the development of methods for solving this problem.

The following exercise is designed to be solved using technology such as a calculator or computer spreadsheet. For assistance, see the technology supplement.

28. A very large dictionary. What size dictionary can be accommodated by a spell checker that uses 33 checks? What size dictionary can be accommodated by a spell checker that uses 100 checks? We know names for numbers like millions, billions, or trillions. Do you have a name for the number of words you found for 100 checks?

CHAPTER SUMMARY

In this chapter, a *graph* is understood to be a collection of points joined by a number of line segments. The points are called *vertices*, and the segments connecting them are called *edges*. A highway map is a typical example of a graph.

Modeling with graphs and finding Euler circuits

Graphs can be helpful in determining efficient routes. One such route, an *Euler circuit*, begins and ends at home and traverses each edge exactly once. It is the type of route a mail carrier would use. A connected graph has an Euler circuit precisely when each vertex has even order. There are fairly straightforward procedures for constructing Euler circuits. One such procedure is simply to combine shorter circuits into longer ones. Another such procedure is *Fleury's algorithm*. If a graph does not have an Euler circuit, a route that backtracks as little as possible is desired. Such routes can often be found if the graph is first *Eulerized* by temporarily adding duplicate edges.

Hamilton circuits and traveling salesmen: Efficient routes

A driver of a delivery truck desires a route that starts and ends at home and visits each destination exactly once. Such a route is an example of a *Hamilton circuit*. In general, it is not easy to determine whether a graph has a Hamilton circuit. Even if a Hamilton circuit is known to exist, finding it can be problematic. There are guides that can help. For example, if a graph has a vertex of order 2, then each edge meeting that vertex must be part of any Hamilton circuit. But in practice, the search for a Hamilton circuit may involve a good deal of trial and error.

Some of the most important routing problems involve minimum-distance circuits that visit each city exactly once. Many such problems can be phrased in terms of *complete graphs*, graphs in which each pair of vertices is joined by an edge. If we assign distances (more generally, values) to each edge, then the *traveling salesman problem* is the problem of finding a shortest Hamilton circuit. This simple-sounding problem turns out to be surprisingly difficult. Mathematicians believe that there is no "quick" way to find a best route. In practice, no quick solution is known, and computers are used in some cases to find the solution. Generally speaking, we must settle for an approximately shortest route. One procedure for finding such a route is the *nearest-neighbor algorithm*: At each step, it travels to the nearest vertex not already visited (except at the final step, where it returns to the starting point). Another method used to find approximate solutions of the traveling salesman problem is the *cheapest link algorithm*. Algorithms that are much better than these two are available, but the better algorithms are more complex.

Trees: Why are spell checkers so fast?

Special counting rules apply to *trees*, which are graphs with no circuits. The most obvious occurrences of trees in popular culture are tournament brackets, such as those for the NCAA basketball tournament. In this context, they serve as useful guides. But more important trees lie just below the surface of much of our daily activity. A spell checker, for example, is a tree functioning inside a computer that saves the authors of this text much embarrassment. A typical spell checker may know 100,000 or more words. A tree structure allows the computer to determine whether a typed word is in its dictionary using no more than 17 comparisons, not 100,000 comparisons. Trees also play important roles in many business and industrial settings.

KEY TERMS

CHAPTER QUIZ

1. For the graph in **Figure 7.147**, find the degree of each vertex.

Answer   *A*: 3; *B*: 2; *C*: 3; *D*: 3; *E*: 2; *F*: 3.

If you had difficulty with this problem, see Example 7.2.

2. The edges in **Figure 7.148** represent electrical lines that must be inspected, and the vertices represent junctions. Is there a route that will not require the inspector to travel along already inspected lines? That is, is there an Euler circuit for the graph in Figure 7.148? Be sure to explain your answer.

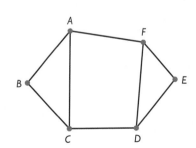

FIGURE 7.147 Finding degrees.

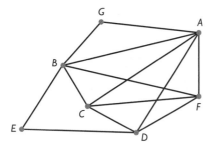

FIGURE 7.148 Inspecting electrical lines.

Answer No Euler circuit exists because some of the vertices have odd order.

If you had difficulty with this problem, see Example 7.3.

3. The edges in **Figure 7.149** represent streets that need to be swept, and the vertices represent intersections. Is there a route for the street sweeper that will not require traversing of already clean streets? In other words, is there an Euler circuit for the graph in Figure 7.149? If an Euler circuit exists, find one.

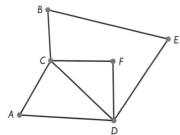

FIGURE 7.149 Street sweepers.

Answer One Euler circuit is *A-C-D-F-C-B-E-D-A*.

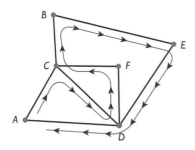

If you had difficulty with this problem, see Example 7.5.

4. For the graph in **Figure 7.150**, either find a Hamilton circuit or explain why no such circuit exists.

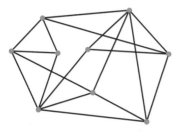

FIGURE 7.150 Hamilton circuit?

Answer One Hamilton circuit is

If you had difficulty with this problem, see Example 7.7.

5. Solve the traveling salesman problem for the map in **Figure 7.151** by calculating the mileage for each possible route. Use Oklahoma City as a starting point.

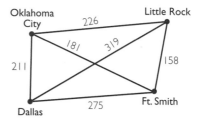

FIGURE 7.151 Traveling salesman.

Answer Oklahoma City - Ft. Smith - Little Rock - Dallas - Oklahoma City; distance: 869 miles.

If you had difficulty with this problem, see Example 7.8.

6. There are 31 members of an organization. The plan for disseminating news requires one member to contact two others with the information. Each member who receives the news contacts two members who have not yet gotten the news. This continues until all members have the information. Use the formula

$$\text{Number of vertices} = 2 \times \text{Number of parents} + 1$$

to determine how many members contact others.

Answer 15.

If you had difficulty with this problem, see Example 7.11.

7. Make a spell-checking tree for the following dictionary:

aardvark, bear, cat, deer, emu, fox, giraffe, hound, impala, javelina, krill

How many checks are needed to test whether "dove" is a valid word?

Answer For "dove": 4 checks.

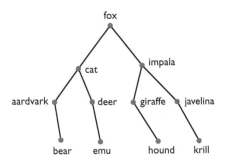

If you had difficulty with this problem, see Example 7.13.

8. We have a dictionary consisting of 300 words. If we use a binary tree as a spell checker, how many checks might be required to test whether "gnu" is a valid word?

Answer 9 checks.

If you had difficulty with this problem, see Example 7.14.

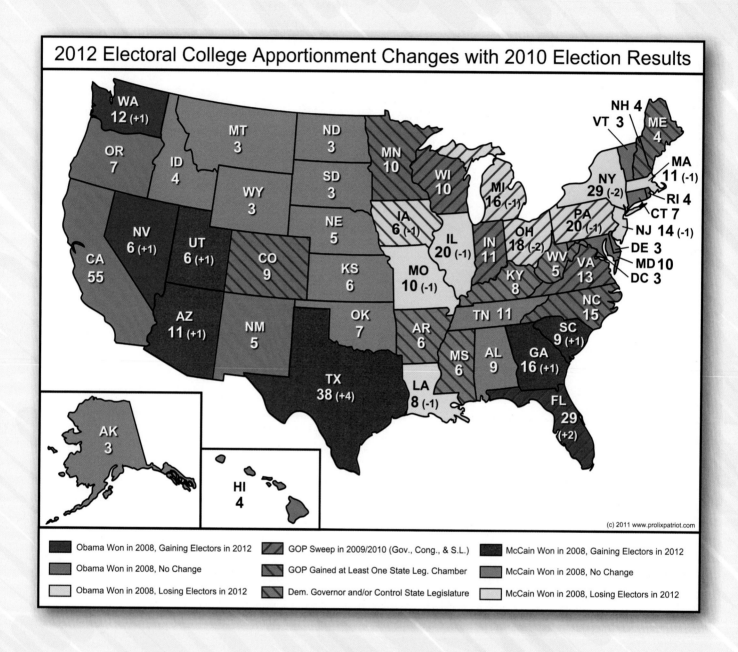

2012 Electoral College Apportionment Changes with 2010 Election Results

WA 12 (+1)
OR 7
MT 3
ND 3
MN 10
NH 4
VT 3
ME 4
WI 10
ID 4
WY 3
SD 3
MI 16 (-1)
NY 29 (-2)
MA 11 (-1)
RI 4
CT 7
NV 6 (+1)
UT 6 (+1)
CO 9
NE 5
IA 6 (-1)
IL 20 (-1)
IN 11
OH 18 (-2)
PA 20 (-1)
NJ 14 (-1)
DE 3
MD 10
DC 3
CA 55
KS 6
MO 10 (-1)
KY 8
WV 5
VA 13
AZ 11 (+1)
NM 5
OK 7
AR 6
TN 11
NC 15
SC 9 (+1)
MS 6
AL 9
GA 16 (+1)
TX 38 (+4)
LA 8 (-1)
FL 29 (+2)
AK 3
HI 4

(c) 2011 www.prolixpatriot.com

Obama Won in 2008, Gaining Electors in 2012	GOP Sweep in 2009/2010 (Gov., Cong., & S.L.)	McCain Won in 2008, Gaining Electors in 2012
Obama Won in 2008, No Change	GOP Gained at Least One State Leg. Chamber	McCain Won in 2008, No Change
Obama Won in 2008, Losing Electors in 2012	Dem. Governor and/or Control State Legislature	McCain Won in 2008, Losing Electors in 2012

VOTING AND SOCIAL CHOICE

Voting rules and their consequences are not always as straightforward as one might first imagine, especially when some participants may have more than one vote. A prime example is the U.S. Electoral College. In electing a president of the United States, each state receives a number of electoral votes equal to the number of representatives and senators from that state. If a candidate wins the majority of the popular vote in a state, the candidate normally receives all the electoral votes from that state (the current exceptions are Maine and Nebraska). The following article from the *New York Times* concerns the controversy arising from this system.

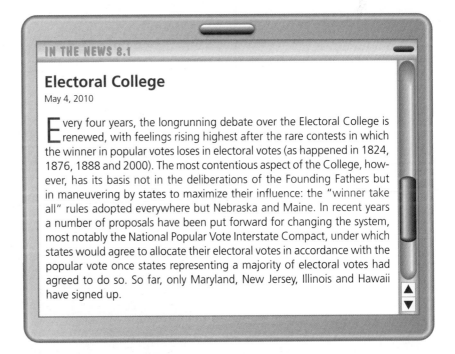

IN THE NEWS 8.1

Electoral College

May 4, 2010

Every four years, the longrunning debate over the Electoral College is renewed, with feelings rising highest after the rare contests in which the winner in popular votes loses in electoral votes (as happened in 1824, 1876, 1888 and 2000). The most contentious aspect of the College, however, has its basis not in the deliberations of the Founding Fathers but in maneuvering by states to maximize their influence: the "winner take all" rules adopted everywhere but Nebraska and Maine. In recent years a number of proposals have been put forward for changing the system, most notably the National Popular Vote Interstate Compact, under which states would agree to allocate their electoral votes in accordance with the popular vote once states representing a majority of electoral votes had agreed to do so. So far, only Maryland, New Jersey, Illinois and Hawaii have signed up.

Questions of how a group should make decisions arise not only in weighty situations such as the election of a president but also in everyday situations, for example, choosing which movie a group of friends will rent for the night. Other situations in which making a decision is difficult come up in the context of equitable division. Typical examples include inheritance and property settlements associated with divorces. Determining how to divide assets fairly is often difficult and sometimes contentious.

A particularly sticky division problem comes up when delegates to the House of Representatives are assigned. The issue of how many representatives each state should get was a problem when our country was founded, and it remains an area of controversy today. We consider all of these topics in this chapter.

8.1 Measuring voting power: Does my vote count?

The following is an excerpt from an article appearing at the Web site Kyivpost.com. This site focuses on Ukraine, a country in Eastern Europe that formerly was part of the Soviet Union.

IN THE NEWS 8.2

Ukraine President Accuses Rivals of "Coup d'Etat"
September 12, 2008

President Viktor Yushchenko accused his two arch rivals on Thursday of plotting a "coup d'etat" by joining forces to cut his powers and replace a pro-Western "orange" coalition in parliament.

A coalition of groups loyal to the president and prime minister, collapsed after Yushchenko's Our Ukraine party walked out last week. It denounced Prime Minister Yulia Tymoshenko's tactical voting alliance in the chamber with the more Russia-friendly party of ex-premier Viktor Yanukovich.

"The de facto formation in parliament of a new coalition is the implementation of a political scenario with a sole aim—to carry out a coup d'etat through a redistribution of powers," the president's press service quoted Yushchenko as saying.

Ukraine President Viktor Yushchenko.

This article refers to political coalitions and to a "tactical voting alliance." One question that arises is how to measure the relative power of voters or voting blocks. Voting alliances or *coalitions* play an important role here. It may seem that a voter's power should be proportional to the number of his or her votes, but that is not the case. Just because one entity (a person, state, or coalition) has more votes than another does not mean that entity has more power. In this section, one way we will measure voting power is in terms of coalitions of voters.

Here's an example to illustrate the importance of voters joining together to form an alliance or coalition. Suppose the board of directors of ACE Computer Corporation has three members. Ann controls 19 shares, Beth controls 18 shares, and Cate controls only 3 shares. They want to elect a chairperson, and the election requires a simple majority of the voting shares. Because there are 40 shares altogether, 21 shares are required for a majority.

How powerful are these voters? It might seem on the surface that Ann is the most powerful but that Cate doesn't have much power at all because she has only 3 shares out of a total of 40. A good case can be made, however, that all three members have equal power. Let's see how to make that case.

None of the members has enough shares (21) to win. The only way for any member to be elected is to convince one of the other members to throw support to her. If Cate can convince Ann to support her, she will have $3 + 19 = 22$ votes, so she will be elected. If Cate can convince Beth to support her, she will have $3 + 18 = 21$ votes, so she will be elected. Thus, if Cate can get the support of either Ann or Beth, then she will be elected, even though she controls only 3 shares. Similarly, if Beth can convince either Cate or Ann to support her, she will win. This observation suggests that all three members have equal power.

This example illustrates that voting power is not simply proportional to the number of votes a person or block has. Coalitions play an important role in voting power, and we now focus on the question of measuring voting power in terms of coalitions.

Voting power and coalitions

One way to measure voting power is the *Banzhaf index*. This index is named after John F. Banzhaf III, an attorney who first published his method in 1965.[1]

A group of voters who vote the same way is called a *voting coalition*. (Each coalition includes at least one voter.) In the example involving the board of ACE Corporation, no one voter could win alone but coalitions consisting of two or three voters could. Banzhaf devised a way to measure the relative power of voters in terms of *winning coalitions*.[2] We will describe his scheme shortly, but first we have to explain the term *quota*, the number of votes necessary to win.

---

[1] "Weighted Voting Doesn't Work: A Mathematical Analysis," *Rutgers Law Review* **19** (1965), 317–343.

[2] In the real-life situation addressed by Banzhaf in his original paper, votes were allocated among six district supervisors, but three of the supervisors had no power because their votes were never necessary to determine the outcome. See Exercises 43 and 44 for the details.

The U.S. Constitution.

Key Concept

In a voting system, the number of votes necessary to win the election is called the **quota**.

In many cases, the quota is just a simple majority, which requires more than half of the total number of votes cast. For instance, in the previous example there were 40 votes altogether, so the quota for a simple majority is 21. In the case of a cloture vote in the U.S. Senate, however, the quota is 60 votes out of 100. To ratify a constitutional amendment in the United States requires (among other things) approval by three-quarters of the states, which makes a quota of 38 states.

In this section, you should assume (unless you are instructed otherwise) that the quota is a simple majority. The *methods* we use are unaffected by the choice of quota. Changing the quota can, however, affect the power of voters. See Exercises 24 through 26.

Winning coalitions and critical voters

We have noted that some voters have more sway in determining outcomes than others and that voters may increase their influence by banding together into coalitions. Banzhaf devised a way to measure the relative power of voters in terms of coalitions. The first step in describing his method is to look more closely at coalitions.

Key Concept

A set of voters with enough votes to determine the outcome of an election is a **winning coalition**; otherwise, it is a **losing coalition**.

In the example involving the ACE Corporation, the winning coalitions consist of any two voters or all three. Here is another example.

EXAMPLE 8.1 Finding the winning coalitions: County convention

Let's suppose there are three delegates to a county convention: Abe has 4 votes from his precinct, Ben has 3 votes, and Condi has 1 vote. A simple majority of the votes wins.

a. What is the quota?

a. Make a table listing all of the coalitions of voters. Designate which of them are winning coalitions.

SOLUTION

a. There are 8 votes, so the quota for a simple majority is 5 votes.

b. The accompanying table shows all possible coalitions together with the votes that each coalition controls. The last column indicates whether or not the coalition controls a majority (at least 5) of the votes and therefore is a winning coalition.

Number of votes			Total votes	Winning coalition?
4	3	1		
Abe	Ben	Condi	8	Yes
Abe	Ben		7	Yes
Abe		Condi	5	Yes
Abe			4	No
	Ben	Condi	4	No
	Ben		3	No
		Condi	1	No

TRY IT YOURSELF 8.1

Suppose that Abe has 5 votes and that Ben and Condi have 3 votes each. A simple majority is required to win. Find the quota and make a table listing all of the coalitions of voters. Designate which of them are winning coalitions.

The answer is provided at the end of this section.

Now we focus attention on the voters who are crucial components of winning coalitions.

Key Concept

A member of a winning coalition is said to be a **critical voter** for that coalition if the coalition is no longer a winning one when that voter is removed.

Let's determine the critical voters for each of the three winning coalitions in Example 8.1. These winning coalitions appear in the top three rows of the table.

Number of votes			Total votes	Winning coalition?
4	3	1		
Abe	Ben	Condi	8	Yes
Abe	Ben		7	Yes
Abe		Condi	5	Yes

Abe, Ben, Condi (8 total votes): If Abe is removed, the remaining coalition controls only 4 votes—not a majority. So, removing Abe changes the coalition from a winning one to a losing one. Therefore, Abe is a critical voter for this coalition.

Removing either Ben or Condi from the coalition leaves at least 5 votes controlled by the remaining coalition, so the remaining coalition is still a winning coalition. Therefore, neither Ben nor Condi is a critical voter in this coalition.

Abe, Ben (7 total votes): Removing either voter makes it a losing coalition, so both Abe and Ben are critical for this coalition.

Abe, Condi (5 total votes): As with the preceding coalition, removing either voter makes it a losing coalition. So, both Abe and Condi are critical for this coalition.

We summarize all of this information in a *coalition table* (see **Table 8.1**). It lists the critical voters.

TABLE 8.1 Coalition Table

Number of votes			Total votes	Winning coalition?	Critical voters
4	3	1			
Abe	Ben	Condi	8	Yes	Abe
Abe	Ben		7	Yes	Abe, Ben
Abe		Condi	5	Yes	Abe, Condi
Abe			4	No	Not applicable
	Ben	Condi	4	No	Not applicable
	Ben		3	No	Not applicable
		Condi	1	No	Not applicable

> **SUMMARY 8.1**   **Winning Coalitions and Critical Voters**
>
> A set of voters with enough votes to determine the outcome of an election is a *winning coalition*. A voter in a winning coalition is *critical* for that coalition if the coalition is no longer a winning one when that voter is removed. We can summarize the essential information about coalitions in a *coalition table*.

Counting coalitions

To make a coalition table, we had to list all the different coalitions of voters. Suppose there are 7 voters. How many different coalitions are possible? Each of the 7 voters is either in the coalition or not in it, so there are 2^7 possibilities for voters to be either in or not in a coalition.[3] This includes all the voters not being in, which isn't a coalition, so there are $2^7 - 1$ coalitions possible. In general, if there are n voters, then there are $2^n - 1$ possible coalitions. In the case of voters Abe, Ben, and Condi that we considered in Example 8.1, the number of voters is $n = 3$, so there should be $2^3 - 1 = 7$ coalitions—just as indicated in the coalition table, Table 8.1. Of course, for large collections of voters, it is impractical to list all the possible coalitions.

> **SUMMARY 8.2**   **Number of Coalitions**
>
> For n voters, there are $2^n - 1$ possible coalitions (each of which includes at least 1 voter).

The Banzhaf power index

Now we define the Banzhaf index for a voter in terms of the number of instances in which that voter is critical.

Key Concept

A voter's **Banzhaf power index** is the number of times that voter is critical in some winning coalition divided by the total number of instances of critical voters. The index is expressed as a fraction or as a percentage.

We illustrate this definition by computing the Banzhaf index for each voter in the situation we have been considering.

EXAMPLE 8.2   Computing the Banzhaf index: County convention

Use the coalition table, Table 8.1, to compute the Banzhaf index for each delegate to the county convention described in Example 8.1.

SOLUTION

From the coalition table, we see that overall there were 5 instances of critical voters. This means that to find the Banzhaf index of a voter, we first find the number of times that voter is critical and then divide by 5.

Abe was critical in 3 of these 5 instances, so he has a Banzhaf index of 3/5 or 60%. Ben was critical in only 1 of these 5 instances, so he has a Banzhaf index of

---

[3] Here, we are using the Counting Principle from Section 3 of Chapter 5.

1/5 or 20%. Condi also was critical in just 1 of these 5 instances, so she also has a Banzhaf index of 1/5 or 20%.

TRY IT YOURSELF 8.2

Make a coalition table for the board of ACE Computer Corporation, where Ann controls 19 shares, Beth controls 18 shares, and Cate controls only 3 shares. Find the Banzhaf index of each voter.

The answer is provided at the end of this section.

Let's make a couple of observations based on the preceding example. First, note that even though Abe has only one more vote than Ben, he has three times the voting power of Ben (as measured by the Banzhaf index). Also note that Ben has 3 votes to Condi's 1 vote, but they have the same voting power. Once again, this indicates that simply knowing how many votes a voter has does not really tell us much about their voting power. The Banzhaf index gives us a way to measure the real power of voters.

Now we compute the Banzhaf index in a historical example involving four voting blocks.

EXAMPLE 8.3 Calculating the Banzhaf index: The 1988 Democratic National Convention

At the 1988 Democratic National Convention, there were four principal candidates: Michael Dukakis with 1401 votes, Jesse Jackson with 1218 votes, Al Gore with 325 votes, and Bruce Babbitt with 197 votes. Although the votes of the delegates are not strictly controlled by the candidates, it is not uncommon for candidates to ask for their delegates to vote for another candidate, effectively forming a coalition. A simple majority is required to win the nomination.

    a. What is the quota?

    b. Determine the winning coalitions.

    c. Determine the critical voters (in this case, candidates) in each winning coalition.

    d. Determine the Banzhaf index of each candidate.

SOLUTION

    a. The total number of votes is 3141, and half of that number is 1570.5. Therefore, a simple majority requires at least 1571 votes, and that is the quota.

Presidential candidate Jesse Jackson addressing the 1988 Democratic Convention.

b. There are $2^4 - 1 = 15$ possible coalitions. The accompanying table lists them along with the total number of votes for each. (We use D for Dukakis, J for Jackson, G for Gore, and B for Babbitt.) The winning coalitions are those with a total of 1571 or more.

Votes				Total	Winning
1401	1218	325	197	votes	coalition?
D	J	G	B	3141	Yes
D	J	G		2944	Yes
D	J		B	2816	Yes
D	J			2619	Yes
D		G	B	1923	Yes
D		G		1726	Yes
D			B	1598	Yes
D				1401	No
	J	G	B	1740	Yes
	J	G		1543	No
	J		B	1415	No
	J			1218	No
		G	B	522	No
		G		325	No
			B	197	No

c. The accompanying table lists only the winning coalitions along with the critical voters in each case. Note that in the first coalition (the one with all four voters), there is no critical voter. Removing any of the voters from this coalition does not cause it to be a losing coalition.

Winning Coalitions Only

Votes				Total	Critical
1401	1218	325	197	votes	voters
D	J	G	B	3141	None
D	J	G		2944	D
D	J		B	2816	D
D	J			2619	D, J
D		G	B	1923	D
D		G		1726	D, G
D			B	1598	D, B
	J	G	B	1740	J, G, B

d. To calculate the Banzhaf index of a voter, we divide the number of times that voter is critical in some winning coalition by the total number of instances of critical voters. In this case, there are 12 instances in which a voter or candidate is critical. This means that to find the Banzhaf index of a candidate, we first find the number of times that candidate is critical and then divide by 12.

Let's do the computation for each candidate:

Dukakis is critical in 6 of these 12 instances. Therefore, his Banzhaf index is $6/12 = 1/2$ or 50%.

Jackson is critical in 2 of these 12 instances. Therefore, his Banzhaf index is $2/12 = 1/6$ or about 16.67%.

Gore is critical in 2 of these 12 instances. Therefore, his Banzhaf index is $2/12 = 1/6$ or about 16.67%.

Babbitt is critical in 2 of these 12 instances. Therefore, his Banzhaf index is $2/12 = 1/6$ or about 16.67%.

The calculations in Example 8.3 show that, as measured by the Banzhaf index, the most powerful candidate by far was Dukakis. They also show that the other three candidates all had the same power—even though Jackson had far more support than Gore and Babbitt.

> **SUMMARY 8.3   Banzhaf Power Index**
>
> A voter's *Banzhaf index* is the number of times that voter is critical in some winning coalition divided by the total number of instances of critical voters. The index is expressed as a fraction or as a percentage.

Swing voters and the Shapley-Shubik power index

In a seminal paper, Lloyd Shapley and Martin Shubik[4] devised a method of measuring voting power that is somewhat different from the Banzhaf index. The Banzhaf index is based on the idea of a voter who is *critical* for a coalition, and the Shapley–Shubik index is based on the concept of a *swing voter*. To describe the Shapley-Shubik index, we assume that the voters vote in order and that their votes are added as they are cast.

We start with the first voter. If that voter has enough votes to decide the election (because her votes meet the quota), she is the swing voter. If not, add the second voter's votes. If the total of the two meets the quota, the second voter is the swing. If not, add a third voter's votes, and so on until the total of the votes cast meets the quota. The deciding voter for this order of voting is the *swing voter*.

Key Concept

Suppose that voters vote in order and that their votes are added as they are cast. The **swing voter** is the one whose votes make the total meet the quota and thus decides the outcome. Which is the swing voter depends on the order in which votes are cast.

Flags of the European Union.

EXAMPLE 8.4 Finding the swing voter: Council of the European Union

Member states of the European Union have votes on the Council assigned by the 2001 Treaty of Nice. The number of votes is roughly determined by a country's population but progressively weighted in favor of smaller countries. Ireland has 7 votes, Cyprus has 4 votes, and Malta has 3 votes. Suppose that these three countries serve as a committee where a simple majority is required to win, so the quota is 8 votes. Make a table listing all the permutations[5] of the voters and the swing voter in each case.

---

[4]Shapley and Shubik are prominent game theorists who were portrayed in the book and movie *A Beautiful Mind*. See their paper "A Method for Evaluating the Distribution of Power in a Committee System," *American Political Science Review* **48** (1954), 787–792.

[5]Recall from Section 4 of Chapter 5 that a permutation of objects is just an arrangement of them in a certain order.

SOLUTION

There are $n!$ different permutations[6] of n objects, where $n! = n \times (n-1) \times \cdots \times 1$. Our three objects are Ireland, Cyprus, and Malta and $3! = 3 \times 2 \times 1 = 6$, so there will be six rows in our table. We let the first row represent the voting order Ireland, Cyprus, Malta. Then Ireland, with 7 votes, doesn't have a majority, but the 4 votes of Cyprus added to Ireland's do give a majority. Therefore, Cyprus is the swing voter for this order, and we indicate this in the last column of the first row.

If we continue in this way for each row, we obtain the following table:

Order of voters			Swing
Ireland (7)	Cyprus (4)	Malta (3)	Cyprus
Ireland (7)	Malta (3)	Cyprus (4)	Malta
Cyprus (4)	Ireland (7)	Malta (3)	Ireland
Cyprus (4)	Malta (3)	Ireland (7)	Ireland
Malta (3)	Ireland (7)	Cyprus (4)	Ireland
Malta (3)	Cyprus (4)	Ireland (7)	Ireland

TRY IT YOURSELF 8.4

Germany has 29 votes on the European Union Council, Spain has 27, and Sweden has 10. Suppose these three countries serve as a committee where a simple majority of 34 votes is required to win. Make a table listing all the permutations of the voters and the swing voter in each case.

The answer is provided at the end of this section.

Note that in Example 8.4, the second voter is not always the swing voter. The third voter is the swing in the fourth and sixth rows of the table. Note, too, that Ireland is the swing voter 4 out of 6 times. This proportion is the key to the measurement we will now introduce.

Definition of the Shapley–Shubik power index

The Shapley–Shubik index measures the distribution of power by observing the proportion of times a given voter is the swing.

Key Concept

The **Shapley–Shubik power index** of a given voter is calculated as the fraction (or percentage) of all permutations of the voters in which that voter is the swing.

EXAMPLE 8.5 Computing the Shapley-Shubik index: Committee of the Council

Compute the Shapley–Shubik power index for the committee of Ireland, Cyprus, and Malta from Example 8.4.

SOLUTION

For each voter, we calculate the fraction of all permutations of the voters in which that voter is the swing. There are six permutations. Ireland is the swing in 4 of the 6 cases, so the index for Ireland is $4/6 = 2/3$ or about 66.67%. Cyprus is the swing in only 1 of the 6 cases, so the index for Cyprus is $1/6$ or about 16.67%. Malta also

---

[6]For more information on permutations, see Section 4 of Chapter 5.

Thus, the Shapley–Shubik index gives at the very least a method of revealing possibly hidden bias or complexity in a system.

> **CALCULATION TIP 8.1** Computer Software
>
> Computing either the Banzhaf index or the Shapley–Shubik index by hand is impractical even for moderately large collections of voters. Computer software that performs the calculations (as long as the number of voters is not terribly large) is freely available. One such program for the Banzhaf index is called ipdirect, and one for the Shapley–Shubik index is called ssdirect. Both programs can be found at www.warwick.ac.uk/~ecaae.

Try It Yourself answers

Try It Yourself 8.1: Finding the winning coalitions: County convention   The quota is 6 votes.

Number of votes			Total votes	Winning coalition?
5	3	3		
Abe	Ben	Condi	11	Yes
Abe	Ben		8	Yes
Abe		Condi	8	Yes
Abe			5	No
	Ben	Condi	6	Yes
	Ben		3	No
		Condi	3	No

Try It Yourself 8.2: Computing the Banzhaf index: ACE Computer Corporation

Number of votes			Total votes	Winning coalition?	Critical voters
19	18	3			
Ann	Beth	Cate	40	Yes	None
Ann	Beth		37	Yes	Ann, Beth
Ann		Cate	22	Yes	Ann, Cate
Ann			19	No	Not applicable
	Beth	Cate	21	Yes	Beth, Cate
	Beth		18	No	Not applicable
		Cate	3	No	Not applicable

Each voter has an index of $2/6 = 1/3$ or about 33.33%.

Try It Yourself 8.4: Finding the swing voter: Council of the European Union

Order of voters			Swing
Germany (29)	Spain (27)	Sweden (10)	Spain
Germany (29)	Sweden (10)	Spain (27)	Sweden
Spain (27)	Germany (29)	Sweden (10)	Germany
Spain (27)	Sweden (10)	Germany (29)	Sweden
Sweden (10)	Germany (29)	Spain (27)	Germany
Sweden (10)	Spain (27)	Germany (29)	Spain

Try It Yourself 8.5: Computing the Shapley–Shubik index: Council of the European Union   Germany: 1/3; Spain: 1/3; Sweden: 1/3.

Exercise Set 8.1

Leave each power index as a fraction unless you are directed to do otherwise.

1. How many coalitions? In a collection of 5 voters, how many different coalitions are possible?

2. How many coalitions? In a collection of 10 voters, how many different coalitions are possible?

3. How many permutations? In a collection of 5 voters, how many different permutations are possible?

4. How many permutations? In a collection of 10 voters, how many different permutations are possible?

5. State coalitions. Each of the 50 states is assigned a certain number of electoral votes. How many possible coalitions (each coalition including at least 1 state) are there?

    a. 2^{50}

    b. $2^{50} - 1$

    c. $50! - 1$

    d. $50!$

    e. 50^2

6. State orders. Each of the 50 states is assigned a certain number of electoral votes. How many vote orders would we need to consider in order to calculate the Shapley–Shubik index for each state?

    a. 2^{50}

    b. $2^{50} - 1$

    c. $50! - 1$

    d. $50!$

    e. 50^2

7. A strategy. Your political opponent has put together a coalition of voters who control sufficient votes to defeat you in the upcoming election. You wish to improve your chances by persuading some of the coalition members to switch their support to you. Should you talk with critical voters in the coalition or with noncritical voters in the coalition? Explain why you gave the answer you did.

8. Two voters. Suppose there are only two voters. Anne controls 152 votes, and the quota is 141 votes. What is Anne's Banzhaf index? What is the Banzhaf index of the other voter?

9. Two voters again. Suppose there are only two voters. Anne controls 152 votes, and the quota is 141 votes. What is Anne's Shapley–Shubik index? What is the Shapley–Shubik index of the other voter?

10. One person, one vote. In a group of 100 voters, each person has 1 vote, and the quota is a simple majority. What is the Banzhaf index for each voter? *Hint:* Each voter should have the same index.

11. One person, one vote again. In a group of 100 voters, each person has 1 vote, and the quota is a simple majority. What is the Shapley–Shubik index for each voter? *Hint:* Each voter should have the same index.

Three delegates Suppose there are three delegates to a county convention. Abe has 100 votes, Ben has 91 votes, and Condi has 10 votes. A simple majority wins. Exercises 12 through 16 refer to this situation.

12. What is the quota?

13. Make a table listing all of the coalitions and the critical voter in each case.

14. Find the Banzhaf index for each voter.

15. Make a table listing all of the permutations and the swing voter in each case.

16. Find the Shapley–Shubik index for each voter.

Four voters Suppose there are four voters: A with 13 votes, B with 6 votes, C with 5 votes, and D with 2 votes. Suppose that a simple majority is required to win. This information will be used in Exercises 17 through 23.

17. What is the quota?

18. How many coalitions are there?

19. Make a coalition table.

20. Find the Banzhaf index for each voter.

21. How many permutations of four voters are there?

22. Make a table listing all of the permutations of the voters and the swing voter in each case.

23. Find the Shapley–Shubik index for each voter.

Ace Solar Suppose there are four stockholders in a meeting of the Ace Solar Corporation and that each person gets as many votes as the number of his or her shares. Assume that Abe has 49 shares, Ben has 48 shares, Condi has 4 shares, and Doris has 3 shares. Exercises 24 through 26 refer to this situation.

24. Assume that a simple majority is required to prevail in a vote. Make a table listing all of the permutations of the voters and the swing voter in each case, and calculate the Shapley–Shubik index for each voter.

25. Assume now that a two-thirds majority is required to prevail in a vote, so the quota is 70. Calculate the Shapley–Shubik index for each voter.

26. Refer to Exercises 24 and 25. Does the voting power of Condi increase or decrease with the change from a simple majority to a two-thirds majority?

Five voters Suppose there are five voters, A, B, C, D, and E, with 6, 5, 4, 3, and 2 votes, respectively. Suppose that a simple majority is required to win. This information will be used in Exercises 27 through 31.

27. What is the quota?

28. How many permutations of five voters are there?

29. How many coalitions are there?

30. Fill out the accompanying coalition table.

		Votes			Total votes	Winning coalition?	Critical voters
6	5	4	3	2			
A	B	C	D	E			
A	B	C	D				
A	B	C		E			
A	B	C					
A	B		D	E			
A	B		D				
A	B			E			
A	B						
A		C	D	E			
A		C	D				
A		C		E			
A		C					
A			D	E			
A			D				
A				E			
A							
	B	C	D	E			
	B	C	D				
	B	C		E			
	B	C					
	B		D	E			
	B		D				
	B			E			
	B						
		C	D	E			
		C	D				
		C		E			
		C					
			D	E			
			D				
				E			

31. Find the Banzhaf index for each voter.

32. Four states. This exercise refers to Example 8.6, which discusses the four swing states in the 2004 U.S. presidential election. Recall that Florida had 27 electoral votes, Michigan had 17, Ohio had 20, and Pennsylvania had 21. Assume again that a simple majority vote from only these four states would determine the election, so the quota is 43. Find the Banzhaf index for each of these four states.

33. Party switch. In 2001, the U.S. Senate had 50 Republicans and 50 Democrats. On May 24 of that year, James Jeffords, a Republican from Vermont, announced his intention to leave the Republican Party and declare himself an Independent. That made the Senate makeup 49 Republicans, 50 Democrats, and 1 Independent. For a simple majority vote, make a table listing all of the permutations of the three voting blocks and the swing voter in each case. What is the Shapley–Shubik index for each voting block?

34. Party switch 2. Repeat Exercise 33, this time making a coalition table. What is the Bahzhaf index for each voting block?

Electoral College Suppose that there are only three hypothetical states with a distribution of popular and electoral votes as shown in the accompanying table. (The number of electoral votes for a state is the size of its congressional representation.)

	State A	State B	State C	Total
Population	100	200	400	700
Senators	2	2	2	6
Representatives	1	2	4	7
Electoral votes	3	4	6	13

This information will be used in Exercises 35 and 36.

35. Find the Shapley–Shubik index for each state using the popular vote and then using the electoral vote. Assume that a simple majority is required in each case.

36. Use your answers to Exercise 35 to explain why states with small populations resist changing the Electoral College system.

The Security Council The United Nations Security Council has five permanent members (China, France, Great Britain, Russia, and the United States) and 10 rotating members. Each member has 1 vote, and 9 votes are required to pass a measure. But each permanent member has a veto. That is, if a permanent member votes "no," the measure fails. Although this is not a true weighted voting system, it is equivalent to one. Exercises 37 through 40 explore this fact.

37. Suppose that each rotating member has 1 vote and that each permanent member has 7 votes, for a total of 45 votes. If the quota is set at 39 votes, explain why a measure cannot pass if a permanent member votes "no."

38. Consider the weighted system in Exercise 37. Explain why a measure will pass if nine members of the council, including all of the permanent members, support it.

39. Explain why the weighted voting system described in Exercise 37 is equivalent to the actual United Nations Security Council voting system. *Note:* The Shapley–Shubik index for this weighted voting system can be calculated using a computer program. The result is 19.63% for each permanent member and 0.19% for each rotating member.

40. In Exercises 37 through 39, we failed to mention something: In reality, a permanent member of the Security Council may abstain, which does not count as a veto. Suppose France abstains and no one else does. Then 9 votes, including those of all four of the remaining permanent members, are required to pass a measure. Devise an equivalent weighted voting system for these 14 members, as was done previously. That is, determine how many votes should be allotted to each of the remaining four permanent members and what the quota should be. (Assign 1 vote to each rotating member.) *Note:* The Shapley–Shubik index for this weighted voting system can be calculated using a computer program. The result is 23.25% for each permanent member and 0.70% for each rotating member.

Exercises 41 and 42 may be suitable for group work.

41. A research project. Voting coalitions are a vital part of parliamentary systems. A nice example is the Israeli Knesset, where there are several parties and coalitions often are necessary for forming a government. Write a report on the Knesset or other parliaments. Focus on coalitions.

42. A research project. At national party conventions, state delegations usually vote for the presidential nominee in alphabetical order. Find other situations in which votes are taken in a particular order.

History When John Banzhaf introduced his index of voting power in 1965, he analyzed the voting structure of the Nassau County Board of Supervisors. (Nassau County is located in the western part of Long Island, New York.) That board had divided 115 votes among the districts based on population. The following table shows the vote allocation based on the population in 1964. In the table, Hempstead #1 represents the presiding supervisor, always from Hempstead, and Hempstead #2 represents the second supervisor from Hempstead.

District	Number of votes
Hempstead #1	31
Hempstead #2	31
Oyster Bay	28
North Hempstead	21
Glen Cove	2
Long Beach	2

Exercises 43 and 44 refer to this voting system.

43. Assume that a simple majority is required to win a vote. Explain why North Hempstead, Glen Cove, and Long Beach have zero power (in the sense that their votes are never necessary to determine the outcome of an election). Do not answer this question by calculating the Banzhaf index.

44. Even though the number of votes was allocated according to the size of a district, clearly voting power was not allocated that way. This voting system was modified several times over the years because of legal challenges. Investigate this history and write a brief summary, including the final resolution.

History Exercises 45 and 46 refer to the article by Shapley and Shubik in which they first defined their index.

45. The following quotation is from the introductory portion of the Shapley–Shubik article. "It is possible to buy votes in most corporations by purchasing common stock. If their policies are entirely controlled by simple majority votes, then there is no more power to be gained after one share more than 50% has been acquired." How does this observation support the importance of the concept of a swing voter?

46. Here is another quotation from the introductory portion of the Shapley–Shubik article. "Put in crude economic terms, …if votes of senators were for sale, it might be worthwhile buying forty-nine of them, but the market value of the fiftieth (to the same customer) would be zero." How do the numbers 49 and 50 in the statement reflect the fact that the paper was published in 1954? How would you update the statement to be true now?

History 2 Exercises 47 and 48 refer to the following quotation from the article by Shapley and Shubik in which they first defined their index.

To illustrate, take Congress without the provision for overriding the President's veto by means of two-thirds majorities. This is now a pure tricameral system with chamber sizes of 1, 97, and 435. The values come out to be slightly under 50% for the President, and approximately 25% each for the Senate and House, with the House slightly less than the Senate.

47. Note that the figure of 97 refers to the Senate. Explain why it was valid in 1954 (when the paper was published). *Hint:* Consider the role of the vice president.

48. Does it make sense that the president's power index is nearly 50%? Explain.

The following exercises are designed to be solved using technology such as calculators or computer spreadsheets. For assistance, see the technology supplement.

49. Three voters.

   a. Produce a spreadsheet that computes the Banzhaf index in the case when there are three voters and the quota is a simple majority. The spreadsheet should allow the user to enter the number of votes controlled by each voter. Do the same for the Shapley–Shubik index.

   b. Use your spreadsheet to check the computations of the Shapley–Shubik index for some of the previous examples and exercises.

   c. Compare the results given by your spreadsheet with those given by the software mentioned in Calculation Tip 8.1.

50. Four voters.

   a. Produce a spreadsheet that computes the Banzhaf index in the case when there are four voters and the quota is a simple majority. The spreadsheet should allow the user to enter the number of votes controlled by each voter. Do the same for the Shapley–Shubik index.

   b. Solve Example 8.3 (on the 1988 Democratic National Convention) using your spreadsheet.

   c. Experiment with different vote totals for Dukakis, Jackson, Gore, and Babbitt in Example 8.3 to see how the power index depends on these totals, and report your results.

   d. Compare the results given by your spreadsheet with those given by the software mentioned in Calculation Tip 8.1.

51. Paradox. Attempting to measure voting power can give rise to paradoxes.[7] Two voting schemes, each with a total of 11 votes, are given. Assume that a two-thirds majority is required to win. Express each value of the Banzhaf index as a percentage rounded to two decimal places.

   a. How many votes are required to win?

   b. Suppose that A has 5 votes; B has 3 votes; and C, D, and E have 1 vote each. Use a computer program to find the Banzhaf index for each voter.

---

[7]See Dan S. Felsenthal and Moshe Machover, *The Measurement of Voting Power: Theory and Practice, Problems and Paradoxes* (Cheltenham, UK: Edward Elgar, 1998).

c. Suppose that, instead of the votes in part b, A gives away 1 vote to B. So A has 4 votes; B has 4 votes; and C, D, and E still have 1 vote each. Use a computer program to find the Banzhaf index for each voter.

d. Did the Banzhaf index for A increase or decrease after giving away 1 vote? Does this make sense?

e. Explain why C, D, and E have no voting power in the voting scheme of part c. Use this observation to compute by hand your answers to part c.

8.2 Voting systems: How do we choose a winner?

TAKE AWAY FROM THIS SECTION

There is no perfect voting system if there are three or more candidates. Many different voting systems are used.

The following excerpt is from a syndicated opinion piece submitted by the Minnesota Voters Alliance.

IN THE NEWS 8.3

Asheville Citizen Times Claims Favoring Instant Runoff Voting Don't Hold Up to Scrutiny

MATT MARCHETTI   September 2, 2008

According to elections expert Steven J. Brams, Ph.D. New York University, with IRV [instant runoff voting] "ranking your favorite candidate first could cause him to lose, whereas ranking him last could cause him to win—just the opposite of what you want the system to do." This is utterly unacceptable. Voters shouldn't need to bring a calculator to the voting booth! . . .

IRV is billed as a "new" idea, which will empower voters, provide more choices and make elections fair. This "new" system, which was devised over a century ago, is touted as eliminating the "Nader effect," and as guaranteeing a majority winner. However, this "spoiler effect" is a legitimate form of political speech which should not be frivolously eliminated. Moreover, IRV doesn't guarantee a majority winner anyway!

A plurality should be acceptable in a representative system, but if some people insist on having majority winners, IRV won't help. A runoff is used only when no candidate wins a majority on the first ballot. Whatever happens after that doesn't change the fact that the eventual winner never got a majority of "first choices." Thus, IRV merely creates the illusion of a majority.

One of the hallmarks of a democracy is that its leaders are chosen by the "people," but there are many voting schemes by which this may be accomplished. This excerpt shows how controversial such schemes can be. The opinion piece refers to instant runoff voting, the spoiler effect, and plurality voting.

If a candidate receives a majority of the votes cast, many would agree that he or she should be the winner. Even this is not as simple as it sounds, as the Electoral College system for the U.S. president shows. And even if we agree that a candidate who receives a majority of the votes should be the winner, it may be that no one achieves that goal. When there are more than two candidates, the problem of determining the "people's choice" is difficult, to say the least. Indeed, a case can be made that for three or more candidates, there is no completely satisfactory system for determining a winner, as we will see.

There are many systems currently in use. In this section, we discuss several popular ones, including the method called instant runoff voting in the preceding excerpt. We also examine the troubling paradox of Condorcet and Arrow's impossibility theorem.

Plurality voting and spoilers

A *voting system* is a set of rules under which a winner is determined. The simplest voting system for three or more candidates is *plurality* voting.

Key Concept

In the system of **plurality** voting, the candidate receiving more votes than any other candidate wins.

On the surface, this system seems quite reasonable, but there are difficulties. One objection to plurality voting is that if there are a number of candidates, the winner could be determined by only a small percentage of the electorate.

EXAMPLE 8.7 Examining plurality voting: Four candidates

Suppose there are four candidates competing for 100 votes. What is the smallest number of votes a candidate could receive and still win a plurality?

SOLUTION

If the four candidates received an equal number of votes, they would have $100/4 = 25$ each. A candidate could have a plurality with as few as 26 votes.

TRY IT YOURSELF 8.7

Suppose there are five candidates competing for 100 votes. What is the smallest number of votes a candidate could receive and still win a plurality?

The answer is provided at the end of this section.

In Example 8.7, note that a candidate could receive only 26% of the votes cast, far less than a majority, and still win a plurality of the votes.

With three or more candidates, there can be a *spoiler*.

Key Concept

A **spoiler** is a candidate who has no realistic chance of winning but whose presence in the election affects the outcome.

Typically, a spoiler splits the vote of a more popular candidate who may otherwise have won a majority. Many argue that this was the case with Ross Perot in the 1992 U.S. presidential race and with Ralph Nader in the 2000 presidential race. (Of course, the president is not determined by popular vote, but this example still illustrates a point.)

Republican George W. Bush narrowly defeated Democrat Al Gore in the 2000 presidential campaign.

EXAMPLE 8.8 Understanding plurality voting and spoilers: The 2000 U.S. presidential election

The state of Florida was decisive in the 2000 U.S. presidential race. Although there are many controversies surrounding the balloting itself, the Florida vote tally reported by the Federal Election Commission was

Candidate	Votes
George W. Bush	2,912,790
Al Gore	2,912,253
Ralph Nader	97,488
Others	40,579

Here, the category "Others" includes five other candidates and write-in votes.

a. What percentage of all votes cast were for Bush? Did any candidate achieve a majority of votes cast?

b. The state of Florida determines the winner of an election by plurality, so Bush was declared the winner. Let's suppose that the vote was conducted with only the top three candidates and that all their supporters continued to vote for them. Also, assume that the votes cast for "Others" were divided somehow among the top three candidates. Would it be possible for any of the three to achieve a majority?

c. Now let's assume that Nader had not entered the race in Florida and that supporters of Bush, Gore, and those in the "Other" category remained loyal to their candidates. Assume that the Nader voters would somehow divide their votes among the remaining three candidates (or not vote at all). In order to change the outcome of the election, Gore needs to pick up more Nader voters than Bush. How many more?

d. Exit polls indicated that 21% of Nader voters would have voted for Bush and 47% of them would have voted for Gore. Political scientists who have studied the election estimate that 14% to 17% of Nader voters would have voted for Bush and 32% to 40% would have voted for Gore.[8] Do these figures indicate that Nader was a spoiler for that election?

SOLUTION

a. Of the 5,963,110 votes cast, Bush received 2,912,790 votes, which is 2,912,790/5,963,110 or about 48.8% of the votes. This is not a majority. Bush won more votes than any other candidate, so no candidate received a majority of votes cast.

b. Half of 5,963,110 votes is 2,981,555, so a majority is 2,981,556 votes. That is 68,766 votes more than were cast for Bush. So, even if all the "Others" votes were cast for Bush (or Gore or Nader), no one would have had a majority.

c. Overall, Bush received 537 more votes than Gore. So, Gore would need 538 more Nader votes than Bush.

d. Let's take the maximum estimate of 21% for Bush and the minimum estimate of 32% for Gore. Now 21% of the Nader votes is 20,472 new votes for Bush, and 32% of the Nader votes is 31,196 new votes for Gore. This gives Bush a total of 2,933,262 votes and Gore a total of 2,943,449 votes. If Nader had not been on the ballot, the outcome of the election would have been different. Therefore, Nader could reasonably be considered a spoiler for the election.

Because of the problems with plurality voting we have mentioned, it is desirable to have an alternative way of choosing a winner.

Runoffs, preferential voting, and the Borda count

The Heisman Trophy.

There are many alternatives to plurality-rule voting. Some of them use runoff elections in case no candidate garners a majority. In one such method, the *top-two runoff system*, there is a second election with the two highest vote-getters. This method is used in many state elections. Another runoff method is the *elimination runoff system*, in which the lowest vote-getter is eliminated and a new vote is tallied among the remaining candidates. This process continues until someone achieves a majority.

In practice, to avoid several rounds of voting, a voter lists candidate preferences on a single *ranked* ballot. Such a ballot is used in the voting for the Heisman Trophy, awarded annually to the most outstanding collegiate football player in the United States: Each ballot allows a first-choice, second-choice, and third-choice vote.

A ranked ballot contains much more information than a ballot that allows only one choice. In particular, a ranked ballot shows all the preferences of the voter, that is, which candidate is preferable to which other candidates. *Preferential voting systems* usually use ranked ballots to avoid the need for repeated votes. Different preferential voting systems use the preferences indicated on the ranked ballots in different ways, sometimes with quite different outcomes.

The elimination runoff system is also called the *Hare system*.

Variations on this voting system are used today in the city of San Francisco and in Australia (as well as other countries). The excerpt opening this section refers to such a system as *instant runoff voting*. If there are only three candidates, the elimination runoff system is equivalent to the top-two runoff system. If there are four or more candidates, these two systems can produce surprisingly different outcomes.

---

[8] C. S. P. Magee, "Third-Party Candidates and the 2000 Presidential Election," *Social Science Quarterly* **84** (2003), 574–595.

Key Concept

In the **top-two runoff system**, if no candidate receives a majority then there is a new election with only the two highest vote-getters.

In the **elimination runoff system**, if no candidate receives a majority then the lowest vote-getter is eliminated and a vote is taken again among those who are left. This process continues until someone achieves a majority.

These two systems are examples of **preferential voting systems**, in which voters express their ranked preferences between various candidates, usually using a **ranked ballot**.

EXAMPLE 8.9 Finding the winner in the elimination runoff system: 10 voters, three candidates

Consider the following ranked ballot outcome for 10 voters choosing among three candidates: Alfred, Betty, and Gabby.

Rank	4 voters	4 voters	2 voters
First choice	Alfred	Gabby	Betty
Second choice	Betty	Alfred	Gabby
Third choice	Gabby	Betty	Alfred

As an example of how to interpret this table, look at the row labeled "First choice." The first-choice votes were 4 for Alfred, 4 for Gabby, and 2 for Betty. The fourth column, labeled "2 voters," indicates that 2 voters chose Betty first, Gabby second, and Alfred third.

Determine the winner under the elimination runoff system. (Because there are only three candidates, this is equivalent to the top-two runoff system.)

SOLUTION

No candidate has a majority of the first-choice votes. Betty has the fewest first-choice votes (2), so she is eliminated in the first round. For the second round, we first remove Betty from the original table:

Rank	4 voters	4 voters	2 voters
First choice	Alfred	Gabby	~~Betty~~
Second choice	~~Betty~~	Alfred	Gabby
Third choice	Gabby	~~Betty~~	Alfred

Now with Betty eliminated, for the 4 voters represented in the second column, Alfred is the first choice and Gabby is the second choice. Similarly, for the 4 voters represented in the third column, Gabby is the first choice and Alfred is the second choice. For the 2 voters represented in the fourth column, Gabby is the first choice and Alfred is the second choice among the remaining candidates. The ballot outcomes for the second round are as follows:

Rank	4 voters	4 voters	2 voters
First choice	Alfred	Gabby	Gabby
Second choice	Gabby	Alfred	Alfred

Now the first-choice votes are 4 votes for Alfred and $4 + 2 = 6$ votes for Gabby. In this runoff, Gabby has a majority of the votes and is the winner.

TRY IT YOURSELF 8.9

Consider the following ranked ballot outcome for 10 voters choosing among three candidates: Alfred, Betty, and Gabby.

Rank	5 voters	3 voters	2 voters
First choice	Alfred	Betty	Gabby
Second choice	Betty	Gabby	Alfred
Third choice	Gabby	Alfred	Betty

Determine the winner under the elimination runoff system.

The answer is provided at the end of this section.

There are other methods that use ranked ballots. In 1770, Jean-Charles de Borda proposed a system of this type, and it bears his name today.

Key Concept

The *Borda count* assigns for each ballot 0 points to the choice ranked last, 1 point to the next higher choice, and so on. The **Borda winner** is the candidate with the highest total Borda count.

Although the Borda count system takes into account strength of feelings about the voting choices, one drawback of this system is that it is possible for a candidate to win a majority of first-place votes and yet lose the election. The following example illustrates this fact.

EXAMPLE 8.10 Using the Borda count: Our first choice loses

Our group needs to make a decision on which food to order. To make the choice, everybody marks a ranked ballot for pizza, burgers, and tacos. The number 2 indicates a first-place vote, the number 1 a second-place vote, and so on. The table below shows the five ballots cast.

	Pizza	Tacos	Burgers
Ballot 1	2	1	0
Ballot 2	2	1	0
Ballot 3	2	1	0
Ballot 4	0	2	1
Ballot 5	0	2	1

a. Did one of the food items receive a majority of first-place votes?

b. Use the Borda count to determine what the group should order.

SOLUTION

a. First-place votes are indicated by the number 2. There were three first-place votes for pizza, which is a majority of the five first-place votes.

b. For pizza, the Borda count is $2 + 2 + 2 + 0 + 0 = 6$. The Borda count for tacos is $1 + 1 + 1 + 2 + 2 = 7$, and for burgers, the count is $1 + 1 = 2$. According to the Borda count, the group should order tacos. This is true in spite of the fact that pizza got a majority of first-choice votes.

We mentioned earlier that the voting for the Heisman Trophy uses a ranked ballot. A first, second, and third choice is indicated on each ballot. In the next example, we consider the race for the 2009 Heisman.

EXAMPLE 8.11 Finding the Borda count and Borda winner: Heisman Trophy

Here are the vote counts of the five finalists for the 2009 Heisman, the closest race in the history of the award to that point.

Player	First-place votes	Second-place votes	Third-place votes
Toby Gerhart (Stanford)	222	225	160
Mark Ingram (Alabama)	227	236	151
Colt McCoy (Texas)	203	188	160
Ndamukong Suh (Nebraska)	161	105	122
Tim Tebow (Florida)	43	70	121

Mark Ingram, 2009 Heisman Trophy winner.

Determine the Borda counts and the Borda winner.[9]

SOLUTION

The Borda count assigns numbers 2, 1, and 0 to the choices. We incorporate this information into the original table:

Player	First-place votes	Second-place votes	Third-place votes
Toby Gerhart (Stanford)	222	225	160
Mark Ingram (Alabama)	227	236	151
Colt McCoy (Texas)	203	188	160
Ndamukong Suh (Nebraska)	161	105	122
Tim Tebow (Florida)	43	70	121
Borda count value	2	1	0

---

[9]The actual system used to award the Heisman is a slight modification of the Borda count. See Exercise 34. The outcome here is not affected.

To find the Borda count for Toby Gerhart, we assign 2 points to each of his 222 first-place votes, 1 point to each of his 225 second-place votes, and 0 points to each of his 160 third-place votes. Then we add to find the total number of points. Here are the results:

$$\text{Borda count for Toby Gerhart} = 222 \times 2 + 225 \times 1 + 160 \times 0 = 669$$
$$\text{Borda count for Mark Ingram} = 227 \times 2 + 236 \times 1 + 151 \times 0 = 690$$
$$\text{Borda count for Colt McCoy} = 203 \times 2 + 188 \times 1 + 160 \times 0 = 594$$
$$\text{Borda count for Ndamukong Suh} = 161 \times 2 + 105 \times 1 + 122 \times 0 = 427$$
$$\text{Borda count for Tim Tebow} = 43 \times 2 + 70 \times 1 + 121 \times 0 = 156.$$

Mark Ingram has the highest Borda count, so he is the Borda winner—and the winner of the Heisman Trophy in 2009.

TRY IT YOURSELF 8.11

Here are the vote counts of the three finalists for the 2008 Heisman:

Player	First-place votes	Second-place votes	Third-place votes
Sam Bradford (Oklahoma)	300	315	196
Colt McCoy (Texas)	266	288	230
Tim Tebow (Florida)	309	207	234

Determine the Borda counts and the Borda winner.

The answer is provided at the end of this section.

The next example shows that the four voting systems we have considered can give four different outcomes for an election.

EXAMPLE 8.12 Differing voting systems give differing outcomes

Consider the following ranked ballot outcome for 100 voters choosing among four options: A, B, C, and D.

Rank	28 voters	25 voters	24 voters	23 voters
First choice	A	B	C	D
Second choice	D	C	D	C
Third choice	B	D	B	B
Fourth choice	C	A	A	A

a. Who wins under plurality voting?

b. Who wins under the top-two runoff system?

c. Who wins under the elimination runoff system?

d. Who wins under the Borda count system?

SOLUTION

a. To determine the winner under plurality voting, we must assume that if voters could choose only one name, they would select their first choice. The result of such a vote would be

Candidate	A	B	C	D
Votes	28	25	24	23

Candidate A has the most first-choice votes, so A wins under plurality voting.

b. As we saw in part a, the top two vote-getters on an unranked ballot are A with 28 votes and B with 25 votes. To determine the outcome of a runoff between A and B, we use the preferences shown on the ballots, but we consider only A and B. That is, we eliminate C and D from the original table showing the voter preferences and look at the resulting preferences as expressed by the ranking on the ballots.

Rank	28 voters	25 voters	24 voters	23 voters
Adjusted first choice	A	B	B	B
Adjusted second choice	B	A	A	A

In this runoff, B has 72 first-choice votes. This is a clear majority, so B is the winner under the top-two runoff system.

c. In the first round, D has the fewest votes (23) and is eliminated. If we remove D from the original table, the combined ballot outcomes are

Rank	28 voters	25 voters	24 voters	23 voters
First choice	A	B	C	C
Second choice	B	C	B	B
Third choice	C	A	A	A

Now the first-choice votes are 28 votes for A, 25 votes for B, and 47 votes for C. Therefore, B (having the fewest votes) is eliminated. If we remove B from the table, the combined ballot outcomes are

Ranks	28 voters	25 voters	24 voters	23 voters
First choice	A	C	C	C
Second choice	C	A	A	A

In this runoff, C has 72 votes and so is the winner under the elimination runoff system.

d. The Borda count assigns numbers 3, 2, 1, and 0 to the choices. We incorporate this information into the original table:

Rank	Borda count value	28 voters	25 voters	24 voters	23 voters
First choice	3	A	B	C	D
Second choice	2	D	C	D	C
Third choice	1	B	D	B	B
Fourth choice	0	C	A	A	A

Here are the Borda counts:

Borda count for A $= 28 \times 3 + 25 \times 0 + 24 \times 0 + 23 \times 0 = 84$
Borda count for B $= 28 \times 1 + 25 \times 3 + 24 \times 1 + 23 \times 1 = 150$
Borda count for C $= 28 \times 0 + 25 \times 2 + 24 \times 3 + 23 \times 2 = 168$
Borda count for D $= 28 \times 2 + 25 \times 1 + 24 \times 2 + 23 \times 3 = 198.$

Hence, D is the winner based on the Borda count.

Example 8.12 is not just a curiosity. It shows some of the difficulties in determining exactly who is the "people's choice." All four of the candidates in this example have a reasonable argument that they deserve to be named the winner.

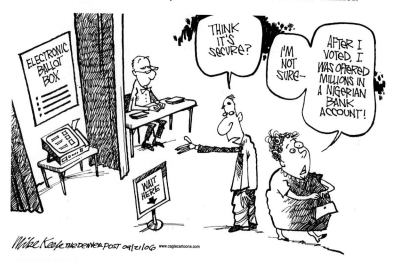

Mike Keefe THE DENVER POST 09/21/06   www.caglecartoons.com

SUMMARY 8.5   Common Preferential Voting Systems

A *voting system* is a scheme for determining the winner of an election when there are three or more candidates. Common systems that are in use today include variants of the following.

Plurality: The candidate with the most votes wins.

Top-two runoff: If no one garners a majority of the votes, a second election is held with the top two vote-getters as the only candidates.

Elimination runoff: Successive elections are held where the candidate with the smallest number of votes is eliminated. This continues until there is a majority winner.

Borda count: Voters rank the candidates first to last. The last place candidate gets 0 points, the next 1 point, and so on. The candidate with the most points wins.

Condorcet's paradox

Normally, an individual's preferences are *transitive*. That is, if I prefer Pepsi over Coke and I prefer Coke over Dr. Pepper, then it seems obvious that, if I am given a choice between Pepsi and Dr. Pepper, I'll choose Pepsi.

It may seem paradoxical, but this transitivity is not necessarily true if the preferences are expressed by groups instead of individuals. We will illustrate this by the following example.

Suppose there are three voters and three candidates, say, A, B, and C. The voters' preferences are shown in the accompanying table.

Preferences	Voter 1	Voter 2	Voter 3
First choice	A	B	C
Second choice	B	C	A
Third choice	C	A	B

Now a majority of voters (voters 1 and 3) prefer candidate A over candidate B, and a majority of voters (voters 1 and 2) prefer candidate B over candidate C. And yet, paradoxically, a majority of voters do not prefer candidate A over candidate C; in fact, a majority of voters (voters 2 and 3) prefer candidate C over candidate A.

In 1785, the Marquis de Condorcet published an essay that described this paradox, which is now named after him. He referred to this phenomenon as the *intransitivity of majority preferences*.

The existence of this paradox suggests that we focus on a candidate who would win in any head-to-head contest with every other candidate. The name given to such candidates honors Condorcet.

Key Concept

In an election, a **Condorcet winner** is a candidate who beats each of the other candidates in a one-on-one election.

Note that in the simple example above, there is no Condorcet winner.

EXAMPLE 8.13 Finding the Condorcet winner: The Gentoo Foundation

The Gentoo Foundation, which distributes the software Gentoo Linux, uses a Condorcet voting process for elections to its board of trustees.[10] Suppose in such an election there are seven voters and three candidates, say, A, B, and C. The voters' preferences are shown in the accompanying table.

Preferences	3 voters	2 voters	2 voters
First choice	A	C	C
Second choice	B	B	A
Third choice	C	A	B

Is there a Condorcet winner? If so, which candidate is it?

SOLUTION

We find the results of each head-to-head contest. Consider a contest between A and B. There are $3 + 2 = 5$ voters who rank A over B and only two voters who rank B over A. Therefore, A wins the one-on-one contest with B. Here are results of the other two head-to-head contests:

$$\text{A and C: C wins by 1 (4 to 3).}$$
$$\text{B and C: C wins by 1 (4 to 3).}$$

There is a Condorcet winner, and it is C.

TRY IT YOURSELF 8.13

In an election, there are seven voters and three candidates, say, A, B, and C. The voters' preferences are shown in the accompanying table.

Preferences	3 voters	2 voters	2 voters
First choice	B	B	C
Second choice	A	C	A
Third choice	C	A	B

Is there a Condorcet winner? If so, which candidate is it?

The answer is provided at the end of this section.

---

[10]See www.gentoo.org/foundation/en/.

Even if there is a Condorcet winner, that candidate may lose the election in all the voting systems we have discussed. This is a further troubling aspect of common voting schemes, and we illustrate it in the following example.

EXAMPLE 8.14 Different voting systems and the Condorcet winner

In a certain election, there are seven voters and four candidates, say, A, B, C, and D. The tally of ranked ballots is as follows:

Rank	Voter 1	Voter 2	Voter 3	Voter 4	Voter 5	Voter 6	Voter 7
First choice	A	A	B	C	D	A	C
Second choice	B	D	A	B	B	D	B
Third choice	C	B	C	A	A	B	A
Fourth choice	D	C	D	D	C	C	D

    a. Who wins in the plurality system?

    b. Who wins in the top-two runoff system?

    c. Who wins in the elimination runoff system?

    d. Who wins the Borda count?

    e. Is there a Condorcet winner? If so, which candidate is it?

SOLUTION

    a. To determine the winner in the plurality system, we must assume that if voters could choose only one name, they would select their first choice. The result of such a vote would be

Rank	Voter 1	Voter 2	Voter 3	Voter 4	Voter 5	Voter 6	Voter 7
First choice	A	A	B	C	D	A	C

The 7 votes are distributed as follows: Candidate A has 3 votes, B has 1 vote, C has 2 votes, and D has 1 vote. Candidate A has the most votes and wins in the plurality system.

    b. The first-place winner is A. Candidate C is second with 2 votes. In a runoff between A and C, A wins 5 votes to 2. So, A wins the top-two runoff.

    c. Because B and D get only 1 vote each, they are eliminated and A wins, just as in part b.

    d. We calculate the Borda count as follows:

$$\text{Borda count for A} = 3 \times 3 + 1 \times 2 + 3 \times 1 = 14$$
$$\text{Borda count for B} = 1 \times 3 + 4 \times 2 + 2 \times 1 = 13$$
$$\text{Borda count for C} = 2 \times 3 + 2 \times 1 = 8$$
$$\text{Borda count for D} = 1 \times 3 + 2 \times 2 = 7.$$

So, A wins the Borda count.

    e. There is a Condorcet winner, and it is B, as the following list of head-to-head outcomes shows:

$$\text{B beats A (4 to 3)}$$
$$\text{B beats C (5 to 2)}$$
$$\text{B beats D (4 to 3).}$$

In conclusion, B wins in any head-to-head contest but loses to A in the top-two runoff, elimination runoff, Borda count, and plurality systems.

> **SUMMARY 8.6**   **Condorcet Winners**
>
> In an election, a *Condorcet winner* is a candidate who would beat all others in a one-on-one election. Some elections do not produce a Condorcet winner.

Arrow's impossibility theorem

Given the problems with the various voting systems described in this section, it's reasonable to ask whether all voting systems for three or more candidates are fraught with such problems. Is there a voting system we haven't considered that in some sense is "the best" or maybe the most "fair"? Of course, this depends on what you mean by best or fair. For example, if it means that the candidate with the most votes wins, then you should be happy with plurality voting.

The notion of "best" in reference to voting procedures has been a subject of debate and research.[11] In general terms, the answer to the question of what is best or most fair lies in the expectation that voting systems have certain properties that can be viewed as reasonable.

One such expectation for a fair voting system is the *Condorcet winner criterion*.

Key Concept

The **Condorcet winner criterion** says that if there is a Condorcet winner—that is, a candidate preferred over all others in one-on-one comparisons—then he or she should be the winner of the election.

Another apparently reasonable property of voting system involves the effect of alternatives that seem irrelevant. Suppose I am asked to choose between hamburgers and hot dogs as my favorite food, and I choose hamburgers. Later pizza is added to the list as a third choice for me to rank. Even if I like pizza far more than either hamburgers or hot dogs, I will still prefer hamburgers over hot dogs. No matter where I rank pizza on my list, it will not change the fact that hamburgers will be ranked over hot dogs. Pizza is an irrelevant alternative—that is, liking or disliking pizza is irrelevant to the question of whether I prefer hamburgers to hot dogs.

Key Concept

The condition of **independence of irrelevant alternatives** says the following: Suppose candidate A wins an election and candidate B loses. Then suppose there is another election in which no voter changes his or her relative preference of A to B. Then B should still lose to A no matter what else may have happened concerning other candidates.

Of course, as we saw with Condorcet's paradox, what is reasonable for an individual may be surprisingly unreasonable for a group. Here is an example where the introduction of an "irrelevant alternative" changes the outcome of an election:

---

[11] Alan Taylor's book *Mathematics and Politics: Strategy, Voting, Power and Proof* (New York: Springer-Verlag, 1995) gives a good overview of the many ideas in this fruitful area of debate.

Suppose that there are 10 voters and that the winner is the one with a plurality of votes. Assume that with candidates A and B in the race, the outcome is as follows:

Rank	6 voters	4 voters
First choice	A	B
Second choice	B	A

Candidate A wins not only the plurality vote but also the majority vote by 6 to 4. Now suppose that candidate C enters the race and that the new 10-vote tally is

Rank	3 voters	3 voters	4 voters
First choice	C	A	B
Second choice	A	B	A
Third choice	B	C	C

This time, candidate B wins with a plurality of 4 votes. Note that none of the voters in the elections changed their preference between A and B. Nonetheless, the introduction of the irrelevant alternative of candidate C changed the election results from A to B.

In this case, we see that the irrelevant alternative was nothing more than a spoiler. So the criterion of independence of irrelevant alternatives perhaps seems a bit less realistic than on first consideration. If the irrelevant alternative is, for example, everyone's first choice, then the irrelevant alternative would win under any system, but that would not violate the independence of irrelevant alternatives. The presence of a spoiler—as in the preceding example, changing the outcome from A to B—can violate the independence of irrelevant alternatives.

The economist Kenneth Arrow studied the properties of voting systems. He was seeking "fair" systems—in particular, systems that would have the reasonable Condorcet winner criterion and also independence of irrelevant alternatives. These two properties are quite different, but surely both seem fair. In point of fact, in 1950, Arrow proved the following startling theorem.[12]

SUMMARY 8.7 **Arrow's Impossibility Theorem**

If there are three or more candidates, there is no voting system (other than a dictatorship) for which both the Condorcet winner criterion and independence of irrelevant alternatives hold.

This is quite surprising for at least two reasons: First, the condition of independence of irrelevant alternatives seems reasonable. Second, any voting system can be adjusted to satisfy the Condorcet winner criterion by saying that, if there is a Condorcet winner, then that candidate wins, and if there isn't, then we carry on with the voting system as originally designed.

Arrow's theorem can be thought of as saying that there is no "best" voting system for three or more candidates. Therefore, we must consider the strengths and weaknesses of any voting system and decide what criteria *seem* the most fair to the voters.

---

[12]Kenneth Arrow, "A Difficulty in the Concept of Social Welfare," *Journal of Political Economy* 58 (1950), 328–346.

Try It Yourself answers

Try It Yourself 8.7: Examining plurality voting: Five candidates   21.

Try It Yourself 8.9: Finding the winner in the elimination runoff system: 10 voters, three candidates   Alfred.

Try It Yourself 8.11: Finding the Borda count and Borda winner: Heisman Trophy Bradford: 915; McCoy: 820; Tebow: 825. Bradford is the Borda winner.

Try It Yourself 8.13: Finding the Condorcet winner: Seven voters, 3 candidates Yes: B.

Exercise Set 8.2

1. The majority loses? In Example 8.10, we saw an outcome involving a ranked ballot in which one candidate received a majority of first-place votes but lost the Borda count. Is it possible to have a two-candidate race in which one candidate receives a majority of first-place votes but loses the Borda count?

2. Condorcet winners. In a ranked ballot tally, is it possible that there is more than one Condorcet winner? Justify your answer.

3. Bush versus Clinton. The popular vote in the 1992 presidential election was

Candidate	Votes
Clinton	44,909,806
Bush	39,104,550
Perot	19,734,821
Other	665,746

a. Conventional wisdom is that Perot took more votes from Bush than from Clinton, but there is no real consensus on the effect of Perot's candidacy. Suppose that 60% of Perot voters had voted for Bush instead and the rest had voted for Clinton. Would Bush have had a plurality of the popular vote? If 65% of Perot voters had voted for Bush instead, would Bush have had a plurality of the popular vote?

b. In your opinion, was Perot a spoiler in this election?

Who wins? The vote tally of ranked ballots for a slate of four candidates is as follows:

Rank	3 voters	3 voters	2 voters	4 voters	4 voters
First choice	A	C	B	C	D
Second choice	B	D	A	B	B
Third choice	C	B	C	A	A
Fourth choice	D	A	D	D	C

Exercises 4 through 8 refer to this information.

4. Who wins a plurality vote?

5. Who wins a top-two runoff?

6. Who wins an elimination runoff?

7. Who wins the Borda count?

8. Is there a Condorcet winner? If so, who?

Who wins? The vote tally of ranked ballots for a slate of four candidates is as follows:

Rank	2 voters	1 voter	3 voters	4 voters	4 voters	1 voter
First choice	C	A	B	D	D	B
Second choice	B	C	C	A	B	A
Third choice	D	B	A	B	A	D
Fourth choice	A	D	D	C	C	C

This information is used in Exercises 9 through 13.

9. Who is the plurality winner?

10. Who is the top-two runoff winner?

11. Who is the elimination runoff winner?

12. Who is the Borda winner?

13. Is there a Condorcet winner? If so, who?

Who wins? The vote tally of ranked ballots for a slate of four candidates is as follows:

Rank	4 voters	5 voters	1 voter	4 voters	4 voters	2 voters
First choice	C	A	B	D	D	B
Second choice	B	C	C	A	B	A
Third choice	D	B	A	B	A	D
Fourth choice	A	D	D	C	C	C

This information is used in Exercises 14 through 18.

14. Who is the plurality winner?

15. Who is the top-two runoff winner?

16. Who is the elimination runoff winner?

17. Who is the Borda winner?

18. Is there a Condorcet winner? If so, who?

Who wins? The vote tally of ranked ballots for a slate of four candidates is as follows:

Rank	4 voters	10 voters	2 voters	4 voters	3 voters
First choice	A	C	B	A	B
Second choice	B	A	D	C	A
Third choice	C	D	C	B	C
Fourth choice	D	B	A	D	D

This information is used in Exercises 19 through 23.

19. Who is the plurality winner?

20. Who is the top-two runoff winner?

21. Who is the elimination runoff winner?

22. Who is the Borda winner?

23. Is there a Condorcet winner? If so who?

Who wins? The tally of ranked ballots is shown below.

Rank	2 voters	2 voters	3 voters	2 voters	2 voters
First choice	A	C	B	A	B
Second choice	B	A	D	C	C
Third choice	C	B	C	B	A
Fourth choice	D	D	A	D	D

This information is used in Exercises 24 through 28.

24. Who is the plurality winner?

25. Who is the top-two runoff winner?

26. Who is the elimination runoff winner?

27. Who is the Borda winner?

28. Is there a Condorcet winner? If so who?

Who wins? The vote tally of ranked ballots for a slate of four candidates is as follows:

Rank	4 voters	5 voters	3 voters	4 voters	6 voters
First choice	A	C	B	A	B
Second choice	B	A	D	C	C
Third choice	C	D	C	B	A
Fourth choice	D	B	A	D	D

This information is used in Exercises 29 through 33.

29. Who is the plurality winner?

30. Who is the top-two runoff winner?

31. Who is the elimination runoff winner?

32. Who is the Borda winner?

33. Is there a Condorcet winner? If so who?

34. **Heisman Trophy.** In Example 8.11, we considered the following vote counts of the five finalists for the 2009 Heisman Trophy:

Player	First-place votes	Second-place votes	Third-place votes
Toby Gerhart (Stanford)	222	225	160
Mark Ingram (Alabama)	227	236	151
Colt McCoy (Texas)	203	188	160
Ndamukong Suh (Nebraska)	161	105	122
Tim Tebow (Florida)	43	70	121

The actual system used to determine the winner of the Heisman assigns 3 points to each first-place vote, 2 points to each second-place vote, and 1 point to each third-place vote. Use this method to determine the point total for each of the finalists and find the winner of the Heisman.

Florida again Let's suppose the voting tally in the state of Florida for the 2000 presidential election represents voting blocks in the state. Let's classify them as follows:

Voting block	Votes
Republican	2,912,790
Democrat	2,912,253
Independent	97,488
Other	40,579

Assume that the quota is determined by a majority vote. This information is used in Exercises 35 through 37. Leave each index as a fraction.

35. **Banzhaf index.** Find the Banzhaf index for each voting block. *Suggestion:* This will be easier if you first observe that the smallest block has no voting power.

36. **Shapley–Shubik index.** Find the Shapley–Shubik index for each block.

37. **Interpretation.** Interpret what your answers to Exercises 35 and 36 say about the relative power of the four blocks. How is this related to the discussion about Ralph Nader in Example 8.8?

Bush versus Gore In the 2000 presidential election, the official popular vote tallies were as follows:

Candidate	Popular vote
Al Gore	51,003,926
George W. Bush	50,460,110
Others	3,898,559

Gore won a plurality of the popular vote, but Bush won the electoral vote and so won the presidency. Let's treat these results as voting blocks. Assume that the quota is determined by a majority vote. This information is used in Exercises 38 through 40. Leave each index as a fraction.

38. **Shapley–Shubik index.** What is the Shapley–Shubik index for each block?

39. **Banzhaf index.** What is the Banzhaf index for each block?

40. **Interpretation.** Interpret what your answers to Exercises 38 and 39 say about the relative power of the three blocks.

Exercises 41 and 42 are suitable for group work.

41. **The tyranny of the majority.** There are critiques of majority rule from Plato,[13] de Tocqueville,[14] religious communities,[15] and contemporary authors,[16] each for quite different reasons.

---

[13] *The Republic*, for example.

[14] *Democracy in America*, 1835.

[15] M. Sheeran, *Beyond Majority Rule*, Annual Meeting of the Religious Society of Friends, Philadelphia, 1983.

[16] For example, Lani Guinier, *The Tyranny of the Majority: Fundamental Fairness in Representative Democracy* (New York: Free Press, 1994).

Winston Churchill is quoted as saying, "Democracy is an abysmal system of government. Its only advantage is that it is infinitely superior to anything else we've come up with so far." Report on the concerns of scholars regarding the concept of majority rule.

42. Constitutional limits. Majority rule does not mean that the majority totally runs the show. For example, in the United States, it is not lawful for a majority to disenfranchise a minority. There are many constitutional limits to what a majority may impose on a minority. Report on a few of these.

43. History. It has happened four times in U.S. history that a presidential candidate lost the popular vote but won the presidency. Report on these four incidents.

44. History. The 1960 presidential race between Richard Nixon and John F. Kennedy was another very close race. In this election, Illinois was crucial. There were allegations of voter fraud in Illinois that, if proven valid, could have changed the results of the election. Report on the 1960 presidential race in Illinois.

45. History. The year 1960 was the first time presidential candidates faced each other in televised debates. The debates stirred controversy for any number of reasons. Report on the 1960 presidential debates.

46. History. In the United States, the winner of a presidential race is the candidate who receives a majority of the electoral vote. If no candidate achieves a majority, the House of Representatives decides the winner of the election. This happened only once, in 1824. Report on the 1824 presidential election.

8.3 Fair division: What is a fair share?

TAKE AWAY FROM THIS SECTION

Fair division of assets can be difficult, but there are mathematically sound ways to divide fairly.

The following article is from the Web site Luxist.

IN THE NEWS 8.4

John Cleese's Big Divorce Settlement

DEIDRE WOOLLARD   August 20, 2009

Monty Python star John Cleese has reached a divorce settlement with his wife and the numbers aren't pretty. Cleese was married to psychotherapist Alyce Faye Eichelberger in 1992. Now he must pay her a total settlement worth $19.7 million. The terms of the divorce give her an apartment in New York, a $3.3 million London home and part ownership of a California beach house. Cleese will be worth less than his former wife. His total net worth will be around $16.5 million. To his credit, Cleese seems to be able to laugh about the situation telling London's *Daily Telegraph* that "I will know in future if I go out with a lady they will not be after me for my money." Cleese, who is 70, hasn't stopped working. He is said to currently be writing a one-man show called My Alyce Faye Divorce Tour. Cleese's settlement looks mild compared to some others. The *Telegraph* rounds up the top ten including Rupert and Anna Murdoch, Michael and Juanita Jordan, and Neil Diamond and Marcia Murphey.

Monty Python star John Cleese and his ex-wife, Alyce Faye Eichelberger.

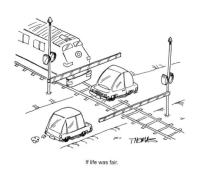

If life was fair.

There are many situations, such as divorces and inheritances, where goods need to be divided equitably. One way of accomplishing this is to liquidate all assets and divide the resulting cash. But in many cases, liquidation is neither practical nor desirable. For example, if a small business is part of an inheritance, one heir may have an interest in taking over the business rather than selling it. There is also the issue of value other than cash. For example, my father's old guitar has little cash value but holds great sentimental value to me. In this section, we will look at ways to divide items so that, as nearly as possible, all parties see the division as fair. In the process we will see that the idea of fairness is difficult to define. See also Exercises 24 and 25, where the idea of envy-free procedures is explored.

Divide-and-choose procedure

Perhaps the simplest, and sometimes most practical, fair-division scheme is the *divide-and-choose* procedure. Suppose you and I wish to share a chocolate chip cookie. I could break the cookie into two pieces and give you one of the pieces, but maybe you wouldn't be happy with the outcome. Maybe you're not happy because I gave you the smaller piece, or maybe the pieces are of equal size but I gave you the half with fewer chocolate chips. One simple solution is that I break the cookie but you choose which piece you want (or *vice versa*—you break it and I choose). If I intentionally break the cookie "unfairly," I am setting myself up to get the less desirable piece.

Key Concept

In the **divide-and-choose procedure**, one person divides the items into two parts and the other person chooses which part he or she wants.

Here's another example: Suppose my sister and I inherit a jewelry collection that we want to divide. She could separate the jewelry into two piles and let me take the pile I like best. While this may not result in an equal division, it is likely that both of us will at least consider it fair. Perhaps my sister is particularly fond of a necklace that Mom wore. When she divides the jewelry, she may intentionally put the necklace in a small pile, knowing that I will choose the larger pile (which has the greater cash value). Sister is happy because she gets to wear Mom's necklace, and I am happy because I got a larger cash value than I expected.

This division procedure can be extended easily to three people. The following extension is the *lone-divider method*. Suppose now that my brother, my sister, and I want to divide fairly a jewelry collection. I divide the jewelry into three piles that I consider to be of equal value. There are piles 1, 2, and 3. My brother and my sister each must surely consider that at least one pile has at least one-third of the total value. Suppose my brother likes pile number 1 and my sister likes pile number 2. Then brother and sister each take the pile they like, and I get pile number 3. Suppose, on the other hand, that each of them likes the same pile, say, pile 1. Then I choose whichever of piles 2 and 3 that I like. The remaining two piles are put back together, and brother and sister use the divide-and-choose method for two persons to divide what's left. Just as in the case of Mom's necklace, there may be various strategies employed to get the desired outcome.

In Exercises 21 and 22, you will be asked to extend this procedure to more than three parties. Related procedures are the *lone-chooser method* (Exercise 19) and the *last-diminisher method* (Exercise 20).

Adjusted winner procedure

In practice, most fair-division problems are much more complex than dividing a cookie. There are a number of ways of resolving them. One that works well when only two people are involved is the *adjusted winner procedure*.

Key Concept

In the **adjusted winner procedure**, two people assign points to bid on each item, assigning a total of 100 points. Initial division of the items gives each item to the person offering the highest bid. The division is then adjusted based on the ratios of the bids for each item so that ultimately each person receives a group of items whose bid totals are the same for each person.

To see how it works, let's suppose my sister and I want to divide an inheritance. The assets consist of a guitar, a jewelry collection, a car, a small library, and a certain amount of cash. To start the procedure, each of us takes 100 points and divides those points among the assets according to how we individually value the items. Suppose our point values are as follows:

My points	Item	Sister's points
35	Guitar	10
10	Jewelry	10
20	Car	40
15	Library	10
20	Cash	30

Initially, each item goes to the person who bid the most for it. So, in the initial round, I get the guitar and the library for a point total of 50. My sister gets the car and the cash for a point total of 70. We give the tied item, the jewelry, to the point leader—my sister. She now has a total of 80 points. This is not a fair division. In a sense, I am only 50% satisfied, but my sister is 80% satisfied. We need to adjust to even the totals.

To make the score come out even, my sister must give me some of her property. Here is how we do it. For each item currently belonging to my sister (jewelry, car, and cash), we calculate the following fraction:

$$\frac{\text{Sister's bid}}{\text{My bid}}.$$

Note that my sister's bid goes on top because she is the leader (in terms of point totals), and we are considering the items she owns. This fraction serves as a measure of the relative value placed by each of us on the items. We calculate these fractions:

$$\text{Ratio for jewelry} = \frac{10}{10} = 1$$

$$\text{Ratio for car} = \frac{40}{20} = 2.0$$

$$\text{Ratio for cash} = \frac{30}{20} = 1.5.$$

We arrange these ratios in increasing order:

Jewelry	Cash	Car
1	1.5	2

Now we begin transferring items from my sister to me in this order until we reach an item that changes the point leader. We first transfer the jewelry. This gives me a point total of 60 and my sister a point total of 70. Transferring the next item, the cash, from my sister to me would make me the point leader. The cash is the *critical item*, and we will transfer just enough of it from my sister to me to make the score come out even. We want to take p percent of the cash from my sister and give it to me. That will reduce her point score by p percent of her bid for the cash and increase my score by p percent of my bid for the cash. We want the resulting point totals to be the same. Expressed as an equation, this is

My score $+ p$ percent of my cash bid $=$ Sister's score $- p$ percent of her cash bid.

Writing p in decimal form gives the equation

$$60 + 20p = 70 - 30p.$$

We solve this equation as follows:

$$60 + 20p = 70 - 30p \quad \text{Add } 30p \text{ to each side.}$$
$$60 + 50p = 70 \quad \text{Subtract 60 from each side.}$$
$$50p = 10 \quad \text{Divide both sides by 50.}$$
$$p = 0.20.$$

So, in the end, I get the guitar, the jewelry, the library, and 20% of the cash. My sister gets the car and the remaining 80% of the cash. Now my sister's total points are $40 + 30 \times 0.80 = 64$, and my total points are $35 + 10 + 15 + 20 \times 0.20 = 64$, so each one of us is 64% satisfied with the outcome. This is a fair division, according to each person's valuation of the items.

> **SUMMARY 8.8** **Adjusted Winner Procedure**
>
> Each of two people makes a bid totaling 100 points on a list of items to be divided.
>
> **Step 1: Initial division of items:** Items go to the person with the highest bid. Tied items are held for the moment. The person with the highest score is the *leader*, and the person with the lowest score is the *trailer*.
>
> **Step 2: Tied items:** Tied items go to the leader.
>
> **Step 3: Calculate leader/trailer ratios:** For each item belonging to the leader, calculate the ratio
>
> $$\frac{\text{Leader's bid}}{\text{Trailer's bid}}.$$
>
> **Step 4: Transference of some items from leader to trailer:** Transfer items from the leader to the trailer in order of increasing ratios as long as doing so does not change the lead. The first item whose transference would change the lead is the *critical item*.
>
> **Step 5: Division of the critical item:** Take p percent of the critical item from the leader and give it to the trailer. Here, p in decimal form is the solution of the equation
>
> $$\text{Trailer's score} + p \times \text{Trailer's bid} = \text{Leader's score} - p \times \text{Leader's bid}.$$

EXAMPLE 8.15 Using the adjusted winner procedure: An inheritance

The adjusted winner procedure can be used to divide an inheritance when items have sentimental value that would not be reflected by simply using dollar values. Use the adjusted winner procedure to divide the listed items fairly when the following bids are made:

John	Item	Faye
50	Vacation condominium	65
20	Red 1962 GT Hawk	15
20	Family silver set	15
10	Dad's Yale cap and gown	5

SOLUTION

Initially, John gets the Hawk, the silver, and the cap and gown, and Faye gets the condominium. John has a score of 50, and Faye has a score of 65. Faye is the leader. We must use Faye's only item, the condo, to even the score. (The condo is the critical item.) We are going to take p percent of the condo from Faye and give it to John. We find the decimal form of p by solving the equation from Step 5 of Summary 8.8:

$$\text{John's score} + p \times \text{John's bid} = \text{Faye's score} - p \times \text{Faye's bid}$$
$$50 + 50p = 65 - 65p \quad \text{Add } 65p \text{ to each side.}$$
$$50 + 115p = 65 \quad \text{Subtract 50 from each side.}$$
$$115p = 15 \quad \text{Divide by 115.}$$
$$p = \frac{15}{115} \quad \text{or about 0.13.}$$

So we take 13% of the condo from Faye and give it to John. In the end, Faye gets 87% ownership in the condo, while John gets the Hawk, the silver, the cap and gown, and 13% ownership in the condo. The shared ownership could be accomplished, for example, by giving John use of the condo for 7 weeks (13% of 52 weeks) and Faye use of it for the remaining 45 weeks each year.

TRY IT YOURSELF 8.15

Redo the preceding example using the following bids:

John	Item	Faye
60	Vacation condominium	45
15	Red 1962 GT Hawk	20
15	Family silver set	20
10	Dad's Yale cap and gown	15

The answer is provided at the end of this section.

There are situations where we transfer more than one item. We look at one of these scenarios in the following example.

EXAMPLE 8.16 Using the adjusted winner procedure: Multiple transfers

Use the adjusted winner procedure to divide the listed items fairly when the following bids are made:

Anne	Item	Becky
50	Stereo	40
5	CDs	5
6	Tapes	5
31	Cabinet	10
8	TV	40

SOLUTION

Initially, Anne gets the stereo, the tapes, and the cabinet for a point total of 87. Becky gets the TV for a point total of 40. Anne is the leader, so she gets the tied item, the CDs. Anne's new point total is 92. Because Anne has the lead, we look at ratios of bids for items she owns, with Anne's bid in the numerator:

$$\text{Ratio for stereo} = \frac{50}{40} = 1.25$$

$$\text{Ratio for CDs} = \frac{5}{5} = 1$$

$$\text{Ratio for tapes} = \frac{6}{5} = 1.2$$

$$\text{Ratio for cabinet} = \frac{31}{10} = 3.1.$$

We arrange these ratios in increasing order:

CDs	Tapes	Stereo	Cabinet
1	1.2	1.25	3.1

Starting from the left, we transfer as many items as we can from Anne to Becky without changing the point leader. We transfer the CDs and the tapes, giving Anne a new score of 81 and Becky a new score of 50. The next item on the list is the stereo. Its transference would change the leader, so this is the critical item. We want to transfer p percent of the stereo from Anne to Becky. We use the equation from Step 5 of Summary 8.8:

$$\text{Becky's score} + p \times \text{Becky's bid} = \text{Anne's score} - p \times \text{Anne's bid}$$
$$50 + 40p = 81 - 50p$$
$$90p = 31$$
$$p = \frac{31}{90} \quad \text{or about } 0.34.$$

In the final division, Becky gets the CDs, the tapes, the TV, and 34% of the stereo, and Anne gets the cabinet and the remaining 66% of the stereo.

TRY IT YOURSELF 8.16

Use the adjusted winner procedure to divide the listed items fairly when the following bids are made:

Alfred	Item	Betty
90	Computer	90
9	Cables	8
1	Printer paper	2

The answer is provided at the end of this section.

The adjusted winner procedure has several advantages. Under the assumption that neither person has a change of heart about the original allocation of points, the following can be shown to be true:

1. Both people receive the same number of points.

2. Neither person would be happier if the division were reversed.

3. No other division would make one person better off without making the other worse off.

This means that the allocation is as fair as anyone knows how to make it. It may or may not result in two happy campers, but from an objective point of view, it is a fair division.

The adjusted winner procedure does have several drawbacks. One drawback is that it is only for division between two people or parties. Another drawback is that the procedure is useful only if there are two or more items to be divided. If there is only one item to be divided, each person must bid all 100 points for the item and, ultimately, each person will be given 50% ownership of that one item. A third drawback is that the procedure may require that ownership of an item be divided in some way between the people involved, or that some similar arrangement be made for one of the items. Such an arrangement may be problematic if, for example, the item has sentimental value and neither side wants to relinquish it.

The Knaster procedure

The *Knaster procedure* is a method for dividing items among several parties, and it doesn't require dividing ownership of items. The procedure is based on having the parties assign monetary value to each item, their *bid*. The bidding is without knowledge of others' bids. In the end, all receive at least their fair share of their own bids. For example, if there are five parties, then each party receives at least one-fifth of its bid on each item. This procedure is also known as the *method of sealed bids*.

Key Concept

In the **Knaster procedure**, or **method of sealed bids**, each of three or more people assigns a dollar value to each item through secret bidding. Each item ultimately goes to the highest bidder while cash is used to equalize the distribution of the items according to each bidder's value of the item.

Sealed bid real estate sales in Dubai.

To illustrate the procedure, let's first assume there is only one item to be divided, an automobile. There are four people involved, Abe, Ben, Caleb, and Dan, and each is entitled to one-fourth of the value (to them) of the car. Each bids the value he places on the car, and each is going to wind up with a value of at least one-fourth of his bid. Suppose the bids are as follows:

Person	Abe	Ben	Caleb	Dan
Bid	$30,000	$23,000	$28,000	$25,000

The highest bidder, Abe, gets the car. But that is the entire estate, and Abe is only entitled to one-fourth of it. He corrects it with cash. The value of the car to him is $30,000, and he is entitled to only one-fourth of that amount, $7500. So he puts the remaining three-fourths, $22,500, into a *kitty*.

Now each of the other three withdraws from the kitty one-fourth of his assigned value of the car. So, Ben takes one-fourth of $23,000, which is $5750. Similarly, Caleb withdraws $7000, and Dan takes $6250. Note that the kitty is not being divided evenly between Ben, Caleb, and Dan. The reason is that they did not each assign the same value to the car. Each person is being given one-fourth of the value he assigned to the car. The total withdrawals from the $22,500 kitty are $5750 + 7000 + 6250 = $19,000. That leaves a kitty of $3500 to be divided equally among the four people, $875 for each. Each person is receiving one-fourth of the value he assigned to the car, plus $875.

The following table summarizes the complete transaction:

Person Bid	Abe $30,000	Ben $23,000	Caleb $28,000	Dan $25,000	Kitty
Car award	Car				0
To kitty	−22,500				22,500
From kitty		5750	7000	6250	3500
From kitty	875	875	875	875	0
Final share	Car less $21,625	$6625	$7875	$7125	$0

One important result of this procedure is that each person gets at least a fourth of his own evaluation of the worth of the car.

If there is more than one item to be divided, the Knaster procedure can be applied to each item sequentially, as illustrated in the example below.

off the mark.com   by Mark Parisi

JAMES DID NOT LEAVE A WILL, BUT HE'LL STILL BE ABLE TO HELP DETERMINE WHICH OF YOU GETS HIS INHERITANCE...

offthemark.com
©2007 MARK PARISI DIST. BY UFS INC.

EXAMPLE 8.17 Employing the Knaster procedure with three heirs: Keeping the family farm

Julie, Anne, and Steve have inherited the family farm. They want the farm to remain in the family, and they want to divvy up the parts of the farm rather than try to share ownership in each part. Using the Knaster procedure, they make sealed bids for the three parts of the farm: the farmhouse, the wheat fields, and the cattle operation. Here are the bids:

Bid	Julie	Anne	Steve
Farmhouse	$120,000	$60,000	$75,000
Wheat fields	210,000	450,000	390,000
Cattle operation	300,000	750,000	600,000

Determine the fair division of the farm using the Knaster procedure.

SOLUTION

Each part of the farm is given to the highest bidder, so Julie will get the farmhouse, Anne will get both the wheat fields and the cattle operation, and Steve will not get any of the farm (although he will get cash). We begin with the farmhouse, which goes to Julie. Because she valued the farmhouse at $120,000 but is entitled to only one-third of that value ($40,000), she puts $80,000 in the kitty. Anne and Steve are each entitled to one-third of their respective bids for the farmhouse, so Anne takes from the kitty one-third of $60,000, or $20,000, and Steve takes one-third of $75,000, or $25,000. That leaves $80,000 − 20,000 − 25,000 = $35,000 in the kitty, as shown below.

Bid	Julie	Anne	Steve	Kitty
Farmhouse	$120,000	$60,000	$75,000	
Wheat fields	210,000	450,000	390,000	
Cattle operation	300,000	750,000	600,000	
Farmhouse award	Farmhouse			0
To kitty	−80,000			80,000
From kitty		20,000	25,000	35,000

Similarly, Anne places the highest bid for the wheat fields, which she gets. Because she valued the fields at $450,000 but is entitled to only one-third of that value ($150,000), she puts $300,000 in the kitty. Julie and Steve are each entitled to one-third of their respective bids for the wheat fields, so Julie takes from the kitty one-third of $210,000, or $70,000, and Steve takes one-third of $390,000, or $130,000. The net effect is to add $300,000 − 70,000 − 130,000 = $100,000 to the $35,000 already in the kitty, as shown below.

Bid	Julie	Anne	Steve	Kitty
Farmhouse	$120,000	$60,000	$75,000	
Wheat fields	210,000	450,000	390,000	
Cattle operation	300,000	750,000	600,000	
Farmhouse award	Farmhouse			0
To kitty	−80,000			80,000
From kitty		20,000	25,000	35,000
Wheat fields award		Wheat fields		35,000
To kitty		−300,000		335,000
From kitty	70,000		130,000	135,000

Anne also has the highest bid for the cattle operation, which she gets. Because she valued the cattle operation at $750,000 but is only entitled to one-third of that value ($250,000), she puts $500,000 in the kitty. Julie and Steve are each entitled to one-third of their respective bids for the cattle operation, so Julie takes from the kitty one-third of $300,000, or $100,000, and Steve takes one-third of $600,000, or $200,000. The net effect is to add $500,000 − 100,000 − 200,000 = $200,000 to the $135,000 already in the kitty, as shown below. Finally, the balance of $335,000 left in the kitty is divided equally among the three heirs, so each gets approximately $111,667, as shown below.

Bid	Julie	Anne	Steve	Kitty
Farmhouse	$120,000	$60,000	$75,000	
Wheat fields	210,000	450,000	390,000	
Cattle operation	300,000	750,000	600,000	
Farmhouse award	Farmhouse			0
To kitty	−80,000			80,000
From kitty		20,000	25,000	35,000
Wheat fields award		Wheat fields		35,000
To kitty		−300,000		335,000
From kitty	70,000		130,000	135,000
Cattle operation award		Cattle operation		135,000
To kitty		−500,000		635,000
From kitty	100,000		200,000	335,000
Kitty distribution	111,667	111,667	111,667	

Summing the columns, we find that Julie gets the farmhouse and cash totaling

$$-\$80,000 + 70,000 + 100,000 + 111,667 = \$201,667.$$

Anne gets the wheat fields, the cattle operation, and cash totaling

$$\$20,000 - 300,000 - 500,000 + 111,667 = -\$668,333;$$

that is, Anne needs to pay $668,333. Steve gets no property, but he does get cash totaling

$$\$25,000 + 130,000 + 200,000 + 111,667 = \$466,667.$$

TRY IT YOURSELF 8.17

Determine the fair division of the farm using the Knaster procedure if the bids are as follows:

Bid	Julie	Anne	Steve
Farmhouse	$120,000	$60,000	$75,000
Wheat fields	210,000	450,000	480,000
Cattle operation	330,000	780,000	690,000

The answer is provided at the end of this section.

SUMMARY 8.9   **Knaster Procedure or Method of Sealed Bids**

Step 1: Each person bids a dollar amount for each item to be distributed. The bidding is without knowledge of others' bids—hence the phrase "sealed bids."

Step 2: For each item separately:

  a. The item is given to the person with the highest bidder, the winner.
  b. The winner contributes to the kitty the difference between that person's bid and his or her share of that item (which is his or her bid divided by the number of bidders).
  c. Those who don't win the item receive from the kitty their share of their bid for the item (which is their bid divided by the number of bidders).

Step 3: The money left in the kitty is distributed equally to each bidder.

One disadvantage of the Knaster procedure is that it may require participants to have a large amount of cash available.

Try It Yourself answers

Try It Yourself 8.15: Using the adjusted winner procedure: An inheritance   John gets 95% ownership in the vacation condominium and Faye gets the red 1962 GT Hawk, family silver set, Dad's Yale cap and gown, and 5% ownership in the vacation condominium.

Try It Yourself 8.16: Using the adjusted winner procedure: Multiple transfers Alfred gets the cables and 46% ownership of the computer, and Betty gets the paper and 54% ownership of the computer.

Try It Yourself 8.17: Employing the Knaster procedure with three heirs: Keeping the family farm   Julie gets the farmhouse plus $205,000, Anne gets the cattle operation and pays $245,000, and Steve gets the wheat fields plus $40,000.

Exercise Set 8.3

In each case, round any answer in percentage form to one decimal place unless you are directed to do otherwise.

1. Strategy. Suppose Abe, Caleb, and Mary wish to divide fairly an inheritance of jewelry using the lone-divider method. Mary is particularly fond of a pair of earrings. If she performs the first division, what strategy should she use to give her the best chance of getting the earrings?

2. Strategy. Suppose Abe, Caleb, and Mary wish to divide fairly an inheritance of jewelry using the lone-divider method. Mary is particularly fond of a pair of earrings. If she does not perform the first division, what strategy should she use to give her the best chance of getting the earrings?

Adjusted winner In Exercises 3 through 9, use the adjusted winner procedure to divide items if the initial bids are as shown.

3. Carla and Daniel divide an art collection.

Daniel	Item	Carla
20	Drawings	25
40	Paintings	30
40	Art supplies	45

4. Mike and Simon divide automobiles.

Mike	Item	Simon
50	BMW	30
25	Mercedes	30
25	Lexus	40

5. Anna and Ashley divide a small library.

Anna	Item	Ashley
10	Math books	10
10	Craft books	30
80	Travel books	60

6. Monica and Angel divide items associated with travel.

Monica	Item	Angel
35	Plane tickets	30
40	Luggage	30
25	Clothes	40

7. Acom divides into two companies, Bcom and Ccom. Bcom and Ccom divide some of the assets of Acom.

Bcom	Item	Ccom
15	Appliance dealership	20
5	Coffeeshop	10
35	Computer outlet	20
20	Bakery	20
25	Cleaning service	30

8. David and Mike divide household items.

David	Item	Mike
40	Dishes	40
10	Cooking utensils	30
20	Crockery	10
25	Furniture	10
5	Linens	10

9. Ben and Ruth divide a Southwest art collection.

Ben	Item	Ruth
30	Hopi kachinas	20
20	Navajo jewelry	30
30	Zuni pottery	20
15	Apache blankets	20
5	Pima sand paintings	10

10. What to do? How would you fairly divide things if the initial bids are as follows?

Person 1	Item	Person 2
50	Item 1	50
30	Item 2	30
20	Item 3	20

11. What to do? How would you fairly divide things if the initial bids are as follows?

Person 1	Item	Person 2
25	Item 1	25
25	Item 2	25
25	Item 3	25
25	Item 4	25

The Knaster procedure In Exercises 12 through 18, use the Knaster procedure to divide fairly property with the given bids.

12. Oscar, Ralph, Mary, and Wanda wish to divide a business property.

Person	Oscar	Ralph	Mary	Wanda
Bid	$33,000	$40,000	$38,000	$32,000

13. J. C., Trevenor, Alta, and Loraine inherit a home and wish to divide it.

Person	J. C.	Trevenor	Alta	Loraine
Bid	$330,000	$340,000	$380,000	$320,000

14. Shirley, Brooke, Tommy, and Jessie inherit a plot of land that they wish to divide.

Person	Shirley	Brooke	Tommy	Jessie
Bid	$45,000	$48,000	$43,000	$48,000

15. Lee, Max, Anne, and Becky wish to divide a watercolor.

Person	Lee	Max	Anne	Becky
Bid	$100	$150	$90	$120

16. Ann, Liz, Jeanie, and Lorie wish to divide an art collection.

Person	Ann	Liz	Jeanie	Lorie
Bid	$33,000	$38,000	$38,000	$32,000

17. James and Gary want to divide three small businesses.

Bid	James	Gary
Business 1	$12,000	$22,000
Business 2	22,000	2000
Business 3	8000	12,000

18. A conglomerate splits into three companies and needs to divide the distribution rights for a clothing line and a food line.

Bid	Company 1	Company 2	Company 3
Clothing line	$12,000,000	$21,000,000	$6,000,000
Food line	21,000,000	6,000,000	6,000,000

19. Lone-chooser method. The *lone-chooser method* is an adaptation of the divide-and-choose procedure for three people. For the purposes of illustration, we will use a pie. The first and second persons use the divide-and-choose procedure to divide the pie into two pieces. Now each person divides his or her share of the pie into three pieces. The third person chooses one slice from each half. The first two people each keep what is left of their half.

   a. Explain why each person believes that he or she has a fair portion of the pie at the end of this procedure.

   b. Propose a division scheme of this sort for four people.

20. Last-diminisher method. The *last-diminisher method* works for items that can be reasonably divided in virtually any way we wish—land or a cake, for example. The method would make no sense for a home. We will describe the method for three people, Anne, Becky, and Alice, who are placed in this order randomly. Let's suppose they want to divide up cookie dough from a mixing bowl. They proceed as follows:

   Step 1: Anne scoops what she believes is a fair portion of the cookie dough into her own bowl.

   Step 2: Becky has two options: She can pass if she thinks that Anne's portion is fair or less than fair. Becky's other option is to take Anne's bowl of dough from her and then return some of the dough to the mixing bowl, thereby reducing the amount to what she considers a fair portion. In this case, we say she *diminishes* the dough.

   Step 3: Alice has the same kind of options as Becky. She may pass or take Becky's bowl, and in the latter case she returns some of the dough to the mixing bowl.

The last person of the three to take and diminish keeps the bowl she is holding. (If no one diminished, then Anne, the one who started, would keep the bowl.) Let's say that person is Becky. The remaining two, Anne and Alice, use the divide-and-choose procedure for two to allocate the rest of the dough.

Explain why each of Anne, Becky, and Alice believes she has a fair share of the cookie dough at the end of this procedure.

21. Divide-and-choose for four. You want to divide a seashell collection among four people. Suggest a divide-and-choose procedure that you think is fair.

22. Five people. Propose a divide-and-choose procedure for five people.

23. History. In 1945, the Polish mathematician Bronislaw Knaster introduced the procedure that bears his name. In fact, the study of fair division and the related mathematical area called *game theory* developed rapidly in a period starting before World War II and extending to the early stages of the Cold War. Research the relevance of game theory to military strategy in general and in particular to the doctrine of mutually assured destruction.

Exercises 24 and 25 are suitable for group work.

Envy A division procedure is *envy-free* if, when all is said and done, no one has reason to wish for someone else's division. In Exercises 24 and 25, we investigate this idea.

24. Divide-and-choose for two people. Explain why the divide-and-choose procedure for two players is envy-free.

25. Lone-divider. The point of this exercise is that the lone-divider method need not be envy-free. Abe, Ben, and Dan are dividing three chocolate chip cookies: cookie 1 with 10 chocolate chips, cookie 2 with 9 chocolate chips, and cookie 3 with no chocolate chips. Abe doesn't care one way or the other about chocolate chips. As far as he is concerned, the three cookies are of equal value. Ben and Dan care only about chocolate chips—they are going to pick the chips out of the dough and eat only the chocolate.

In the rest of the exercise, we assume the following: Abe, Ben, and Dan use the lone-divider method to divide the three cookies. Abe starts. The three cookies have equal value as far as he can see, so he just lays out the three cookies for Ben and Dan. There are various possible outcomes after this.

   a. Let's suppose that Ben and Dan both prefer cookie 1. (This seems reasonable.) According to the procedure, Abe can take either cookie 2 or cookie 3. Suppose he takes cookie 2. Then Ben and Dan divide cookie 1 and cookie 3 between themselves using the divide-and-choose procedure for two people. (The cookies can be broken!) Explain why either Ben or Dan will envy Abe.

   b. In part a, we sketched one possible outcome. There may be reasons for Ben and Dan not to state a preference for cookie 1. List all of the possible preferences the two of them can state. Show that if they state different preferences from each other, then one of them will envy the other. Also show that if they state the same preference as each other, then Abe can settle on a choice that will make Ben or Dan envy him.

The following exercise is designed to be solved using technology such as a calculator or computer spreadsheet. For assistance, see the technology supplement.

26. Knaster procedure. Make a spreadsheet that executes the Knaster procedure.

8.4 Apportionment: Am I represented?

The following excerpt from the *Huffington Post* discusses changes in congressional representation due to the results of the 2010 census.

IN THE NEWS 8.5

New Census Numbers Portend Significant Latino Role

ARTURO VARGAS   December 20, 2010

For the Latino community in particular, Tuesday December 21 is a very big day, and it has nothing really to do with the holidays but rather with numbers and more numbers. The U.S. Census Bureau today releases its official 2010 population figures for the nation, and congressional apportionment totals for each state (reapportionment is the process of dividing all the seats in the U.S. House among the 50 states based on population figures gathered from the U.S. Census).

The numbers let states know if they gain, lose or maintain their congressional seats. Several states are expected to experience gains in the number of House seats, and the Latino population played a significant role in those gains. Furthermore, Latino growth across the country is expected to help certain states maintain their number of congressional seats, or minimize their losses. In other words, it is a win-win for the Latino community specifically and for the country as a whole.

You may be surprised to learn that the U.S. Constitution lays out some basic rules for the makeup of the House of Representatives, but leaves both the size of the House and the exact number of representatives for each state to be determined by Congress. This is the problem of *apportionment* and, as this article shows, the issue remains important to this day. The Constitution gives these rules in Article 1, Section 2: "Representatives and direct taxes shall be apportioned among the several states which may be included within this union, according to their respective numbers...." The numbers—that is, the population of each state—is established by the census: "The actual enumeration shall be made within three years after the first meeting of the Congress of the United States, and within every subsequent term of ten years, in such manner as they shall by law direct." The history of the census is fascinating in itself. The nation has addressed questions such as how to count slaves, people overseas, those in the military, and noncitizens. In particular, before the Civil War, only three-fifths of the population of slaves was counted for apportionment purposes.

The first Congress met in Congress Hall in Philadelphia.

The Constitution specifies a maximum size for the House of Representatives: "The number of Representatives shall not exceed one for every thirty thousand, but each state shall have at least one Representative." It specifies the initial size (65 members) and distribution of the House. Congress regularly increased the size of the House to account for increasing population until 1911, when the size was fixed at 435. The size has remained at this level since then.[17]

There has been considerable political controversy about these issues since the beginning of the republic. We highly recommend the comprehensive treatment of apportionment in *Fair Representation* by Michel Balinski and H. Peyton Young[18]

---

[17]It was temporarily increased to 437 in 1959 when Alaska and Hawaii were admitted to the Union.

[18]Michel Balinski and H. Peyton Young, *Fair Representation: Meeting the Ideal of One Man, One Vote* (New Haven, CT, and London: Yale University Press, 1982).

and the 2001 Congressional Research Service Report for Congress on proposals for change in the House of Representatives apportionment formula.[19]

Apportionment: The problem

To understand some of the difficulties with apportionment, we look at the 1810 House of Representatives. The census of 1810 gave the population of the United States for apportionment purposes as 6,584,255. The size of the House was set at 181 members. To find the ideal size of a congressional district, we divide the U.S. population by the size of the House. Then we divide the population of each state by the size of a congressional district to determine that state's proper share of the House, its *quota*.

Key Concept

To find the size of the ideal district, we use the formula

$$\text{Ideal district size} = \frac{\text{U.S. population}}{\text{House size}}.$$

The ideal district size is also known as the **standard divisor**. We use this number to determine each state's quota:

$$\text{State's quota} = \frac{\text{State's population}}{\text{Ideal district size}}.$$

The United States in 1810.

For the population in 1810 with the House at 181, we find an ideal district size of

$$\text{Ideal district size} = \frac{6,584,255}{181}$$

---

[19]David C. Huckabee, *The House of Representatives Apportionment Formula: An Analysis of Proposals for Change and Their Impact on States*, Congressional Research Service, 2001, http://usinfo.org/enus/government/branches/docs/housefrm.pdf.

or about 36,377.099. For each state, we divide the population by the ideal district size to calculate the quota. For example, Connecticut had a population of 261,818, so its quota is

$$\text{Quota for Connecticut} = \frac{261,818}{36,377.099}$$

or about 7.197. Here are the results of the 1810 census and the quota (to three decimal places) for each state with the ideal district size of 36,377.099. We have added a column giving the quota rounded to the nearest whole number.

State	Population	Quota	Rounded quota
Connecticut	261,818	7.197	7
Delaware	71,004	1.952	2
Georgia	210,346	5.782	6
Kentucky	374,287	10.289	10
Maryland	335,946	9.235	9
Massachusetts	700,745	19.263	19
New Hampshire	214,460	5.895	6
New Jersey	241,222	6.631	7
New York	953,043	26.199	26
North Carolina	487,971	13.414	13
Ohio	230,760	6.344	6
Pennsylvania	809,773	22.261	22
Rhode Island	76,888	2.114	2
South Carolina	336,569	9.252	9
Tennessee	243,913	6.705	7
Vermont	217,895	5.990	6
Virginia	817,615	22.476	22
Total	**6,584,255**	**181**	**179**

The question is what to do with the fractional parts. For example, North Carolina's quota is 13.414. Should North Carolina get 13 representatives, or should it get 14? This is no small issue—the number of representatives allocated to a state makes a big difference in its congressional voting power.

The obvious solution is simply to round each quota to the nearest whole number, as we have done in the last column of the table. One difficulty with this solution is shown in the total. Rounding each state's quota to the nearest whole number gives a total of only 179 representatives. There are two representatives not assigned to any state. This shows why the easy solution of rounding to the nearest whole number by itself would not be a consistently reliable method for apportionment. In general terms, the problem of apportionment is how to deal with the fractions.

Since 1790, several apportionment methods for the U.S. House of Representatives have been developed and either used to allocate seats after various censuses or cited in court cases, even as recently as 2001 (after the 2000 census). These methods fall into two categories: those that rank the fractions, and those that round the fractions.

Hamilton's solution: Ranking the fractions

The first apportionment bill was passed by Congress in 1792. It was based on a method developed by Alexander Hamilton. This method calculates the quotas for each state and then makes an initial allocation to each state by rounding down to the whole number part of the quota.[20] The sum of these whole numbers is less than

---

[20]In practice, this number is never allowed to be 0.

the size of the House, so the leftover members are allocated, one per state, in order from the highest fractional part down. For this reason, this method is also called the *method of largest fractional remainders.*[21]

Let's see the results given by this method for the 1810 Congress.

State	Quota	Quota rounded down: Initial seats	Fractional part of quota	Added seats	Final seats
Connecticut	7.197	7	0.197		7
Delaware	1.952	1	**0.952**	1	2
Georgia	5.782	5	**0.782**	1	6
Kentucky	10.289	10	0.289		10
Maryland	9.235	9	0.235		9
Massachusetts	19.263	19	0.263		19
New Hampshire	5.895	5	**0.895**	1	6
New Jersey	6.631	6	**0.631**	1	7
New York	26.199	26	0.199		26
North Carolina	13.414	13	**0.414**	1	14
Ohio	6.344	6	0.344		6
Pennsylvania	22.261	22	0.261		22
Rhode Island	2.114	2	0.114		2
South Carolina	9.252	9	0.252		9
Tennessee	6.705	6	**0.705**	1	7
Vermont	5.990	5	**0.990**	1	6
Virginia	22.476	22	**0.476**	1	23
Total	181	173		8	181

Using the whole number part of the quota for initial allocation, we see that 173 seats are allocated. This leaves $181 - 173 = 8$ leftover seats. These seats are given to the 8 states for which the fractional parts of the quotas are the greatest. These states correspond to the boldface numbers in the column giving the fractional parts. The final allocations are shown in the last column of the table.

SUMMARY 8.10   **Hamilton's Method for Apportionment**

Step 1: We calculate the quota for each state by dividing its population by the ideal district size.

Step 2: We give to each state the number of representatives corresponding to the whole number part of its quota.

Step 3: We allocate the leftover seats, that is, the difference between the House size and the total given in Step 2, as follows: Rank the states by the size of the fractional part of the quota, from greatest to least, and give one leftover member to each state in that order until the leftovers are exhausted.

EXAMPLE 8.18 Apportioning by Hamilton's method: City councilors

Our city has four districts: North, East, West, and South. There are required to be 10 city councilors allocated according to population, with at least one from each district. Given the populations below, calculate the quota for each district. (You need keep

---

[21] Hamilton's method is also known as the Vinton method or Hamilton–Vinton method because Representative Samuel Vinton was a strong advocate for using this method for the 1850 census.

only one place beyond the decimal point.) Then use Hamilton's method to determine the number of councilors to represent each district.

District	Population
North	3900
East	4800
West	18,000
South	3300
Total	**30,000**

SOLUTION

We have 10 seats to fill and a total population of 30,000, so the ideal district size is

$$\text{Ideal district size} = \frac{\text{Total population}}{\text{Number of seats}}$$
$$= \frac{30,000}{10}$$
$$= 3000.$$

We divide the population in each district by this district size, and the resulting quota is shown in the second column of the accompanying table.

District	Quota	Quota rounded down: Initial seats	Fractional part of quota	Added seats	Final seats
North	1.3	1	0.3		1
East	1.6	1	0.6	1	2
West	6.0	6	0.0		6
South	1.1	1	0.1		1
Total	**10**	**9**		**1**	**10**

The third column of the table shows that the whole number parts sum to 9, so there is one leftover councilor to allocate. Because the quota of 1.6 has the largest fractional part (0.6), the East district gets that councilor. The final result: North and South should each get 1 councilor, East should get 2 councilors, and West should get 6 councilors.

TRY IT YOURSELF 8.18

For the same city, districts, and populations, use Hamilton's method to apportion a total of 20 councilors. You need keep only one place beyond the decimal point.

The answer is provided at the end of this section.

The congressional bill of March 26, 1792, that allocated the U.S. House of Representatives according to Hamilton's method was vetoed by President George Washington. Indeed, this was the very first presidential veto. Some of the reasons for Washington's veto are investigated in Exercise 8. Some additional problems with Hamilton's method are investigated in Exercises 9 and 10. Hamilton's method was not used until it was revived for the 1850 census. Instead, a method devised by Thomas Jefferson was used beginning with the 1790 census.

Jefferson's solution: Adjusting the divisor

Thomas Jefferson, disturbed by the fractions, proposed a different solution. The first step is to round down (but not less than 1, because each state is guaranteed at least

On April 5, 1792, George Washington cast the first presidential veto.

one representative) using the whole number part of the quota just as with the Hamilton method. But rather than rank the fractional parts, Jefferson proposed that the divisor be adjusted until the sum of the rounded-down quotas is the correct size—the legislated size of the House of Representatives. There are complicated formulas that can tell us how to adjust the divisor but, with the aid of a spreadsheet, trial-and-error is a suitable method. Let's consider again the allocation of seats for 1810. We start with the divisor as the ideal district size of 36,377.099 and round each quota down as our first allocation effort. As the fourth column of the accompanying table shows, this allocates only 173 seats. In the fifth column, we decrease the divisor to 34,000 to increase the number of seats allocated. For example, with a divisor of 34,000, Connecticut's adjusted quota is $261{,}818/34{,}000 = 7.701$ to three decimal places, which rounds down to 7. The overall result is a total of 187 seats allocated—too many. So, we increase the divisor to 35,000 and try again. The result is a sum of 181, as is shown in the last column of the table. Because this is the correct number of seats, we have found the final allocation.

State	Population	Quota using divisor 36,377.10	First try: Quota rounded down	Second try: Decrease divisor to 34,000, round down	Third try: Increase divisor to 35,000, round down
Connecticut	261,818	7.197	7	7	7
Delaware	71,004	1.952	1	2	2
Georgia	210,346	5.782	5	6	6
Kentucky	374,287	10.289	10	11	10
Maryland	335,946	9.235	9	9	9
Massachusetts	700,745	19.263	19	20	20
New Hampshire	214,460	5.895	5	6	6
New Jersey	241,222	6.631	6	7	6
New York	953,043	26.199	26	28	27
North Carolina	487,971	13.414	13	14	13
Ohio	230,760	6.344	6	6	6
Pennsylvania	809,773	22.261	22	23	23
Rhode Island	76,888	2.114	2	2	2
South Carolina	336,569	9.252	9	9	9
Tennessee	243,913	6.705	6	7	6
Vermont	217,895	5.990	5	6	6
Virginia	817,615	22.476	22	24	23
Total	**6,584,255**	**181**	**173**	**187**	**181**

Comparing this apportionment with Hamilton's, we see that they are very similar. But Jefferson's method gives one less delegate each to New Jersey, North Carolina, and Tennessee and gives one additional delegate to each of Massachusetts, New York, and Pennsylvania.

As we noted, trial-and-error (with the help of a spreadsheet) is an effective way to adjust divisors to find an allocation that yields the correct House size. Finding the proper divisor is usually not difficult because there is often a wide range of workable answers. In the table above, any divisor between 34,856 and 35,037 will produce the correct size.

SUMMARY 8.11   **Jefferson's Method for Apportionment**

Step 1: We start using the ideal district size as the divisor.

Step 2: We calculate the quota for each state by dividing its population by the divisor.

Step 3: We round down each quota to the nearest whole number (but not less than 1) and sum the rounded quotas.

Step 4: If the sum from Step 3 is larger than the size of the House, we increase the divisor and repeat Steps 2 and 3. If the sum is too small, we decrease the divisor and repeat Steps 2 and 3. We continue this process until a divisor is found for which the sum of the rounded-down quotas (but not less than 1) is the House size. The result is the apportionment given by Jefferson's method.

EXAMPLE 8.19 Apportioning by Jefferson's method: City councilors

Recall the situation studied in Example 8.18: Our city has the four districts North, East, West, and South. There are required to be 10 city councilors allocated according to population, with at least one from each district. Given the populations below, apportion the number of councilors to represent each district using Jefferson's method. Quotas may be reported using one place beyond the decimal point.

District	Population
North	3900
East	4800
West	18,000
South	3300
Total	**30,000**

SOLUTION

We have 10 councilors and a total population of 30,000, so our ideal district size is $30,000/10 = 3000$. We use this as our first divisor. Dividing gives quotas of 1.3, 1.6, 6, and 1.1, which round down to 1, 1, 6, and 1 (see the accompanying table). These sum to 9, so we need to adjust the divisor down.

District	Population	Quota using divisor 3000	First try: Quota rounded down	Second try: Decrease divisor to 2500, round down
North	3900	1.3	1	1
East	4800	1.6	1	1
West	18,000	6.0	6	7
South	3300	1.1	1	1
Total	30,000	10	9	10

Trying a divisor of 2500 gives quotas of 1.6, 1.9, 7.2, and 1.3, which round down to 1, 1, 7, and 1. These sum to 10, so North, East, and South should each get 1 councilor, and West should get 7 councilors. Note that this distribution differs from that of Hamilton's method in Example 8.18.

TRY IT YOURSELF 8.19

For the same city, districts, and populations, use Jefferson's method to apportion a total of 8 councilors. Quotas may be reported using one place beyond the decimal point.

The answer is provided at the end of this section.

Jefferson's method was adopted by Congress for the Third Congress (1793) and used with each census through 1830. Even from the beginning, it was clear that small states may well be disproportionately underrepresented by this method. For 1800, Jefferson's method awards Delaware only a single representative based on its apportionment population[22] of 61,812—far more than the ideal district size of 34,680 at that time. As Jefferson's method continued to be applied for the censuses through 1830, it became more and more evident that the method favored larger states over smaller states. Note for the 1810 results that, compared with Hamilton's method, Jefferson's method gives extra representatives to the larger states of Massachusetts, New York, and Pennsylvania, and takes them away from the smaller New Jersey, North Carolina, and Tennessee.

In addition to the problem of bias in favor of larger states, there was the question of *staying within the quota*. In Exercise 32, you will be asked to verify that in the 1820 census New York had a quota of 32.503. We would thus expect New York to get either 32 or 33 representatives. In fact, Jefferson's method gives New York 34 representatives, thus *violating quota*. Exercise 23 gives a simple situation where quota is violated by Jefferson's method. By contrast, Hamilton's method always stays within quota.

Key Concept

A desirable trait of any apportionment method is to **stay within quota**, meaning that the final apportionment for each state would be within one of the quota. That is, the final apportionment should be the quota either rounded down or rounded up. Apportionment methods that don't stay within quota are said to **violate quota.**

---

[22]The actual population of Delaware was enumerated in the census as 64,273; for apportionment purposes, however, each of Delaware's 6153 slaves was counted as three-fifths of a person, as noted at the beginning of this section.

Return to Hamilton's method: Paradoxes

From 1850 through the end of the nineteenth century, Congress nominally required that Hamilton's method be used to apportion representatives. In practice, Congress would begin with apportionment according to Hamilton's method and then add members not following the method. Because presidential elections are determined by the Electoral College, whose membership by state depends on apportionment, Samuel Tilden would have won the presidency over Rutherford B. Hayes in 1876 had Hamilton's method been consistently used. In three situations, however, Hamilton's method presents troubling and paradoxical results.

The first problem is an apparent inconsistency in allocating members: It is possible for a state to lose a member if the House size is increased by one member! A similar observation was made based on the census of 1870, but it was the 1880 census that brought the issue to the political forefront. While doing some calculations to determine the House size, the chief clerk of the Census Office noticed the *Alabama paradox*: For a House size of 299, Alabama would be allocated 8 representatives. For a House size of 300, Alabama would be allocated only 7 representatives. The census data for 1880 and the calculations made by the clerk appear in **Table 8.2**. They show that if the House size increased from 299 to 300, then Illinois and Texas would each gain a seat and Alabama would lose a seat.

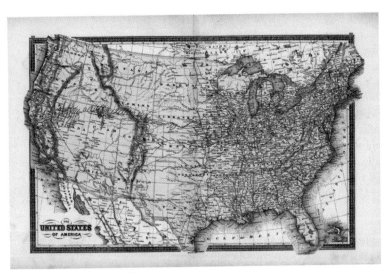

The United States in 1880.

EXAMPLE 8.20 Exploring the Alabama paradox: 1880 census

    a. For a House size of 299, determine the ideal district size and verify the quotas for Alabama and Illinois.

    b. How do the fractional parts of the quotas for Alabama and Illinois compare for 299 members?

    c. How do the fractional parts of the quotas for Alabama and Illinois compare for 300 members?

    d. Use parts b and c to give a plausible explanation for the paradox.

SOLUTION

    a. The table shows that the total population is 49,371,340. For a House size of 299, the ideal district size is the total population divided by 299, which is

TABLE 8.2 Alabama Paradox: 1880 Census, Hamilton's Method

State	Population	House size of 299				House size of 300			
		Quota	Initial seats	Added seats	Total seats	Quota	Initial seats	Added seats	Total seats
Alabama	1,262,505	7.646	7	1	8	7.671	7		7
Arizona	802,525	4.860	4	1	5	4.876	4	1	5
California	864,694	5.237	5		5	5.254	5		5
Colorado	194,327	1.177	1		1	1.181	1		1
Connecticut	622,700	3.771	3	1	4	3.784	3	1	4
Delaware	146,608	0.888	1		1	0.891	1		1
Florida	269,493	1.632	1		1	1.638	1		1
Georgia	1,542,180	9.340	9		9	9.371	9		9
Illinois	3,077,871	18.640	18		18	18.702	18	1	19
Indiana	1,978,301	11.981	11	1	12	12.021	12		12
Iowa	1,624,615	9.839	9	1	10	9.872	9	1	10
Kansas	996,096	6.033	6		6	6.053	6		6
Kentucky	1,648,690	9.985	9	1	10	10.018	10		10
Louisiana	939,946	5.692	5	1	6	5.711	5	1	6
Maine	648,936	3.930	3	1	4	3.943	3	1	4
Maryland	934,943	5.662	5	1	6	5.681	5	1	6
Massachusetts	1,783,085	10.799	10	1	11	10.835	10	1	11
Michigan	1,636,937	9.914	9	1	10	9.947	9	1	10
Minnesota	780,773	4.728	4	1	5	4.744	4	1	5
Mississippi	1,131,597	6.853	6	1	7	6.876	6	1	7
Missouri	2,168,380	13.132	13		13	13.176	13		13
Nebraska	452,402	2.740	2	1	3	2.749	2	1	3
Nevada	62,266	0.377	1		1	0.378	1		1
New Hampshire	346,991	2.101	2		2	2.108	2		2
New Jersey	1,131,116	6.850	6	1	7	6.873	6	1	7
New York	5,082,871	30.783	30	1	31	30.886	30	1	31
North Carolina	1,399,750	8.477	8		8	8.505	8		8
Ohio	3,198,062	19.368	19		19	19.433	19		19
Oregon	174,768	1.058	1		1	1.062	1		1
Pennsylvania	4,282,891	25.938	25	1	26	26.025	26		26
Rhode Island	276,531	1.675	1	1	2	1.680	1	1	2
South Carolina	995,577	6.029	6		6	6.050	6		6
Tennessee	1,542,359	9.341	9		9	9.372	9		9
Texas	1,591,749	9.640	9		9	9.672	9	1	10
Vermont	332,286	2.012	2		2	2.019	2		2
Virginia	1,512,565	9.160	9		9	9.191	9		9
West Virginia	618,457	3.745	3	1	4	3.758	3	1	4
Wisconsin	1,315,497	7.967	7	1	8	7.993	7	1	8
Total	49,371,340	299	279	20	299	300	282	18	300

49,371,340/299 or about 165,121.539. To verify the quotas, we divide the state population by the ideal district size. For Alabama, the quota is therefore 1,262,505/165,121.539 = 7.646 to three decimal places. For Illinois, we have a quota of 3,077,871/165,121.539 or about 18.640.

b. With 299 members, Alabama's fractional part is 0.646, and the fractional part for Illinois is 0.640. Alabama has the larger fractional part.

c. The table shows that with 300 members, the fractional part for Alabama is 0.671 and the fractional part for Illinois is 0.702. Illinois has the larger fractional part.

d. From parts b and c, we see that increasing the size of the House from 299 to 300 causes the fractional part of Alabama's quota to become smaller than the fractional part for Illinois. Because Hamilton's method allocates leftover seats according to the size of that fractional part, it seems reasonable that Illinois could gain a seat and Alabama could lose a seat in the process.

TRY IT YOURSELF 8.20

For a House size of 300, determine the ideal district size and verify the quotas for Alabama and Illinois.

The answer is provided at the end of this section.

Another troubling consequence of Hamilton's method is the *population paradox*. In the 1900 census, Virginia's population was 1,854,184 and growing at 1.07% per year, whereas Maine's population was 694,466 and growing at 0.67% per year. The quota for Virginia was 9.599, which Hamilton's method rounded up to 10, and Maine's quota was 3.595, which was rounded down to 3. Note how close the fractional parts are: 0.599 and 0.595. Because the nation as a whole was growing faster than either state and Virginia had a larger population, after one year Virginia's fractional part of 0.599 would slip below that of Maine. Then Maine's quota could be rounded up and Virginia's rounded down. This could happen despite the fact that Virginia was growing faster than Maine.

A third problematic consequence of Hamilton's method is the *new states paradox*. When Oklahoma became a state in 1907, 5 seats were added to the House. This was the number of seats to which Oklahoma would be entitled, based on its population then. Oklahoma was indeed awarded 5 seats, and one would assume that the number of representatives for the other states would be unchanged. If Hamilton's method had been used, however, it would have changed the number of representatives for Maine and New York.

These paradoxes were sufficiently troubling to Congress that in time new methods were sought that would avoid real and perceived problems with Hamilton's method.

More adjusted divisor methods: Adams and Webster

Problems with Hamilton's method led Congress to revive Jefferson's method and modify it in the hopes of making it better. Any system for apportionment that rounds the quotas by some method and then adjusts the divisor until the House is of the correct size is an *adjusted divisor method*. Jefferson's method is one of several adjusted divisor methods. All of the methods follow the same steps as in Summary 8.11, except that in Step 3 they vary in the method by which they round. Jefferson's method rounds all the quotas down (but not less than 1).

In the 1820s and 1830s, the U.S. population was moving westward, adding states west of the Appalachians, and the proportion of the population living in New England decreased. These and other factors inspired John Quincy Adams to reexamine Jefferson's method. Adams believed that Jefferson's rounding down of quotas left some people unrepresented and failed to meet the constitutional intent that all the people be represented. He suggested an adjusted divisor method in which all quotas are rounded *up*. We noted earlier that Jefferson's method favored larger states, and it comes as no surprise that Adams's method favors smaller states. Like Jefferson's method, Adams's method could violate quota. Adams's method was never formally used by Congress.

Jefferson's method always rounds down and Adams's method always rounds up. A third adjusted divisor method was proposed by Daniel Webster. Webster's method rounds to the nearest whole number (rather than always up or always down) before adjusting the divisor. This method was used by Congress with the 1840 census and revived for use in the censuses of 1900, 1910, and 1930.

"How do we know we can't fool all the people all the time if we don't give it a try?!"

SUMMARY 8.12   Three Adjusted Divisor Methods for Apportionment

1. We calculate ideal district size by dividing the total population by the size of the House. This is used as the first divisor.

2. We calculate the quota for each state by dividing its population by the divisor.

3. We round each quota to a whole number as follows:

Jefferson's method: We round down (but not less than 1).

Adams' method: We round up.

Webster's method: We round to the nearest whole number (up if the fractional part is 0.5 or greater and down otherwise, but not less than 1).

Then we sum the resulting rounded quotas.

4. If the sum from Step 3 is larger than the size of the House, we increase the divisor and repeat Steps 2 and 3. If the sum is too small, we decrease the size of the divisor and repeat Steps 2 and 3. We continue this process until a divisor is found for which the sum of the rounded quotas is the House size. This is the final apportionment.

EXAMPLE 8.21 Comparing methods: Montana's loss is Florida's gain

According to the 2010 census, this was the population of four states:

State	Population
Montana	994,416
Alaska	721,523
Florida	18,900,773
West Virginia	1,859,815
Total	**22,476,527**

Congress allocated 32 seats in total to these states.

a. Calculate the ideal district size and each state's quota. Keep three places beyond the decimal point.

b. Calculate the apportionment of the 32 seats according to Hamilton's method.

c. Calculate the apportionment of the 32 seats according to Jefferson's method.

d. Calculate the apportionment of the 32 seats according to Adams's method.

e. Calculate the apportionment of the 32 seats according to Webster's method.

SOLUTION

a. The ideal district size is the total population divided by the number of seats, or $22{,}476{,}527/32 = 702{,}391.469$. The quota for each state is calculated by dividing the state's population by this divisor:

State	Population	Quota
Montana	994,416	1.416
Alaska	721,523	1.027
Florida	18,900,773	26.909
West Virginia	1,859,815	2.648
Total	22,476,527	32

b. For Hamilton's method, we make an initial allocation by rounding the quotas down. That gives an initial allocation of 30 seats. The remaining 2 seats are given to the states having the largest fractional parts of their quotas— Florida and West Virginia. The final allocation is shown in the last column of the accompanying table.

State	Population	Quota	Quota rounded down: Initial seats	Fractional part of quota	Added seats	Final seats
Montana	994,416	1.416	1	0.416		1
Alaska	721,523	1.027	1	0.027		1
Florida	18,900,773	26.909	26	0.909	1	27
West Virginia	1,859,815	2.648	2	0.648	1	3
Total	22,476,527	32	30		2	32

According to Hamilton's method, Montana should get 1 seat, Alaska 1 seat, Florida 27 seats, and West Virginia 3 seats.

c. For Jefferson's method, we begin as in part b by rounding the quota down as an initial allocation. Because the total from the initial allocation is less than 32, the total number of seats, we adjust the divisor to a smaller number. Using an adjusted divisor of 700,000 and calculating the corresponding quotas and then rounding down, we obtain 31 seats, so we try an even smaller divisor of 675,000.

State	Population	Quota using divisor 702,391.469	First try: Quota rounded down	Second try: Decrease divisor to 700,000, round down	Another try: Decrease divisor to 675,000, round down
Montana	994,416	1.416	1	1	1
Alaska	721,523	1.027	1	1	1
Florida	18,900,773	26.909	26	27	28
West Virginia	1,859,815	2.648	2	2	2
Total	22,476,527	32	30	31	32

According to Jefferson's method, Montana should get 1 seat, Alaska 1 seat, Florida 28 seats, and West Virginia 2 seats.

d. For Adams's method, we make an initial allocation by rounding the quotas up. Because the total from the initial allocation is more than 32 (see the accompanying table), the total number of seats, we adjust the divisor to a larger number. Using an adjusted divisor of 750,000, calculating the corresponding quotas, and then rounding up, we obtain 32 seats, the desired number. The last column of the table below shows this result.

State	Population	Quota using divisor 702,391.469	First try: Quota rounded up	Another try: Increase divisor to 750,000, round up
Montana	994,416	1.416	2	2
Alaska	721,523	1.027	2	1
Florida	18,900,773	26.909	27	26
West Virginia	1,859,815	2.648	3	3
Total	22,476,527	32	34	32

According to Adams's method, Montana should get 2 seats, Alaska 1 seat, Florida 26 seats, and West Virginia 3 seats.

e. For Webster's method, we make an initial allocation by rounding the quotas to the nearest whole number, that is, we round down if the fractional part of the quota is less than 0.5 and round up otherwise. Because the total from the initial allocation is 32 (see the accompanying table), no further adjustments are needed: According to Webster's method, Montana should get 1 seat, Alaska 1 seat, Florida 27 seats, and West Virginia 3 seats.

State	Population	Quota using divisor 702,391.469	First try: Quota rounded to nearest whole number
Montana	994,416	1.416	1
Alaska	721,523	1.027	1
Florida	18,900,773	26.909	27
West Virginia	1,859,815	2.648	3
Total	22,476,527	32	32

Note that the four methods used in Example 8.21 give three different answers.

TABLE 8.3 Geometric Means

Whole number	Geometric mean	Whole number	Geometric mean	Whole number	Geometric mean
1	1.414	11	11.489	21	21.494
2	2.449	12	12.490	22	22.494
3	3.464	13	13.491	23	23.495
4	4.472	14	14.491	24	24.495
5	5.477	15	15.492	25	25.495
6	6.481	16	16.492	26	26.495
7	7.483	17	17.493	27	27.495
8	8.485	18	18.493	28	28.496
9	9.487	19	19.494	29	29.496
10	10.488	20	20.494	30	30.496

The Huntington–Hill method

The Huntington-Hill method is an adjusted divisor method that deserves special attention because it is the one that is in use today. Huntington–Hill follows the steps of the other adjusted divisor methods (Jefferson, Adams, Webster), but rounding is done using the *geometric mean*: If n is the whole number part of the quotient (initially the quota), the geometric mean we use is $\sqrt{n(n+1)}$. We round up if the quotient is at least this large; otherwise, we round down. **Table 8.3** shows some values of the geometric mean.

For example, for $n = 1$, the geometric mean is about 1.414. This means that a quotient of 1.433 (whose whole number part is 1) is rounded up to 2 by Huntington–Hill. On the other hand, the geometric mean for 5 is 5.477. So a quotient of 5.433 is rounded down to 5 by Huntington–Hill. Both quotients have the same fractional part, 0.433, but they are rounded differently. Note that Huntington–Hill gives a benefit to smaller states because it rounds the small quotient 1.433 up, but it rounds the larger quotient 5.433 down.

EXAMPLE 8.22 Using Huntington–Hill: Four states

Use the Huntington–Hill method to allocate 32 seats to the four states in Example 8.21.

SOLUTION

We begin by showing the quota using the ideal district size of 702,391.469 and the corresponding geometric mean (from Table 8.3). We round down if the quota is less than the geometric mean; otherwise, we round up.

State	Population	Quota using divisor 702,391.469	Geometric mean	First try: Rounded quota
Montana	994,416	1.416	1.414	2
Alaska	721,523	1.027	1.414	1
Florida	18,900,773	26.909	26.495	27
West Virginia	1,859,815	2.648	2.449	3
Total	22,476,527	32		33

In the final column of the table, we see that we have allocated a total of 33 seats—too many. So we use a larger divisor of 710,000.

State	Population	Quota using divisor 702,391.469	Geometric mean	First try: Rounded quota	Second try: Quotient using divisor of 710,000	Rounded quotient
Montana	994,416	1.416	1.414	2	1.401	1
Alaska	721,523	1.027	1.414	1	1.016	1
Florida	18,900,773	26.909	26.495	27	26.621	27
West Virginia	1,859,815	2.648	2.449	3	2.619	3
Total	22,476,527	32		33		32

The last column of the table shows that this allocates a total of 32 seats, which is the required number. So Huntington–Hill awards 1 seat to Montana, 1 to Alaska, 27 to Florida, and 3 to West Virginia.

TRY IT YOURSELF 8.22

Use the Huntington-Hill method to allocate the 10 councilors from Example 8.18.

The answer is provided at the end of this section.

Figure 8.1 illustrates the differing effects of the five methods from Examples 8.21 and 8.22 on allocation of seats to Montana, Alaska, Florida, and West Virginia. Since each of these methods has historically been used at various times by Congress, these differing effects raise the question, "Which is fair?"

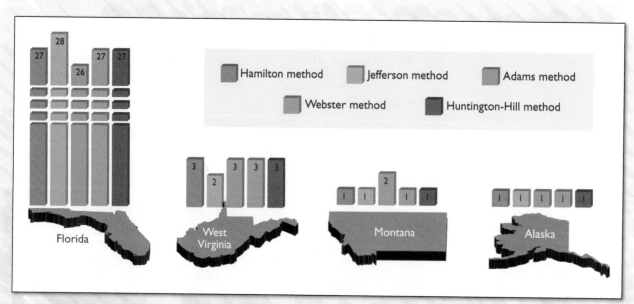

FIGURE 8.1 Results of various apportionment methods.

Is it fair? What is the perfect method?

We've seen five different methods for apportionment: Hamilton's method and four adjusted divisor methods. Currently, Congress uses the Huntington–Hill method, which rounds according to the geometric mean. Is this method fair? The answer surely depends on what criteria are used to define "fair." Of these methods, only Hamilton's method never violates quota, and yet it suffers from the Alabama, population, and new states paradoxes. Only adjusted divisor methods avoid the population paradox, but all adjusted divisor methods violate quota.[23] In fact, it can be shown that no method both avoids the population paradox and never violates quota. There is no perfect method.

All adjusted divisor methods violate quota, but by how much? Of all the methods, Webster's violates quota the least: The probability of violating quota is only 0.61 time per 1000 times, that is, about once per 1640 apportionments or once per 16,400 years on average (because each apportionment occurs once every 10 years).[24]

From the very beginning, there has been tension between smaller states and larger states because the different methods differ in their effect, depending on the size of the state. Michel Balinski and H. Peyton Young have analyzed the period from 1790 to 1970 and calculated the bias of the adjusted divisor methods toward smaller states.[25] In **Figure 8.2**, the vertical axis indicates percentage bias toward smaller states.

As we can see from Figure 8.2, the Huntington–Hill method (called the Hill method in the graph) has a slight bias of about 3% in favor of smaller states, whereas Webster's method is the closest to being unbiased. Congress continues to use the Huntington–Hill method, despite lawsuits filed by states after both the 1990 and 2000 censuses. Although the Supreme Court has determined that Congress has the discretion to choose an apportionment method, we can anticipate attempts to change apportionment methods in the future.

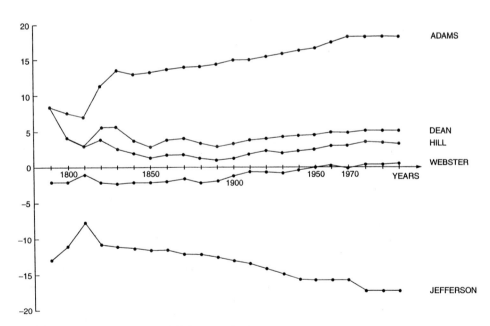

FIGURE 8.2 Biases of adjusted divisor methods.

---

[23] See Balinski and Young, *op. cit.*, p. 79.

[24] See Balinski and Young, *op. cit.*, p. 81.

[25] See Balinski and Young, *op. cit.*, p. 75.

Try It Yourself answers

Try It Yourself 8.18: Apportioning by Hamilton's method: City councilors   North and East each get 3, West gets 12, and South gets 2.

Try It Yourself 8.19: Apportioning by Jefferson's method: City councilors   Using a divisor of around 3500, we see that North, East, and South each get 1 councilor and West gets 5 councilors.

Try It Yourself 8.20: Exploring the Alabama paradox: 1880 census   The ideal district size is 49,371,340/300 or 164,571.133. The Alabama quota is 1,262,505/164,571.133 = 7.671 to three places. The Illinois quota is 3,077,871/164,571.133 = 18.702 to three places.

Try It Yourself 8.22: Using Huntington–Hill: City councilors   1 each to North and South, 2 to East, and 6 to West.

Exercise Set 8.4

1. District size in 1790. The U.S. population in 1790 was 3,615,920, and the House size was set at 105. What was the ideal congressional district size in 1790? Use three decimal places in your answer.

2. District size in 2000. In 2000, there were 435 members of the House of Representatives. The U.S. population in 2000 was 291,424,177. What was the ideal congressional district size in 2000? Use three decimal places in your answer.

3. New York's quota in 1790. The district size in 1790 was 34,437. The population of New York in 1790 was 331,589. What was New York's quota in 1790? Use two decimal places in your answer.

4. Virginia's quota in 1790. The district size in 1790 was 34,437. The population of Virginia in 1790 was 630,560. What was Virginia's quota in 1790? Use two decimal places in your answer.

5. California's quota in 1990. In 1990, the U.S. population was 248,709,873. There were 435 members of the House of Representatives. The population of California in 1990 was 29,839,250. What was California's quota in 1990? Use two decimal places for your answer.

6. Minnesota's quota in 1990. In 1990, the U.S. population was 248,709,873. There were 435 members of the House of Representatives. The population of Minnesota in 1990 was 4,387,029. What was Minnesota's quota in 1990? Use two decimal places for your answer.

7. An interesting House size. The U.S. Constitution states that the minimum size of a congressional district is 30,000 people. The U.S. population in 2010 was 308,400,408. If each district size were 30,000 people, how many House members would we have had in 2010? Round your answer to the nearest whole number.

8. Washington's veto. President Washington vetoed the Hamilton apportionment bill. Part of his reasoning involved the state of Delaware. Hamilton's plan would have awarded Delaware 2 House seats. The population of Delaware in 1790 was 55,540.

a. With 2 House seats, what would have been the size of a congressional district in Delaware?

b. Compare your answer to part a with the constitutional minimum for the number of people in a congressional district.

9. Hamilton's method and ties. One problem with Hamilton's method could arise if there are ties in the quotas. (This would be highly unlikely to happen in practice.) Suppose a country consisted of State 1 with population 140, State 2 also with population 140, and State 3 with population 720. That is a total population of 1000. Suppose the House size is 10.

a. What is the ideal district size?

b. Calculate the quota for each of the three states.

c. Explain the difficulty Hamilton's method has in allotting the 10 House seats.

10. Hamilton's method and small states. When a state's population is too small, Hamilton's method (if unaltered) may not assign it any representatives. Suppose, for example, that the country consists of two states: State 1 with a population of 9 and State 2 with a population of 91. There are 5 seats to be assigned. How does Hamilton's method assign these 5 seats? It should be noted that Hamilton's method was later modified to account for this difficulty.

A small town   A small town elects a total of 12 board members from four districts, North, South, East, and West. The populations of each district are shown below.

District	Population
North	100
South	150
East	200
West	700

This information is used in Exercises 11 through 16.

11. Ideal district size for small town. Find the ideal district size for this small town. Use two decimal places in your answer.

12. Small town board using Hamilton's method. Find the distribution of the 12 board members using Hamilton's method.

13. Small town board using Jefferson's method. Find the distribution of the 12 board members using Jefferson's method.

14. Small town board using Adams's method. Find the distribution of the 12 board members using Adams's method.

15. Small town board using Webster's method. Find the distribution of the 12 board members using Webster's method.

16. Small town board using Huntington–Hill method. Find the distribution of the 12 board members using the Huntington–Hill method.

A larger town A town elects a total of 15 board members from four districts, North, South, East, and West. The populations of each district are shown below.

District	Population
North	145
South	150
East	326
West	735

This information is used in Exercises 17 through 22.

17. Ideal district size for larger town. What is the ideal district size for this town? Use one decimal place in your answer.

18. Larger town board using Hamilton's method. Find the distribution of the 15 board members using Hamilton's method.

19. Larger town board using Jefferson's method. Find the distribution of the 15 board members using Jefferson's method.

20. Larger town board using Adams's method. Find the distribution of the 15 board members using Adams's method.

21. Larger town board using Webster's method. Find the distribution of the 15 board members using Webster's method.

22. Larger town board using Huntington–Hill. Find the distribution of the 15 board members using the Huntington–Hill method.

23. Violating quota. This exercise shows how Jefferson's method can violate quota. The following table shows the population for a nation consisting of State 1, State 2, and State 3. The quota for a House size of 25 is also given.

State	Population	Quota
State 1	200	3.571
State 2	200	3.571
State 3	1000	17.857
Total	1400	25

   a. Show that 52 works as a divisor for Jefferson's method.

   b. For which state is quota violated?

The trouble with a new state Suppose that Oklahoma, Texas, Arkansas, and New Mexico form a Board of Representatives to attract business to the region. There are to be 125 members of the board, and the membership from each state was determined by its population in 2009. The populations of these states are given in the table below.

State	Population
Oklahoma	3,687,050
Texas	24,782,302
Arkansas	2,889,450
New Mexico	2,009,971
Total	33,368,773

This information is used in Exercises 24 through 26.

24. Ideal district size for the coalition. Find the ideal district size for the coalition. Use three decimal places in your answer.

25. Hamilton's method for the coalition. Determine the distribution of board members among the four states if Hamilton's method is used.

26. Kansas joins the coalition. Kansas, which has a population of 2,805,747, joins the coalition. Because the population of Kansas is close to that of Arkansas, it is decided that 11 new board members should be added, for a new total membership of 136.

   a. Determine the distribution of board members among the five states if the 136-member board is selected using Hamilton's method.

   b. Do Kansas and Arkansas have the same number of board members?

   c. Which of the four original states gained a board member with the addition of Kansas to the coalition?

27. Dean's method. Dean's method is an adjusted divisor method. It follows the steps of the other adjusted divisor methods, but rounding is done using the *harmonic mean*: If n is the whole number part of the quotient (initially the quota), the harmonic mean we use is $\dfrac{n(n+1)}{n+1/2}$. We round up if the quotient is at least this large; otherwise, we round down. The following table provides a few helpful values for the harmonic mean:

Whole number	Harmonic mean
1	1.333
2	2.400
26	26.491

What allocation does Dean's method give for the four states in Example 8.21?

The following exercises are designed to be solved using technology such as calculators or computer spreadsheets. For assistance, see the technology supplement.

Exercises 28 through 32 are suitable for group work.

The 1790 Congress The table below gives state populations in 1790, when the size of the House was 105. The ideal district size in 1790 was 34,437, and the table shows the quotas to three decimal places. This information will be used in Exercises 28 through 31.

State	Population	Quota
Connecticut	236,841	6.878
Delaware	55,540	1.613
Georgia	70,835	2.057
Kentucky	68,705	1.995
Maryland	278,514	8.088
Massachusetts	475,327	13.803
New Hampshire	141,822	4.118
New Jersey	179,570	5.214
New York	331,589	9.629
North Carolina	353,523	10.266
Pennsylvania	432,879	12.570
Rhode Island	68,446	1.988
South Carolina	206,236	5.989
Vermont	85,533	2.484
Virginia	630,560	18.311
Total	3,615,920	105

28. Hamilton's solution for 1790. Find each state's representation for 1790 using Hamilton's apportionment method.

29. Jefferson's solution for 1790. What allocation would Jefferson's method give for 1790?

30. Adams's solution for 1790. What allocation would Adams's method give for 1790?

31. Webster's solution for 1790. What allocation would Webster's method give for 1790?

32. Jefferson's method for 1820: violating quota. The population figures for 1820 are given in the table below.

State	Population	State	Population
New York	1,368,775	Connecticut	275,208
Pennsylvania	1,049,313	New Jersey	274,551
Virginia	895,303	New Hampshire	244,161
Ohio	581,434	Vermont	235,764
North Carolina	556,821	Indiana	147,102
Massachusetts	523,287	Louisiana	125,779
Kentucky	513,623	Alabama	111,147
South Carolina	399,351	Rhode Island	83,038
Tennessee	390,769	Delaware	70,943
Maryland	364,389	Missouri	62,496
Maine	298,335	Mississippi	62,320
Georgia	281,126	Illinois	54,843
		Total	8,969,878

a. Use Jefferson's method to calculate the apportionment of 213 representatives for 1820.

b. Verify that quota is violated for New York.

c. For what other states is quota violated?

CHAPTER SUMMARY

This chapter revolves around questions of how a group should make decisions. How to measure voting power and how to evaluate voting systems are two topics from the political process. Other situations in which making a decision is difficult arise in the context of equitable division, such as dividing an inheritance. One important division problem is the apportionment of delegates to the U.S. House of Representatives among the 50 states. In this chapter, we consider all of these topics.

Measuring voting power: Does my vote count?

In many situations, voters or voting blocks control multiple votes. Intuitively, it would seem that in such situations a voter's power is directly proportional to the number of votes controlled. But *voting power* is more complicated than may appear at first blush. Voting power is hard to define, but we discuss two common methods of measuring voting power.

The number of votes required to decide the outcome of an election is the *quota*. Often the quota is a simple majority, which requires more than half of the total number of votes cast.

Consider the case in which Voter A has 20 votes, Voter B has 20 votes, and Voter C has 1 vote. It appears on the surface that Voters A and B have more power than Voter C. But if a simple majority of 21 votes is required, then any pair of the three voters can combine to win an election regardless of how the third votes. One might reasonably claim that the three voters actually have equal power.

One measure of voting power is the *Banzhaf power index*. A coalition of voters is a *winning coalition* if it controls sufficient votes to determine the outcome of an election. A voter in a winning coalition is *critical* for that coalition if his or her absence would prevent the coalition from being a winner. We summarize the information for the preceding example in a *coalition table*.

Number of votes			Total votes	Winning coalition?	Critical voters
20	20	1			
A	B	C	41	Yes	None
A	B		40	Yes	A, B
A		C	21	Yes	A, C
A			20	No	Not applicable
	B	C	21	Yes	B, C
	B		20	No	Not applicable
		C	1	No	Not applicable

The Banzhaf index of a voter is the number of times that voter is critical in a winning coalition divided by the total number of instances of critical voters. So the Banzhaf index of each of the three voters in the example is 2/6 = 1/3 or about 33.33%. This indicates that they, indeed, have equal voting power.

The *Shapley–Shubik power index* is another measure of voting power. If voters cast their votes in order, then one specific voter, the *swing voter*, may cast a vote that lifts the total to the quota. The Shapley–Shubik power index for a voter is the fraction of all possible orderings in which that voter is the swing. For the preceding example, we list the possible orderings of the three voters and the swing voter.

Voting order			Swing
A	B	C	B
A	C	B	C
B	A	C	A
B	C	A	C
C	A	B	A
C	B	A	B

Each voter is the swing voter in 2 of the 6 orderings, so the Shapley–Shubik index of each is 2/6 = 1/3 or about 33.33%. This indicates once again that the three voters have equal power.

The Banzhaf and Shapley–Shubik indices are not always equal. But each can point out bias or complexity in a system.

Voting systems: How do we choose a winner?

There are several ways in which a winner is selected in elections that involve more than two candidates, and there may be different winners depending on the voting system used. Among these systems are *plurality* voting, the *top-two runoff*, the *elimination runoff*, and the *Borda count*.

Suppose, for example, that the ranked ballot total for a certain election is

Rank	28 votes	25 votes	24 votes	23 votes
First choice	A	B	C	D
Second choice	D	C	D	C
Third choice	B	D	B	B
Fourth choice	C	A	A	A

For this ballot total, A wins the plurality, B wins the top-two runoff, C wins the elimination runoff, and D wins the Borda count. Each of the four candidates has a reasonable claim to be the "people's choice."

A *Condorcet* winner of an election is a candidate who would beat all others in a head-to-head election. Some elections have a Condorcet winner, and some don't. Even if there is a Condorcet winner, none of the systems described above can guarantee that the Condorcet winner will win the election.

If the voters prefer one candidate over another, the introduction of a third candidate should not reverse the result. That is the principle of *independence of irrelevant alternatives. Arrow's impossibility theorem* tells us that no voting system (other than a dictatorship) that always selects the Condorcet winner can have this property. One interpretation of this result is that no voting system is perfect, and we should be aware that the voting system may play as large a role as the vote total in selecting the winner.

Fair division: What is a fair share?

The problem of fair division arises in many settings, including divorces and inheritances. The simplest kind of fair division scheme is the *divide-and-choose* procedure for two people. One participant divides and the other chooses the share he or she likes best. Fair division does not necessarily mean equal division.

When several items are to be divided between two parties, the *adjusted winner procedure* may be used. In this scheme, each party assigns a value to each item so that the total value for each party is 100 points. The procedure uses the point distribution to arrive at a fair division. The adjusted winner procedure has some advantages:

1. Both parties receive the same point value.

2. Neither party would be happier if the division were reversed.

3. No other division would make one party better off without making the other worse off.

One difficulty with this scheme is that it may require sharing of items that are not easily divisible.

A division scheme that can be used for multiple parties is the *Knaster procedure*. It is normally applied to one item at a time. In this procedure, each party bids on the item. The highest bidder gets the item but must then contribute cash to be divided among the other parties. The division of the cash takes into account each party's valuation. One advantage of this procedure is that each party gets at least a fixed fraction of its valuation of each item. For example, if there are five parties, then each party gets at least one-fifth of its valuation of each item. On the other hand, this procedure may require each party to have relatively large amounts of cash on hand.

In general, fair division is a complex issue with no simple solutions that work in all cases.

Apportionment: Am I represented?

The problem of *apportionment* is concerned with the allocation of representatives for a governing body among several groups. The main focus of this section is on historical approaches to apportionment of seats in the U.S. House of Representatives among the 50 states.

Congress sets the size of the House of Representatives. To find the ideal size of a congressional district, we divide the U.S. population by the size of the House. Then we divide the population of each state by the size of a congressional district to determine that state's proper share of the House, its *quota*. In general, the quota will not be a whole number, and the varying approaches to apportionment differ in the way they treat the fractional parts.

Hamilton's method initially gives each state the number of representatives corresponding to the whole number part of its quota. (In practice, this number is never allowed to be 0.) If the sum of the resulting quotas is less than the House size, the leftover seats are distributed, one per state, in order from the highest fractional part down.

For *Jefferson's method* of apportionment, we round each quota down (but not less than 1). If the sum of the initial rounded quotas is larger than the size of the House, we increase the divisor, recalculate the quotas, round, and check the sum again. If the sum of the initial rounded quotas is too small, we decrease the divisor and repeat those steps. We continue this process until a divisor is found for which the sum of the rounded-down quotas (but not less than 1) is the House size.

Other methods of apportionment are similar to Jefferson's in that they are *adjusted divisor methods*. They differ only in the way rounding is done. *Adams's method* rounds the quotas up, *Webster's method* rounds to the nearest whole number, and the *Huntington–Hill method* rounds in terms of the geometric mean. Currently, Congress uses the Huntington–Hill method for apportionment.

There is no perfect method of apportionment. Hamilton's method suffers from the *Alabama, population,* and *new states* paradoxes. The final apportionment for a state should be the quota either rounded down or rounded up, thus *staying within quota*. But all adjusted divisor methods fail this test: They *violate quota*.

KEY TERMS

quota, p. 480
winning coalition, p. 480
losing coalition, p. 480
critical voter, p. 481
Banzhaf power index, p. 482
swing voter, p. 485
Shapley–Shubik power index, p. 486
plurality, p. 494
spoiler, p. 494
top-two runoff system, p. 497

elimination runoff system, p. 497
preferential voting systems, p. 497
ranked ballot, p. 497
Borda winner, p. 498
Condorcet winner, p. 503
Condorcet winner criterion, p. 505
independence of irrelevant alternatives, p. 505
divide-and-choose procedure, p. 510

adjusted winner procedure, p. 511
Knaster procedure (or method of sealed bids), p. 516
ideal district size, p. 524
standard divisor, p. 524
state's quota, p. 524
stay within quota, p. 530
violate quota, p. 530

CHAPTER QUIZ

1. Suppose there are three delegates to a small convention. Alfred has 10 votes, Betty has 7 votes, and Gabby has 3 votes. A simple majority wins. Find the quota, make a table listing all of the coalitions and the critical voter in each case, and find the Banzhaf index for each voter.

Answer Quota: 11 votes. We use A for Alfred, B for Betty, and G for Gabby.

Number of votes			Total votes	Winning coalition?	Critical voters
10	7	3			
A	B	G	20	Yes	A
A	B		17	Yes	A, B
A		G	13	Yes	A, G
A			10	No	Not applicable
	B	G	10	No	Not applicable
	B		7	No	Not applicable
		G	3	No	Not applicable

Alfred: 3/5; Betty: 1/5; Gabby: 1/5.

If you had difficulty with this problem, see Example 8.2.

2. There are three voters: A with 10 votes, B with 8 votes, and C with 3 votes. A simple majority wins. Find the quota, make a table listing all of the permutations of the voters and the swing voter in each case, and calculate the Shapley–Shubik index for each voter.

Answer Quota: 11 votes.

Order of voters			Swing
A (10)	B (8)	C (3)	B
A (10)	C (3)	B (8)	C
B (8)	A (10)	C (3)	A
B (8)	C (3)	A (10)	C
C (3)	A (10)	B (8)	A
C (3)	B (8)	A (10)	B

A: 1/3; B: 1/3; C: 1/3.

If you had difficulty with this problem, see Example 8.5.

3. Consider the following ranked ballot outcome for 14 voters choosing among three candidates: Alfred, Betty, and Gabby.

Rank	7 voters	4 voters	3 voters
First choice	Alfred	Betty	Gabby
Second choice	Betty	Gabby	Alfred
Third choice	Gabby	Alfred	Betty

Determine the winner under the elimination runoff system and the winner using the Borda count.

Answer Elimination runoff: Alfred; Borda count: Alfred.

If you had difficulty with this problem, see Example 8.9.

4. In an election, there are 15 voters and three candidates, say, Abe, Ben, and Condi. The voters' preferences are shown in the accompanying table.

Preferences	7 voters	5 voters	3 voters
First choice	Abe	Ben	Condi
Second choice	Ben	Condi	Abe
Third choice	Condi	Abe	Ben

Is there a Condorcet winner? If so, which candidate is it?

Answer No.

If you had difficulty with this problem, see Example 8.13.

5. Use the adjusted winner method to divide the listed items fairly when the following bids are made:

Michael	Item	Juanita
40	Lakeside cottage	30
30	Penthouse apartment	25
30	Ferrari	45

Answer Michael gets the cottage and 55% of the apartment. Juanita gets the Ferrari and 45% of the apartment.

If you had difficulty with this problem, see Example 8.15.

6. Ann, Beth, Cate, and Dan wish to divide a computer. Use the Knaster procedure to divide fairly the computer with the following bids:

Person	Ann	Beth	Cate	Dan
Bid	$1200	$2000	$1200	$1600

Answer Ann: $425; Beth: computer less $1375; Cate: $425; Dan: $525.

If you had difficulty with this problem, see Example 8.17.

7. Our city has four districts: North, East, West, and South. There are required to be 15 city councilors allocated according to population, with at least one from each district. Given the populations below, calculate the quota for each district. (You need keep only one place beyond the decimal point.) Then use Hamilton's method to determine the number of councilors to represent each district.

District	Population
North	4200
East	13,500
West	7600
South	9200
Total	34,500

Answer

District	Quota	Quota rounded down: Initial seats	Fractional part of quota	Added seats	Final seats
North	1.8	1	0.8	1	2
East	5.9	5	0.9	1	6
West	3.3	3	0.3		3
South	4.0	4	0.0		4
Total	15	13		2	15

North should get 2 councilors, East should get 6, West should get 3, and South should get 4.

If you had difficulty with this problem, see Example 8.18.

8. Our city has four districts: North, East, West, and South. There are required to be 15 city councilors allocated according to population, with at least one from each district. Given the populations below, use Jefferson's method to determine the number of councilors to represent each district.

(You need keep only one place beyond the decimal point in each quotient.)

District	Population
North	4200
East	13,500
West	7600
South	9200
Total	34,500

Answer

District	Quota using divisor 2300	First try: Quota rounded down	Second try: Decrease divisor to 2000, round down
North	1.8	1	2
East	5.9	5	6
West	3.3	3	3
South	4.0	4	4
Total	15	13	15

North should get 2 councilors, East should get 6, West should get 3, and South should get 4.

If you had difficulty with this problem, see Example 8.19.

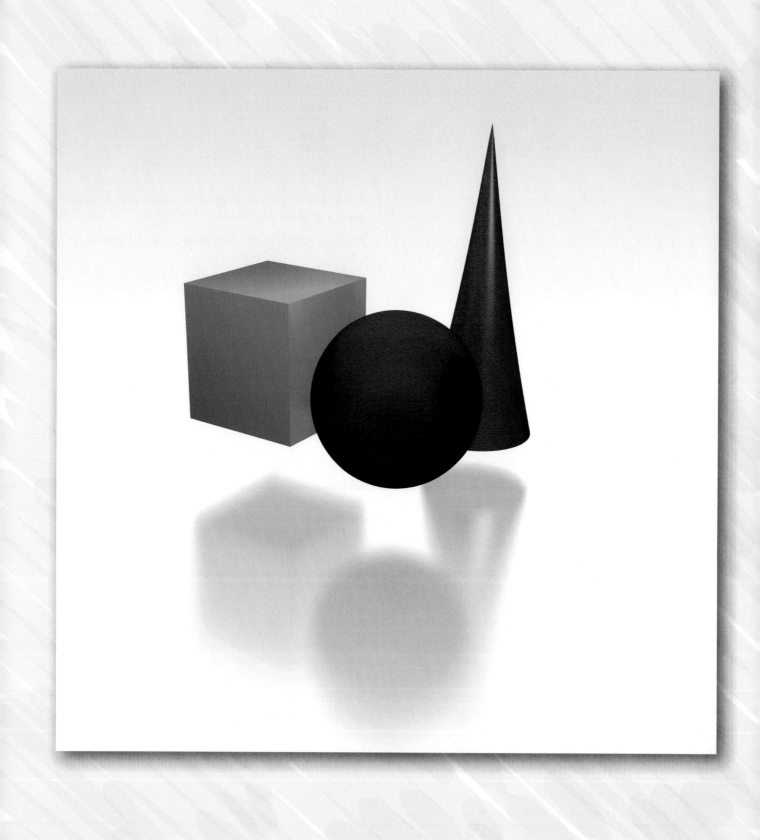

GEOMETRY

Y ou might be surprised by the following news report, which appears on the Web site of MSNBC. It indicates that all of us may have an innate knowledge of geometry, even though we are not conscious of it.

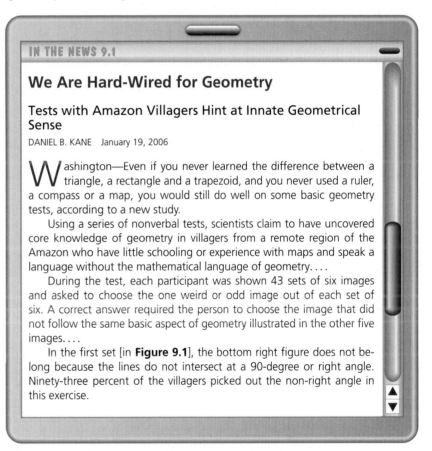

IN THE NEWS 9.1

We Are Hard-Wired for Geometry

Tests with Amazon Villagers Hint at Innate Geometrical Sense

DANIEL B. KANE   January 19, 2006

W ashington—Even if you never learned the difference between a triangle, a rectangle and a trapezoid, and you never used a ruler, a compass or a map, you would still do well on some basic geometry tests, according to a new study.

Using a series of nonverbal tests, scientists claim to have uncovered core knowledge of geometry in villagers from a remote region of the Amazon who have little schooling or experience with maps and speak a language without the mathematical language of geometry....

During the test, each participant was shown 43 sets of six images and asked to choose the one weird or odd image out of each set of six. A correct answer required the person to choose the image that did not follow the same basic aspect of geometry illustrated in the other five images....

In the first set [in **Figure 9.1**], the bottom right figure does not belong because the lines do not intersect at a 90-degree or right angle. Ninety-three percent of the villagers picked out the non-right angle in this exercise.

In this chapter, we examine topics from geometry that are encountered in everyday life. In the first section, we study the measurement of geometric figures. In the next section, we examine how the notion of geometric similarity is related to the

FIGURE 9.1 Which figure is the odd one?

idea of proportionality. In the last section, we consider symmetries of shapes and the notion of tiling the plane.

9.1 Perimeter, area, and volume: How do I measure?

The following excerpt is from the *New York Times*. It highlights the use of geometry in fashion.

IN THE NEWS 9.2

Designers Outline a New Geometry

SUZY MENKES   March 5, 2010

Paris—New geometry—linear and circular cutting—rules the runways for the fall/winter Paris season. From the Issey Miyake math lesson through the Rick Owens asymmetrics to Vionnet's folded squares, sharp cutting and soft folds are the power pieces.

The interlacing black nets at the Issey Miyake show on Friday might have been a set for a sports event—but the designer Dai Fujiwara had a more complex answer. The structure, like the collection's colorful knits, were designed according to the geometric theories of William Thurston, the U.S. mathematician who took a bow with the designer at the end of the show. . . .

The compass may have played a bigger role than the ruler in this show, for everything seemed to work in the round, even a black leather coat with whorls of stitching and hose covered in a pattern of watery globules.

Citing everything from the Milky Way to a doughnut, Mr. Fujiwara made straight angle tailoring credible, as well as the rounded version of the wind coats that Mr. Miyake invented three decades ago. And you did not need a top grade in math to understand the fundamentals of this thought-provoking Issey Miyake show: clean geometric lines with imaginative embellishment, like the squares of organza appliqued with stars for a celestial moment.

The geometric shapes mentioned in this article are our concern in this section. Here, we see how to apply in practical settings some basic formulas for measuring figures in the plane and three-dimensional objects.

Finding the area: Reminders about circles, rectangles, and triangles

For most of us geometry begins with some basic formulas that have applications in everyday life.

Circles   A circle is a figure where all points are a fixed distance, the *radius*, from a fixed point, the *center*. For a circle of radius r (see **Figure 9.2**):

$$\text{Area} = \pi r^2$$

$$\text{Circumference} = 2\pi r.$$

Recall that the number π is approximately 3.14159.

The formula for the circumference of a circle can be rewritten as

$$\frac{\text{Circumference}}{\text{Diameter}} = \pi.$$

This tells us that the ratio of circumference to diameter is always the same no matter which circle you study. The ratio is the same for the equator of Earth and the equator of a baseball.

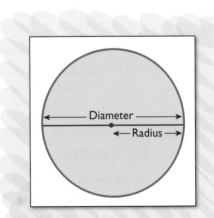

FIGURE 9.2 Radius and diameter of a circle.

Circumference/Diameter is the same for Earth and a baseball.

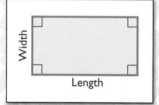

FIGURE 9.3 Opposite sides of a rectangle are of equal length.

Rectangles　A rectangle has four right angles, and opposite sides are equal. (See **Figure 9.3**.) Here are two basic formulas:

$$\text{Area} = \text{Length} \times \text{Width}$$
$$\text{Perimeter} = 2 \times \text{Length} + 2 \times \text{Width}.$$

Triangles　A triangle is a figure with three sides. To find the area of a triangle, we select any one of the three sides and label it the *base*. Then we find the *height* by starting at the vertex opposite the base and drawing a line segment that meets the base in a right angle. In **Figure 9.4**, we have chosen the base to be *AB*, and the resulting height is "inside" the triangle. In **Figure 9.5**, we have chosen the base to be the side *AC*, which results in a height "outside" the triangle. No matter which side we choose for the base, the area is always given by the formula

$$\text{Area} = \frac{1}{2} \text{ Base} \times \text{Height}.$$

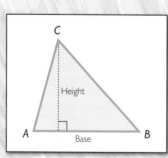

FIGURE 9.4 Base with height inside the triangle.

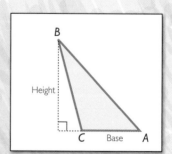

FIGURE 9.5 Different base with height outside the triangle.

Key Concept

The **perimeter** of a geometric figure is the distance around it. In the case of a circle, the perimeter is referred to as its **circumference** and equals 2π times the radius. In the case of a rectangle, the perimeter is the sum of the lengths of its four sides. In the case of a triangle, the perimeter is the sum of the lengths of its three sides.

The **area** of a geometric figure measures the region enclosed by the figure. In the case of a circle, the area is π times the radius squared. In the case of a rectangle, the area is the product of the length and the width. In the case of a triangle, the area is one-half the product of the length of the base times the length of the height.

Right triangles   A *right triangle* is a triangle with one 90-degree (or right) angle. The traditional names for sides of a right triangle are illustrated in **Figure 9.6**. One of the most familiar of all facts in geometry is the famous theorem of Pythagoras.

Pythagorean theorem   For the right triangle shown in Figure 9.6,

$$a^2 + b^2 = c^2.$$

In this formula, c always represents the *hypotenuse*, which is the side opposite the right angle. The other two sides of a right triangle are the *legs*. The converse of the Pythagorean theorem is true: If a triangle has three sides of lengths a, b, and c such that $a^2 + b^2 = c^2$, then that triangle is a right triangle with hypotenuse c.

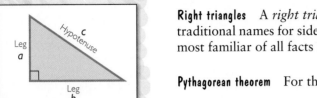

FIGURE 9.6 The sides of a right triangle.

Key Concept

The **Pythagorean theorem** states that for a right triangle, the square of the length of the hypotenuse c equals the sum of the squares of the lengths of the two legs, a and b, so $a^2 + b^2 = c^2$.

Applications of basic geometry formulas

Now we see some ways these formulas are used in practice.

EXAMPLE 9.1 Using circumference and diameter: Runners

Suppose two runners A and B are side-by-side, 2 feet apart, on a circular track, as seen in **Figure 9.7**. If they both run one lap, staying in their lanes, how much farther did the outside runner B go than the inside runner A? (Note that no information was given about the diameter of the track.)

FIGURE 9.7 Two running tracks.

SOLUTION

Let r denote the radius of the inside lane, where A is running. Then $r + 2$ is the radius of the outside lane, where B is running. The distance covered by runner A is

$$\text{Length of inside track} = 2\pi r.$$

The distanced covered by B is

$$\text{Length of outside track} = 2\pi (r + 2)$$
$$= 2\pi r + 2\pi \times 2$$
$$= \text{Length of inside track} + 4\pi.$$

Therefore, the distances covered by the two runners differ by 4π or about 12.6 feet. At first glance, it may appear that a longer inside track would cause a greater difference in the distance the runners travel. But the difference is the same, 12.6 feet, whether the inside track is 100 yards in diameter or 100 miles in diameter.

TRY IT YOURSELF 9.1

Suppose the runners are side-by-side, 3 feet apart. How much farther does the outside runner go?

The answer is provided at the end of this section.

Next we use the formula for the area of a circle.

EXAMPLE 9.2 Using area and diameter: Pizza

At a local Italian restaurant a 16-inch-diameter pizza costs \$15 and a 12-inch-diameter pizza costs \$10. Which one is the better value (i.e., costs less per square inch)?

SOLUTION

First we find the area of each pizza in terms of its radius r using the formula

$$\text{Area} = \pi r^2.$$

The radius of the 16-inch-diameter pizza is $r = 8$ inches. Hence, its area is

$$\text{Area of 16-inch-diameter pizza} = \pi \times (8 \text{ inches})^2 = 64\pi \text{ square inches}$$

(which is about 201 square inches). The radius of the 12-inch-diameter pizza is $r = 6$ inches, so its area is

$$\text{Area of 12-inch-diameter pizza} = \pi \times (6 \text{ inches})^2 = 36\pi \text{ square inches}$$

(which is about 113 square inches).

The larger pizza costs[1]

$$\frac{\$15}{64\pi \text{ square inches}}$$

or about $0.075 per square inch, and the smaller pizza costs

$$\frac{\$10}{36\pi \text{ square inches}}$$

or about $0.088 per square inch. Thus, the larger pizza is the better value.

TRY IT YOURSELF 9.2

Pizza from this restaurant comes with a thin cheese ring around its perimeter. How long are the cheese rings of the two pizzas in the example above?

The answer is provided at the end of this section.

Now we apply the Pythagorean theorem.

EXAMPLE 9.3 Applying the Pythagorean theorem: Slap shot

Hockey player A is located 5 vertical yards and 7 horizontal yards from the goal, and hockey player B is 8 vertical yards and 3 horizontal yards from the goal. (See **Figure 9.8**.) Which player would have the longer shot at the goal?

SOLUTION

The distance for player A is marked a in the figure. It represents the hypotenuse of a right triangle. Similarly, b is the distance for player B, and it represents the hypotenuse

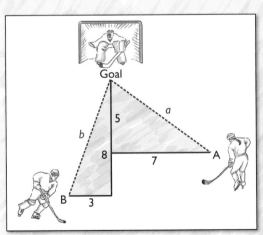

FIGURE 9.8 Two hockey players.

---

[1]On your calculator, enter this as $15/(64\pi)$.

of a right triangle. We use the Pythagorean theorem:

$$a^2 = 5^2 + 7^2 = 74$$

$$a = \sqrt{74} \text{ or about 8.6 yards.}$$

$$b^2 = 8^2 + 3^2 = 73$$

$$b = \sqrt{73} \text{ or about 8.5 yards.}$$

Player A has the longer shot.

TRY IT YOURSELF 9.3

Which player has the longer shot if they are positioned as in **Figure 9.9**?

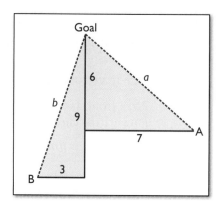

FIGURE 9.9 Positions for Try it Yourself 9.3.

The answer is provided at the end of this section.

Heron's formula

We know how to find the area of a triangle if we know the lengths of the base and height. Heron's formula gives the area of a triangle if we know the lengths of all three sides. Referring to the triangle shown in **Figure 9.10** with sides a, b, and c, we calculate one-half of the perimeter, the *semi-perimeter*, $S = \frac{1}{2}(a + b + c)$. Herons formula says that the area is given by

$$\text{Area} = \sqrt{S(S - a)(S - b)(S - c)}.$$

For the triangle in **Figure 9.11**, which has sides 5, 6, and 9, we find that $S = \frac{1}{2}(5 + 6 + 9) = 10$. Thus, the area is

$$\text{Area} = \sqrt{10(10 - 5)(10 - 6)(10 - 9)} = \sqrt{200}$$

or about 14.1 square units.

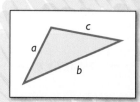

FIGURE 9.10 A triangle with given sides.

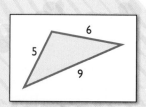

FIGURE 9.11 Applying Heron's formula.

EXAMPLE 9.4 Applying geometric formulas: Getting ready for spring

My lawn has the shape shown in **Figure 9.12**, where the dimensions are in feet.

a. I want to fence my lawn this spring. How many feet of fencing will be required?

b. I want to fertilize my lawn with bags of fertilizer that cover 500 square feet each. How many bags of fertilizer do I need to buy?

SOLUTION

a. The lawn has an irregular shape, but by adding the line segment shown in **Figure 9.13**, we can divide it into three pieces: a right triangle, a rectangle, and an oblique triangle (a triangle with no right angle).

The length of fence we need is the perimeter. To find the perimeter, we need to find the hypotenuse (marked H in Figure 9.13) of the right triangle. The legs

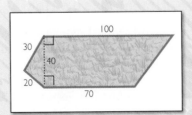

FIGURE 9.12 My lawn.

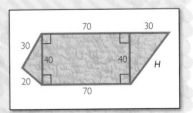

FIGURE 9.13 Dividing into a rectangle and two triangles.

have length 30 feet and 40 feet. We find H using the Pythagorean theorem:

$$c^2 = a^2 + b^2$$
$$H^2 = 30^2 + 40^2$$
$$H^2 = 2500$$
$$H = \sqrt{2500}$$
$$H = 50 \text{ feet.}$$

Now we calculate the perimeter of the lawn as

$$\text{Perimeter} = 30 + 20 + 70 + 50 + 30 + 70 = 270 \text{ feet.}$$

b. Look again at Figure 9.13. The area of the lawn is the area of the rectangle plus the area of the right triangle plus the area of the oblique triangle. For the rectangle, we find

$$\text{Area of rectangle} = \text{Length} \times \text{Width}$$
$$= 70 \times 40$$
$$= 2800 \text{ square feet.}$$

For the right triangle, we find

$$\text{Area of right triangle} = \frac{1}{2} \text{ Base} \times \text{Height}$$
$$= \frac{1}{2} \times 30 \times 40$$
$$= 600 \text{ square feet.}$$

We use Heron's formula to calculate the area of the oblique triangle. The semi-perimeter is $\frac{1}{2}(40 + 30 + 20) = 45$, so the area is

$$\text{Area of oblique triangle} = \sqrt{S(S - a)(S - b)(S - c)}$$
$$= \sqrt{45(45 - 40)(45 - 30)(45 - 20)}$$
$$= \sqrt{84{,}375}.$$

This is about 290 square feet.

Thus, the total area of the lawn is $2800 + 600 + 290 = 3690$ square feet. Each bag of fertilizer covers 500 square feet. Dividing 3690 by 500 gives about 7.4. Therefore, we will need to buy eight bags of fertilizer.

Three-dimensional objects

So far we have discussed geometric objects lying in a plane, but many objects that we want to measure are not flat. Now we turn to the study of three-dimensional objects. First we consider volume.

The simplest volume to compute is that of a box. Labeling the dimensions of a box as *length, width,* and *height* as shown in **Figure 9.14**, we calculate the volume using

$$\text{Volume of a box} = \text{Length} \times \text{Width} \times \text{Height.}$$

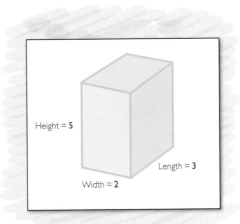

FIGURE 9.14 Dimensions of a box.

For example, the volume of the box in Figure 9.14 is

$$2 \text{ units} \times 3 \text{ units} \times 5 \text{ units} = 30 \text{ cubic units.}$$

Note that the volume of a box can also be expressed as the area of the base times the height:

$$\text{Volume} = \text{Area of base} \times \text{Height.}$$

This formula also holds for any three-dimensional object with uniform cross sections, such as a cylinder. A cylinder has uniform circular cross sections, and a box has uniform rectangular cross sections.

For the cylinder shown in **Figure 9.15**, the radius is r, so the area of the base is πr^2. Using h for the height, we obtain

$$\text{Volume} = \text{Area of base} \times \text{Height} = \pi r^2 h.$$

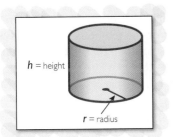

FIGURE 9.15 A cylinder.

EXAMPLE 9.5 Calculating the volume of a cylinder: A swimming pool
What is the volume of a cylindrical wading pool that is 6 feet across and 15 inches high? Report your answer to the nearest cubic inch.

SOLUTION
Because the pool is 6 feet across, the diameter of the circular base is 6 feet. The diameter is given in feet and the height is given in inches, so we need to convert units for one of them. It doesn't matter which one we choose. Let's change the diameter of

6 feet to 72 inches. Then the radius is $r = 36$ inches, and the height is $h = 15$ inches. The volume of the pool is

$$\text{Volume} = \text{Area of base} \times \text{Height}$$
$$= \pi r^2 h$$
$$= \pi (36 \text{ inches})^2 \times 15 \text{ inches}.$$

This is about 61,073 cubic inches.

TRY IT YOURSELF 9.5

Suppose you want to fill the same pool to a depth of 12 inches and your hose puts out 475 cubic inches per minute (about 2 gallons per minute). How long will it take you to fill the pool? Report your answer to the nearest minute.

The answer is provided at the end of this section.

Now we apply the formula for the volume of a box.

EXAMPLE 9.6 Calculating the volume of a box: Cake batter

Cake batter rises when baked. The best results for baking cakes occur when the batter fills the pan to no more than two-thirds of the height of the pan. Suppose we have 7 cups of batter, which is about 101 cubic inches. We have a pan that is 2 inches high and has a square bottom that is 9 inches by 9 inches. Is this pan large enough?

SOLUTION

This pan forms the shape of a box, and we want to find the volume based on two-thirds of the full height of 2 inches. Thus we use a height of

$$\text{Height} = \frac{2}{3} \times 2 = \frac{4}{3} \text{ inches.}$$

We find

$$\text{Volume} = \text{Length} \times \text{Width} \times \text{Height}$$
$$= 9 \text{ inches} \times 9 \text{ inches} \times \frac{4}{3} \text{ inches}$$
$$= 108 \text{ cubic inches.}$$

This pan will easily hold batter with a volume of 101 cubic inches.

TRY IT YOURSELF 9.6

We have a round pan that is 10 inches across and 2 inches high. To bake a cake, we need to fill it to only two-thirds of the height. Is this pan large enough for baking 101 cubic inches of batter?

The answer is provided at the end of this section.

The volumes of other geometric objects are addressed in the exercises.

Now we consider surface area. To find the surface area of a cylinder of radius r and height h (excluding the top and bottom), think of the cylinder as a can that we split lengthwise and roll out flat. See **Figure 9.16**. This gives a rectangle with width h and length equal to the circumference of the circular base. The circumference of the base is $2\pi r$, so the surface area of the cylinder (excluding the top and bottom) is

$$\text{Surface area of cylinder} = 2\pi rh.$$

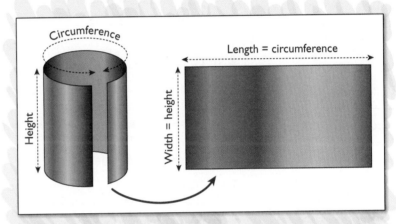

FIGURE 9.16 Splitting the cylindrical side of a can and rolling it out flat.

EXAMPLE 9.7 Calculating volume and surface area: A can

A tin can has a radius of 1 inch and a height of 6 inches.

   a. How much liquid will the can hold? Round your answer in cubic inches to one decimal place.

   b. How much metal is needed to make the can? Round your answer in square inches to one decimal place.

SOLUTION

   a. The base is a circle of radius 1, so its area is

$$\text{Area of base} = \pi \times 1^2 = \pi \text{ square inches.}$$

The height is 6 inches, so the volume is

$$\text{Volume} = \text{Area of base} \times \text{Height} = \pi \times 6 \text{ cubic inches.}$$

This is about 18.8 cubic inches.

   b. The metal needed to make the can consists of the top, bottom, and cylindrical side. We already found that the base has area π square inches, so

$$\text{Area of top and bottom} = 2\pi \text{ square inches.}$$

To find the area of the cylindrical side, we use the formula $2\pi rh$ for the surface area of a cylinder:

$$\text{Area of side} = 2\pi \times 1 \times 6 = 12\pi \text{ square inches.}$$

The total area includes the top and bottom of the can, a total area of $2\pi + 12\pi = 14\pi$ or about 44.0 square inches.

TRY IT YOURSELF 9.7

A can of tuna fish has a diameter of $3\frac{1}{4}$ inches and a height of $1\frac{3}{8}$ inches. Find its volume. (Round your answer in cubic inches to one decimal place.) How much metal is needed to make the can? (Round your answer in square inches to one decimal place.)

The answer is provided at the end of this section.

SUMMARY 9.1   **Volumes and Surface Areas**

- The volume of a box is

$$\text{Volume of a box} = \text{Length} \times \text{Width} \times \text{Height.}$$

- The volume of a cylinder is

$$\text{Area of base} \times \text{Height.}$$

If the cylinder has radius r and height h, this equals $\pi r^2 h$.

- The surface area of a cylinder (excluding the top and bottom) of radius r and height h is $2\pi r h$.

Try It Yourself answers

Try It Yourself 9.1: Using circumference and diameter: Runners   The difference in the distances is 6π or about 18.8 feet.

Try It Yourself 9.2: Using area and diameter: Pizza   The perimeter of the 12-inch pizza is 37.7 inches. The perimeter of the 16-inch pizza is 50.3 inches.

Try It Yourself 9.3: Applying the Pythagorean theorem: Slap shot   Player B has the longer shot.

Try It Yourself 9.5: Calculating the volume of a cylinder: A swimming pool 103 minutes, or 1 hour and 43 minutes.

Try It Yourself 9.6: Calculating the volume of a box: Cake batter   Multiplying the height by 2/3 gives a volume of about 105 cubic inches, so the pan is large enough.

Try It Yourself 9.7: Calculating volume and surface area: A can   The volume is 11.4 cubic inches; the amount of metal needed is 30.6 square inches.

Exercise Set 9.1

1. Pool fence. A circular swimming pool is 10 feet across and is enclosed by a fence. How long is the fence?

2. Pool cover. A circular swimming pool is 15 feet across and has a cover. How much material is in the pool cover?

3. Finding perimeter. Find the perimeter of the figure in **Figure 9.17**.

4. Finding the area. Find the area of a right triangle whose legs have lengths 8 and 15.

5. Finding the hypotenuse. Find the length of the hypotenuse of a right triangle whose legs have lengths 3 and 7.

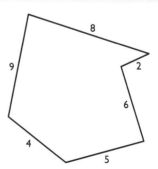

FIGURE 9.17

6. Finding a leg. The hypotenuse of a right triangle has length 7, and a leg has length 3. Find the length of the other leg.

7. A basketball problem. In basketball, free-throws are taken from the free-throw line, which is 13 feet from a point directly below the basket. The basket is 10 feet above the floor. These dimensions are shown in **Figure 9.18**. How far is it from where the free-throw shooter stands at the free-throw line to the basket?

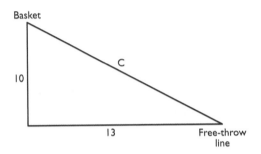

FIGURE 9.18

8. Distance between towns. Town B is 5 miles due north of town A, and town C is due east of town A. It is 9 miles from town B to town C. The distance from town A to town C is marked *b* in **Figure 9.19**. Find *b*, the distance from town A to town C.

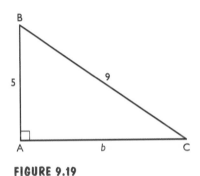

FIGURE 9.19

9. Area of a right triangle. Verify that the triangle with sides 3, 4, and 5 shown in **Figure 9.20** is a right triangle, and find its area. This is the *3-4-5 right triangle*.

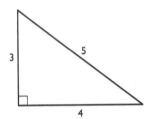

FIGURE 9.20 The 3-4-5 right triangle.

10. Whole number sides. Exercise 9 introduced the 3-4-5 right triangle. Can you find other right triangles with whole number sides? *Hint*: You are looking for whole numbers *a*, *b*, and *c* so that $a^2 + b^2 = c^2$.

11. Finding the area. The hypotenuse of a right triangle has length 11, and a leg has length 4. Find the area of the triangle.

Heron's formula Exercises 12 through 15 use Heron's formula.

12. Find the area of a triangle with sides 7, 8, and 9.

13. A garden. Find the area of a garden that is a triangle with sides 4 feet, 5 feet, and 7 feet.

14. Find the area of a triangle with sides 6 yards, 6 yards, and 7 yards.

15. One triangular plot has sides 4 yards, 5 yards, and 6 yards. Another has sides 3 yards, 6 yards, and 7 yards. Which plot encloses the larger area?

16. Making drapes. A window in your home has the shape shown in **Figure 9.21**. The top is a semi-circle (half of a circle). In order to buy material for drapes, you need to know the area of the window. What is it?

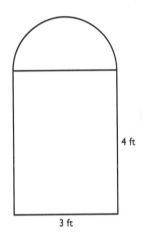

FIGURE 9.21 A window.

17. Trimming a window. *This is a continuation of Exercise 16.* How many feet of molding do we need in order to make a border for the window in Figure 9.21?

18. Carpet for a ramp. A ramp is 3 feet wide. It begins at the floor and rises by 1 foot over a horizontal distance of 10 feet. (See **Figure 9.22**.) How much carpet is needed to cover the (top of the) ramp?

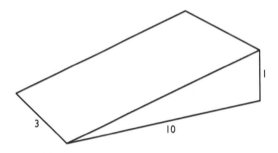

FIGURE 9.22 A ramp.

19. Isosceles triangles. An *isosceles* triangle has two equal sides. In an isosceles triangle, consider the line segment starting at the vertex that touches both of the equal sides and meeting the opposite side in a right angle. It is a fact that this segment cuts the opposite side (the base) in half. Use this fact to find the area and perimeter of the isosceles triangle in **Figure 9.23**.

(The base is 10 units, and the height is 20 units.) *Suggestion:* First use the Pythagorean theorem to find the lengths of the unknown sides of the triangle.

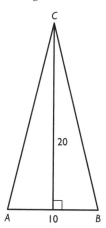

FIGURE 9.23 Isosceles triangle.

Perimeter and area of a triangle In Exercises 20 through 23, use the Pythagorean theorem to find the missing lengths and then find the area and perimeter. (Do not include the height in the perimeter.)

20. Refer to the triangle in **Figure 9.24**.

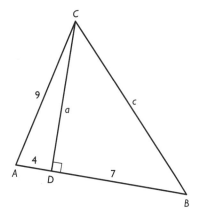

FIGURE 9.24

21. Refer to the triangle in **Figure 9.25**.

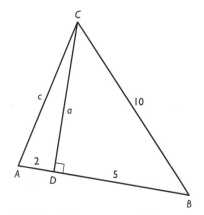

FIGURE 9.25

22. Refer to the triangle in **Figure 9.26**.

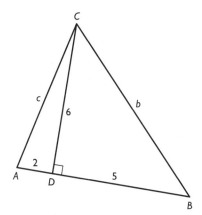

FIGURE 9.26

23. Refer to the triangle in **Figure 9.27**.

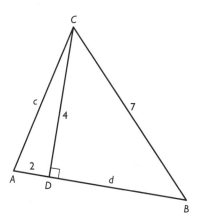

FIGURE 9.27

24. Pitcher foam. A restaurant serves soda pop in cylindrical pitchers that are 5 inches in diameter and 8 inches tall. If the pitcher has a 1-inch head of foam, how much soda is lost as a result?

25. Soda and soup. *This is a continuation of Exercise 24.* A typical soup can is about 2.5 inches in diameter and 3.5 inches tall. Compare the amount of soda lost in Exercise 24 with the volume of a soup can.

A soup can A soup can has a diameter of $2\frac{5}{8}$ inches and a height of $3\frac{3}{4}$ inches. Exercises 26 through 29 refer to this can.

26. Find the volume of the soup can.

27. How many square inches of paper are required to make the label on the soup can?

28. How many square inches of metal are required to make the soup can?

29. When you open the soup can, how far does the can opener travel?

30. Distance from Dallas to Fort Smith. Dallas, Texas, is 298 miles due south of Oklahoma City, Oklahoma. Fort Smith, Arkansas, is 176 miles due east of Oklahoma City. How many miles long is a plane trip from Dallas directly to Fort Smith?

31. Houston to Denver. Houston, Texas, is 652 miles due south of Kansas City, Missouri. Denver, Colorado, is 558 miles due west of Kansas City. How many miles long is a plane trip from Houston directly to Denver? Round your answer to the nearest mile.

32. Big-screen TVs. The size of a television set is usually classified by measuring its diagonal. Suppose that one giant television measures 5 feet long by 3 feet high, and another is a 4.5-foot square.

   a. Which television has the larger diagonal?

   b. Which television has the larger screen in terms of area?

Monitor sizes The size of a computer monitor (also called a display) is usually measured in terms of the diagonal of the rectangle. Exercises 33 through 35 explore the implications of this fact.

33. One monitor. A monitor measures 12.6 inches by 16.8 inches. What is the length of the diagonal?

34. Another monitor. A monitor measures 10.3 inches by 18.3 inches. What is the length of the diagonal?

35. Same size? Calculate the areas of the rectangles described in Exercises 33 and 34. (Round your answers to the nearest whole number.) In light of your answers to those exercises, does measurement of a monitor in terms of the diagonal give complete information about the size?

36. Pizza coupon. At your local pizza place, a large pizza costs $10.99. You have a coupon saying you can buy two medium pizzas for $12.99. A medium pizza has a diameter of 12 inches, and a large pizza has a diameter of 14 inches. Which is the better value (measured in dollars per square inch), one large pizza or two medium pizzas with the coupon?

Taxes and square footage Property taxes on a home are often based on its valuation, determined by the number of square feet of living space. Incorrect measurements of this area can significantly affect your property taxes. Exercises 37 and 38 concern square footage of a home.

37. A home. Suppose the floor plan for a home is shown in Figure 9.28. (The curved part is a half circle.) How many square feet of floor space does this home have?

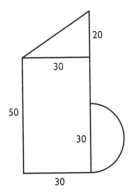

FIGURE 9.28 Floor plan.

38. Another home. The floor plan for a home is shown in Figure 9.29. How many square feet of floor space does this home have?

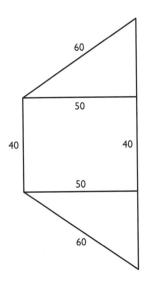

FIGURE 9.29 Floor plan.

Getting ready for spring My lawn has the shape shown in Figure 9.30, where the dimensions are in feet. Exercises 39 through 43 concern my lawn.

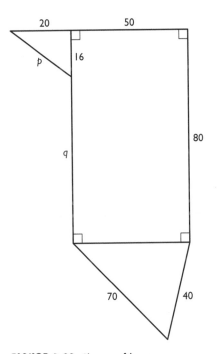

FIGURE 9.30 Shape of lawn.

39. Find p. Find the length marked p.

40. Find q. Recall that opposite sides of a rectangle are equal. Use this fact to find the length marked q.

41. Fencing needed. I want to fence my lawn this spring. Use your answers to Exercises 39 and 40 to determine how many feet of fencing will be required.

42. Area of triangle. Find the area of the triangle at the bottom of the figure. (This is not a right triangle.)

43. Fertilizer needed. I want to fertilize my lawn with bags of fertilizer that cover 500 square feet each. Use your answer to Exercise 42 to determine how many bags of fertilizer I need to buy.

44. Area of Oklahoma. Like many states, Oklahoma has an irregular shape, as shown in **Figure 9.31**. This makes its area difficult to calculate exactly. The following table shows distances between certain towns in Oklahoma.

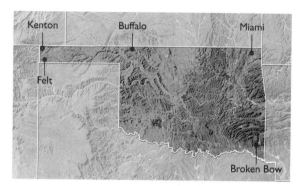

FIGURE 9.31 Map of Oklahoma.

From	To	Miles
Miami	Broken Bow	193
Miami	Buffalo	256
Buffalo	Kenton	187
Kenton	Felt	20

Use the given information to estimate the area of the state of Oklahoma.

45. Perimeter of Oklahoma. *This is a continuation of Exercise 44.* Suppose we wished to set up a string of cameras every 10 miles around the border of Oklahoma to monitor the migration of birds across state lines. Use the information given in Exercise 44 to estimate the number of cameras needed.

46. More on the area of Oklahoma. *This is a continuation of Exercise 44.* The actual area of the state of Oklahoma is 69,960 miles. Explain how you would make a more accurate calculation than was possible in Exercise 44 if you had a detailed map of the state.

47. Area of Nevada. Like many states, Nevada has an irregular shape, as shown in **Figure 9.32**. This makes its area difficult to calculate exactly. The following table shows distances between certain towns in Nevada.

From	To	Miles
Carson City	Ely	258
Carson city	Needles, CA	426
Ely	Jackpot	166
Ely	Needles, CA	339

Use the given information to estimate the area of the state of Nevada.

FIGURE 9.32 Map of Nevada.

48. Perimeter of Nevada. *This is a continuation of Exercise 47.* Suppose we wished to set up a string of cameras every 10 miles around the border of Nevada to monitor the migration of birds across state lines. Use the information given in Exercise 47 to estimate the number of cameras needed.

49. More on area of Nevada. *This is a continuation of Exercise 47.* The actual area of the state of Nevada is 110,567 square miles. Explain how you would make a more accurate calculation than was possible in Exercise 47 if you had a detailed map of the state.

50. Area of your state. Locate a good map of your state and use it to estimate its area.

51. Perimeter of your state. Locate a good map of your state and use it to estimate its perimeter.

52. Cement for a sidewalk. You need to order cement to make a concrete sidewalk 3 feet wide and 100 feet long. Assume that the depth of the concrete is 4 inches. How much cement is needed, in cubic feet? Cement is usually ordered in cubic yards. How many cubic yards of cement do you need?

53. Gravel for a driveway. You need to order gravel for your driveway, which is 8 feet wide and 230 feet long. Assume you would like the gravel to be 6 inches deep. How much gravel is needed, in cubic feet? Gravel is ordered in cubic yards. How many cubic yards of gravel do you need?

54. Water for an in-ground pool. Your in-ground pool is rectangular and has a depth of 5 feet. If the width of the pool is 15 feet and the length is 25 feet, how much water does the pool hold?

Balls The surface area and volume of a ball of radius r are given by

$$\text{Volume} = \frac{4}{3}\pi r^3$$

$$\text{Surface area} = 4\pi r^2.$$

These formulas are used in Exercises 55 through 58.

55. A baseball. The diameter of a baseball is 2.9 inches. What is the volume of a baseball? A baseball is covered in cowhide. How many square inches are needed to make the cover of a baseball?

56. A bigger baseball. *This is a continuation of Exercise 55.* If you had twice the leather needed to cover a baseball, could you cover a ball with twice the radius? *Suggestion:* Calculate the surface area of a ball of radius 2.90 inches.

57. A basketball. The diameter of a standard basketball is 9.39 inches. Find the volume. An NBA basketball is made of "composite leather," a synthetic material designed to withstand slam dunks. How many square inches are required to make a basketball?

58. A bigger basketball. *This is a continuation of Exercise 57.* If you had twice the composite leather needed to cover a basketball, could you cover a ball with twice the radius? *Suggestion:* Calculate the surface area of a ball of radius 9.39 inches.

Cones. To measure a *cone* as shown in **Figure 9.33**, we need to know its height h and the radius r of its top. For a cone of these dimensions, we have

$$\text{Volume} = \frac{1}{3}\pi r^2 h$$

$$\text{Surface area} = \pi r \sqrt{r^2 + h^2}.$$

Note that the surface area does not include the top circle of the cone. Exercises 59 through 62 use these formulas.

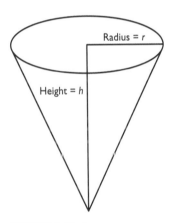

FIGURE 9.33 A cone.

59. A cone for ice cream has a top diameter of 2.25 inches and a height of 4.5 inches. How much ice cream will this cone hold?

60. A (solid) conical spire is made of concrete. The spire has a radius of 3 feet and a height of 10 feet. How much concrete is needed to make the spire?

61. How many square feet of paint are needed to paint a cone with a top diameter of 5 feet and a height of 6 feet? You will paint the entire cone including the top circle.

62. A waffle cone has a top diameter of 2.25 inches and a height of 4.5 inches. What is the area of the dough needed to make the cone?

63. A demonstration of the Pythagorean theorem. This exercise uses a puzzle to demonstrate why the Pythagorean theorem is true. Trace the square shown in **Figure 9.34** onto your own paper. Cut out the square and then cut the square along the three interior lines shown. You should have three pieces of a puzzle. Reassemble the three pieces into the figure shown in **Figure 9.35**. Explain how this demonstrates why the Pythagorean theorem is true.

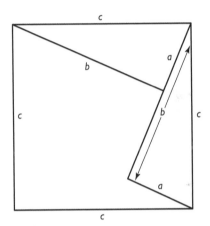

FIGURE 9.34 Trace and cut to make a Pythagorean puzzle.

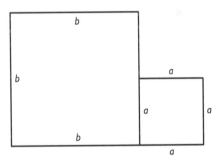

FIGURE 9.35 Reassemble the puzzle to make this picture.

9.2 Proportionality and similarity: Changing the scale

The subject of proportion is of interest not only to mathematicians but also to artists. The mix of the two disciplines can occasion controversy. Consider the following article from *Al-Ahram*, an Egyptian journal. The author is a professor of mathematics at Cairo University and Misr University for Science and Technology.

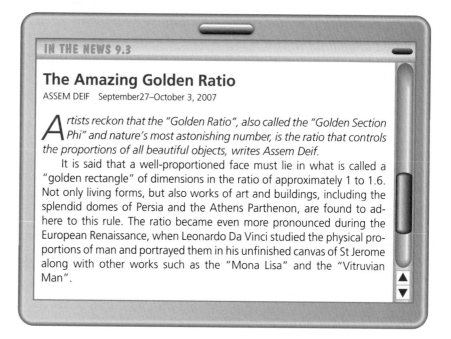

IN THE NEWS 9.3

The Amazing Golden Ratio

ASSEM DEIF   September 27–October 3, 2007

*A*rtists reckon that the "Golden Ratio", also called the "Golden Section Phi" and nature's most astonishing number, is the ratio that controls the proportions of all beautiful objects, writes Assem Deif.

It is said that a well-proportioned face must lie in what is called a "golden rectangle" of dimensions in the ratio of approximately 1 to 1.6. Not only living forms, but also works of art and buildings, including the splendid domes of Persia and the Athens Parthenon, are found to adhere to this rule. The ratio became even more pronounced during the European Renaissance, when Leonardo Da Vinci studied the physical proportions of man and portrayed them in his unfinished canvas of St Jerome along with other works such as the "Mona Lisa" and the "Vitruvian Man".

The reader is encouraged to look at Exercise 52 for a somewhat different point of view.

The opening article is concerned with ratio and proportion in a geometric setting. Proportionality relationships are quite common in many settings, both numerical and geometrical. For example, they arise when we convert from one unit of measurement to another. In this section, our purpose is to examine proportionality and similarity relationships that arise in geometry. In the process we will look more closely at the golden ratio, which is at the heart of much controversy regarding the interplay of art and mathematics.

"I got it on eBay."

Definition and examples

To say that one variable quantity is proportional to another simply means that one is always a constant multiple of the other.

Key Concept

One variable quantity is **(directly) proportional** to another if it is always a fixed (nonzero) constant multiple of the other. Expressed in terms of a formula, this says that the quantity A is proportional to the quantity B if there is a nonzero constant c such that $A = cB$. The number c is the **constant of proportionality.**

Note that if $A = cB$, we can turn the relationship around and write $B = \frac{1}{c}A$, which says that B is proportional to A with constant of proportionality $\frac{1}{c}$.

Another way to write the equation $A = cB$ is $\frac{A}{B} = c$. Expressed in this way, the proportionality formula tells us that two quantities are proportional if their ratio is constant.

For example, if you are driving at a constant speed of 60 miles per hour, then the distance you travel is proportional to the time you spend driving. If distance is measured in miles and time is in hours,

$$\text{Distance} = 60 \times \text{Time}.$$

The constant of proportionality is your speed, 60 miles per hour. Note that distance traveled increases by the same amount, 60 miles, for each extra hour of travel. We can also write the relationship as

$$\frac{\text{Distance}}{\text{Time}} = 60.$$

Note that distance traveled is proportional to time only when your speed is constant. If your speed varies, distance is not proportional to time.

You can probably think of lots of quantities that are not proportional to each other. One example would be your height and your age. Although your height increases as you get older, it is not proportional to your age because you don't grow by the same amount each year.

One of the most important instances of proportionality in geometry is the familiar formula for the circumference of a circle:

$$\text{Circumference} = \pi \times \text{Diameter}.$$

This formula tells us that a circle's circumference is proportional to its diameter and that the number π is the constant of proportionality. Expressed as a ratio, this relationship is

$$\frac{\text{Circumference}}{\text{Diameter}} = \pi.$$

It was pointed out in the preceding section that if we divide the circumference of a very small circle by its diameter, we get exactly the same value, π, as we would if we divided the length of the equator of Earth by its diameter. This is another way of saying that the circumference is proportional to the diameter.

On the other hand, the radius r and area A of a circle are related by

$$A = \pi r^2.$$

Here, A is not a constant multiple of r, so the area is not proportional to the radius. Because the diameter is twice the radius, the area of a circle is not proportional to its diameter either.

EXAMPLE 9.8 Applying proportionality: Forestry

Biologists and foresters use the number of the trees and their diameters as one measure of the condition and age of a forest. Measuring the diameter of a tree directly is not easy, especially for a large tree. On the other hand, the circumference is easy to measure simply by running a tape measure around the tree. Is the diameter of a tree proportional to its circumference? (Assume that the cross section is circular in shape, as in **Figure 9.36**.) If so, what is the constant of proportionality?

FIGURE 9.36 Foresters find the diameter of trees by measuring the circumference.

SOLUTION

We know that circumference is proportional to diameter, and the relationship is

$$\text{Circumference} = \pi \times \text{Diameter}.$$

We can rearrange this formula to find

$$\text{Diameter} = \frac{1}{\pi} \times \text{Circumference}.$$

Thus, the diameter is proportional to the circumference, and the constant of proportionality is $1/\pi$.

TRY IT YOURSELF 9.8

We saw down a tree, and the top of the resulting stump is circular in shape. The area of this circle is the cross-sectional area of the base of the tree. Is the cross-sectional area of the base proportional to the circumference?

The answer is provided at the end of this section.

Properties of proportionality

We can discover an important property of proportionality by further consideration of tree measurement, as in the previous example. Foresters determine the diameter of a tree by measuring its circumference. What happens to the circumference of a tree if the diameter is doubled? We can find out by multiplying both sides of the equation

$$\text{Circumference} = \pi \times \text{Diameter}$$

by 2. The result is

$$2 \times \text{Circumference} = \pi \times (2 \times \text{Diameter}).$$

Thus, doubling the diameter causes the circumference to double as well. Conversely, a tree with twice the circumference of another has twice the diameter. The same is true for any multiple—tripling the diameter causes the circumference to triple, and halving the diameter cuts the circumference in half. It turns out that this fact is true for any proportionality relation, and it is another characterization of proportionality.

SUMMARY 9.2   **Properties of Proportionality**

1. Two variable quantities are proportional when one is a (nonzero) constant multiple of the other. This multiple is the constant of proportionality. Another way to say this is that two quantities are proportional when their ratio is always the same.

2. If a variable quantity A is proportional to another quantity B with proportionality constant c, then B is proportional to A with proportionality constant $1/c$.

3. If one of two proportional quantities is multiplied by a certain factor, the other one is also multiplied by that factor. That is, if y is proportional to x, then multiplying x by k has the result of multiplying y by k. In particular, if one quantity doubles, then the other will double as well.

EXAMPLE 9.9 Comparing volumes using proportionality: Two cans

It is a fact that the volume of a cylindrical can of diameter 3 inches is proportional to the height of the can. One such can is 4 inches high, and another is 12 inches high. How do their volumes compare?

SOLUTION

If one of two proportional quantities is multiplied by a certain factor, the other one is also multiplied by that factor. A change from 4 inches to 12 inches is a tripling of the height. Therefore, the volume of the taller can is three times that of the shorter one.

TRY IT YOURSELF 9.9

Two cans have diameter 3 inches. If one can has a height of 10 inches and the second can has a height of 2 inches, how does the volume of the second can relate to that of the first?

The answer is provided at the end of this section.

Another example involves a square. Recall that the perimeter of a square of side S is given by $P = 4S$. This formula says that the perimeter of a square is proportional to the length of one side, with constant of proportionality 4. Assuming that the cost of a square wooden picture frame is proportional to the length of wood used, we can say that the cost of wood for framing the picture is proportional to the length of a side.

Euclid's Elements.

The golden ratio φ

The article at the beginning of this section refers to a quantity called the "golden ratio." This special number, denoted by the Greek letter ϕ (phi, pronounced "fie"), is approximately 1.62. As one can see from that article, discussion of the golden ratio arises in a variety of contexts, including art and architecture.

Some claim that awareness of the golden ratio can be found in the Egyptian pyramids and in the writings of Pythagoras, but the first known direct mention of the number comes from Euclid (around 300 B.C.), who wrote

> If a straight line is cut in extreme and mean ratio, then as the whole is to the greater segment, the greater segment is the lesser segment.

The great pyramids of Egypt.

This somewhat confusing statement can be restated to give a definition of the golden ratio.

Key Concept

Cut the segment in **Figure 9.37** so that the ratios $\dfrac{\text{Longer}}{\text{Shorter}}$ and $\dfrac{\text{Whole}}{\text{Longer}}$ are equal. This common ratio is the **golden ratio**.

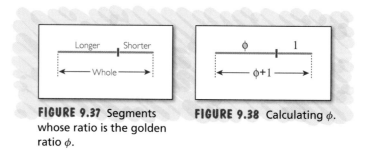

FIGURE 9.37 Segments whose ratio is the golden ratio ϕ.

FIGURE 9.38 Calculating ϕ.

This definition can be used to derive the formula

$$\phi = \frac{1 + \sqrt{5}}{2},$$

which is about 1.62. A derivation is given in Algebraic Spotlight 9.1.

ALGEBRAIC SPOTLIGHT 9.1 Formula for the Golden Ratio

A formula for the golden ratio can be found with a bit of algebra. If we let the shorter length be 1 unit and the longer be ϕ units, as shown in Figure 9.37 and **Figure 9.38**, we find:

$$\frac{\text{Longer}}{\text{Shorter}} = \frac{\text{Whole}}{\text{Longer}}$$

$$\frac{\phi}{1} = \frac{\phi + 1}{\phi}$$

$$\phi^2 = \phi + 1$$

$$\phi^2 - \phi - 1 = 0.$$

We can use the *quadratic formula* from algebra to solve this equation for ϕ. Recall that the quadratic formula says that if $ax^2 + bx + c = 0$, then $x = \dfrac{-b \pm \sqrt{b^2 - 4ac}}{2a}$.

Our equation is $\phi^2 - \phi - 1 = 0$, so in the quadratic formula we put ϕ in place of x and use $a = 1$, $b = -1$, and $c = -1$. The quadratic formula allows us to conclude that

$$\phi = \frac{1 \pm \sqrt{(-1)^2 - 4 \times 1 \times (-1)}}{2 \times 1} = \frac{1 \pm \sqrt{5}}{2}.$$

We discard the negative solution. The positive solution is

$$\phi = \frac{1 + \sqrt{5}}{2},$$

which is about 1.62.

The number ϕ arises in a number of settings, but perhaps its best-known occurrence is in the context of a *golden rectangle*.

Key Concept

A golden rectangle is a rectangle for which the ratio of the length to the width is the golden ratio ϕ. That is, among golden rectangles the length is proportional to the width with constant of proportionality ϕ.

One golden rectangle is shown in **Figure 9.39**.

EXAMPLE 9.10 Constructing a golden rectangle: 3-feet wide

The width (the shorter side) of a rectangle is 3 feet. What would the length have to be to make this a golden rectangle?

SOLUTION

Because

$$\frac{\text{Length}}{\text{Width}} = \phi,$$

we have

$$\text{Length} = \text{Width} \times \phi.$$

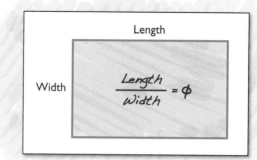

FIGURE 9.39 A golden rectangle: $\dfrac{\text{Length}}{\text{Width}} = \phi$.

Because the width is 3 feet, the length is $3 \times \phi$ feet. This is about 3×1.62 or roughly 4.9 feet.

TRY IT YOURSELF 9.10

The longer side of a golden rectangle is 5 feet. How long is the shorter side?

The answer is provided at the end of this section.

Golden rectangles in art and architecture

Much has been written concerning the aesthetic qualities of ϕ. The relative dimensions of golden rectangles are thought by many to make them the most visually pleasing of rectangles.

In the 1860s, the German physicist and psychologist Gustav Theodor Fechner conducted the following experiment. Ten rectangles varying in their length-to-width ratios were placed in front of a subject, who was asked to select the most pleasing one. The results showed that 76% of all choices centered on the three rectangles having ratios of 1.75, 1.62, and 1.50, with a peak at 1.62. These results, together with many other measurements of picture sizes, book sizes, and architecture, were reported by Fechner in his book *Vorschule der Aesthetik* (School for Aesthetics). He claimed that his observations supported the proposition that the golden rectangle is pleasing to the human eye.

Comparing the article that opens this section with the article in Exercise 52 makes clear the divide that exists regarding people's perceptions of the importance of ϕ. The article opening this section shows that some people claim the golden rectangle is evident in some of Leonardo da Vinci's paintings, including the *Mona Lisa*. But the second article disputes that claim. We will not settle the issue here, but we can say that there is some evidence that the famous Spanish artist Salvador Dali expressly used the golden rectangle in his painting *The Sacrament of the Last Supper*. And more recently, the number ϕ played a role in Dan Brown's popular novel *The Da Vinci Code*.

Many claim to see evidence of the golden ratio in the works of da Vinci.

Similar triangles

Unlike golden rectangles, *similar triangles* are not likely to become part of the popular culture. But they are examples of proportionality and form the basis of a great deal of important mathematics, including trigonometry. They are also very important in applications of mathematics.

Key Concept

Two triangles are **similar** if all corresponding angles have the same measure.

Figure 9.40 shows similar triangles. For comparison, **Figure 9.41** shows triangles that are not similar.

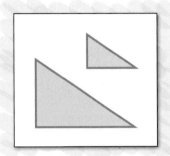

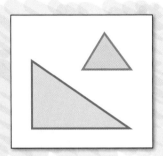

FIGURE 9.40 Similar triangles: Same angle measures.

FIGURE 9.41 Triangles that are not similar: Different angle measures.

As is made clear from these figures, similar triangles have the same shape. The most important feature of similar triangles is that the ratios of corresponding sides are the same. That is, the lengths of corresponding sides are proportional. This key fact is illustrated in **Figure 9.42**. In the notation of that figure, the similarity relationship can be expressed using the equations

$$\frac{A}{a} = \frac{B}{b} = \frac{C}{c}.$$

A typical use of similarity can be illustrated using **Figure 9.43**. Let's find the missing lengths B and c in the figure. First we find B:

$$\frac{A}{a} = \frac{B}{b}$$
$$\frac{12}{3} = \frac{B}{2}$$
$$4 = \frac{B}{2}$$
$$8 = B.$$

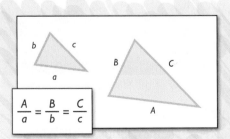

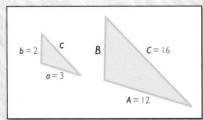

FIGURE 9.42 In similar triangles, ratios of corresponding sides are equal.

FIGURE 9.43 An example of similarity.

In a similar fashion, we find the length c:

$$\frac{A}{a} = \frac{C}{c}$$
$$4 = \frac{16}{c}$$
$$4c = 16$$
$$c = 4.$$

An alternative way of thinking about two similar triangles is to think of them as having sides marked using the same numbers but with different units of length. For example, a triangle with sides of length 3 inches, 4 inches, and 5 inches is similar to a triangle with sides of length 3 feet, 4 feet, and 5 feet. Both are similar to a triangle with sides of length 3 miles, 4 miles, and 5 miles.

It is a fact that the three angles of a triangle must add up to 180 degrees. Consequently, if two corresponding angles of a pair of triangles have the same measure, so does the third, and the triangles are similar.

> **SUMMARY 9.3**   **Similar Triangles**
>
> **1.** Triangles are similar if all corresponding angles have the same measure.
>
> **2.** For similar triangles, ratios of corresponding sides are the same. Therefore, the lengths of corresponding sides are proportional.
>
> **3.** If two of the three pairs of corresponding angles of a triangle have the same measure, the triangles are similar.

EXAMPLE 9.11 Using similar triangles: Shadows

Suppose we find that a tall building casts a shadow of length 25 feet. At the same time, a 6-foot-tall man casts a shadow 2 feet long. How tall is the building?

Shadow of the Empire State Building.

SOLUTION

We can represent the man and the building using vertical lines. Thus, the two triangles in **Figure 9.44** each have a right angle at their base. Because we are measuring the shadows at the same time of day, the angle of sunlight with the vertical (marked α

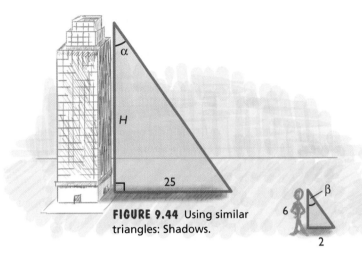

FIGURE 9.44 Using similar triangles: Shadows.

and β) is the same for both the man and the building. Because the two triangles have two pairs of equal angles, the triangles are similar. Hence, the ratios of corresponding sides are the same. In the figure, we have marked the height of the building as H. Therefore,

$$\frac{H}{6} = \frac{25}{2}$$

$$H = 6 \times \frac{25}{2}$$

$$H = 75.$$

The building is 75 feet tall.

TRY IT YOURSELF 9.11

A 5-foot-tall woman casts a shadow 3 feet long. What will be the length of the shadow of a 12-foot-tall tree at the same time of day?

The answer is provided at the end of this section.

Other types of proportionality

We have already noted that the area A of a circle is *not* proportional to the radius r. Because $A = \pi r^2$, the area is a constant multiple of the *square* of the radius, however. Therefore, we may say that A is proportional to r^2. We express this relationship verbally by saying that *the area of the circle is proportional to the square of the radius, with proportionality constant π*. Another way to look at this relationship is to note that the ratio A/r^2 is a constant, the number π, no matter how large or small the circle might be.

EXAMPLE 9.12 Comparing areas: Radius doubled

If the radius of a circle is doubled, what happens to the area?

SOLUTION

If the radius is r and the area is A, the relationship is given by the formula $A = \pi r^2$. If the radius is doubled, we replace r in the expression πr^2 by $2r$. This gives a new area of $\pi(2r)^2 = 4\pi r^2 = 4A$. The result is that the area is multiplied by 4.

TRY IT YOURSELF 9.12

Suppose the diameter of a pizza is reduced by half. What happens to the amount of pizza?

The answer is provided at the end of this section.

In the next example, we consider proportionality relationships involving volume and surface area.

EXAMPLE 9.13 Applying proportionality relationships: Mailing boxes

Suppose we have several boxes of different sizes, each of which is in the shape of a cube.

a. Is the volume of a box proportional to the length of one side of the box, the square of the length, or the cube of the length? Find the constant of proportionality.

b. One of our boxes has sides that are twice as long as those of another box. How much more does the larger box hold than the smaller box?

c. Is the surface area of a box proportional to the length of one side of the box, the square of the length, or the cube of the length?

d. The cost of wrapping paper for a box is proportional to the surface area. What happens to the cost of wrapping paper if the sides of the cube are doubled in length?

SOLUTION

a. Let x denote the length of one side of the cube, and let V denote the volume. Because the box is cubical in shape, the length, width, and height are all the same, x. Therefore,

$$V = \text{Length} \times \text{Width} \times \text{Height}$$
$$V = x \times x \times x$$
$$V = x^3.$$

Thus, the volume is proportional to the cube of the length, and the constant of proportionality is 1.

b. As in part a, we denote the length of a side by x. If the sides are doubled, we replace x by $2x$. This gives a new volume of

$$(2x)^3 = 2^3 x^3 = 8x^3.$$

The other volume was x^3. Thus, the larger volume is 8 times the smaller volume.

c. The amount of paper needed is the surface area of the box. The surface of a cube has six square "faces." If x represents the length of a side, the area of one face is $x \times x = x^2$. Therefore, the area of the surface of the cube is $6x^2$. This means that the surface area is proportional to the square of the length, with constant of proportionality 6.

d. We know that the cost is proportional to the surface area. To find the surface area of the larger cube, we replace x by $2x$ in the formula we used in part c:

$$\text{New surface area} = 6 \times (2x)^2$$
$$= 4(6x^2).$$

The larger surface area is 4 times the other surface area. So 4 times as much wrapping paper will be needed for the larger cube as for the smaller cube. Therefore, the cost of wrapping paper is multiplied by 4 when the sides are doubled in length.

The idea of proportionality occurs often in biology. For example, as a simple model, biologists may say that the weight of an animal is proportional to the cube of its height. Models like this show that animals grown to an enormous size (as often represented in science fiction) would collapse under their own weight. That is the point of the next example.

Classic poster of King Kong with Fay Wray (from the original movie released April 7, 1933).

EXAMPLE 9.14 Applying proportionality relationships: Overgrown apes

In a science fiction movie, an ape has grown to 100 times its usual height.

a. The weight of an ape is taken to be proportional to the cube of its height. How does the weight of the overgrown ape compare to its original weight?

b. The cross-sectional area of a limb is taken to be proportional to the square of the height. How does the cross-sectional area of a limb of the overgrown ape compare to the original area?

c. The pressure on a limb is the weight divided by the cross-sectional area. Use parts a and b to determine how the pressure on a limb of the overgrown ape compares to the original pressure.

SOLUTION

a. Let h denote the height of an ape and W the weight. Because the weight is proportional to the cube of the height, we have the formula

$$W = ch^3.$$

Here, c is the constant of proportionality. If the height is multiplied by 100, we replace h by $100h$. This gives a new weight of

$$c(100h)^3 = 100^3 \times ch^3.$$

The original weight was ch^3. Thus, the new weight is 100^3 times the original weight. In other words, the weight increases by a factor of $100^3 = 1,000,000$ (one million).

b. Let A denote the cross-sectional area of a limb and, as in part a, let h denote the height. Because the cross-sectional area is proportional to the square of the height, we have the formula

$$A = Ch^2.$$

Here, C is the constant of proportionality. If the height is multiplied by 100, we replace h by $100h$. This gives a new area of

$$C(100h)^2 = 100^2 \times Ch^2.$$

The original area was Ch^2. Thus, the new area is 100^2 times the original area. In other words, the area increases by a factor of $100^2 = 10,000$ (ten thousand).

c. Because the pressure on a limb is the weight divided by the cross-sectional area, the pressure will increase by a factor of

$$\frac{\text{Factor by which weight increases}}{\text{Factor by which area increases}} = \frac{100^3}{100^2} = 100.$$

The pressure on a limb of the overgrown ape is 100 times the original pressure.

The tremendous increase in pressure on a limb means that the overgrown ape would collapse under its own weight. Science fiction aside, such *scaling arguments* are used by biologists to study the significance of the size and shape of organisms. Exercise 49 provides one example.

Try It Yourself answers

Try It Yourself 9.8: Applying proportionality: Forestry   No.

Try It Yourself 9.9: Comparing volumes using proportionality: Two cans   The volume of the second can is one-fifth that of the first can.

Try It Yourself 9.10: Constructing a golden rectangle: 5 feet long $5/\phi$ or about 3.1 feet.

Try It Yourself 9.11: Using similar triangles: Shadows   7.2 feet.

Try It Yourself 9.12: Comparing areas: Radius halved   The area is one-fourth of what it was, which means we get one-fourth the pizza.

Exercise Set 9.2

1. Luggage size. The Web site of Southwest Airlines says that the airline allows a maximum weight of 70 pounds and a maximum size of 62 inches per checked piece of luggage. By "size," it means the sum Length + Width + Height. Is the size proportional to the length? Is the size proportional to the volume of the suitcase?

Which are proportional?   Exercises 2 through 11 ask about proportionality relationships. For those that describe proportionality relationships, give the constant of proportionality. In some cases, your answer may depend on your interpretation of the situation.

2. If each step in a staircase rises 8 inches, is the height of the staircase above the base proportional to the number of steps?

3. Suppose that hamburger meat costs $2.70 per pound. Is the cost of the hamburger meat proportional to its weight?

4. Is the volume of a balloon proportional to the cube of its diameter?

5. Is the volume of a balloon proportional to its diameter?

6. Is the area of a pizza proportional to its diameter?

7. Is the volume of a pizza with diameter 12 inches proportional to its thickness?

8. Is the area of land in a square lot proportional to the length of a side?

9. Is the area of a rectangular TV screen proportional to the length of its diagonal?

10. Consider a box with a square base and a height of 2 feet. Is the volume proportional to the area of the base?

11. The surface area S of a sphere of radius r is given by $S = 4\pi r^2$. Is the surface area of a balloon proportional to the square of its diameter?

12. Circle. Recall that the area of a circle of radius r is given by the formula $A = \pi r^2$. Your friend says that this formula shows that the area is proportional to the radius with proportionality constant πr because the area is πr times r. Is your friend correct? Explain.

The golden ratio   Exercises 13 through 23 concern the golden ratio ϕ.

13. Making golden rectangles. The shorter side of a golden rectangle is 5 inches. How long is the longer side? Round your answer to one decimal place.

14. More golden rectangles. The longer side of a golden rectangle is 10 inches. How long is the shorter side? Round your answer to one decimal place.

15. TV screens. Measure your television screen. What is the ratio of the longer side to the shorter? Is the screen a golden rectangle?

16. Computer screens. Measure your computer screen. What is the ratio of the longer side to the shorter? Is the screen a golden rectangle?

17. This text. Is this textbook a golden rectangle?

18. Arm measurements. The distance from your shoulder to your fingertips divided by the distance from your elbow to your fingertips is thought to be close to ϕ. Take these measurements of your own arm and calculate the ratio. Is your answer close to ϕ?

19. Body measurements. Your height divided by the distance from your belly button to your toes is thought to be close to ϕ. Take these measurements of your own body and calculate the ratio. Is your answer close to ϕ?

20. Leg measurements. The distance from your hip to your toes divided by the distance from your knee to your toes is thought to be close to ϕ. Take these measurements of your own body and calculate the ratio. Is your answer close to ϕ?

21. Bees in The Da Vinci Code. An excerpt from the book *The Da Vinci Code* by Dan Brown goes as follows:

Plants, animals and even human beings all possessed dimensional properties that adhered with eerie exactitude to the ratio of PHI [sic, ϕ] to 1.... So the ancients assumed PHI must have been preordained by the Creator....

Langdon grinned. "Ever study the relationship between female and males in a honeybee community?"

"Sure. The female bees always outnumber the male bees."

"And did you know that if you divide the number of female bees by the number of male bees in any beehive in the world, you always get the same number?"

"You do?"

"Yup. PHI."

The girl gasped. "No way!"

There are various resources, including the Web, available for doing research about bees. Does your research bear out the claim that bees "adhered with eerie exactitude to the ratio of ϕ to 1"?

22. Golden rectangle in art. Do some research on the use of the golden rectangle in art and write a report.

23. Golden rectangle in architecture. Do some research on the use of the golden rectangle in architecture and write a report.

Similar triangles Exercises 24 through 28 deal with similarity of triangles.

24. Finding missing sides. The triangles in **Figure 9.45** are similar. Find the sides labeled A and B.

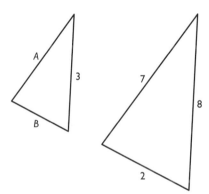

FIGURE 9.45

25. More missing sides. The triangles in **Figure 9.46** are similar. Find the sides labeled A and b.

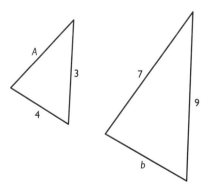

FIGURE 9.46

26. Shadows. A pole casts a 4-foot shadow, and at the same time of day a 6-foot-tall man casts a 1-foot shadow. How tall is the pole?

27. More shadows. A 10-foot-tall building casts a 4-foot shadow, and at the same time of day a child casts a 2-foot shadow. How tall is the child?

28. Tennis. A tennis player stands 39 feet behind the net, which is 3.5 feet tall. He extends his racket to a height of 9 feet and hits a serve. The ball passes 6 inches above the net. How far beyond the net does the tennis ball strike the court? *Suggestion:* If d is the distance past the net, the distance from the server to where the ball strikes the court is $39 + d$.

29. Equilateral triangles. It is a fact that the area of an equilateral triangle is proportional to the square of a side. If all sides of an equilateral triangle are doubled, what happens to the area?

30. Box. Suppose a box has a fixed height of 9 feet and a square base. Explain why the length of one of the sides of the base is proportional to the square root of the volume. What is the constant of proportionality?

Volume of a sphere The volume V of a sphere of radius r is $V = \frac{4}{3}\pi r^3$, and the surface area S is given by $S = 4\pi r^2$. Exercises 31 through 34 make use of these formulas.

31. A balloon. The amount of gas in a balloon of radius r is proportional to:

  a. r **b.** r^2 **c.** r^3 **d.** π

What is the constant of proportionality?

32. A larger balloon. A child wants a balloon of twice the radius of the offered balloon. How does the increase in radius affect the amount of gas in the balloon?

33. Painted styrofoam balls. The amount of paint needed to color a styrofoam ball of radius r is proportional to:

  a. r **b.** r^2 **c.** r^3 **d.** π

What is the constant of proportionality?

34. Baseballs and softballs. A softball has a radius 1.33 times that of a baseball. How many times as much material is required to cover a softball than a baseball?

A cylinder Recall that a cylinder with radius r and height h has volume $V = \pi r^2 h$ and surface area $S = 2\pi rh$ (not counting the top and bottom). Exercises 35 through 43 make use of these formulas.

35. A can. If the height of a cylindrical can is held constant, the volume is proportional to:

  a. r **b.** r^2 **c.** h **d.** π

What is the constant of proportionality?

36. Larger can. Two soup cans have a height of 4 inches. One can is twice the radius of the other. How does the amount of soup in the cans compare?

37. Volume of a cylinder. If the radius is held constant, the volume of a cylindrical can is proportional to:

  a. r **b.** r^2 **c.** h **d.** π

What is the constant of proportionality?

38. More on larger cans. Two soup cans have a radius of 2 inches. One can is twice the height of the other. How does the amount of soup in the cans compare?

39. Surface area of a cylinder. If the cylinder is a can of tuna and the radius is held constant, the area of the label is proportional to:

   a. r **b.** r^2 **c.** h **d.** π

What is the constant of proportionality?

40. The label of a can. Two tuna cans have a radius of 2 inches. One is twice as tall as the other. How do the sizes of the labels on the cans compare?

41. Surface area of a cylinder II. If the height is held constant, the surface area of a can is proportional to:

   a. r **b.** r^2 **c.** h **d.** π

What is the constant of proportionality?

42. More about the label. Two soup cans have a height of 4 inches. One can has 3 times the radius of the other. How do their labels compare in area?

43. Tin for a can. If we include the top and bottom of a cylindrical can, the total surface area is $A = 2\pi rh + 2\pi r^2$. If the height is held constant, the total surface area is proportional to:

   a. r **d.** π
   b. r^2 **e.** none of these
   c. h

A cone Suppose a cone has height h and the radius of its top circle is r. Then the volume is given by $V = \frac{1}{3}\pi r^2 h$. The surface area (not counting the top circle) is given by $A = \pi r\sqrt{r^2 + h^2}$. Exercises 44 through 48 make use of these formulas.

44. Varying radius. If the height of a cone is held constant, the volume is proportional to:

   a. r **d.** π
   b. r^2 **e.** none of these
   c. h

What is the constant of proportionality?

45. Ice cream. Two ice cream cones have the same height. One's radius is twice the radius of the other. Compare the amounts of ice cream the cones can hold.

46. Volume of a cone. If the radius of a cone is held constant, the volume is proportional to:

   a. r **d.** π
   b. r^2 **e.** none of these
   c. h

What is the constant of proportionality?

47. More ice cream. Two ice cream cones have the same radius. One has twice the height of the other. Compare the amounts of ice cream the cones can hold.

48. Surface area of a cone. If the radius of the top circle is held constant, the surface area is proportional to:

   a. r **d.** π
   b. r^2 **e.** $\sqrt{r^2 + h^2}$
   c. h **f.** none of these

49. Basal metabolic rate. The *basal metabolic rate* (BMR) of an animal is a measure of the amount of energy it needs for survival. An animal with a larger BMR will need more food to survive. One simple model of the BMR states that among animals of a similar shape, the rate is proportional to the square of the length. (This model assumes that the BMR is proportional to the surface area.) If one animal is 10 times as long as another, how do the BMR values compare?

50. History. The idea of proportionality has a long history in geometry. Explain how the idea of proportion is defined by Eudoxus and used in Euclid's *Elements*.

51. History. Scaling arguments such as those used in Example 9.14 were given by Galileo in his 1638 book *Two New Sciences*. Summarize the arguments used by Galileo and explain how he applied them.

52. A different point of view. The following article, which paints a somewhat different picture from the article that opens this section, appears at the Web site of the Mathematical Association of America.

IN THE NEWS 9.4

Good Stories, Pity They're Not True
KEITH DEVLIN June 2004

The enormous success of Dan Brown's novel *The Da Vinci Code* has introduced the famous Golden Ratio . . . to a whole new audience. Regular readers of this column will surely be familiar with the story. The ancient Greeks believed that there is a rectangle that the human eye finds the most pleasing. . . .

Having found this number, the story continues, the Greeks then made extensive use of the magic number in their architecture, including the famous Parthenon building in Athens. Inspired by the Greeks, future generations of architects likewise based their designs of buildings on this wonderful ratio. Painters did not lag far behind. The great Leonardo Da Vinci is said to have used the Golden Ratio to proportion the human figures in his paintings—which is how the Golden Ratio finds its way into Dan Brown's potboiler.

It's a great story that tends to get better every time it's told. Unfortunately, apart from the fact that Euclid did solve the line division problem in his book *Elements*, there's not a shred of evidence to support any of these claims, and good reason to believe they are completely false. . . .

Find other sources that report on the role of the golden ratio in art and architecture. Write a report on your findings.

9.3 Symmetries and tilings: Form and patterns

TAKE AWAY FROM THIS SECTION

Various types of symmetries appear in art, architecture, and nature. *Rotational* symmetry and *reflectional* symmetry are two of these types.

The chief forms of beauty are order and symmetry and definiteness, which the mathematical sciences demonstrate in a special degree.
—Aristotle

Symmetry is among the most important properties of a geometric object. It may not be easy to define symmetry, but the idea is intuitive and has an aesthetic appeal. The following article appeared on the Web site of *ScienceNews*.

IN THE NEWS 9.5

Forms of Symmetry

JULIE REHMEYER   April 21, 2007

Symmetry attracts us. Studies comparing people's reactions to different faces have shown, for example, that they find highly symmetrical faces more attractive than less symmetrical faces. The symmetry of faces is simple and bilateral, but other three-dimensional objects can be symmetric in complex ways, leading to different kinds of beauty.

Bathsheba Grossman, a sculptor in Santa Cruz, California, mines subtle forms of symmetry for inspiration. Her results are swirling and proportional. They may be simple or complex, but they always come together into a precise, intriguingly symmetrical pattern.

When Grossman dreams up a new sculpture, she starts by contemplating what its symmetries will be. In mathematical terms, she thinks about its symmetry group. "I have to pick a group before I can come up with an idea," she says.

Three-dimensional objects can be symmetrical in a variety of ways. Mathematicians have identified all of the different symmetry groups that such objects can have. The two most basic kinds of symmetry are "reflective" and "rotational." . . .

Grossman's favorite form of symmetry is remarkably simple: It contains 180-degree rotations around its three perpendicular coordinate axes. "Three rotations and no reflections; what could be finer?" she muses.

This article alludes to "rotational" and "reflective" symmetry in art. In this section, we examine these symmetries and also take a look at tilings of the plane.

Rotational symmetries in the plane

We will examine the two types of symmetry suggested by the above article: rotational symmetry and reflectional symmetry.

Key Concept

A planar figure has **rotational symmetry** about a point if it remains exactly the same after a rotation about that point of less than 360 degrees.

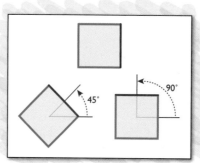

FIGURE 9.47 A square has 90-degree rotational symmetry but not 45-degree rotational symmetry.

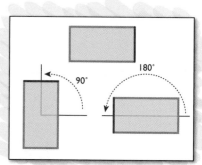

FIGURE 9.48 A rectangle has 180-degree rotational symmetry but not 90-degree rotational symmetry.

Let's agree at the outset that when we speak of a "rotation," we always mean it to be in a clockwise direction. If we rotate a square about its center by 90 degrees, the result is a square that is an exact copy of the one we started with. Therefore, we say that a square has 90-degree *rotational symmetry*. In fact, a rotation of a square returns an exact copy of the square if and only if we rotate it by a multiple of 90 degrees. For example, if the square is rotated by 45 degrees, it is tilted and does not remain the same. These facts are illustrated in **Figure 9.47**.

Now consider a rectangle that is *not* a square. This object does not have 90 degrees rotational symmetry because rotation by 90 degrees does not leave the rectangle exactly the same as it was originally. It does have 180-degree rotational symmetry about its center, however. These facts are illustrated in **Figure 9.48**.

Another example of rotational symmetry is provided by a circle. A circle has *complete rotational symmetry* because if we rotate it any number of degrees about the center of the circle, the result is exactly the same as the original circle. By way of comparison, the triangle in **Figure 9.49** has no rotational symmetries at all.

FIGURE 9.49 This triangle has no rotational symmetries.

EXAMPLE 9.15 Finding rotational symmetries: Pentagram

Find the rotational symmetries of the five-pointed star, or pentagram, in **Figure 9.50**.

SOLUTION

In order to preserve the star, we must rotate through an angle that takes one star point to another. The star points divide 360 degrees into five equal angles, so a rotation of 360 degrees/5 = 72 degrees moves one star point to the next. The pentagram has 72-degree rotational symmetry about its center. The other rotational symmetries are multiples of 72 degrees, namely 2 × 72 degrees = 144 degrees, 3 × 72 degrees = 216 degrees, and 4 × 72 degrees = 288 degrees.

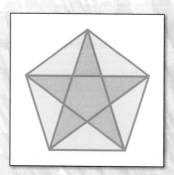

FIGURE 9.50 A pentagram.

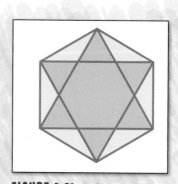

FIGURE 9.51 A six-pointed star.

TRY IT YOURSELF 9.15

Find the rotational symmetries of the six-pointed star shown in **Figure 9.51**.

The answer is provided at the end of this section.

Many familiar objects have rotational symmetries, as is shown in the following photos.

The pictured starfish has 60-degree rotational symmetry.

The lunaria has 90-degree rotational symmetry, and the dahlia has many more rotational symmetries.

Reflectional symmetry of planar figures

When we reflect a figure in the plane about a line in the plane, we can think of letting the line serve as an axis and flipping the plane about that line. If we imagine a mirror placed along the line, the reflection of the figure about the line is the mirror image of the figure. **Figure 9.52** illustrates the reflection of a shape about the line *L*.

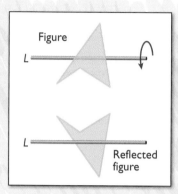

FIGURE 9.52 Reflection is the same as flipping the plane about a line.

FIGURE 9.53 This shape has reflectional symmetry about the line shown.

FIGURE 9.54 This star has reflectional symmetry about five different lines.

"You're a nice guy and all that, but I'm looking for somebody more *symmetrical.*"

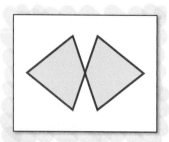

FIGURE 9.55

Key Concept

A planar figure has **reflectional symmetry about a line** L if the figure is identical to its reflection through L.

For example, the dart shape in **Figure 9.53** has reflectional symmetry about the line shown. Similarly, the triangle in Figure 9.49 has reflectional symmetry about the horizontal line that bisects the angle at the left-hand side of the picture. Note that neither of these figures shows any rotational symmetry. The five-pointed star in **Figure 9.54** has reflectional symmetry about each of the five lines shown there.

Reflectional symmetries are shown in many familiar settings, including the faces of animals.

EXAMPLE 9.16 Finding symmetries: A given shape

What are the rotational and reflectional symmetries of the shape in **Figure 9.55**?

SOLUTION

There is reflectional symmetry about both horizontal and vertical lines. The shape has a rotational symmetry of 180 degrees.

Faces show reflectional symmetry.

Many flags such as the Macedonian flag show reflectional symmetry through both horizontal and vertical lines.

TRY IT YOURSELF 9.16

What are the rotational and reflectional symmetries of the shape in **Figure 9.56**?

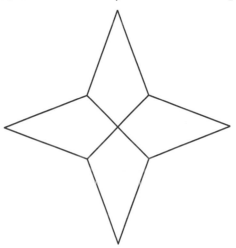

FIGURE 9.56

The answer is provided at the end of this section.

Regular tilings of the plane

Now we consider ways of filling the plane with geometric figures that don't overlap.

Key Concept

A tiling or **tessellation** of the plane is a pattern of repeated figures that cover up the plane.

Simple examples are the tilings shown in **Figure 9.57** on the following page, which uses one square tile, and **Figure 9.58**, which uses one square and two different rectangular tiles. Perhaps one of these matches the configuration in your bathroom floor.

FIGURE 9.57 A tiling by squares.

FIGURE 9.58 A tiling by squares and two rectangles.

Now we consider one special kind of tiling. A *regular polygon* is a polygon (a closed figure made of three or more line segments) in which all sides are of equal length and all angles have equal measure.

Key Concept

A regular tiling is a tiling of the plane that consists of repeated copies of a single regular polygon, meeting edge to edge, so that every vertex has the same configuration.

The tiling in Figure 9.57 is regular, but the tiling in Figure 9.58 is not regular because different vertices have different configurations. The fact that each vertex of a regular tiling must have the same configuration lets us quickly narrow the list of polygons that might regularly tile the plane. At each vertex, the polygons must fit together, and their angles must add up to 360 degrees. This is illustrated in **Figure 9.59**, where the vertices of six polygons fit together and have angles adding up to 360 degrees. For example, if the tile is a square, then it takes four 90-degree angles to add up to 360 degrees. That means that Figure 9.57 shows the only way to make a regular tiling of the plane by squares. If the angle of a regular polygon is greater than 120 degrees, it cannot make a regular tiling of the plane. This is illustrated in **Figure 9.60**, which shows that if two angles are greater than 120 degrees, then there isn't room for a third angle of that size because the sum of three or more such angles would exceed 360 degrees.

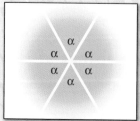

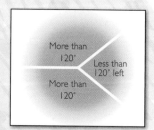

FIGURE 9.59 In a regular tiling, the angles that meet at a vertex must sum to 360 degrees. If there are six polygons, the angles must be 60 degrees.

FIGURE 9.60 If a polygon has an angle of more than 120 degrees, a regular tiling cannot be created.

To see which other regular polygons might be used to make a regular tiling of the plane, let's first list the sizes of angles in some regular polygons.

Regular polygon	Angle size
Equilateral triangle	60°
Square	90°
Pentagon	108°
Hexagon	120°

Regular polygons with more than six sides have angles larger than 120 degrees. Hence, they cannot regularly tile the plane.

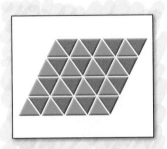

FIGURE 9.61 A regular tiling with equilateral triangles.

EXAMPLE 9.17 Finding regular tilings: Equilateral triangle

Show how to use an equilateral triangle to make a regular tiling of the plane.

SOLUTION

The tiling is shown in **Figure 9.61**.

TRY IT YOURSELF 9.17

Find a regular tiling of the plane using regular hexagons.

The answer is provided at the end of this section.

As a contrast to the example above, you are asked in Exercise 16 to show that there is no regular tiling of the plane by a regular pentagon.

Irregular tilings

Regular tilings must satisfy two conditions: They use a single regular polygon, and the same configuration of edges must occur at each vertex. Tilings that are not regular

Irregular tilings are common in art and architecture.

fail to meet one of these two conditions. One classic example of a tiling that is not regular is shown in the accompanying photo of floor tiles. This tiling does satisfy the second condition, so it is *semi-regular*.

Figure 9.62 shows how to cut a regular hexagon into three (nonregular) pentagons. We saw in Try It Yourself 9.17 that the plane can be tiled by regular hexagons. Then such a tiling automatically gives an irregular tiling by pentagons. See **Figure 9.63**.

The idea, suggested by Figure 9.63, that we can make new tilings by subdividing old ones is fruitful. We will use it to show how to use *any* triangle to tile the plane. We begin with the triangle shown in **Figure 9.64**. The first step is to note that we can put together two copies of this triangle to make a parallelogram, as shown in **Figure 9.65**. Now we use parallelograms to tile the plane as shown in **Figure 9.66**, and this gives the required tiling by triangles.

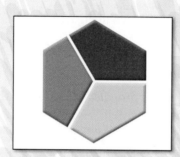 

FIGURE 9.62 Cutting a regular hexagon into three (irregular) pentagons gives an irregular tiling.

FIGURE 9.63 Using a hexagon tiling to make a pentagon tiling.

FIGURE 9.64 A triangle.

FIGURE 9.65 Two copies of a triangle make a parallelogram.

FIGURE 9.66 Using the parallelogram to make a triangle tiling.

EXAMPLE 9.18 Making new tilings from old: Three given pieces

Show that the three pieces in **Figure 9.67** can be used to tile the plane.

SOLUTION

Note that the three pieces go together to make the regular hexagon shown in **Figure 9.68**. We know that regular hexagons tile the plane, and that gives a tiling using the three pieces shown.

FIGURE 9.67 Pieces for Example 9.18.

FIGURE 9.68 Pieces go together to make a regular hexagon.

TRY IT YOURSELF 9.18

Show that the three pieces in **Figure 9.69** can be used to tile the plane.

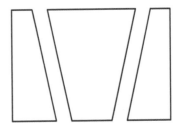

FIGURE 9.69

The answer is provided at the end of this section.

Escher tilings

Some of the best known tilings of the plane are those of M. C. Escher. Several of his famous tilings are displayed here. They are interesting from both a mathematical and an artistic point of view. They are much more elaborate than any tilings we have produced, but they are constructed using similar ideas. Let's show how we can make tilings using unusual shapes.

Some famous Escher prints.

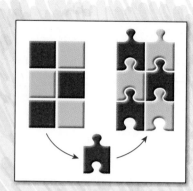

FIGURE 9.70 A regular tiling by squares changed into an irregular tiling.

Let's start with the regular tiling by squares shown on the left in **Figure 9.70**. Now from each square, let's remove a puzzle tab from the bottom and add it to the top of the square. The resulting tiling piece is the irregular shape shown at the bottom of Figure 9.70. We get the new tiling of the plane shown on the right in Figure 9.70.

A little thought can lead one to produce truly unusual tiling pieces. Many of Escher's prints use this idea as well as more sophisticated ones. Some start with hexagons rather than squares.

Try It Yourself answers

Try It Yourself 9.15: Finding rotational symmetries: Six-pointed star   It has 60-degree rotational symmetry (as well as 120-degree, 180-degree, 240-degree, and 300-degree).

Try It Yourself 9.16: Finding symmetries: A given shape   90-degree rotational symmetry (as well as 180-degree and 270-degree). Reflectional symmetry through vertical and horizontal lines as well as two 45-degree lines.

Try It Yourself 9.17: Finding regular tilings: Hexagon

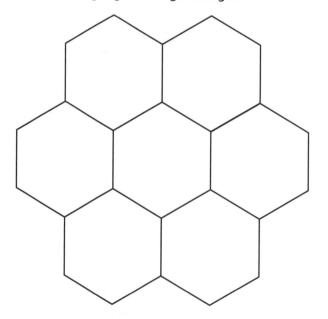

Try It Yourself 9.18: Making new tilings from old: Three given pieces   The three pieces fit together to make a rectangle, which can tile the plane.

Exercise Set 9.3

1. Complete rotational symmetry. Find a planar figure other than a circle that has complete rotational symmetry.

Making figures with given symmetry   In Exercises 2 through 7, find an example of a figure with the given symmetry.

2. 45-degree rotational symmetry

3. 30-degree rotational symmetry

4. n-degree rotational symmetry

5. 180-degree rotational symmetry but no reflectional symmetries

6. Reflectional symmetries through exactly one line

7. Reflectional symmetries through exactly two lines

Many rotations   If R_θ denotes a clockwise rotation of the plane through an angle of θ degrees, we use R_θ^n to mean that we make a clockwise rotation of the plane through an angle of θ degrees n times. For example, R_{45}^2 means a clockwise rotation of 45 degrees followed by a second clockwise rotation of 45 degrees. The result is a rotation through 90 degrees. This means

$$R_{45}^2 = R_{90}.$$

Exercises 8 through 12 make use of this idea.

8. What is the result of R_{45}^8?

9. Find θ between 0 and 360 degrees so that $R_{60}^8 = R_\theta$.

10. Find θ between 0 and 360 degrees so that $R_{30}^{15} = R_\theta$.

11. For what value of n will R_2^n result in leaving each point in the plane fixed?

12. Find θ between 0 and 360 degrees so that $R_{10}^{255} = R_\theta$.

13. A hexagon. What are the reflectional and rotational symmetries of a regular hexagon?

14. An octagon. What are the reflectional and rotational symmetries of a regular octagon?

15. Tiling with a rhombus. Show how to make a tiling of the plane with a rhombus, which is a square pushed over (as shown in **Figure 9.71**).

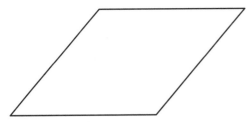

FIGURE 9.71 A rhombus.

16. Tiling with a pentagon. Explain why there is no regular tiling of the plane using regular pentagons. *Suggestion:* Each angle of a regular pentagon is 108 degrees. How many of these would need to fit together at a vertex?

17. Tiling with hexagons and triangles. Show how to tile the plane using regular hexagons and triangles. *Suggestion:* Put the hexagons together so that triangular spaces are left.

18. Tiling with unusual shapes. Show how to tile the plane using the shape in **Figure 9.72**.

FIGURE 9.72

19. Tiling the plane. Show how to tile the plane using the shape in **Figure 9.73**.

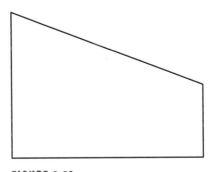

FIGURE 9.73

20. Tiling the plane. Show how to tile the plane using the shape in **Figure 9.74**.

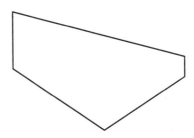

FIGURE 9.74

Penrose tiles   The tilings of the plane that are most familiar to us are *periodic* in that they are made up of the repetition of a single pattern. For example, the tiling in **Figure 9.75** on the following page is periodic, and the repeated pattern is outlined in that figure. But the tiling in **Figure 9.76** is not periodic. In 1974, Roger Penrose found a pair of tiles, *darts* (shown in **Figure 9.77**) and *kites* (shown in **Figure 9.78**). (The distance ϕ is the golden ratio discussed in the preceding section and is approximately equal to 1.62.) These tiles, if assembled according to some simple rules, always make nonperiodic tilings. These tiles can be found as commercially available puzzles, and many people find them fun and challenging. Exercises 21 through 24 deal with *Penrose tilings*.

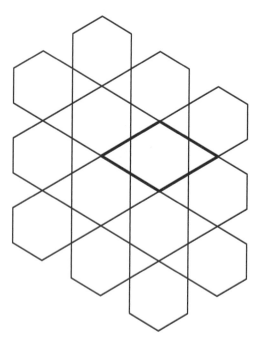

FIGURE 9.75 A periodic tiling.

21. Why it is not periodic. Our goal here is to explain why the tiling in Figure 9.76 is nonperiodic.

    a. How many points of rotational symmetry does the tiling in Figure 9.76 have? That is, how many points can we take for the center of a rotation less than 360 degrees that leaves the figure unchanged?

    b. Can a tiling with exactly one point of rotational symmetry be periodic?

    c. Explain using parts a and b why the tiling in Figure 9.76 is not periodic.

FIGURE 9.76 A nonperiodic tiling.

22. Angles of kites and darts. Find the angles marked α, β, γ, and δ in Figures 9.77 and 9.78. Recall the following facts from geometry:

    • The angle sum of every triangle is 180 degrees.

    • Recall that an isosceles triangle has two equal sides. It is a fact that the "base angles" of an isosceles triangle have the same degree measure. (See **Figure 9.79.** The angles marked "1" and "2" are base angles because they are opposite the equal sides.)

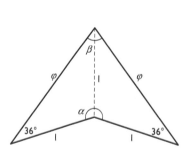

FIGURE 9.77 A dart.

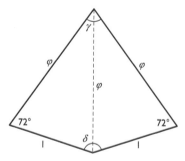

FIGURE 9.78 A kite.

FIGURE 9.79 Base angles of an isosceles triangle.

23. A periodic tiling with kites and darts. Show that a kite and a dart can be fitted together to make a rhombus, each side of which is ϕ. Explain why this gives a periodic tiling of the plane using kites and darts.

24. Experimenting with kites and darts. Kites and darts come with assembly rules, which are as follows:

    Assembly Rule 1: Pieces must be assembled along matching edges.

    Assembly Rule 2: The kites and darts in **Figures 9.80** and **9.81** have certain angles marked with dots. Dotted angles must be adjacent. **Figure 9.82** shows an assembly that is allowed because the dots match. **Figure 9.83** shows an assembly that is not allowed because the dots do not match.

Fit together a number of kites and darts to begin a tiling of the plane. It is a fact that any such tiling you make will be nonperiodic. An example of such a nonperiodic tiling is shown in **Figure 9.84.**

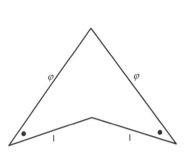

FIGURE 9.80 A dart with dots.

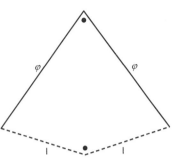

FIGURE 9.81 A kite with dots.

FIGURE 9.84 A nonperiodic tiling using kites and darts by Roger Penrose.

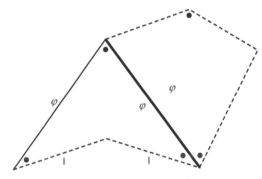

FIGURE 9.82 An allowable assembly.

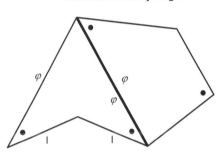

FIGURE 9.83 An assembly that is not allowed.

CHAPTER SUMMARY

In this chapter, we consider geometric topics that are encountered in everyday life. We study how to measure geometric objects and how the notion of geometric similarity is related to the idea of proportionality. We also consider symmetries of shapes and the notion of tiling the plane.

Perimeter, area, and volume: How do I measure?

Two-dimensional geometric objects can be measured in terms of perimeter and area. The most common figures are circles, rectangles, and triangles. For each of these objects, there are standard formulas for computing area and perimeter. The area and circumference of a circle of radius r are given by the formulas

$$\text{Area} = \pi r^2$$
$$\text{Circumference} = 2\pi r.$$

For a rectangle, the formulas are

$$\text{Area} = \text{Length} \times \text{Width}$$
$$\text{Perimeter} = 2 \times \text{Length} + 2 \times \text{Width}.$$

One way to find the area of a triangle is to pick one side as the *base* and then find the *height* by starting at the opposite vertex and drawing a line segment that meets the base in a right angle. The area is given by the formula

$$\text{Area} = \frac{1}{2} \text{ Base} \times \text{Height}.$$

Another way to find the area of a triangle is to use *Heron's formula*:

$$\text{Area} = \sqrt{S(S-a)(S-b)(S-c)}.$$

Here, a, b, and c are the sides of the triangle, and S is the *semi-perimeter*, defined by $S = \frac{1}{2}(a+b+c)$.

The *Pythagorean theorem* relates the legs a, b of a right triangle with the hypotenuse c. It states that

$$a^2 + b^2 = c^2.$$

Conversely, if any triangle has three sides of lengths a, b, and c such that $a^2 + b^2 = c^2$, then that triangle is a right triangle with hypotenuse c.

Three-dimensional geometric objects can be measured in terms of volume and surface area. For a box, we have the formula

$$\text{Volume} = \text{Length} \times \text{Width} \times \text{Height}.$$

For a cylinder with radius r and height h, the volume is

$$\text{Volume} = \pi r^2 h,$$

and the surface area (excluding the top and bottom) is given by

$$\text{Surface area} = 2\pi r h.$$

There is a general formula for the volume of any three-dimensional object with a uniform cross section:

$$\text{Volume} = \text{Area of base} \times \text{Height}.$$

Proportionality and similarity: Changing the scale

Proportionality relationships are common in both numerical and geometrical settings. One quantity is *proportional* to another if it is always a fixed (nonzero) constant multiple of the other. This multiple is the *constant of proportionality*. An alternative definition of proportionality requires the ratio of the two quantities to be constant.

If one of two proportional quantities is multiplied by a certain factor, the other one is also multiplied by that factor. For example, if one quantity doubles, then the other doubles as well. This property is useful when we compare the measurements (such as area) of geometrically similar objects.

The *golden ratio* is a special number that arises in geometry. It is denoted by the Greek letter ϕ and is approximately equal to 1.62. Here is one way to define the golden ratio: Divide a line segment into two pieces, one longer and one shorter, so that the ratios $\dfrac{\text{Longer}}{\text{Shorter}}$ and $\dfrac{\text{Whole}}{\text{Longer}}$ are equal. This common ratio is by definition the golden ratio.

The number ϕ appears in geometry in *golden rectangles*, which are rectangles for which the ratio of the length to the width is ϕ. The golden ratio appears in other ways as well. The significance of the golden ratio and golden rectangles in science and art is the subject of some debate.

In geometry, one important application of proportionality involves similar triangles. Two triangles are *similar* if all corresponding angles have the same measure. The most important feature of similar triangles is that the ratios of corresponding sides are the same.

Sometimes a quantity is proportional to the square or some other power of another quantity. For example, the area of a circle is proportional to the square of the radius, with constant of proportionality π.

Symmetries and tilings: Form and patterns

One of the most important properties a geometric object can have is symmetry. We consider two types of symmetry, rotational and reflectional. A planar figure has *rotational symmetry* if it remains exactly the same after a rotation of less than 360-degrees. For example, a square has 90-degree rotational symmetry, and a rectangle has 180-degree rotational symmetry.

A planar figure has *reflectional symmetry* about a line if the figure is identical to its reflection through that line. If we imagine a mirror placed along the line, the mirror image of the figure is the same as the original figure.

A *tiling* of the plane is a pattern of repeated figures that cover up the plane. Such a tiling is *regular* if it consists of repeated copies of a single regular polygon, meeting edge to edge, so that every vertex has the same configuration. For example, squares can be used to make a regular tiling of the plane.

Irregular tilings are often encountered as well. They can be obtained by subdividing existing tilings. The tilings of M. C. Escher are interesting from both a mathematical and an artistic point of view.

KEY TERMS

perimeter, p. 552
circumference, p. 552
area, p. 553
Pythagorean theorem, p. 553
directly proportional, p. 569

constant of proportionality, p. 569
golden ratio, p. 572
golden rectangle, p. 597
similar, p. 575
rotational symmetry, p. 583

reflectional symmetry about a
   line, p. 586
tiling or tessellation, p. 587
regular tiling, p. 588

CHAPTER QUIZ

1. The net on a tennis court is 3.5 feet high. The baseline is 39 feet from the bottom of the net. How far is the baseline from the top of the net?

Answer   39.2 feet.

If you had difficulty with this problem, see Example 9.3.

2. Use Heron's formula to find the area of a triangle with sides 11, 14, and 16.

Answer   75.5 square units.

If you had difficulty with this problem, see Example 9.4.

3. A trunk is 2 feet by 3 feet by 4 feet. What is the volume of the trunk?

Answer   24 cubic feet.

If you had difficulty with this problem, see Example 9.6.

4. The volume of concrete needed to make a sidewalk is proportional to the length of the sidewalk. (Assume that the sidewalk has a fixed width and depth.) Six cubic yards of concrete are required to make a sidewalk that is 100 feet long. How much concrete is needed to make a sidewalk that is 25 feet long?

Answer   1.5 cubic yards.

If you had difficulty with this problem, see Example 9.9.

5. A 5-foot-tall girl casts a 3-foot shadow at the same time that a building casts a 12-foot shadow. How tall is the building?

Answer   20 feet tall.

If you had difficulty with this problem, see Example 9.11.

6. The volume of a cylindrical water tank of height 30 foot is proportional to the square of the diameter. Two tanks have height 30 feet, and the diameter of one is 3 times the diameter of the other. How do the volumes compare?

Answer   The larger tank has 9 times the volume of the smaller.

If you had difficulty with this problem, see Example 9.12.

7. Find the rotational symmetries of an equilateral triangle.

Answer   It has 120-degree rotational symmetry (as well as 240-degree).

If you had difficulty with this problem, see Example 9.15.

8. Find the reflectional symmetries of an equilateral triangle.

Answer   It has reflectional symmetry about three lines: the lines from a vertex to the midpoint of the opposite side.

If you had difficulty with this problem, see Example 9.16.

APPENDIX 1 Unit Conversion

In this appendix, we discuss conversion of physical units (e.g., feet to inches) and currency conversion.

Physical units

The most common system of weights and measures in the United States is informally called the "U.S. Customary System" or the "English system." Most of the rest of the world uses what we usually call the "metric system." The official name of this system of weights and measures is Le Système International d'Unités, or "SI" for short.

Some basic relationships for unit conversions are contained in Tables A1.1 and A1.2.

TABLE A1.1 Conversion Among Familiar English Units

Length	1 foot	12 inches
	1 yard	3 feet
	1 (statute) mile	5280 feet
	1 nautical mile	1.15 (statute) miles
	1 lightyear	5.88×10^{12} miles
	1 parsec	3.26 lightyears
Area	1 acre	43,560 square feet
Volume	1 tablespoon	3 teaspoons
	1 fluid ounce	2 tablespoons
	1 cup	8 fluid ounces
	1 pint	2 cups
	1 quart	2 pints
	1 gallon	4 quarts
Weight	1 pound	16 ounces
	1 ton	2000 pounds

TABLE A1.2 English/Metric Conversions

Length	1 inch	2.54 centimeters
	1 foot	30.48 centimeters
	1 meter (100 centimeters)	3.28 feet
	1 centimeter (10 millimeters)	0.39 inch
	1 kilometer (1000 meters)	0.62 mile
Volume	1 liter	1.06 quarts
	1 quart	0.95 liter
	1 gallon	3.79 liters
Weight/mass	1 ounce	28.35 grams
	1 pound	453.59 grams
	1 kilogram (1000 grams)	2.20 pounds

The following tips may be helpful in using these tables.

Tip 1: Conversion factors: To convert one unit to another, we multiply by a number called the *conversion factor*. For example, to convert feet to inches, we multiply length in feet by 12 because there are 12 inches in a foot. Here, the conversion factor is 12 inches per foot.

Informally, the conversion factor from unit *A* to unit *B* is the number of *B*'s in a single *A*. For example, from Table A1.1 we find that there are 5280 feet in a mile. Thus, 5280 feet per mile is the conversion factor for changing miles to feet. For example, 5 miles is $5 \times 5280 = 26{,}400$ feet.

To convert feet to miles, the conversion factor is the reciprocal, $\dfrac{1}{5280}$ miles per foot. For example, 2400 feet is $2400 \times \dfrac{1}{5280}$ or about 0.45 mile. Tip 3 below elaborates on this point.

Tip 2: Dimensional analysis: Sometimes it's easy to get confused when doing unit conversions—do I multiply or divide? The units themselves can help you decide. To illustrate this approach, called *dimensional analysis*, let's consider how we convert yards to feet.

The word "per" always means division, so we can write the conversion factor for yards to feet, 3 feet per yard, as $3\,\dfrac{\text{feet}}{\text{yard}}$. Let's see how this observation helps in converting yards to feet. A football field is 100 yards long. Here's how to convert that length to feet:

$$100 \text{ yards} = 100 \text{ yards} \times 3\,\frac{\text{feet}}{\text{yard}}$$

$$= 100 \times 3 \, \cancel{\text{yards}}\,\frac{\text{feet}}{\cancel{\text{yard}}}$$

$$= 300 \text{ feet.}$$

Note that the "yards" units cancel, leaving "feet" as desired.

This dimensional analysis is even more helpful when we do complex unit conversions. For example, let's convert 60 miles per hour to feet per second. We can perform the conversion in steps: First change miles to feet and then change hours to seconds.

To convert miles to feet, we use the conversion factor of 5280 feet per mile from Table A1.1:

$$60 \text{ miles} = 60 \text{ miles} \times 5280\,\frac{\text{feet}}{\text{mile}}$$

$$= 60 \times 5280 \, \cancel{\text{miles}} \times \frac{\text{feet}}{\cancel{\text{mile}}}$$

$$= 60 \times 5280 \text{ feet.}$$

Note how, after we cancel, the final units come out to be "feet," which is what we intended. The fact that this happened confirms that we're on the right track.

We proceed similarly to convert 1 hour to seconds:

$$1 \text{ hour} = 1 \text{ hour} \times 60\,\frac{\text{minutes}}{\text{hour}} \times 60\,\frac{\text{seconds}}{\text{minute}}$$

$$= 1 \times 60 \times 60 \, \cancel{\text{hour}} \times \frac{\cancel{\text{minutes}}}{\cancel{\text{hour}}} \times \frac{\text{seconds}}{\cancel{\text{minute}}}$$

$$= 60 \times 60 \text{ seconds.}$$

Note again how, after we cancel "minutes" and "hours," the final units come out to be "seconds," which is what we intended.

Therefore,

$$60 \, \frac{\text{miles}}{\text{hour}} = \frac{60 \times 5280 \text{ feet}}{60 \times 60 \text{ seconds}}$$

$$= 88 \, \frac{\text{feet}}{\text{second}}.$$

Tip 3: Reversing conversions: Table A1.2 tells us that there are 2.2 pounds in a kilogram, so to convert from kilograms to pounds, we multiply by 2.2. To convert from pounds to kilograms, we use the reciprocal $\frac{1}{2.2}$:

$$\text{Conversion factor for kilograms to pounds} = 2.2 \, \frac{\text{pounds}}{\text{kilogram}},$$

so

$$\text{Conversion factor for pounds to kilograms} = \frac{1}{2.2} \, \frac{\text{kilograms}}{\text{pound}}.$$

(This is about 0.45 kilogram per pound.) For example, to convert 180 pounds to kilograms, we proceed as follows:

$$180 \text{ pounds} = 180 \text{ pounds} \times \frac{1}{2.2} \, \frac{\text{kilograms}}{\text{pound}}$$

$$= 180 \times \frac{1}{2.2} \, \text{pounds} \, \frac{\text{kilograms}}{\text{pound}}$$

$$= \frac{180}{2.2} \, \text{kilograms}.$$

This is about 82 kilograms.

In general, we find the conversion factor for B units to A units by taking the reciprocal of the conversion factor for A units to B units.

Currency conversion

Unit conversion involving currency arises in connection with currency exchange. For example, as of this writing the value of one U.S. dollar is 0.7773 euro. Hence, the *exchange rate* for converting dollars to euros is 0.7773 euro per dollar. For example, to convert $25 to euros, we proceed as follows:

$$25 \text{ dollars} = 25 \text{ dollars} \times 0.7773 \, \frac{\text{euro}}{\text{dollar}}$$

$$= 25 \times 0.7773 \, \text{dollars} \times \frac{\text{euro}}{\text{dollar}}$$

$$= 25 \times 0.7773 \text{ euros}.$$

This is about 19.43 euros.

To find the exchange rate for conversion in the other direction, we use reciprocals. For example, under an exchange rate of 0.7773 euro per dollar, the exchange rate for converting euros to dollars is $\frac{1}{0.7773}$ or about 1.29 U.S. dollars per euro.

Practice Exercises

In these exercises, round all answers to one decimal place unless you are instructed otherwise.

1. Quarter pounders. How many grams of meat are in a "quarter pounder"?

2. Speed limit. In Canada the speed limit is measured in kilometers per hour (abbreviated km/h), and it's common to see 100 km/h speed limit signs in Canada. How fast is 100 km/h in miles per hour?

3. Cola. You can get a quart bottle of cola for $1.50 or a liter bottle for $1.55. Which is the better buy in terms of price per volume?

Cans and bottles   Soft drinks are often sold in six-packs of 12-ounce cans and in 2-liter bottles. A liter is about 33.8 fluid ounces.

4. Which is the greater volume: a six-pack or 2 liters?

5. A store offers a 2-liter bottle of soft drink for $1.15 and a six-pack of 12-ounce cans for $1.20. Which is the better value (based on the price per ounce)?

6. Currency conversion. As of this writing, 1 U.S. dollar is worth 12.72 Mexican pesos. What is the exchange rate for converting dollars to pesos? What is the exchange rate for converting pesos to dollars? Round your answer to three decimal places.

7. Currency conversion. In the summer of 2010, 1 U.S. dollar was worth 1.04 Canadian dollars. An automobile in Toronto sells for 23,000 Canadian dollars. The same car sells in Detroit for 22,000 U.S. dollars. Which is the better buy?

8. Conversion factors. What is the conversion factor for kilometers to miles? What is the conversion factor for miles to kilometers? Round your answer to two decimal places.

9. Conversion factors. What is the conversion factor for parsecs to lightyears? What is the conversion factor for lightyears to parsecs? Round your answer to two decimal places.

10. Star Wars. In the 1977 film *Star Wars Episode IV: A New Hope*, the character Han Solo says that a certain starship "made the Kessel Run in less than 12 parsecs." Do the units here make sense?*

11. Cubic centimeters. The quantity of fluid administered in an inoculation is usually measured in cubic centimeters, abbreviated cc. One liter is 1000 cc, and 1 liter is about 33.8 fluid ounces. How many cubic centimeters are in 1 fluid ounce?

12. Football field. How many meters long is a football field? (A football field is 100 yards long.)

13. Meters and miles. Which is longer, a 1500-meter race or a 1-mile race? By how much?

14. Meters and yards. Which is longer, a 100-meter dash or a 100-yard dash? By how much? Round your answer to the nearest yard.

15. Records. In 1996 Ato Boldon of UCLA ran the 100-meter dash in 9.92 seconds. In 1969 John Carlos of San Jose State ran the 100-yard dash in 9.1 seconds. Which runner had the faster average speed? What was his speed in yards per second? Round your answer to two decimal places.

16. Gasoline. My car gets 23 miles per gallon, and gasoline costs $2.85 per gallon. If we consider only the cost of gasoline, how much does it cost to drive each mile? Round your answer in dollars to two decimal places.

17. Speed. A car is traveling at 70 miles per hour. What is its speed in feet per second?

18. Gravity. In the metric system the acceleration due to gravity near Earth's surface is about 9.81 meters per second per second. What is the acceleration due to gravity as measured in the English system, that is, in feet per second per second?

19. Speed of light. The speed of light is 300 million meters per second. What is the speed of light measured in miles per hour? (Round your answer to the nearest million miles per hour.) How long does it take light to travel the distance of 93 million miles from the sun to Earth? (Round your answer to the nearest minute.)

Lightning   The speed of sound in air on a certain stormy day is about 1130 feet per second. This fact is used in Exercises 20 and 21.

20. What is the speed of sound measured in miles per second? In miles per hour? Round the former answer to three decimal places and the latter answer to the nearest whole number.

21. You see a flash of lightning striking some distance away, and five seconds later you hear the clap of thunder. How many miles away did the lightning strike? Round your answer to the nearest mile. *Note*: The speed of light is so much faster than the speed of sound that you can assume that the strike occurs at the instant you see it.

A drug bust   The following is an excerpt from the *Roanoke Times*:

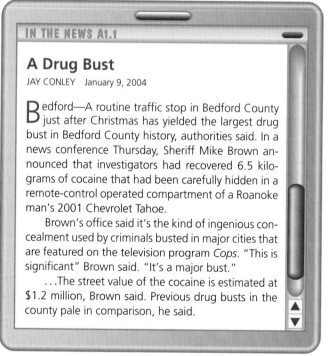

IN THE NEWS A1.1

A Drug Bust

JAY CONLEY   January 9, 2004

Bedford—A routine traffic stop in Bedford County just after Christmas has yielded the largest drug bust in Bedford County history, authorities said. In a news conference Thursday, Sheriff Mike Brown announced that investigators had recovered 6.5 kilograms of cocaine that had been carefully hidden in a remote-control operated compartment of a Roanoke man's 2001 Chevrolet Tahoe.

Brown's office said it's the kind of ingenious concealment used by criminals busted in major cities that are featured on the television program *Cops*. "This is significant" Brown said. "It's a major bust."

…The street value of the cocaine is estimated at $1.2 million, Brown said. Previous drug busts in the county pale in comparison, he said.

This information is used in Exercises 22 through 24.

22. How many pounds of cocaine were seized in this bust?

---

*Various explanations of this can be found on the Web.

23. What is the street value of a pound of cocaine? What is the street value of an ounce of cocaine? In each case, round your answer in dollars to the nearest whole number.

24. Assume that a typical dose contains half a gram of pure cocaine. How much would a dose cost? Round your answer to the nearest dollar.

25. A project, the Gimli glider. There are several examples of errors involving unit conversion that had serious consequences.

One such story with a happy ending involved the *Gimli glider*, a commercial aircraft that ran out of fuel at an altitude of about 28,000 feet. The crew glided the aircraft to a landing, and no one was seriously injured. Do a research project on the Gimli glider. Pay special attention to how unit conversion played a part in the incident.

Brief Answers to Practice Exercises

1. 113.4 grams.

2. 62 miles per hour.

3. The liter bottle.

4. A six-pack.

5. A six-pack.

6. 12.72 pesos per dollar; 0.079 dollar per peso.

7. The car in Detroit.

8. 0.62 mile per kilometer; 1.61 kilometers per mile.

9. 3.26 lightyears per parsec; 0.31 parsec per lightyear.

10. No, a parsec is a measure of distance.

11. 29.6 cc.

12. 91.5 meters.

13. A 1-mile race is 360 feet longer.

14. A 100-meter dash is about 9 yards longer.

15. Ato Boldon; 11.02 yards per second.

16. $0.12 dollar.

17. 102.7 feet per second.

18. 32.2 feet per second per second.

19. 671 million miles per hour; 8 minutes.

20. 0.214 mile per second; 770 miles per hour.

21. 1 mile.

22. 14.3 pounds.

23. $83,916; $5245.

24. $92

25. Answers for this research project will vary.

APPENDIX 2 Exponents and Scientific Notation

The following is a summary of the basic rules for exponents.

Exponents

Negative exponents: a^{-n} is the reciprocal of a^n.

Definition	Example
$a^{-n} = \dfrac{1}{a^n}$	$2^{-3} = \dfrac{1}{2^3} = \dfrac{1}{8}$

Zero exponent: If a is not 0, then a^0 is 1.

Definition	Example
$a^0 = 1$	$2^0 = 1$

Basic properties of exponents:

Property	Example
$a^p a^q = a^{p+q}$	$2^2 \times 2^3 = 2^{2+3} = 2^5 = 32$
$\dfrac{a^p}{a^q} = a^{p-q}$	$\dfrac{3^6}{3^4} = 3^{6-4} = 3^2 = 9$
$(a^p)^q = a^{pq}$	$\left(2^3\right)^2 = 2^{3 \times 2} = 2^6 = 64$

Powers of 10

Powers of 10 provide an easy way to represent both large and small numbers. If n is a positive whole number, we find 10^n by starting with 1 and moving the decimal point n places to the right. For example,

For 10^3 *3 tells how to move the decimal point*   1.000   *right shift by 3 places results in*   1000.

A negative power of 10 causes the decimal point to be moved to the *left*. If n is a positive whole number, we find 10^{-n} by starting with 1 and moving the decimal point n places to the left. For example,

For 10^{-3} *−3 tells how to move the decimal point*   0001.0   *left shift by 3 places results in*   0.001

Scientific notation

When we multiply by a power of 10, the effect is to move the decimal point, just as when powers of 10 stand alone. For example,

For 3.45×10^3 *3 tells how to move the decimal point*   3.450   *right shift by 3 places results in*   3450.

For 3.45×10^{-3} *−3 tells how to move the decimal point*   003.45   *left shift by 3 places results in*   0.00345

This observation leads to the idea of expressing numbers in *scientific notation*.

Key Concept

A positive number is said to be expressed in **scientific notation** if it is written as a power of 10 multiplied by a number that is less than 10 and greater than or equal to 1.

For example, to write 24,500 in scientific notation, we start with 2.45 because that is between 1 and 10. To get 24,500 from 2.45, we need to move the decimal place four places to the right. That is the power of 10 that we use. The result is

24,500 expressed in scientific notation is 2.54×10^4.

To write 0.000245 in scientific notation, we start once more with 2.54. This time we need to move the decimal point four places to the left. That means we use an exponent of -4:

0.000245 expressed in scientific notation is 2.45×10^{-4}.

CALCULATION TIP A2.1

Calculators often resort to scientific notation when they encounter very large or very small numbers. They usually do this by expressing the power of 10 with the letter "E" or "e". For example, if I enter $2,000,000^{10}$ in my Microsoft Windows© calculator, I see in the display

$$1.024e+63$$

This means 1.024×10^{63}, which is scientific notation.

If I enter 2^{-100} in my calculator, I see in the display

$$7.8886090522101180541172856528279e{-}31$$

Rounding off to one place, this means 7.9×10^{-31}, which is scientific notation.

Scientific notation is useful in dealing with very large and very small numbers. For example, according to the U.S. Treasury Department, the national debt as of January 22, 2009, was \$10,618,718,703,374.78. This is over 10.6 *trillion* dollars. To write 10.6 trillion or 10,600,000,000,000 in scientific notation, we start with 1.06 because that number is between 1 and 10. To get 10,600,000,000,000 from 1.06, we must move the decimal point 13 places to the right. This is the power of 10 we use:

$$10.6 \text{ trillion} = 10,600,000,000,000 = 1.06 \times 10^{13}.$$

In chemistry there is a very large number called *Avogadro's number*. It is the number of atoms in 12 grams of carbon-12 and is about 602,000,000,000,000,000,000,000 (21 zeros in a row). To express this number in scientific notation, we start with 6.02 because it is between 1 and 10. We need to move the decimal point 23 places to the right, so this is the power of 10 that we use. The result is 6.02×10^{23}.

As another example, the wavelength of blue light is 0.000 000 475 meters (6 zeros after the decimal place). To express this number in scientific notation, we begin with 4.75 because it is between 1 and 10. We need to move the decimal seven places to the left, so we use -7 for the power of 10. The result is 4.75×10^{-7}.

Scientific notation not only gives us an easy way to express very large and very small numbers, but also makes it easier for us to do calculations based on the rules of exponents. For example, light travels about 186 thousand miles per second. Let's figure out how far it will travel in a year (i.e., the number of miles in a lightyear). In scientific notation, 186 thousand is 1.86×10^5. There are $60 \times 60 \times 24 \times 365$ or about 31.5 million seconds in a year. That is 3.15×10^7. So in a year, light will travel about

$$1.86 \times 10^5 \times 3.15 \times 10^7 = (1.86 \times 3.15) \times (10^5 \times 10^7)$$
$$= (1.86 \times 3.15) \times 10^{5+7}$$
$$= 5.859 \times 10^{12} \text{ miles.}$$

That is almost 6 trillion miles.

Practice Exercises

1. Earth to moon. The distance from the center of Earth to the center of the moon is about 239,000 miles. Express this number in scientific notation.

Scientific notation In Exercises 2 through 9, express the given number in scientific notation.

2. Two hundred million

3. 953,000

4. Twenty-eight billion

5. Twenty and a half trillion

6. 0.12 (twelve hundredths)

7. 0.0003 (three ten-thousandths)

8. Fifteen millionths

9. One-hundred-and-eight billionths

World population As of this writing, the world's population was about 6.8 billion, and the population of Canada was 34 million. This information is used in Exercises 10 through 12.

10. Write in scientific notation the world population.

11. Write in scientific notation the population of Canada.

12. What percentage of world population is the population of Canada?

Brief Answers to Practice Exercises

1. 2.39×10^5

2. 2×10^8

3. 9.53×10^5

4. 2.8×10^{10}

5. 2.05×10^{13}

6. 1.2×10^{-1}

7. 3×10^{-4}

8. 1.5×10^{-5}

9. 1.08×10^{-7}

10. 6.8×10^9

11. 3.4×10^7

12. 0.5%

APPENDIX 3 Calculators, Parentheses, and Rounding

Parentheses

Difficulties with calculators often stem from improper use of parentheses. When they occur in a formula, their use is essential, and sometimes additional parentheses must be supplied.

For example, expressions like $(1 + 0.06/12)^{36}$ arise in Chapter 4 when we consider compound interest. In such expressions, the parentheses are crucial. If you omit them, you may get the wrong answer. Let's see what happens. The correct calculator entry is

$$\boxed{(}\ 1\ \boxed{+}\ .06\ \boxed{\div}\ 12\ \boxed{)}\quad\boxed{\wedge}\ 36.$$

The calculator's answer is 1.196680525. If the parentheses are omitted, the expression becomes $1 + 0.06/12^{36}$, which we enter as

$$1\ \boxed{+}\ .06\ \boxed{\div}\ 12\ \boxed{\wedge}\ 36.$$

The calculator's answer is 1. As we can see, the parentheses make a big difference.

Sometimes when parentheses do not appear, we must supply them. For example, no parentheses appear in the expression $\dfrac{1.24 + 3.75}{1.62}$. But when we enter this expression, we must enclose $1.24 + 3.75$ in parentheses to tell the calculator that the entire expression goes in the numerator. The correct entry is

$$\boxed{(}\ 1.24\ \boxed{+}\ 3.75\ \boxed{)}\ \boxed{\div}\ 1.62.$$

The calculator gives an answer of 3.080246914.

If we forget to use parentheses and enter $1.24\ \boxed{+}\ 3.75\ \boxed{\div}\ 1.62$, the calculator thinks only 3.75 (not the entire expression $1.24 + 3.75$) goes in the numerator. The result is a wrong answer: 3.554814815.

EXAMPLE A3.1 Using parentheses

Supply parentheses for proper entry of the following expressions into a calculator.

a. $\dfrac{2}{3.6 + 5.7}$.

b. $2^{1.7 \times 3.9}$

SOLUTION

a. We need to enclose $3.6 + 5.7$ in parentheses to indicate that the entire expression is included in the denominator of the fraction. The proper entry is

$$2\ \boxed{\div}\quad\boxed{(}\ 3.6\ \boxed{+}\ 5.7\ \boxed{)}.$$

The result is 0.2150537634.

b. We need to enclose 1.7×3.9 in parentheses to indicate that the entire expression is the exponent. The proper entry is

$$2\ \boxed{\wedge}\quad\boxed{(}\ 1.7\ \boxed{\times}\ 3.9\ \boxed{)}.$$

The result is 99.04415959.

Rounding

It is often unnecessary to report all the digits given as an answer by a calculator, so it is common practice to shorten the answer by *rounding*.

First we need to decide how many digits to keep when we round. There is no set rule for determining this number. Generally, in this text we keep one digit beyond the decimal point. In certain applications, though, it is appropriate to keep fewer or more digits. For example, when we report the balance of an account in dollars, we usually keep two digits beyond the decimal point, that is, we round to the nearest cent. Another consideration in rounding is the accuracy of the data on which we base our calculation. We should not report more digits in our final result than the number of digits in the original data.

The usual method for rounding a number begins with the digit to be rounded. That digit is increased by 1 if the digit to its right is 5 or greater and is left unchanged if the digit to its right is less than 5. All the digits to the right of the rounded digit are dropped.

For example, let's round 23.684 to one decimal place. That means we are going to keep only one digit beyond the decimal point. We focus on the 6. Since the next digit to its right (8) is 5 or larger, we increase the 6 to 7 and drop all the digits to the right of it. This results in 23.7.

If we choose to round 23.684 to two decimal places, we are going to keep two digits beyond the decimal point. We focus on the 8. The next digit is a 4, which is less than 5. So we leave the 8 unchanged and delete the remaining digits. This results in 23.68.

Although it is often appropriate to round final answers, rounding in the middle of a calculation can lead to errors. We were cautioned about this in Calculation Tip 4.1 in Section 4.1. Let's look a bit more closely at this situation.

Here is a typical financial calculation. It gives the balance in dollars of an account after 5 years of daily compounding:

$$\text{Balance} = 10{,}000 \left(1 + \frac{0.06}{365}\right)^{1825}.$$

The right way to calculate it: Use parentheses and enter the formula as it reads. Here is the sequence of keystrokes:

$$10000 \;\boxed{\times}\; \boxed{(}\; 1 \;\boxed{+}\; .06 \;\boxed{\div}\; 365 \;\boxed{)}\; \boxed{\wedge}\; 1825\; .$$

What you should see in the calculator display is

```
10000*(1+.06/365)^1825
        13498.25527
```

Rounding to the nearest cent yields 13,498.26 dollars.

Warning! The wrong way to calculate it: It is not uncommon for students to calculate $0.06/365 = 0.00016438\cdots$, round this number to 0.00016, and plug that result into the formula. This is a bad idea. Here's what happens if you do this: You are calculating $10{,}000(1.00016)^{1825}$. This gives 13,390.7174, which rounds to 13,390.72 dollars. The difference between this answer and the correct answer above is $107.54. The reason for this discrepancy is that your calculator stores a lot of decimal places that can have a big effect on the final answer. It's the rounding of $0.06/365 = 0.00016438\cdots$ to 0.00016 that makes the $107.54 error.

The moral of the story is that it is best to avoid rounding numbers in the middle of a calculation. Avoiding rounding not only improves accuracy but also is less work. Next we see one way to avoid rounding in the middle.

Chain calculations

For complicated formulas, it is possible to make calculations in steps without intermediate rounding. To illustrate the procedure, let's recall the monthly payment formula from Section 4.2:

$$\text{Monthly payment} = \frac{\text{Amount borrowed} \times r(1+r)^t}{((1+r)^t - 1)}.$$

Here, t is the term in months and $r = \text{APR}/12$ is the monthly interest rate as a decimal. Let's do the calculation when the amount borrowed is $10,000, $t = 30$ months, and the monthly rate is $r = 0.05/12$. With these values, the formula gives

$$\text{Monthly payment} = \frac{10000 \times 0.05/12 \times (1+0.05/12)^{30}}{\left((1+0.05/12)^{30} - 1\right)}.$$

If we choose to enter the entire expression all at once, we must be very careful to get all the parentheses right:

10000 $\boxed{\times}$ .05 $\boxed{\div}$ 12 $\boxed{\times}$ $\boxed{(}$ 1 + .05 $\boxed{\div}$ 12 $\boxed{)}$ $\boxed{\wedge}$ 30

$\boxed{\div}$ $\boxed{(}$ $\boxed{(}$ 1 $\boxed{+}$ .05 $\boxed{\div}$ 12 $\boxed{)}$ $\boxed{\wedge}$ 30 $\boxed{-}$ 1 $\boxed{)}$.

It is difficult to enter this expression without making a mistake. Here is a method that makes entering a bit easier. First calculate $(1 + .05/12)^{30}$ using

$\boxed{(}$ 1 $\boxed{+}$ .05 $\boxed{\div}$ 12 $\boxed{)}$ $\boxed{\wedge}$ 30.

The calculator displays 1.132854218, and it automatically stores the result. We can recall it using the keystrokes $\boxed{\text{2nd}}$ [ANS]. So we can now enter the formula as

10000 $\boxed{\times}$ .05 $\boxed{\div}$ 12 $\boxed{\times}$ $\boxed{\text{2nd}}$ [ANS] $\boxed{\div}$ $\boxed{(}$ $\boxed{\text{2nd}}$ [ANS] $\boxed{-}$ 1 $\boxed{)}$.

The answer, rounded to the nearest cent, is $355.29.

Practice Exercises

Using parentheses Exercises 1 through 7 focus on the use of parentheses. In each case, show how parentheses should be used to calculate the given expression and then find the value rounded to one decimal place.

1. $\dfrac{3.1 + 5.8}{2.6}$

2. $\dfrac{3.1}{5.8 - 2.6}$

3. $\dfrac{3.1 - 1.3}{5.8 + 2.6}$

4. $3^{1.1-2.2}$

5. $3 - \dfrac{1.6 + 3.7}{4.3}$

6. Calculate the following three expressions. The only difference is placement of parentheses. Compare the results.
   a. $2 * 3 + 4$
   b. $2 * (3 + 4)$
   c. $(2 * 3) + 4$

7. Calculate the following four expressions. The only difference is placement of parentheses. Compare the results.
   a. $2 * 3 + 4^2$
   b. $(2 * 3 + 4)^2$
   c. $2 * (3 + 4)^2$
   d. $(2 * 3) + 4^2$

8. Round the number 14.2361172 to one decimal place.

9. Round the number 14.2361172 to two decimal places.

10. Round the number 14.2361172 to three decimal places.

11. **Intermediate rounding.** This exercise shows problems that can arise using improper calculation techniques.
   a. Calculate $100,000(1 + 0.07/12)^{100}$ to the nearest cent.
   b. Observe that $0.07/12 = 0.0058333333\cdots$. Round this to 0.006, and use that result to calculate $100,000(1 + 0.07/12)^{100}$ to the nearest cent.
   c. What is the difference between your two answers?

Brief Answers to Practice Exercises

1. $(3.1 + 5.8) \div 2.6 = 3.4$
2. $3.1 \div (5.8 - 2.6) = 1.0$
3. $(3.1 - 1.3) \div (5.8 + 2.6) = 0.2$
4. $3 \wedge (1.1 - 2.2) = 0.3$
5. $3 - (1.6 + 3.7) \div 4.3 = 1.8$
6. a. 10
   b. 14
   c. 10
7. a. 22

b. 100

c. 98

d. 22

8. 14.2

9. 14.24

10. 14.236

11. a. $178,896.73

b. $181,885.50

c. $2988.77

BRIEF ANSWERS TO ODD-NUMBERED EXERCISES

Exercise Set 1.1

1.

Fair condition	
	% who lived
Mercy	98.3%
County	96.6%

3. Answers will vary.

5. 75%

7. 10%

9. Answers will vary.

11. Answers will vary.

13.

Below poverty line	
	% passing
School A	52.5%
School B	36.2%

15. 90.5%

17. 36.2%

19. Answers will vary.

21.

All participants		
	Total	% with insurance
Not traced	1174	17%
5-year group	416	11%

23.

White participants		
	Total	% with insurance
Not traced	126	83%
5-year group	12	83%

25. They are the same in both cases.

27. Answers will vary.

Exercise Set 1.2

1. The premises are that all dogs go to heaven and my terrier is a dog. The conclusion is that my terrier will go to heaven.

3. The premise is that the president was caught in a lie. The conclusion is that nobody believes the president anymore.

5. The premise is that people have freedom of choice. The conclusion is that they will choose peace.

7. The premise is that bad times have a scientific value. The conclusion is that these are occasions a good learner would not miss.

9. The argument is invalid.

11. The argument is invalid.

13. The argument is invalid.

15. Dismissal based on personal attack

17. False cause

19. False dilemma

21. Appeal to common practice

23. Appeal to ignorance

25. Straw man

27. False dilemma

29. Circular reasoning

31. Straw man

33. Answers will vary.

35. Answers will vary.

37. The number of blocks is the cube (the third power) of the size of a side. 125,000 blocks fit in a cube 50 inches on each side.

39. These are the positive odd integers in order.

41. Answers will vary.

Exercise Set 1.3

1. (NOT p) → (NOT q)

3. (France OR England) AND (eighteenth OR nineteenth century)

5. If I don't buy this MP3 player, then I can't listen to my favorite songs. No.

7. Answers will vary.

9. q → (NOT p)

11. q → (NOT p)

13. p: "We pass my bill"; r: "The economy will recover"; $p \rightarrow r$

15. c: "I want cereal for breakfast"; e: "I want eggs for breakfast"; c AND e

17. c: "You clean your room"; f: "I'll tell your father"; (NOT c) → f

19. d: "I'll go downtown with you"; m: "We can go to a movie"; r: "We can go to a restaurant"; (m OR r) → d

21. "It is not the case that he's American or Canadian." "He's not American and he's not Canadian."

23. False   **25.** True   **27.** True

31. False

33. a: True; b: False; c: True

35. If you clean your room, I won't tell your father.

37. If you don't agree to go to a movie or a restaurant, then I won't go downtown with you.

39. If it is not a math course, then it is not important.

41. If you don't take your medicine, you won't get well.

43. If I told your father, then you didn't clean your room.

45. If I went downtown with you, then you agreed to go to a movie or a restaurant.

47. All important courses are math courses.

49. If you get well, then you took your medicine.

51. If you do not get arrested, then you did not drink and drive.

53. If I didn't tell your father, then you didn't clean your room.

55. If I did not go downtown with you, then you did not agree to go to a movie or a restaurant.

57. If you do not drive a Porsche, then you are not my friend.

59. If radar does not see thunderstorms, then they are not approaching.

61.

p	q	(p OR q)	$\rightarrow$
T	T	T	T
T	F	T	T
F	T	T	T
F	F	F	T

63.

p	q	(p OR q)	AND
T	T	T	T
T	F	T	T
F	T	T	F
F	F	F	F

65. Their truth values are both (F, T, T, T).

67. 0

69. 1

~~   ~~ s will vary.

k, l, m, n, p, q, r, s, t, v, w, x, y, z}

s of *A*. *B* and *C* are disjoint.

7. *A* is a subset of *B*, *A* is a subset of *C*, and *B* is a subset of *C*.

9.

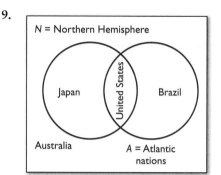

11.

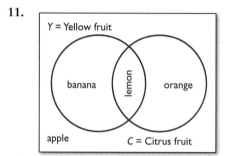

13. 68

15. 851

17. 991

19. 1989

21.

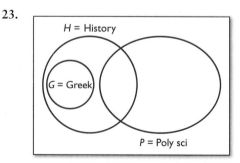

23.

25. Answers will vary.

27. True positives: 780

29.

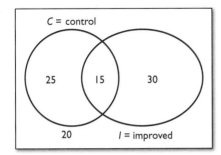

31. 84 blue cars

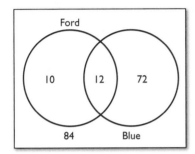

33.

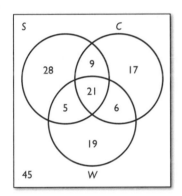

35. 24

37. 27

39. 64

41. 140

43. 88

45. Answers will vary.

Exercise Set 1.5

1. The bacterium is 50 times as large as the virus.

3. $10,000

5. 10 million

7. 96,907 miles. That's nearly four times around the world.

9. 39,878 cubic feet

11. 9.5 years

13. 2.0 blogs each second

15. About 25 minutes

17. Answers will vary.

19. $100

21. Estimate 1%; exact figure 1.15%

23. The carpet costing $12 per square yard

25. 25,592 people per square mile

Exercise Set 2.1

1. One choice: Let the year be the independent variable and the number of home runs the dependent variable. Then the function gives the number of home runs in a given year.

3. 63.1%

5. 9487

7. 10,473; answers will vary

9. About 0.7 percentage point per year (Some variation in answers is acceptable.)

11. The percentage grew more slowly after 2005 than it did during the first part of that decade.

13. 1975: 38,979 million dollars; 1991: 144,869 million dollars

15. 573,108 million dollars

17. 57.5%

19. 1996: 2.9%; 1997: 3.0%; 1998: 3.2%; 1999: 3.3% (Some variation in answers is acceptable.)

21. $98

23. $454 per student

25. −0.07 dollar per student for each additional student enrolled

27. The cost per student decreases more at the low end of the scale, for consolidations of smaller schools.

29. Answers will vary.

31. Enrollment in French is declining.

33.

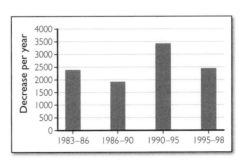

35. German

37. 33.5%

Exercise Set 2.2

1. Bar graph:

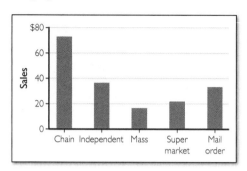

Scatterplot:

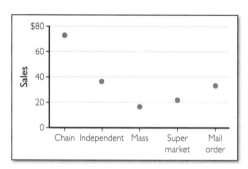

Line graph:

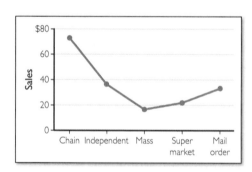

Answers will vary.

3. One possibility is

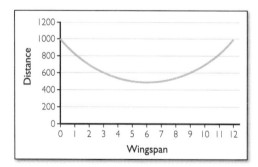

5. One possibility is

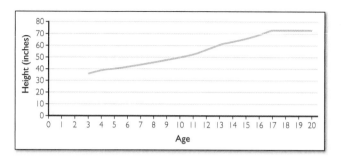

7. One possibility is

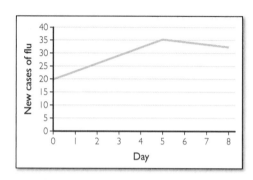

9.

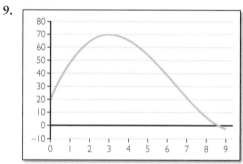

11. One possibility is

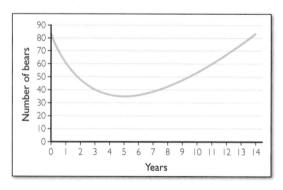

13. Answers will vary.

15. No. Answers will vary.

17. Mystery curve 1: growth; Mystery curve 2: sales. Answers will vary.

19. Answers will vary.

Exercise Set 2.3

1. A narrow range on the vertical axis should be used. One possibility is

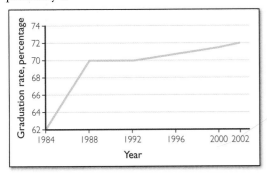

3. We can emphasize the changes by choosing to display a range of numbers on the vertical axis that is much closer to the data values. In the following figure, we have used a range on the vertical axis of 170 to 215.

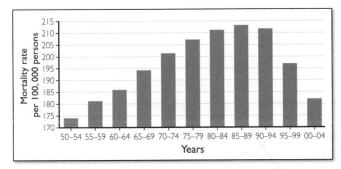

5. The fact that nearly 60% of employees are in the un-known category makes the graph of questionable value.

7. The lines connecting the dots do not represent any conceivable data. A table, or perhaps a bar graph or scatterplot, would be better.

9. Answers will vary. Knowing the number of drivers in each category would be important.

11.

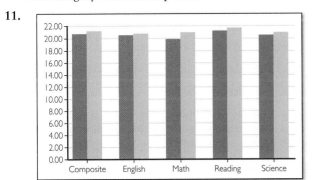

13. a. From 1945 to 1990
   b. 0.29 thousand dollars per year, or 290 dollars per year
   c. 1.14 thousand dollars per year, or 1140 dollars per year
   d. Answers will vary.

15. A justification is difficult to find. Perhaps it seems natural to predict growth using a straight line.

17. Answers will vary.

19.

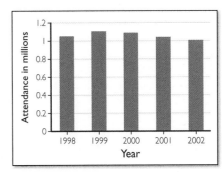

Answers will vary.

21.

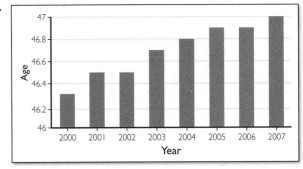

23. About 1.5%

25.

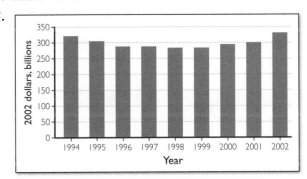

Answers will vary.

27. a.

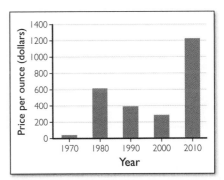

b.

Year	2010 dollars
1970	$198.03
1980	$1555.90
1990	$628.96
2000	$351.68
2010	$1224.53

c.

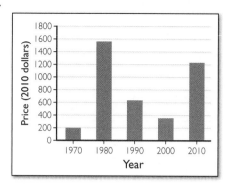

29. a.

Year	Price (2010 dollars)
1970	$82.65
1980	$76.20
1990	$205.00
2000	$409.50
2010	$500.00

b.

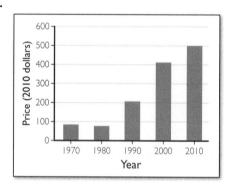

c. Tickets were about 6 times as expensive in 2010.

31. The vertical axis represents percentage of the total by weight.

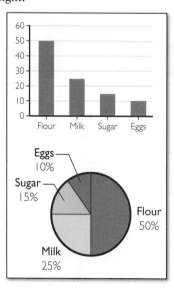

1. Yes

3. No

5. No

7. No

9. Yes. If M is the number of miles and g is the number of gallons, then $M = 25g$.

11. No

13. It is linear. $T = 0.25I + 4393$

15. The growth rate is 200 feet per mile. The formula is $E = 200h + 3500$.

17. About $-\$10$ per month

19. The formula predicts $-\$1$.

21. The slope is 17 meters per second per kilometer. The initial value is 1534 meters per second. Answers will vary.

23. 1568 meters per second

25. 1568 meters per second

27. -0.746 mile per hour per degree; answers will vary

29. 35.07, or about 35, miles per hour

31. If C is the amount collected (in trillions of dollars) and t is the time in years since 2003, then $C = 0.178t + 1.952$.

33. If N is the net income and s is the total sales, both in dollars, then $N = 0.03s - 15{,}257$. The commission is 3%.

35. There is a constant growth rate of 14 dollars in tax per 50 dollars in taxable income.

37. $0.28

39. There is a constant growth rate of 6 patients diagnosed per 5 days, or 1.2 patients diagnosed per day.

41. $F = 1.2t + 35$ if F denotes the number of diagnosed flu cases and t the time in days.

43. No. The data do not appear to be linear.

45. 2007; answers will vary.

47. From $67,500 through $67,800, there is $15 additional tax due for each span. From $67,900 through $68,600, there is $25 additional tax due for each span.

49. 15% and 25%

51. Answers will vary.

53. 2004

55. About 130.6; no

1. Mary

3. Answers will vary.

5. 0.2%; answers will vary

7. If we use H for the number of hosts (in millions) and t for the time in years since 1995, the formula is $H = 8.2 \times 1.43^t$.

9. $2.64 billion; $6.06 billion

11. 164 degrees

13. 0.67

15. Answers will vary. If N is the population and t is time in years since the start, then $N = 500 \times 1.02^t$. After 5 years: 552.

17. Answers will vary; 336.23 million

19. 305%

21. Measure the thickness of a stack and divide by the number of sheets.

23. $0.0000001 \times 2^{50} = 112,589,991$ kilometers

25. $180.61

27. $23.42

29. 8192 million, or 8.192 billion. Probably not—it's comparable to the world population.

31. 16,777,216 million

33. At age 55: $531,759.45; at age 65: $3,174,243.86. At age 55: option 1; at age 65: option 2

35. 27.1 pounds

37. 1.95 parts per million

39. $10^5 = 100,000$; 10^{10}

41. 24%

43. Answers will vary; 228%

45. 21.2 years after 2010, or early 2032

Exercise Set 3.3

1. Answers will vary.

3. 100 times

5. The one in Alaska was about twice as intense.

7. $1.76 \times 10^{16} = 17,600,000,000,000,000$

9. The Indonesian quake released 31.6 times as much energy as the San Francisco quake.

11. 83 decibels

13. 75 decibels

15. 1.16

17. 1000 times

19. After 153 months, or 12 years and 9 months

21. 2.32 half-lives; 69.6 years

23. $C = 100 \times 1.025^t$

25. 56 years

27. $348.68

29. 26 trillion dollars

31. Answers will vary.

33. Answers will vary.

35. Answers will vary.

37. Answers will vary.

39. Answers will vary.

Exercise Set 4.1

1. $50; $2050

3.

Quarter	Interest	Balance
1	$60.00	$3060.00
2	$61.20	$3121.20
3	$62.42	$3183.62
4	$63.67	$3247.29

5.

Quarter	Interest	Balance
1	$50.00	$2050.00
2	$51.25	$2101.25
3	$52.53	$2153.78
4	$53.84	$2207.62

7. Simple interest would yield $200 in interest. Semi-annual compounding would yield $205 in interest.

9. For simple interest, the balance is $2200.00. For monthly compounding, the balance is $3310.20.

11. $657.50

13. Option 2 is better if you retire at age 65. Option 1 is better if you retire at age 55.

15. $7607.64

17. First quarter: $2040.00. Second quarter: $2080.80. Third quarter: $2122.42. Fourth quarter: $2164.87. Total interest $164.87. The APY is 8.24%.

19. 3.91%, a bit larger than the table

21. a. $412.16; 8.24%
    b. $5400. We know that compounding is taking place because, according to the table, the APY is higher than the APR.
    c. The APY is 4.05%, which agrees with the value in the table.

23. True

25. False

27. False

29. 7.50%

31. $3313.57

33. $2451.36. This is how much the investment will be worth after 10 years.

35. It doubles in 8 years and doubles again in 8 more years.

37. 11 years and 7 months. This is 5 months less than the estimate of 12 years given by the Rule of 72.

39. 8.19%

Exercise Set 4.2

1. $386.66; $3199.60

3. The rule of thumb says that the payment is at least $1041.67. The monthly payment is $1096.78.

5. $15,014.57

7. Rule of Thumb 4.2 gives the correct answer of $750.00 per month.

9. $14.17

11. $227.20

13. $1904.90

15. $664.81; $955.44

17. Answers will vary.

19. We omit the column showing the monthly payment of $8.56.

No.	Applied interest	Applied balance	Outstanding balance
			$100.00
1	$0.42	$8.14	$91.86
2	$0.38	$8.18	$83.68
3	$0.35	$8.21	$75.47

21. 63.57%

23. $583.74

25. You will have paid $120,146.40 in interest. It is more than twice as much as with the 15-year mortgage.

27. After 90 months, your equity is $36,856.30. That is 36.9% of the principal. Answers will vary.

29. $3536.06

31. You still owe $1963.94.

33. $801.83

35. Yes, this is too good to be true: By Rule of Thumb 4.2, the payment must be at least $277.78.

37. Answers will vary.

39. $3333.33

41. $683.33

43. We can afford to borrow $132,675. 74, or about $132,676, and this is enough to purchase the home.

45. Answers will vary.

47. 3 years

Exercise Set 4.3

1. $5027.42

3. $116.91

5. Both give $202.26.

7. $33,394.34

9. The formula gives $301.97; the table gives $301.98.

11. $31,733.38

13. The total amount is $132.07. Benny is wrong by part 2 of Rule of Thumb 4.4.

15. $10,012.57

17. $3295.02

19. $597,447.22

21. $497,872.68

23. $625,357.24

25. $279,161.54

27. $14,479.30

29. $4000.00

31. $3000.00

33. $960,000.00

35. $750,000.00

37. $137,799.63

39. Starting at age 40 requires a monthly deposit of nearly 4 times that for starting at age 20. Starting at age 20 requires $27,000. Starting at age 40 requires $59,655, over twice as much.

41. 4 years

43. Answers will vary.

45.

Age	Balance
60	$498,648.60
61	$533,721.61
62	$570,957.83
63	$610,490.71
64	$652,461.89
65	$697,021.76
66	$744,329.98
67	$794,556.07
68	$847,880.00
69	$904,492.83
70	$964,597.41

Exercise Set 4.4

1. We list only previous balance, finance charge, and new balance.

Previous balance	Charge	New balance
$800.00	$9.33	$709.33
$709.33	$13.46	$1022.79

3. We list only previous balance, finance charge, and new balance.

Previous balance	Charge	New balance
$1000.00	$24.70	$1324.70
$1324.70	$32.77	$1757.47

5. We present an abbreviated table.

Previous balance	Payment	Charge
$4500.00	$112.50	$97.75
$4985.25	$124.63	$103.21

7. $4500 \times (1.0175 \times 0.975)^t$ dollars

9. $1729.40

11. $9531.25

13. 13,244 months, or 1103 years and 8 months

15. 172 months, or 14 years and 4 months

17. 135 months, or 11 years and 3 months

19. $3000 \times (1.01 \times 0.95)^t$ dollars

21. $21.01

23. 49 months, or 4 years and 1 month

25. $2500

27. $70.21

29. $681.75

31. $1414.00

33. The balance would increase each month if you made only the minimum payment, even if you made no other charges.

35. We present only a portion of the table.

Month	Previous balance	Payment	Finance charge
21	$437.42	$21.87	$4.16
22	$419.71	$20.99	$3.99
23	$402.71	$20.14	$3.83

The next payment is $19.32.

Exercise Set 4.5

1. 1979; 13.3%. There was no CPI figure in 1944 from which to compute a percent change.

3. $3120.00

5. 39.0%

7.

Year	CPI	Rate
1936	20	100%
1937	40	100%
1938	80	100%

9. 300%; no

11. 23.08%, or about 23.1%

13. $107.51

15. $4081.25

17. $5100.00

19. Answers will vary.

21. $17,883.25

23. Answers will vary.

25. $41,300; $5357.50

27. $127.50; a 2.7% increase

29. The DJIA increases by 15.12 points.

31. 1996; 1999

33. The stock is bought on day 3 and sold on day 5. The profit is $1200.

35. $1500

37. Answers will vary.

39. Answers will vary.

41. $1834.41

43. $19,925

Exercise Set 5.1

1. a. 0.5; 1/2
   b. 0.25; 25%
   c. 1/20; 5%
   d. 0.65; 65/100 = 13/20
   e. 0.875; 87.5%
   f. 37/100; 37%

3. Opinion

5. Empirical

7. Opinion

9. They are not always equally likely outcomes. It depends on the weather conditions.

11. A fraction is 0 exactly when the numerator is 0.

13. 90.8%

15. Answers will vary.

17. 7/8

19. 3/4

21. Those outcomes are not equally likely.

23. 13/15

25. 5/15 = 1/3

27. 25/144

29. 125/1728

31. 8

33. 3/8

35. 7/8

37. 66%

39. 1/2

41. Women: 15.4%; men: 19.4%

43. More than twice as high

45. $\dfrac{34}{52,250,000}$; empirical

47. Answers will vary.

49. 1/3; 9/19

51. 1/2; 1 to 1

53. 7 to 3

55. 5/7; 2/7

57. 6/11

59. Answers will vary.

61. Answers will vary.

63. Answers will vary.

Exercise Set 5.2

1. 100%; 0%

3. 20.0%

5. Answers will vary.

7. Sensitivity is 75.0%. Specificity is 85.0%

9. NPV is higher if prevalence is low.

11. PPV: 50.4%; NPV: 96.6%

13. 0.41/0.65 is about 0.631.

15. PPV: 55.6%; NPV: 93.2%

17. 95%

19. Exactly 50%

21. 99.7%

23. PPV: 82.6%; NPV: 98.7%

25.

	Ill	Healthy	Total
Positive	495	95	590
Negative	5	9405	9410
Total	500	9500	10,000

27. 99.9%

29. 69% of those without prostate cancer test negative.

31. PPV: 62.7%; NPV: 82.5%

33. a. Exactly 50%
    b. 100.0%
    c. 999

35. Answers will vary.

37. 59.9%; NPV

39. 1/3

41. 7.1%

43. a. 1/52
    b. 1/4
    c. 1/13
    d. 1/52 = 1/4 × 1/13

45. The probabilities are not the same.

47. a. 1/2
    b. 1/3
    c. 2/5 or 0.40

49. Answers will vary.

Exercise Set 5.3

1. You can make many more plates using numerals and letters than with only numerals.

3. 40

5. 10,000,000

7. 52

9. a. $21^4 = 194,481$
   b. 1/194,481
   c. 50/194,481

11. Just over 28 years

13. No

15. Yes

17. Answers will vary.

19. 1/4

21. Yes; 4/9

23. 1/9

25. 8/9

27. (10 × 5)/(15 × 14) = 5/21

29. 25/144

31. $4^2/52^2 = 1/169$

33. $13^2/52^2 = 1/16$

35. $(13 \times 12)/(52 \times 51) = 1/17$

37. $50/95 = 10/19$

39. 90.8%

41. 80.2%

43. 1/36

45. 1/9

47. Answers will vary.

49. Answers will vary.

51. 72.9%

53. 5

55. 80%

57. 11%

59. $1/16^4 = 1/65{,}536$; 46

61. $31/16^2 = 31/256$; 11.7% or about 12%

63. Probability of at least one double six: $1 - (35/36)^{24}$ or 0.49

65. These numbers are so large that only specialized mathematical software can calculate them exactly. For 50 people, the correct answer is 30,414,093,201,713, 378,043,612,608,166,064,768,844,377,641,568,960, 512,000,000,000,000 or about 3.041×10^{64}. For 200, the answer is too large to reproduce here. The answer is approximately 7.887×10^{374}.

Exercise Set 5.4

1. 56

3. Yes

5. No

7. No

9. Yes

11. $26^3 = 17{,}576$; $26^4 = 456{,}976$

13. $12! = 479{,}001{,}600$

15. 366!

17. 56

19. 35

21. 21

23. 2002

25. 715

27. Yes

29. 0.049

31. 0.031

33. 60

35. 0.1176

37. 0.0002

39. 0.0003

41. 9828

43. When we counted the number of ways to select 2 days from 7, we used combinations, which do not consider order. But when we counted the number of ways to choose 2 birthdays from 31, we used permutations, which do take order into account. The correct answer is 21/465.

45. $1 - \dfrac{490!/(10!480!)}{500!/(10!490!)}$

47. 2359/117,999 or about 0.02

49. Answers will vary.

51. Exercise 45: 0.18; Exercise 46: 0.51; Exercise 48: 0.00 (less than 4.6×10^{-44})

Exercise Set 5.5

1. They would not make a profit.

3. $255 billion

5. −25%

7. Larry, by $25.00

9. Answers will vary.

11. Answers will vary.

13. The strategy assumes you have an essentially infinite amount of money so that you can double your wager many times. Also, casinos place a limit on how large a wager they will accept. If you lose too many times in a row, you will not be able to place a large enough wager.

15. $1000

17. $1.17

19. −23.1%

21. 0%

23. −30%

25. 1.8; 72

27. $86.50

29. $15,092.50

31. −5%; no

33. The expected value is 0%.

35. −5.3%

37. 18/37; 48.6%

39. 18/37, or 48.6%

41. No. The numbers are chosen randomly.

43. $\dfrac{59!}{5!54!} \times 39 = 195{,}249{,}054$

45. Answers will vary.

47. Answers will vary.

Exercise Set 6.1

1. Mean: 86.9; median: 87; mode: none

3. The mean is 75 for both. Medians: 75 and 100, respectively. According to the grading scale, both students receive a "C."

5. For May, mean: $15,495; median: $15,000; mode: $15,000. For June, mean: $15,026; median: $13,750; mode: $12,500. For July, mean: $14,974; median: $12,500; mode: $12,500.

7. Minimum: about 71; first quartile: 90; median: 110; third quartile: about 140; maximum: about 154.

9. The scores show improvement. No downside is evident.

11. a. The vertical axis is the salary in dollars.

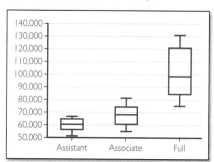

b. Answers will vary.

13. Mean: 79.5; standard deviation: 10.16

15. Mean: 85%; standard deviation: 0

17. 3.20

19. Mean: 10; standard deviation: 2.5; number of points: 6

21. a. $\mu = 26.1$
b. $\sigma = 6.2$
c. 39 bulbs

23. Answers will vary.

25. Answers will vary.

27. Sample standard deviation = $22,712.58. Population standard deviation = $21,413.62.

29. Answers will vary.

Exercise Set 6.2

1. Distributions 2 and 3 are normal.

3. Yes, the temperatures are approximately normally distributed, though the curve is not perfectly symmetric—it's somewhat skewed.

5. 68% of ridge counts are between 90 and 190, 95% are between 40 and 240, and 99.7% are between 0 and 290.

7. 340

9. −1.3

11. 0.3

13. z-score 1.9; 97.1%

15. z-score 2.7; 0.3% or 0.4% score 140 or higher

17. 93.3%

19. 0.6

21. 1%

23. 25.3 and 34.5 degrees Celsius

25. 14%

27. 2%

29. A: 6.7%; B: 24.2%

31. a. Mean 70%, standard deviation 2.3
b. 30.9%

33. 1; 2

35. 2.3%

37. 72.6%

39. Our town is extremely unusual. It's almost 20 standard deviations above the mean.

41. a. Mean: 20%; standard deviation: 1.3.
b. Extremely unusual. This is almost 4 standard deviations above the mean.

43. Extremely unusual; no

45. Answers will vary.

47. Answers will vary.

49. Answers will vary.

Exercise Set 6.3

1. 2.6%

3. Margin of error is about 3.3%. Perhaps the author of the article simply rounded 3.3 up to 3.5.

5. Answers will vary. Yes: At a 95% confidence level, we can say that between 56% and 64% support the Patriot Act.

7. The rule of thumb gives the same margin of error, 3.5%.

9. Answers will vary. About 816 people were sampled.

11. Answers will vary.

13. Doubling the sample size multiplies the margin of error by a factor of $1/\sqrt{2}$ or about 0.7. Cutting the margin of error in half requires quadrupling the sample size.

15. 2.6%

17. With 90% confidence, between 45.4% and 50.6% like salad. The majority level is within the margin of error, so we cannot say with 90% confidence that a majority like salad.

19. a. 87%
    b. 500

21. Answers will vary.

23. Answers will vary.

25. 3%

27. 95% of the polls represented the voting population with an error of at most 3 percentage points.

29.

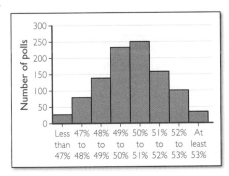

31. Answers will vary.

33.

% error	Size
2.0	2500
2.1	2268
2.2	2066
2.3	1890
2.4	1736
2.5	1600
2.6	1479
2.7	1372
2.8	1276
2.9	1189
3.0	1111

Exercise Set 6.4

1. Answers will vary.

3. Small p-values, not large ones, indicate statistical significance.

5. The p-value is 0.184, which is not normally considered statistically significant.

7. Half of the women would be studied for 5 or more years, and half of the women would be studied for 5 or fewer years.

9. Because the study was terminated, the researchers must have considered the difference to be statistically significant.

11. This study indicates that they should.

13. Answers will vary.

15. 14

17. a. These variables are negatively correlated: As distance increases, brightness decreases.
    b. These variables are uncorrelated.
    c. These variables are positively correlated: Earthquakes with larger readings on the Richter scale generally cause more damage.
    d. These variables are negatively correlated: A golfer wins by getting the lowest score. So the lower a golfer's score, the higher her ranking.

19. No

21. Yes

23. 80 million should remember

25. Answers will vary.

27. Answers will vary

29. Answers will vary.

31. Answers will vary.

33. Answers will vary.

35. The trend line is $y = -240x + 564$, where y is the number sold and x the price in dollars. The correlation coefficient is -0.97. This represents a strong negative linear correlation.

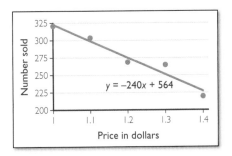

37. 0.028

Exercise Set 7.1

1. Answers will vary.

3.
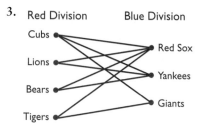
Red Division   Blue Division

5.
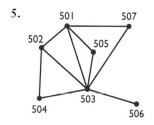

7.
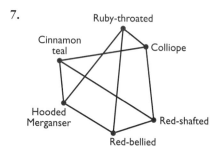

9. Each vertex has degree 4.

11. The worker would like to have a route traversing each edge (line) exactly once.

13. One Euler circuit is *A-B-F-C-D-E-A-D-B-C-A*.

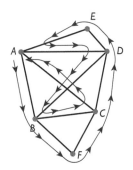

15. No Euler circuit exists.

17. One Euler circuit is Post Office-*F-B-E-D*-Post Office-*C-B*-Post Office.

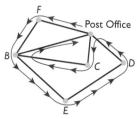

19. No Euler circuit exists.

21. No Euler circuit exists

23. No Euler circuit exists.

25. No Euler circuit exists.

27. No Euler circuit exists.

29. One correct answer is

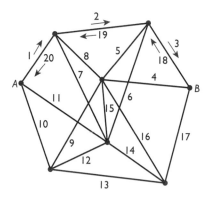

31. One correct answer is

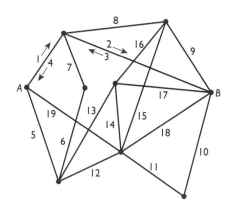

33. **a.** The vertices of odd degree are marked *A* and *B*.

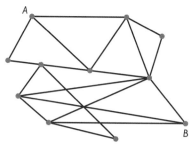

b.

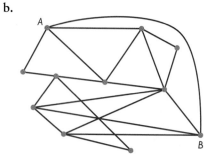

c. One Euler circuit is

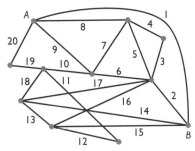

d. Delete the first edge from the Euler circuit to get the Euler path.

35.

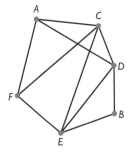

37. No such graph exists.

39. Answers will vary.

Exercise Set 7.2

1. Euler circuit

3. Yes

5. The highlighted edges show one Hamilton circuit.

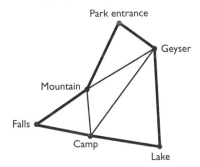

7. No Hamilton circuit exists.

9. The highlighted edges show one Hamilton circuit.

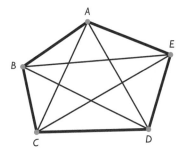

11. The highlighted edges show one Hamilton circuit.

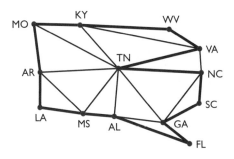

13. The highlighted edges show one Hamilton circuit.

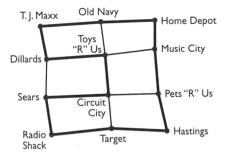

15. The highlighted edges show one Hamilton circuit.

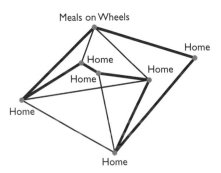

17. The highlighted edges show one Hamilton circuit.

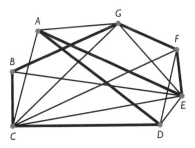

19. No Hamilton circuit exists.

21. Ft. Wayne - Cleveland - Richmond - Louisville - Ft. Wayne: Distance 1496 miles

23. Provo - Colorado Springs - Santa Fe - Tucson - Provo: Distance 2244 miles

25. *A-D-E-C-B-A* has length 15.

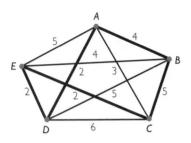

27. *A-D-E-B-C-A* has length 14.

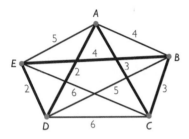

29. *A-D-E-B-C-A* has length 14.

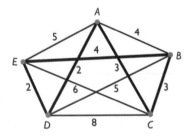

31. *A-D-E-C-B-A* has length 15.

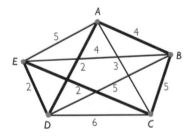

33. *A-D-E-B-C-A* has length 14.

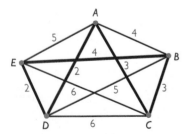

35. *A-D-E-B-C-A* has length 14.

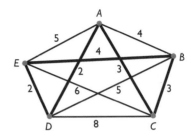

37. a. *A-D-E-B-C-A* has length 14.
   b. *B-C-A-D-E-B* has length 14.
   c. *C-A-D-E-B-C* has length 14.
   d. *D-E-B-C-A-D* has length 14.
   e. *E-D-A-C-B-E* has length 14.
   f. The shortest of the routes found start at *A*, *B*, *C*, *D*, or *E*. Answers will vary.

39. 14

41. 18

43. Answers will vary.

45. Answers will vary.

Exercise Set 7.3

1. One solution is

3.

5. No

7. *D, G, H, I, K*

9.

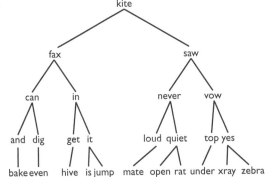

11. 7 checks

13. 16 checks

15. 501

17. 1024 leaves; 176 show at least 7 passes.

19. 7

21. 255

23. 240

25.

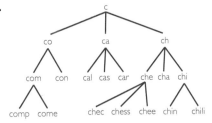

27. Answers will vary.

Exercise Set 8.1

1. $2^5 - 1 = 31$

3. $5! = 120$

5. (b) $2^{50} - 1$

7. Critical voters

9. Anne 1, other voter 0

11. 1/100

13. We use *A* for Abe, *B* for Ben, and *C* for Condi.

Votes			Total votes	Winning coalition?	Critical voters
100	**91**	**10**			
A	*B*	*C*	201	Yes	None
A	*B*		191	Yes	*A, B*
A		*C*	110	Yes	*A, C*
A			100	No	not applicable
	B	*C*	101	Yes	*B, C*
	B		91	No	not applicable
		C	10	No	not applicable

15.

Voters			Swing
Abe	Ben	Condi	Ben
Abe	Condi	Ben	Condi
Ben	Abe	Condi	Abe
Ben	Condi	Abe	Condi
Condi	Abe	Ben	Abe
Condi	Ben	Abe	Ben

17. 14

19. We list only winning coalitions.

Coalition	Votes	Critical voters
A B C D	26	A
A B C	24	A
A B D	21	A
A B	19	A, B
A C D	20	A
A C	18	A, C
A D	15	A, D

21. 24

23. *A*: 18/24 = 3/4; *B, C, D* each 2/24 = 1/12

25. Abe: 1/2, Ben: 1/2, Condi: 0, Doris: 0

27. 11

29. 31

31. *A*: 1/3, *B*: 7/27, *C*: 5/27, *D*: 1/9, *E*: 1/9

33.

R	D	I	D
R	I	D	D
D	R	I	R
D	I	R	I
I	R	D	D
I	D	R	D

Democrats: 4/6 = 2/3, Republicans: 1/6, Independent: 1/6

35. Using the popular vote, the indices are State A: 0/6 = 0, State B: 0/6 = 0, State C: 6/6 = 1. Using the electoral vote, the index is 1/3 for each state.

37. A "no" vote leaves only 38 possible "yes" votes, which is less than the quota.

39. By the two preceding exercises, a measure passes under this weighted system if, and only if, it passes under the actual system.

41. Answers will vary.

43. They are not critical voters in any winning coalition.

45. Answers will vary.

47. Answers will vary.

49. Answers will vary.

51. a. 8 votes
  b. A: 47.37%, B: 36.84%, C: 5.26%, D: 5.26%, E: 5.26%
  c. A: 50%, B: 50%, C: 0%, D: 0%, E: 0%
  d. The index for A increased. Answers will vary.
  e. The 8-vote quota will be reached exactly when both A and B are in support. The votes of the remaining voters don't matter. In calculating the Banzhaf index, we can discard voters with no power.

Exercise Set 8.2

1. No

3. a. No; yes
  b. Answers will vary.

5. C

7. B

9. D

11. D

13. D

15. D

17. B

19. C

21. C

23. C

25. A

27. B

29. B

31. A

33. No

35. Republican: 1/3, Democrat: 1/3, Independent: 1/3, Other: 0

37. The first three blocks have equal power and the fourth has no power. Answers will vary.

39. 1/3 for each

41. Answers will vary.

43. Answers will vary.

45. Answers will vary.

Exercise Set 8.3

1. Put the earrings in a small pile.

3. Daniel gets the paintings and 35.3% of the art supplies. Carla gets the drawings and 64.7% of the art supplies.

5. Anna gets 71.4% of the travel books. Ashley gets the math books, craft books, and 28.6% of the travel books.

7. Bcom gets the computer outlet, the bakery, and 9.1% of the cleaning service. Ccom gets the appliance dealership, the coffee shop, and 90.9% of the cleaning service.

9. Ben gets the Hopi kachinas and the Zuni pottery. Ruth gets the Navajo jewelry, the Apache blankets, and the Pima sand paintings. There is no need to divide any of the items.

11. Various answers are possible. One answer is: Person 1 gets items 1 and 2 and person 2 gets items 3 and 4.

13. J.C.: $91,875; Trevenor: $94,375; Alta: home less $275,625; Loraine: $89,375

15. Lee: $33.75; Max: watercolor less $103.75; Anne: $31.25; Becky: $38.75

17. James: Business 2 and $7500; Gary: Business 1 and Business 3 less $7500

19. a. Answers will vary.
  b. Answers will vary.

21. Answers will vary.

23. Answers will vary.

25. a. No matter how they divide cookie 1 and cookie 3, at least one will wind up with fewer chocolate chips than Abe.
  b. Answers will vary.

Exercise Set 8.4

1. 34,437.333

3. 9.63

5. 52.19

7. 10,280

9. a. 100
  b. States 1 and 2 have a quota of 1.4. State 3 has a quota of 7.2.
  c. Hamilton's method gives 2 seats to each of States 1 and 2 and 7 seats to State 3. That is a total of 11 seats.

11. 95.83

13. North 1, South 1, East 2, West 8

15. North 1, South 2, East 2, West 7

17. 90.4

19. North 1, South 1, East 4, West 9

21. North 2, South 2, East 3, West 8

23. a. A divisor of 52 gives an allocation of 3 seats for each of States 1 and 2 and 19 seats for State 3.
  b. State 3

25. Oklahoma 14, Texas 93, Arkansas 11, New Mexico 7

27. Montana should get 2 seats, Alaska 1 seat, Florida 26 seats, and West Virginia 3 seats, using Dean's method.

29.

State	Seats
Connecticut	7
Delaware	1
Georgia	2
Kentucky	2
Maryland	8
Massachusetts	14
New Hampshire	4
New Jersey	5
New York	10
North Carolina	10
Pennsylvania	13
Rhode Island	2
South Carolina	6
Vermont	2
Virginia	19

31.

State	Seats
Connecticut	7
Delaware	2
Georgia	2
Kentucky	2
Maryland	8
Massachusetts	14
New Hampshire	4
New Jersey	5
New York	10
North Carolina	10
Pennsylvania	13
Rhode Island	2
South Carolina	6
Vermont	2
Virginia	18

Exercise Set 9.1

1. 31.4 feet

3. 34 units

5. $\sqrt{58}$ or about 7.6 units

7. $\sqrt{269}$ or about 16.4 feet

9. Answers will vary; 6 square units

11. 20.5 square units

13. 9.8 square feet

15. For the 4-5-6 triangle, the area is about 9.9 square yards. For the 3-6-7 triangle, the area is about 8.9 square yards. The 4-5-6 triangle is larger.

17. 15.7 feet

19. Area is 100 square units. Perimeter is 51.2 units.

21. Area is $7\sqrt{75}/2$ or about 30.3 square units. Perimeter is $17 + \sqrt{79}$ or about 25.9 units.

23. Area is $4(2 + \sqrt{33})/2$ or about 15.5 square units. Perimeter is $9 + \sqrt{20} + \sqrt{33}$ or about 19.2 units.

25. The can holds 17.2 cubic inches. Therefore, the amount of lost soda is more than that.

27. Area: 30.9 square inches

29. Circumference: 8.2 inches

31. 858 miles

33. 21 inches

35. 212 square inches and 188 square inches; no

37. 2153.4 square feet

39. $\sqrt{656}$ or about 25.6 feet

41. 349.6 feet

43. 11

45. Answers will vary.

47. Answers will vary.

49. Answers will vary.

51. Answers will vary.

53. 920 cubic feet, 34.1 cubic yards

55. Volume: 12.8 cubic inches; Area: 26.4 square inches

57. Volume: 433.5 cubic inches; Area: 277.0 square inches

59. Volume: 6.0 cubic inches

61. 70.7 square feet

63. Answers will vary.

Exercise Set 9.2

1. No; no

3. Yes; 2.70 dollars per pound if the cost is measured in dollars and the weight is measured in pounds

5. No

7. Yes; the surface area of the pizza

9. No

11. Yes; π

13. 8.1 inches

15. Answers will vary.

17. Answers will vary.

19. Answers will vary.

21. Answers will vary.

23. Answers will vary.

25. $A = 7/3$; $b = 12$

27. 5 feet tall

29. The area is multiplied by 4.

31. c. r^3; $4\pi/3$

33. b. r^2; 4π

35. b. r^2; πh

37. c. h; πr^2

39. c. h; $2\pi r$

41. a. r; $2\pi h$

43. e. None of these

45. One can hold 4 times that of the other.

47. One can hold twice as much as the other.

49. The BMR of the longer animal is 100 times that of the shorter.

51. Answers will vary.

Exercise Set 9.3

1. Answers will vary.

3. Answers will vary.

5. Answers will vary.

7. Answers will vary.

9. $\theta = 120$ degrees

11. $n = 180$

13. Reflectional symmetries through lines through opposite angles and lines through the mid-points of opposite sides (6 lines altogether). Rotational symmetry of 60 degrees (as well as 120, 180, 240, and 300 degrees).

15. One tiling is:

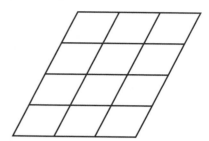

17. Answers will vary.

19. Two copies go together to make a rectangle.

21. Answers will vary.

23.

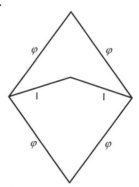

Answers will vary.

CREDITS

In the News, Illustrations, and Tables

Ch. 1

p. 3: (In the News 1.1) "21st Century Skills: Rethinking How Students Learn," *OttawaStart*, October 27, 2010, http://ottawastart.com/story/12251.php.

p. 4: (In the News 1.2) Douglas E. Hall, "Warning: Why Average Isn't Average: Simplistic Statistics Can Be Misleading in Measuring for Accountability in Education," New Hampshire Center for Public Policy Studies, June 1998, http://www.nhpolicy.org/reports/average.html.

p. 13: (In the News 1.3) Sam Hudzik, "Testiness, accusations in final gov debate," WBEZ, October 29, 2010, http://www.wbez.org/story/bill-brady/snippy-exchanges-guilt-association-charges-mark-final-debate-illinois-governor#.

p. 25: (In the News 1.4) Belle Dumé, "Quantum logic gate lights up," PhysicsWorld.com, August 8, 2003, http://physicsworld.com/cws/article/news/18020.

p. 40: (In the News 1.5) Joshua Glenn, "Venn and the art of diagrams," *Boston Globe*, May 2, 2004, http://www.boston.com/news/globe/ideas/articles/2004/05/02/venn_and_the_art_of_diagrams/.

p. 56: (In the News 1.6) Naftali Bendavid, Elizabeth Williamson, & Sudeep Reddy, "Stimulus Package Unveiled," January 16, 2009, http://online.wsj.com/article/SB123202946622485595.html.

Ch. 2

pp. 71–72: (In the News 2.1, Figure 2.1) Owen Thomas, "Facebook Slashes Its Growth Rate," *Gawker*, October 24, 2007, http://gawker.com/314772/facebook-slashes-its-growth-rate?tag=valleywag.

p. 72: (In the News 2.2) Steve Wieberg, "NCAA Football Grad Rates at All-Time High, but Top Schools Falter," *USA Today*, October 27, 2010, http://www.usatoday.com/sports/college/2010-10-27-ncaa-graduation-rates-study_N.htm.

p. 86: (table) Data from *National Vital Statistics Reports* 51, no. 1 (December 2002).

p. 88: (Exercises 3–4 table) Data from U.S. Census Bureau, *Current Population Reports*, 2010; (Exercises 5–7 table) Data from the U.S. Department of Transportation, *Air Travel Consumer Report*, 2010; (Figure 2.7) Data from the annual *CIRP Freshman Survey* by the Higher Education Research Institute at UCLA, for the given year; (Exercises 12–15 table) Data from the U.S. National Center for Education Statistics, *Digest of Education Statistics*, 2010; (Exercises 16–18 table) Data from The Higher Education Research Institute, UCLA, *The American Freshman: National Norms*, 2005.

p. 89: (Exercises 23–27 table) John Riew, "Economies of Scale in High School Operation," *Review of Economics and Statistics* 48 (August 1966): 280–287. These figures have not been adjusted for inflation; (Exercises 31–33 table) Data from Association of

Departments of Foreign Languages, New York, *ADFL Bulletin* 31, no. 2 (Winter 2002); (Exercises 34–35 table) Data from The College Board, *National Summary Report, 2005*.

p. 90: (In the News 2.3) Fred Pearce, "Climate: The Great Hockey Stick Debate," *New Scientist* 2543 (March 18, 2006): 40.

pp. 93–94: (Figures 2.11–2.13 and Example 2.11 table) Adapted from J. T. Bonner, *Size and Cycle* (Princeton, NJ: Princeton University Press, 1965).

p. 94: (Figure 2.14) From http://cnx.org/content/m10927/latest.

p. 95: (Figures 2.15–2.16 and Try It Yourself 2.11 table) Data from Harlow Shapley, "Note on the Thermokinetics of Dolichoderine Ants," *Proceedings of the National Academy of Science* 10 (1924): 436–439.

p. 96: (Figure 2.17) Data through 2004 are from http://ratecalc.cancer.gov. The data for 1995–99 and 2000–04 come from the authors' personal correspondence with the National Institutes of Health, National Cancer Institute, Division of Cancer Epidemiology and Genetics.

p. 102: (Exercise 1 table) Data from *2003 Statistical Abstract of the United States*, Table 135, data for 2002; (Figure 2.26) Data from www.sifry.com/alerts/archives/000436.html.

p. 104: (In the News 2.4) Lawrence K. Altman, "Study Says Drug Ads in Medical Journals Frequently Mislead," *The New York Times*, June 1, 1992.

p. 107: (In the News 2.5) Beth Gollob, "State ACT Scores Improve, Still Low," *The Oklahoman*, August 16, 2006.

p. 109: (Try It Yourself 2.16 table) Data from www.act.org/news/releases/2000/08-17-00.html.

p. 110: (Figures 2.42–2.43) Data from Frederick Klein, "Americans Hold Increasing Amounts in Cash Despite Inflation and Other Drawbacks," *Wall Street Journal*, July 5, 1979.

p. 115: (Figure 2.50 and Example 2.18 tables) Data from www.news.ucdavis.edu/search/news_detail.lasso?id=8360.

p. 116: (Fuel economy) Based on a figure from Edward R. Tufte, *The Visual Display of Quantitative Information* (Cheshire, CT: Graphics Press, 2001): p. 57.

pp. 117–118: (Figures 2.51–2.52) Data from OSU Human Resources, as of December 31, 2002.

p. 118: (Figure 2.55) Data from New York State Department of Motor Vehicles, as presented in Robert R. Johnson, *Elementary Statistics* (North Scituate, MA: Duxbury Press, 1976), 80–82; (Figure 2.56) Data from www.theweddingreport.com.

p. 119: (Figure 2.59) Data from Tulsa County Public Facilities Authority; (Figure 2.60 and Exercises 21–24 table) Data from *Statistical Abstract of the United States, 2003*.

p. 121: (Figure 2.61) Data from http://usgovernmentrevenue.com/yearrev2002_0.html.

p. 424: (In the News 7.2) Martin Tolchin, "Pilots Given Leeway to Pick Efficient Routes," *New York Times*, May 23, 1993, http://www.nytimes.com/1993/05/23/us/pilots-given-leeway-to-pick-efficient-routes.html.

p. 443: (In the News 7.3) Gina Kolata, "Math Problem, Long Baffling, Slowly Yields," *New York Times*, March 12, 1991, http://www.nytimes.com/1991/03/12/science/math-problem-long-baffling-slowly-yields.html.

p. 460: (In the News 7.4) "How Did That Chain Letter Get to My Inbox?" National Science Foundation, May 16, 2008, http://www.nsf.gov/news/news_summ.jsp?cntn_id=111580.

Ch. 8

p. 477: (In the News 8.1) "Electoral College," *New York Times*, May 4, 2010, http://topics.nytimes.com/topics/reference/timestopics/subjects/e/electoral_college/index.html.

p. 478: (In the News 8.2) "Ukraine President Accuses Rival of 'Coup d'Etat,'" *New York Times*, September 12, 2008, http://www.kyivpost.com/news/nation/detail/29681.

pp. 483–485: (Example 8.3) We would like to acknowledge Prof. Tom Quint of the University of Nevada for bringing to our attention the example of the 1988 Democratic convention used in Example 8.3.

p. 493: (In the News 8.3) Matt Marchetti, "*Asheville Citizen Times* Claims Favoring Instant Runoff Voting Don't Hold Up to Scrutiny," September 2, 2008, http://irvbad4nc.blogspot.com/2008/09/asheville-citizen-times-claims-favoring.html.

p. 495: (Example 8.8 table) Data from the Federal Election Commission, http://www.fec.gov/pubrec/2000presgeresults.htm.

p. 509: (In the News 8.4) Deidre Woollard, "John Cleese's Big Divorce Settlement," *Luxist*, August 20, 2009, http://www.luxist.com /2009/08/20/john-cleeses-big-divorce-settlement.

p. 522: (In the News 8.5) Arturo Vargas, "New Census Numbers Portend Significant Latino Role," *Huffington Post*, December 20, 2010, http://www.huffingtonpost.com/arturo-vargas/new-census-numbers-porten_b_799454.html.

Ch. 9

p. 549: (In the News 9.1) Daniel B. Kane, "We Are Hard-Wired for Geometry," MSNBC, January 19, 2006, http://www.msnbc.msn.com/id/10925120/.

p. 550: (In the News 9.2) Suzy Menkes, "Designers Outline a New Geometry," *New York Times*, March 5, 2010, http://www.nytimes.com/2010/03/06/fashion/06iht-rrick.html.

p. 568: (In the News 9.3) Assem Deif, "The Amazing Golden Ratio," *Al-Ahram* 864(September 27–October 3, 2007), http://weekly.ahram.org.eg/2007/864/heritage.htm.

p. 582: (In the News 9.4) Keith Devlin, "Good Stories, Pity They're Not True," Mathematical Association of America, June 2004, http://maa.org/devlin/devlin_06_04.html.

p. 583: (In the News 9.5) Julie Rehmeyer, "Forms of Symmetry," *Science News*, April 21, 2007. http://sciencenews.org/articles/20070421/mathtrek.as.

Appendix

p. A-4: (In the News A1.1) Jay Conley, "A Drug Bust," *Roanoke Times*, January 9, 2004.

Photographs and Cartoons

Ch. 1

p. 2: © Fancy/Alamy; **p. 4:** Bill Watterson, Universal Uclick; **p. 5:** Tim Sloan/AFP/Getty Images; **p. 7:** Justin Sullivan/Getty Images; **p. 8:** AP Photo/Tony Dejak; **p. 14:** AP Photo/Ron Edmonds; © Mike Baldwin/Cornered; **p. 17:** Photographer_s Mate 1st Class John Lill/U.S. Navy; 20: Mandel Ngan/AFP/Getty Images; **p. 22:** © Steve Marcus/Las Vegas Sun/Reuters/Corbis; **p. 25:** © Piyaphanta/Dreamstime.com; **p. 27:** © 2011 Google; **p. 28:** Getty Images/PhotoAlto; **30:** DILBERT: © Scott Adams/Distributed by United Feature Syndicate, Inc.; **p. 31:** AP Photo/Susan Walsh; **p. 34:** Omar Torres/Newscom; **p. 35:** Alamy; **p. 37:** cartoonstock.com; **p. 40:** The Print Collector/Alamy; **p. 44:** © 2000–2010 Richard Stevens 3/dieselsweeties.com; **p. 55:** Michael Marcovici; **p. 57:** (Milky Way) NASA, ESA and The Hubble Heritage Team (STScI/AURA), (Earth) NASA Goddard Space Flight Center/http://visibleearth.nasa.gov, (San Francisco Bay) NASA, (flowers) Photodisc, (bee) Getty Images RF/National Geographic, (DNA) Scott Camazine/Alamy; **p. 60:** Harley Schwadron/cartoonstock.com; **p. 61:** Goddard/cartoonstock.com.

Ch. 2

p. 70: Urbanmyth/Alamy; **p. 75:** AP Photo/Matthias Schrader; **p. 77:** Ed Fischer/Cartoonstock.com; **p. 78:** The U.S. National Archives and Records Administration; **p. 83:** Krista Kennell/Sipa via AP Images; **p. 84:** AP Photo/Richard Vogel; **p. 90:** *Climate Change 2001: The Scientific Basis*, Contribution of Working Group I to the Third Assessment Report of the Intergovernmental Panel on Climate Change; **p. 92:** Photodisc; **p. 94:** Mark Parisi/offthemark.com; **p. 97:** Schwadron/cartoonstock.com; **p. 99:** AP Photo/East

Liverpool Review, Wayne Maris; **p. 102:** Courtesy of David L. Sifry (www.sifry.com/alerts) and Technorati; **p. 104:** MPI/Getty Images; **p. 105:** cartoonstock.com; **p. 107:** Blend Images/Alamy; **p. 108:** © 2006, The Oklahoma Publishing Company; **p. 111:** cartoonstock.com; **p. 114:** AP Photo/Tom Strattman; **p. 116:** Courtesy of HowStuffWorks.com; **p. 121:** National Geographic Stock.

Ch. 3

p. 126: Alamy; **p. 127:** © INTERFOTO/Alamy; **p. 129:** AP Photo; **p. 131:** cartoonstock.com; **p. 134:** NASA; **p. 136** (top photo) VIN MORGAN/AFP/Getty Images; (bottom photo) © Peter Arnold, Inc. Alamy; **p. 140:** Schwadron/cartoonstock.com; **p. 146:** Department of Energy; **p. 149:** CDC; **p. 150:** cartoonstock.com; **p. 152:** Christopher Weyant/The New Yorker Collection/www.cartoonbank.com; **p. 157:** Dreamstime.com; **p. 159:** Alamy; **p. 160:** cartoonstock.com; **p. 164:** © 2004, California Institute of Technology; **p. 167:** USGS; **p. 168:** Walter Mooney, U.S. Geological Survey; **p. 171:** © Elvinstar | Dreamstime.com; **p. 173:** © Zimmytws | Dreamstime.com; **p. 179:** G. William Schwert, 2000–2004.

Ch. 4

p. 184: KRT/Newscom; **p. 185:** © Arenacreat | Dreamstime.com; **p. 188:** cartoonstock.com; **p. 189:** © Karenr | Dreamstime.com; **p. 192:** cartoonstock.com; **p. 193:** Carr/Newscom; **p. 195:** cartoonstock.com; **p. 200:** © Route66 | Dreamstime.com; **p. 203:** DILBERT © 2008 Scott Adams. Used by permission of UNIVERSAL UCLICK. All rights reserved; **p. 204:** Mike Baldwin/cartoonstock.com; **p. 207:** liquid library; **p. 211:** © Draghicich |

INDEX